Quaternary of the Thames

THE GEOLOGICAL CONSERVATION REVIEW SERIES

The comparatively small land area of Great Britain contains an unrivalled sequence of rocks, mineral and fossil deposits, and a variety of landforms that span much of the earth's long history. Well-documented ancient volcanic episodes, famous fossil sites and sedimentary rock sections used internationally as comparative standards, have given these islands an importance out of all proportion to their size. The long sequences of strata and their organic and inorganic contents have been studied by generations of leading geologists, thus giving Britain a unique status in the development of the science. Many of the divisions of geological time used throughout the world are named after British sites or areas, for instance, the Cambrian, Ordovician and Devonian systems, the Ludlow Series and the Kimmeridgian and Portlandian stages.

The Geological Conservation Review (GCR) was initiated by the Nature Conservancy Council in 1977 to assess, document and ultimately publish accounts of the most important parts of this rich heritage. Since 1991, the task of publication has been assumed by the Joint Nature Conservation Committee on behalf of the three country agencies, English Nature, Scottish Natural Heritage and the Countryside Council for Wales. The GCR series of volumes will review the current state of knowledge of the key earth-science sites in Great Britain and provide a firm basis on which site conservation can be founded in years to come. Each GCR volume will describe and assess networks of sites of national or international importance in the context of a portion of the geological column, or a geological, palaeontological or mineralogical topic. The full series of approximately 50 volumes will be published by the year 2000.

Within each individual volume, every GCR locality is described in detail in a self-contained account, consisting of highlights (a précis of the special interest of the site), an introduction (with a concise history of previous work), a description, an interpretation (assessing the fundamentals of the site's scientific interest and importance), and a conclusion (written in simpler terms for the non-specialist). Each site report is a justification of a particular scientific interest at a locality, of its importance in a British or international setting and ultimately of its worthiness for conservation.

The aim of the Geological Conservation Review series is to provide a public record of the features of interest in sites being considered for notification as Sites of Special Scientific Interest (SSSIs). It is written to the highest scientific standards but in such a way that the assessment and conservation value of the site is clear. It is a public statement of the value given to our geological and geomorphological heritage by the earth-science community which has participated in its production, and it will be used by the Joint Nature Conservation Committee, English Nature, the Countryside Council for Wales and Scottish Natural Heritage in carrying out their conservation functions. The three country agencies are also active in helping to establish sites of local and regional importance. Regionally Important Geological/Geomorphological Sites (RIGS) augment the SSSI coverage, with local groups identifying and conserving sites which have educational, historical, research or aesthetic value, enhancing the wider earth science conservation perspective.

All the sites in this volume have been proposed for notification as SSSIs; the final decision to notify, or renotify, lies with the governing Councils of the appropriate country conservation agency.

Information about the GCR publication programme may be obtained from:

Earth Science Branch,
Joint Nature Conservation Committee,
Monkstone House,
City Road,
Peterborough PE1 1JY.

Titles in the series

1. **Geological Conservation Review**
   An introduction
2. **Quaternary of Wales**
   S. Campbell and D.Q. Bowen
3. **Caledonian Structures in Britain**
   South of the Midland Valley
   Edited by J.E. Treagus
4. **British Tertiary Volcanic Province**
   C.H. Emeleus and M.C. Gyopari
5. **Igneous Rocks of South-West England**
   P.A. Floyd, C.S. Exley and M.T. Styles
6. **Quaternary of Scotland**
   Edited by J.E. Gordon and D.G. Sutherland
7. **Quaternary of the Thames**
   D.R. Bridgland

Quaternary of the Thames

D.R. Bridgland

Bridgland Earth Science Consultancy,
Darlington, UK.

Scientific Editor: D. Q. Bowen
GCR Editor: W. A. Wimbledon

SPRINGER-SCIENCE+BUSINESS MEDIA, B.V.

First edition 1994

Originally published by Chapman & Hall in 1994
Softcover reprint of the hardcover 1st edition 1994

Typeset in 10pt ITC Garamond Book by Herrington Geoscience and Exe Valley Dataset Ltd, Exeter

ISBN 978-94-010-4303-8 ISBN 978-94-011-0705-1 (eBook)
DOI 10.1007/978-94-011-0705-1

A catalogue record for this book is available from the British Library

Library of Congress Cataloging-in-Publication data available

Contents

Contents

Contributors

T. Allen, 30 Fourth Avenue, Romford, Essex, RM7 0UB.

Dr D.R. Bridgland, Earth Science Consultancy, 41 Geneva Road, Darlington, County Durham, DL1 4NE.

Dr D.A. Cheshire, Division of Environmental and Earth Science, University of Hertfordshire, College Lane, Hatfield, Hertfordshire, AL10 9AB.

Professor G.R. Coope, Department of Geological Sciences, Birmingham University, P.O. Box 363, Birmingham, B15 2TT.

Dr P.L. Gibbard, Sub-department of Quaternary Research, Botany School, Downing Street, Cambridge, CB2 3EA.

P. Harding, Wessex Archaeology, Portway House, South Portway Estate, Old Sarum, Salisbury, Wiltshire, SP4 6EB.

R. Wrayton, 9 Barnfield, Wickford, Essex, SS11 8HP.

Acknowledgements

D.R. Bridgland wishes to acknowledge the assistance and encouragement of a great many colleagues since the initiation of his GCR work on the Thames in 1980. He is particularly indebted to Dr W.A. Wimbledon, who was an inspiration in the early years of the project and provided much valuable discussion on stratigraphy. Professor J. Rose contributed much constructive advice and criticism as referee, which helped guide the project towards publication. Invaluable discussion and advice has also been supplied by Mr T. Allen, Professor D.J. Briggs, Dr S. Campbell, Dr D.A. Cheshire, Professor G.R. Coope, Dr P.L. Gibbard, Mr P. Harding, Dr D.T. Holyoak, Dr D.H. Keen, Mr J. MacNabb, Dr D. Maddy, Dr A.J. Stuart, Dr A.J. Sutcliffe, Dr R.M. Tipping and, in particular, by Dr P. Allen, Dr D.A. Cheshire, Dr R.A. Kemp, Dr R.C. Preece, Dr C.B. Stringer, Mr F.F. Wenban-Smith, and Mr J.J. Wymer, who have all read and commented on parts of the text. He would like to thank those who have contributed photographs (these are individually attributed in their captions) and unpublished work to the volume (these are: D.A. Cheshire – Westmill Quarry and Ugley Park Quarry; P. Harding – Globe Pit and Lion Pit tramway cutting; T. Allen, G.R. Coope, P.L. Gibbard and R. Wrayton – Great Totham), as well as those who have supplied personal communications, namely P. Allen, D.Q. Bowen, J.A. Catt, A.P. Currant, M.H. Field, P.L. Gibbard, R.W. Hey, D.T. Holyoak, R.A. Kemp, M.P. Kerney, J. McNabb, R.J. MacRae, D. Maddy, R.C. Preece, H.M. Rendell, J.E. Robinson, H.M. Roe, J. Rose, A.J.R. Snelling, A.J. Sutcliffe, C. Turner, A.H. Weir, F.F. Wenban-Smith, C.A. Whiteman and J.J. Wymer.

Assistance with the bibliography, provided by Paula Bridgland and Caroline Mee, is gratefully acknowledged. Paula Bridgland also helped with word processing and proof reading at various stages of the project. Thanks are also due to the GCR Publication Production Team: Dr D. O'Halloran, Project Manager; Neil Ellis, Publications Manager; Nicholas D.W. Davey, Scientific Officer (Editorial Assistant) and Valerie Wyld, GCR Sub-editor. Diagrams were drafted initially by Paul Butler and Nicholas D.W. Davey and were subsequently completed by Lovell Johns Ltd in collaboration with the author.

Access to the countryside

This volume is not intended for use as a field guide. The description or mention of any site should not be taken as an indication that access to a site is open or that a right of way exists. Most sites described are in private ownership, and their inclusion herein is solely for the purpose of justifying their conservation. Their description or appearance on a map in this work should in no way be construed as an invitation to visit. Prior consent for visits should always be obtained from the landowner and/or occupier.

Information on conservation matters, including site ownership, relating to Sites of Special Scientific Interest (SSSIs) or National Nature Reserves (NNRs) in particular counties or districts may be obtained from the relevant country conservation agency headquarters listed below:

English Nature,
Northminster House,
Peterborough PE1 1UA.

Scottish Natural Heritage,
12 Hope Terrace,
Edinburgh EH9 2AS.

Countryside Council for Wales,
Plas Penrhos,
Ffordd Penrhos,
Bangor,
Gwynedd LL57 2LQ.

Preface

The principal aim of this volume is to provide descriptions of Sites of Special Scientific Interest, selected as part of the Geological Conservation Review, that yield evidence for the Pleistocene history of the River Thames and its tributaries. Although defined thematically, the volume covers all Pleistocene GCR sites in the Thames valley. A number of sites in southern East Anglia are also included because they provide important evidence bearing on the history of the Thames system. The justification for a GCR volume devoted to the Thames lies in the special importance of the river in the British Quaternary. The volume is concerned with the history of Britain over approximately the last two million years of geological time, during the repeated glacial phases of the Quaternary 'Ice Age' and the warmer intervals between them. Many such climatic fluctuations are recorded in the deposits of the Thames, which can be shown to have existed as the predominant west to east drainage line in south-eastern England throughout the Pleistocene.

The Thames has left a detailed record of its earlier presence in the form of deposits, which comprise fluvially aggraded floodplain sediments (predominantly gravels) preserved on the valley sides, where they form geomorphological terrace features. The oldest deposits are at the highest levels, forming a terrace 'staircase' that records successive stages in the evolution of the valley. Thames deposits are also found in areas no longer drained by the river and are a testament to the dramatic changes that have occurred in the catchment and courses of the Thames and its tributaries through time. The deposits contain, at certain localities and stratigraphical levels, the remains of plants and animals. These have been used as the main indicators of past climates, which varied from extreme Arctic cold to conditions at least as temperate as today. In addition, the fluvial and associated sediments have yielded prolific evidence, in the form of flint tools and very rare human bones, for the activities of early humans in southern Britain. There is considerable overlap between the research by geologists and that by archaeologists on the Thames deposits laid down during the last 500,000 years, many of which yield important archaeological evidence that is of considerable value to the geologist as a potential means of dating.

The Thames terrace deposits represent a less complete record than, for example, the considerable accumulation of sediment that underlies The Netherlands, laid down at the edge of the subsiding North Sea Basin by the Rhine, but they provide what is probably the most detailed terrestrial Pleistocene sequence in the British Isles. The neighbouring East Anglia region, better established as a rich source of Quaternary information (much of which is marine) and the home of several Pleistocene stratotypes, has a less complete record of the cold stages, which are represented by the gravel terraces of the Thames. Correlation of the Thames sequence with those of East Anglia and The Netherlands is in its early stages, but promises major advances in the near future.

The Thames sequence, although based on a terrace system in which down-cutting events and periods of non-deposition account for a significant (but unknown) proportion of the time represented, nevertheless benefits from a degree of continuity, in that it derives from a single continuously existing source, the river itself. It therefore promises, amongst the British terrestrial sequences, to be the most readily correlated with the global oceanic record. In recent years an oxygen isotope stratigraphy, providing a detailed record of climatic fluctuations during the Pleistocene, has been reconstructed from the more or less continuous sequence of deep-sea sediments. Its use as a chronostratigraphical and 'climatostratigraphical' standard provides a basis for correlating the often widely separated interglacial and glacial deposits of terrestrial sections throughout the world. Relating the major terrestrial sequences, such as that in the Thames basin, to this standard thus offers the best prospect for global correlation within the Pleistocene. An improved understanding of the biostratigraphical and lithostratigraphical evidence yielded by the terrace sediments, as well as the application of geochemical dating methods, has made it possible to place the Thames sequence more precisely within the emerging global geochronological framework.

A pragmatic approach to the task of selecting Thames Pleistocene sites has been employed. Important sites that allow the demonstration of stratigraphical relations have been selected, as have sites with fossiliferous sediments that provide evidence for dating the Thames sequence, as well as sites yielding significant Palaeolithic evidence. To these have been added a number of sites that are important for tracing the early evolution of the river. No attempt has been made to grade sites according to their relative significance.

By the very nature of the geological record, it is clear that some sites will represent unique occurrences for which no parallels exist, while others are merely the best available sites in deposits of extensive and widespread interest, for which alternative localities could be substituted at a later date. It is important for those engaged in geological conservation to recognize the difference between these two types of site. On the one hand careful management and maximum protection are required, while on the other hand there is a need to recognize when a designated locality is no longer a key site and has been superseded. The latter calls for a degree of flexibility in the provision of geological sites, so that the coverage can be updated to keep abreast with both advances in the science (from continuing research) and with the creation of new sites by continued quarrying activity.

Accounts of the evolution of the Thames, as evidenced by its terrace sequence, have been required reading for generations of students. Many treatises on the Thames deposits and their fossils have become classics of the scientific literature; the Swanscombe Skull, for example, has been of particular fascination for scientists and lay readers alike. The story of the blocking and diversion of the Thames by an ice sheet some 450,000 years ago is fundamental to an understanding of the shaping of southern Britain. This volume brings together, for the first time, all the themes and debates that have figured in the many attempts to reconstruct the history of this important river.

A general introduction to the Quaternary, aimed at the less specialist reader, but also providing a succinct overview for the earth scientist, is given elsewhere in this series (Campbell and Bowen, 1989).

D.R. Bridgland and W.A. Wimbledon

Chapter 1

The Pleistocene of the Thames

THE QUATERNARY RECORD OF THE RIVER THAMES

Few rivers of so modest a size have had so much attention devoted to their deposits and geological history as the Thames. The present size of the river belies its importance to British Quaternary geology, however, since Thames sediments provide a framework for this the latest part of the geological record in Britain. Indeed, there is abundant evidence to show that the modern Thames is a mere shadow of its Pleistocene forebear. Not only did it once flow from the London Basin out across East Anglia to north Norfolk, but there are indications that its headwaters may once have drained a large part of the West Midlands and even North Wales (Fig. 1.1). During Middle and Late Pleistocene cold episodes, when sea level was much lower and Britain was joined to the continent, the river extended for many kilometres over areas that now lie offshore; during these episodes it was a tributary of the Rhine system. Thus Thames deposits provide a potential means for correlation between the London Basin and other areas of Britain, as well as with the North Sea Basin and surrounding parts of the European continent.

The river has undergone a number of significant changes during the Pleistocene, some as a result of glaciation. It appears initially to have been a 'consequent stream', flowing along the approximate centre of the London syncline and receiving tributaries from the north and south. Many of the latter would have been of even greater antiquity, having originated on the newly formed Weald and Chiltern uplands and draining into the sea that covered the London Basin during the Palaeogene. The central drainage of the western end of the London Basin is the province of the River Kennet, the Thames upstream from Reading probably having originated as an early river flowing southwards into the basin down the northern limb of the syncline. The presence of quartz pebbles in the Palaeocene deposits of Buckinghamshire suggests that a fluvial route through the Chilterns, tapping pre-Cretaceous strata beyond, was in existence by that time (see Chapter 3, Part 1). The relative lowering of sea level that brought about marine regression in the late Neogene was responsible for the initiation of the Middle and Lower Thames as the main drainage line along the emergent sea-floor.

It appears that, by the Early Pleistocene, the Thames had acquired a much more extensive catchment, with Midlands and Welsh detrital material being brought into the London Basin via the Upper Thames system. There is still controversy as to whether the Thames once drained these areas directly, or whether glacial transport carried this 'exotic' material into its catchment during the Early Pleistocene. It is apparent that this supply ended by the Middle Pleistocene and that the Upper Thames was at that time confined to the south-east of the Cotswold escarpment (see Chapter 2).

THE THAMES TERRACE SYSTEM

Research on the Thames terrace system first began over a hundred years ago, although there was relatively little progress until the early years of this century. The description by Trimmer (1853) of a gravel terrace 150 ft (46 m) above tidal level at Dartford is one of the earliest such records from the Thames. But the 'type area' for the Thames terrace sequence was established at about the turn of the century as the region in the Middle Thames around Maidenhead, Slough and Beaconsfield, where the lower parts of the succession were recognized and defined by officers of the Geological Survey (see Chapter 3, Part 3). The Survey applied this scheme throughout the Middle and Lower Thames valleys, classifying the terrace gravels as Boyn Hill, Taplow or Floodplain Gravel. These names replaced what had become a confusion of terms based on height above river level, relative elevation (High, Middle and Low terraces) or numerical schemes (First, Second etc. terraces) in which numbering from high to low or vice versa was employed by different authors (for summary, see Gibbard, 1985).

The system of named terraces accounted only for the 'valley gravels', which extend up to *c.* 40 m above the modern floodplain. Deposits above this level were generally mapped as 'Glacial Gravel' or, higher still, 'Pebble Gravel', the latter being regarded by most authors as marine rather than fluviatile (see Chapter 3). The view that many of these high-level deposits are the remains of older terraces of the Thames was already in existence by the turn of the century and has progressively gained widespread acceptance, although there is still a degree of controversy over whether the highest gravels may be marginal marine deposits (see Chapter 3, Parts 1 and 2). The full terrace sequence, as recognized at present, has been built up over the last sixty years by additions to the original Geological Survey scheme, the most important work being by S.W. Wooldridge, F.K. Hare, R.W. Hey and P.L. Gibbard (see Chapter 3). Prior to the 1960s, work on the Middle Thames was generally based on geomorphological mapping, which led to the recognition of

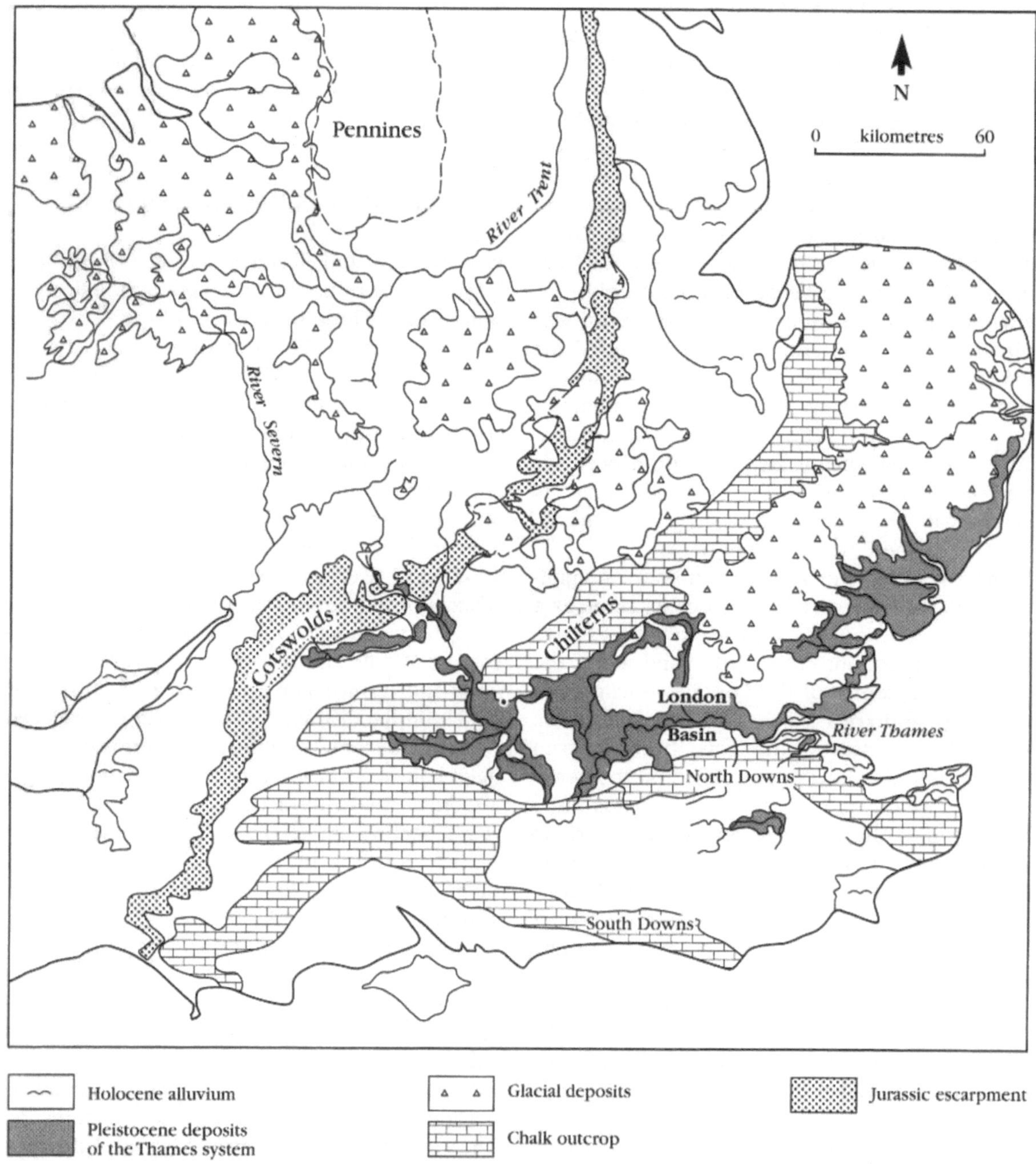

Figure 1.1 Map of southern and central England, showing the division into the catchments of the modern Thames, Severn and Trent rivers. As is described in the text, the area to the north-west of the Cotswolds escarpment was probably drained by the Thames in the Early Pleistocene. In the early Middle Pleistocene it was drained by the Trent system (the proto-Soar of Shotton, 1953).

further terraces both between and above those of the original scheme. These were (in stratigraphical order) the Harefield, Rassler, Winter Hill and Black Park Terraces, all above the Boyn Hill, and the Lynch Hill Terrace, between the Boyn Hill and the Taplow (Hare, 1947; Sealy and Sealy, 1956; see below and Fig. 3.2). Additional, less well-developed terrace features have been described at various times, but their existence as products of fluvial aggradation has not been confirmed.

Later studies have concentrated on the aggradational deposits themselves, rather than the terrace 'flats' formed by their upper surfaces. The latter are geomorphological features and thus may be recognized independently from the sediment bodies underlying them, which formed as a result of aggradation by the river from an earlier, lower base level to the level of the terrace surface. The study of gravel remnants, as opposed to terrace geomorphology, allowed several earlier Thames aggradations to be recognized, despite the fact that the deposits laid down during these events are now highly degraded and fragmentary. Thus the highest and earliest elements of the sequence, the Nettlebed, Stoke Row, Westland Green, Satwell, Beaconsfield and Gerrards Cross Gravels, have all been defined in recent years (see Chapter 3). A refined classification of terrace deposits, based on lithostratigraphy, was pioneered in East Anglia (Rose *et al.*, 1976; Rose and Allen, 1977) and has been applied to the sequences in the Middle Thames (Gibbard, 1985; Chapter 3), in the Lower Thames (Bridgland, 1988a; Gibbard *et al.*, 1988; Chapter 4) and in Essex (Bridgland, 1988a; Chapter 5).

Work in the Upper Thames basin (see Chapter 2) has generally been carried out entirely in isolation from that downstream of the Chiltern escarpment. The Upper Thames represents an important link between the London Basin and the Midlands, both of which have complex Quaternary sequences. Correlation of Pleistocene deposits between these areas has been hampered by the difficulty in tracing terrace aggradations between the Upper and Middle Thames valleys through the constriction of the Goring Gap, where terrace preservation is minimal. Biostratigraphical evidence from fossiliferous sites in the Upper Thames valley has provided some grounds for correlation, but the interpretation of all the key localities has been the subject of controversy (see Chapter 2). An important marker within the Upper Thames sequence is provided by the first appearance of glacially derived material, mainly fresh flint, which was introduced through gaps in the Cotswolds escarpment. The glaciation that carried flint to the Cotswolds has been attributed to the Wolstonian (Saalian) Stage (Shotton, 1973a). However, recent work in the Midlands has raised important questions about the established chronology in that area, suggesting in particular that the glaciation of the Cotswolds occurred during the Anglian Stage (see below; Table 1.1; Chapter 2).

Several important review articles have described the Thames sequence, from different viewpoints, in recent years. The most significant and useful of these are by Evans (1971), Brown (1975), Hey (1976a), Clayton (1977), Green and McGregor (1980) and Gibbard (1983). Several major works have described and discussed the formation of the Thames terraces (Wooldridge and Linton, 1939, 1955; Zeuner, 1945, 1959; Wymer, 1968; Gibbard, 1985).

THE DIVERSION OF THE THAMES

The suggestion that the Thames formerly flowed to the north of London, through the Vale of St Albans, dates back to the turn of the century, when this old drainage route was recognized by Salter (1905). Sherlock and Noble (1912) considered that glaciation of the Vale of St Albans ended Thames drainage by this northern route. Wooldridge (1938, 1960), however, recognized two previous courses of the Thames, the first being that recognized by Salter and the second an intermediate route through the Finchley area (see Chapter 3, Part 2). He suggested that separate glacial advances were responsible for the demise of both routes, the first as a result of a 'Chiltern Drift' ice advance and the second in response to the main 'Chalky Boulder Clay' (Anglian Stage) glaciation. Recent workers have found little evidence to support the existence of a pre-Anglian glaciation in the London Basin (Moffat and Catt, 1982; Avery and Catt, 1983; Green and McGregor, 1983), but have determined that the Thames flowed through the Vale of St Albans until blocked by ice during the Anglian Stage (Gibbard, 1977; Green and McGregor, 1978a; Cheshire, 1986a; see Chapter 3, Part 2). In addition, Gibbard (1979) has demonstrated that the Thames was diverted from the Vale of St Albans directly into its modern valley without using Wooldridge's intermediate route, which in fact represents the pre-Anglian valley of the Mole-Wey tributary (see Chapter 3).

Gibbard (1977) showed that the Anglian glacial deposits of the Vale of St Albans provide a stratigraphical marker that enables correlation of the Thames terrace sequence with the established British Pleistocene chronology. He demonstrated that the Winter Hill Gravel of the Middle Thames was accumulating at the time of the glacial advance, thus implying an Anglian age for that formation. The next terrace aggradation (topographically lower) in the sequence, the Black Park Gravel, is also regarded as contemporaneous with the occupation of the Vale of St Albans by ice, but this formation has been traced into the modern Thames valley through London, whereas the Winter Hill Formation can be traced downstream into the Westmill Lower Gravel of the Vale of St Albans, which underlies Anglian glacial deposits. This indicates that the diversion of the Thames occurred between the aggradation of the Winter Hill and Black Park Terrace deposits, at the time of the Anglian glacial maximum (see Chapter 3, Part 2).

THE CONTINUATION OF THE THAMES INTO ESSEX

The first evidence for the route of the pre-diversion Thames downstream from the Vale of St Albans was derived from sub-drift contour mapping by Wooldridge and Henderson (1955), who recognized in their 'Mid Essex Depression' a buried valley system extending eastwards from Ware to Bishops Stortford, Chelmsford and Colchester. Detailed sub-drift mapping of parts of this area was subsequently published in various Geological Survey (Mineral Assessment Unit) reports, which include exhaustive borehole records but no regional interpretation. Green *et al.* (1982) suggested correlations of isolated gravel samples from this area with the Middle Thames sequence on the basis of 'clast-lithological analysis' (which is the identification of the clasts that make up a deposit and the calculation of their relative proportions – see Bridgland, 1986a), but their work lacked the added control of geological mapping. The first attempt to combine both types of evidence and to trace terrace aggradations through this area by reconstructing their downstream profiles has recently been made by Whiteman (1990; see below and Chapter 5).

Confirmation of the extension of the pre-diversion Thames system into East Anglia came with the recognition of early Thames gravels in Essex, Suffolk and Norfolk, important early contributions to this work being made by S.H. Warren, R.W. Hey, J. Rose and P. Allen (see Chapter 5). The distribution of these various aggradations, which make up the Kesgrave Sands and Gravels (Rose *et al.*, 1976), reflects a progressive southward migration of the pre-diversion Thames valley (see Chapter 5, Part 1). This migration culminated in a course through Chelmsford and Colchester, recognized in the lowest part of the 'Mid Essex Depression'. Gravel representing this final pre-diversion floodplain has been traced to the Clacton area, where it forms the lowest of a sequence of Kesgrave Group formations and is in close juxtaposition with post-diversion Thames deposits (Bridgland, 1988a; see Chapter 5, Part 1). The distinction in this area between pre- and post-diversion gravels is based on differences in their clast composition, assisted by the recognition of an influx of distal outwash from the Anglian glaciation at the end of the last pre-diversion aggradational phase. Using the stratigraphical evidence for the glaciation and the diversion of the Thames, the sequence in the Clacton area can be correlated directly with that in the Middle Thames and the Vale of St Albans (Bridgland, 1988a; see Chapter 5).

The post-diversion terrace formations of the Lower Thames continue into eastern Essex, where they make up the Low-level East Essex Gravel Subgroup (Bridgland, 1988a). These formations extend northwards from the Southend district, running approximately parallel to the coast (see Chapter 5, Part 2), but only the oldest (immediately post-dating the diversion) reaches the Clacton area. Earlier terrace deposits are preserved in south-eastern Essex, however, representing formations within a High-level East Essex Gravel Subgroup, but these are products of the River Medway that pre-date the diversion

Table 1.1 Correlation of Quaternary deposits within the Thames system. Rejuvenations that have occurred since the Anglian glaciation are indicated.

The continuation of the Thames into Essex

Age (in thousands of years)	Upper Thames	Middle Thames	Lower Thames	Essex	Stage	^{18}O
	– – – – – – – Recent floodplain and channel deposits: Holocene alluvium of floodplain and coast – – – – – – – – –				Holocene	1
10						
	Northmoor Gravel	Shepperton Gravel	Submerged	Submerged	late Devensian	2-4
	– – – – – – – – – – – – *Rejuvenation event* – – – – – – – – – – – –					
71						
		Temperate climate deposits at South Kensington (Ismaili centre), Isleworth and Kempton Park	Submerged	Submerged	early/mid-Devensian? interstadial(s)	5a &/or 5c?
?						
	Cold climate gravels above Eynsham Gravel	**Reading area** U. levels of Taplow Gravel; **Slough area** Kempton Park Gravel	East Tilbury Marshes Gravel	Submerged	early-mid-Devensian	5d–2
122						
	Eynsham Gravel	Within Taplow Formation; Trafalgar Square and Brentford deposits	Below floodplain	Submerged	Ipswichian (*sensu* Trafalgar Square)	5e
128						
	Stanton Harcourt Gravel	Taplow Gravel; Basal Kempton Pk Gravel - incl. Spring Gardens Gravel of Gibbard (1985)	Basal East Tilbury Marshes Gravel	Submerged	late Saalian	6
		– – – – *Rejuvenation event* – – – – – – – – – – – –				
		Taplow Gravel	Mucking Gravel			
186						
	Stanton Harcourt Channel Deposits, interglacial Magdalen Grove, Summertown etc.	Interglacial deposits at Redlands Pit, Reading	Interglacial deposits at Aveley, Ilford (Uphall Pit), West Thurrock, Crayford and Northfleet	Submerged	Intra-Saalian temperate episode	7
245						
	Basal Summertown-Radley Formation at some sites?	Basal Taplow Gravel?	Basal Mucking Gravel	Submerged		
	– – – – – – – – – – – – *Rejuvenation event* – – – – – – – – – – – –				mid-Saalian	8
	Wolvercote Gravel at some sites?	Lynch Hill Gravel	Corbets Tey Gravel	Barling Gravel		
303						
	Wolvercote Channel Deposits		Interglacial deposits at Ilford (Cauliflower Pit), Belhus Park, Purfleet and Grays	Shoeburyness Channel interglacial deposits	Intra-Saalian temperate episode	9
339						
	Basal Wolvercote Gravel	Basal Lynch Hill Gravel?	Basal Corbets Tey Gravel	Shoeburyness Channel - basal gravel		
	– – – – – – – – – – – – *Rejuvenation event* – – – – – – – – – – – –				early Saalian	10
	? Moreton Drift (Arkell, 1947a)					
	Hanborough Gravel	Boyn Hill Gravel	Orsett Heath Gravel	Southchurch/Asheldham/ Mersea Island/Wigborough Gravel		
?						
	Reworked mammalian fauna in Hanborough Gravel		Swanscombe deposits	Southend/Asheldham/ Cudmore Grove/Clacton Channel Deposits	Hoxnian (*sensu* Swanscombe)	11
423						
	Basal Hanborough Gravel?	Basal Boyn Hill Gravel?	} Basal Orsett Heath Gravel (incl. Basal Gravel at Swanscombe)	Southend/Asheldham/ Cudmore Grove/Clacton Channel - basal gravel		
	– – – – – – – – – *Rejuvenation event* – – – – – – – – –					
	Freeland Formation	Black Park gravel				
	Moreton Drift?	Anglian glacial deposits	Hornchurch Till	U.St Osyth/U. Holland Gravel	Anglian	12
	Freeland Formation	Winter Hill/Westmill Gravel	Valley did not exist as a Thames course prior to this	St Osyth/Holland Formation		
478						
	Sugworth Channel Deposits	Rassler Gravel?		Wivenhoe/Cooks Green Fm Ardleigh/St Osyth Formation Waldringfield Gravel	Cromerian Complex	21–13
?						
	Combe Formation	Gerrards Cross Gravel		Bures Gravel*	Early Pleistocene	pre–21
	Higher divisions of the Northern Drift Group	Beaconsfield Gravel Satwell Gravel ?gravel at Chorleywood Westland Green Gravels Stoke Row Gravel Nettlebed Gravel Nettlebed interglacial deposits		Moreton Gravel* Stebbing Gravel*		

* Nomenclature for High-level Kesgrave Subgroup in Essex follows Whiteman (1990).

of the Thames. The sequence in eastern Essex therefore indicates that the Medway formerly flowed northwards across this area, towards its confluence with the pre-diversion Thames. This early Medway valley was never glaciated and continued to operate throughout the Anglian glaciation. It is apparent that the newly diverted Thames took over the former Medway route between Southend and Clacton, where it rejoined its former valley (Bridgland, 1988a). In the southern North Sea, Anglian (Elsterian) ice is thought to have ponded the waters of the Thames-Medway and the continental Rhine system, forming a huge proglacial lake (see Gibbard, 1988a). This lake eventually spilled over the watershed that then existed between North Sea drainage and English Channel drainage, to form a through valley, flowing southwestwards, in the position of the present-day Straits of Dover. After the Anglian Stage it appears that, during phases of low sea level, the Rhine and Thames united in the area of the southern North Sea and flowed south-westwards through the English Channel to the Atlantic Ocean (Gibbard, 1988a).

Later onshore terrace formations record the continued southward migration of the Thames. This migration culminated in the valley system, submerged by the Holocene transgression, that has been mapped beneath modern marine sediments on the floor of the southern North Sea using seismic techniques (D'Olier, 1975; Bridgland and D'Olier, 1989). Following the diversion, and as the Thames migrated southwards, the modern drainage system of Essex became established, rivers such as the Blackwater and Crouch being initiated as left-bank tributaries of the Thames. Later Pleistocene deposits in the areas of northern, central and eastern Essex formerly drained by the Thames are invariably the products of these tributary rivers (see Chapter 5, Part 3).

CLASSIFICATION OF THAMES TERRACES AND DEPOSITS

Two types of classification are used in parallel in this volume; one for the series of morphological features, the terrace surfaces, and the other for the deposits that the Thames and its tributaries have laid down during the Pleistocene. The terrace surfaces are geomorphological features formed by alternating periods of fluvial aggradation and downcutting, the surfaces representing former floodplains that were abandoned as a result of rejuvenation. Terrace surfaces can also be formed by erosion, but features of this type are almost entirely unrepresented in the Thames system. Thus, wherever morphological terrace features are recognized in the Thames valley, the parallel geological classification of the deposits can also be applied. In contrast, remnants of fluvial deposits can be recognized in areas where severe post-depositional modification, such as erosion or burial by later sediments, prevents the recognition of original terrace surfaces. Consequently, the geological (lithostratigraphical) method of classification is generally more useful.

Geomorphological classification

The system of naming terraces after type localities is favoured here, rather than numbering or other alternatives (see above). The established sequence of Thames terrace features is as follows (see also Fig. 3.2):

Terrace	**Underlying lithostratigraphical formation**
9. Lower Floodplain	Shepperton Gravel
8. Upper Floodplain	Kempton Park Gravel
7. Taplow	Taplow Gravel
6. Lynch Hill	Lynch Hill Gravel
5. Boyn Hill	Boyn Hill Gravel
4. Black Park	Black Park Gravel
3. Winter Hill	Winter Hill Gravel
2. Rassler	Rassler Gravel
1. Harefield	Gerrards Cross Gravel

Lithostratigraphical classification

Formal lithostratigraphical classification was first applied to the type-sequence of Middle Thames terrace deposits by Gibbard (1983, 1985), although a number of earlier workers had classified the sediments rather than the terrace surfaces, particularly when describing older, dissected aggradations (Wooldridge, 1938; Hey, 1965). Gibbard adopted the scheme for lithostratigraphical nomenclature recommended by the stratigraphical guides (Hedberg, 1976;

Holland *et al.*, 1978), in which the following hierarchical divisions are applied:

Group	Two or more formations
Formation	The primary unit, into which the entire stratigraphical column is divided
Member	Named unit within a formation
Bed	Named distinctive layer within a formation or member

Gibbard considered individual terrace aggradations to be of member status and grouped these into a 'Middle Thames Valley Gravel Formation'. He sought to differentiate between his various members on the basis of clast-lithological differences, applying techniques of statistical evaluation (Gibbard, 1985, 1986). A similar method was adopted by McGregor and Green (1986), who also regarded the study of clast lithology as a lithostratigraphical method but did not advocate formal nomenclature. Hey (1986) adopted the Gibbard model in subdividing the Northern Drift of the Upper Thames (see Chapter 2), again on the basis of clast-lithological differences. However, Bridgland (1988b, 1990a), citing the various stratigraphical guides, observed that lithostratigraphical classification should be based on gross lithological properties rather than on laboratory techniques such as clast-lithological analysis. The former properties may include major breaks in sedimentary continuity (unconformities), such as occur at the base and top of each individual terrace aggradation. These allow the individual aggradations to be separated using basic techniques of geological mapping (including geomorphological mapping), thus making them primary units. For this reason Bridgland proposed that individual terrace aggradations should be classified as formations, rather than members. In some cases several of these can be collected together to form a group, for which no single type locality is necessary, although other lithostratigraphical units must be defined at a type locality. Members and beds may be defined within some of the primary units: for example, the Swanscombe Lower Gravel and other Swanscombe units (see Chapter 4) are members or beds within the Boyn Hill/Orsett Heath Gravel Formation, since they fall within the body of sediment that can be mapped as the Boyn Hill/Orsett Heath Gravel (see Chapter 4).

Although the Middle Thames is regarded as the type area for the Thames sequence, recent work in Essex has revealed that there are a number of terrace formations in that area for which no Middle Thames equivalent is recognized (Bridgland, 1988a; Whiteman, 1990; see Chapter 5). It is therefore necessary to refer to type localities outside the Thames valley to define these formations. In fact, separate nomenclature has been established in recent years for the Lower Thames, Essex and southern East Anglia and has been in existence since the earliest research in the Upper Thames (where the nomenclature used in the London Basin has never been applied). The overriding reason for this proliferation of names is that correlation between these various areas has been problematic, usually because of breaks in the continuity of recognizable terrace remnants. The most important breaks of this type coincide with the Goring Gap, central London and the Essex till sheet; in the first of these there has been little preservation of terrace deposits because of the constriction of the valley, whereas in the other two the evidence is preserved but largely inaccessible. Once correlation is on a sounder footing, however, it would be desirable to suppress synonyms and use a single nomenclature, probably that established in the Middle Thames (see Table 1.1).

THE STRATIGRAPHICAL FRAMEWORK: PLEISTOCENE CHRONOSTRATIGRAPHY AND CORRELATION

The interpretation of the floral and faunal content of temperate deposits, taken together with geological evidence for the deposition of other sediments under intensely cold or even glacial conditions, has been used as a basis for 'climato-stratigraphical' subdivision of Pleistocene time. During the middle part of this century a scheme for climato-stratigraphical (relative) dating of the British Pleistocene succession was established using palynology (West, 1963, 1968). This scheme, which is still in use, distinguishes different interglacial episodes on the basis of their distinctive patterns of vegetational development, as determined by the analysis of pollen assemblages from successive horizons within depositional sequences. Moreover, climatic fluctuation has been accepted as a guide for the division of Pleistocene time (Shotton, 1973b), so that time periods corresponding with interglacials and glacials have been defined as chronostratigraphical stages. Thus palynological analyses allowed pollen-bearing sequences to be allocated to particular stages, enabling a chronostratigraphy for the British Pleistocene to be developed, as summarized by Mitchell *et al.* (1973). The following stages were recognized by these authors in the sequence post-dating the marine crags of East Anglia (in stratigraphical order):

Holocene = Flandrian (warm)
Devensian (cold)
Ipswichian (warm)
Wolstonian (cold)
Hoxnian (warm)
Anglian (cold)
Cromerian (warm)
Beestonian (cold)
Pastonian (warm)
Baventian (cold)*

* most recent stage within the Norwich Crag, the youngest marine crag.

Mitchell *et al.* recognized two interglacials between the Anglian Stage and the Holocene. These correspond with the Hoxnian Stage (type locality at Hoxne in Suffolk) and the Ipswichian Stage (type locality at Bobbitshole, near Ipswich, Suffolk). The Ipswichian and the Holocene were separated by the last glaciation, within the Devensian Stage. The time interval between the Hoxnian and Ipswichian Stages was ascribed by Mitchell *et al.* (1973) to the Wolstonian Stage, which replaced the Gipping Stage of West (1963). This stage, regarded by Mitchell *et al.* as a cold episode, was defined at Wolston, Warwickshire, where a detailed sequence of gravels, sands, clay and till occurs, entirely ascribed to the Wolstonian (Shotton, 1973a, 1973b). The absence of earlier or later sediments at Wolston underlines the acknowledged difficulties in relating the type-Wolstonian deposits to the local Pleistocene sequence, let alone the British climatostratigraphical scheme (Bowen, 1978). In recent years it has been suggested that the glacial sequence at Wolston is in fact equivalent to the Anglian Stage deposits of East Anglia, therefore invalidating the term Wolstonian (Perrin *et al.*, 1979; Sumbler, 1983a, 1983b; Rose, 1987, 1988, 1989, 1991). Furthermore, there is mounting evidence from other research that more than a single climatic cycle separates the Hoxnian and Ipswichian interglacials (see below and Table 1.1). As no redefinition of this time interval has been forthcoming, the corresponding continental term, Saalian, is used in this volume. This is not without some problems, as the definition and subdivision of the European Saalian Stage is also under scrutiny at present. It is widely accepted, however, that this stage incorporates a series of climatic fluctuations separating the continental equivalents of the Hoxnian and Ipswichian, the Holsteinian and Eemian Stages respectively (Zagwijn, 1985, 1986; Bowen *et al.*, 1986b; Sibrava, 1986a, 1986b; de Jong, 1988; see below). Whatever the result of any review of the continental Saalian, the term is used in this volume for the time interval between the Swanscombe interglacial (Hoxnian *sensu* Swanscombe) and the Ipswichian *sensu* Trafalgar Square (for definitions of these, see below and Table 1.1).

The pollen-based 'climato-stratigraphical' model has formed the basis for British Pleistocene studies since its inception, other biostratigraphical evidence generally being related to the palynological sequence. Within the Thames system, well-documented interglacial sites at Swanscombe and Trafalgar Square have been attributed to the Hoxnian and Ipswichian respectively, providing a biostratigraphical framework for Thames terrace stratigraphy that continues to be used. However, the relative dating of various sites within the Lower Thames terrace sequence at intermediate heights between Swanscombe (23–30 m O.D.) and Trafalgar Square (around Ordnance Datum) has proved controversial (see Chapter 4). Misgivings have been expressed about the above model on the basis of discrepancies between the palynological record and the evidence from Pleistocene mammals (Sutcliffe, 1964, 1975, 1976, 1985; Shotton, 1983; Green *et al.*, 1984). At several sites, including examples in the Lower Thames (see Chapter 4, Aveley), different mammalian faunas have been found in deposits that appeared, from their palynology, to represent the Ipswichian Stage. This led to the suggestion,

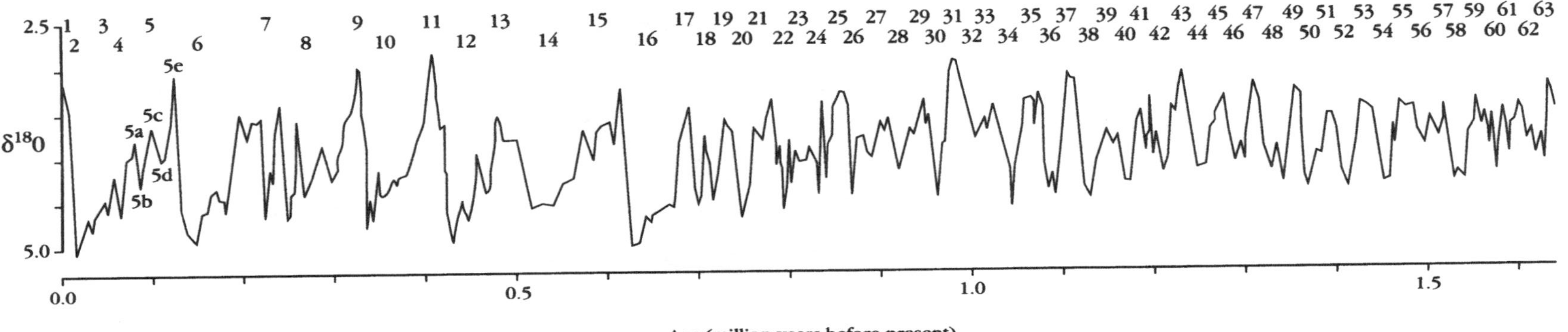

Figure 1.2 The oxygen isotope record, as represented in a borehole (Site 607) in the mid-Atlantic at latitude *c.* 41°N. Numbered stages are shown at the top; even-numbered ones are relatively cold (more ice) and odd-numbered ones relatively warm (less ice). Note that the amplitude and wavelength of the curve increases at around 0.7 million years ago (the $\delta^{18}O$ scale is a ratio obtained by comparing the proportion of ^{18}O to ^{16}O in samples to that in a mean sea-water standard). Compiled from data published by Ruddiman *et al.* (1989).

first made by A.J. Sutcliffe, that there had been two separate interglacials since the Hoxnian. These had similar patterns of vegetational development and are therefore indistinguishable on the basis of palynology. The morphological separation of apparent Ipswichian sites in the Lower Thames within different terraces has been cited in support of this suggestion (see Chapter 4). Corroboration has come from radiometric dating of bone and travertine (Szabo and Collins, 1975; Green *et al.*, 1984), from studies of molluscan palaeontology (Allen, 1977; see Chapter 4, Purfleet) and from the analysis of the amino acid content of these same fossils (Miller *et al.*, 1979; Bowen *et al.*, 1989; see below). Stratigraphical evidence from the Warwickshire Avon (Bridgland *et al.*, 1989; Maddy, 1989; Maddy *et al.*, 1991a) goes even further, in suggesting that two additional climatic cycles occurred between the Anglian and Ipswichian Stages.

With the current uncertainty about the chronological significance of the palynological framework, although pollen analyses remain critically important in the study of Quaternary sites, the search for an alternative scheme has occupied many workers in the past few years. An obvious basis for a stratigraphical standard is the record provided by ocean-bed sediments, since these are from environments where sedimentation during the Pleistocene is presumed to have been uninterrupted. The pattern of climatic fluctuations during the Pleistocene has been recognized in these sediments from the study of oxygen isotopes (^{16}O and ^{18}O) in the remains of foraminifera (Emiliani, 1955, 1957; Shackleton, 1969; Shackleton and Opdyke, 1973). Changes in the relative frequencies of the isotopes reflect parallel changes in the isotopic composition of sea-water in direct response to fluctuations of global climate through time. The lighter isotope (^{16}O) is relatively common in water evaporated from the oceans, so during cold episodes, when much of this water goes to form long-term accumulations of ice, the sea water becomes enriched in the heavy isotope (^{18}O). Fluctuations in the relative frequencies of these two isotopes have been plotted against time, as represented by the oceanic sedimentary record, to provide a graphic representation of climatic change through the Pleistocene, the 'oxygen isotope curve' (Fig. 1.2). The climatic fluctuations observed in this curve also form the basis for 'oxygen isotope stratigraphy', in which a scheme of 'oxygen isotope stages' is recognized. These stages represent alternate cold and warm episodes and are numbered in reverse stratigraphical sequence from Stage 1, the equivalent of the Holocene (Fig. 1.2 and Table 1.1; see Bowen, 1978, for explanation). Thus even-numbered ^{18}O stages represent cold episodes and odd-numbered stages represent warm intervals. It must be emphasized that the oxygen isotope record can only indicate an episode of ice-cap depletion and gives no direct indication of global climate or vegetational development. Nevertheless, the oceanic record affords a truly global framework for Pleistocene chronology, since the same sequence can be recognized in oceans throughout the world. More problematic, however, is the correlation of the deep-sea record with discontinuous terrestrial sequences, for which radiometric dating methods are rarely available. On land, relative dating methods are generally used, relying on land-based fauna and/or flora that cannot be related directly to the oceanic stratigraphy.

One approach to this problem that has been pursued in recent years is an attempt to establish a stratigraphy based on a relative dating method in which the progressive alteration of amino acids in mollusc shells is measured. The amino acid L-isoleucine epimerizes to form D-alloisoleucine progressively through time, although the rate of change is dependent on temperature. During life only L-isoleucine is present in a shell, but after death and incorporation in sediments, progressive epimerization causes increasing amounts of D-alloisoleucine to be present, until an equilibrium level is reached. Therefore the ratio between these two amino acids gives an indication of relative age, which can be calibrated using radiometric dating where available (Wehmiller, 1982; Bowen *et al.*, 1985, 1989). This method has been applied to shells from a number of important sites in the Thames system (Miller *et al.*, 1979; Bowen *et al.*, 1989).

The desirability of relating terrestrial sequences such as that of the Thames terraces to the oxygen isotope record is obvious. However, apart from a general consensus that the Ipswichian interglacial represents Oxygen Isotope Substage 5e (Gascoyne *et al.*, 1981; Shotton, 1983; Bowen *et al.*, 1985; Stringer *et al.*, 1986; Bowen and Sykes, 1988; Campbell and Bowen, 1989), there has been little agreement about correlation between the British and European sequences and the deep-sea cores. Of critical importance is the position, in relation to the oxygen isotope record, of the Anglian Stage, which is well-established in the British stratigraphy and particularly important within the Thames sequence. The Anglian Stage has generally been correlated with Oxygen Isotope Stage 12, primarily because this appears to represent one of the most severe cold episodes within the Middle Pleistocene deep-sea record (Shackleton and Opdyke, 1973; Shackleton, 1987; Bowen *et al.*, 1986a, 1986b; Bowen and Sykes, 1988; Campbell and Bowen, 1989). Support for this view is provided in this volume from the number of post-diversion (post-Anglian) climatic fluctuations recognized within the Lower Thames sequence (see Chapter 4).

Evidence from amino acid ratios suggests that the Hoxnian Stage interglacial, as represented at Swanscombe and Clacton (see respective site descriptions), equates with Oxygen Isotope Stage 11 (Bowen *et al.*, 1986a, 1989; Table 1.1), but similar evidence from Hoxne suggests correlation with Stage 9 (Bowen *et al.*, 1989). If these ratios provide an accurate indication of age, the Thames sites that have traditionally been regarded as Hoxnian may prove to belong to a hitherto undefined temperate episode that followed immediately after the Anglian glaciation, whereas later temperate deposits within the Thames sequence might be true equivalents of the sequence at Hoxne (see Chapter 4). However, the type-Hoxnian sequence has been interpreted as a kettle-hole infill, formed immediately following deglaciation late in the Anglian Stage (West, 1956). Thus, if the Anglian equates with Stage 12, it is difficult to envisage the type-Hoxnian sediments post-dating Stage 11. For the deposits at Swanscombe and Clacton, which unequivocally post-date the diversion of the Thames, to equate with Stage 11 and those at Hoxne with Stage 9, the till underlying the latter would have to post-date the glaciation that effected the diversion. As the East Anglian glacial deposits have been attributed to a single pre-Hoxnian glaciation (since the post-Hoxnian Gipping glaciation was disproved – see Chapter 5), there is a clear conflict between the interpretation of the Hoxne sequence as a kettle-hole infill and the amino acid ratios from the site (see, however, below and Chapter 2 for discussion of the possibility that a post-Stage 12 glaciation is represented in the Cotswolds). It is perhaps appropriate to question whether the origin of the lake-beds at Hoxne in a kettle-hole can be demonstrated unequivocally. Lake basins have been formed in East Anglia in other ways, as is indicated by the occurrence of several large closed depressions containing modern lakes and/or substantial infills of Holocene sediments. The formation of these features, the 'meres' of central East Anglia, is attributed to solution of bedrock Chalk beneath the cover of Anglian till (Bennett *et al.*, 1991).

It is possible that the amino acid ratios obtained from Hoxne are misleading and that the type-Hoxnian sequence equates with Oxygen Isotope Stage 11 (and is therefore the same age as the Swanscombe deposits). Correlation of the British Hoxnian and continental Holsteinian sequences on the basis of palynology is considered by many workers to be highly convincing, it being possible to match important aspects of vegetational evolution in both (Turner, 1975). If the Hoxnian is taken to equate with Oxygen Isotope Stage 11, Stage 9 remains to be identified within the British terrestrial succession, falling within the interval called 'Wolstonian' by Mitchell *et al.* (1973), and here referred to as the Saalian Stage. The Stage 9 temperate episode is believed to be represented by deposits in the Upper and Lower Thames sequences and in eastern Essex (Table 1.1). It is notable that shells from sediments at Grays and Purfleet in the Lower Thames, considered from stratigraphical evidence to represent Stage 9, have produced amino acid ratios suggestive of greater antiquity (see Chapter 4). This may indicate that amino acid analyses of shells from deposits older than 200,000 years in the Lower Thames are as yet problematic, although these ratios do provide important evidence in support of the pre-Ipswichian age of the Grays and Purfleet sediments, which remains a subject of some controversy. Shells from a site a few kilometres upstream from Purfleet, at Belhus Park (see Chapter 4, Purfleet), have recently yielded amino acid ratios consistent with a Stage 9 age (Bowen, 1991).

An undefined interglacial, palynologically similar to the type-Ipswichian, is now well-established from several sites in southern and Midland England as representing Oxygen Isotope Stage 7 (Shotton, 1983; Bowen *et al.*, 1989). Many deposits formerly attributed to the Ipswichian may represent this earlier interglacial, which (like the Stage 9 temperate episode) falls within the Saalian (Wolstonian) Stage. Deposits now attributed to Stage 7 include those previously distinguished from Ipswichian sediments on the basis of mammals (Sutcliffe, 1975, 1976, 1985; Green *et al.*, 1984). Separation may also be possible using molluscs (see Chapter 2, Stanton Harcourt and Magdalen Grove). The age of Oxygen Isotope Stage 7 has been estimated at 245,000–186,000 years (Martinson *et al.*, 1987), largely on the basis of dating by the uranium-series method at a number of important sites. Worthy of mention in this respect is Pontnewydd Cave, North Wales, where a range of uranium-series dates, with corroboration using the thermoluminescence dating technique, suggests an age of 225,000–160,000 years for the fossiliferous Lower Breccia (Green, 1984; Campbell and Bowen, 1989). At Marsworth, Buckinghamshire, travertine clasts containing leaf-impressions from interglacial tree species have provided three dates, pointing to an age between 200,000 and 140,000 years (Green *et al.*, 1984). Other uranium-series dates attributed to Stage 7 are from Stoke Goldington, Buckinghamshire, where an age of 200,000–180,000 years is suggested (Green *et al.*, in Bridgland *et al.,* 1989). These dates contrast with indications, again using the uranium-series technique, of an age of 130,000–100,000 years for Ipswichian deposits at Victoria Cave, Settle, North Yorkshire (Gascoyne *et al.*, 1981) and in Minchin Hole and Bacon Hole caves, South Wales (Bowen *et al.*, 1985; Stringer *et al.*, 1986; Bowen and Sykes, 1988; Campbell and Bowen, 1989). A 'marine stratotype' for Oxygen Isotope Stage 7 has been proposed at Minchin Hole (Bowen *et al.,* 1985). At this site, raised beach deposits that are attributed to Stage 7 underlie cold-climate cave deposits. The last-mentioned are in turn overlain by further raised beach sediments from which uranium-series dates indicative of the Ipswichian Stage have been obtained (see above). Temperate-climate sediments attributed to Stage 7 are recognized at several sites within the Thames system (see Chapter 2, Stanton Harcourt and Magdalen Grove; Chapter 4, Lion Pit, Aveley and Baker's Hole). There is support for a Stage 7 age for Thames deposits at Aveley, Crayford and Stanton Harcourt, and for their separation from sites representing the Ipswichian Stage (Substage 5e), from amino acid geochronology (Bowen *et al.*, 1989).

This complex stratigraphical scheme, in which four separate interglacials are recognized between the glaciations of the Anglian and Devensian Stages, receives considerable support from some of the more complete Pleistocene sequences on the continent. There is a long-standing correlation of the British Ipswichian and the continental Eemian Stages and of the British Hoxnian and continental Holsteinian Stages (Mitchell *et al.*, 1973). In recent years 'climatostratigraphical' schemes in a number of European countries have been claimed to include one or more additional temperate episodes between the Holsteinian and Eemian. This interval is classified on the continent as the Saalian, which is the equivalent of the Wolstonian Stage as defined by Mitchell *et al.* (1973). The Saalian succession, named after the River Saale in north Germany, comprises the deposits of two glaciations, the Drenthe and Warthe (for English summaries, see Evans (1971), Bowen (1978), Sibrava (1986a, 1986b) and Gibbard (1988a)). Many German authors have regarded these as separate advances within a single glacial episode (Duphörn *et al.*, 1973; Ehlers, 1981), the only early suggestions to the contrary being based on equivocal geomorphological evidence such as the relative 'freshness' of depositional features (moraines) related to the two ice sheets (Bowen, 1978). There have, however, been recent claims that an intra-Saalian interglacial cycle separated these glaciations, on the basis of evidence from biostratigraphy, sea levels and cycles of soil formation in thick loess sequences (Wiegank, 1972; Kukla, 1975, 1977; Turner, 1975; Brunnacker, 1986; Cepek, 1986; Grube *et al.*, 1986; Sarnthein *et al.*, 1986; Sibrava, 1986a, 1986b). Kukla (1975), Sarnthein *et al.* (1986) and Sibrava (1986a, 1986b) have also presented arguments for a second additional temperate episode between the Holsteinian and Eemian Stages.

Intra-Saalian temperate episodes have been recognized in several areas of northern Europe, predominantly on the grounds of biostratigraphy or from the study of soil horizons within loess sequences. Evidence for a single additional climatic cycle between the Holsteinian and Eemian has been described in Germany, from the Middle Rhine valley and from the type area of the Holsteinian Stage in Schleswig-Holstein (Brunnacker *et al.*, 1982; Brunnacker, 1986; Sarnthein *et al.*, 1986). However, a significant number of areas have now produced evidence for two intra-Saalian temperate episodes. Kukla (1975, 1977) based his recognition of two post-Holsteinian but pre-Eemian interglacials on soils within the loess sequence of central Europe. A similar sequence of palaeosols has been described from Normandy, where four temperate-climate soils are recognized within a thick loess succession at St-Pierre-les-Elbeuf. The lowest of these is correlated with the Holsteinian and the highest with the Eemian (Lautridou *et al.*, 1974, 1983; Lautridou, 1982; Sarnthein *et al.*, 1986). Evidence for two intra-Saalian temperate half-cycles is also recognized in east Germany (Cepek and Erd, 1982; Cepek, 1986; Sarnthein *et al.*, 1986). In their summary of Pleistocene correlation in Europe, Bowen *et al.* (1986b) also recognized three post-Elsterian and pre-Eemian interglacials in Poland, Russia and the Carpathians. The names that have been most commonly applied to the two additional temperate episodes are 'Domnitz' or 'Wacken' for the earlier and 'Treene' for the later interval (Bowen *et al.*, 1986b).

In northern Holland, palynological studies have revealed two temperate episodes within the Saalian sequence, both pre-dating the Drenthe glaciation. These have been named the Hoogeveen and Bantega Interstadials (Zagwijn, 1973, 1986; de Jong, 1988). In the south, at a site near Maastricht, a fully temperate molluscan fauna was described by Meijer (1985), who ascribed it to an intra-Saalian interglacial episode. Although classifying the Hoogeveen and Bantega episodes as interstadials, de Jong (1988) admitted that the former has many of the features of a full interglacial and is difficult to distinguish from the Holsteinian when incomplete fragments of vegetational sequences are studied using palynology. This implies that the fauna described by Meijer might relate to this same episode. Zagwijn (1973) had previously suggested a correlation between the Hoogeveen Interstadial and the 'Wacken Warmzeit' as recognized in Schleswig-Holstein.

The same problems in correlating the terrestrial stratigraphy with the oxygen isotope chronology apply on the continent as in Britain; the absence of reliable dating methods and the frequency of gaps within the sequences on land mean that all correlation schemes of this type are tentative at present. Notwithstanding these reservations, it has been widely agreed that the Eemian Stage and Oxygen Isotope Substage 5e are equivalent (see, for example, Sibrava, 1986b; de Jong, 1988), supporting correlation with the British Ipswichian. However, there is considerable doubt about the oceanic equivalent to the Holsteinian, with stages 7, 9, 11 and 13 all being contenders (Kukla, 1975, 1977; Bowen, 1978; Zagwijn, 1978; Linke *et al.*, 1985; Sarnthein *et al.*, 1986 (provides summary); de Jong, 1988; Grün *et al.*, 1988; Schwarcz and Grün, 1988). De Jong (1988) favoured correlation of the Holsteinian with Stage 9 in a tentative scheme in which he linked the Elsterian glaciation to Stage 10. He considered both the Hoogeveen and Bantega temperate intervals to fall within Oxygen Isotope Stage 7. However, Sarnthein *et al.* (1986) argued from biostratigraphical evidence, supported by geochronometric dating, for a correlation between the Holsteinian Stage of northwest Europe and Oxygen Isotope Stage 11. This is more easily reconciled with the evidence from central Europe, Germany and Normandy, where two full climatic cycles appear to separate the Holsteinian and Eemian. The Dutch record could also be accommodated within such a scheme if the Hoogeveen and Bantega episodes were separately correlated with Stages 9 and 7 respectively. The attribution of the type-Holsteinian to Stage 11 has been questioned, however, by Schwarcz and Grün (1988), who favoured correlation with Stage 7.

The British sequence can be accommodated readily in the scheme of Sarnthein *et al.* (1986). The Anglian Stage, which is generally regarded as the equivalent of the Elsterian, has been equated by most authorities with Stage 12 (Wymer, 1985a; Bowen *et al.*, 1986a, 1986b). This would imply a Stage 11 age for the Holsteinian, which follows the Elsterian, and for the Hoxnian, which is regarded as being immediately post-Anglian in age (Turner, 1973). Thus the correlation of the Hoxnian-dated Thames sites at Clacton and Swanscombe with the Holsteinian is confirmed; these have been ascribed on stratigraphical grounds to Stage 11 (see Chapter 4). However, doubts remain about the Hoxnian type locality. If suggestions that this represents Stage 9 are correct (see above), correlation of the Hoxnian *sensu* Hoxne with a post-Holsteinian temperate episode in Europe, such as the 'Wacken (Domnitz) Warmzeit', might be implied. Because of this uncertainty, subsequent references to the Hoxnian in this volume will distinguish, where appropriate, between Hoxnian *sensu* Hoxne and Hoxnian *sensu* Swanscombe, the latter equating with Oxygen Isotope Stage 11 and the former of uncertain age.

Some discussion of the pre-Anglian sequence is also required here, as it is now clear that the earliest Thames deposits date back to the Early Pleistocene or even the Pliocene, although biostratigraphical evidence for precise dating is extremely limited. Recent research, particularly on the Norwich Crag Formation, has led to considerable revision of the pre-Anglian stratigraphical scheme of Mitchell *et al.* (1973). West (1980) added a series of climatic cycles to the scheme, under the names Pre-Pastonian a–d. From comparisons with the continental record it has become apparent that the British succession, which is effectively confined to East Anglia, contains several important hiatuses. The following revised scheme was outlined in the latest summary of the East Anglian sequence by Zalasiewicz and Gibbard (1988), with an additional modification from Gibbard *et al.* (1991):

Anglian
Cromerian
  HIATUS
Beestonian
  HIATUS (added by Gibbard *et al.*, 1991)
Pastonian
Pre-Pastonian d }
Pre-Pastonian c } Substages
Pre-Pastonian b }
  HIATUS
Pre-Pastonian a
  HIATUS
Baventian

A number of workers have argued in recent years that the British sequence includes evidence for a further climatic cycle between the type-Cromerian and the Anglian (Bishop, 1982; Currant, *in* Roberts, 1986). This is based on a single but important difference in the mammalian faunas from some sites that would otherwise be classified as Cromerian. Most Cromerian small-mammal assemblages, including those from the stratotype and from sites within the Thames system at Sugworth and Little Oakley, contain the extinct water vole *Mimomys savini* (Hinton). Sites at Ostend, Norfolk (Stuart and West, 1976; Stuart, 1982a), Boxgrove, Sussex (Roberts, 1986), and Westbury-sub-Mendip, Somerset (Bishop, 1982), have yielded similar faunas, but with the vole *Arvicola cantiana* (Hinton) instead of *M. savini*. The former species has been interpreted as the evolutionary descendant of the latter (von Koenigswald, 1973; Sutcliffe and Kowalski, 1976; Stuart, 1982a, 1988). Not all authors accept this faunal change as evidence for an additional climatic cycle, however. An alternative explanation that has been proposed is that the sites with *A. cantiana* represent the later part of the Cromerian *sensu* West Runton, whereas the assemblages with *M. savini* date from the early part of the same temperate episode (Stuart and West, 1976; Stuart, 1982a, 1988).

The same controversy exists on the continent, where mammalian assemblages of Cromerian aspect containing both *M. savini* and *A. cantiana* occur. For example, freshwater deposits beneath Elsterian till at Voigstedt, Germany, have yielded an assemblage of small mammals remarkably similar to that of the West Runton freshwater bed, including *M. savini* (Kahlke, 1965; Stuart, 1981). Pollen spectra from overlying clays suggest that the Voigtstedt mammalian assemblage may relate to Cromerian pollen biozones CrII or early CrIII. *Mimomys savini* also occurs in the sparse fauna of small mammals from the upper part of a sequence capped by Elsterian till at Süssenborn, Germany (Kahlke, 1969). The rich assemblage of large mammals from this locality probably dates partly from the Cromerian and partly from the previous cold episode (Stuart, 1982a, p. 118). Early Middle Pleistocene sites yielding faunas with *A. cantiana* include Mosbach (Germany), Vértesszöllós (Hungary) and Stránská Skala (Czechoslovakia) (Kahlke, 1975; Janossy, 1975, 1987).

Comparison with the lower Middle Pleistocene sequence in The Netherlands (Zagwijn *et al.*, 1971; Zagwijn, 1985, 1986; de Jong, 1988), which is more complete than that in Britain, indicates that the hiatus between the Cromerian and Beestonian Stages corresponds to several climatic cycles. It has been suggested that the British Beestonian and Anglian Stages correlate with the Dutch Menapian and Elsterian Stages respectively (Zalasiewicz and Gibbard, 1988). Further reappraisal of correlation between The Netherlands and Britain has recently indicated that the Beestonian is much older than hitherto believed; an age of over 1.5 million years has been suggested (Gibbard *et al.*, 1991). Dutch geologists currently recognize a sequence of at least six interglacial and five cold oscillations between the Menapian and Elsterian (Zagwijn, 1986; de Jong, 1988), as follows:

Elsterian Stage

'Cromerian Complex'	Interglacial IV (Noordbergum) Glacial C Interglacial III (Rosmalen) Glacial B Interglacial II (Westerhoven) Glacial A Interglacial I (Waardenburg)
'Bavelian Complex'	Dorst Glacial Leerdam Interglacial Linge Glacial Bavel Interglacial

Menapian Stage (upper boundary *c.* 1,000,000 years BP)

The four interglacials of the 'Cromerian Complex' are defined on palynology alone. However, evidence from palaeomagnetism enables the first of these to be separated from the remainder of the complex, since the Matuyama–Brunhes magnetic reversal roughly coincides with the transition from Interglacial I to Glacial A (de Jong, 1988). Deposits from the Waardenburg interglacial are therefore the most recent sediments to show a reversed geomagnetic polarity. The Matuyama–Brunhes magnetic boundary, dated at around 780,000 years BP (Shackleton *et al.*, 1990), has been suggested as the base of the Middle Pleistocene (Richmond and Fullerton, 1986), which would therefore fall within the 'Cromerian Complex'.

At Noordbergum, the most recent of the 'Cromerian Complex' interglacials (IV) has yielded the vole *A. cantiana*, suggesting that it post-dates the type-Cromerian of West Runton (van Kolfschoten, 1988). On the other hand, Bridgland *et al.* (1990) have presented a palynological argument against correlating the West Runton and Little Oakley sites with the three earliest interglacials from the Dutch 'Cromerian Complex'. If the suggestion (above) that faunas with *A. cantiana* represent a later temperate episode than the type-Cromerian is correct, there would appear to be no equivalent of the West Runton interglacial yet recognized in The Netherlands, implying that the Dutch sequence is also incomplete (Bridgland *et al.*, 1990).

Pre-Cromerian (*sensu* West Runton) interglacials may be represented within the Kesgrave Group at Ardleigh (Chapter 5) and Broomfield, both in Essex, and these may equate with one or more of the post-Menapian temperate episodes recognized in The Netherlands (Gibbard, 1988b).

TERRACE FORMATION

There has been considerable debate about the possible correlation between the formation of river terraces and climatic fluctuation during the Pleistocene (Zeuner, 1945, 1959; Wymer, 1968; Clayton, 1977; Rose, 1979; Green and McGregor, 1980, 1987). Zeuner (1945, 1959) considered depositional fluviatile terraces to fall into two groups, climatic and thallassostatic. In the lower reaches of rivers, he believed that thallassostatic terraces were predominant, formed by aggradation in response to rises in sea level. In the higher and middle reaches of rivers, remote from the effects of sea-level change, he envisaged aggradation in response to climatic deterioration (and its effect on river energy and sediment supply), forming climatic terraces. Since sea-level changes during the Pleistocene were climatically controlled, both types of terrace formation are potentially related to climatic fluctuation. Recognizing these two types of terrace, many authors have regarded terrace aggradations in the higher reaches of river valleys as the product of cold-climate environments, whereas those in the lower reaches have frequently been attributed to aggradation in response to relative rises in sea level during interglacials (see, for example, Evans, 1971). Direct correlation between these two types of terrace in the lower and higher reaches of river valleys should not be possible, although interdigitation of the two sets of deposits would be expected, especially as the cold-climate aggradations were laid down when sea level was much lower, so that the present lower reaches of rivers would then have been a considerable distance inland from the contemporary coast.

Zeuner believed that the major downcutting events (rejuvenations) in all parts of river valleys occurred in response to falls in sea level. In the case of the Thames, he recognized a series of erosional 'benches', overlain by sheets of cold-climate gravel, that he attempted to correlate with phases of low sea level. The mechanism of rejuvenation is poorly understood, however. Investigations in the North Sea have indicated that fluvially formed valley floors and terraces continue offshore beneath Holocene marine deposits (D'Olier, 1975; Bridgland and D'Olier, 1987, 1989). It is apparent that the Holocene sea-level rise brought about a considerable accumulation of estuarine and marine alluvium in the lower reaches and estuaries of rivers such as the Thames; presumably a return to a low sea level similar to that of the Devensian Stage would cause these deposits to be dissected as the rivers returned to their pre-Holocene floodplain levels. However, there is no reason why a fall in sea level should cause rivers to cut down to a lower level than they have previously occupied, as has happened at each 'rejuvenation' between different terraces of the Thames. Instead, their valleys would merely be extended further and further beyond the interglacial coastline with the progressive decline in eustatic sea level during the onset of 'glacial' conditions, reoccupying the channels (like those beneath the southern North Sea) that were submerged at the end of the previous cold episode.

Evans (1971) presented a model for the chronostratigraphical interpretation of the Thames terrace sequence, in an early attempt at correlation with the deep-sea record. In Evans' scheme, cycles of river aggradation and rejuvenation were superimposed against the background of a progressive decline in relative sea level since the Pliocene, which is indicated by evidence for successive interglacial sea levels from raised beaches and shorelines both in Britain and abroad. Despite doubts in recent years about the marine origin of some of the high-level features used by Evans to formulate his views, it remains clear that fluvial base levels in most areas have been progressively lowered during the Pleistocene, irrespective of cycles of river aggradation and rejuvenation. This may reflect a gentle tectonic (isostatic) adjustment to the redistribution of material by rivers over this period; an uplift of terrestrial areas in response to erosion and a downwarping of marine areas under the weight of fluvially derived sediment. The reconstruction of Early Pleistocene floodplain levels of the Thames indicates that a vast amount of material has been removed from the land area of Britain since that time. Since land masses can be shown to rise over thousands of years following the removal, by deglaciation, of the weight of ice sheets (isostatic rebound), it is clear that the removal of solid rock over hundreds of thousands of years must result in a similar, if more gradual, tectonic adjustment.

This process of tectonic adjustment provides a possible explanation for the progressive lowering of fluvial base levels that is required for the formation of a terrace sequence. Given this progressive uplift of land areas, the initiation of a cycle of aggradation and downcutting, either in response to climatic or thallassostatic factors, would bring about terrace formation. The evidence from the Thames allows some insight into this process. Firstly, it is clear that during cold episodes the river was a considerably more active agent of both deposition and erosion than during temperate intervals, such as the present interglacial (the Holocene). This is partly because (1) there would have been a greater discharge during the spring melt season than would have occurred under temperate conditions and (2) the paucity of vegetation during cold intervals would have allowed more ready transfer of sediment to river channels. Secondly, it is apparent from the common preservation of temperate sediments in the Upper Thames that aggradation during interglacials was not restricted to the lower reaches of the valley. In all parts of the Thames system the preservation of interglacial sediments seems to be restricted to minor channel-fills or lenses within much larger bodies of cold-climate sand and gravel. In the Lower Thames, more extensive sheets of interglacial sediment appear to have accumulated under estuarine conditions, providing the only evidence in support of the theory of thallassostatic terrace formation. In fact, rises in relative sea level during interglacials appear to have drowned the lower reaches of river valleys, leading to accumulations of predominantly fine-grained estuarine sediment, such as those preserved within the Lower Thames sequence (see Chapter 4).

It is also apparent, from the stratigraphical position of many of the occasional sedimentary remnants from temperate episodes that occur within terrace aggradations, that more than one cold-climate episode is often represented by a single aggradational sequence. Even when interglacial sediments occur close to the base of a sequence, they are usually underlain by a basal gravel that is suggestive of a colder climate (examples are the basal gravels at Swanscombe and Clacton and the basal gravels of buried channels in eastern Essex (Bridgland, 1988a)). It is therefore possible to suggest a modified climatic model for terrace formation, which may be directly related to the climatic cycles recorded in the oxygen isotope curve:

Phase 1 Downcutting by rivers during a time of high discharge, under cold climatic conditions. The limits of this rejuvenation would be controlled by base level.

Phase 2 Aggradation of sand and gravel and the formation of floodplains at the new level; energy levels remain high, but sedimentation now exceeds erosion, leading to a vertical accumulation of sediment (final part of cold half cycle).

Phase 3 Limited deposition by less powerful rivers under temperate conditions (interglacial). This usually takes place in single-thread channels covering only small areas of floodplains, although overbank deposits may be more extensive. Estuarine sediments accumulate above phase 2 deposits (often overlapping these) in the lower reaches of valleys.

Phase 4 Climatic deterioration results in increases in discharge coupled with enhanced sediment supplies, brought about by the decline of interglacial vegetation and increases in erosion and mechanical weathering. This causes the removal and/or reworking of existing floodplain deposits and the renewed aggradation of sand and gravel.

Then Discharge exceeds sediment supply, causing renewed downcutting (repeat of phase 1).

In the above model, the aggradation of sands and gravels occurs both at the beginning (phase 4) and end (phase 2) of cold climatic episodes. Of these, the latter is probably the principal aggradational phase, represented by most of the classic terrace gravel accumulations within the Thames system. Deposits from the three depositional phases may occur in superposition, but are likely also to be variously represented in different parts of former floodplains, so that later deposits may be banked laterally against earlier ones. The gravels of phases 2 and 4 are, however, indistinguishable without the recognition of interglacial sediments (phase 3) within the system. This presents an important stratigraphical problem, because these two phases would have been separated by an entire warm climatic half-cycle and are of different geochronological ages, yet they are represented beneath a single terrace surface, often at identical elevations. This makes it particularly important to fully assess the complex aggradational sequences that underlie terrace surfaces, as previously emphasized by Green and McGregor (1980, 1987) and Gibbard (1985).

The above model resembles the scheme for climatic terrace formation outlined by Zeuner (1945, 1959), although Zeuner considered that downcutting, once initiated in cold episodes by a fall in sea level, continued to work upstream during the subsequent temperate period. He therefore attributed downcutting in the higher reaches of valleys to interglacials. Similar climatic terrace models have been proposed by Wymer (1968) and Green and McGregor (1980, 1987). Wymer followed Zeuner in linking downcutting phases to falls in sea level, whereas Green and McGregor considered that both downcutting and aggradation may be triggered by hydrological changes as well as changes in base level.

Most of the depositional sequences within the Thames system can be interpreted according to the model outlined above. The main exceptions occur where sediments from more than a single interglacial episode are recorded from a particular terrace aggradation (for example, the Summertown-Radley sequence of the Upper Thames – see Chapter 2). In these cases rejuvenation appears not to have occurred between the two temperate episodes represented. No temperate deposits have yet been found within the Winter Hill or Black Park aggradations (excepting, perhaps, the Sugworth deposits – see below). These aggradations are correlated with the glaciation (Anglian Stage) of parts of the Thames catchment and the rejuvenation that separated them appears to have directly resulted from the diversion of the river (see Chapter 3, Part 2).

THE STRATIGRAPHY OF THE THAMES SEQUENCE

The earliest stratigraphical scheme for the interpretation of the Thames terraces, in which a complex sequence of climatic fluctuations was envisaged, was proposed by King and Oakley (1936). Evidence from biostratigraphy and Palaeolithic artefact assemblages formed the principal bases for this scheme, in which numerous alternating phases of downcutting and aggradation by the river were recognized, far greater in number than the mapped terraces. Phases of downslope movement of soliflucted colluvium, which is locally interbedded with the Thames terrace deposits, were also used as an indication of periglacial episodes. Considerable complexity was necessary in order to reconcile similarities between the archaeological and fossil content of deposits within different terraces, as well as the occurrence of different Palaeolithic industries within single terraces. The scheme therefore reflected what at that time was regarded as a progressive typological evolution of Palaeolithic implements through the Middle and Late Pleistocene. It also sought to correlate the Thames sequence with the limited Pleistocene chronology then recognized. The great complexity of King and Oakley's scheme offered little prospect for extending the stratigraphy from areas in which both interglacial and artefact-bearing deposits are commonly preserved, such as the Lower Thames (where the scheme was established), to other less informative parts of the catchment.

Repeated rejuvenation and aggradation without a progressive lowering of base level, as envisaged by King and Oakley, is necessary to accommodate the post-Anglian Thames sequence within the Pleistocene climato-stratigraphical scheme of Mitchell *et al.* (1973), since the latter recognizes fewer climatic cycles than the number of post-diversion Thames terraces. Perhaps because of this, King and Oakley's scheme remained in favour until quite recently. However, the realization that there is little basis for the Palaeolithic typological succession favoured in the middle part of this century has coincided with the recognition, from the oceanic record, that a greater number of climatic fluctuations has occurred since the Anglian Stage. Furthermore, a stratigraphical reappraisal of the terrace system suggests that the sequence can be reconstructed satisfactorily using a new, simpler model that relates downcutting and aggradation to palaeoclimatic fluctuations and which can be correlated with the oxygen isotope record.

The only previous attempt to relate the Thames sequence to the oceanic record was by Evans (1971), who based his correlation on estimates for successive interglacial sea levels, many of which were extrapolated from terrace projections. Evans assumed that each major terrace aggradation could be correlated with a particular interglacial sea level, the successive fall in fluviatile terrace levels corresponding to a parallel decline in sea level maxima during the Pleistocene. He attempted to correlate this sequence of levels with the warm cycles of the oxygen isotope curve. Despite the fact that the Thames gravel aggradations are now recognized as the products of cold-climate episodes, and the demonstrable invalidity of his conclusions regarding sea level, the correlations of interglacial sites with the oceanic record suggested by Evans show remarkable similarities with those outlined below. This fact suggests that there is a clear correlation between terrace formation and climatic fluctuation, so that simply counting backwards from the present interglacial and river level provides at least an approximate means for correlation.

It has been shown in recent years that the Anglian glaciation and the resultant diversion of the river provides an important stratigraphical marker within the Thames succession, both in the form of changes in sediment constituents (the addition of glacial 'erratics') and in the obvious change in drainage routes (Gibbard, 1977; Bridgland, 1980, 1988a; Cheshire, 1986a). Biostratigraphical evidence, derived from the occasional preservation of fossiliferous sediments within the terrace sequence, has proved valuable for relative dating with reference to the standard British Pleistocene stratigraphy (see above). Important early interglacial remnants have come to light in recent years at Sugworth, in the Upper Thames (see Chapter 2), at Nettlebed on the Chiltern dip slope (see Chapter 3, Part 1) and at a number of sites within the Kesgrave Sands and Gravels of Essex (see Chapter 5, Part 1). Post-Anglian Stage interglacial sites are more numerous, occurring in the Upper Thames (see Chapter 2), Lower Thames (see Chapter 4) and in Essex (see Chapter 5, Parts 2 and 3).

In the present volume, suggested correlations of the post-diversion Thames sequence with the oxygen isotope curve (see Table 1.1) are based on a number of lines of evidence. Firstly, the sequence of cold-climate terrace aggradations has been traced throughout the Thames system, primarily by reconstructing the three-dimensional form of the now-dissected sediment bodies, particularly their downstream profiles (Fig. 1.3). Biostratigraphical evidence provides important support for this scheme, which is described in detail below. Secondly, interglacial sediments (representing phase 3 of the terrace model described above) preserved within the primarily cold-climate gravel sequences provide evidence of climatic events that can be related to the standard British Pleistocene chronology. Correlation with the oxygen isotope record can then be attempted with reference to climato-stratigraphical markers, such as the presumed equivalence of (1) the Anglian Stage and Oxygen Isotope Stage 12 and (2) the Ipswichian Stage with Oxygen Isotope Substage 5e. Interglacial deposits at Aveley and Stanton Harcourt, widely accepted as correlating with Stage 7 (see Chapters 2 and 4, respectively), provide important evidence for correlation between different parts of the Thames valley and between the terrace system and the oxygen isotope record. Four post-diversion (post-Anglian) temperate episodes are recognized within the Thames sequence (see Table 1.1).

Many of the deposits ascribed here to Oxygen Isotope Stage 7 have previously been attributed to the Ipswichian Stage. For this reason, when reference is made in this volume to Substage 5e, the term 'Ipswichian (*sensu* Trafalgar Square)' will be used. Similarly, there is some doubt whether the Hoxne type sequence and deposits in the Thames system at Swanscombe and Clacton, attributed to the Hoxnian Stage, are of equivalent age (see above). As the deposits at Swanscombe and Clacton are considered here to represent Oxygen Isotope Stage 11, the term 'Hoxnian (*sensu* Swanscombe)' will be applied in this volume to that stage. The term 'Swanscombe interglacial' (Bowen *et al.*, 1989) will also be used. If the indications, from amino acid stratigraphy, that the type-Hoxnian correlates with Oxygen Isotope Stage 9 were to be confirmed (Bowen *et al.*, 1989), the implication would be that the Swanscombe and Clacton sediments are pre-Hoxnian (*sensu* Hoxne) and that the true correlatives of the Hoxne sediments within the Thames system are the deposits at Wolvercote, Purfleet and Grays (see Table 1.1). Although the Wolvercote and Grays deposits have been claimed to be Hoxnian by some authors (see Chapters 2 and 4), their age has remained controversial.

Considerable advances have been made in recent years in the dating of pre-diversion Thames deposits. The first evidence for the climato-stratigraphical dating of such deposits came from southern East Anglia, where the gravels of the Kesgrave Group were recognized as products of the pre-Anglian Thames (Rose *et al.*, 1976; Rose and Allen, 1977; see Chapter 5). The upper levels of the Kesgrave Group sands and gravels were shown to have been subjected to pedogenesis during both warm and cold episodes, before burial by Anglian Stage glacial sediments. These processes resulted in the formation of the Valley Farm Soil, a temperate-climate palaeosol, superimposed upon which was the Barham Soil, formed under intensely cold conditions. Rose and Allen concluded that the Kesgrave Group gravels were laid down during the Beestonian (at that time considered to immediately pre-date the Cromerian) and that the temperate Valley Farm Soil was developed during the Cromerian Stage.

This interpretation has since been shown to be oversimplified. Subsequent work has allowed the subdivision of the Kesgrave Group (Hey, 1980; Allen, 1983, 1984; Bowen *et al.*, 1986a; Bridgland, 1988a; Whiteman, 1990), which is now regarded as a series of individual formations dating from numerous periglacial episodes within the Early and early Middle Pleistocene (Bowen *et al.*, 1986a; Zalasiewicz and Gibbard, 1988). Studies of the Valley Farm Soil have shown that at many sites it is highly complex, with evidence for repeated climatic fluctuation, particularly where developed on higher and older formations within the Kesgrave Group (Kemp, 1985a, 1987a, 1987b; Zalasiewicz and Gibbard, 1988). A discussion of the current stratigraphical scheme for the pre-Anglian Thames sequence appears below.

Hey (1980) correlated early Kesgrave Group deposits in Norfolk with the 'Pre-Pastonian a' Stage, as defined by West (1980; see above), on the basis of the first appearance in the East Anglian marine sequence of a clast assemblage including quartzites from the Midlands and Greensand chert from the Weald, which he considered to have been introduced by the Kesgrave Thames. This is of considerable significance, as the recent reappraisal of the correlation between the East Anglian and Dutch stratigraphies suggests that the Pastonian Stage, hitherto placed near the end of the Early Pleistocene, is in fact equivalent to the European Tiglian C5–6 (Gibbard *et al.*, 1991). Since this part of the Tiglian has an estimated age of over 1,600,000 years, considerable antiquity is implied for the earliest formations of the Kesgrave Group. Hey classified the deposits in

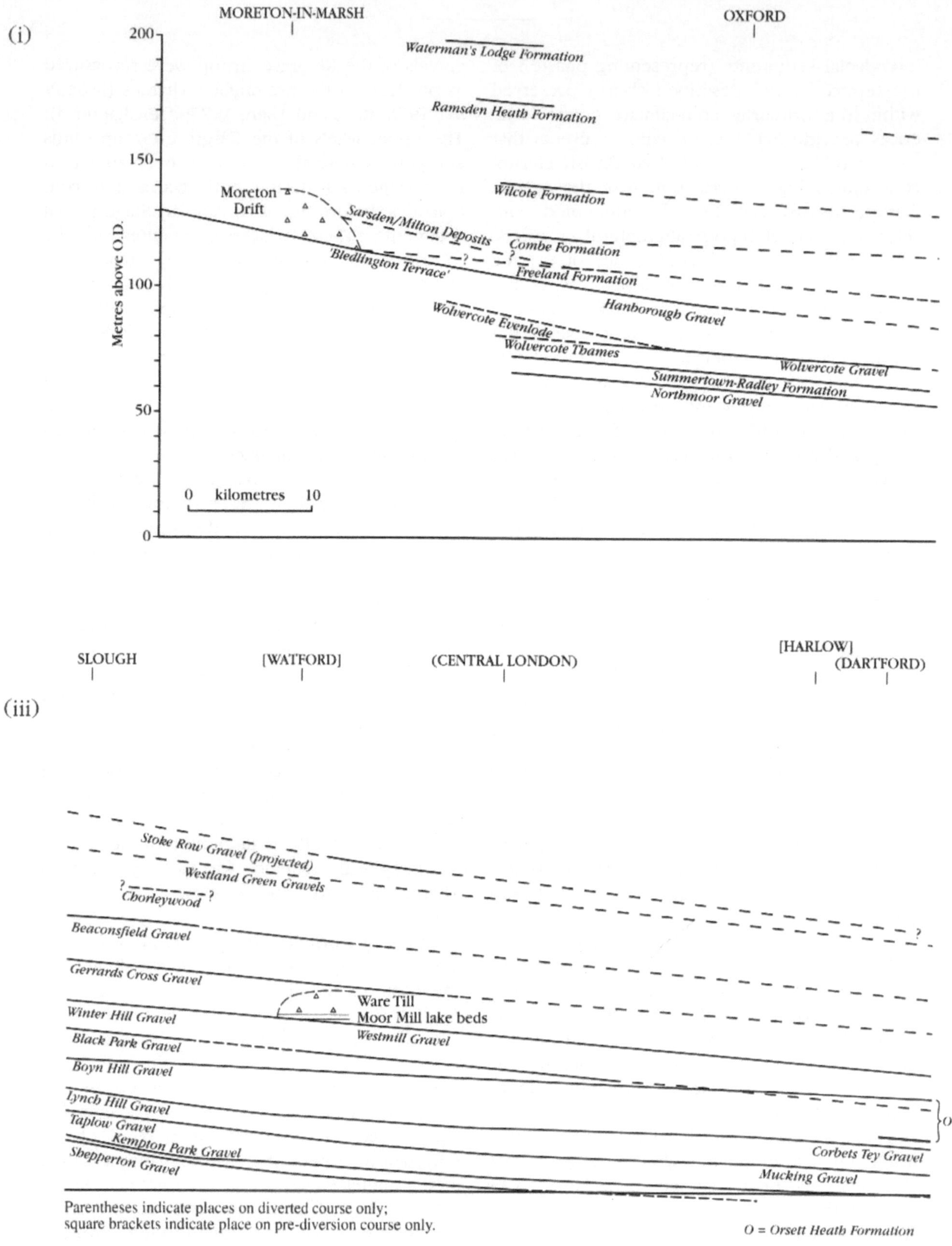

Figure 1.3 Longitudinal profiles of Thames terrace surfaces throughout the area covered by the present volume. The main sources of information used in the compilation of this diagram are as follows: Arkell (1947a, 1947b), Briggs and Gilbertson (1973), Briggs *et al.* (1985), Evans (1971) and Sandford (1924, 1926) for the

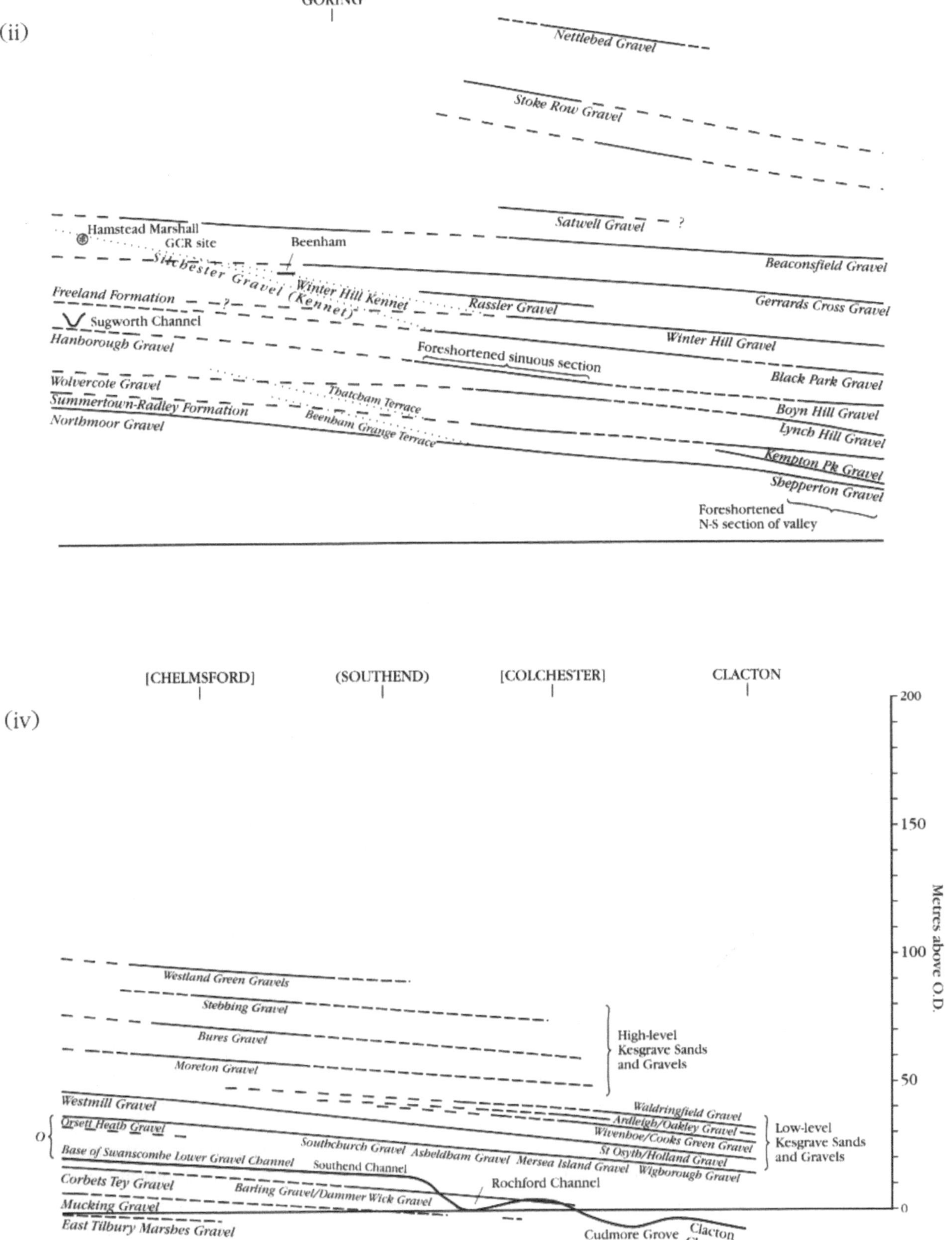

Upper Thames; Gibbard (1985) and Sealy and Sealy (1956) for the Middle Thames; Bridgland (1983a, 1988a) and Bridgland *et al.* (1993) for the Lower Thames and eastern Essex; Whiteman (1990) for central Essex.

question as Westland Green Gravels, which he differentiated from lower-level Kesgrave Group formations on the basis of clast lithology. Allen (1983, 1984) recognized two divisions of this high-level gravel in Suffolk, which he termed Baylham Common Gravel and Westland Green Gravels (in stratigraphical sequence).

Allen (1983, 1984) found that a lower formation, below the Westland Green Gravels, could be distinguished within the Kesgrave Group in south-eastern Suffolk, on the basis of clast lithology and mapping. He named this the Waldringfield Gravels. Bridgland (1988a) described three lower formations within the Kesgrave Group on the Tendring Plateau, northern Essex (see Chapter 5, Part 1). Therefore the full sequence of formations within the Kesgrave Group of pre-diversion Thames deposits (and equivalent Thames-Medway deposits, marked *) is as follows:

St Osyth/Holland* Gravel Wivenhoe/Cooks Green* Gravel Ardleigh/Oakley* Gravel Waldringfield Gravel	Low-level Kesgrave Subgroup
Westland Green Gravels (*sensu* Allen) Baylham Common Gravels	High-level Kesgrave Subgroup

As shown, this group of six formations can be divided, on the basis of clast lithology and geomorphological criteria, into two subgroups (see Chapter 5, Part 1). The lowest formation, the St Osyth/Holland Gravel, has been correlated (Bridgland, 1988a) with the (Anglian Stage) Winter Hill/Westmill Gravel of the Vale of St Albans, on the grounds that it is overlain by deposits representing an unglaciated Medway, formed while the Thames was blocked by Anglian ice (see Chapter 5, St Osyth and Holland-on-Sea). In north-eastern Essex, interglacial sediments are interbedded with gravels of the Ardleigh/Oakley and Wivenhoe/Cooks Green Formations (Bridgland *et al.*, 1988; Chapter 5, Part 1). Assessment of biostratigraphical evidence from these suggests that the two formations were laid down in the interval between the Beestonian and Anglian Stages, which is otherwise poorly represented in Britain (see above).

Recent re-evaluation of the correlation of pre-diversion Thames gravels between the Middle Thames and Essex by Whiteman (1990; see also Chapters 3 and 5) has major stratigraphical implications. By collating all available borehole information, Whiteman compiled a three-dimensional picture of the various Thames gravels buried beneath the Anglian till sheet of central Essex. He found that these deposits could be divided into several well-preserved terrace formations and that these could be traced both upstream to the Vale of St Albans and Middle Thames and downstream to the areas around Colchester and Ipswich, where the subdivisions of Allen (1983, 1984) had already been established. Whiteman found that the unit classified as Westland Green Gravels in Suffolk is, in fact, a continuation of the Gerrards Cross Gravel of the Middle Thames, the youngest pre-Winter Hill formation according to Gibbard (1983, 1985). From Whiteman's evidence, the Baylham Common Gravel of Suffolk appears to equate with the Beaconsfield Gravel, leaving older formations in the Middle Thames without equivalents in East Anglia. Therefore only formations within the older High-level Kesgrave Subgroup would appear to have equivalents upstream, with the exception of the St Osyth/Holland Gravel (see above). The name Westland Green Gravels, as used in Suffolk and Norfolk, is invalidated by this reappraisal, which implies that the gravels correlated by Hey with the 'Pre-Pastonian a' Stage are equivalents of the Gerrards Cross or Beaconsfield Formations. This has major implications for the ages of higher parts of the Middle Thames sequence (see Chapter 3).

Whiteman has argued that the older High-level Kesgrave Subgroup formations, and their upstream equivalents, reflect the maximum extent of the Thames catchment, when the river drained a considerable region beyond the Cotswolds. In contrast, the Low-level Kesgrave Subgroup represents a series of aggradations reflecting a much reduced catchment, a change that gave rise to the differences in composition between the two subgroups, effectively a reduction in far-travelled material. In the Middle Thames the absence of any significant gravel formations (with the possible exception of the Rassler Formation – see below) in the interval between the Gerrards Cross Gravel and the Winter Hill Gravel (in effect, a hiatus in the sequence in that area), implies that, between the 'Pre-Pastonian a' and Anglian stages, terrace generation was restricted to southern East Anglia. The sequence of just four formations representing this interval in the latter area, one of which is of Anglian age, is clearly fewer than the corresponding number of climatic cycles now recognized. Thus the climatic model for terrace formation breaks down during this interval. It may be that the beheaded river, with a much reduced discharge and flowing in an oversized valley, was less effective at both erosion and aggradation, so that rejuvenations affected only its lower reaches. This would explain the selective preservation in Essex of aggradation products representing the interval between the 'Pre-Pastonian a' and the Anglian. Only after its diversion by Anglian ice into a new course did the river resume the aggradation of terraces in all parts of its basin. The severity of climatic fluctuation during the post-Anglian period, as opposed to the earlier part of the Pleistocene, may have had an important influence on terrace formation, as climatic change is seen as the driving force in the model for terrace generation outlined above.

Whiteman's reinterpretation of the sequence in East Anglia has major implications throughout the Thames catchment. The most important of these is that the Gerrards Cross Gravel and all earlier formations are very much older than was hitherto believed. They are all Early Pleistocene or older and so, therefore, are their upstream equivalents recognized by Hey (1986) within the 'Northern Drift' of the Upper Thames (see Chapter 2). Prior to Whiteman's work, the Gerrards Cross Gravel was held to be of early Anglian age or only slightly older (Gibbard, 1983, 1985). This posed difficulties in reconciling Hey's interpretation of the 'Northern Drift' with the recently revised Middle Pleistocene sequence in the Midlands proposed by Sumbler (1983a, 1983b) and Rose (1987, 1989). Rose's work, in particular, suggests that a major eastward-draining valley existed in the area north of the Cotswolds for an unknown interval of time leading up to the Anglian. This is the proto-Soar valley first recognized by Shotton (1953), containing the Baginton Gravel and Baginton Sand of the type 'Wolstonian' sequence, originally defined in the Coventry area. Rose concluded that the Wolston sequence represents the Anglian Stage rather than equating with the continental Saalian, as envisaged by Shotton (1953, 1973b). It was difficult to see how the 'Severn-Thames', the river considered to have deposited the 'Northern Drift', could have co-existed with the 'proto-Soar' in the immediate pre-Anglian and early Anglian period. According to Rose (1987, 1989), the 'proto-Soar' deposited gravel and sand in the Stratford-upon-Avon district to an elevation of *c.* 100 m O.D. at this time. This is approximately 60 m lower than an upstream projection of the Freeland Formation (the lowest division of the 'Northern Drift' – see below) would allow a contemporary Severn-Thames valley floor to have traversed the area of the modern Avon valley. By erecting a scheme in which the Thames ceased to flow from the Midlands before the Middle Pleistocene, Whiteman appears to have resolved these difficulties. It is still evident (from their gradients in the Upper Evenlode) that the earlier Middle Pleistocene formations in the Evenlode valley reflect a larger Thames catchment than at present, but the Thames may have been beheaded by the proto-Soar well before the deposition of the Freeland Formation. A comparable upstream projection of the Gerrards Cross Gravel, the lowest Thames formation considered by Whiteman to reflect a catchment beyond the Cotswolds, crosses the Stratford area only slightly higher than the projection of the Freeland Formation, largely because there is no indication for a marked upstream steepening of the older formation (perhaps because of the scarcity of data points) as there is with the Freeland. The considerable early Middle Pleistocene erosion by the proto-Soar that must have occurred between Gerrards Cross Gravel and Anglian times (if the re-interpretation of the Wolston sequence as Anglian is correct) appears to have coincided with a relative quiescence of the Thames, as described above. Further consideration of the important evidence for correlation between the Midlands and the Thames basin through the area of the Evenlode headwaters is given in Chapter 2.

EVIDENCE FROM PALAEOLITHIC ARTEFACTS IN THAMES DEPOSITS

Support for the revised chronostratigraphical interpretation of the Thames sequence outlined in this chapter is derived from the distribution of Palaeolithic artefacts within the sequence, although the role of archaeology in this scheme is very much reduced in comparison with many earlier models. The occurrence of Lower

Palaeolithic artefacts in the lower terraces of the Thames sequence (within the Black Park Gravel and all later formations) is extremely well documented, most material of this type being collected, before mechanization, by gravel diggers in commercial pits. For this reason large collections of Palaeolithic tools were assembled, but there is little information available about their exact provenance. Much has been learned, however, from the few sites that have been systematically excavated in recent years. These include the type site of the Clactonian Industry (see Chapter 5), itself part of the Thames story, and Swanscombe (see Chapter 4), the only site in England to have produced Lower Palaeolithic human remains.

Assemblages of Lower Palaeolithic artefacts are broadly divisible into industries with formal tools (hand-axes) and those in which only flaking from cores was carried out. Early assemblages comprising only cores and flakes are assigned to the Clactonian Industry, in which flaking was simplistic, involving minimal prior shaping of the core, which might also have been serviceable as a crude 'chopper'. More advanced flake-core industries appear later in the sequence; these involved the use of a more advanced flint knapping method, known as the Levallois technique, in which cores were carefully prepared in order to yield flakes of a desired size and shape. Although some assemblages showing evidence of the Levallois flaking technique lack hand-axes, many include large numbers of such implements.

Hand-axe industries, both with or without evidence for the use of the Levallois technique, are collectively termed Acheulian. Until recently the Acheulian Industry was subdivided using a scheme in which collections of cruder implements were regarded as being earlier than assemblages showing more skilful or painstaking workmanship. This industry was also held to make its first appearance later in the British stratigraphical record than the Clactonian, largely on the basis of the evidence from Swanscombe (Wymer, 1968, 1974). The recent discovery at Boxgrove, West Sussex, of skilfully made hand-axes in conjunction with a late Cromerian fauna (Roberts, 1986) has led to the realization that typological subdivisions within the Acheulian Industry, and other hand-axe classifications such as the 'Abbevillian' and 'Chellean', have no chronological significance. On biostratigraphical grounds, the Boxgrove industry is now widely believed to pre-date the Clactonian assemblages from Clacton and Swanscombe. However, a flake/core industry from High Lodge, Suffolk, regarded as Clactonian (J. McNabb, pers. comm.), may be of similar antiquity. The evidence from Boxgrove and High Lodge would appear to indicate that both the Clactonian and Acheulian industries were in operation in Britain before the Anglian Stage glaciation. Wymer (1988) regarded both the Boxgrove and High Lodge industries as pre-Anglian and included more controversial collections from Kent's Cavern, Devon, and Westbury-sub-Mendip, Somerset, in the same category.

Some doubt remains about the age of the Boxgrove artefacts. They are associated with a 'late Cromerian' mammalian fauna (Roberts, 1986; see above), but amino acid ratios from marine shells from the site suggest a post-Anglian age, in Oxygen Isotope Stage 11 (Bowen and Sykes, 1988; Bowen, 1991).

The Palaeolithic content of the Thames terrace deposits continues to yield important stratigraphical information, however, in that the first appearance of artefacts in the sequence of gravels provides a marker. The first appearance in the terrace sequence of artefacts showing the Levallois flaking technique is also stratigraphically significant. Evidence from the Lower Thames suggests that this technique was first being used on a large scale during Oxygen Isotope Stage 8, since it appears in the uppermost levels of the Corbets Tey Formation. It is also well-represented in the basal deposits of the Mucking Formation, also attributed to Stage 8 (see Table 1.1). Deposition of the latter formation continued into the Stage 7 temperate episode (see Chapter 4, Lion Pit). It is interesting to note that the Levallois technique is recognized in the Palaeolithic assemblage from Pontnewydd Cave, North Wales, in deposits that have recently been ascribed to Stage 7 (Green, 1984; Campbell and Bowen, 1989).

CORRELATION OF THAMES TERRACES

The last systematic attempt at correlation between the terraces of the different parts of the Thames catchment was that by Evans (1971). Gibbard (1985) traced individual terrace formations in detail within the Middle Thames valley and suggested correlations with both the Upper and Lower Thames, but he failed to provide a comparison of his data for different terraces by plotting all the information on a single long-profile diagram. Without this comparison the degree of separation of terraces and differences in their downstream gradient are difficult to assess.

The long-profile diagram provided here (Fig. 1.3) forms an important part of the evidence for correlating terraces across the principal zones of demarcation that separate the various parts of the Thames basin; namely, the Goring Gap, the urban area of London and the till sheet of Essex. The main problems in providing a composite long-profile diagram of this type arise from differences in the courses followed by different formations. In particular, the later gravels appear to follow markedly more sinuous courses, although this may be a reflection of the fact that they are better preserved and it is therefore possible to reconstruct their routes in greater detail. In general it is possible to overcome this problem by plotting gravel heights against a simplified Thames course, but this has the effect of greatly shortening the more sinuous parts of some formations, causing apparent increases in gradient that are artefacts of the cartographic technique (see Fig. 3.3).

The Middle Thames terrace sequence as plotted in Figs 1.3 and 3.3 differs from that outlined by Gibbard (1985) in a number of important ways. The first of these concerns the important Winter Hill Gravel Formation, which represents the final phase of aggradation by the Thames prior to its diversion. The Winter Hill Gravel can be traced downstream into the Westmill Lower Gravel of the Vale of St Albans, which is overlain by Anglian glacial deposits (see Chapter 3, Part 2). Sealy and Sealy (1956) recognized upper and lower divisions of the Winter Hill Terrace, an elevation in status of two facets of the terrace previously mapped by Hare (1947). Gibbard (1983, 1985) reinterpreted this double terrace feature in the area north of Slough as the dissected remnants of a sequence, formerly in superposition, of fluvial deposits (Winter Hill Lower Gravel) overlain by deltaic deposits (Winter Hill Upper Gravel). He believed the deltaic gravels to be restricted to this area and to have prograded into an ice-dammed lake that formed in the Vale of St Albans when the Thames route north of London was blocked by the Anglian glaciation (see Chapter 3, Part 2). Upstream from the Marlow area, the Winter Hill Formation is poorly preserved, except in the region of the 'Ancient Channel', which represents an abandoned course of the Thames last used during the aggradation of the Black Park Gravel (see Chapter 3, Highlands Farm Pit). Gibbard (1985), in his reconstruction of the Winter Hill (Lower) Gravel, included in this formation deposits to the north and south of the 'Ancient Channel' that had been attributed by Sealy and Sealy (1956) to their Rassler Terrace. This correlation requires a marked steepening of the gradient of the formation upstream from Winter Hill.

However, an alternative interpretation can be presented. The Sealys mapped the gravel flooring the 'Ancient Channel', now correlated with the Black Park Gravel, as Lower Winter Hill Terrace, but also recognized a few small remnants of an additional terrace formation, intermediate between this and the Rassler Terrace. The two most significant remnants of this additional formation, which the Sealys classified as their Upper Winter Hill Terrace, are at Bellehatch Park (SU 748803) and Crowsley Park (SU 725804), at 86.5 and 88 m O.D. respectively. If the Winter Hill/Westmill Gravel of the Slough–Watford area is projected upstream to the vicinity of the 'Ancient Channel', its elevation conforms closely to that of the Sealys' Upper Winter Hill Terrace remnants. This suggests that the latter are in fact part of the Winter Hill Formation, whereas the higher gravel correlated by Gibbard with this unit represents a separate, earlier formation. As the Rassler Wood type locality (SU 822854) of the Rassler Terrace is included in that earlier formation, the name Rassler Gravel is here given to this previously undefined unit (see Fig. 1.3 and Table 1.1).

This reinterpretation of the Winter Hill Gravel and adjacent deposits in the Reading area is not based solely on the longitudinal projection of formation levels. There are stratigraphical arguments, based on the correlation of the Middle and Upper Thames terraces (see Fig. 1.3), in support of the revised stratigraphy outlined immediately above. The steep gradient of the Winter Hill Gravel implied in Gibbard's reconstruction requires the contemporaneous floodplain level in the Upper Thames valley to have been in excess of 110 m O.D., apparently converging upstream with the Gerrards Cross Gravel (see Fig. 3.3). This is problematic, since the Sugworth Channel interglacial deposits, ascribed by most authors to the Cromerian Stage, occupy a lower altitudinal position within the Upper Thames terrace 'staircase' than this. The interglacial sediments at Sugworth lie at under 90 m O.D. and are overlain by a decalcified gravel attributed to the Freeland Formation, which has been correlated with the late Anglian Black Park Gravel (see Chapter 2, Sugworth; Fig. 1.3). According to Gibbard's reconstruction, the earlier of the two Anglian Stage formations, the Winter Hill Gravel, forms a higher terrace than the late Anglian Freeland Formation. This interpretation would appear to imply that the Winter Hill Gravel is also older than the (Cromerian) Sugworth channel-fill, which underlies the Freeland Formation. Such a conclusion is clearly untenable. The revised interpretation outlined above, in contrast, suggests that the Winter Hill aggradation and the (lower) Black Park Gravel converge upstream and that both can be correlated with the Freeland Formation (Fig. 1.3). This allows a more satisfactory interpretation of the sequence at Sugworth, with decalcified Anglian Stage gravel overlying Cromerian deposits. The convergence of the Winter Hill and Black Park formations can also be readily explained. It has already been noted that the rejuvenation that occurred between the deposition of these two units was the direct consequence of the diversion of the Thames. The Black Park Gravel was laid down while ice still occupied the Vale of St Albans (see Chapter 3, Part 2). The two gravels are therefore closely associated and the time interval between them was short.

The second important difference between the scheme presented here and that of Gibbard (1985) is that the latter author recognized an additional aggradation between the Taplow Gravel and the Kempton Park Gravel in the Reading area, his Reading Town Gravel. Reassessment of the altitudinal distribution of gravel remnants in the Middle Thames associated with the Taplow aggradation (see Chapter 3, Fern House Pit) suggests a rather different interpretation, however. Projection of the Taplow Gravel upstream from the type area indicates that the Reading Town Gravel of Gibbard (1985) is the true upstream continuation of the Taplow Formation (confirming the geomorphological interpretation of Sealy and Sealy (1956)). It is suggested in Chapter 3 that interglacial sediments underlying the Taplow (Reading Town) Gravel at Redlands Pit, Reading, ascribed by Gibbard (1985) to the Ipswichian Stage, may instead be correlatives of an intra-Saalian interglacial correlated with Oxygen Isotope Stage 7 (see Chapter 3, Fern House Pit).

This re-evaluation of the terrace stratigraphy in the Reading area has major repercussions for correlation with the Upper Thames. Gibbard (1985) correlated the Summertown-Radley aggradation of the Upper Thames, which he considered to have continued until the early Devensian, with his Reading Town Gravel; this correlation appears highly plausible on altitudinal grounds. However, the Summertown-Radley aggradation is seen by many workers to be a highly complex succession of temperate- and cold-climate deposits, the main part of which was deposited during the latter part of the Saalian Stage. At Stanton Harcourt the main cold-climate gravel of the Summertown-Radley aggradation overlies a temperate channel-fill that has been claimed as representative of Oxygen Isotope Stage 7 (Shotton, 1983; Chapter 2, Stanton Harcourt and Magdalen Grove), although other workers, including Gibbard (1985), have regarded it as Ipswichian. The Stage 7 age is supported by amino acid ratios (Bowen *et al.*, 1989) and correlations based on mammalian faunas (Shotton, 1983).

Correlations based on this revised terrace stratigraphy have implications for the dating of another important site upstream from the Reading area, but this time in the tributary Kennet valley, at Brimpton (see Chapter 3, Part 3). At Brimpton a sequence of gravels incorporates fossiliferous silt and clay lenses at several stratigraphical levels. These have been attributed, on biostratigraphical grounds, to a series of interstadial episodes within the Devensian Stage (Bryant *et al.*, 1983). However, projection of the Middle Thames terrace formations up the Kennet valley, even allowing for the probable increased gradient in the tributary, strongly indicates correlation of the Brimpton deposits with the Taplow Gravel (see Fig. 3.3). This would appear to imply a pre-Devensian age, as the deposition of the Taplow Formation in the Slough area is considered here to have occurred between Oxygen Isotope Stages 8 and 6 (Table 1.1). In the Middle Thames the main post-interglacial phase of aggradation (phase 4 of the climatic terrace model) of the Taplow Formation is therefore ascribed to early Stage 6. In the Summertown-Radley Formation of the Upper Thames, which is also correlated with the Taplow Gravel, the main Stage 6 gravel aggradation is overlain by sediments attributed to the Ipswichian (*sensu* Trafalgar Square) and, probably, the early part of the Devensian (see Chapter 2). This unusually long sequence is believed to be repeated in the Reading area (see Chapter 3, Fern House Pit). This is thought to indicate that the Stage 6 rejuvenation (phase 1 of the terrace model) only occurred in the Thames valley below Reading. It is therefore possible for the Taplow Formation above Reading to include sediments ranging in age from late in Oxygen Isotope Stage 8 to the mid-Devensian Stage (Table 1.1; see Chapter 3, Part 3).

The results of the correlation between the Upper and Middle Thames described above have considerable significance for the correlation of glacial events between the London Basin and the Midlands. As has been noted, the Thames was diverted, between the Winter Hill and Black Park aggradational phases, by a glacial advance into Hertfordshire during the Anglian (see Chapter 3, Part 2). This stage is widely regarded as the equivalent of Oxygen Isotope Stage 12 of the deep-sea record (see above). Glacial deposits also impinge on the sequence in the headwaters of the Upper Thames around Moreton-in-Marsh, reputedly overlying deposits correlated with the Hanborough Gravel and supplying outwash material to the Wolvercote Gravel (Arkell, 1947a, 1947b; Briggs and Gilbertson, 1974; Chapter 2). The correlation scheme between the Upper and Middle Thames shows that the Hanborough Gravel is an upstream equivalent of the Boyn Hill Gravel and the Wolvercote of the Lynch Hill Gravel. It is suggested (see Chapter 2) that the glacial input into the Wolvercote Gravel may have occurred during the pre-interglacial aggradational phase (phase 2 in the scheme for terrace formation outlined earlier in this chapter) of the Wolvercote/Lynch Hill aggradation. The combination of terrace correlation and climatostratigraphy suggests that the interglacial represented within the Wolvercote/Lynch Hill aggradation, as seen in the Wolvercote Channel, equates with Oxygen Isotope Stage 9 (see Chapter 2). The Hanborough/Boyn Hill Formation is believed to include remnants of deposits laid down during Oxygen Isotope Stage 11 (Hoxnian *sensu* Swanscombe), although in the Upper Thames this interglacial is represented only by derived mammalian remains in the dominant post-interglacial (phase 4) aggradation. The Hanborough/Boyn Hill Formation clearly post-dates the diversion of the Thames and, therefore, the Anglian (Oxygen Isotope Stage 12) glaciation of Hertfordshire. The glaciation of the Cotswolds thus appears to post-date the phase 4 aggradation of the Hanborough Gravel, which, according to the terrace model outlined above, occurred earlier in the same cold episode as the phase 2 aggradation of the Wolvercote Gravel. The glaciation must have taken place during this same cold episode, since it fed outwash material into the Wolvercote Gravel. The implication of the correlations proposed here is that this cold episode is equivalent to Oxygen Isotope Stage 10 and not Stage 12 (see Chapter 2).

This is of great significance, because it implies that glacial deposits that have been ascribed to the Anglian may in fact be the products of two separate glacial episodes (equivalent to Oxygen Isotope Stages 12 and 10). This interpretation relies heavily, however, on Arkell's account of the stratigraphy of the Moreton-in-Marsh area and on the recognition of a glacial input into the Wolvercote Gravel. In a recent review of the stratigraphical significance of glacially derived flint in the Wolvercote Terrace deposits, Maddy *et al.* (1991b) have challenged the notion that such material appears for the first time in this formation (see Chapter 2). It is interesting to note that Arkell (1943, 1947b) was initially inclined to correlate the Moreton glaciation with his Freeland Terrace, which would allow correlation with both Stage 12 Anglian terrace formations in the London Basin, the Winter Hill and the Black Park Gravels (see Fig. 1.3 and Table 1.1). He rejected this correlation following his assessment of the stratigraphy in the watershed area between the Evenlode and Stour valleys. Kellaway *et al.* (1971) recorded high-level quartzite-rich sands and gravels at Sarsden, well above the level of the Hanborough Gravel (Figs 2.1 and 2.3), which they interpreted as outwash of the 'Northern Drift' glaciation. Although they reported that flint was absent from these deposits, Tomlinson (1929) had previously described a gravel at a similar height (Fig. 2.3) on the opposite side of the Evenlode valley, at Milton-under-Wychwood (*c.* SP 264182; Fig. 2.1), in which flint was present. Tomlinson suggested that this might represent an upstream continuation of the Hanborough Terrace, but it occurs in association with Northern Drift (mapped as till) and was mapped as 'Glacial Gravel' on the Geological Survey map (Sheet 236). Both the Sarsden and Milton gravels are at elevations that could suggest correlation with the Freeland Formation (Fig. 2.3), providing a possible link between that deposit and the glaciation of the Cotswolds. However, Briggs (1973) described gravels, which he attributed to glacial outwash, overlying the Hanborough Formation in the Upper Evenlode, thus supporting Arkell's later interpretation.

It is worth noting that there are breaks in the downstream continuity of all the Evenlode gravels and miscorrelation between the lower and upper parts of the valley cannot be ruled out. This could mean that the gravels underlying the Moreton glacial deposits, the 'Bledlington Terrace' of Arkell (1947b), are older than the Hanborough Formation. The limestone content of these deposits would seem to preclude correlation with any pre-Hanborough formation, but burial by till could have effected the preservation of calcareous clasts in an older deposit. If the longitudinal profile of the Freeland Formation downstream from the type area in the Lower Evenlode was to be continued upstream at the same gradient, instead of at the increased gradient needed to take it up to the level of the Sarsden and Milton deposits described above (see Figs 1.3 and 2.3), it would pass through the Moreton-in-Marsh area at exactly the altitude of the Bledlington gravels. Thus the Bledlington and Paxford gravels (see Fig. 2.3) could be upstream equivalents of the Freeland Formation rather than the Hanborough Formation, which would allow correlation of the Cotswolds and Vale of St Albans glaciations. The steepness of the floodplain gradient in the Upper Evenlode during Freeland Formation times is dependent to a large extent on the size of the catchment at that time. The older Northern Drift formations are believed to have shallow gradients throughout the area, reflecting a catchment that extended well beyond the present Stour–Evenlode watershed, whereas the later Evenlode terraces are much steeper, being confined to the modern catchment (Fig. 2.3). The obvious importance of the interpretation of the Upper Thames terraces and their relation to the glacial deposits of the Moreton-in-Marsh area, both for studies of the Thames sequence and for British glacial and interglacial stratigraphy in general, indicates that a reappraisal of the evidence in the Evenlode valley is urgently required.

CONCLUSIONS

In this volume a radically new scheme for terrace stratigraphy in the Thames basin is proposed. This results from the incorporation of the latest evidence for Pleistocene chronostratigraphy and geochronology in a critical review of the evidence from which the Thames sequence has been reconstructed. Although much of the latter derives from previously published work spanning more than a century, new investigations, carried out as part of the GCR project, are described here for the first time.

The proposed new scheme for Thames terrace stratigraphy attempts to correlate this fragmentary, but complex, terrestrial sequence with the more continuous sedimentary and climatic record from deep-sea cores. This is based on biostratigraphical evidence, with support from amino acid geochronology, and on assumptions about the correlation between cycles of terrace formation, as represented in the Thames sequence, and climatic fluctuation during the Pleistocene. Although many of the new interpretations based on these premises will be regarded as controversial, the task of correlating the Thames sequence with the oceanic record should have the highest priority, as it promises to improve greatly the resolution of Pleistocene terrestrial stratigraphy in Britain. The terrace sequence of the Thames is particularly suited to this type of approach, since the terraces themselves provide a range of deposits that extend from the Early Pleistocene to the last glaciation. This can be regarded as a genuine 'long sequence' of the type needed to provide a land-based framework for correlating with the global deep-sea record. Although it is punctuated with numerous gaps (rejuvenations between different terrace levels), the Thames sequence benefits from the continuity of its formative agent, the river itself. With King and Oakley's (1936) model involving repeated

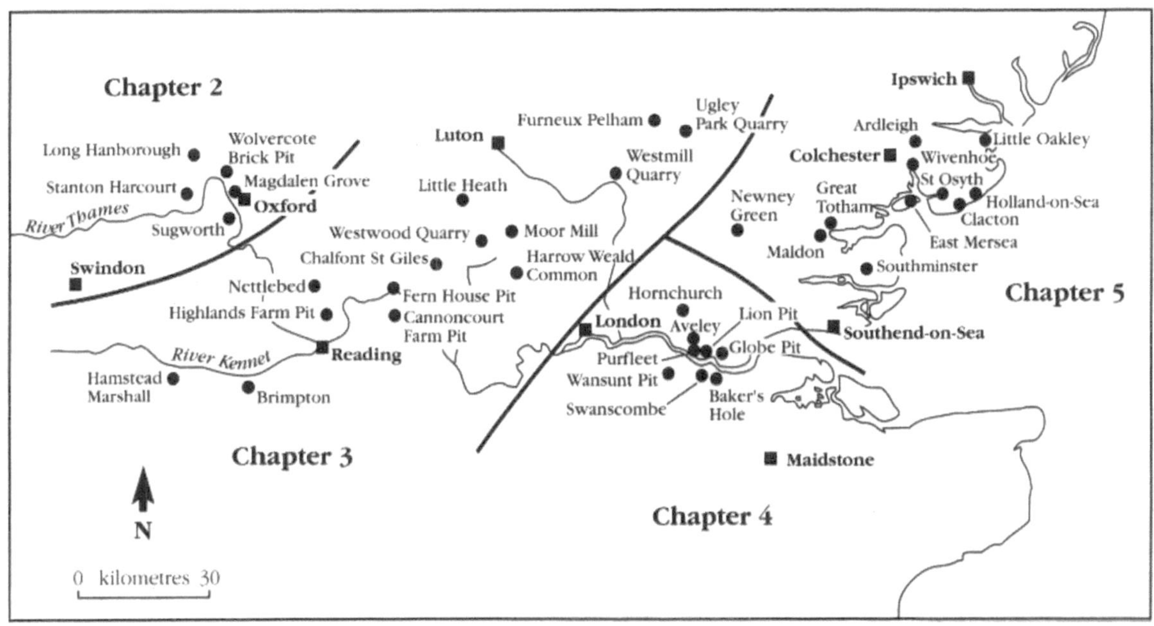

Figure 1.4 Map showing the locations of the GCR sites described in this volume.

occupation of similar terrace levels now considered obsolete, and with many more climatic cycles now recognized than the chronology of Mitchell *et al.* (1973) allowed, there is a real possibility that the terrace sequence matches closely the sequence of climatic fluctuations recognized in the oxygen isotope record.

In the remaining chapters the various GCR sites (Fig. 1.4) are described in detail, with further consideration given, where relevant, to wider issues such as correlation with the oceanic record.

Chapter 2

The Upper Thames basin

INTRODUCTION

This chapter is devoted to the first of three geographical divisions of the present Thames drainage system, the Upper Thames basin. This comprises the catchment areas of the various streams that combine to form the Thames upstream from the Chiltern escarpment (Fig. 2.1). In many ways the Upper Thames is most easily treated as a separate system from the valley downstream of the Goring Gap. Its Quaternary terrace record is extensive, particularly in the valley of the River Evenlode, which appears to have been the main stream until late in the Middle Pleistocene. Unfortunately, there is little preservation of terrace deposits in the area of the Goring Gap and, therefore, no continuity with the succession in the Middle Thames. This has contributed to a long-standing uncertainty over correlation between the Upper Thames terraces and those in the London Basin. The problem of correlation between the Upper and Middle Thames has been addressed in this volume using projections of sediment-body longitudinal profiles between the two areas, reinforced with biostratigraphical evidence for the age of surviving interglacial deposits (see Chapter 1 and Fig. 1.3).

THE PLEISTOCENE SEQUENCE IN THE UPPER THAMES

Classification of the Pleistocene sequence in the Upper Thames basin has been somewhat haphazard, with some deposits being named as lithological units and others named after the geomorphological terrace features to which they give rise. Following the trend towards the former method, established in other parts of the Thames catchment (Chapter 1), geological names were applied to parts of the terrace sequence by Briggs *et al.* (1985) and formal lithostratigraphy was used to classify the high-level Northern Drift deposits by Hey (1986). The nomenclature of these and earlier authors is incorporated in the lithostratigraphical scheme shown in Table 2.1, the rationale for which is discussed in Chapter 1.

The Pleistocene succession in the Upper Thames basin is readily divisible into two parts: (1) the older, high-level deposits of the Northern Drift Group (Table 2.1), which are primarily preserved parallel to the Evenlode valley and are devoid of calcareous clasts, and (2) more recent, lower-level terrace gravels composed largely of local limestones (see Figs 2.1, 2.2 and 2.3).

Table 2.1 Lithostratigraphical classification of Upper Thames deposits.

Group	Formation	Member
	Northmoor Gravel (= Floodplain Terrace)	
	Summertown-Radley	Eynsham Gravel Stanton Harcourt Gravel Stanton Harcourt Channel Deposits
	Wolvercote Gravel	Wolvercote Channel Deposits
	Hanborough Gravel	
Northern Drift	Freeland Combe Wilcote Ramsden Heath Waterman's Lodge	Sugworth Channel Deposits

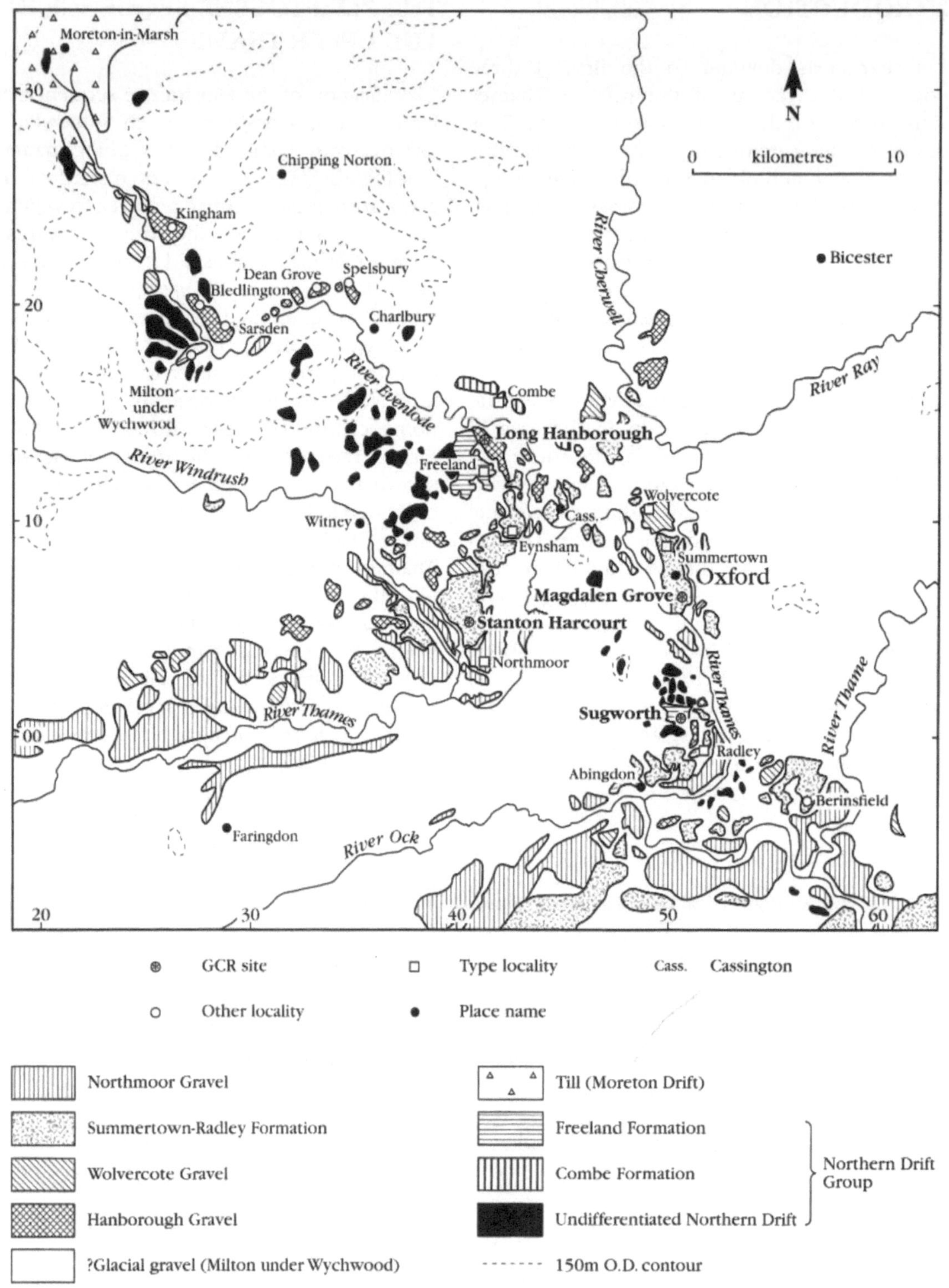

Figure 2.1 The gravels of the Upper Thames catchment.

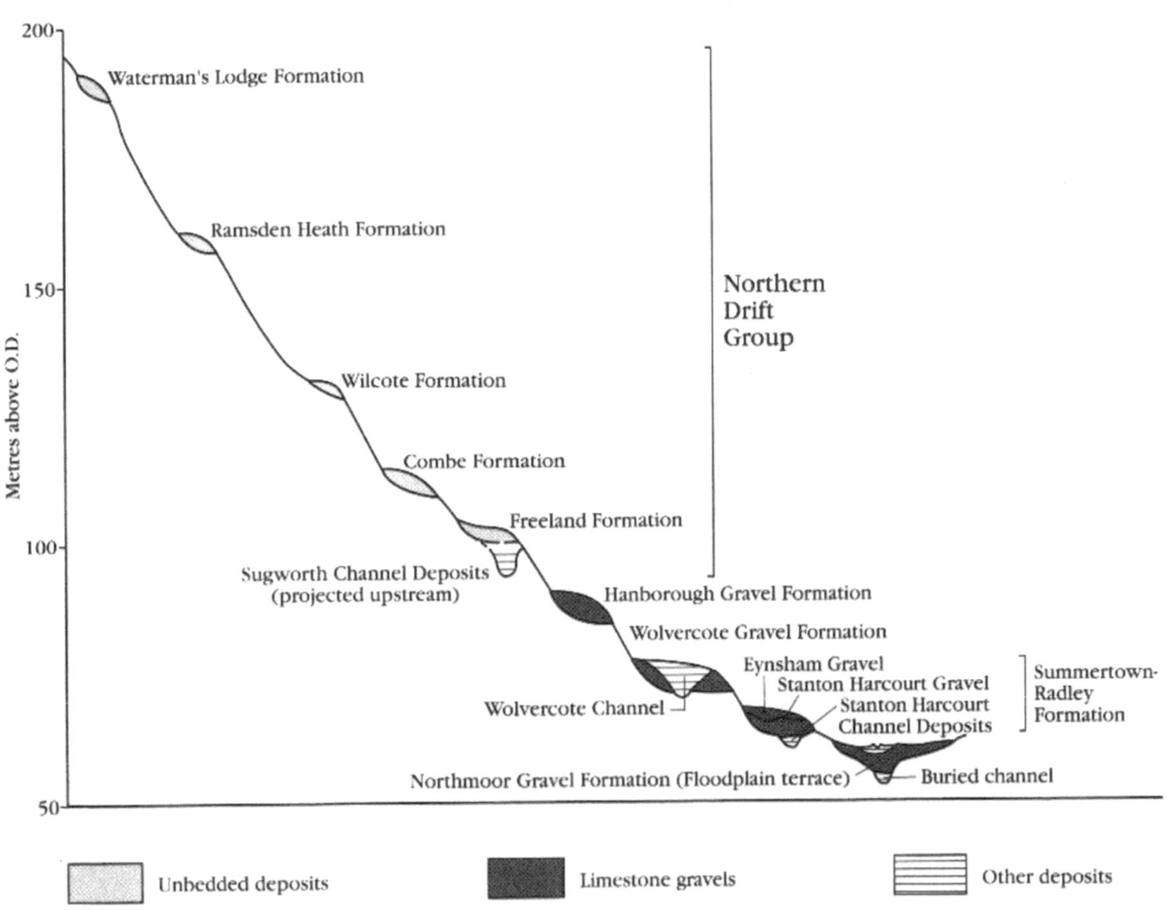

Figure 2.2 Idealized transverse section through the terrace deposits of the Upper Thames (Evenlode).

The Northern Drift remnants contain pebbles of quartz, quartzite, chert and flint derived from the area beyond and to the north of the present Thames catchment, hence the name, which was first applied to these particular deposits by Hull (1855). These high-level deposits are generally unbedded and contain large proportions of silt and clay. For this reason they have, until recently, been regarded as the decalcified remains of till, usually attributed to the Anglian glaciation (Briggs and Gilbertson, 1974). Recent discoveries at Sugworth, near Abingdon, have not only led to the rejection of this interpretation, but have somewhat undermined the distinction between the older and younger Pleistocene deposits. At Sugworth, an interglacial fluviatile channel-fill (see below) was found beneath a spread of high-level unbedded clayey gravel that has been attributed to the Freeland Formation (Hey, 1986). These interglacial sediments yielded biostratigraphical evidence for a Cromerian age and contained pebbles of limestone as well as rock-types characteristic of the Northern Drift (Shotton *et al.*, 1980). This discovery, which was the first occasion on which calcareous gravel had been observed in association with the high-level deposits, demonstrated that Northern Drift material was present in the Upper Thames basin by Cromerian times.

Since the discovery of the Sugworth site, an alternative interpretation of the Northern Drift has gradually gained credence, following the work of Hey (1986). This holds that the high-level material is the degraded and decalcified remains of old terrace gravels of the Upper Thames river system. Some workers continue to believe that glaciation was responsible for the introduction of the Northern Drift pebbles into the Upper Thames basin (Bowen *et al.*, 1986a), but others consider that the Thames catchment was once considerably more extensive,

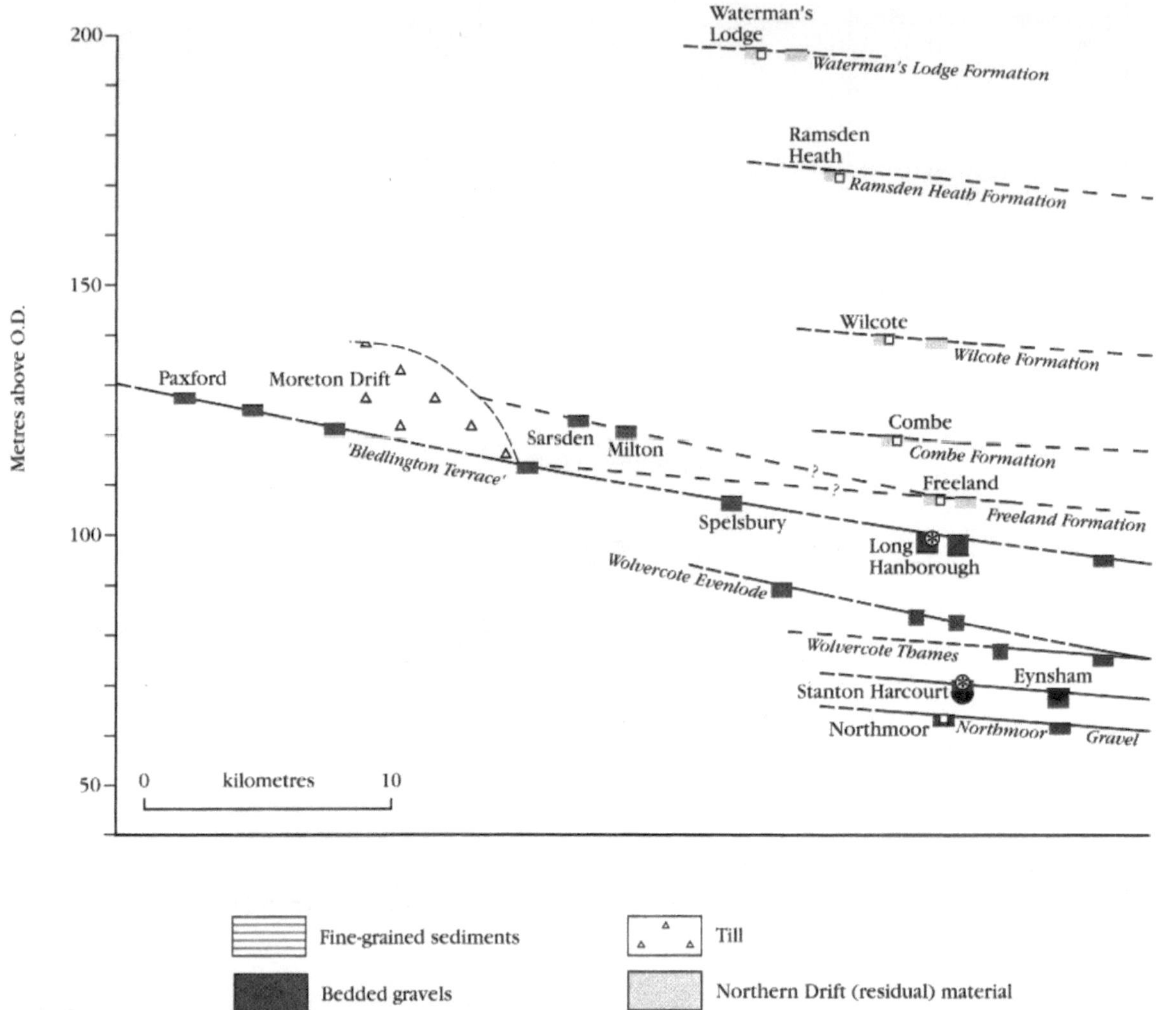

Figure 2.3 Longitudinal profiles of the Upper Thames terrace deposits. Compiled from the following sources: Arkell (1947a, 1947b); Bishop (1958); Briggs and Gilbertson (1973); Briggs *et al.* (1985); Evans (1971); Kellaway *et al.* (1971); Sandford (1924, 1926); Tomlinson (1929).

encompassing the source areas of at least some of these rock-types (Bridgland, 1988c). The distribution of the Northern Drift suggests that the Evenlode was the main drainage line at this time.

The highest and oldest of the various limestone-rich terrace gravels, the Hanborough Gravel Formation, differs from the others in that it is, like the Northern Drift, preferentially preserved in the Evenlode valley, suggesting that this river continued as the main drainage line within the Upper Thames system. This view is supported by the reconstruction of the long-profile of this formation, which extends into the present watershed area between the Evenlode and the Stour (a left-bank tributary of the Warwickshire–Worcestershire Avon) and has a gentle gradient, contrasting with those of later formations (Fig. 2.3). The Hanborough Gravel has yielded the remains of temperate-climate mammals, on the basis of which it was believed until recently to have formed during an interglacial. Most of the bones have come from the base of the deposits, however. In contrast, molluscan remains from silts interbedded with higher parts of the formation indicate cool conditions (Briggs and Gilbertson, 1973). The aggradation is therefore now attributed to a cold episode, probably within the early part of the Saalian Stage. Arkell (1947a, 1947b) considered

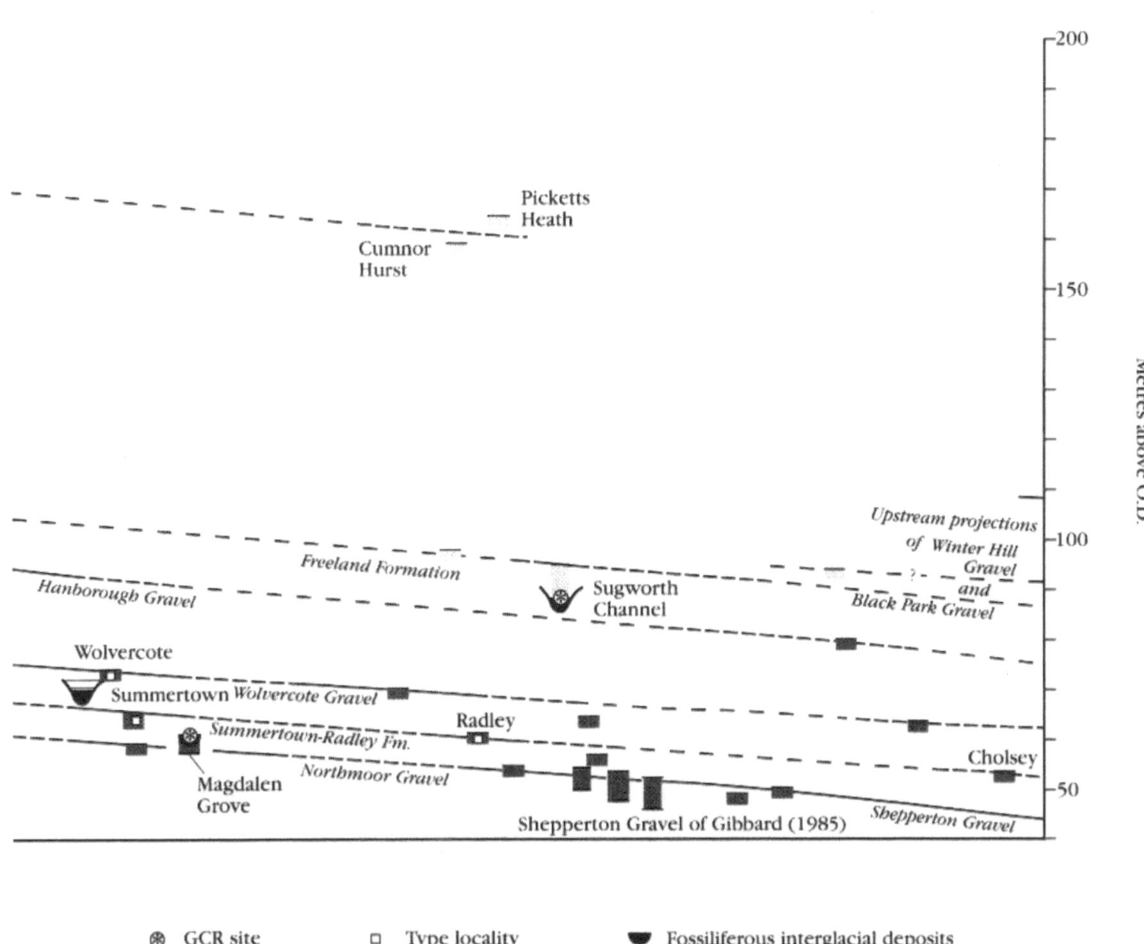

the Hanborough Gravel to pass upstream beneath the deposits of the Cotswolds glaciation (Moreton Drift), an interpretation of great importance in correlating the Pleistocene sequences of the Thames basin and the Midlands (see below and Chapter 1).

The next formation in the Upper Thames sequence, the Wolvercote Gravel, is of considerable stratigraphical significance, in that it has long been held to be the first to contain material introduced into the basin by the glaciation of the Cotswolds. This interpretation is generally attributed to Bishop (1958), who recognized the relatively high flint content of the terrace deposits at Wolvercote and traced the formation upstream in the Cherwell valley, linking it with the glaciation of the Fenny Compton area. However, a succession of later authors have contrasted this deposit with the (higher) Hanborough Gravel and concluded that the glaciation of the Cotswolds occurred between the aggradation of the two formations (Briggs and Gilbertson, 1973; Briggs *et al.*, 1985). Moreover, the Wolvercote Gravel has frequently been interpreted as outwash derived directly from this glaciation, an idea that originated with Tomlinson (1929), who noted the importance of the input of fresh flint many years before Bishop. The Wolvercote Formation is best developed in the Cherwell and Evenlode valleys, both of which flow from gaps in the Cotswold escarpment through which outwash is believed to have escaped from the Midlands ice sheet (Bishop, 1958; Briggs, 1973). This glaciation has generally been correlated with the Saalian Stage (Shotton, 1973a, 1973b) and the associated Wolvercote Gravel has been correlated with the Taplow Gravel of the Middle Thames (Gibbard, 1985), which has been firmly dated in the late part of that stage (see Chapter 3). However, recent re-evaluation of the sequence in the Midlands (Sumbler, 1983a, 1983b; Rose, 1987, 1989) has suggested that the ice advance that reached the Cotswolds may have occurred during the earlier Anglian Stage, initiating a major and continuing controversy over the age of this glaciation. The conclusion in this volume (Chapter 1), that the Wolvercote Gravel equates with the Lynch Hill Gravel of the Middle Thames, in no way resolves this problem, as the latter formation is also attributed to the Saalian Stage (Gibbard, 1985; Table 1.1; see below, Wolvercote). The veracity of the input of flint into the Upper Thames system between the deposition of the Hanborough and Wolvercote Gravels has been called into question in a recent review of the evidence from the Cherwell and Evenlode valleys (Maddy *et al.*, 1991b; see below, Long Hanborough and Wolvercote).

The most important site associated with the Wolvercote Formation, at Wolvercote itself, revealed a large channel cut through the terrace deposits and filled with a sequence of fossiliferous gravels and sands followed by peat and silty clay. The gravels at the base of the channel-fill proved to be the richest source of Palaeolithic artefacts discovered to date in the Upper Thames. Unfortunately, there have been no extensive exposures in these important sediments for many years, so relatively little of their palaeontological potential has been realized. It is known that they represent a cooling sequence following fully temperate conditions, possibly reflecting the end of an interglacial. There has been considerable controversy over the identity of the interglacial represented (see below, Wolvercote), but the correlation of the Thames sequence with the deep-sea record proposed in Chapter 1 suggests that the channel was infilled during Oxygen Isotope Stage 9. This revised correlation raises the possibility that neither of the traditionally recognized post-Anglian interglacials (corresponding with the Hoxnian and Ipswichian Stages) is represented at Wolvercote. Although no conservable remnant of Wolvercote Channel sediments has yet been located, it is hoped that a GCR site in these critically important deposits will be identified in due course.

The next terrace in the Upper Thames sequence, the Summertown-Radley Terrace, is the most extensively preserved. The terrace surface is underlain by a complex aggradational sequence (here classed as a formation) apparently spanning two temperate episodes and one full cold interval, with parts of two others. A complex stratigraphy was recognized in early research on the Summertown-Radley deposits, a cold-climate mammal fauna being identified in the lower part of the sequence, with richer fossiliferous sediments in the upper part containing mammals and molluscs indicative of temperate conditions (Sandford, 1924, 1926). These upper (temperate-climate) sediments, recently given the name Eynsham Gravel (Briggs *et al.*, 1985), have generally been attributed to the last (Ipswichian) interglacial (see below).

The cold-climate Stanton Harcourt Gravel, underlying the Eynsham Gravel, probably constitutes the largest part of the Summertown-Radley Formation. In a large working quarry at Stanton Harcourt, a channel cut in Oxford Clay and filled with richly fossiliferous sediments was found beneath typical cold-climate gravels of the Stanton Harcourt Member. The fauna and flora from this channel is of fully interglacial character, but differs significantly from the assemblages from the Eynsham Gravel. The Stanton Harcourt Channel Deposits have thus been interpreted as the product of an earlier temperate event, possibly one that is as yet undefined, chronostratigraphically, in Britain. This event has been correlated with Oxygen Isotope Stage 7 of the deep-sea record (Shotton, 1983; Bowen *et al.*, 1989; Chapter 1). It is suggested below that many of the historical records of interglacial sediments within the Summertown-Radley Formation that have previously been ascribed to the Eynsham Gravel (Briggs *et al.*, 1985) should be equated instead with the Stanton Harcourt Channel Deposits (see below, Stanton Harcourt and Magdalen Grove). Included amongst these reinterpreted sites is a pit in the grounds of Magdalen College, Oxford, which was re-excavated in 1984 as part of the GCR programme (Briggs *et al.*, 1985).

The Pleistocene sequence in the Upper Thames is completed by the 'Floodplain Terrace' (Sandford, 1924, 1926), otherwise known as the Northmoor Terrace (Arkell, 1947a). This aggradation, here termed the Northmoor Gravel Formation, reaches only *c.* 2–3 m above the alluvium of the modern valley floor and lies 6–10 m below the level of the Summertown-Radley Terrace (Fig. 2.2). Recent mapping has suggested that the floodplain surface is a composite morphological feature (Robson, 1976; Harries, 1977). The aggradational sequence forming the Northmoor Terrace comprises well-bedded limestone gravels, with abundant evidence of a contemporaneous periglacial climate, and is considered to have been laid down during the Devensian (last glacial) Stage (for example, Briggs *et al.*, 1985). Beds and lenses of organic material, widely preserved within the formation, have been found by radiocarbon dating to represent three distinct periods. Plant macrofossils, pollen and the remains of large mammals, molluscs and insects have been found in these various organic sediments. The oldest and lowest occurrences are attributed to cold-climate phases within the latter part of the mid-Devensian, with no evidence of deposition much earlier than 40,000 years BP (these sediments appear to post-date the mid-Devensian (Upton Warren) interstadial of Coope *et al.* (1961)). A further group, higher within the sequence, relates to the Late Devensian late-glacial interstadial; the remainder, from channels cut into the terrace surface, are of early Holocene age (Briggs *et al.*, 1985). The last mentioned coincides with a sedimentary change from (predominantly planar-bedded) gravels to fine-grained silts and peat, thought to reflect an adjustment from a braided gravelly floodplain to a single channel river. A site has come to light recently at Cassington, in the valley of the main Thames, where earlier Devensian sediments may be represented. Early indications, principally from molluscan remains, suggest that channel-fills of early or early mid-Devensian age are represented here within a sequence that is mapped as Northmoor Gravel (D. Maddy, pers. comm.; see below, Stanton Harcourt and Magdalen Grove).

Numerous commercial workings have operated in and continue to exploit the deposits of the Northmoor Formation, but the sections are consistently below the water table, so that abandoned workings rapidly become flooded. This makes the conservation of sections a difficult proposition and no sites in the Northmoor Formation have been selected in the GCR. Devensian and Holocene fluvial deposits are invariably situated at low levels in valleys, so that this problem applies quite generally. New exposures are, however, frequently opened, allowing temporary access to these widely distributed Upper Pleistocene sediments. Devensian fossiliferous deposits are represented within the GCR Pleistocene coverage of the Thames basin at Great Totham, in Essex (see Chapter 5).

SUGWORTH ROAD CUTTING (SP 513007)

Highlights

This is the only Cromerian site discovered to date in the modern Thames valley. It provides critical evidence for interpreting terrace stratigraphy in the Upper Thames basin, as well as an important record of pre-Anglian fauna and flora.

Introduction

The Sugworth interglacial site, discovered during the construction of the A34 Abingdon by-pass in 1972–3, provides important information about the Quaternary evolution of the Upper Thames basin. The sequence at Sugworth comprises fossiliferous clays, silts, sands and gravels in a channel cut into Kimmeridge Clay, capped by an unstratified sandy clayey gravel. The site is remarkable in that, unlike all previously described interglacial sediments in the Upper Thames, it lies at a higher level than the various limestone-rich terrace gravels, apparently falling within the Northern Drift Group (Briggs *et al.*, 1975b; Goudie and Hart, 1975; Shotton *et al.*, 1980; see Introduction to this Chapter).

Synonymous terms for the Northern Drift are 'Triassic Drift' (Harmer, 1907), 'Plateau Drift' (Sandford, 1924, 1926; Goudie, 1976; Briggs *et al.*, 1985) and 'Unbedded Drift', (Dines, 1928). These deposits have generally been interpreted as weathered till (Callaway, 1905; Tomlinson, 1929; Dines, 1946; Briggs and Gilbertson, 1974; Goudie, 1976), although Arkell (1947a, 1947b) considered the lower-level elements of the Northern Drift to include degraded terrace gravels (of proglacial origin). Shotton *et al.* (1980) concluded that much of the Northern Drift might be of fluvial origin, but that glacial transport must have originally carried the material into the Upper Thames basin. Hey (1986) divided the entire Northern Drift into individual fluvial aggradations on the joint bases of elevation and clast composition. He disputed the glacial origin of the material, concluding instead that the Evenlode branch of the Upper Thames once drained a considerably larger catchment, including much of the West Midlands, and that the Northern Drift was the decalcified remains of terrace gravels laid down by this river. Hey's conclusions have been supported by Bowen *et al.* (1986a), Bridgland (1988c) and Whiteman (1990).

The elevation of the Sugworth deposit in relation to the terrace sequence of both the Upper Thames and the Thames system in general, according to the correlation scheme outlined in Chapter 1 (Fig. 1.3), indicates that it is one of the earliest interglacial remnants to survive. This is confirmed by biostratigraphical evidence, particularly from molluscs (Gilbertson, 1980; Preece, 1989) and vertebrates (Stuart, 1980, 1982a), which points to a Cromerian age. Problems of terrace correlation between the Upper and Middle Thames have led to this evidence being questioned (Gibbard, 1983, 1985; Hey, 1986; Chapter 1), but the re-interpretation of the Middle Thames sequence in the Reading area proposed in this volume (Chapter 1) appears to resolve these difficulties.

Description

In 1972–3, the excavation of a cutting for the A34 Abingdon by-pass revealed a number of channels cut into the Kimmeridge Clay and infilled with Pleistocene sediments to a height of 40 m above present river level (Fig. 2.4). These sediments are predominantly medium-coarse orange sand with seams of gravel. One of the channel-fills, in a section cut during realignment of Sugworth Lane (Fig. 2.4), also includes a sequence of grey organic sands, silts and clays (Fig. 2.5). These were found to be richly fossiliferous, yielding beetles, molluscs, ostracods, vertebrates and macroscopic plant remains, as well as poorly preserved pollen (Briggs *et al.*, 1975b; Gibbard and Pettit, 1978; Shotton *et al.*, 1980; Holman, 1987). Other channel features (Fig. 2.4) to the north, at Bagley Wood, and to the south, at Lodge Hill, contain only unfossiliferous sand. Overlying these channel-fills and overlapping them onto the Kimmeridge Clay is a layer of unbedded clayey gravel, 1–2 m thick, containing a similar assemblage of pebbles to the Northern Drift of the district. This is apparently part of a large spread of such 'Plateau Drift' mapped by the Geological Survey in the area (New Series, Sheet 236). Both the fossiliferous channel sediments and the overlying unbedded clayey gravel contain the range of clast-types characteristic of the Northern Drift: predominantly quartz and quartzites, with Carboniferous chert, patinated flint and occasional igneous and metamorphic rocks.

The fossiliferous deposits were concentrated within the central section of the Sugworth Channel, the organic sediments passing laterally into unfossiliferous sands like those filling the Bagley Wood and Lodge Hill channels. Part of this central area was well-exposed in excavations for the Sugworth Lane bridge abutments (Fig. 2.5). Material was collected here from both sides of the A34 cutting, but most of the work was carried out in the excavation for the eastern bridge abutment. Unfortunately, this section is now sealed beneath the concrete bridge structure and its immediate continuation eastwards presumably lies beneath the new alignment of Sugworth Lane.

Interpretation

According to the model for terrace formation and the stratigraphical scheme for terrace correlation presented in this volume (Chapter 1), the sequence at Sugworth may be ascribed to the Freeland Formation and the Sugworth Channel Deposits regarded as a member of that formation. The latter are therefore also included within the Northern Drift Group (Fig. 2.2). They represent deposition under temperate-climate conditions (phase 3 of the terrace model – see Chapter 1), during a temperate episode within the 'Cromerian Complex' (see Chapter 1), whereas the overlying clayey gravel is regarded as the decalcified remains of the subsequent cold-climate (phase 4) aggradation, which is correlated with the Winter Hill and Black Park Gravel Formations of the Middle Thames, both attributed to the Anglian Stage.

Implications of the Sugworth deposit for terrace stratigraphy

The apparent occurrence of Northern Drift material overlying the Sugworth deposit was taken as evidence for the considerable antiquity of the latter in the preliminary reports of the site. The consensus of opinion at that time held that the Northern Drift was a glacial deposit of Anglian age (Briggs and Gilbertson, 1973, 1974; Shotton, 1973a), implying a broadly pre-Anglian age for the Sugworth Channel Deposits. Biostratigraphical evidence from

(A)

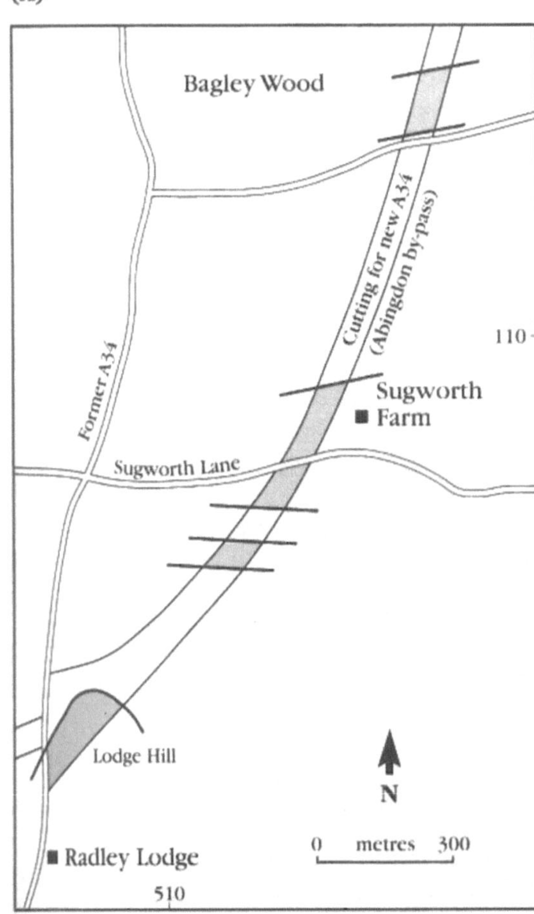

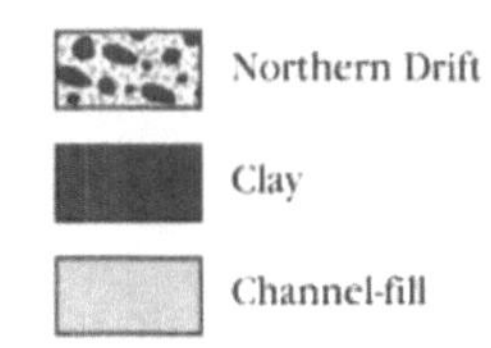

Figure 2.4 Map (A) and section (B) showing the location of the Sugworth channel and its relation to other channel features revealed during the construction of the A34 road cutting (after Shotton *et al.*, 1980).

(B)

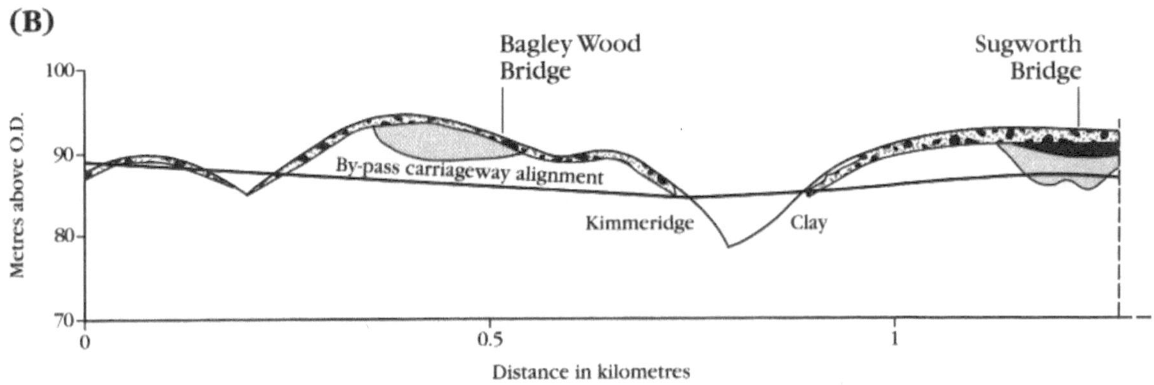

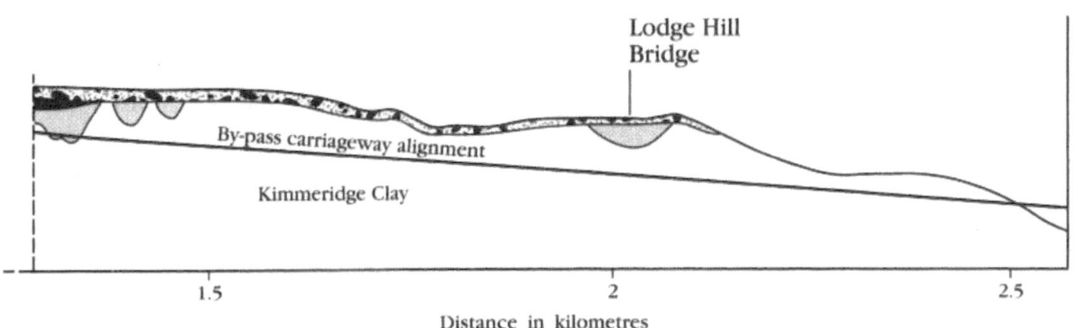

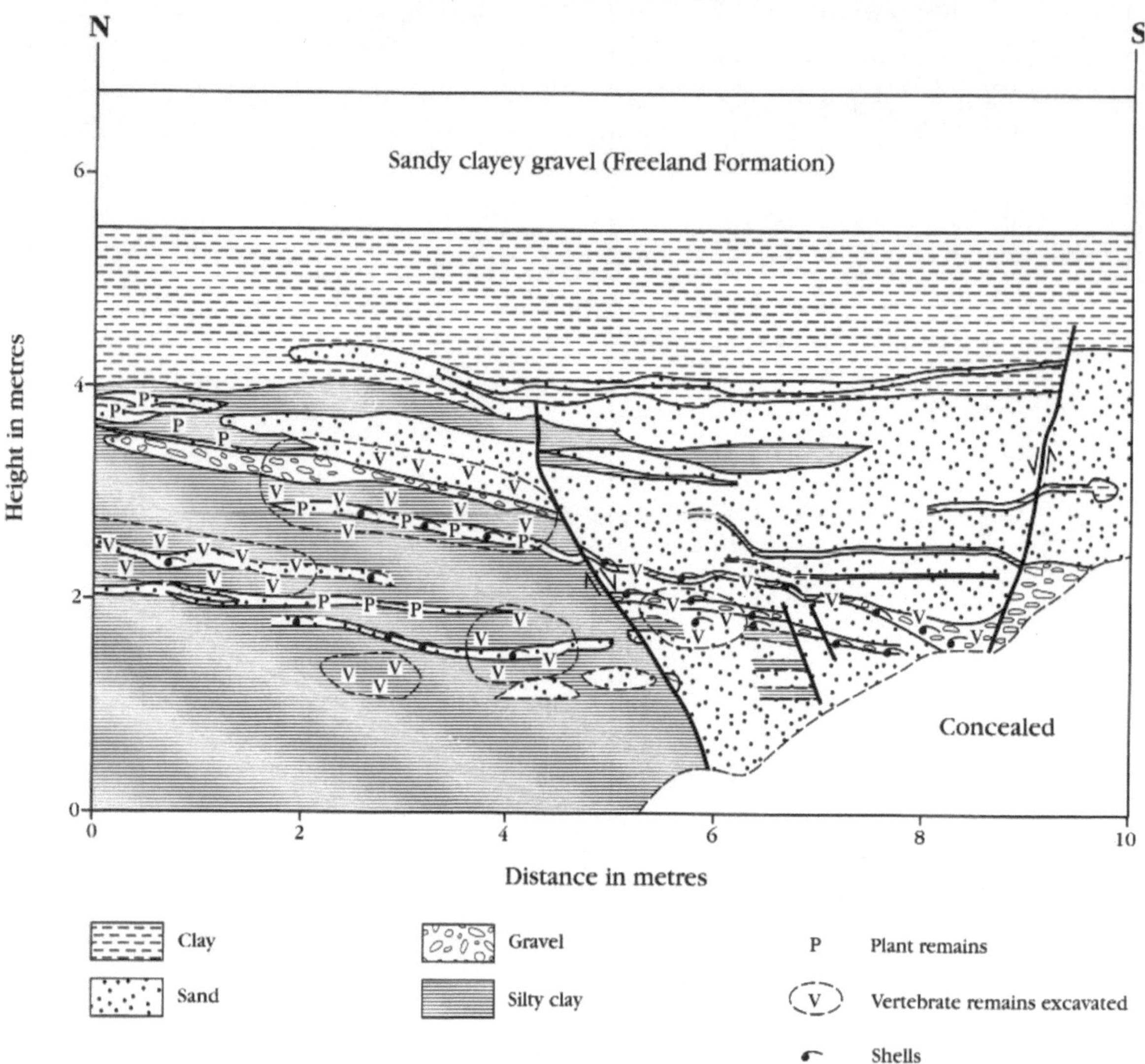

Figure 2.5 Section through the Sugworth Channel Deposits in the east abutment for the Sugworth Lane overbridge (after Shotton *et al.*, 1980).

Sugworth appeared to confirm this view by pointing to deposition during the Cromerian (Briggs *et al.*, 1975b; Goudie and Hart, 1975). It was, however, pointed out by these authors that clasts of lithologies characteristic of the Northern Drift occurred within the Sugworth Channel Deposits, posing difficulties for the existing stratigraphical model, which hinged upon the acceptance of a single period of Northern Drift deposition, during the Anglian Stage.

The results of more extensive analyses of the Sugworth deposits have since been published, including detailed reports on the palaeobotany (Gibbard and Pettit, 1978) and faunal content (Shotton *et al.*, 1980). The assemblages of mammal remains (Stuart, 1980 and in Shotton *et al.*, 1980) and Mollusca (Gilbertson, in Shotton *et al.*, 1980) provide particularly good evidence in support of deposition during the Cromerian (see below).

Osmaston (in Shotton *et al.*, 1980) concluded, from its size and the sedimentology of the deposits filling it, that the channel at Sugworth was the product of the Thames and not of a tributary stream. He calculated that the river responsible for cutting the Sugworth Channel had a considerably larger discharge than the present Thames south of Oxford, probably around seven times the size. The three channels exposed by the A34 cutting have been interpreted as related features, probably repeated intersections with the same meandering course of an early Thames channel (Briggs *et al.*, 1975b; Goudie and Hart, 1975; Goudie, 1976).

Shotton *et al.* (1980), elaborating on the earlier observations of Briggs *et al.* (1975b) and Goudie and Hart (1975), concluded that the simple explanation of the Northern Drift as an Anglian deposit, glacial or otherwise, is untenable. They found that the Sugworth Channel contains both calcareous Jurassic material and typical Northern Drift erratics from beyond the Thames catchment. They interpreted the presence of the latter as an indication that Northern Drift deposits were already in existence during the Cromerian, to provide a source for the erratics in the Sugworth Channel. They therefore concluded that some or all of the Northern Drift was pre-Cromerian.

Heavy-mineral analysis has revealed differences between the unbedded deposit overlying the channel sediments at Sugworth and samples of typical Northern Drift from elsewhere in the Oxford area, despite the fact that all were mapped as 'Plateau Drift' by the Geological Survey. In fact, the unbedded deposit at Sugworth has a heavy-mineral content similar to the underlying channel sediments (Shotton *et al.*, 1980). The main difference is the paucity of garnet in the typical Northern Drift, despite the richness of this mineral in the Triassic of the West Midlands, the main source of Northern Drift pebbles. Garnet is relatively abundant in the channel sediments and overlying unbedded clayey gravel at Sugworth. This difference can probably be attributed to the weathering of garnet over a lengthy period. It therefore argues for the greater antiquity of the Northern Drift in general, and particularly those remnants at higher levels, in comparison with the channel sediments and overlying deposits of Northern Drift type at Sugworth. If this is the case, a pre-Anglian age for the higher-level Northern Drift is clearly indicated.

Shotton (1981) adopted a polygenetic interpretation of the Northern Drift. He argued that much of this material is probably the de-calcified upstream continuation of the various pre-Anglian gravels of the Middle Thames, but that the Triassic-derived clasts must have originally been carried into the area by glacial means, since the soft clays of the Lias would have always given rise to a north-east/south-west-trending strike valley separating the Cotswolds from the source of these rocks in the West Midlands (Shotton *et al.*, 1980; Shotton, 1981, 1986). This argument hinges on the effectiveness of geologically controlled strike valleys as barriers to through-drainage. However, it is only necessary to look as far as the Weald to observe rivers habitually crossing outcrops of non-resistant rock-types, such as the Weald Clay and the Gault Clay, that have given rise to strike valleys.

By comparing the relative proportions of their quartz and quartzite components, Hey (1986) demonstrated that Northern Drift remnants from different altitudes represent distinct fluvial aggradations. He sought to correlate these with the pre-Anglian gravels of the Middle Thames (see Chapter 1). According to Hey, the unbedded gravel capping the Sugworth Channel is a decalcified and degraded remnant of the fluvial deposit defined as his Freeland Member (here classed as a formation). It is apparent, therefore, that the Northern Drift represents a series of decalcified terrace deposits of various ages, the oldest, highest units presumably dating from the Early Pleistocene, whereas the youngest, the Freeland Formation, is post-Cromerian (*sensu* Sugworth). This interpretation explains the occurrence of Northern Drift material both in and above the Sugworth Channel Deposits and also explains the preservation within the deposits at Sugworth of an unusually large amount of semi-durable garnet. The Freeland Formation, as the lowest (and youngest) within the Northern Drift Group, would be expected to be the least weathered. This is not only confirmed by the relative abundance of garnet at Sugworth, but also by the presence of a few surviving limestone clasts (Shotton *et al.*, 1980).

The distribution of substantial Northern Drift remnants alongside the course of the River Evenlode suggests that this river had a much larger catchment in pre-Anglian times than at present. It is therefore possible to envisage the early Evenlode draining the source areas of the Northern Drift erratics directly, obviating the need to invoke glacial transport (Bridgland, 1986b, 1988c; Hey, 1986; Fig. 2.6). This idea was, in fact, very popular amongst earlier workers, many of whom considered a 'Severn-Thames' river to have existed prior to the Middle Pleistocene (Ellis, 1882; Davis, 1895, 1899, 1909; Buckman, 1897, 1899a, 1899b, 1900; White, 1897).

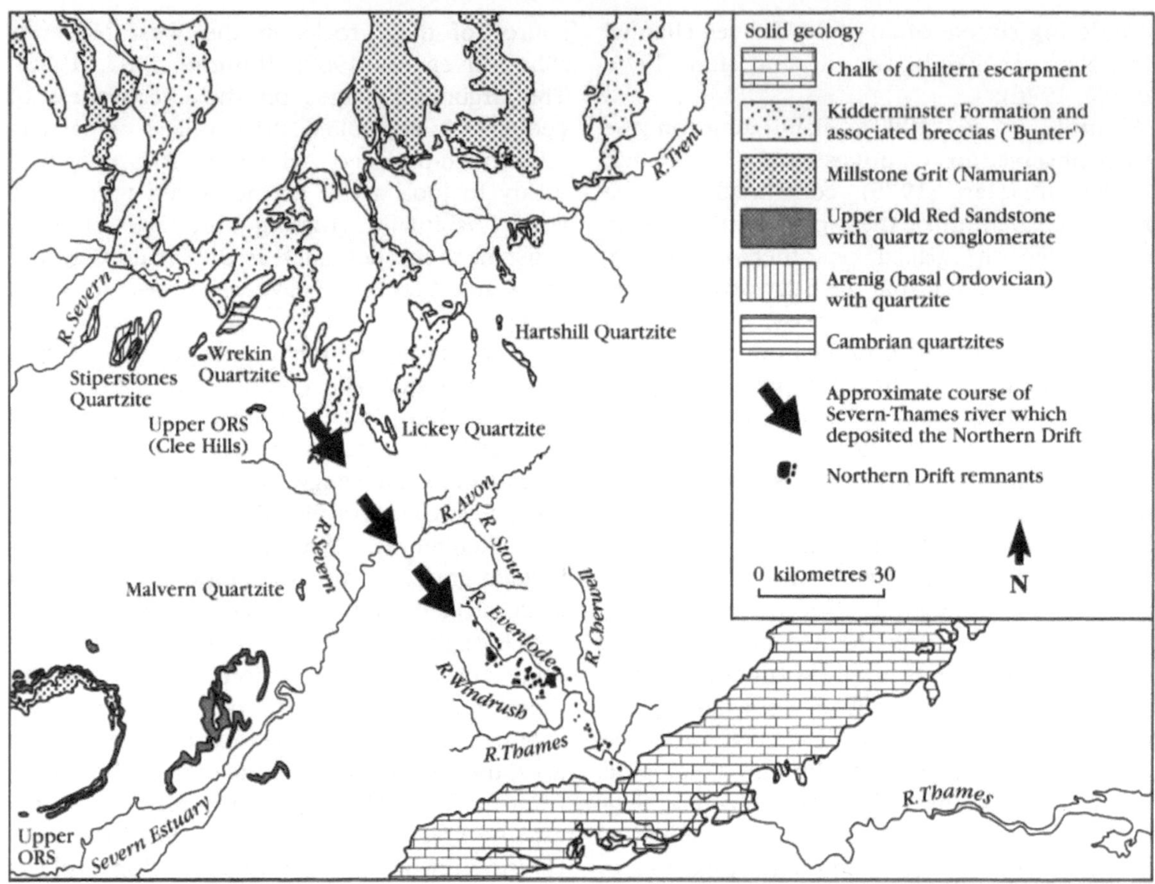

Figure 2.6 Map showing the course of the hypothetical Severn–Thames river of the Early Pleistocene.

Biostratigraphical implications of the Sugworth deposit

The most critical palaeontological evidence at Sugworth, and the strongest indication of a Cromerian age, comes from the vertebrate and molluscan remains. Large mammals from the site include *Dicerorhinus etruscus* (Falconer), restricted in Britain to the Cromerian and Pastonian Stages. Amongst the small mammals present is the water vole *Mimomys savini* (Hinton), which is unknown after the Cromerian. This species was replaced by *Arvicola cantiana* (Hinton), either during the latter part of the Cromerian Stage (*sensu* West Runton) or between it and an additional temperate episode that has been claimed, on the basis of this faunal change, to have occurred between the Cromerian and the Anglian Stage (Stuart, 1988; see Chapter 1). Teeth comparable with those of the extinct shrew *Sorex savini* (Hinton) from the type-Cromerian at West Runton were also present (Stuart, *in* Goudie and Hart, 1975; Stuart, 1980, 1982a).

The Mollusca included a number of species that first appear in the British record in the Cromerian Stage, namely *Valvata naticina* (Menke), *Bithynia inflata* (Hansén), *Marstonniopsis scholtzi* (Schmidt) and *Unio crassus* (Philipson). These occur with *Valvata goldfussiana*, which disappears from the record after the Cromerian, and *Tanousia runtoniana* (Reid) (= *Nematurella runtoniana* auctt.), which is unique to that stage. The dominance of taxa indicative of fully temperate conditions, coupled with the consistent fall in species number from the base to the top of the channel-fill (which implies climatic deterioration), suggests correlation with biozone CrIII (late temperate) of the interglacial, somewhat after the climatic optimum (Gilbertson, in Goudie and Hart, 1975; Gilbertson, 1980). Additional molluscan species have recently been recognized by Preece (1989).

Other aspects of the fauna and flora are less conclusive. The ostracods include abundant *Scottia browniana* (Jones), which limits the age to Hoxnian or earlier, together with occasional species of early Middle Pleistocene or late Early Pleistocene affinities (Robinson, 1980). The beetle fauna from Sugworth indicates a temperate climate and the presence of deciduous woodland (Osborne, in Shotton *et al.*, 1980). However, very few sites of Cromerian age in Britain have yielded insect remains, so stratigraphical comparisons cannot yet be made on this basis. The palynological evidence from Sugworth is insufficiently detailed to distinguish between the Hoxnian and Cromerian stages, although biozone III of one or the other is indicated (Gibbard and Pettit, 1978; Pettit and Gibbard, in Shotton *et al.*, 1980). The consensus view of Shotton *et al.*, based on consideration of the various biostratigraphical evidence, was that the Sugworth Channel was infilled during biozone CrIII of the Cromerian Stage.

Wider significance and correlation

Recent attempts at correlation between high-level terraces in the Middle Thames and the various decalcified aggradations of the Northern Drift Group have led to a reconsideration of the stratigraphical position and age of the Sugworth deposits. Gibbard (1985) suggested a correlation between the late Anglian Black Park Gravel of the Middle Thames and the Freeland Terrace, on the basis of elevation and projected long-profile reconstructions. The same author correlated the Winter Hill Gravel of the Middle Thames, which he held to be contemporaneous with the Anglian Stage glaciation of the Vale of St Albans, with the Combe Terrace of Arkell (1947a, 1947b), the division of the Northern Drift immediately higher in the terrace sequence than the Freeland Formation (see Fig. 2.2; Chapter 1). These correlations were supported by Hey (1986). However, as both these authors recognized, there are inherent difficulties with this interpretation. According to Hey (1986), the channel-fill at Sugworth is, on the basis of standard terrace stratigraphy, intermediate in age between the Combe and Freeland aggradations. This would appear to imply that the Sugworth Channel Deposits are themselves of Anglian age, a correlation that is clearly untenable. Gibbard's interpretation differed from that of Hey in that he did not include the upper unbedded gravel at Sugworth in the Freeland Formation. On the basis of his terrace correlations, Gibbard (1985) considered the Sugworth deposits to be too low (at 85.5–92 m O.D.) to be related to either of the Anglian aggradations, the Combe and Freeland Formations (the latter he projected to *c.* 96 m O.D. in the Sugworth area). Instead, he placed them between the late Anglian Freeland Formation and the early Saalian Hanborough Gravel. Gibbard was therefore forced into the conclusion, in spite of the biostratigraphical evidence outlined above, that the Sugworth sediments were deposited during the late Hoxnian (biozone HoIII), thus explaining his earlier inclusion (Gibbard, 1983, p. 23) of the site in the Hoxnian Stage.

This conclusion has received little support. Preece (1989) reiterated the biostratigraphical case for assigning the site to the Cromerian. He pointed out, however, that the continental record indicates considerable stratigraphical complexity for the Lower Pleistocene to lower Middle Pleistocene period. In The Netherlands, in particular, a series of glacial-interglacial cycles is known from this period, giving rise to the term 'Cromerian Complex' (Zagwijn, 1986; de Jong, 1988; see Chapter 1). Preece therefore accepted that the Sugworth sediments might represent a different temperate episode within the Middle Pleistocene to the type-Cromerian at West Runton. Despite this, he was adamant that the deposits at Sugworth could not be post-Cromerian. A similar view has been adopted by Bridgland *et al.* (1990), in a three-way biostratigraphical comparison of the sites at Sugworth, Little Oakley (Essex) and the West Runton stratotype. All three sites were interpreted as 'broadly Cromerian', although it was recognized that difficulties existed in correlation with the complex Dutch sequence. It is therefore possible that different episodes within the 'Cromerian Complex' may be represented by the three British sites (see also Chapter 5, Little Oakley).

Further support for the correlation of the Sugworth Channel Deposits with the Cromerian Stage (*sensu lato*) has recently been provided by amino acid geochronology. Preece (1989) quoted similar D-alloisoleucine : L-isoleucine ratios from *Valvata piscinalis* (Müller) shells from Sugworth (mean 0.286 ± 0.016, n = 2) and West Runton (mean 0.283 ± 0.037, n = 5). These compare with significantly lower ratios from the same species from Hoxnian sites (from Hoxne, for example, a mean of 0.243 ± 0.023, n = 3). These ratios were measured in the

INSTAAR laboratories of the University of Colorado. Rather different results were produced by Bowen *et al.* (1989), using the same technique. These authors obtained a comparable mean ratio from *Valvata goldfussiana* (Wüst) from Sugworth (0.296 ± 0.008, n = 4), but this is significantly lower than *V. piscinalis* ratios obtained by this laboratory from specimens from West Runton (0.348 ± 0.011, n = 5) and from Little Oakley (0.324 ± 0.004, n = 2 and 0.336 ± 0.027, n = 4) (Bridgland *et al.*, 1990; see Chapter 5, Little Oakley). The closest match for the Sugworth ratio amongst the Bowen *et al.* data for *Valvata* species is with Clacton (0.299 ± 0.002, n = 3) and Ingress Vale, Swanscombe (mean 0.297 ± 0.009, n = 5), both of which they ascribed to Oxygen Isotope Stage 11. Bowen *et al.* considered the Hoxnian (*sensu* Hoxne) to correlate with Oxygen Isotope Stage 9, so the above interpretation still holds that Sugworth is pre-Hoxnian. However, Bowen *et al.* (1986b) correlated the Anglian Stage with Oxygen Isotope Stage 12, which would therefore place Sugworth after the Anglian.

The above correlation, which implies the separation of the Anglian and Hoxnian by an extra interglacial-glacial cycle, provides some support for Gibbard's (1985) interpretation of the Sugworth Channel Deposits as Hoxnian. However, even if the Swanscombe deposits are attributed, on the basis of amino acid ratios, to a separate temperate episode, earlier than the Hoxnian (*sensu* Hoxne), the palaeontological evidence from the Swanscombe deposits strongly resembles that from other Hoxnian sites and provides little support for a correlation with Sugworth (see Chapter 4, Swanscombe). Moreover, the mammalian and molluscan assemblages from the two sites strongly indicate that Sugworth, with clear Cromerian affinities, is older than Swanscombe.

The revised scheme for terrace correlation between the Upper and Middle Thames, outlined in Chapter 1 (Fig. 1.3), allows a more satisfactory interpretation of the Sugworth Channel Deposits and their relation to British Pleistocene chronostratigraphy. This revised scheme holds that the Freeland Formation is the upstream equivalent of both the Winter Hill and the Black Park Gravels of the Middle Thames, thus indicating a much lower elevation for the Upper Thames floodplain during the earlier part of the Anglian Stage than was envisaged by Gibbard (1985). No Anglian aggradation therefore occupies a higher terrace position than the sequence at Sugworth. This interpretation allows the biostratigraphical evidence for a Cromerian age for the channel deposits to be reconciled with the relative dating of the various terrace formations, derived from their stratigraphical relation to the glaciation of the Vale of St Albans.

One of the few other Cromerian sites in Britain outside the type area (the Cromer Forest Beds of north Norfolk), that at Little Oakley (north-east Essex), also occurs within the Thames system (see Chapter 5). The deposits at Little Oakley appear to fall within the Low-level Kesgrave Group (see Chapters 1 and 5). It was suggested in Chapter 1 that during the early Middle Pleistocene, phases of downcutting (rejuvenations) by the Thames had little effect in the valley upstream from Essex. This view is supported by the Upper Thames sequence, where the Sugworth deposits, believed on palaeontological grounds to be of similar age to those at Little Oakley (Bridgland *et al.*, 1988, 1990), appear to underlie the upstream equivalent of the (Anglian) Winter Hill and Black Park aggradations, implying that no rejuvenation took place in this area between the Cromerian and the end of the Anglian.

Summary

The Sugworth GCR site is of major importance to British Pleistocene stratigraphy. The oldest interglacial remnant within the Upper Thames sequence is found at this locality. Its interpretation has been the subject of some controversy, resulting largely from difficulties in correlating terrace deposits between the areas north and south of the Chilterns. There appears, however, to be overwhelming palaeontological evidence that it is of a broadly Cromerian age. Sugworth therefore represents one of only a handful of Cromerian localities in this country. Cromerian deposits that can be related to fluvial terrace sequences are extremely rare and Sugworth is the only such site in the valley of the modern Thames. It has important implications both for Thames terrace stratigraphy and for correlation with the neighbouring Midlands region.

Satisfactory correlation of British Cromerian sites with the more complex Dutch sequence has yet to be achieved. This correlation would be an essential preliminary to matching the British terrestrial stratigraphy with the oceanic early Middle Pleistocene oxygen isotope record. Evidence from fossiliferous sites such as Sugworth will provide the basic data for such correlation. Early results from amino acid geochronology suggest that correlation with the oceanic record will be possible in the near future.

Conclusions

The richly fossiliferous channel-fill deposits at Sugworth (containing fossil molluscs, beetles, ostracods, vertebrates and plant remains, including pollen) are believed to have been deposited during an interglacial (one of the warmer, non-glacial phases of the Pleistocene 'Ice Age'), about half a million years ago.

These interglacial sediments, deposited by an ancestral Thames that was larger than the present river, are overlain by an enigmatic, clayey, gravelly deposit. This later deposit appears to form part of a river terrace that accumulated during a much colder climatic phase, about 450,000 years ago (the Anglian Stage), when ice sheets covered most of Britain. Prior to the discovery of the Sugworth interglacial channel, it was widely believed that such clayey and gravelly sediment, found covering high ground in the Upper Thames, was a true glacial till (boulder clay). This Northern Drift, as it was called (because it contains numerous pebbles of rocks derived from the area north of the Cotswolds, the present limit of the Thames catchment), is now considered to represent the remnants of a series of early river terraces, now weathered and degraded, dating from a period during the early Pleistocene when the Thames drained a far larger catchment than that of the modern river.

LONG HANBOROUGH GRAVEL PIT (SP 418136)

Highlights

This pit is the type locality for the Hanborough Gravel Formation, the oldest of the limestone terrace gravels of the Upper Thames. Sedimentary features, periglacial structures and snail faunas in the deposits at Long Hanborough have been fundamental in demonstrating that these and, by comparison, many other Thames terrace gravels are largely the products of fluvial deposition under intensely cold conditions.

Introduction

Long Hanborough Gravel Pit provides sections in the gravel that forms the Hanborough Terrace of the River Evenlode. This deposit, the Hanborough Gravel Formation, is aggraded to *c.* 100 m O.D. in its type area and is the highest of the limestone-dominated terrace gravels of the Oxford region (Sandford, 1924, 1926; Arkell, 1947a, 1947b; Briggs and Gilbertson, 1973, 1974). Higher and older formations, with the exception of the Sugworth Channel sediments, are preserved only as decalcified Northern Drift deposits, dominated by quartz and quartzites (see Introduction to this Chapter and the Sugworth report). The Hanborough Formation is also recognized in the Cherwell and Thames valleys (Fig. 2.1).

The largest remnant of Hanborough Terrace deposits is that on which the villages of Church Hanborough and Long Hanborough are built (Fig. 2.7). Since the Long Hanborough pit is situated on the 'type outcrop' of the Hanborough Gravel, it may be considered to represent the type section for this formation, there being no other exposures surviving in the area. Although earlier pits have been described (Sandford, 1924; Arkell, 1947a, 1947c), much important recent work on the Hanborough Terrace sediments has been carried out at the GCR site, notably the description of a cold-climate molluscan fauna from silty horizons within the gravel sequence (Briggs and Gilbertson, 1973, 1974). This did much to overturn old ideas that the aggradation of terrace deposits (here and elsewhere) took place during temperate periods. The Hanborough Gravel is now believed to have been laid down under cold (periglacial) conditions, but the identification of the cold episode during which this took place is the subject of controversy, particularly as a result of recent work in the Midlands (Rose, 1987, 1989), to the north-west of the Evenlode catchment (see below and Chapter 1).

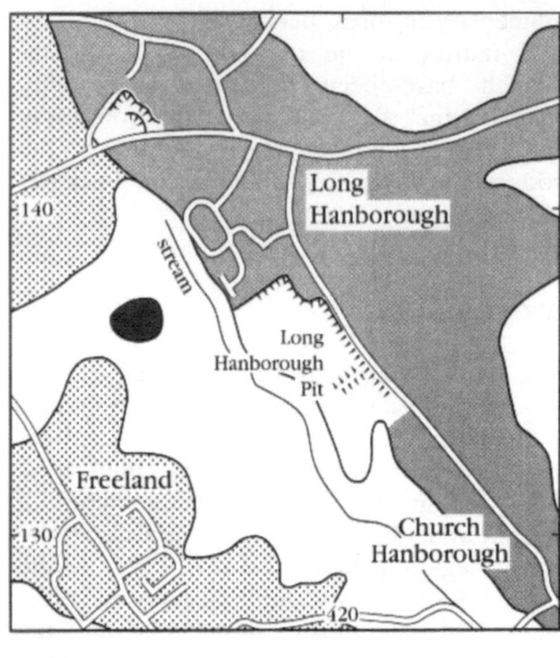

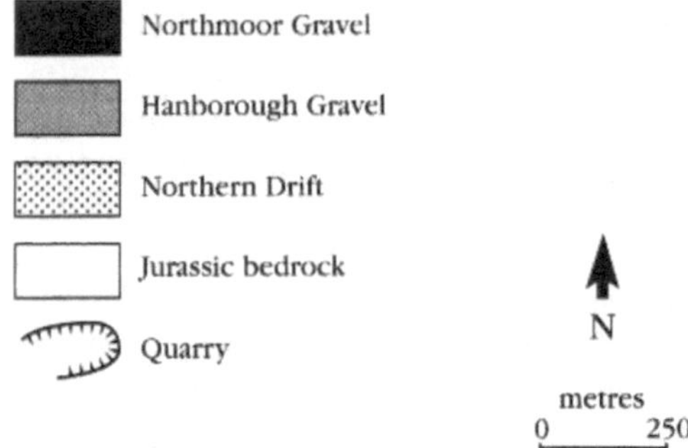

Figure 2.7 Map of the Long Hanborough area, showing the location of the GCR site. The quarry in the top left part of the diagram is Duke's Pit.

Description

Early descriptions of Hanborough Terrace deposits record a variety of different exposures. According to Arkell (1947a, p. 202), 'a dozen large pits at Long Hanborough and Church Hanborough' had existed (or did exist) by 1947. The most famous site was Duke's Pit (SP 415143), 0.5 km to the north-west of the present site, which was described by Sandford (1924), Richardson (1935) and Arkell (1947a, 1947c). Sandford (1924, 1925) also described a working known as Lay's Pit, Long Hanborough, near the northern edge of the gravel spread, and two pits in the Hanborough Terrace of the Cherwell at Kirtlington. Numerous less important exposures were described in the second Oxford memoir (Sandford, 1926), but the most extensive list of pits in the Hanborough Terrace was provided by Dines (1946) in the Witney memoir. In more recent years attention has been centred on the present site, to the south of Long Hanborough village, as well as on exposures at Dean Grove, *c.* 10 km upstream (Briggs and Gilbertson, 1973, 1974), and at Cassington (Kellaway *et al.*, 1971).

Sandford (1924) established that a number of mammalian species occurred in the gravel of the Hanborough Terrace, namely ***Palaeoloxodon antiquus*** (Falconer and Cantley), ***Mammuthus primigenius*** (Blumenbach), ***Dicerorhinus*** sp. (probably ***D. hemitoechus*** (Falconer)), ***Bos primigenius*** (Bojanus), ***Equus ferus*** (Bodaent) and ***Cervus elaphus*** (Falconer) (Sandford, 1924, 1925, 1926). Arkell (1947c) reported the discovery of an Acheulian hand-axe in Duke's Pit, Long Hanborough. This is the only palaeolith that can be attributed with certainty to the Hanborough Terrace gravel (Arkell, 1947c; Wymer, 1968; Roe, 1976; Briggs *et al.*, 1985), although a previous record exists of an early find that probably came from these deposits (Anon., 1908; Manning and Leeds, 1921; Wymer, 1968).

The most recent detailed description of the Long Hanborough GCR site was by Briggs and Gilbertson (1973), who recorded the following sequence at NGR SP 418138:

	Thickness
3. Loam, red-brown, unbedded. Contains scattered non-calcareou s pebbles and is typically up to 1 m thick. This deposit fills pipes in the gravel down to a depth of 2.5 m from the surface	up to 1 m
2. Gravel, pale-coloured (pinkish), with cross-bedding and localized channelling; dominantly composed of limestone clasts, with quartz, quartzite and sandstone, plus rare flint and igneous or metamorphic rocks. Occasional seams and thin beds of silt occur, locally containing non-marine Mollusca (see Fig. 2.8).	1–3 m
1. Oxford Clay	

The recovery by Briggs and Gilbertson of a gastropod assemblage from silts within bed 2, comprising *Pupilla muscorum* (L.), *Oxyloma pfeifferi* (Rossmässler), *Catinella arenaria* (Bouchard-Chantereaux), *Planorbis* spp., *Columella columella* (Martens), *Agriolimax* spp., *Pisidium vencentianum* (Woodward), *P. nitidum* (Jenyns) and *Trichia hispida* (L.) provided important additional evidence for palaeoenvironmental conditions during the aggradation of the Hanborough Terrace deposits. The origin of the 'pipes' filled with red brown loam (bed 3) has been the subject of considerable debate (see below).

Interpretation

The Hanborough Formation has been recognized as a separate, early division of the limestone gravels of the Upper Thames for over a century. In early Geological Survey memoirs both Hull (1859) and Green (1864) noted that the gravel at Long Hanborough and Church Hanborough lay at a greater elevation than the bulk of the terrace deposits in the area. The Hanborough Terrace, which has also been referred to as the '100 ft Terrace', the 'High Terrace' and the 'Fourth Terrace' (Pocock, 1908), is better preserved in the Evenlode valley than in the other branches of the Upper Thames system.

Sandford (1924), in the first of his major papers on the Upper Thames, introduced the term 'Handborough Terrace' (using the old spelling). He interpreted the various remnants of this formation as the product of a temperate-climate environment, on the basis of the mammalian remains found in them (see above). Arkell (1947a, 1947b) interpreted the Hanborough Terrace as an interglacial gravel 'delta', remarking on the evidence in the Upper Thames basin for a single period of major rock-disintegration and terrace formation, represented by the aggradation at Hanborough. This he correlated with the comparable 'Silchester Gravel' of the Kennet valley, assigning both to the 'First Interglacial' (this would now be equated with the Cromerian).

Arkell was the first to claim that the Hanborough Gravel has a gentler downstream gradient than later terrace deposits in the Evenlode valley, including the floodplain sediments. It therefore intersects with the projected profiles of successively more recent formations as it is traced upstream. Upstream from Kingham, where the Hanborough Gravel is only *c.* 10 m above the modern floodplain, Arkell believed it to underlie glacial deposits, the 'Moreton Drift'. These flint-rich glacial sediments are believed to result from the glaciation that deposited the 'Chalky Boulder Clay' ('Chalky Till') of the Midlands, this col in the Cotswolds escarpment representing its furthest encroachment into the modern Thames catchment (Arkell, 1947a, 1947b; Bishop, 1958; Kellaway *et al.*, 1971; Briggs, 1973). To the north, beyond the present Stour–Evenlode watershed, a limestone gravel similar to the Hanborough Formation occurs beneath the Moreton Drift in the area of Great Wolford and Stretton-on-Fosse. Arkell believed that this deposit, which he called Paxford Gravel, was an upstream continuation of the Hanborough Gravel, laid down by a formerly more extensive Evenlode.

This interpretation is of considerable importance: if the Paxford Gravel is correctly identified as an upstream equivalent of the Hanborough Formation, the latter must pre-date the 'Chalky Till' glaciation of the Cotswolds. The same conclusion has also been reached from the analysis of gravel composition. It has been widely reported that the Hanborough Gravel of the Evenlode valley does not contain the typical north-eastern erratic components, fresh flint in particular, that were introduced into the Stour–Evenlode watershed area by the 'Chalky Till' glaciation; it has been claimed instead that such material first appears in the Wolvercote Gravel (Bishop, 1958; Kellaway *et al.*, 1971; Briggs, 1988; see below, Wolvercote). This view has been challenged recently by Maddy *et al.* (1991), who have suggested that there is no significant difference between the gravel contents of these two formations in the Evenlode.

Kellaway *et al.* (1971) and Briggs and Gilbertson (1973) have supported Arkell's reconstruction of the Hanborough Terrace long-profile and its convergence upstream with the valley floor. However, Tomlinson (1929) and Kellaway *et al.* (1971) recorded gravels of Hanborough type from the Wychwood area at levels that are too high to be consistent with this interpretation. If Arkell's correlation of the Paxford and Hanborough Gravels is correct, these higher calcareous deposits may be non-decalcified remnants of the lowest Northern Drift formation(s) or they may be locally derived 'fan gravels'. However, it has been shown in

Chapter 1 that miscorrelation may have occurred between the Upper and Lower Evenlode valleys, with the implication that the gravels underlying the Moreton Drift might be older than the Hanborough Formation. Further investigation of the gravel remnants in the Evenlode valley is required in order to assess the relations of the Hanborough Formation to the glacial deposits of the Cotswolds.

More recent studies in the south Midlands have challenged Arkell's conclusion that the Hanborough Gravel dates from the Cromerian. Tomlinson (1963) placed it in the 'Great Interglacial' (Hoxnian) on the grounds of its mammalian fauna and its relation, established by Arkell, to the Cotswolds glaciation, which she attributed to the Saalian. Wymer (1968) pointed out that no elements of the fauna found at Long Hanborough are diagnostic of the Cromerian as opposed to the Hoxnian Stage and that a deposit as old as the Cromerian might be expected to be more decalcified. He also considered substantial accumulations of gravel of the type represented at Long Hanborough to be more typical of cold-climate than of interglacial conditions.

Kellaway *et al.* (1971) thought that the relative abundance of Triassic quartzites and the paucity of flint in the Hanborough Gravel implied formation during the interval between the 'Northern Drift glaciation' and the 'Chalky Till' glaciation. According to Shotton (1973a), this would place the Hanborough aggradation between the Anglian and Saalian (Wolstonian) glaciations. The rejection, in recent years, of the Northern Drift as a glacial deposit renders this view obsolete (see above, Sugworth); it is now clear that the Northern Drift is a complex group of deposits spanning the Lower and lower Middle Pleistocene (Chapter 1). Another problem in assessing the age of the Hanborough Formation is the dispute that has arisen in recent years as to which glacial episode is represented by the 'Chalky Till' of the West Midlands (see below).

An important challenge was made by Briggs and Gilbertson (1973, 1974) to the long-held belief that the Hanborough Terrace deposits were of interglacial origin. This view, based on the occurrence of occasional mammal bones in the gravels (see above), had been questioned by Wymer (1968), but reiterated by Kellaway *et al.* (1971). Briggs and Gilbertson presented new evidence from the Long Hanborough pit, from studies of sedimentology and molluscan faunas, pointing to deposition under cold-climate conditions. The sedimentological analyses, based on samples from temporary exposures at Milton-under-Wychwood and Kingham, as well as the pits at Dean Grove and Long Hanborough, revealed a considerable downstream increase in coarse-grade local material. This would have been incorporated in the gravel as the river flowed through the Evenlode Gorge, suggesting a period of highly active local erosion. In addition to this, the sediment type and the range of sedimentary structures occurring in the deposits suggest deposition by a braided river. The full picture is therefore one of a braided, gravel-bed river in an open environment with sparse vegetation.

Assemblages of non-marine Mollusca obtained by Briggs and Gilbertson (1973) from two silt/fine-sand horizons in the Long Hanborough pit (Fig. 2.8) provide further important evidence. The lower of these horizons yielded species indicative of subarctic conditions, with *Pupilla muscorum*, a xerophilous land-snail, dominating an assemblage otherwise typical of marshlands and small streams. In the higher silty horizon *P. muscorum* was replaced by *Oxyloma pfeifferi*, a snail found in marshland or damp habitats, in a general assemblage hinting at slightly milder conditions than that from the lower horizon. Briggs and Gilbertson likened the assemblages from both horizons to those from the various Devensian interstadials. They concluded that the Hanborough Gravel accumulated under cold (periglacial) conditions and that the interglacial mammalian remains previously recorded, which appear to have been concentrated near the base of the aggradation, were probably reworked from earlier deposits and/or land surfaces. They suggested an early Wolstonian (Saalian) age as the most probable, taking into account (1) the occurrence of the derived temperate-climate mammalian fauna, which they considered likely to be of Hoxnian age, and (2) the apparent link with the pre-Moreton Drift ('Chalky Till') Paxford Gravel of the Stour–Evenlode watershed area, first suggested by Arkell (1947a, 1947b). At that time the Saalian age of the Moreton Drift, now in doubt, was unchallenged. If this glaciation is of Anglian age, as suggested by more recent authors (see above), and the correlation of the Hanborough and Paxford Gravels is correct, the latest possible age for the Hanborough Gravel would be Anglian. The implication of this would be a return to the interpretation of the derived mammalian fauna as being of Cromerian age, a view that has no support from biostratigraphy and is contrary to evidence from terrace stratigraphy in the Thames basin as a whole (see below).

Biostratigraphical evidence for the age of the Hanborough Terrace is rather sparse. The Mollusca obtained by Briggs and Gilbertson (1973) are of considerable palaeoenvironmental significance, but yield no information of significance to relative dating (although amino acid ratios from these shells might prove informative at some future time). Many of the teeth of ***Palaeoloxodon antiquus*** from the Hanborough Terrace deposits were described as having primitive or archaic affinities (Sandford, 1925), but recent reassessment of the material has not confirmed this potential indication of antiquity (Lister, 1989). These derived mammalian remains are of potential significance, however. Both mammoth and straight-tusked elephant appear for the first time in the Pleistocene of north-west Europe in the Elsterian (Anglian) Stage and have not been recorded from pre-Elsterian interglacials (Lister, 1989). Unless this fauna represents an intra-Anglian interstadial, it would not be expected beneath Anglian glacial deposits. All the mammalian species recorded from the Hanborough Gravel have, in contrast, been recognized from deposits ascribed to the Hoxnian Stage (*sensu* Swanscombe and *sensu* Hoxne). Therefore this derived (temperate-climate) fauna provides support for the chronostratigraphical scheme for the Thames terraces

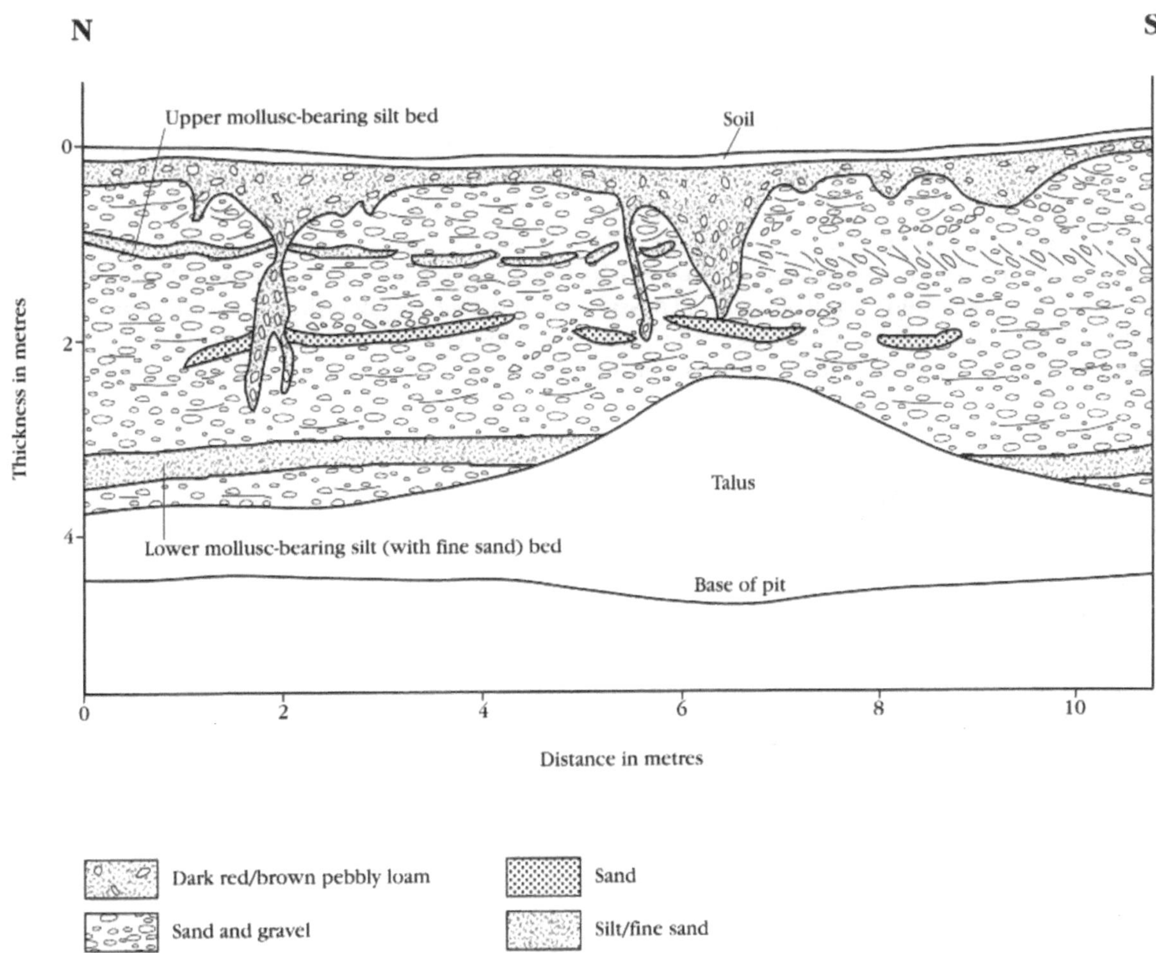

Figure 2.8 The east face at Long Hanborough Gravel Pit (after Briggs and Gilbertson, 1973). This shows 'frost boils'/decalcification pipes, filled with pebbly loam. Note the effects of frost heaving in the adjacent bedding.

proposed in Chapter 1, in which it was considered to represent Oxygen Isotope Stage 11 (Hoxnian *sensu* Swanscombe), although reworked into a Stage 10 gravel aggradation.

Arkell's correlation of the Paxford and Hanborough Gravels has been questioned by a number of authors. Shotton (1953) correlated the Paxford Gravels with the early Wolstonian (*sensu* Wolston) Baginton-Lillington Gravels of the Leamington area, whereas Richardson and Sandford (1963) described a mammalian assemblage of cold-climate affinities from the Paxford deposits. At this time the Hanborough Formation was confidently attributed to the Hoxnian, on the basis of its mammalian fauna, and regarded as interglacial in origin. Arkell's interpretation was also challenged by Kellaway *et al.* (1971), who noted compositional similarities between the Hanborough and Paxford Gravels, but regarded them as the products of separate river basins. Briggs and Gilbertson (1973, 1974) considered that their new evidence for cold conditions during the aggradation of the Hanborough Terrace eliminated the objections of Shotton (1953) and Richardson and Sandford (1963) and were inclined to accept Arkell's hypothesis. Strong additional evidence for a pre-Cotswolds glaciation age for the Hanborough Gravel was cited by Briggs (1973; Briggs and Gilbertson, 1974). He recognized quartzite-rich gravels (outwash) and a purple clay (Triassic-rich till?) in the upper Evenlode valley and attributed these to an ice sheet moving from the north-west, independent of, and probably slightly earlier than, that which deposited the (flinty) Moreton Drift. According to Briggs, the quartzite-rich outwash overlies the Hanborough Gravel upstream of Kingham, but has a steeper downstream gradient and is incised below the Hanborough Terrace level further downstream. However, Maddy *et al.* (1991b) have recently pointed to descriptions by Gray (1911) and Tomlinson (1929) of significant quantities of flint in deposits at Bledlington and Milton-under-Wychwood that Arkell (1947a, 1947b) included in the Hanborough Terrace. These latter observations suggest that, even if the Hanborough Gravel can be demonstrated to underlie the Moreton Drift, flint from the Cotswolds glaciation was already being fed into the Upper Thames by the time of Hanborough Gravel deposition. As has been discussed in Chapter 1, there is a distinct possibility that the gravels underlying the glacial deposits in the Upper Evenlode, the 'Bledlington Terrace' of Arkell (1947b), might be older than the Hanborough Gravel of the type area, perhaps the equivalent of the Freeland Formation (see Fig. 1.3).

Solution and periglacial features

The various large-scale post-depositional structures with v-shaped cross-sections that disrupt the deposits at Long Hanborough have attracted considerable interest. Periglacial and solution features have been noted in the Hanborough Terrace deposits by many workers (Sandford, 1924, 1926; Dines, 1946; Arkell, 1947a; Briggs and Gilbertson, 1973, 1974; Briggs, 1976a). According to Sandford (1924, p. 124), '... the top of the (Duke's) pit is marked by a remarkable series of so-called solution-pipes to a depth of 10 ft or more, filled with brown gravelly clay'. Sandford noted root remains in some of these features and suggested that the roots of trees might have been responsible for their formation, pointing out that the Hanborough area was formerly part of Wychwood Forest. He observed similar features in the Hanborough Gravel of the Cherwell valley at Kirtlington. Sandford reported that bedding could be traced through the 'pipes' and that it also sagged into them. Dines (1946), however, noted that the bedding was deformed upwards on either side of these features, which he believed to be infilled with material foreign to the Hanborough Gravel. These observations were confirmed by Arkell (1947a).

The 'pipes' in the gravel at Long Hanborough are striking features, picked out by their dark red-brown, clayey infill, which stands out against the surrounding pale limestone gravels (Fig. 2.9). They were described in some detail by Kellaway *et al.* (1971) as part of a study of various Pleistocene structures in the Upper Thames and Cotswold regions. These authors confirmed Arkell's view: 'clearly the "wedges" are long tapering pipes or inverted cones, not wedge shaped masses arranged on a polygonal plan' (Kellaway *et al.*, 1971, p. 12). However, they did not believe the features to be simple solution pipes. They noted that the immediately adjacent calcareous gravel was disturbed and showed no sign of decalcification. Furthermore, they suggested that the dark infill was not the residue of gravel that had been decalcified, but was derived from a formerly continuous superficial deposit that had fallen in from above. They established that the 'pipes' were widely distributed at about 3 m intervals on the flat part of the terrace remnant, but were absent from the peripheral slopes, which they believed to have been cambered. They suggested that the features were the degraded bases of 'frost boils', formed in a periglacial environment. Kellaway *et al.* also noted the occurrence of calcrete and small-scale normal faulting at Long Hanborough, as well as 'gulls' near the edge of the gravel spread at Church Hanborough, where cambering has occurred.

Briggs and Gilbertson (1973, 1974) recognized various ground-ice pseudomorphs (ice-wedge casts and frost-cracks) at several places in association with the solution pipes, as well as small-scale festooning (involutions) within the top metre of sediment. They accumulated evidence that ice-wedge formation had been contemporaneous with gravel deposition (intraformational), claiming this as support for their conclusion that the Hanborough Terrace deposits were laid down under periglacial conditions (Briggs and Gilbertson, 1973). They later reserved judgement on the matter, considering it possible that the observed features might be 'dip and fault structures', related to cambering, rather than ice-wedge casts (Briggs and Gilbertson, 1974). Briggs (1976a, 1988) suggested that the characteristic Hanborough Terrace 'pipes' might have originated as cylindrical growths of ice within the upper part of the gravel, perhaps associated with festooning, followed, under warmer conditions, by solution at points where the gravel fabric was aligned vertically. This two-stage hypothesis of 'pipe' formation explains the fact that the upward deformation of adjacent bedding is confined to the top part of the pipes; these parts were presumably initiated by periglacial processes. The hypothesis also explains why the widest pipes show the least

Figure 2.9 Sections at Long Hanborough photographed early in this century. This view was first published in the Witney Geological Survey memoir, where it was attributed to Duke's Pit. Note the prominent 'frost boil'/pipe features, filled with darker material. Upward deformation of the gravel bedding adjacent to the left-hand pipe is clearly shown. Photograph reproduced by courtesy of the British Geological Survey (A3188).

deformation (deformed zones having been removed by solution), the opposite to what might be expected in the best-developed features if they were simple periglacial structures.

Correlation with the sequence in the London Basin

Correlation of the Hanborough Terrace with the Pleistocene sequence in the London Basin has previously proved somewhat problematic (see Chapter 1 and Fig. 1.3). Previous attempts at correlation between the Upper and Middle Thames have usually associated the Hanborough Gravel with the Boyn Hill (Sandford, 1924, 1926; King and Oakley, 1936) or the Winter Hill Formations (Sandford, 1932; Wooldridge, 1938). Arkell's (1947a) correlation of the Hanborough and Silchester Gravels (see above) would, at the time, have implied parity with the Winter Hill Terrace, but the Silchester (Hampstead Marshall Terrace) Gravel has recently been correlated with the Black Park Gravel of the Thames (Gibbard, 1985; see Chapter 3, Hamstead Marshall). Lower facets within what was recognized as the Winter Hill Terrace when Sandford and King and Oakley were active have subsequently been attributed to the Black Park Formation (see Chapters 1 and 3).

The palaeolith from Duke's Pit (see above) has also been interpreted as evidence for relative antiquity, having been regarded as 'Early Acheulian' or Abbevillian (Arkell, 1947c). This interpretation was not supported by Wymer (1968) or Roe (1976), both of whom advised against drawing conclusions from single finds. In any case, the separate existence of earlier phases of the Acheulian Industry, characterized by less skilfully made implements, is no longer widely accepted (Chapter 1).

Arkell's interpretation has remained a minority view, most authors having correlated the Hanborough and Boyn Hill Gravels, largely on the basis of the similarity of the mammalian faunas at Hanborough and Swanscombe. This view was widely accepted prior to the work of Briggs and Gilbertson (1973, 1974), both aggradations being assigned to the Hoxnian Stage (Mitchell *et al.*, 1973). Briggs and Gilbertson, in showing the Hanborough Formation to have accumulated under periglacial conditions, cast doubt on this correlation. The basis for this doubt has subsequently been negated by the work of Gibbard (1985), who has shown that the Boyn Hill Gravel of the Middle Thames is also largely the result of aggradation under periglacial conditions and that interglacial sediments such as those at Swanscombe are atypical. Gibbard placed the Boyn Hill Gravel in the early 'Wolstonian' (Saalian) Stage and considered the Hanborough Gravel to be its direct upstream equivalent. The remains of interglacial mammals found in these gravels were considered by Gibbard to have been reworked during the destruction of Hoxnian sediments similar to those preserved at Swanscombe.

Recent reappraisal of the glacial stratigraphy of the Midlands (Perrin *et al.*, 1979; Sumbler, 1983a; Rose, 1987, 1989, 1991; Chapter 1) has important implications for this correlation. In East Anglia it has been generally accepted, since the work of Bristow and Cox (1973) and Perrin *et al.* (1973), that the local 'Chalky Till' is entirely of Anglian age, but in the Midlands till of the same general type has been placed in the Saalian (Wolstonian) (Shotton, 1953, 1973a; Kelly, 1964). Recent research has challenged this last view, however, with the implication that the glaciation of the Vale of Moreton may, in fact, be of Anglian age (Sumbler, 1983a; Rose, 1987, 1989). Such an interpretation would seem to support Arkell's (1947a, 1947b) view, in which the Hanborough Terrace was regarded as earlier than the Catuvellaunian (= Anglian) glaciation. However, the scheme for terrace correlation between the Upper and Middle Thames valleys, outlined in Chapter 1 (Fig. 1.3), has an important bearing on this discussion. A pre-Anglian glaciation age for the Hanborough Gravel would rule out correlation with the Boyn Hill Gravel of the Middle Thames, which post-dates the Anglian glaciation of the Vale of St Albans (Gibbard, 1979, 1985). Correlation with the Black Park Gravel would also prove difficult, since this formation is believed to have been deposited in the Middle Thames valley while Anglian ice sheets occupied parts of the Vale of St Albans (Gibbard, 1977; see Chapter 3).

The conclusion that the Hanborough Gravel may be older than the Anglian glaciation therefore poses major problems for terrace correlation between the Upper and Middle Thames. The Black Park Gravel (formerly the Lower Winter Hill Terrace – see Chapter 3, Highlands Farm Pit) is aggraded to over 90 m O.D. in the district immediately downstream from the Goring Gap (Evans, 1971, fig. 53; Gibbard, 1985, fig. 10), whereas the Hanborough Terrace has already fallen below 80 m O.D. immediately upstream of the gap (Fig. 2.3). These facts strongly indicate that the Hanborough Terrace must correlate with a formation lower, and therefore later, than the Black Park Gravel. It must also, therefore, post-date the glaciation of the Vale of St Albans. In fact, altitudinal evidence unequivocally indicates correlation between the Hanborough and Boyn Hill Gravels (Chapter 1 and Fig. 1.3), as many past authors have claimed (Sandford, 1924, 1926; King and Oakley, 1936; Gibbard, 1985).

The Arkell (1947a, 1947b) interpretation of the Hanborough Gravel underlying the glacial deposits of the Moreton-in-Marsh area is thus of considerable significance. The downstream equivalent of the Hanborough Formation, the Boyn Hill Gravel, is ascribed to the early Saalian (Gibbard, 1985), so a broadly Saalian (post-Swanscombe interglacial) age appears to be implied for the Cotswolds glaciation. It should be noted, however, that the records of flints in deposits ascribed to the Hanborough Formation (Maddy *et al.,* 1991b; see above) may indicate that at least part of the Hanborough aggradation incorporates outwash from the Cotswolds glaciation. Maddy *et al.* (1991b) did not address the question of whether Arkell's interpretation of the stratigraphical relations between the Hanborough Gravel and the glacial deposits is correct. This question, which is discussed above and in Chapter 1, remains of paramount importance and is an issue requiring urgent attention.

In Chapter 1, schemes for correlating the Upper and Middle Thames terraces and for relating these to the oceanic oxygen isotope sequence were described. According to these, and if Arkell's interpretation is correct, the relation of the Hanborough Formation to Moreton Drift implies that the Cotswolds glaciation was later than that which diverted the Thames from the Vale of St Albans. If the latter is correctly correlated with Oxygen Isotope Stage 12 (Bowen *et al.*, 1986b), the former may equate with Stage 10 (Chapter 1). If Bowen *et al.* (1989) are correct in correlating the type-Hoxnian with Oxygen Isotope Stage 9, it is possible that the Moreton glaciation is both post-Anglian and pre-Hoxnian (*sensu* Hoxne), although it appears that it might post-date the interglacial sediments at Swanscombe (see Chapter 1).

It remains possible, in the light of the complexity of sedimentary sequences beneath single terrace surfaces implied by the model for terrace formation advocated in Chapter 1, that late Anglian (Oxygen Isotope Stage 12) outwash is included within a range of deposits making up the Hanborough Formation. This would allow the Moreton Drift to be ascribed to the same glacial event as the tills of the London Basin, although there is still the major problem that the glaciation of the Cotswolds appears to occur (or at least persist) after the downcutting to the Hanborough/Boyn Hill terrace level. The flinty gravels described by Gray (1911) and Tomlinson (1929) at Milton and Bledlington (see above) would, according to this view, equate with the gravels underlying the Moreton Drift, but the Hanborough deposits containing the mammalian bones must presumably date from the subsequent cold episode (Stage 10), if the fauna is correctly attributed to Stage 11 (Hoxnian *sensu* Swanscombe). The two sets of gravels would thus represent phases 2 (Stage 12) and 4 (Stage 10) of the model for terrace formation, with phase 3 represented by the *remanié* fauna. Further discussion of the age and stratigraphical relations of the Moreton Drift will be found below, in the Wolvercote report.

Conclusions

The gravels at Long Hanborough were deposited by a braided, gravel-bed river that flowed across a sparsely vegetated, treeless landscape during one or more of the cold-climate phases of the Pleistocene. Since the gravels at Long Hanborough contain the bones of large mammals, of species that preferred warm climates, the deposits were formerly thought to have accumulated during a temperate (interglacial) phase. The discovery of the remains of cold-tolerant snails within the gravels, however, has shown that the deposits were laid down under harsh, periglacial conditions. It seems likely that the bones of the temperate-climate mammals were derived from older, interglacial sediments. It appears that the gravels at Long Hanborough were laid down by an early River Evenlode, which drained a larger area than the present catchment. The precise age of the deposits is difficult to establish. On the basis of correlations with sequences in the English Midlands and elsewhere in the Thames basin, it is likely that the Long Hanborough gravels date back to around 300,000 to 400,000 years BP, somewhat later than the most intense of the Pleistocene glaciations (the Anglian glaciation). The interest of the site is enhanced by a series of infilled pipe structures that penetrate down into the gravels. These are thought to have formed by a combination of periglacial processes and the solution of carbonate-rich sediments, but their precise origin is uncertain.

THE WOLVERCOTE GRAVEL AND WOLVERCOTE CHANNEL DEPOSITS

Although no GCR site exists there, it is necessary, because of its importance to the Quaternary history of the Upper Thames, to discuss the evidence from Wolvercote (type locality of the Wolvercote Gravel and the Wolvercote Channel Deposits) at this point in the text.

During the latter part of the 19th century, a large channel was revealed in a brick pit at Wolvercote (SP 498105), apparently cut through the gravel of the Wolvercote Terrace into Oxford Clay (Figs 2.10 and 2.11). This channel was found to contain a sequence of Pleistocene sediments that have yielded molluscan and mammalian remains, plant macrofossils and the largest Palaeolithic assemblage from the Upper Thames basin (Bell, 1894a, 1904; Sandford, 1924, 1926; Arkell, 1947a; Wymer, 1968; Roe, 1981; Tyldesley, 1986a, 1986b, 1988). The pit is

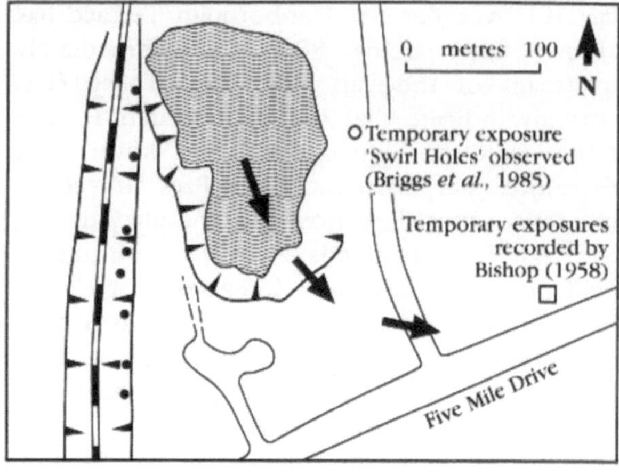

Figure 2.10 Map of Wolvercote brick pit and the surrounding area, showing the possible alignment of the Wolvercote Channel.

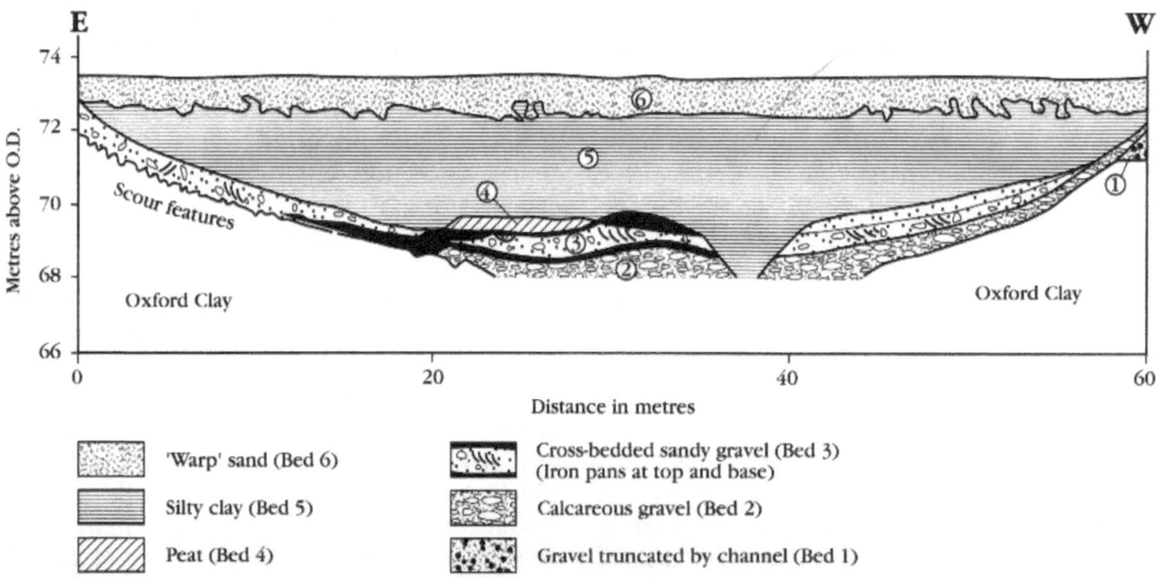

Figure 2.11 Section through the Wolvercote Channel (after Sandford, 1924).

now an ornamental lake surrounded by residential development, making reinvestigation difficult. Temporary exposures in the channel deposits were recently observed on the eastern side of the pit (Briggs *et al.*, 1985; Tyldesley, 1986b), but attempts to locate the channel in a railway cutting immediately to the west revealed only Oxford Clay with pockets of gravel at the surface (Bridgland and Harding, 1986). Work is continuing in open areas close to the brick pit to locate further remnants of the fossiliferous sediments, if any exist. So far no GCR site has been identified at Wolvercote, but it is hoped that future investigations will reveal Wolvercote Channel Deposits at a potentially conservable location.

Sections were open in the Wolvercote brick pit until the 1930s, but the lack of later opportunities to study the site has led to contrasting interpretations in the ensuing years, with both Hoxnian and Ipswichian ages being proposed for the channel deposits (Bishop, 1958; Wymer, 1968; Evans, 1971; Shotton, 1973a; Roe, 1981; Briggs *et al.*, 1985).

Many of the Wolvercote palaeoliths are of a highly distinctive, technologically advanced type, a fact that has caused some authors to argue for a Late Pleistocene age (for example, Roe, 1981), irrespective of the position of the Wolvercote Formation within the Upper Thames terrace sequence. The Wolvercote Terrace deposits have generally been attributed to the Saalian Stage (Bishop, 1958; Tomlinson, 1963; Shotton, 1973a; Briggs and Gilbertson, 1974, 1980), an interpretation based on the supposed first appearance in these gravels of material (in particular, fresh flint) from the 'Chalky Till' glaciation of the Cotswolds (Bishop, 1958; Goudie and Hart, 1975). It has been suggested recently that the 'Chalky Till' of the Cotswolds area might be of Anglian age (Bowen *et al.*, 1986a; Rose, 1987), raising the possibility that the Wolvercote Gravel is pre-Saalian and adding further fuel to the controversy over the age of the Wolvercote Channel.

The sequence at Wolvercote

It is difficult to determine the precise location of the Wolvercote Channel deposits from the early descriptions of the brick pit (Bell, 1894a, 1904; Pocock, 1908; Sandford, 1924, 1926), although they were clearly present in the southern and eastern faces. However, Bell (1894a, 1894b, 1904) provided detailed descriptions of the sections, including several illustrations, and Sandford (1924) provided an east-west section through the channel that has been repeatedly reproduced in subsequent publications (Sandford, 1926; Dines, 1946; Wymer, 1968; Roe, 1981; Tyldesley, 1986a) and which forms the basis for Fig. 2.11. Recent summaries have been provided by Wymer (1968), Roe (1981) and Tyldesley (1986a). The sequence at Wolvercote can be summarized as follows:

		Thickness (where known)
'warp sand'	6. Clayey gravelly sand, yellow-brown, in involutions	1–2 m
Wolvercote Channel Deposits	5. Silt and clay, laminated	up to 4.5 m
	4. Peat, thin and localized within base of bed 5	
	Iron-pan	
	3. Sandy gravel, current-bedded, with shells	
	Iron-pan	
	2. Calcareous gravel containing bones and artefacts	
Wolvercote Gravel	1. Bedded gravel, truncated by the channel deposits	
Oxford Clay		

Fresh flint, derived from the Cotswolds glaciation, is claimed to be present in beds 1–3.

According to the published records, the channel deposits overlie, at their margins, a gravel (bed 1) attributed by Sandford (1924) to the Wolvercote Terrace, although Bishop (1958) considered this to be part of an older aggradation (see below). The basal channel deposits comprise calcareous gravels (bed 2), containing bones and artefacts, that fill and overlie small hollows in the surface of the Oxford Clay. First described by Bell (1904), these features have been interpreted as potholes or 'swirl-holes' (Sandford, 1924; Arkell, 1947a). This appears to be an early record of scour features similar to those described recently beneath Pleistocene gravels (1) overlying London Clay at Stoke Newington (Harding and Gibbard, 1984), (2) overlying Pleistocene silt (Lower Loam) at Swanscombe and (3) overlying Thanet Sand at Globe Pit (see Chapter 4). Sandford (1924) noted that many of the large vertebrate bones were obtained from gravel within these scour features.

Above the basal gravel a series of ferruginous, cross-bedded sands and gravels was recorded (bed 3). Certain layers, predominantly at the base and top of the bed, were cemented into iron-pans. Between the lower and upper iron-pans, Sandford (1924, 1926) recorded shelly sands and a lens of clay containing shells and organic material. In the most comprehensive record, Kennard and Woodward (1924) recorded seventeen molluscan species from these beds. These gravelly deposits were separated by an erosive contact from the main infill of the channel, which comprised laminated silty clay (bed 5). A peat horizon (bed 4), which occurred locally at the base of this deposit, yielded plant fossils and beetles (Bell, 1894a, 1904; Reid, 1899; Blair, 1923; Duigan, 1956).

The channel sequence was capped by an upper sand (bed 6), up to 2 m thick, which, with the top of bed 5, was deformed by cryoturbation. This sand, with only a small clay component (Spiller, in Sandford, 1924), has generally been called 'warp sand'. It has also been recognized above the Wolvercote Terrace gravel away from the channel. Most authors have attributed this bed, directly or by implication, to solifluction, although Sandford (1925) thought it might result partly from the decalcification of the underlying sediments. It is also possible that an input of wind-blown, originally fluviatile, sands occurred. Spiller (in Sandford, 1924) showed from its heavy-mineral content that the 'warp sand' has closer affinities to the local fluvial deposits than to the bedrock.

Most of the palaeoliths from the Wolvercote site came from the basal gravel (bed 2) of the channel-fill or from the overlying cross-bedded ferruginous gravel. Some artefacts are considerably stained by iron, suggesting derivation from in or near one of the iron-pans (Sandford, 1924). The assemblage comprises mainly pointed hand-axes of a particularly well-made and characteristic type, although there is a fair representation of both primary and finishing flakes. About two-thirds of the material is in an unpatinated sharp or mint condition, suggesting that it was knapped at the site (Wymer, 1968). Palaeoliths have also been found in the Wolvercote Terrace gravel, both here and at Pear Tree Hill (SP 494111) to the north (Bell, 1904; Wymer, 1968). These are predominantly heavily abraded and patinated hand-axes, of the pointed type.

The Wolvercote Gravel (bed 1) is preserved, in the area of the Wolvercote brick pit, predominantly in pockets in the surface of the Oxford Clay, described as 'somewhat flask-shaped holes' by Bell (1904, p. 126). These were noted particularly on the western side of the pit (Pocock, 1908) and have been observed at the top of sections in the adjacent railway cutting (Bridgland and Harding, 1986). According to Sandford (1926), the terrace gravel was better represented in temporary sections to the south and east than in the brick pit. Bishop (1958) described temporary sections to the south-east of the pit in deposits that he ascribed to the Wolvercote Terrace and from which he obtained molluscs and ostracods.

Stratigraphy, age and correlation of Wolvercote deposits

Bell (1894a, p. 198) noted that the gravel of the Wolvercote area was older than the 'Summertown and Oxford gravel'. Pocock (1908) classified the former under the title 'Third Terrace', part of a sequence of four terraces numbered topographically upwards. Sandford (1924, 1926) redefined this as his Wolvercote Terrace. The deposits forming this terrace are here classified as the Wolvercote Gravel Formation, of which the Wolvercote Channel Deposits are a member.

The stratigraphical position of the Wolvercote Channel, particularly in relation to the Wolvercote Terrace gravel, has been a topic of a prolonged controversy. According to Sandford (1924, 1926), the Wolvercote Channel cuts through gravel of the Wolvercote Terrace, a view accepted by most authors. As the Wolvercote Gravel has generally been correlated with the Saalian, the temperate-climate channel deposits have been ascribed to the Ipswichian Stage

(Shotton, 1973a; Roe, 1981). A conventional interpretation of terrace stratigraphy in the Upper Thames basin would place the Wolvercote Terrace and Channel earlier than the Summertown-Radley Terrace (Figs 2.2 and 2.3), which has also yielded temperate-climate fauna and flora (see below, Stanton Harcourt and Magdalen Grove). A similar conclusion was reached, following the interpretation of molluscan faunas from various sites in the Upper Thames, by Kennard and Woodward (in Sandford, 1924). Despite this, Sandford (1925, 1926, 1932) concluded that the Wolvercote Channel Deposits post-date the temperate-climate sediments representing the upper part of the Summertown-Radley aggradation (the Eynsham Gravel – see below, Stanton Harcourt and Magdalen Grove), although he considered that both originated in the same interglacial. His correlation table (Sandford, 1932, p. 10) makes it clear that he considered this warm interval to immediately pre-date the last glaciation, implying correlation with the last interglacial (Ipswichian), but he also suggested that the lake beds at Hoxne, later to become the type site of an earlier interglacial in Britain (Mitchell *et al.*, 1973), were of equivalent age. This led some later authors, notably Bishop (1958), to claim that Sandford regarded the channel as Hoxnian.

Sandford's view, that the Wolvercote Channel Deposits post-dated the upper Summertown-Radley aggradation, was later consolidated by Dines (1946) and Arkell (1947a), although the latter placed both in the 'Great Interglacial' (Hoxnian Stage). Most subsequent authors who have concurred with Sandford's stratigraphical interpretation (Shotton, 1973a; Briggs and Gilbertson, 1974) have considered the Wolvercote Channel sequence to belong to the Ipswichian Stage. Evidence for climatic cooling, from plant macrofossils in the peat (see below), implies that the latter part of an interglacial is represented. The stratigraphical relations of the Wolvercote Channel have become a major point of controversy, however, since Bishop (1958) suggested that the channel deposits were older, rather than younger, than the Wolvercote Gravel. He believed the channel-fill to be of late Hoxnian to early Saalian age, an interpretation that also found favour with Wymer (1968).

Briggs (1976b) presented a summary of the various possible stratigraphical interpretations of the Wolvercote Channel Deposits. These are: (1) the channel post-dates the Wolvercote Gravel but pre-dates the Summertown-Radley Formation (the view of Kennard and Woodward (in Sandford, 1924)); (2) the Wolvercote Channel Deposits immediately post-date the upper Summertown-Radley aggradation (the later suggestion by Sandford (1925, 1926, 1932)); (3) the Wolvercote Channel is a pre-Wolvercote Terrace feature (the interpretation of Bishop (1958)) and (4) the Wolvercote Channel Deposits are contemporaneous with the upper Summertown-Radley aggradation, but laid down in a steeply sloping tributary valley and, therefore, at a greater elevation. The last view corresponds with the idea of formation in a 'hanging' tributary valley, possibly an early River Ray, suggested by Arkell (1947a). Whereas the earlier reconstructions (1 and 2) implied an Ipswichian age, Bishop (1958) interpreted the channel as Hoxnian. Briggs (1976b) favoured the first or third of the above hypotheses, considering that deposition of the Wolvercote Channel Deposits before the Summertown-Radley sediments, in line with conventional terrace stratigraphy, was inherently more likely than alternative models requiring complex sequences of erosion and aggradation. The larger number of climatic fluctuations now recognized in the late Middle Pleistocene allows a more straightforward interpretation of the Wolvercote sediments, as in option 1 above; they do not necessarily correlate with any other interglacial sediments recorded from the Upper Thames basin (see below, correlation).

Palaeontological evidence

Information on the faunal content of the Wolvercote Gravel and Wolvercote Channel Deposits is rather sparse. The Wolvercote Gravel has yielded scanty remains of mammals; the only such records appear to be horse, from Spelsbury (Sandford, 1924, 1926, 1939), and a wolf's tooth from Pear Tree Hill (Bell, 1904), where the gravel was regarded by Sandford (1924) as part of the Wolvercote Terrace (although the provenance of the tooth must be regarded as dubious). The fauna from the Wolvercote Channel Deposits comprises *Palaeoloxodon antiquus*, *Dicerorhinus hemitoechus*, *Bos primigenius, Cervus elaphus* and *Equus caballus* L., with bison, reindeer and bear recorded with less certainty (Sandford, 1924, 1925, 1926). Molluscan faunas were recorded from both the channel deposits (Kennard and

Woodward in Sandford, 1924) and the Wolvercote Gravel (Bishop, 1958), in each case dominated by *Trichia hispida*. None of the species recognized give a useful indication of climate (Bishop, 1958), although Gilbertson (1976) claimed that the fauna from the channel showed greater temperate affinities than that from the terrace gravel; however, he considered it unlikely that either represented exceptionally cold conditions. The source of the molluscan fauna attributed by Bishop (1958) to the Wolvercote Terrace lies less than 200 m from the brick pit and, although the deposits recorded there by Bishop were dominated by gravel, it seems likely that they represent a continuation of the channel-fill rather than the cold-climate terrace deposits (Fig. 2.10).

Plant macrofossils (Bell, 1894a, 1904; Reid, 1899; Duigan, 1956) and beetle remains (Blair, 1923) from the peaty horizon (bed 4) at the base of the silty clay (Fig. 2.11) indicate cool-temperate conditions (Duigan, 1956), probably colder than during the deposition of the gravels (beds 1–3) (Briggs *et al.*, 1985; Briggs, 1988). This interpretation is based on the presence of the arctic-alpine plant *Draba incana* L. (Duigan, 1956), the northern weevil *Notaris aethiops* (Fabricius) and a number of mosses of cold-climate affinities (Bell, 1904). Confirmation of progressive cooling during the infilling of the Wolvercote Channel was recently obtained from sparse pollen, probably from bed 5, from temporary sections near the eastern edge of the pit (Briggs *et al.*, 1985; Briggs, 1988). These showed a change from pine-dominated forest to open conditions, a sequence suggestive of the later part of an interglacial cycle. None of the palaeontological data provides any clear indication of the age of the channel-fill, except that the mammalian and molluscan species indicate deposition during the late Middle Pleistocene or Late Pleistocene.

The archaeological evidence

Both Wymer (1968) and Roe (1981) singled out Wolvercote brick pit as the most important Palaeolithic site in the Upper Thames valley. It is, in fact, the only locality in this region to have yielded a large collection of well-made Lower Palaeolithic tools, fashioned from good quality flint (presumably imported from the Chilterns), in a condition suggesting the proximity of a working site. Moreover, the site is of great Palaeolithic and Pleistocene significance as the source of an industry that is possibly unique in Britain. This claim is based on the characteristic and unusual form of certain of the best-made artefacts within the assemblage, which have been compared with some of the most recent hand-axe industries on the continent (Roe, 1981; Tyldesley, 1986a). Not all of the Wolvercote material is flint; artefacts made from quartzite and greywacke are included in the collections (Wymer, 1968; Roe, 1976, 1981; Tyldesley, 1988).

Summaries of the Palaeolithic assemblage from Wolvercote have been provided by Sandford (1924), Roe (1964, 1976, 1981), Tyldesley (1986a) and Wymer (1968). The typological harmony of the unabraded implements suggests that a single industry is represented, supporting the notion that a working site existed in the vicinity (Wymer, 1968; Roe, 1981). These hand-axes are characteristically large, finely made tools, showing evidence for soft-hammer working and often with a markedly plano-convex cross-section. This type of implement from Wolvercote, sometimes referred to as 'tongue-shaped' (Evans, 1897) or 'slipper-shaped' (Sandford, 1924, 1926), has been compared to the continental Micoquian industries, which are generally attributed to the last glaciation or the last interglacial (Sandford, 1924, 1925; Roe, 1981). Tyldesley (1986a) noted that classic 'Wolvercote Channel style' hand-axes formed a small proportion of the assemblage from the brick pit, eight specimens in all, but that there was a considerable cluster of other implements sharing many of the 'classic' features.

In his analysis of British hand-axe assemblages on the basis of implement typology, Roe (1964, 1968a) allocated the Wolvercote collection to a group of its own, it being the only British assemblage dominated by plano-convex bifacial tools, which Roe thought likely to be of relatively late inception. Wymer (1968), on the other hand, considered the Wolvercote industry to be broadly comparable to that from the Swanscombe Middle Gravel, both sites lacking ovate hand-axes and Levallois flakes and cores. Wymer recognized that the large plano-convex implements from Wolvercote were exceptional, but was prepared to accept the conclusion of Bishop (1958) that the Wolvercote Channel was pre-Saalian (see above). Roe (1981) disputed Wymer's view both of the affinities of the implement assemblage and of the likely age of the channel deposits. He cited the occurrence of typologically comparable Micoquian hand-axe assemblages at continental sites such as La Micoque in France and Bocksteinschmiede in Germany, both attributed to the last interglacial/glacial cycle. For the Wolvercote Channel Deposits to be as late as these continental industries, they would have to post-date much, if not all, of the Summertown-Radley Formation. This view, advocated by Roe (1981) and tentatively supported by Tyldesley (1986a, 1986b), would appear to conform with the later stratigraphical interpretation of Sandford (1925, 1932).

A detailed study of the Wolvercote Palaeolithic collections has been completed recently by Tyldesley (1986a, 1986b, 1988). Amongst her observations, she noted that the characteristic Wolvercote implements could be the work of a single craftsman who, given the limited size of the assemblage, could have made all the surviving artefacts in a single day. She also noted that tools similar to the characteristic Wolvercote types occurred within some French Micoquian industries, although they formed a less important part of these assemblages than at Wolvercote. She found that German Micoquian industries generally lacked such forms, however, and concluded (1986a) that the similarity of the Wolvercote hand-axes to the French material could have resulted from coincidence.

Briggs *et al.* (1985) questioned the value of typological refinement as an indication of a relatively recent age for the Wolvercote Channel Palaeolithic assemblage. This reflects recent thinking amongst archaeologists, which results largely from the recognition, at sites such as Boxgrove (Sussex), that relatively advanced knapping techniques were used in Britain in the early Middle Pleistocene (see Chapter 1). It is therefore apparent that the Palaeolithic assemblage from Wolvercote brick pit, though forming an important part of the scientific interest at the site, is of little value for relative dating.

Correlation

Since they lack diagnostic biostratigraphical evidence, determination of the age of the Wolvercote deposits relies heavily on the interpretation of the Upper Thames sequence as a whole. It is now widely believed that the deposits underlying the Summertown-Radley Terrace contain evidence for two separate interglacials, correlated with Oxygen Isotope Stages 7 and 5 (Briggs and Gilbertson, 1980; Shotton, 1983; Briggs *et al.*, 1985; Bowen *et al.*, 1989; Chapter 1; see also below, Stanton Harcourt and Magdalen Grove). These deposits are lower within the Upper Thames terrace sequence, and therefore younger, than the Wolvercote Gravel. According to the climatic model for terrace formation favoured in this volume, this implies that the Wolvercote Formation represents an earlier climatic cycle (cold-temperate-cold) than any part of the Summertown-Radley Formation (see Chapter 1). This model holds that the time interval represented by a typical terrace aggradation straddles a temperate climatic half-cycle, so that interglacial sediments, where preserved, are commonly underlain and overlain by deposits representing different cold episodes. The later, overlying, cold-climate sediments (phase 4 of the climatic model) are usually dominant, if only because they are the last to be deposited prior to rejuvenation. At Wolvercote, however, the pre-interglacial (phase 2) cold-climate aggradation appears to dominate (according to Sandford's interpretation of the relations between the Wolvercote Channel and Wolvercote Gravel). It is likely that a post-interglacial (phase 4) part of the Wolvercote aggradation was deposited elsewhere, although it will be impossible to distinguish it from the earlier phase 2 gravels in the absence of intervening interglacial (phase 3) sediments.

It is apparent that the 'Chalky Till' glaciation of the Cotswolds occurred prior to the deposition of the Wolvercote Channel sediments, since it supplied the fresh flint clasts that occur in the underlying Wolvercote Gravel. The glaciation therefore provides a maximum age for the Wolvercote Channel Deposits (it does not, however, indicate that the Wolvercote Gravel is of similar age to the Cotswolds glaciation – see Long Hanborough and Chapter 1). If the reinterpretation of this glaciation as an Anglian event (Rose, 1987, 1989) is accepted, it is possible to accommodate the Wolvercote Channel in one of two temperate episodes between the Anglian (Oxygen Isotope Stage 12) and the older of the two Summertown-Radley interglacials (Oxygen Isotope Stage 7). Correlation of the Wolvercote Channel Deposits is possible, on this basis, with either Stage 11 or Stage 9 of the oxygen isotope record. However, mammalian bones reworked into the Hanborough Gravel have been attributed to the Hoxnian Stage *sensu* Swanscombe, which is correlated with Oxygen Isotope Stage 11 (Table 1.1), and the Hanborough Gravel itself to Stage 10. Since the Wolvercote Formation clearly post-dates the rejuvenation event that followed the deposition of the Hanborough Gravel, correlation of the Wolvercote Channel Deposits with Oxygen Isotope Stage 9 (rather than 11) is strongly indicated. The stratigraphical correlations advocated here are summarized in Table 2.2.

The correlation of the Wolvercote Formation with the terrace sequence in the London Basin has been attempted by relatively few authors. Both Sandford (1932) and Arkell (1947a) suggested a correlation of the silty infill of the Wolvercote Channel (bed 5) with the Crayford 'brickearth', implying correlation with the Taplow aggradation. Evans (1971) similarly correlated the Wolvercote Terrace and Wolvercote Channel with the Taplow Terrace, considering them to have aggraded in his cycle 4W (equivalent to Oxygen Isotope Stage 7). This correlation was largely based on projection of the terraces through the Goring Gap. Gibbard (1985), using the same method, proposed the same correlation. However, the deposits in the Reading area, immediately downstream from the gap, that were ascribed by Gibbard to the Taplow Formation are reinterpreted in this volume (Chapter 1) as degraded Lynch Hill Gravel, thus implying a correlation between the Wolvercote and Lynch Hill Formations. In the scheme for terrace correlation presented in Chapter 1, the equivalence of the Wolvercote and Lynch Hill Formations was proposed (Fig. 1.3). This provides further support for correlation of the Wolvercote Channel sediments with Oxygen Isotope Stage 9, since the Lower Thames equivalent of the Lynch Hill Formation, the Corbets Tey Gravel, incorporates bodies of temperate-climate sediment at a number of sites that are also ascribed to this stage (Table 1.1; Chapter 4).

Attribution of the Wolvercote Gravel to the Anglian Stage, as suggested by Bowen *et al.* (1986b), is rendered untenable by the correlation of the Wolvercote and Lynch Hill Formations. The projection of the long-profile of this formation downstream into the London Basin (Chapter 1 and Fig. 1.3) shows the Wolvercote/Lynch Hill Gravel to be considerably lower (and therefore later) within the terrace sequence than either of the two Anglian formations, the Winter Hill and Black Park Gravels. Bowen *et al.*'s suggestion was based on the assumption that the Wolvercote Gravel was fed by flint-rich outwash from the Cotswolds

Table 2.2 Stratigraphical interpretation of the Upper Thames deposits advocated in this volume.

Temperate	Stage 7	Stanton Harcourt Channel Deposits
Cold	Stage 8	Basal Stanton Harcourt Channel Deposits and equivalents Wolvercote Gravel (phase 4) Uppermost Wolvercote Channel Deposits
Temperate	Stage 9	Wolvercote Channel Deposits
Cold	Stage 10	Wolvercote Gravel (phase 2) Hanborough Gravel Moreton-in-Marsh glaciation?
Temperate	Stage 11	Mammalian fauna (derived) in Hanborough Gravel
Cold	Stage 12	Hanborough Gravel (pre-bones)? Moreton-in-Marsh glaciation?

glaciation, as suggested by Tomlinson (1929) and Bishop (1958). This assumption, although almost universally accepted in recent years, has now been seriously challenged by Maddy *et al.* (1991b). They have reviewed the published clast-lithological data from the Wolvercote and Hanborough Gravels, which is in any case rather scanty, and concluded that there is no unequivocal indication of an input of flint into the Upper Thames system between these two formations. They found that, in comparison with other material foreign to the modern catchment, the highest percentage of flint actually occurred in a sample of Hanborough Gravel. This led them to suggest that the observed paucity of flint in the Hanborough Gravel, in comparison with the Wolvercote Formation, is the result of the greater incorporation of local limestone material in the older gravel. This evidence, as well as historical records of fresh flint in gravels later ascribed to the Hanborough Formation (Gray, 1911; Tomlinson, 1929; see above, Long Hanborough), suggests that the Cotswolds glaciation occurred during or before the aggradation of the Hanborough Gravel and, therefore, significantly earlier than the deposition of the Wolvercote Gravel. As stated above (see Long Hanborough), the outstanding problem is the relation of the Hanborough Gravel to the Moreton-in-Marsh glacial deposits; however, the dating and correlation of the Wolvercote Formation is in no way affected by the continuing dispute over this relation and the age of the Cotswolds glaciation.

It is widely agreed that a modern study of the Wolvercote deposits is urgently required before a complete understanding of the site's chronostratigraphical position within the Pleistocene can be achieved. Given that Wolvercote brick pit has yielded the largest collection of artefacts from any site in the Upper Thames, and it is the only locality in that area to have yielded Palaeolithic material in association with a temperate fauna, the importance of the site cannot be questioned. The interpretation, presented here, of the temperate-climate deposits at Wolvercote and their included fossils and palaeoliths as representing Oxygen Isotope Stage 9 has been argued almost entirely from stratigraphical evidence derived from other sites in the Thames sequence. A new opportunity for an examination of the Wolvercote Channel Deposits themselves must be awaited in order to test this hypothesis.

STANTON HARCOURT GRAVEL PIT (SP 415052) and MAGDALEN GROVE DEER PARK (SP 520065)

Highlights

These are important sites for interpreting the stratigraphically complex sequence of deposits forming the Summertown-Radley Terrace of the Upper Thames basin. This formation shows evidence of having aggraded during two full climatic cycles, thereby encompassing two major warm episodes, as well as one complete cold episode and parts of two others. Faunal evidence from Stanton Harcourt has been critical in demonstrating that the earlier of these temperate intervals represents an additional, as yet unnamed interglacial in the British Pleistocene – a terrestrial correlative of Oxygen Isotope Stage 7 (*c.* 200,000 years BP) of the deep-sea record. Formerly attributed to the last interglacial (Ipswichian Stage), the Magdalen Grove site now appears to represent further evidence for this newly recognized Stage 7 interglacial.

Introduction

The Summertown-Radley Terrace of the Upper Thames basin is formed by a complex sequence of sediments (the Summertown-Radley Formation) representing periods of both temperate and cold climate. The terrace is widely preserved in the Thames valley downstream from the confluence with the Windrush, but is not well represented in any of the tributaries (Fig. 2.1). The deposits forming this terrace have been well exposed in the past, notably at Eynsham, Summertown, Radley, Cassington, Stanton Harcourt and a number of exposures within the urban area of Oxford (Pocock, 1908; Sandford, 1924, 1926; Dines, 1946). Aggraded to 6–10 m above river level in the Oxford area, the Summertown-Radley Terrace was first named by Sandford (1924), although it had previously been described as the '2nd' or '20 ft' Terrace (Prestwich, 1882; Pocock, 1908).

Sandford (1924, 1926) considered that gravels forming the upper part of the Summertown-Radley aggradation contained the remains of interglacial mammals and molluscs, whereas the underlying lower part of the sequence yielded a mammalian fauna of cold-climate character. In recent years the occurrence of an interglacial deposit clearly stratified beneath Sandford's lower, cold-climate Summertown-Radley gravel has been revealed in a pit to the south of Stanton Harcourt, in the main valley of the Thames (Briggs and Gilbertson, 1980; Briggs *et al.*, 1985; Briggs, 1988). This discovery led to the recognition of a tripartite sequence of deposits beneath this terrace, indicative of a succession of climatic episodes, from temperate to cold (periglacial) and returning again to temperate (Fig. 2.12). The earlier of these two temperate episodes has been widely attributed to a hitherto unrecognized interglacial between the Hoxnian and Ipswichian Stages (Briggs and Gilbertson, 1980; Shotton, 1983; Briggs *et al.*, 1985; Bowen *et al.*, 1989). Reappraisal of the stratigraphy of the Summertown-Radley Formation suggests, however, that the sequence is even more complex, perhaps representing five separate climatic episodes (Table 1.1).

It has generally been assumed that the cold-climate gravels at Stanton Harcourt are equivalent to those described beneath the upper fossiliferous (temperate-climate) Summertown-Radley deposits in early records. Briggs *et al.* (1985) proposed the name Stanton Harcourt Gravel for this unit. However, the upper interglacial sediments have never been observed at Stanton Harcourt, so the full tripartite stratigraphy of the Summertown-Radley aggradation has never been recorded in superposition. Recent new exposures in this terrace have consistently failed to reveal temperate deposits overlying cold-climate gravels, in the relationship described by Sandford, and none of the sites featuring in the early records have been available for study for some years. Attempts were therefore made as part of the GCR site selection programme to relocate some of the early sections, with the result that the deposits at Magdalen Grove (Fig. 2.13), originally described by Sandford (1924, 1926), were re-excavated (Bridgland, 1985a; Briggs *et al.*, 1985). Briggs *et al.* (1985) followed Sandford in attributing the deposits at Magdalen Grove to the (interglacial) upper part of the Summertown-Radley Formation, which they termed the Eynsham Gravel. Detailed consideration of the altitudinal position and fauna from this and other fossiliferous sites in the Summertown-Radley Formation raises serious doubts about this interpretation, suggesting instead that the Magdalen Grove deposits (and certain others formerly attributed to the Eynsham Gravel) were deposited during the earlier (Stage 7) temperate episode.

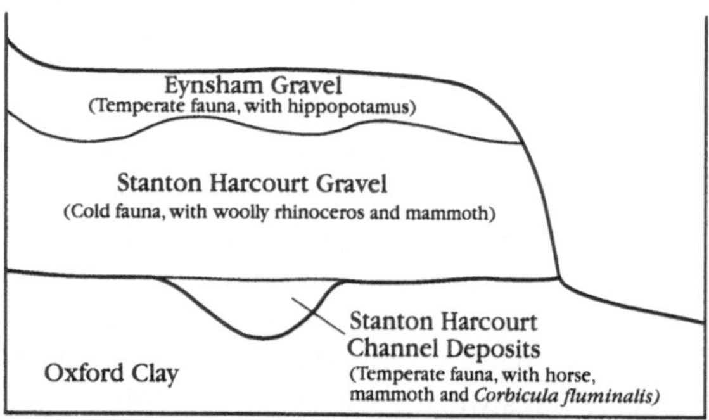

Figure 2.12 Idealized section through the Summertown-Radley Formation.

Description

Lithostratigraphical classification of the Summertown-Radley terrace deposits has already been established in Chapter 1 (Table 1.1), as follows:

Formation	Member	Climate
Summertown -Radley	Unnamed upper gravel at Eynsham	Cold
	Eynsham Gravel	Temperate
	Stanton Harcourt Gravel	Cold
	Stanton Harcourt Channel Deposits	Temperate
	Unnamed lower gravel at Summertown	Cold

This sequence has been determined from records of exposures over a lengthy period. At sites revealing only cold-climate gravels, individual members cannot be identified and the term Summertown-Radley Gravel Formation should be applied.

Descriptions of Summertown-Radley Terrace

deposits appeared in the literature long before the definition of the terrace by Sandford (1924). Prestwich (1882) and Pocock (1908) both recorded various fossils from this formation, but it was Sandford (1924, 1926) who first noted that the fauna allowed a bipartite division of the aggradation. According to Sandford, bones from the lower part of the sequence (= Stanton Harcourt Gravel) were of cold-climate mammals, dominated by *Mammuthus primigenius*, with *Coelodonta antiquitatis* (Blumenbach) and *Bison priscus* (Bojanus). *Equus ferus* and *Ursus* sp. (cited as *Ursus anglicus*) were added to the faunal list by Sandford (1954).

Stratigraphically above the lower, mammoth-rich gravel, Sandford described deposits, usually finer and less ferruginous, that yielded *Palaeoloxodon antiquus*, *Dicerorhinus hemitoechus*, *Hippopotamus amphibius* L., *Bos primigenius*, *Cervus elaphus*, *Equus ferus* and other species typical of temperate environments, as well as molluscan faunas that frequently included the southern bivalve species *Corbicula fluminalis* (Müller). If the records of Sandford (1924) and Dines (1946) are combined, this upper temperate-climate deposit has been recorded at ten

(A)

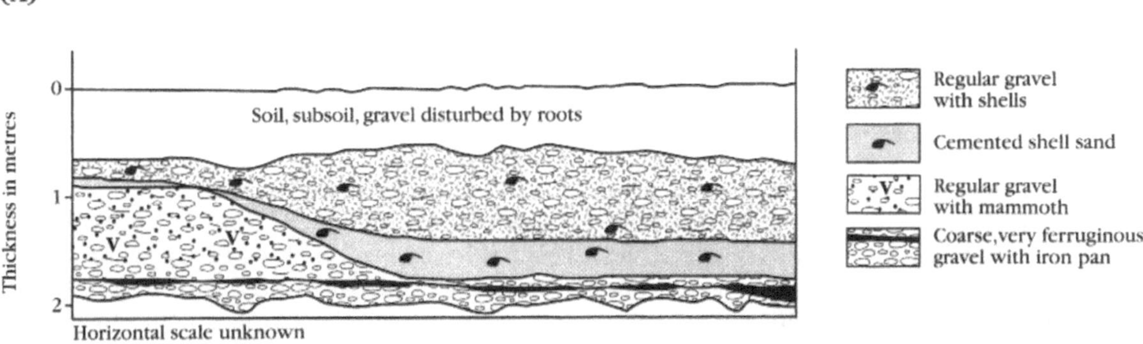

(B)

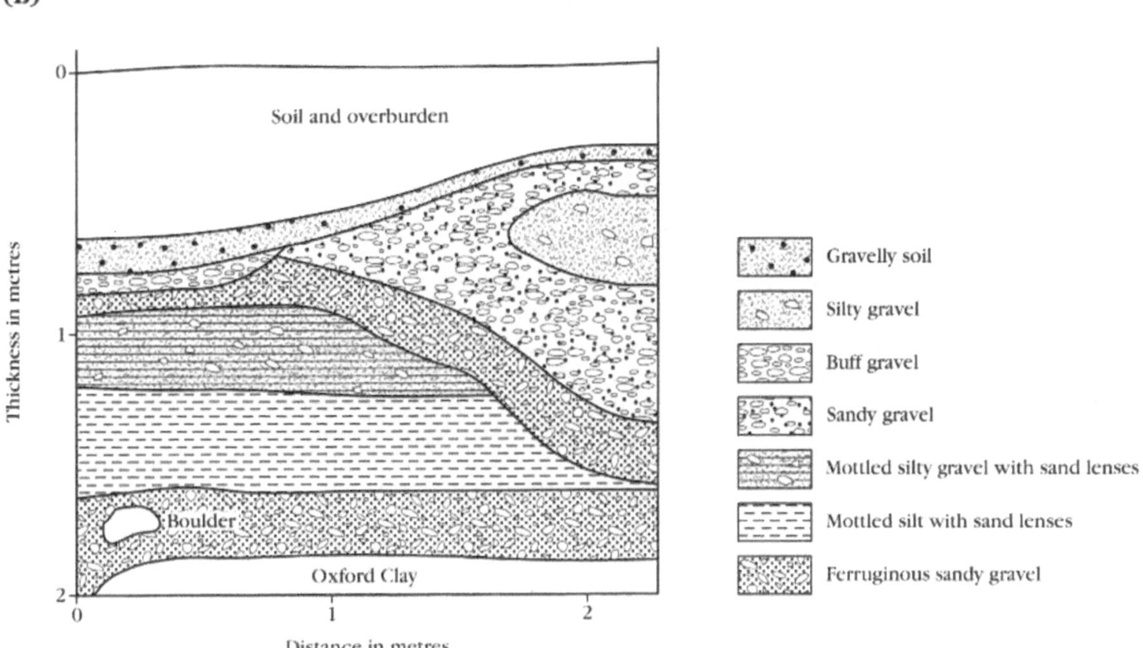

Figure 2.13 A comparison of sections recorded at Magdalen Grove, by (A) Sandford (1924) and (B) Briggs *et al.* (1985).

sites (Briggs *et al.*, 1985), of which the most important were Eynsham (Station Pit, SP 429088), Magdalen College (Magdalen Grove GCR site), Summertown (Webb's Pit, SP 503086) and Radley (Silvester's Pit, precise location uncertain). These were assigned by Briggs *et al.* (1985) to their Eynsham Gravel.

Stanton Harcourt

Erstwhile workings at Stanton Harcourt have been recorded in the various Geological Survey memoirs for the area (Pocock, 1908; Sandford, 1926; Dines, 1946). None of these authors provided any detailed description of the site, although all noted that the gravel here is locally ferruginously cemented into a hard rock. This was used to construct the nearby megaliths, the Devil's Quoits, and appears in the stonework of Stanton Harcourt church. Arkell (1947a) referred to the present pit, to the south of the village. The modern workings, Dix's Pit, exploit an area known as Linch Hill. To avoid confusion with the type locality of the Lynch Hill Terrace of the Middle Thames, it is preferable to refer to this Upper Thames site as Stanton Harcourt.

The Stanton Harcourt site shows two of the five members now recognized within the Summertown-Radley Formation: the Stanton Harcourt Channel Deposits and the Stanton Harcourt Gravel. Although the overlying Eynsham Gravel is absent throughout the large workings here, its type locality, at Station Pit, Eynsham (Briggs *et al.*, 1985), is only 4 km downstream (Sandford, 1924, 1926; Fig. 2.1).

The importance of the Stanton Harcourt site lies in the occurrence there of interglacial sediments underlying the cold-climate Stanton Harcourt Gravel (Briggs and Gilbertson, 1980; Briggs *et al.*, 1985). This temperate-climate deposit appears, from its restricted distribution, to fill a 'single thread' channel cut into the Oxford Clay, which floors the pit (see Fig. 2.12). It is composed of gravels, sands, silts and organic sediments and is highly fossiliferous, yielding mammals, molluscs and beetles, together with wood and plant remains (Fig. 2.14). The rich molluscan fauna, comprising at least 34 aquatic and terrestrial species of warm-climate aspect, includes the important bivalve *Corbicula fluminalis* (Briggs *et al.*, 1985). The mammals include *Mammuthus primigenius*, *Equus ferus* and *Panthera leo* (L.) The first two of these species were also found *in situ* near the base of the overlying Stanton Harcourt Gravel, in company with a small horn core and skull fragment from a bovid, probably *Bison* sp. (Seddon and Holyoak, 1985).

The exposures of Stanton Harcourt Gravel at the type locality (Dix's Pit) have been described in varying detail on a number of occasions (Briggs, 1973, 1976a; Goudie and Hart, 1975; Gilbertson, 1976; Briggs and Gilbertson, 1980; Bryant, 1983; Briggs *et al.*, 1985; Seddon and Holyoak, 1985). Two divisions have been recognized at Stanton Harcourt (Fig. 2.15) within this cold-climate member (Stanton Harcourt Gravel). The lower is coarser and has a maximum thickness of 1.5 m. The higher, finer-grained division is separated from the lower by a well-developed zone of cryoturbation. Capping the sequence is a deposit that has been described as 'coverloam' and which is thought to postdate the formation of the Summertown-Radley Terrace (Briggs *et al.*, 1985). Both divisions of the Stanton Harcourt Gravel contain thin, intermittent sand and silt beds from which molluscan remains have been obtained (Briggs *et al.*, 1985; Seddon and Holyoak, 1985). In addition to the important interglacial channel deposits already described, the Stanton Harcourt Gravel is locally underlain by a basal sand/silt unit, often with ripple drift or parallel lamination, that contains plant remains and occasional Mollusca (Fig. 2.15). Pollen and other plant remains were also obtained from within a single silt-filled channel structure, at a height of *c.* 1 m above the Oxford Clay (Seddon and Holyoak, 1985). These various molluscan assemblages, from minor beds within the Stanton Harcourt Gravel, are in keeping with other evidence, from periglacial structures, implying deposition of this member under cold conditions.

Zones of cryoturbation occur at the top of both divisions of the Stanton Harcourt Gravel (Fig. 2.15). Ice-wedge casts occur at many different levels within the member (Seddon and Holyoak, 1985; Fig. 2.16).

Magdalen Grove

A small pit in Magdalen College Grove (now a deer park) was described by Sandford (1924). He noted that the site had a low elevation within the Summertown-Radley aggradation, but that it nevertheless showed both parts of the bipartite sequence he had established, with a later warm-climate gravel channelled into an earlier deposit

Figure 2.14 Section in the Stanton Harcourt Channel Deposits, showing ***in situ*** mammoth remains. Note that both Pleistocene and Jurassic bivalves are visible, the latter (***Gryphaea*** sp.) scattered on the exhumed Oxford Clay surface as well as within the gravel. (Photo: S. Campbell.)

containing mammoth remains (Fig. 2.13). He regarded this as 'probably the most instructive section as yet opened' (1924, p. 142), an opinion that was supported by Arkell (1947a).

A section at the Magdalen Grove GCR site, re-excavated in 1984, appeared similar to that illustrated by Sandford, revealing silty sands and gravels channelled into an earlier sequence of gravels, silts and sands (Fig. 2.13). However, whereas in Sandford's section the earlier deposit, below the channel feature, was identified as the cold-climate (Stanton Harcourt Gravel) aggradation (on the basis of its fauna, which comprised abundant *Mammuthus primigenius*), the beds in the same stratigraphical position in the 1984 section yielded temperate-climate pollen and faunal remains (Fig. 2.13). These deposits contained molluscs and pollen, as well as a humerus of lion (*Panthera leo*). The molluscan fauna (27 species), comprising land snails, freshwater snails and bivalves (Briggs *et al.*, 1985), included *Corbicula fluminalis* in abundance (see below). The pollen assemblage was dominated by arboreal and herb taxa, with a high incidence of broad-leaved trees; it provided confirmation of an interglacial origin for the silts and sands. These temperate-climate deposits have been attributed to the Eynsham Gravel Member (Briggs *et al.*, 1985), but re-evaluation of the stratigraphy of the Summertown-Radley Formation suggests that this may be incorrect (see below).

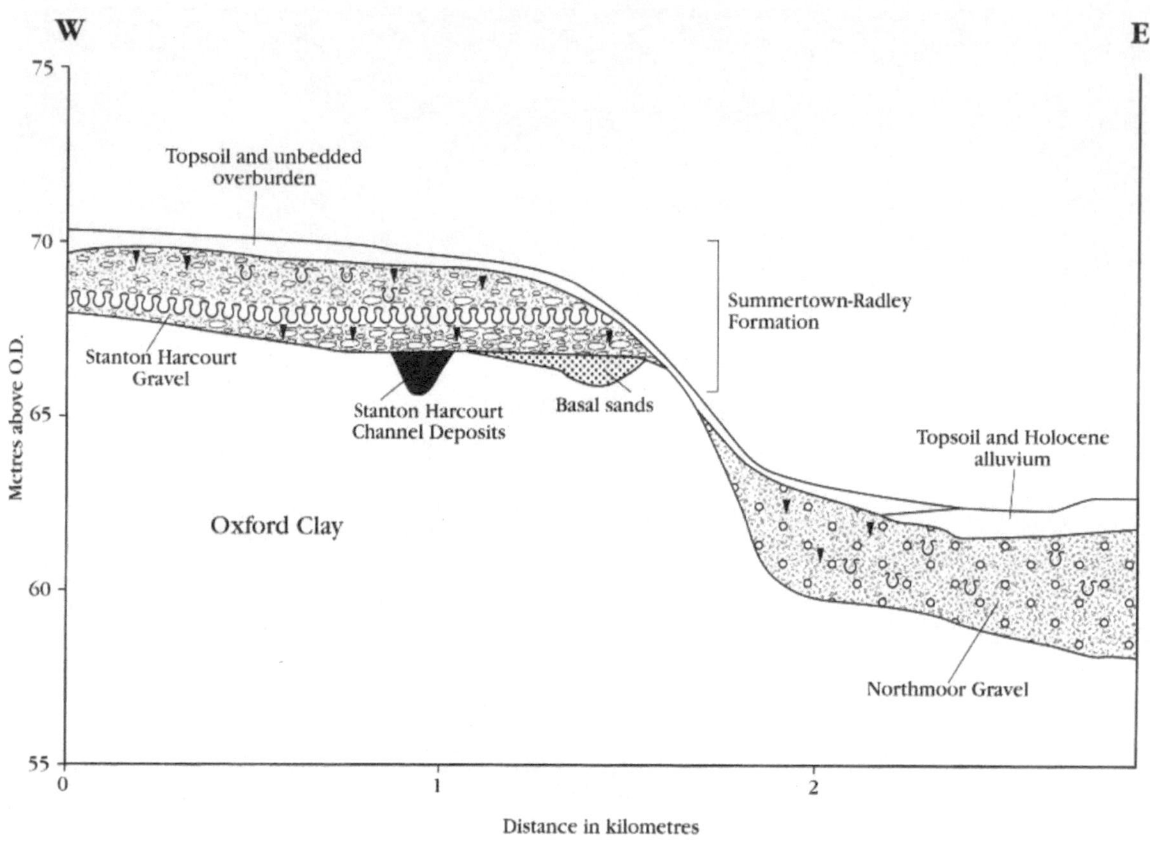

Figure 2.15 Composite section through the deposits at Stanton Harcourt (after Briggs *et al.*, 1985).

Interpretation

The combination of the geological evidence from sites such as Stanton Harcourt and Magdalen Grove, in conjunction with mapping of the distribution of the Summertown-Radley deposits and observations of their stratigraphy throughout their outcrop, has allowed a tripartite sequence to be established (Briggs *et al.*, 1985; Fig. 2.12), as follows:

3. Eynsham Gravel (temperate)
2. Stanton Harcourt Gravel (cold)
1. Stanton Harcourt Channel Deposits (temperate)

Before assessing this sequence, it is necessary to examine the bipartite division of the Summertown-Radley Terrace deposits, proposed by Sandford (1924, 1926), in which only members 2 and 3 (above) were recognized. It is important to establish that the cold-climate gravel that underlies the Eynsham Gravel is the same deposit that overlies the more recently discovered interglacial channel-fill at Stanton Harcourt. This stratigraphical relationship is fundamental to the interpretation of the Stanton Harcourt Channel Deposits as the product of a post-Hoxnian but pre-Ipswichian temperate episode.

Sandford's (1924) distinction of a lower, cold-climate aggradation (the Stanton Harcourt Gravel of Briggs *et al.*, 1985) was further underlined by the discovery, at the base of Summertown-Radley Terrace deposits at Dorchester (SU 569948), of an organic layer that yielded fossil plants indicative of a cold climate (Duigan, 1955).

One of the more significant advances stemming from recent research in the Upper Thames is the realization that the various gravels have predominantly accumulated under periglacial conditions. Earlier models tended to regard cold intervals as periods of erosion and to attribute the bulk of the deposits to interglacials, a view backed up in this region by the relatively

Figure 2.16 Section at Stanton Harcourt, showing an intraformational ice-wedge structure in the Stanton Harcourt Gravel. The complexity of this feature is typical of such structures at this site (see text). (Photo: D.J. Briggs.)

frequent occurrence of warm-climate faunas in the gravels. A significant move away from this view was promoted by Briggs and Gilbertson (1973), who demonstrated that the gravels underlying the Hanborough Terrace were laid down under cold conditions and concluded that the warm-climate mammalian fossils they contained were reworked (see above, Long Hanborough). Following this work, it was suggested by Goudie and Hart (1975) that a large part of the Summertown-Radley Formation was deposited under a 'fluvio-periglacial regime'. This view was based on observations at Standlake and at the present Stanton Harcourt site, which showed that at least the upper 3.5 m of the gravel is of similar character to the Hanborough Formation, suggesting a braided river origin, and that intraformational ice-wedge casts are common. Briggs (1976a) noted that the large complex ice-wedge casts that characterize the gravel at Stanton Harcourt (see Seddon and Holyoak, 1985; Fig. 2.16) are associated with festooning and that, when observed on cleared horizontal surfaces, they form a polygonal 'patterned-ground' effect. Briggs also noted that the ferruginous cementation of the gravels (see above) is typically localized within these ice-wedge cast/festoon features. The braided river origin of the gravels at Stanton Harcourt was confirmed in a sedimentological study by Bryant (1983).

Further support for the interpretation of this gravel as the product of a periglacial environment is provided by a molluscan fauna obtained from an extensive silt band near the base of the succession. This fauna was dominated by the land snail *Oxyloma pfeifferi*, with significant numbers of *Pupilla muscorum* (Gilbertson, 1976; Briggs and Gilbertson, 1980). None of the species present is indicative of particularly cold conditions, but *P. muscorum* requires an exposed, open habitat. This, and the absence of woodland species, leads to the conclusion that this silt accumulated under a climatic regime significantly cooler than that of the present. Seddon and Holyoak (1985) found similar faunas in several silty beds within the Stanton Harcourt Gravel and noted that, although the species present are not good climatic indicators, the low diversity of the assemblages strongly suggests a cold climate. These authors also described pollen and plant assemblages dominated by herb species, mostly of arctic-alpine affinities.

It has become apparent in recent years that the cold-climate Stanton Harcourt Gravel Member very much dominates the Summertown-Radley Formation, certainly in the areas of recent exposure. However, Sandford (1924) considered the upper deposit (the Eynsham Gravel) to be dominant, believing that erosion had removed much of the lower, cold-climate gravel (Stanton Harcourt Member) before the warm-climate deposit was laid down. He noted, however, that the upper deposit was frequently superimposed upon the lower with little sign of any break in the sequence, although occasionally there was evidence of channelling at its base. He illustrated channelling on a large scale at the Magdalen College site (see Fig. 2.13).

A critical reappraisal of the altitudinal relations and faunal content of the various deposits attributed to the uppermost (temperate) Summertown-Radley gravel raises a number of problems that have hitherto largely escaped attention. It appears that many attributions to this member have been erroneous. The original descriptions of the sites (Sandford, 1924, 1926; Dines, 1946) were made long before it was suspected that a further, older interglacial might be represented within the Summertown-Radley Formation (that subsequently identified in the channel deposits at Stanton Harcourt). Faunal records from the various sites in this terrace were merely combined to produce lengthy lists (for example, Kennard and Woodward, 1924) and recent authors have tended to assume, on the basis of the convincing stratigraphical descriptions by Sandford, that these assemblages were entirely derived from the upper deposit, the Eynsham Gravel of Briggs *et al.* (1985).

Eynsham Station Pit is no longer available for study, although degraded faces remain. Excavations here by the GCR Unit in 1984 failed to recover faunal remains (Bridgland, 1985a). Sandford (1924, p. 140) recorded *c.* 3 m of gravel, which was 'noteworthy for the very common occurrence of *Hippopotamus*'. He also noted that teeth of mammoth occurred at the base of the sequence, implying that his 'lower Summertown-Radley gravel' (Stanton Harcourt Gravel) was also represented at the site, making it 'one of the most important in the district' (1924, p. 141). There was faunal evidence for a cold-climate deposit beneath the hippopotamus-bearing levels at Radley and Iffley, as well as at Eynsham. Sandford (1924) also recorded a higher unit at Eynsham, separated from the Eynsham Gravel by an erosional contact, comprising *c.* 1 m of fine pebbly cross-bedded sand. This unit, devoid of fossils, may represent a post-interglacial aggradation (phase 2 of the terrace model – Chapter 1), superimposed upon the Eynsham Gravel during the early Devensian Stage (Table 1.1).

Hippopotamus amphibius is a most important species in the Late Pleistocene of Britain. It is regarded as an indicator for the Ipswichian Stage (*sensu* Trafalgar Square) since, despite the recognition of additional climatic cycles in recent years (Chapter 1), it is thought not to have lived in Britain between the Cromerian and the Ipswichian (Sutcliffe, 1960, 1964; Stuart, 1974, 1982a, 1982b; Gascoyne *et al.*, 1981). This species has been recorded from six sites in the Summertown-Radley Terrace: Eynsham, Wytham, Iffley, Radley, Abingdon and Dorchester (Sandford, 1924, 1965; Briggs *et al.*, 1985).

Many of the sites in the Summertown-Radley Formation, previously ascribed to the Eynsham Gravel Member, have yielded another species of possible stratigraphical significance, the bivalve *Corbicula fluminalis*. It has long been recognized that this mollusc has not lived in Britain during the Holocene, but it is now suggested that it was also absent during the Ipswichian Stage (*sensu* Trafalgar Square), although it has been recorded from many sites formerly regarded as Ipswichian but now considered to be older (Keen, 1990). The principal basis for this view is that there is no reliable record of *C. fluminalis* and hippopotamus occurring together in the same deposit. The Eynsham Gravel would appear to be an exception, but close scrutiny of the records reveals that, of the six sites listed above that have yielded hippopotamus, *C. fluminalis* is recorded only from Radley, where its presence was based on a single abraded fragment (Sandford, 1924; Kennard and Woodward, 1924). It seems highly likely that this was reworked. At other sites, such as Summertown, the species was found in abundance

(Prestwich, 1882; Sandford, 1924), as it is at both the Magdalen Grove and Stanton Harcourt GCR sites. None of these sites has yielded hippopotamus, although numerous other mammalian remains were found. The occurrence of abundant *C. fluminalis* and the absence of hippopotamus from these sites suggests that they represent the earlier of the two temperate intervals recognized from the Summertown-Radley aggradation, that which was first identified from the Stanton Harcourt channel. The true Eynsham Gravel yields hippopotamus but not *Corbicula* (except for derived specimens, as at Radley, reworked from the earlier interglacial member) and is therefore attributable to the Ipswichian Stage (*sensu* Trafalgar Square). It has been recognized unequivocally only at the six sites, listed above, from which hippopotamus is recorded.

Sandford (1924) recognized a lower cold-climate gravel beneath the interglacial beds with *Corbicula* at Summertown and Magdalen College. In both cases this interpretation was based on the occurrence below the *Corbicula* levels of *Mammuthus primigenius*, which is also the most common vertebrate fossil in the Stanton Harcourt Channel Deposits. At Summertown the mammoth was accompanied by horse and ox, which are also present in the Stanton Harcourt Channel Deposits, but neither site yielded woolly rhinoceros *Coelodonta antiquitatis*, the only species that appears, within the Summertown-Radley Formation, to be exclusive to the cold-climate Stanton Harcourt Gravel Member. The elevation of the Summertown site was not recorded, but that at Magdalen Grove (58.8 m O.D.) is at a low level within the altitudinal range of the formation, again suggestive of the lower, rather than the higher, interglacial member. The recorded location of the Summertown pit (see above), in the western edge of the gravel outcrop, suggests that the section there was also relatively low within the formation. These facts seemingly lead to the conclusion that the sites yielding *C. fluminalis* but not hippopotamus represent the earlier of the two Summertown-Radley interglacials and are older than both the Stanton Harcourt and Eynsham Gravels. They also suggest that faunal records from the Summertown-Radley Formation require detailed scrutiny, to determine whether cited upper or lower stratigraphical levels are correct.

The Stanton Harcourt Channel Deposits

The Stanton Harcourt Channel Deposits are of major importance in the British Pleistocene, as they have been repeatedly cited in recent years as providing critical evidence for the recognition of an additional post-Hoxnian/pre-Ipswichian interglacial (Shotton, 1983; Briggs *et al.*, 1985; Bowen *et al.*, 1989; Bridgland *et al.*, 1989). With comparable deposits at Aveley (see Chapter 4), Marsworth (Buckinghamshire) and Stoke Goldington (Buckinghamshire), they have been attributed to a temperate interval at about 180,000 years BP, which has been correlated with Oxygen Isotope Stage 7 (Chapter 1; Shotton, 1983; Bowen *et al.*, 1989).

All these sites have yielded mammalian faunas with mammoth and horse, but lacking hippopotamus, and have insect faunas dominated by the beetle *Anotylus gibbulus* (Shotton, 1983). The sediments at Aveley, Stoke Goldington and Stanton Harcourt have yielded molluscan assemblages that include *Corbicula fluminalis*. A site in the catchment of the Warwickshire–Worcestershire Avon, at Ailstone near Stratford-upon-Avon, has recently been added to this list (Bridgland *et al.*, 1989).

There has not been total agreement with the interpretation of the Stanton Harcourt Channel Deposits as being of pre-Ipswichian age, however. Gibbard (1985) did not accept that they occupy a lower stratigraphical position within the Summertown-Radley Formation. He suggested instead, following Goudie and Hart (1975), that the thick periglacial gravel at Dix's Pit post-dates the interglacial beds recognized by Sandford in the upper part of the formation. The principal reasoning behind this interpretation is that Gibbard, closely adhering to the chronology outlined by Mitchell *et al.* (1973), considered both the Stanton Harcourt Channel Deposits and the Eynsham Gravel to represent the Ipswichian Stage. He claimed support for this conclusion from thermoluminescence dating of silts within the gravel at Stanton Harcourt, which suggested an early Devensian age. This would require the gravel at Stanton Harcourt to be younger than the Eynsham Gravel. Details of two dates obtained by this method, of 91,000 (± 8000) and 93,000 (± 9000) years BP, were provided by Seddon and Holyoak (1985). These authors also cited a radiocarbon date from a similar stratigraphical level, however, that suggested a mid-Devensian age, in the region of 35,000 years BP. This alternative interpretation of the Summertown-Radley sequence raises doubts about the validity of the tripartite sequence established by Briggs and Gilbertson (1980). In support of the view that the Stanton Harcourt Channel Deposits are pre-Ipswichian, in addition to the biostratigraphical arguments already presented, are amino acid ratios recently obtained from specimens of *Corbicula* and *Valvata*. These are comparable to ratios from other sites ascribed to Oxygen Isotope Stage 7 (Bowen *et al.*, 1989).

It is ironical that, whereas the reality of extra climatic cycles within this part of the Pleistocene is now accepted by most workers, the interpretation of the Stanton Harcourt site remains controversial.

The Magdalen Grove deposits

The sediments at Magdalen Grove have provided a wealth of palaeontological information, both from the early collections and from the 1984 re-excavation. The latter provided new information from palynological studies, which have rarely been applied to deposits in the Upper Thames basin. The abundance of broad-leaved trees in the pollen assemblage (see above) argues for a temperate climate, with the occurrence of *Carpinus* and virtual absence of *Betula* suggesting the late-temperate zone (biozone III) of an interglacial (Hunt, 1985). According to Hunt, the absence of *Abies* and the presence of *Picea*, *Acer* and abundant *Carpinus* are features of the Magdalen Grove assemblage that resemble pollen biozone IpIII of the Ipswichian Stage. However, many sites that were once considered Ipswichian, but which are now widely attributed to older, post-Hoxnian but pre-Ipswichian temperate intervals, share these features. As was outlined above, the abundance of *C. fluminalis*, the absence of hippopotamus and the elevation of the deposits at Magdalen Grove suggest that they belong to the earlier of the two interglacials represented within the Summertown-Radley Formation, that correlated with Oxygen Isotope Stage 7.

If this reinterpretation can be corroborated, perhaps by the application of geochronological dating techniques (such as the analysis of the amino acid content of shells – see Chapter 1), Magdalen Grove would be the best reference site for the lower temperate-climate member (the Stanton Harcourt Channel Deposits), since at Stanton Harcourt these are buried beneath several metres of gravel and are therefore inaccessible.

Palaeolithic artefacts from Stanton Harcourt

Palaeolithic implements have been discovered in some numbers in the Summertown-Radley Terrace deposits and offer a possible insight into the approximate age of the aggradation. Sporadic finds of Acheulian hand-axes in these deposits were recorded in the early literature (Evans, 1897; Manning and Leeds, 1921; Almaine, 1922; Smith, 1922; Sandford, 1924). It later became clear that, by Upper Thames standards, a considerable wealth of artefacts occurred throughout the formation, although with the highest concentrations in the cold-climate Stanton Harcourt Gravel (Sandford, 1932, 1939, 1954; Arkell, 1945, 1947a; Wymer, 1968; MacRae, 1991). Both Sandford and Arkell originally considered this to be the earliest appearance of Acheulian implements in the Upper Thames sequence; both thought the prolific Wolvercote Channel to be later than the lower part, if not the whole, of the Summertown-Radley Formation, despite its association with the higher Wolvercote Terrace (see above, Wolvercote). However, one or two hand-axes were later recorded from the Hanborough Gravel (Arkell, 1947c; Wymer, 1968), which is therefore the oldest source of artefacts in the area. Arkell (1945) believed that the numerous worn Acheulian hand-axes found in the Summertown-Radley Terrace deposits were manufactured during a warm period prior to deposition of the cold-climate gravels identified by Sandford. Since no earlier interglacial terrace with such an abundance of artefacts had been identified in the region, he suggested that there might be a third, hitherto undiscovered, earlier division of the Summertown-Radley sequence, of interglacial origin. This suggestion by Arkell pre-empted by more than 30 years the discovery of the lower interglacial channel at Stanton Harcourt and the recognition of the tripartite climatic sequence.

According to Wymer (1968, p. 85) a hand-axe, the largest at that time from west of Oxford, was found in 1962 in the vicinity of the Stanton Harcourt GCR site. This was one of only two discoveries at the locality prior to 1982 (MacRae,

1987). Recent exposures, at Dix's Pit and other nearby workings, have been repeatedly visited by the local archaeologist R.J. MacRae, who has built up a collection of artefacts that includes nearly 50 hand-axes, including (by 1989) 13 made from quartzite (MacRae, 1988, 1989, 1991; MacRae and Moloney, 1988). In the Upper Thames, only the Wolvercote Channel and pits at Berinsfield have yielded larger numbers; those from the latter locality include finds from both the Summertown-Radley and the Northmoor Formations (MacRae, 1982).

The pits at Stanton Harcourt are therefore the most important source of palaeoliths from the Summertown-Radley Terrace. In 1986 a very large hand-axe was discovered there, which, at 26.9 cm long, surpasses the 1962 discovery by 5 cm and is claimed as the third-largest from Britain (MacRae, 1987; Fig. 2.17). Forty-two hand-axes were found at the site between 1984 and the end of 1987, but in the following two years the number retrieved greatly diminished (R.J. MacRae, pers. comm.). According to MacRae, the rate of discovery of mammoth teeth and tusks, which was high from 1984 to 1987, has declined since then in parallel with the artefacts. He reports that this change has occurred as the working faces migrated away from the area of the Stanton Harcourt Channel. This fact, as well as the association of palaeoliths

Figure 2.17 Hand-axe from Stanton Harcourt, discovered by Mr V. Griffin, the excavator driver. Some 26.9 cm long, this is believed to be the third-largest hand-axe discovered to date in Britain (MacRae, 1987). (Photo: R.J. MacRae.)

with mammoth remains (the most abundant vertebrate fossil in the channel deposits), suggests that the incidence of artefacts might be related in some way to the channel. Until recently, all the finds from Dix's Pit were from the Stanton Harcourt Gravel (MacRae, 1985, 1988; MacRae and Moloney, 1988), but a few hand-axes have now been found in the channel deposits (MacRae, 1991). The reinterpretation of the stratigraphy of the Summertown-Radley aggradation, presented above, implies that sediments accumulated during the earlier temperate interval to considerable thicknesses, such as was recorded at Summertown. Subsequently there must have been considerable erosion of these deposits in those areas where the Summertown-Radley Formation is now represented only by later sediments. At Stanton Harcourt, deposits representing the early temperate interval are preserved only as channel deposits incised into the Oxford Clay; these are truncated by the Stanton Harcourt Gravel (Fig. 2.15). The erosive nature of this contact suggests that there may have been considerable reworking of material from the channel deposits into the gravel, perhaps including mammoth remains and artefacts. This might explain a concentration of both in the Stanton Harcourt Gravel in the vicinity of the channel.

Correlation

Attempts at correlating the Upper Thames sequence with the British Pleistocene chronology were, until the new framework based on the deep-sea oxygen isotope record was established (Chapter 1), impeded by the fact that only two interglacials were recognized between the Anglian and Devensian Stages (Mitchell *et al.*, 1973), whereas post-Cromerian interglacial faunas are recorded at four stratigraphical levels within the Upper Thames terraces:

4. Upper Summertown-Radley Formation	(Eynsham Gravel)
3. Lower Summertown-Radley Formation	(Stanton Harcourt Channel Deposits)
2. Wolvercote Channel Deposits	
1. Hanborough Gravel	(derived fauna)

The similarity of the upper Summertown-Radley and Wolvercote Channel interglacial assemblages was stressed by two of the leading authorities on the Upper Thames, Sandford (1924, 1932) and Arkell (1947a). Both authors concluded that these deposits were of comparable ages, despite the fact that they occurred in association with different terraces (see above, Wolvercote). Both also considered the lower cold-climate Summertown-Radley aggradation (Stanton Harcourt Gravel) to pre-date the Wolvercote Channel. Sandford (1924) was impressed with the similarity of the fauna in the upper Summertown-Radley sediments with that from Crayford, implying a correlation with the Taplow Terrace (see Chapter 4, Lion Pit). Sandford (1932) later changed his opinion, correlating Crayford with the Wolvercote Channel Deposits. In this later view, which was supported by Arkell (1945, 1947a), he favoured a correlation of the Eynsham Gravel with the Boyn Hill Terrace at Swanscombe, implying a Hoxnian age. Both Sandford and Arkell were strongly influenced by the first appearance of palaeoliths in the Upper Thames at this level (see above). However, Breuil (Appendix to Sandford, 1932) attributed the upper Summertown-Radley (hippopotamus-bearing) sediments to the 'Riss-Würm interglacial' (= Ipswichian Stage), again proposing a correlation with Crayford. With the recognition, in more recent years, that *Hippopotamus amphibius* was absent from Britain during the Hoxnian, most subsequent writers placed the upper Summertown-Radley deposits in the Ipswichian (for instance, Bishop, 1958; Tomlinson, 1963). Bishop (1958) regarded the Wolvercote Channel as Hoxnian and therefore considerably older. This model was adopted by Wymer (1968), who was then able to claim the Wolvercote Channel as a possible source for the Summertown-Radley implements.

Two recently suggested correlations are worthy of note, being based on attempts to trace the terraces from the Oxford area to the lower reaches of the river. In the first, Evans (1971) linked the Summertown-Radley Terrace with the Upper Floodplain Terrace of the Middle and Lower Thames, claiming support from the occurrence of *Hippopotamus amphibius* in the former; this species occurs beneath the Upper Floodplain Terrace (Kempton Park Formation) at Trafalgar Square (Table 1.1). Gibbard (1985) considered that the Summertown-Radley deposits equated with a variety of deposits downstream of the Goring Gap, ranging in age from late Saalian to Early Devensian. He correlated the terrace not with the Upper Floodplain (Kempton Park Gravel) aggradation, but with his Early Devensian Reading Town Gravel. It is argued in Chapters 1 and 3 that the latter is, in fact, based on a misidentification by Gibbard of the Taplow Gravel in the Reading area. Therefore Gibbard's conclusions actually imply correlation between the Summertown-Radley and Taplow Formations.

The Summertown-Radley sequence provides critical evidence in support of the correlation schemes presented in this volume (Chapter 1) for relating the sequence in the Upper Thames to that elsewhere in the basin and to the deep-sea oxygen isotope record (Fig. 2.18). In these schemes the Stanton Harcourt Channel Deposits are taken as an important stratigraphical marker for Oxygen Isotope Stage 7 and the Eynsham Gravel is assigned to the Ipswichian (*sensu* Trafalgar Square – Substage 5e). Correlation between the Upper and Middle Thames, based on the reconstruction of terrace long-profiles (Fig. 1.3), indicates that the Summertown-Radley Formation is the upstream equivalent of the Taplow Gravel. The latter unit also incorporates temperate-climate sediments attributed to Stage 7, particularly in its Lower Thames equivalent, the Mucking Gravel (see Chapters 1 and 4). As stated above, the Trafalgar Square Ipswichian sediments are found beneath a later aggradation, the Kempton Park Gravel (Fig. 2.18b), that is ascribed to the Devensian Stage (Gibbard, 1985). The Kempton Park Gravel, which gives rise to the 'Upper Floodplain Terrace' of Dewey and Bromehead (1921), cannot be traced further upstream than Marlow (Gibbard, 1985; Chapter 3; Fig. 1.3). This suggests that the rejuvenation event that separated the aggradation of the Kempton Park and Taplow Gravels was restricted to that part of the valley downstream from Reading (Fig. 2.18b). It is interesting to note that Sandford (1965) claimed that the Goring Gap was the site of a major 'nick point', which is the situation now envisaged during the deposition of the Kempton Park Gravel.

This interpretation explains the unusual incidence, in the Summertown-Radley Formation, of two full climatic cycles represented within a single aggradational sequence. The uppermost, unfossiliferous gravel at Eynsham, recorded by Sandford (1924; see above) may represent aggradation at the beginning of the cold half-cycle that followed the Ipswichian (*sensu* Trafalgar Square). The lower gravel at Summertown, underlying deposits containing *Corbicula*, may also represent a cold-climate deposit, presumably representing Oxygen Isotope Stage 8. In terms of the model for terrace formation outlined in Chapter 1, the full Summertown-Radley sequence would therefore be as shown in Table 2.3 (see also Fig. 2.18a).

Rejuvenation to the Northmoor Gravel level occurred during the mid-Devensian (see above, Introduction), perhaps as part of the incision that led to the separation of the Kempton Park and Shepperton Formations in lower reaches of the Thames catchment (Table 1.1). Information regarding the timing of this rejuvenation in the Upper Thames may be obtained in the near future from a recently discovered site in the Northmoor Gravel, at Cassington. Preliminary investigations suggest that organic and shelly channel-fills from the lower part of the sequence here may represent one or more of the interstadial episodes of the early/middle Devensian

Table 2.3 Stratigraphical subdivisions of the Summertown-Radley Formation.

Lithostratigraphical unit	Terrace model (Chapter 1)	Oxygen Isotope Stage
Uppermost gravel at Eynsham	Phase 4	5d-2?
Eynsham Gravel Member	Phase 3	5e
Stanton Harcourt Gravel Member	Phases 2 and 4	6
Stanton Harcourt Channel Deposits	Phase 3	7
Gravel underlying interglacial deposits at sites with *Corbicula*	Phase 2	8

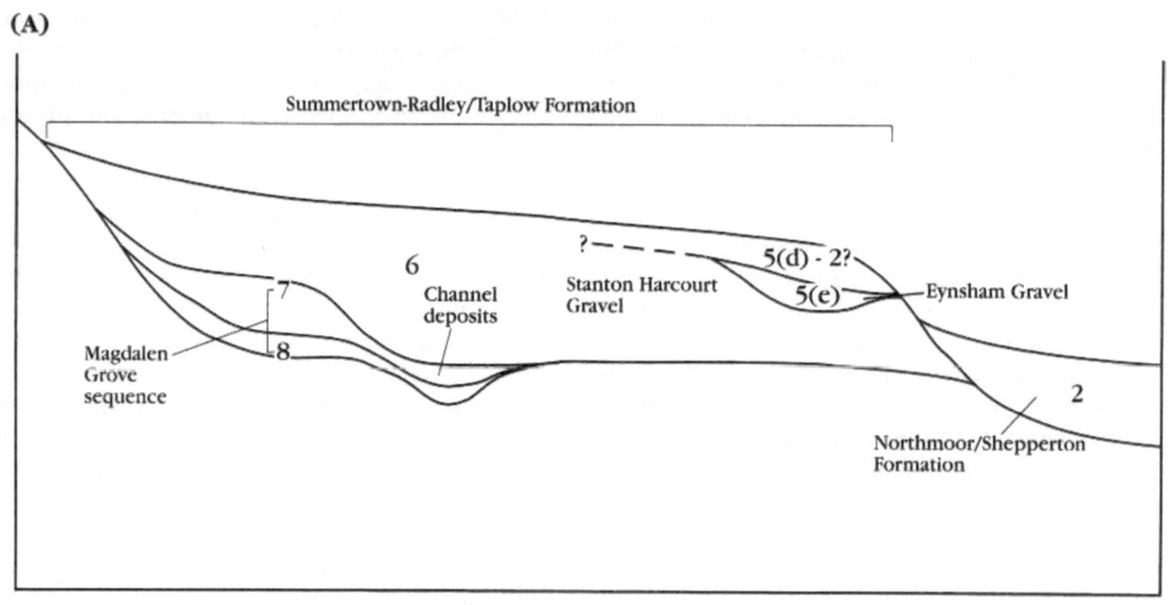

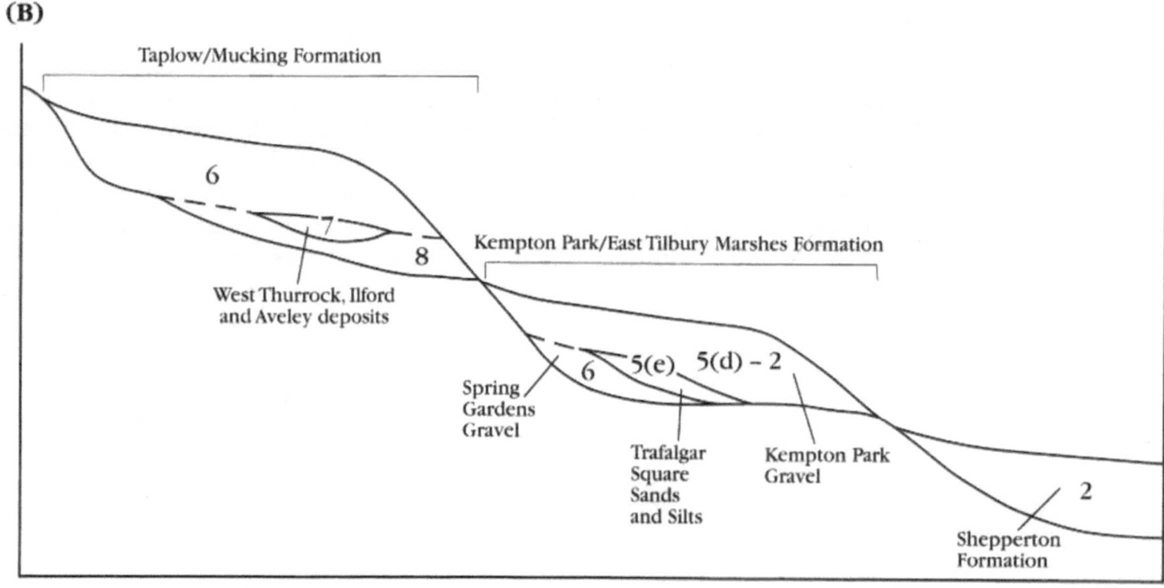

Figure 2.18 Comparison of terrace stratigraphy upstream (A) and downstream (B) from the limit of the Kempton Park Formation. Numbers 2–8 indicate oxygen isotope stages.

(D. Maddy, pers. comm.). This may prove difficult to reconcile with the Middle Thames sequence, in which these temperate episodes are represented within the Kempton Park Formation. It appears that rejuvenations may have occurred at different times in the Upper Thames and Middle Thames valleys.

Summary

The GCR sites at Stanton Harcourt and Magdalen Grove combine to provide coverage of the complex of deposits that underlies the Summertown-Radley Terrace of the Upper Thames. Considerable controversy surrounds the interpretation of this sequence, to the extent that there is disagreement on the subdivision of the deposits and the number of climatic cycles represented. The resolution of these problems is fundamental to an improved understanding of the Upper Thames sequence, which is clearly of considerable importance, given its position between the London Basin and the Midlands. Further research is required to address these difficulties. In particular, additional attempts to date the various sediments would be of value; the cold-climate Mollusca from the Stanton Harcourt Gravel could be suitable for geochronological dating by amino acid analysis. Similar analyses of shells from Magdalen Grove might give support (or otherwise) to the view that the deposits here are of the same age as those from the Stanton Harcourt channel-fill.

Conclusions

The Stanton Harcourt and Magdalen Grove sites provide an insight into a complex sequence of deposits forming a major terrace in the Upper Thames basin (the Summertown-Radley Terrace). The gravels that form the bulk of this sequence were laid down under intensely cold (periglacial) conditions, as is demonstrated at Stanton Harcourt both by structures such as ice-wedge casts and by the shells of cold-climate snails, which occur in silt bands interbedded with the gravels. It has long been recognized, from a number of sites in the region, that parts of the sequence yield interglacial faunas. Early records suggested that they were generally from the upper part of the gravel. They have been widely attributed to the last (Ipswichian) interglacial (around 125,000 years BP), largely on the basis of the occurrence of hippopotamus, which is highly characteristic of that interval.

The more recent discovery of a fossiliferous channel-fill beneath the gravel at Stanton Harcourt has provided critical evidence to show that two separate interglacials are represented within the Summertown-Radley Terrace sequence. This has also been a key site in establishing that the earlier of these interglacials is additional to the sequence of climatic events recognized hitherto in Britain and for its correlation with the oceanic record of Oxygen Isotope Stages, in which it is believed to represent Stage 7 (*c.* 200,000 years BP). The Magdalen Grove deposit has been widely regarded as an example of the upper, last interglacial element of the Summertown-Radley sequence. However, a reconsideration here of the evidence from this and other associated sites now suggests that it is also of Stage 7 age. It is further concluded that five different climate episodes are represented within the Summertown-Radley sequence, respectively cold, warm, cold, warm and cold.

Chapter 3

The Middle Thames

INTRODUCTION

The term 'Middle Thames' refers to that part of the river's course between the Goring Gap and London. The most complete record of fluvial deposition to be found in the Thames catchment is preserved on the dip slope of the Chilterns escarpment, making the Middle Thames basin a classic area for Thames research. Indeed, on the northern side of the valley between Maidenhead and Staines, an impressive sequence from the pre-Anglian Beaconsfield Gravel to the post-Hoxnian Taplow Gravel is present (see Figs 3.1 and 3.2). In this same area, the Devensian Kempton Park Gravel (Upper Floodplain Terrace) is preserved on the south side of the valley. The stratigraphical sequence can be extended upwards, in the areas both upstream and downstream, by way of dissected remnants of older gravels preserved on the Chilterns dip slope (Figs 3.1, 3.2 and 3.3).

Downstream from this, the type area for the Middle Thames terraces, the older sediments (Winter Hill Gravel and above) can be traced north-eastwards along the abandoned Thames route that follows the Chilterns dip slope. This provides an indication of the former course of the river through the Vale of St Albans, before it was blocked by ice during the Anglian glaciation. Thereafter gravels were deposited along the present route of the Thames towards London (see Chapter 1). This change in the course of the Thames was the direct consequence of the glacial blocking of the St Albans valley, but is set against a background of slow but progressive southward migration by the river since its first appearance as the main agent of west to east drainage in the London Basin.

The London Basin is a large synclinal structure stretching from the Savernake Forest in the west to the southern North Sea in the east. It is bounded on its southern side by the North Downs and on its northern side by the Chilterns, opening out to the north-east as that escarpment fades into the broad Chalk outcrop of East Anglia (Fig. 1.1). It is drained principally by streams flowing southwards from the Chilterns dip slope and northwards from the North Downs, but whereas a number of rivers flow through the latter escarpment from the Weald, only the Thames has a substantial catchment to the north of the Chilterns. The role of the Thames as a 'strike' river, flowing along the approximate centre of the syncline, is continued westwards by the Kennet tributary (Fig. 1.1).

The formation of the basin was initiated in pre-Cenozoic times and it was infilled during the Palaeocene and Eocene by a series of mostly marine sediments. Evidence for drainage evolution in south-east England at this time can be gained from the content of pebble beds within these Palaeogene strata. The overwhelming majority of the pebbles are flints, derived from the up-folded margins of the basin. However, a gravelly facies of the Reading Beds on the northern side of the basin, particularly around Lane End, Buckinghamshire, contains considerable quantities of quartz and 'lydite' (pre-Cretaceous chert) pebbles, thought by some to have been derived by way of an early Goring Gap (White, 1906; Wooldridge and Gill, 1925). These unusual beds are considered to be of fluvial origin and may represent the earliest evidence of ancestral Thames drainage. The Upper Bagshot (formerly Barton) Pebble Beds of west Surrey contain significant quantities of Lower Greensand chert and quartz pebbles (Dewey and Bromehead, 1915). The presence of the chert implies that southern tributaries of the Thames, draining the Weald, were established by this time and that denudation in that area had by the Eocene uncovered the Lower Greensand (despite the fact that the principal phase of wealden uplift occurred later, in the Miocene).

The early Neogene is represented in the London area by a few problematic outliers of probable marine sediments preserved on the high ground of the North Downs and Chilterns. These include the Lenham Beds and the Netley Heath, Headley Heath and Little Heath deposits (see Part 1 of this chapter). The next set of deposits in the stratigraphical sequence of the London Basin is the Pebble Gravel. This is a term that has been applied to any deposit

Figure 3.1 (Following two pages) Map showing the gravels of the Middle Thames, the Vale of St Albans and the Kennet valley. Compiled, with reinterpretation as indicated in the text, from the following sources: Cheshire (1986a), Gibbard (1985), Green and McGregor (1978a), Hare (1947), Hey (1965, 1980), Sealy and Sealy (1956), Thomas (1961), Wooldridge (1927a) and the Geological Survey's New Series 1:50,000 and 1:63,360 maps. GCR sites and type localities are shown.

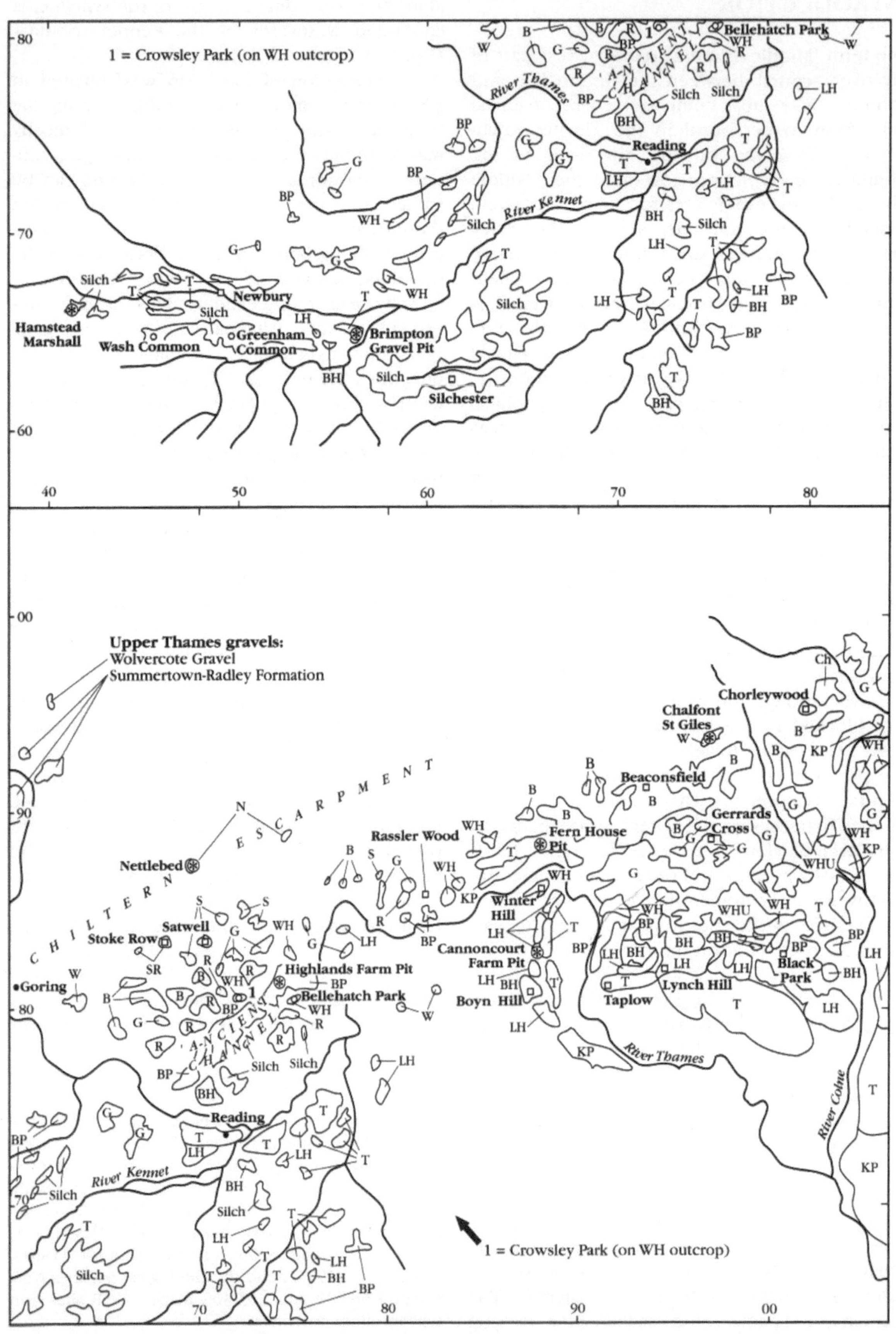
1 = Crowsley Park (on WH outcrop)
Bellehatch Park
River Thames
ANCIENT CHANNEL
Reading
River Kennet
Newbury
Hamstead Marshall
Wash Common
Greenham Common
Brimpton Gravel Pit
Silchester
Upper Thames gravels:
Wolvercote Gravel
Summertown-Radley Formation
CHILTERN ESCARPMENT
Nettlebed
Stoke Row
Satwell
Goring
Highlands Farm Pit
Rassler Wood
Fern House Pit
Winter Hill
Cannoncourt Farm Pit
Boyn Hill
Beaconsfield
Chalfont St Giles
Chorleywood
Gerrards Cross
Taplow
Lynch Hill
Black Park
River Colne
1 = Crowsley Park (on WH outcrop)

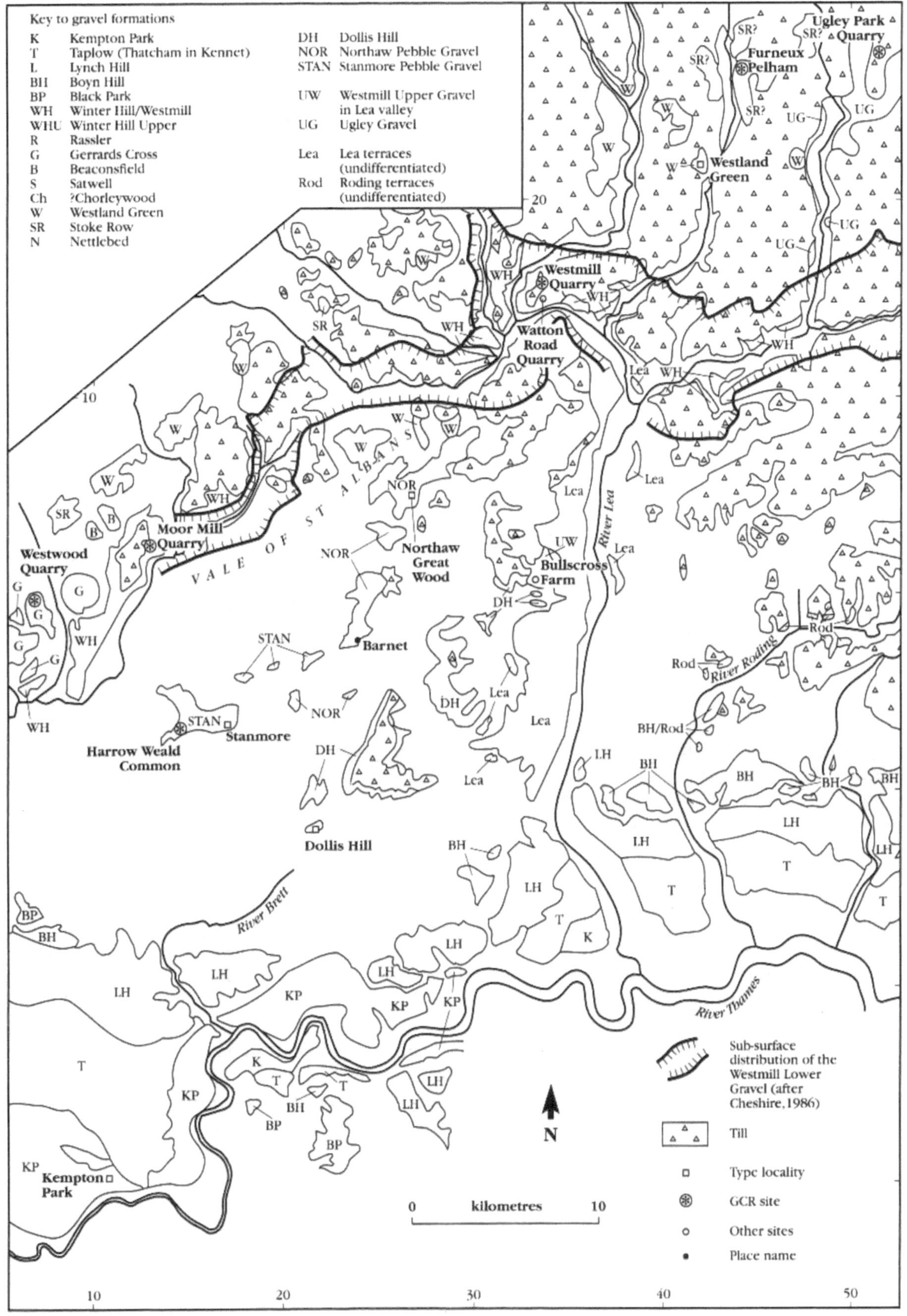
Key to gravel formations
K Kempton Park
T Taplow (Thatcham in Kennet)
L Lynch Hill
BH Boyn Hill
BP Black Park
WH Winter Hill/Westmill
WHU Winter Hill Upper
R Rassler
G Gerrards Cross
B Beaconsfield
S Satwell
Ch ?Chorleywood
W Westland Green
SR Stoke Row
N Nettlebed
DH Dollis Hill
NOR Northaw Pebble Gravel
STAN Stanmore Pebble Gravel
UW Westmill Upper Gravel in Lea valley
UG Ugley Gravel
Lea Lea terraces (undifferentiated)
Rod Roding terraces (undifferentiated)
Ugley Park Quarry
Furneux Pelham
Westland Green
Westmill Quarry
Watton Road Quarry
VALE OF ST ALBANS
Moor Mill Quarry
Westwood Quarry
Northaw Great Wood
Bullscross Farm
Barnet
River Lea
River Roding
Harrow Weald Common
Stanmore
Dollis Hill
River Brent
River Thames
Kempton Park
Sub-surface distribution of the Westmill Lower Gravel (after Cheshire, 1986)
Till
Type locality
GCR site
Other sites
Place name
N
0 kilometres 10
10
20
30
40
50

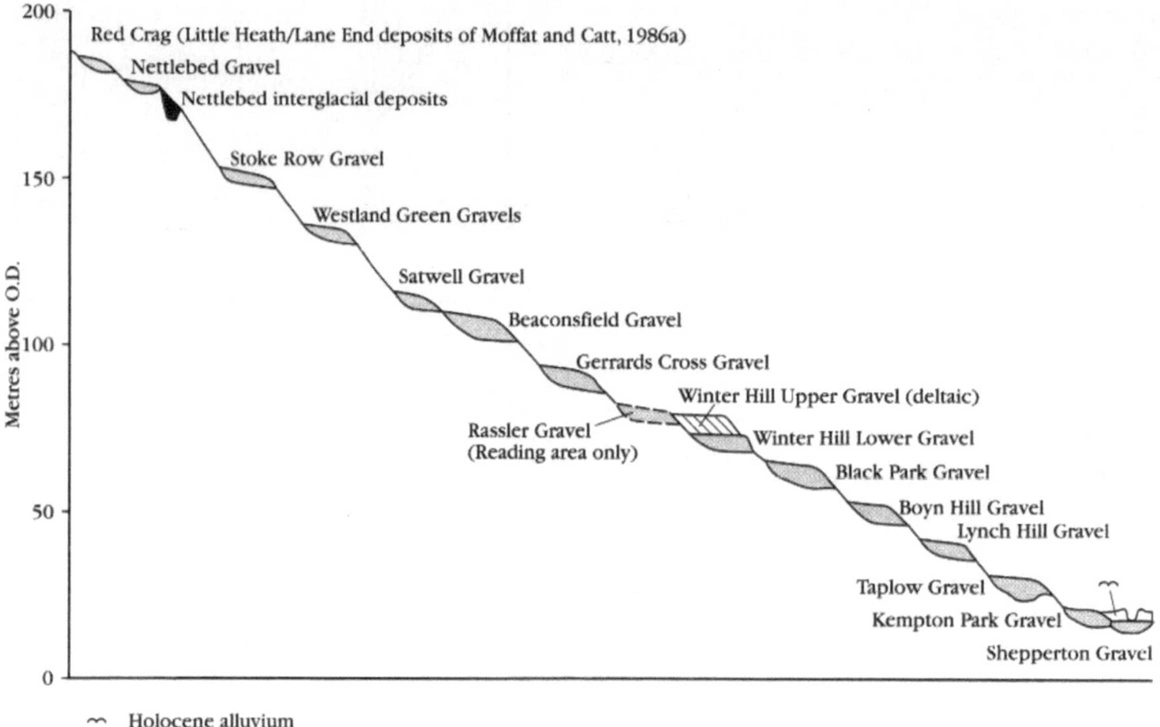

Figure 3.2 Idealized transverse section through the classic Middle Thames sequence of the Slough–Beaconsfield area. The stratigraphical position of the Rassler Gravel, not preserved in this area, is shown.

comprising mainly rounded flint pebbles reworked from the Palaeogene pebble beds. In its stricter sense it refers to high-level plateau gravels in the London Basin, believed to be the oldest drift deposits of the district (Whitaker, 1864, 1889; Wooldridge, 1927a, 1927b; Wooldridge and Linton, 1939, 1955). These deposits do not contain material unequivocally derived from beyond the Cotswold escarpment, such as is characteristic of the later Thames gravels. The initiation of a link with the Midlands, as indicated by the Northern Drift of the Upper Thames basin (described in the preceding chapter), is signalled by the first gravels containing abundant quartzite from the Triassic pebble beds. Such material is an important component of all later Thames gravels, but in particular it is ubiquitous in the Lower Pleistocene and lower Middle Pleistocene pre-diversion gravels. These deposits occur on the northern side of the Middle Thames valley and in the Vale of St Albans, but have generally been classified on Geological Survey maps as 'Glacial Gravels'. Their interpretation as early terrace gravels of the Thames system was only securely established in the late 1930s. This part of the Thames sequence is described in Part 2 of this chapter.

In the Vale of St Albans the lowest of these early Thames gravels interdigitates with Anglian glacial deposits, including till and proglacial lake sediments. It can be demonstrated at a number of sites along this former route that the river was ponded and then diverted as a result of this glaciation (see below, Moor Mill and Westmill).

The last part of this chapter (Part 3) charts the development of the Middle Thames from immediately after the diversion of the river to the Devensian Stage, when the limits of the present floodplain were established. Fossiliferous deposits are extremely rare in the Middle Thames valley, the main palaeontological evidence coming from the occasional bones of large mammals found in the gravels. However, many of the sites described in Part 3 have yielded important assemblages of Palaeolithic artefacts; these take on extra significance, in the absence of biostratigraphical evidence, as a potential means of relative dating.

Two of the sites to be described in Part 3, Hamstead Marshall and Brimpton gravel pits, lie in the tributary Kennet valley. Information from these localities can be readily assimilated into the story of the Thames since, as noted above, the Kennet is essentially a westward extension of the Middle Thames valley along the centre of the London syncline.

Part 1

PLIOCENE/LOWER PLEISTOCENE DEPOSITS IN THE LONDON BASIN
D.R. Bridgland

Introduction

High-level deposits, often capping Palaeogene outliers, are sporadically preserved on the Chilterns and North Downs, the Chalk escarpments bordering the London Basin to the north and south. The highest of these deposits have usually been attributed to Late Pliocene/Early Pleistocene marine episodes, although the incidence of corroborative palaeontological evidence is very rare. In addition, many of the hills north of London, both to the north and south of the Vale of St Albans (Fig. 3.1), are covered by gravelly deposits composed predominantly of rounded flint pebbles of the type that make up the various Palaeogene pebble beds. This type of deposit is commonly known as Pebble Gravel.

Early studies of Pliocene/Lower Pleistocene deposits were concentrated on the North Downs. Most early descriptions of the fossiliferous Lenham Beds on these hills in Kent attributed them to the Pliocene (Prestwich, 1858a; Lyell, 1865; Geikie and Reid, 1866; Reid, 1890; Newton, 1916; Wooldridge, 1927a; see Little Heath), although they are now considered to be Miocene (Curry *et al.*, 1978). Similar deposits were recognized on the North Downs of Surrey at Netley Heath and Headley Heath (Whitaker, 1862; French, 1888) containing, at the former site, fossil molluscs (Stebbing, 1900; Davies, 1917; Chatwin, 1927; Dines and Edmunds, 1929; John and Fisher, 1984). These are poorly preserved moulds in ferruginously indurated horizons and/or clasts that provide evidence of correlation, not with the Lenham Beds, but with the Red Crag of East Anglia. Although formerly regarded as Pliocene (Reid, 1890; Harmer, 1902), the Red Crag has for much of this century been considered to be basal Pleistocene (Baden-Powell, 1950; Boswell, 1952). Later appraisal suggested that at least part of the Red Crag belongs to the Upper Pliocene Series (Cambridge, 1977; West, 1977). Recent estimates of the age of this formation and its suggested correlation with the Praetiglian and Upper Reuverian Stages of The Netherlands imply that the Red Crag is wholly pre-Pleistocene, dating from between 3.5 and 2 million years BP (Zalasiewicz and Gibbard, 1988).

There are very few deposits south of the main Red Crag outcrop and north of the Thames that contain biostratigraphical evidence for a Pliocene age. However, the various high-level outliers rich in well-rounded flint pebbles, the Pebble Gravel, have been recognized as deposits of considerable antiquity, possibly as old as Pliocene, since the middle of the last century. The rounded nature of the majority of its clasts has led numerous authors to regard the Pebble Gravel as a series of marine or littoral deposits (Hughes, 1868; Wood, 1868; Prestwich, 1881, 1890a, 1890b; Whitaker, 1889). In addition to the rounded flints, these deposits also contain small amounts of subangular flint, together with quartz, quartzites and sarsens. The term Pebble Gravel was first used by Whitaker (1864, 1875, 1889), who regarded this type of material as the oldest 'drift' deposit, distinguishing it from the fossiliferous Pliocene beds that occur in comparable situations capping high ground, but which he regarded as part of the 'solid' geology. There has, however, been considerable confusion as to whether certain high-level deposits lacking fauna are unfossiliferous Pliocene marine beds or part of the Pebble Gravel.

In one of the earliest descriptions of the Pebble Gravel, Hughes (1868) referred to it as 'Gravel of the Upper (or Higher) Plain', forming the higher of two dissected gravel-covered plateaux recognized by him in Hertfordshire. According to Hughes: 'there are just enough subangular flints and large partly worn pieces of quartz etc. to show that this gravel derives its pebbly character from the waste of older pebble beds, with which the unworn fragments got mixed, and not that they were all worn together into pebbles along the shingly shore of the Higher-Plain Gravel-sea' (Hughes, 1868, p. 285). Like Hughes, Whitaker (1889) also realized that the rounded nature of the flints in the Pebble Gravel was a feature inherited from the Palaeogene, although both he and Hughes nevertheless favoured a marine origin. Prestwich (1881, 1890b) correlated the Pebble Gravel of the area north of London with his 'Mundesley and Westleton Beds' (later simply 'Westleton Beds'), which he sought to trace southwards and westwards, at increasing altitudes, from East Anglia. Prestwich interpreted the deposits as evidence for a Late Pliocene/Early Pleistocene marine incursion into the London Basin. He realized that these beds dated from near the Pliocene/Lower Pleistocene boundary and considered them to be 'the base of the Quaternary Series' (Prestwich, 1890a, p. 85).

Salter (1896) attempted to subdivide the Pebble Gravel on the basis of composition, recognizing four types. These were:

a. Barnet Gate type – strongly dominated by flint and occurring widely on the highest land in Hertfordshire.

b. Hampstead type – of more restricted occurrence, characterized by smaller pebbles than (a) and by the presence of Greensand chert.

c. High Barnet type – distinguished from (a) and (b) by a sandier matrix, a considerably greater complexity of pebble composition and with a more widespread distribution.

d. Bell Bar type – differing from the others in that it occupies lower levels and has an even greater complexity of composition than (c), including plentiful quartzites.

Within all these compositional divisions, the elevations of the various outliers generally decline from west to east, a factor that led Salter to reject the marine hypothesis in favour of an interpretation of the deposits as reflecting the onset of 'Glacial' conditions. He believed that the various types of Pebble Gravel were the 'first deposits of the Glacial Series' (Salter, 1896, p. 404), but he later emphasized the role played by rivers in their deposition (Salter, 1898, 1901, 1905). In many ways Salter's work, combining altitude and clast-lithological composition as criteria for classification, was far ahead of its time; in the first half of the twentieth century theoretical geomorphology and denudation-chronology prevailed and there was little progress in the interpretation of the Pliocene/Lower Pleistocene deposits in the London Basin.

White (1906) proposed a possible solution to one of the major puzzles presented by the Pebble Gravel, that of the relatively large proportion of quartz it contains, especially in the finer gravel fractions. This material is foreign to the London Basin, and is generally absent in the Palaeocene and Eocene pebble beds. However, White reported the discovery of a markedly different facies of the Palaeocene Reading Beds, at Lane End in Buckinghamshire. Reading Beds gravel at this locality contains abundant quartz as well as subangular flint and 'lydite'. White suggested that this facies was once widespread to the north-west of the present Chiltern escarpment, prior to the erosional recession of the latter during the Pleistocene. This quartzose facies of the Reading Beds was seen by White as a potential source for the non-flint component of the Pebble Gravel.

There followed several years of controversy over the precise age of these deposits at Lane End. Barrow (1919a) questioned White's assertion that they belonged to the Reading Beds. A major factor in favour of White's interpretation was that the quartzose gravel is overlain by what appears to be basal London Clay. Barrow suggested that this clay could have been redeposited and that the quartzose gravel belonged itself to the Pliocene Pebble Gravel. He cited the deposit at Little Heath (see below, Little Heath) as a typical example of Pebble Gravel with a comparable composition. Barrow suggested a mechanism of 'static washing', whereby the mixture of flint and quartz pebbles surrounded by a clayey matrix was left as a residue capping some of the hills in the area north of London. He regarded this type of residual material as the normal Pebble Gravel, quite distinct from a series of deposits occurring at around 400 ft (120 m) O.D. on the Chilterns dip slope, which he believed to be true marine Pebble Gravel. He interpreted the slightly higher Pebble Gravel of the Stanmore area (Stanmore Pebble Gravel – see below, Harrow Weald Common), at up to 500 ft (155 m) O.D., as a beach deposit of equivalent age to the 120 m deposits, regarding the latter as sea-floor sediments, formed near the margins of a Pliocene sea that covered all but the highest parts of the Chalk escarpment.

Some discussion of the age and origin of the Pebble Gravel was undertaken in the various local Geological Survey memoirs published during the following few years (Sherlock, 1922; Sherlock and Noble, 1922; Sherlock and Pocock, 1924; Bromehead, 1925). Sherlock (1924) produced a more detailed argument, in which he proposed that the deposits should be assigned to the Pleistocene rather than the Pliocene and that they were formed by locally nourished glaciers which merely rearranged the local Tertiary materials.

Discussion of the Pliocene/Lower Pleistocene deposits of the London Basin was dominated for the next thirty years by S.W. Wooldridge and his

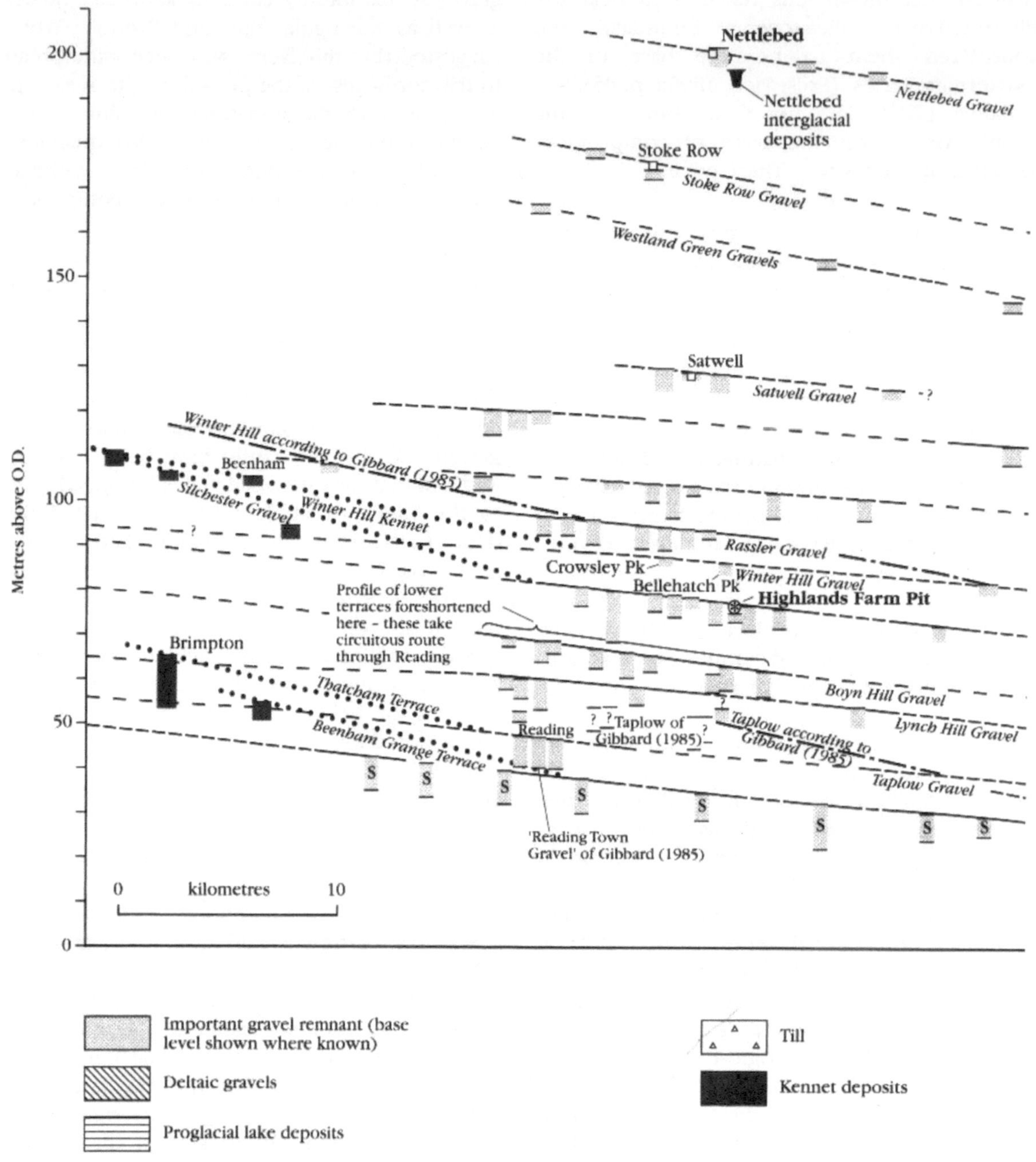

Figure 3.3 Long-profiles of terrace formations in the Middle Thames. Compiled predominantly from data provided by Gibbard (1985), with subordinate information from Sealy and Sealy (1956) and Thomas (1961). Modifications to the source information are described in the text.

co-workers. Wooldridge and Gill (1925) re-investigated the quartzose gravels at Lane End and confirmed their Reading Beds age. They noted that a later Pebble Gravel also occurred at the locality, capping the Palaeogene outlier. They suggested that this younger deposit was of Late Pliocene to Early Pleistocene age and reaffirmed White's view that the quartzose component of the Pebble Gravel might, in the main, have been secondarily derived from the quartzose facies of the Reading Beds. Indeed, they went further and suggested that the Greensand chert that occurs locally in the Pebble Gravel might also have been reworked from the Palaeogene.

In a much-quoted sequence of publications,

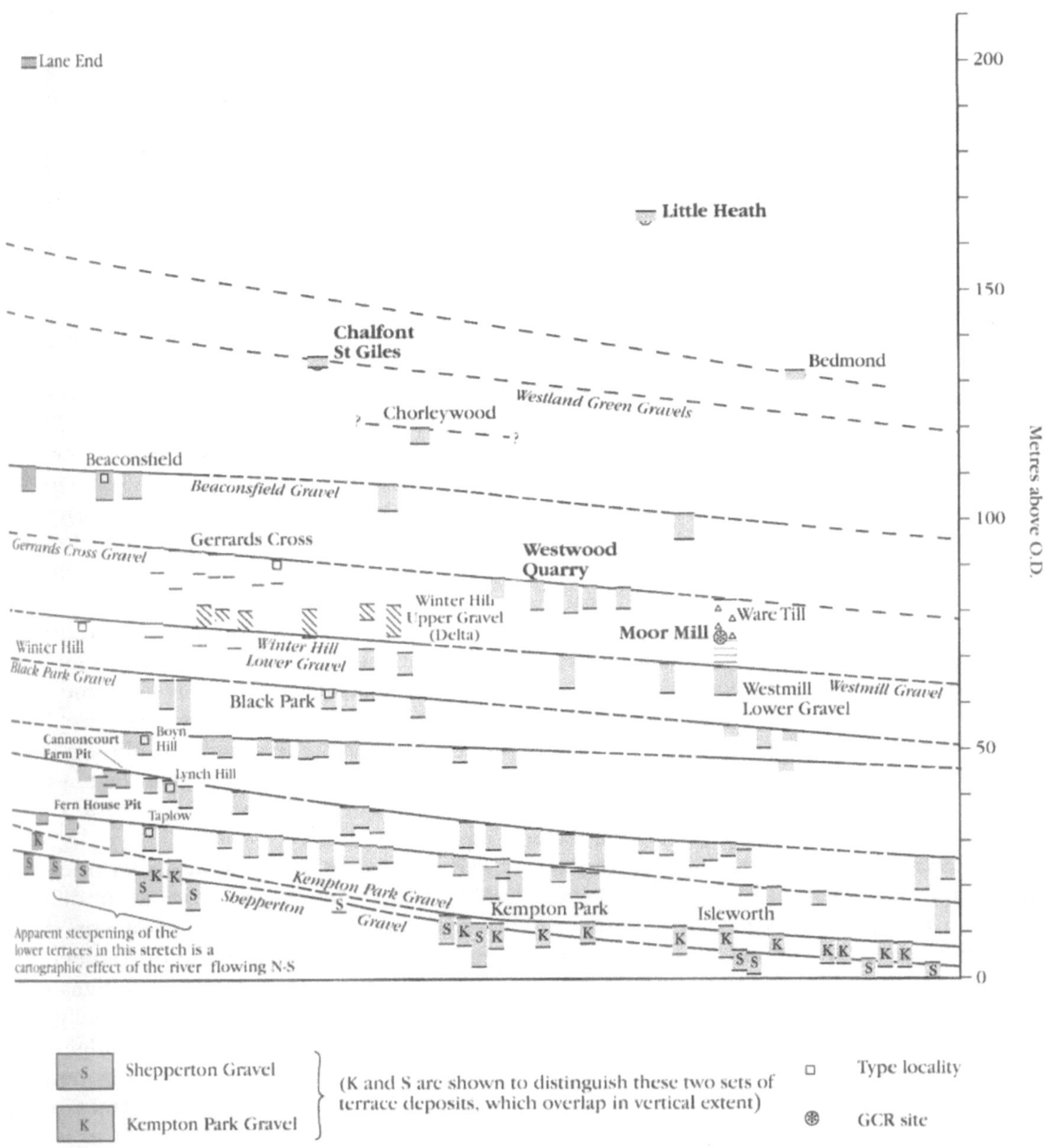

Wooldridge (1927a, 1957, 1960; Wooldridge and Linton, 1939, 1955) redefined the term Pebble Gravel to apply only to the deposits concentrated at around 122 m (400 ft) O.D., Barrow's sea-floor sediments. These deposits, which occupy an area known as the 'South Hertfordshire Plateau' (more or less coincident with Hughes (1868) 'Upper Plain'), have become known as the 'Hertfordshire', 'Lower' or '400 ft Pebble Gravels'. Wooldridge recognized that the higher-level deposits of the Stanmore area contain less non-flint material than the '400 ft' gravels (he regarded the latter as Pebble Gravel *sensu stricto*), thus confirming Salter's observations. Wooldridge ascribed these higher-level gravels to the Late Pliocene marine incursion recognized on the North Downs (at Netley Heath), considering the deposits at Little Heath and Stanmore to be lateral equivalents (see below, Little Heath and Harrow Weald Common). He recognized that chert from the Lower Greensand is locally abundant in the '400 ft Pebble Gravels', reflecting input from Wealden rivers.

Wooldridge (1928) described regional variants of his Pebble Gravel *sensu stricto* in south-west Essex, around Brentwood, Billericay and Laindon. At the last of these, a southern tributary is indicated by the presence of Lower Greensand chert. Wooldridge attributed the fluvial Pebble Gravel (400 ft) to the Late Pliocene or Early Pleistocene. The localization of Greensand chert in north-south trending zones, reflecting south-bank tributaries of an ancestral Thames, led Wooldridge (1927a, 1928) to conclude that the '400 ft Pebble Gravels' were entirely of fluviatile origin. However, he later revised this view and suggested that the chert was distributed along the courses of streams from the Weald across an emergent sea-floor during the Early Pleistocene (Wooldridge, 1957, 1960). Wooldridge was able to delimit, in the area to the north of the modern Lower Lea valley, a confluence area, where, in Pebble Gravel times, a major stream from the south, emanating from the area of the present Mole catchment (Fig. 3.4), met with another from the west, presumably the ancestral Thames.

In recent years parts of the Pebble Gravel (*sensu lato*) have been lithostratigraphically redefined. The first work of this type was by Hey (1965), who found that subdivision was possible in much the way that Salter (1896) had suggested. In particular, Hey recognized in Salter's fourth and lowest category, his 'Bell Bar Group', a compositionally distinctive gravel that could be traced from the Goring Gap to Hertfordshire. This deposit, named the Westland Green Gravels by Hey (1965), was regarded at that time as the highest within the Middle Thames sequence to contain material from the Midlands. It was therefore interpreted by Hey as the first true Thames gravel (see Part 2 of this Chapter). Hey *et al.* (1971) conducted a study of the surface textures of sand grains from all the various types of Pebble Gravel, using scanning electron microscopy. They found a progressive decrease in grain-surface features indicative of a marine or beach environment between the higher-level Pebble Gravel (the Stanmore type – see below, Harrow Weald Common) and the Westland Green Gravels. They concluded from this evidence that the high-level deposits were true littoral gravels, deposited following the maximum Early Pleistocene transgression as 'a single series of regressional marine beaches' (Hey *et al.*, 1971, p. 381). These results confirmed the Westland Green Gravels as fluvial deposits, but the '400 ft Pebble Gravels' appeared to yield conflicting grain-surface evidence, with some support for a partly marine hypothesis, though with most indicators suggesting a dominantly fluvial origin.

Gibbard (1983, 1985) further subdivided the Pebble Gravel, recognizing two divisions higher and older than the Westland Green Gravels, also attributable to the Thames. He applied the names Nettlebed Gravel and Stoke Row Gravel to these (see Fig. 3.2 and below, Nettlebed). It is apparent from a comparison of the clast composition of these two formations that a connection was first made between the headwaters of the Upper Thames system and the Midlands in the interval between the deposition of the Nettlebed and Stoke Row Gravels, although there is some suggestion that material from the Midlands may be present in the Nettlebed Gravel (see below, Nettlebed). Sites associated with deposits laid down after this connection was made are described in Part 2 of this chapter. Descriptions of sites at Little Heath, Harrow Weald Common and Nettlebed appear below. These provide sections in sediments that have all at some time been loosely termed Pebble Gravel. The Nettlebed site is part of the type outcrop of the Nettlebed Gravel, but is of enhanced significance in that a Lower Pleistocene interglacial deposit is also preserved there.

Recent studies have shown that the term Pebble Gravel has been applied to different types of deposit whose only connection is that they contain a high proportion of rounded flint reworked from the Palaeogene. Three main categories can be recognized: (1) high-level gravels on the Chilterns, of possible marine origin, as at Little Heath; (2) high-level remnants north of the Vale of St Albans, of early Thames origin, such as at Nettlebed (these form the highest elements of the terrace 'staircase' that is preserved on the dip slope of the Chilterns) and (3) hill-capping gravels in North London and south Hertfordshire, of fluvial origin, but laid down by south-bank tributaries of the Thames. Only deposits of the third category occur in the type area of the Pebble Gravel, where they were described and classified by Barrow (1919a) and Wooldridge (1927a), although both of these authors suggested correlations with remnants on the Chilterns. The marine gravels, if correctly interpreted, are essentially unrelated to the evolution of the fluvial system, so they should be distinguished from the Pebble Gravel and the latter term restricted to the early fluvial

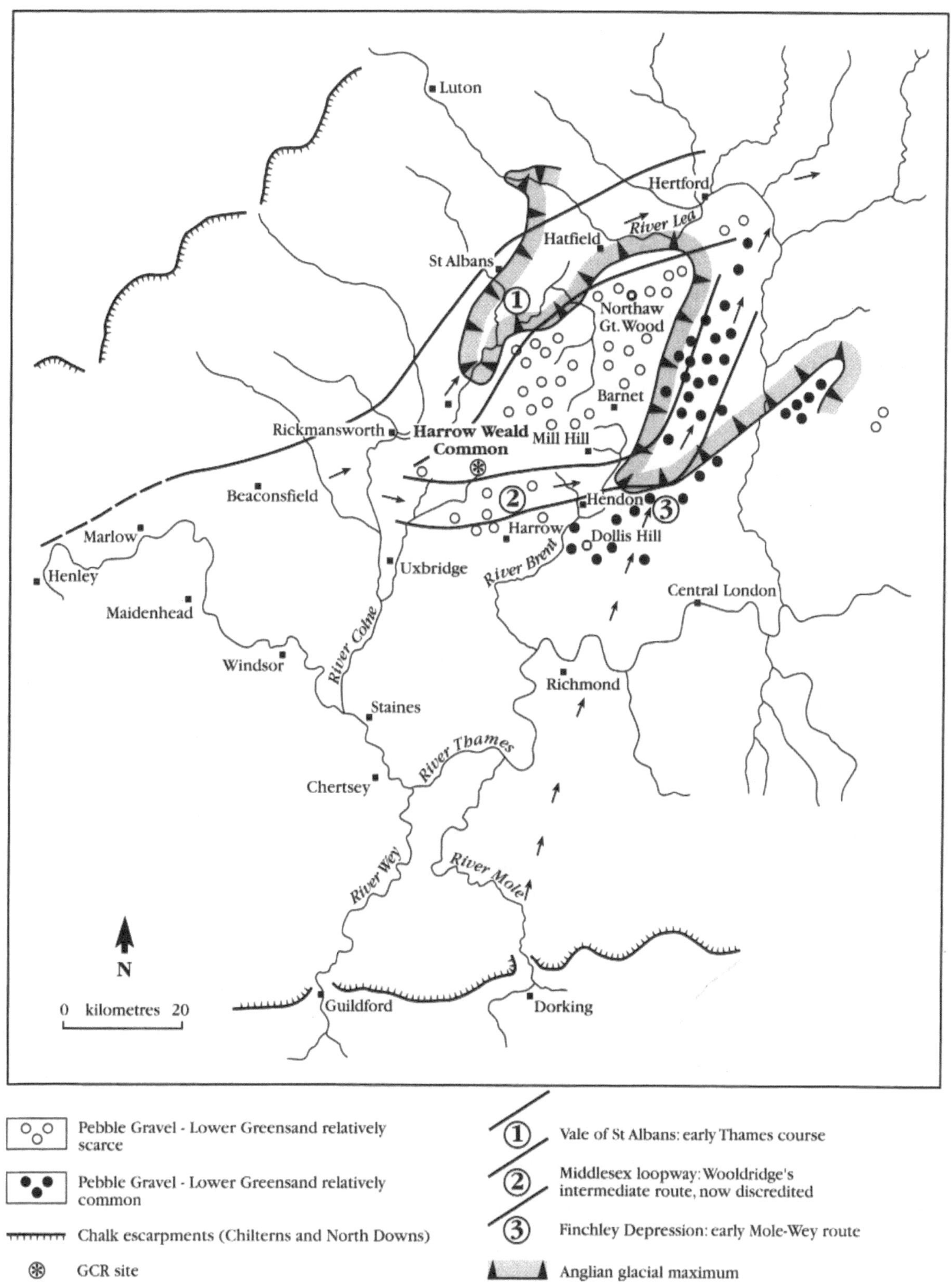

Figure 3.4 Map showing Wooldridge's reconstructed courses of the Thames and its tributary, the Mole-Wey. The distribution of Pebble Gravel remnants is also shown; those remnants in which Greensand chert is scarce are distinguished from those in which it is relatively common.

deposits. Two types of Pebble Gravel remain, representing sub-groups within a Pebble Gravel Group. Firstly, the high-level Thames gravels on the northern side of the Vale of St Albans represent a Chilterns Pebble Gravel Subgroup; it includes the Nettlebed, Stoke Row and Westland Green Formations (Table 3.1). The second subgroup comprises the deposits in North London and south Hertfordshire. Two formations are recognized within this North London Pebble Gravel Subgroup – a higher-level 'Stanmore (Pebble) Gravel' (type locality: Harrow Weald Common, the GCR site) and a lower-level 'Northaw (Pebble) Gravel' (type locality: Northaw Great Wood, TL 281040).

LITTLE HEATH (TL 017083)

D.R. Bridgland

Highlights

A controversial site revealing deposits of shallow marine or fluviatile origin, Little Heath is critical for the understanding of the Late Pliocene/Early Pleistocene evolution of the London Basin and the Middle Thames catchment.

Introduction

The existence at Little Heath of high-level superficial deposits of considerable antiquity was first noted by Prestwich (1890b). Sediments at this site, attributed to a Pliocene marine phase, were subsequently described in detail by Gilbert (1919a). Since that time there has been considerable controversy about the age and origin of unfossiliferous deposits of this type (Barrow, 1919a, 1919b; Sherlock, 1919, 1922, 1924, 1929; Wooldridge, 1927a, 1957, 1960; Wooldridge and Linton, 1939, 1955). They occupy similar topographical positions to both the fossiliferous Pliocene outliers of the North Downs (Stebbing, 1900; Davies, 1917; Chatwin, 1927; Dines and Edmunds, 1929; John and Fisher, 1984) and Chilterns (Dines and Chatwin, 1930) and the earliest fluvial deposits, the Pebble Gravel, which may be of Late Pliocene or Early Pleistocene age (Whitaker, 1864, 1889; Hughes, 1868; Prestwich, 1881, 1890a, 1890b; White, 1895, 1897; Salter, 1896, 1898; Wooldridge, 1927a, 1957, 1960; Wooldridge and Linton, 1939, 1955; see above, Introduction to Part 1). It remains unclear, furthermore, whether deposits such as those at Little Heath are of shallow-marine/littoral or of fluviatile origin. These enigmatic, degraded remnants provide the only indication

Table 3.1 Correlation of tributary and main Thames formations within the Pebble Gravel Group and other pre-diversion gravels in the Middle Thames and Vale of St Albans regions.

Thames formations	Tributary formations (Mole-Wey catchment?)	Group
Winter Hill Gravel	Dollis Hill Gravel	
Gerrards Cross Gravel	Equivalents may be represented within undifferentiated gravels west of Lower Lea valley (Gerrards Cross Gravel to Chorleywood Gravel)	
Rassler Gravel		
Beaconsfield Gravel		
Satwell Gravel		
Chorleywood Gravel		Pebble Gravel (Chorleywood Gravel to Nettlebed Gravel)
Westland Green Gravel	Northaw Pebble Gravel (400 ft)	
Stoke Row Gravel	Stanmore Pebble Gravel (500 ft)	
Nettlebed Gravel		

of the palaeoenvironment in Britain during the Pliocene/earliest Pleistocene.

Prestwich (1890b, p. 139) referred to a Tertiary outlier at Little Heath, preserved in a large depression in the Chalk some 550 ft (168 m) above O.D., associated with an indistinct development of his 'Westleton Shingle'. Barrow (1919a) considered this to represent Pebble Gravel of a comparable composition to the Reading Beds outlier at Lane End, which he also regarded as Pliocene (see above). Wooldridge (1927a) considered the Little Heath Gravel to be a lateral extension of his 'High-level (500 ft) Pebble Gravel' of North London, which he correlated with the Pliocene marine transgression of the North Downs. Recently the deposits at Little Heath have been central to a reappraisal of the evidence for a Pliocene/Lower Pleistocene marine platform on the Chilterns (Moffat, 1980; Moffat and Catt, 1983, 1986a), much of the evidence for which has been questioned in recent years (Pinchemel, 1954; Catt and Hodgson, 1976).

Description

The earliest detailed record of the deposits at Little Heath was by Gilbert (1919a, 1919b), who described a sequence of clay, gravel, sand and 'pebbly clay' overlying the Chalk. The site received little subsequent attention until a new section some 10 m deep was excavated as part of a recent reappraisal of the evidence for Neogene marine deposits on the Chilterns (Moffat, 1980; Moffat and Catt, 1983). The following summary of the sequence at Little Heath is derived from the descriptions by Gilbert and Moffat:

	Thickness
9. Soil with Palaeogene flint pebbles	1 m
8. Pebbly clay, highly variable in composition and thickness, but only 0.4 m in the GCR section	0.4–6 m
7. Stratified loamy sand with clay laminations, showing sun-cracks, rain-spots, ripples, etc., and small-scale normal and reverse faults	2.6 m
6. Stratified coarse gravel, with an undulating surface. This comprises rounded flints reworked from the Palaeogene, together with water-worn flints and occasional pieces of pudding-stone, small quartz and chert pebbles showing beach hammering. The upper 0.3 m, separated from the remainder by a thin clay band, comprises alternations of gravel and sand, including a highly glauconitic sand lens	5.6 m
5. Brown sands, coarse-grained and clearly bedded. Separated from bed 4 by a gravel seam. Not recognized by Gilbert	0.35 m
4. Grey sands, well sorted and faintly bedded. Not recognized by Gilbert	0.4 m
3. Greenish-brown clay with unworn nodular and tabular flints, broken flints and Palaeogene flint pebbles, mixed with quartz, chert and silicified *Inoceramus* shells. The clasts have a black or green surface coloration typical of the lag horizon at the base of the Palaeogene (Bullhead Bed)	*c.* 0.7 m
2. Brown clay (stoneless). Differs in particle-size distribution and mineralogy from bed 3. Not recognized by Gilbert	trace
1. Chalk	

In the recent re-excavation, Chalk was proved at 159 m O.D. in the base of the excavation; it and the brown clay were reached only by augering. Mineralogical and mechanical analyses suggest that beds 2–4 belong to the Reading Beds (Moffat, 1980, 1986; Moffat and Catt, 1983). The remainder of the sequence at Little Heath appears to be of later, probably Pliocene or earliest Pleistocene age. Some evidence of disturbance, probably resulting from solution of the underlying Chalk, was noticed by Moffat (1980).

Interpretation

Gilbert (1919a) based his interpretation of the gravel and loamy sand (beds 6 and 7) at Little

Heath as marine deposits on a number of observations. Firstly, these deposits fine upwards, a feature that he likened to Cambrian marine conglomerates, but which would now be regarded as equally typical of fluvial sedimentation. They possess a grain-supported structure, which Gilbert interpreted as a feature of marine, rather than fluviatile or glacial gravels. There is a gradation between unworn and well-rounded pebbles, which he considered similar to that found on modern beaches in which the materials are 'immediately derived'. Gilbert noted that large clasts showed evidence of beach hammering and that the surface morphology of the gravel body was suggestive of a beach. Finally, laminations and evidence for periodic exposure (sun-cracks, rain-spots, etc.) in the sand (bed 7) were interpreted by Gilbert as evidence of an intertidal environment.

According to Gilbert, the Little Heath gravels were not confined to a depression in the Chalk, as Prestwich (1890b) had claimed, but also capped the Chalk plateau as a horizontal undisturbed sheet. Gilbert compared them to the high-level deposits capping the North Downs in Surrey, attributing both to a Pliocene marine episode. Barrow (1919a) agreed with this assessment, likening the Little Heath gravel to the Pebble Gravel of the area north of London, which contains only local materials plus quartz and 'lydite' pebbles from the Lower Greensand.

Barrow (1919b) divided the Pebble Gravel into two types: one composed of smaller pebbles and occupying a platform at 400 ft, the 'Upper Plain' of Hughes (1868), and the other, a higher, coarser type, typically occupying land at *c.* 500 ft (152 m) O.D. in the Stanmore area. It was with the latter type that Barrow (1919a) classified the Little Heath gravel, noting that it lay even higher, 50 ft (15 m) above the Stanmore deposit, an increase in elevation that was accompanied by a further increase in pebble size. Barrow believed that the coarser, higher (500 ft) Stanmore Pebble Gravel was a beach deposit associated with the same marine episode as the 400 ft (Northaw) Pebble Gravel.

Sherlock (1919) was not convinced of the Pliocene age of the Little Heath deposits. He pointed to the occurrence of gravel in the Reading Beds at Lane End, Buckinghamshire (White, 1906), of similar composition to that at Little Heath, containing quartz, 'lydite' and subangular flints. Sherlock suggested that the various high-level gravels of the Chilterns might also be *in situ* Reading Beds; indeed, in the Aylesbury Geological Survey memoir (Sherlock, 1922), he included all the stratified deposits at Little Heath in the Reading Beds. In an extensive review he argued strongly against the existence of Barrow's '400 ft Platform' and against there having been a submergence in Pliocene times (Sherlock, 1924). He interpreted the Pebble Gravel (*sensu lato*) as 'a Glacial Drift derived from local materials' (Sherlock, 1924, p. 9).

Of considerable significance to this dispute was the allied controversy over White's (1906, 1908a) interpretation of the quartzose gravel at Lane End, Buckinghamshire, as *in situ* Reading Beds (see above, Introduction to Part 1). Although White had interpreted a deposit overlying these gravels as basal London Clay, Barrow (1919a) did not accept that they were Palaeogene. The similarity of the gravels at Lane End and Little Heath, cited by Barrow as evidence that the former was of Pliocene age, was equally indicative to Sherlock (1919, 1922, 1924) that the Little Heath deposits were part of the Reading Beds. However, the Lane End deposits were reinvestigated by Wooldridge and Gill (1925), who fully confirmed White's findings.

Wooldridge (1927a) recognized numerous outliers at high levels on the North Downs and South Downs, all of which he regarded as Diestian (Pliocene) in age, on the basis of their heavy-mineral content. A characteristic heavy-mineral suite, with an abundance of garnet and a relative abundance of monazite and coarse andalusite, had been recognized in the Lenham, Netley Heath and Headley Heath Beds of the North Downs by Davies (1915, 1917). Wooldridge attributed these deposits, together with a number of isolated outliers of high-level gravel on the North Downs and a few in the Chilterns, to a Diestian marine transgression, proposing that they be collectively referred to as Lenham Beds. According to Wooldridge (1927a, p. 80) 'the only locality at which the Lenham Beds are well-developed on the Chiltern plateau is at Littleheath near Berkhampstead'. He claimed that the heavy-mineral assemblage from Little Heath was of Diestian character and quite different from Reading Beds mineral suites (Wooldridge, 1927a, 1927b, in Sherlock, 1929). He also pointed out that the Little Heath sands, being marine, are finer grained and better sorted than the typical (fluviatile) Reading Beds sands of the district. Wooldridge distinguished the various Diestian outliers from the true

Pebble Gravel, although he regarded the latter as little if any younger than the Lenham Beds, and interpreted the highest Pebble Gravel of the Stanmore area as a 'true Diestian shingle'.

The correlation, based on heavy minerals, between the Netley Heath Beds and the Lenham Beds was disproved by the discovery of Late Pliocene (Red Crag) fossils in the former (Chatwin, 1927), indicating that they post-date the Lenham Beds. As a result, Sherlock (1929) argued strongly against the stratigraphical usefulness of heavy-mineral analysis and remained unconvinced that the Little Heath Gravel was of Pliocene age. Red Crag fossils were later discovered in sandstone blocks within high-level drift at Rothamsted, apparently confirming the existence of a Pliocene marine episode in Hertfordshire (Dines and Chatwin, 1930).

Wooldridge and Ewing (1935) carried out a reappraisal of the Palaeocene and alleged Pliocene deposits of the Chilterns, including a detailed petrographical examination. As a result, they concluded that the Little Heath gravel, together with many other patches of high-level deposits in the area, were of Pliocene age. Wooldridge and Linton (1939, p. 55) emphasized the relation of the Little Heath outlier to 'sharply rising ground behind the escarpment edge at Ivinghoe Beacon'. They claimed this to be a fossil shoreline, more conspicuous in the Chiltern area than on the North Downs, being traceable from near the Goring Gap to beyond Luton.

Doubts about this type of geomorphological evidence have been expressed in more recent accounts, however, both in the Chilterns (Pinchemel, 1954; Jones, 1974; Catt and Hodgson, 1976; Moffat, 1980; Moffat and Catt, 1986a, 1986b; Moffat *et al.*, 1986) and south of the Thames (Docherty, 1967; John, 1980; Fisher, 1982). In particular, Moffat (1980; Moffat *et al.*, 1986) noted that the alleged marine 'bench' on the Chiltern dip slope corresponds closely with monoclinal folding of the Chalk and suggested that the feature results from this geological structure rather than from marine erosion. Similar explanations of alleged Pliocene/Early Pleistocene 'benches' in north-west Kent (Docherty, 1967, 1971) and Surrey (John, 1980) have been put forward in recent years.

Moffat (1980; Moffat and Catt, 1983) concluded, on mineralogical grounds, that the lowest three units (beds 2–4) at Little Heath were *in situ* Reading Beds, differentiating them from the overlying gravel and sands on the same basis. He interpreted the upper gravel at Little Heath (bed 6) as a beach deposit, principally on sedimentary grounds, although the presence of *c.* 25% glauconite in the lens of sand within this unit (see above) provides strong evidence for a marine influence, as this mineral is formed only in marine conditions (Porrenga, 1967) and there is no obvious source from which it could be derived secondarily in such quantities. Moffat also considered the overlying sand to be of marine origin, the main evidence being the presence in it of *c.* 5% glauconite. He found no evidence of a glacial origin for the uppermost unbedded gravel (bed 8); the matrix of this unit was found to closely resemble the Reading Beds and to differ from the underlying sand and gravel (beds 5–7). Moffat therefore concluded that the deposit was a colluvial accumulation of bedrock and drift material, presumably derived from upslope areas now removed or modified by erosion.

Although concluding that the geomorphological case for a Pliocene shoreline on the Chilterns was unfounded, Moffat (1980; Moffat and Catt, 1983) supported Wooldridge's (1927a) interpretation of the Little Heath deposits as being of marine origin. He cited the similarity between the particle-size properties of this deposit and the Headley Formation of Surrey as supporting evidence, in addition to similarities in fine-sand mineralogy between both of the above and the Rothamsted Red Crag (Moffat, 1980; Moffat and Catt, 1983, 1986a).

Only one other high-level outlier was accepted by Moffat and Catt (1986a) as an *in situ* Pliocene or Early Pleistocene marine deposit. This is a remnant of Pebble Gravel (*sensu lato*) at Lane End (SU 817917), *c.* 203 m O.D., above the quartzose Reading Beds and basal London Clay (White, 1906; Wooldridge and Gill, 1925; see above, Introduction to Part 1). This remnant was found to be similar to the Little Heath gravel in both its clast and heavy-mineral composition. It lies 28 km to the south-west of the Little Heath outlier and over 30 m higher. Moffat and Catt considered this difference in altitude to be the result of differential tectonic movement in south-eastern Britain since the Pliocene (see below). Moffat (1986) demonstrated that the clast-size distribution of quartz pebbles in the two widely separated outliers was distinct from those in other high-level gravels, but similar to that from quartzose Reading Beds, suggesting derivation solely from that source (see Harrow Weald Common). Other gravel remnants at Gustardwood (TL 175160) and Cowcroft (SP 983017) were interpreted as possibly derived, at least in part, from earlier gravels of the Little Heath type (Moffat, 1980). Moffat considered the matrix of the gravel at Gustardwood to include Late Pleistocene loessic material, implying that the deposit in its present form originated comparatively recently, although *in situ* pedogenic mixing may have caused the introduction of much later (Upper Pleistocene) loess into a Pliocene or Lower Pleistocene gravel.

Moffat and Catt (1986a) demonstrated the degree of post-Pliocene tectonic activity using a projection of the basal surface of the Red Crag from ordnance datum in north-east Essex to almost 200 m O.D. in Surrey and south Buckinghamshire (Fig. 3.5). They attributed the inclination of this surface to subsidence of the north-eastern part of the London Basin during the Pleistocene. It seems likely, however, that uplift of the western area may also have occurred, perhaps as an isostatic response to erosion (Chapter 1), as well as subsidence of the North Sea Basin. The outliers on the North Downs at Netley Heath and Headley Heath have been used to reconstruct this surface, as well as the residual blocks of fossiliferous Red Crag sand at Rothamsted (Fig. 3.5). The Little Heath and Lane End deposits are at suitable elevations to suggest a correlation on this basis with the Red Crag, despite their lack of fossils (Moffat and Catt, 1986a).

However, as Fig. 3.6 shows, the relations

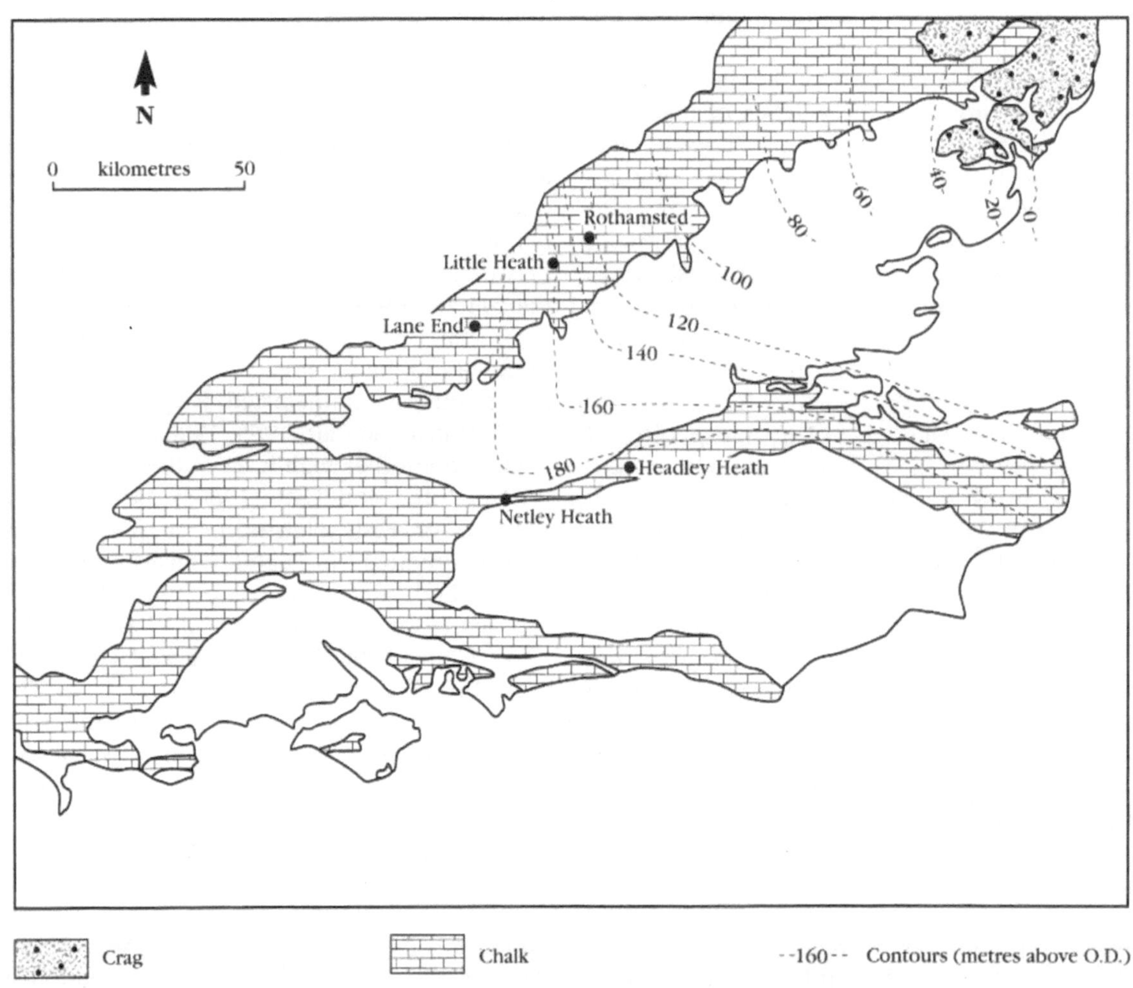

Figure 3.5 Contours on the base of the Red Crag. The positions of sites on the Chilterns and North Downs that may correlate with the Red Crag are shown (after Moffat and Catt, 1986b).

between these two outliers, if they are contemporaneous, imply a north-eastward gradient comparable to that observed in the highest Thames terrace deposits. In the case of the latter, the gradient at least partly reflects an original downstream slope. A fluvial interpretation of the Lane End and Little Heath deposits, which would make them the earliest Neogene river deposits to have been preserved, must therefore be considered. As in the various types of Pebble Gravel, sublittoral/littoral characteristics such as clast shape may reflect derivation from older Palaeogene or Neogene strata. The clast-lithological composition of the gravel at Little Heath (above and Table 3.2) is very much what might be expected for the earliest fluviatile gravels of the Thames system. Thus interpretation as a fluvial aggradation would satisfy the limited evidence that exists for the origin of these deposits, with the possible exception of their high glauconite content at Little Heath. This suggests that it might be premature to consider the littoral origin of the gravel at Little Heath to be settled. The question of whether such remnants truly result from a marine episode is likely to remain controversial for the foreseeable future.

The Little Heath outlier thus represents one of only two gravel remnants that can be interpreted as littoral deposits of Late Pliocene or Early Pleistocene age, out of many such outliers once recognized. It is possible that this deposit broadly correlates with the Red Crag of East Anglia. The evolution of the London Basin at the time of the Pliocene/Lower Pleistocene transition is poorly understood, there being little remaining evidence from which to reconstruct the contemporary coastal or drainage lines. The very existence of high-level littoral deposits *in situ* is a subject of controversy. The GCR site at Little Heath provides an important facility for the study of this enigmatic phase in the early development of the Thames catchment.

Conclusions

The sequence of clay, gravel, sand and pebbly clay deposits overlying the Chalk at Little Heath has long been controversial; even today there is much uncertainty over the age and origin of these sediments. An early interpretation, formerly widely accepted and still with some advocates, held that the Little Heath deposits

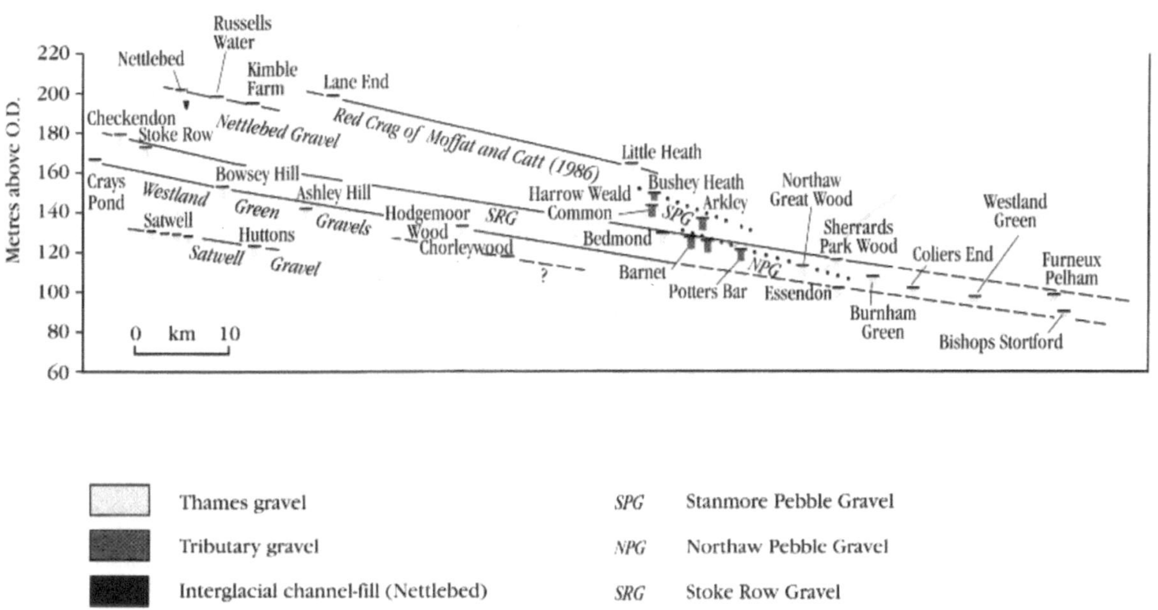

Figure 3.6 Long-profile diagram of higher deposits in the Middle Thames and the Vale of St Albans, showing the North London Pebble Gravels, attributed in this volume to deposition by a south-bank tributary of the early Thames.

Table 3.2 Clast-lithological data (in percentage of total count) from the Middle Thames and Vale of St Albans (compiled from various sources). The data concentrates on key sites, GCR sites and localities mentioned in the text. Note that many different size ranges are included and that these yield strikingly different data (this can be observed where results from different fractions from the same deposits have been analysed). As in Tables 4.2, 5.1 and 5.3, the igneous category includes metamorphic rocks (very rarely encountered) and the quartzite category includes durable sandstones. The Tertiary flint category comprises rounded pebbles (sometimes subsequently broken) reworked from the Palaeogene (see glossary with Table 4.2).

				Flint		**Chalk**	**Southern**		**Exotics**								
Gravel	**Site**	**Sample**	**Size range**	**Tertiary**	**Total**	**Chalk**	**Gnsd chert**	**Total**	**Quartz**	**Quartzite**	**Carb chert**	**Rhax chert**	**Igneous**	**Total**	**Ratio (qtz:qtzt)**	**Total count**	**Source**
Shepperton	Shepperton	1	8-32	4.2	95.4		1.1	1.1	2.3	1.2				3.5	1.85	569	Gibbard (1985)
Gravel	Bray		8-32	3.4	95.5		0.6	0.8	2.7	1.1				3.7	2.43	642	Gibbard (1985)
Kempton Park Gravel	Kempton Park	1	8-32	11.1	62.5		1.5	1.5	28.2	7.8				37.5	2.21	397	Gibbard (1985)
Taplow	Taplow	1	8-32	2.5	91.4		2.1	2.1	5.0	1.5				6.5	3.26	525	Gibbard (1985)
Gravel	Fern House Pit	1	8-32	5.3	93.4		1.5	1.5	1.8	3.2				5.0	0.57	778	Gibbard (1985)
Lynch Hill	Lynch Hill	1	8-32	8.3	91.3		1.9	1.9	4.1	2.7				6.8	0.57	635	Gibbard (1985)
Gravel	Cannoncourt	12	11.2-16	12.7	86.1		4.8	5.1	4.2	2.9	0.9	0.2	0.4	8.8	1.46	454	Harding *et al.* (1991)
	Farm Pit	13	11.2-16	9.7	90.5		4.6	4.6	1.8	1.8	1.0	0.2		4.8	1.00	611	Harding *et al.* (1991)
	Switchback Pit	14	11.2-16	12.6	87.6		4.0	4.4	3.8	3.3	0.2	0.2	0.2	8.0	1.13	452	Harding *et al.* (1991)
Cannoncourt (tributary) **gravel**	Cct Fm Pit	15	11.2-16	1.9	96.1		2.8	3.1	0.2		0.7			0.9		459	Harding *et al.* (1991)
Boyn Hill Gravel	Boyn Hill	1	8-32	6.6	88.9		2.5	2.5	6.6	2.1				8.7	3.21	401	Gibbard (1985)
Black Park	Black Park	1	8-32	9.1	84.6		3.0	3.0	8.5	3.7				12.4	2.27	507	Gibbard (1985)
Gravel	Highlands Fm	1	8-32	5.1	86.4		0.8	0.8	6.3	5.9	0.5		0.2	12.9	1.08	662	Gibbard (1985)
Ugley	Ugley Park	1	16-32	32.5	68.1	13.3			2.7	0.6	1.8	0.2	1.5	20.3	4.50	671	Bridgland (1983a)
Gravel	Quarry	2	16-32	2.2	53.6	26.4			1.6	1.0	1.9	1	1.2	19.9	1.57	683	Bridgland (1983a)
		3	8-16	1.6	50.8	28.0			1.6	3.4	0.5	1	0.8	21.2	1.63	1139	Cheshire (1986a)
		5	8-16	0.6	44.7	38.1	0.1	0.1	0.5	1.6	0.2	0.9	0.9	17.1	0.33	1261	Cheshire (1986a)
	Westmill	11	8-64	•	61.8	21.0	•	•	4.5	8.5	•	•	•	•	0.53	440	Gibbard (1974, 1977)
	Westmill	14	8-64	•	63.2	19.8	•	•	11.6	3.8	•	•	•	•	3.02	444	Gibbard (1974, 1977)
Hoddesdon	Westmill	5	8-64	•	80.5	4.0	•	•	11.2	0.7	•	•	•	•	15.72	367	Gibbard (1974, 1977)
Gravel	Westmill	6	8-64	•	69.6	13.2	•	•	12.9	0.5	•	•	•	•	28.73	441	Gibbard (1974, 1977)
	Westmill	10	8-64	•	74.4	4.0	•	•	13.6	1.2	•	•	•	•	11.83	349	Gibbard (1974, 1977)
	Bullscross Farm	23	8-16	4.2	90.1		0.5	0.5	2.1	3.7	0.2	0.9	0.4	9.4	0.57	968	Cheshire (1986a)
Smug Oak	Moor Mill	1	8-64	•	80.6	1.2	•	•	16.5	1.6	•	•		•	10.31	428	Gibbard (1974)
Gravel		2	8-64	•	80.6	0.7	•	•	17.8	0.7	•	•		•	25.43	738	Gibbard (1974)
Westmill	Westmill	1	8-64	•	88.3		•	•	9.3	1.1	•	•	•	•	8.45	451	Gibbard (1974, 1977)
Lower		4	8-64	•	82.0	1.4	•	•	15.8	0.5	•	•	•	•	31.60	315	Gibbard (1974, 1977)
Gravel	Moor Mill	1	8-64	•	87.7	0.4	•	•	9.6	2.1	•	•		•	4.57	507	Gibbard (1974)
		2	8-64	•	89.6	0.4	•	•	11.3	2.0	•	•		•	5.65	455	Gibbard (1974)
Winter Hill	Winter Hill	1	8-32	6.4	81.0		6.6	6.6	9.8	2.4	0.2			12.4	4.08	500	Gibbard (1985)
Gravel	Mapledurham	1	8-32	5.9	72.8				17.1	8.5			0.6	27.2	2.01	340	Gibbard (1985)
	Stoke Common	1	8-32	8.0	82.8		1.4	1.4	9.9	4.4	0.3		0.8	15.8	2.25	667	Gibbard (1985)
Gerrards	Gerrards Cr.	1	8-32	18.8	66.7		1.8	1.8	23.6	7.3			0.2	31.5	3.24	508	Gibbard (1985)
Cross		2	8-32	18.9	57.0		1.4	1.4	31.7	8.1				41.6	3.92	507	Gibbard (1985)
Gravel		19	11.2-16	•	37.7		1.5	1.5	33.6	25.6	•		•	60.8	1.31	•	Green & McG. (1978a)
		21	11.2-16	•	41.9		3.4	3.4	17.6	34.4	•		•	54.7	0.51	•	Green & McG. (1978a)
	Westwood Qu.	1	8-32	15.3	57.6		1.4	1.4	27.2	10.4			0.3	41.0	2.61	595	Gibbard (1985)
		2	8-32	13.0	59.7		1.5	1.5	27.6	9.0	0.7		0.5	38.8	3.06	744	Gibbard (1985)
		25	11.2-16	•	51.8		2.8	2.8	21.2	22.0	1.0?		0.4	45.4	0.96	•	Green & McG. (1978a & c)
	Upper colluvial gravel	26	11.2-16	•	89.5		0.4	0.4	5.7	3.5				10.1	1.63	•	Green & McG. (1978a & c)
Beacons-	Flackwell Hth	1	8-32	10.5	86.5		0.3	0.3	7.5	5.4	0.3			13.2	1.39	334	Gibbard (1985)
field	Beaconsfield	1	8-32	9.8	47.4		1.0	1.0	42.9	8.7				52.6	4.94	622	Gibbard (1985)
Gravel		2	8-32	11.1	62.5		1.5	1.5	28.2	7.8				37.5	2.21	397	Gibbard (1985)
Satwell Gravel	Satwell	1	8-32	6.0	32.3				42.9	11.0			1.0	67.7	3.90	514	Gibbard (1985)
Additional	Chorleywood	1	8-16	•	35.2		•	•	44.2	18.0	•		•	•	2.46	•	Moffat (1986)
formation?		1	16-32	•	44.8		•	•	27.9	22.3	•		•	•	1.25	•	Moffat (1986)
		10	11.2-16	•	52.1		2.1	2.1	28.5	15.4	•		•	45.8	1.85	•	Green & McG. (1978a)

				Flint		Chalk	Southern		Exotics								
Gravel	**Site**	**Sample**	**Size range**	**Tertiary**	**Total**	**Chalk**	**Gnsd chert**	**Total**	**Quartz**	**Quartzite**	**Carb chert**	**Rhax chert**	**Igneous**	**Total**	**Ratio (qtz:qtzt)**	**Total count**	**Source**
Westland	Cray's Pond	1	8-32	13.2	88.2				7.0	4.6			0.2	11.8	1.50	517	Gibbard (1985)
Green	Chalfont	1	8-16	•	37.6		•	•	46.1	14.6	•		•	•	3.15	•	Moffat (1986)
Gravels	St Giles	1	16-32	•	56.5		•	•	22.1	15.9	•		•	•	1.33	•	Moffat (1986)
		2	8-16	•	33.7		•	•	42.6	19.3	•		•	•	2.21	•	Moffat (1986)
		2	16-32	•	51.1		•	•	20.1	16.9	•		•	•	1.19	•	Moffat (1986)
	Hodgemoor Wood	1	16-32	42.0	66.0		0.5	0.5	22.0	10.0	1.5		•	33.5	2.20	•	Hey (1965)
		86	11.2-16	•	52.2		0.7	0.7	26.4	18.8	•		•	47.0	1.40	•	Green & McG. (1978a)
	Westland Green	1	16-32	53.0	75.0		0.8	0.8	16.0	6.5	1.3		•	24.2	2.46	•	Hey (1965)
Stoke Row	Stoke Row	1	16-32	29.0	66.0				21.0	6.7	4.7		•	34.0	3.13	•	Hey (1965)
Gravel		2	8-32	15.4	46.1				49.9	3.8	0.3			53.9	3.16	369	Gibbard (1985)
	Bedmond	1	8-16	•	41.6		•	•	42.1	10.3	•		•	•	4.09	•	Moffat (1986)
		1	16-32	•	72.3		•	•	14.3	8.9	•		•	•	1.61	•	Moffat (1986)
	Sherrardspark Wood	1	8-16	•	36.4		•	•	42.8	10.1	•		•	•	4.24	•	Moffat (1986)
		1	16-32	•	74.8		•	•	14.2	6.7	•		•	•	2.12	•	Moffat (1986)
		2	8-16	29.8	40.7		0.1	0.1	42.0	12.9	2.9			59.2	3.26	•	Cheshire (1986a)
	Furneux Pelham	1	16-32	52.9	74.6		0.2?	0.2?	18.7	6.0	0.5			25.1	3.12	418	Hey (pers. comm.)
		7	8-16	23.1	45.9		0.3	0.3	37.5	10.3	2.8		•	53.8	3.64	•	Cheshire (1986a)
		7	16-32	56.6	69.0		0.9	0.9	16.8	10.6	1.8		•	30.1	1.58	•	Cheshire (1986a)
Nettlebed	Nettlebed	1	8-16	•	83.9				11.9	0.6	1.8			16.1	19.83	•	Moffat (1986)
Gravel		1	16-32	•	96.2				1.4	0.3	1.7			3.8	4.67	•	Moffat (1986)
	Windmill Hill	C	8-32	32.0	85.0				10.6	4.2	0.2			15.0	2.55	719	Gibbard (1985)
		B	8-32	37.5	81.3				15.1	3.4	0.3			18.7	4.49	416	Gibbard (1985)
	GCR site, above organics	D	8-32	39.5	81.7				14.0	4.1	0.1			18.2	3.41	729	Gibbard (1985)
	Kimble Farm	1	8-32	32.8	92.7				3.9	3.0	0.5			7.3	1.30	281	Gibbard (1985)
Little	Little Heath	1	8-16	•	99.5				0.2		0.3			0.5		•	Moffat (1986)
Heath		1	16-32	•	95.0				3.1	0.2	1.7			5.0	15.50	•	Moffat (1986)
Deposit		2	8-16	•	99.8				0.1	0.1	0.1			0.2	1.00	•	Moffat (1986)
		2	16-32	•	95.6				3.2	0.1	1.1			4.4	32.00	•	Moffat (1986)
Tributary gravels (Mole-Wey system)																	
Dollis Hill	Dollis Hill	1	8-64	44.3	91.3		7.0	7.0	•	•	•			1.7		368	Gibbard (1979)
Gravel	Cockfosters	1	8-64	22.0	90.6		6.3	6.3	•	•	•			3.2		464	Gibbard (1979)
and	Nursery Grove	1	8-16	45.6	74.8		11.3	11.3	7.4	2.4	•			13.9	3.08	1509	Cheshire (1986a)
similar	Bullscross Fm	11	8-16	56.0	86.1		9.1	9.3	1.8	0.4	1.1			4.6	4.50	1072	Cheshire (1986a)
Northaw	Northaw	1	16-32	78.0	92.0		0.7	0.7	4.4	2.2	0.7			8.0	2.00	•	Hey (1965)
Pebble	Great Wood	5	8-16	50.0	62.3		0.2	0.2	24.9	7.1	0.6			37.5	3.51	•	Cheshire (1986a)
Gravel		5	16-32	77.1	90.9		0.5	0.5	2.5	3.1				8.6	0.81	•	Cheshire (1986a)
Stanmore	Harrow	1	8-16	•	87.7		•	•	9.0	0.6	•			•	15.00	•	Moffat (1986)
Pebble Gr.	Weald Com.	1	16-32	•	98.9		•	•	0.3	0.3	•			•	1.00	•	Moffat (1986)

• Information not provided by the source cited.

were part of a complex spread of sands and gravels deposited by a Pliocene sea that covered the London Basin and extended on to the North Downs and the Chilterns (around and before 2 million years ago). This view was supported by mineralogical analyses, which showed that comparable deposits extended over quite large areas. Others have suggested correlation with the Reading Beds (*c.* 60 million years old) or the marine Pliocene Red Crag of East Anglia. Recent work suggests that deposition by a river during the Pliocene or early Pleistocene is a further possibility, which would make the sediments at Little Heath the earliest known Neogene deposits of this type. The marine origin so widely suggested, however, cannot yet be discounted, so it is certain that the Little Heath site will remain controversial in the future.

HARROW WEALD COMMON (TQ 147929)

D.R. Bridgland

Highlights

One of the earliest 'drift' deposits preserved in the Thames catchment, the High-level or Stanmore Pebble Gravel, is exposed at Harrow Weald Common. Although a shallow-marine or beach origin has long been favoured, it is suggested here that this gravel was deposited by a south-bank tributary of the early Thames, probably in Pliocene or Early Pleistocene times.

Introduction

Old gravel workings on Harrow Weald Common provide a rare opportunity to study the Stanmore Pebble Gravel of North London. This deposit aroused interest around the turn of the century (Prestwich, 1890c; Monckton and Herries, 1891; Salter, 1896, 1901; Barrow, 1919a, 1919c). Terms such as '500 ft', 'High-level' or 'Higher Pebble Gravel' have been widely used (Wooldridge, 1927a, 1957, 1960; Wooldridge and Linton, 1939, 1955; Hey *et al.*, 1971) to distinguish the gravels of the Stanmore–Bushey plateau from the lower deposits, now termed Northaw Pebble Gravel (see above). Confusingly, the term 'Higher Pebble Gravels' has been used by Green and McGregor (1978a; McGregor and Green, 1978; Green *et al.*, 1982) in references to Pebble Gravel generally, to distinguish it from lower-level Thames formations.

Most early workers interpreted the Stanmore Gravel as marine, but a recent reinvestigation (Moffat, 1980; Moffat and Catt, 1986a) has suggested that it represents one of the oldest fluvial deposits to be preserved within the Thames Basin.

Description

Shallow pits covering the area of Harrow Weald Common mark the sites where gravelly material has been removed from beneath the plateau surface, which reaches *c.* 145 m O.D. Very little undug ground remains, but faces in the GCR site, at the edge of a residual causeway, reveal *c.* 2 m of gravel, composed predominantly of rounded flint pebbles set in a sandy and loamy matrix.

Whitaker (1864, p. 69) first reported that Pebble Gravel was spread 'over the whole of the hill-top' at Stanmore, in a quotation from the notebook of R. Trench. Prestwich (1890c) provided one of the earliest descriptions of the Stanmore Pebble Gravel at a brick pit that, according to Bromehead (1925), lay to the south of the present GCR site. Prestwich illustrated a section in which *c.* 2 m of unstratified gravel overlay what is now classified as Claygate Beds (then included in the London Clay), the contact being deformed by the loading or involution of the gravel into the underlying silty clay. Salter (1896) recorded the occurrence of 2 m of gravel, composed predominantly of rounded flint pebbles, at 'Harrow Weald', some 146 m (480 ft) above sea level. Unfortunately he did not provide details of its precise location and, in a later paper, listed the same locality, but at 122 m (400 ft) O.D., one mile to the south of Stanmore. The last-mentioned outlier is not identified on the New Series Geological Survey map (Sheet 256), which does, however, show a large spread of Pebble Gravel covering Bushey Heath, Stanmore Common and Harrow Weald Common. Barrow (1919c) described in some detail a section on Harrow Weald Common seen by a Geologists' Association excursion (in May, 1919) close to the Kilns, where he noted that a rare undug remnant of gravel survived. From the description given, the locality visited in 1919 seems likely to have been the GCR site. Barrow observed *c.* 2 m of unbedded gravel composed of large rounded flints with smaller clasts and sand filling the interstices. In the most recent investigation of the site, Moffat (1980) considered the gravel to be vaguely horizontally stratified. He carried out clast-lithological analysis of the material, confirming the preponderance of flint, particularly in the coarser gravel fractions (Moffat, 1980, 1986; Moffat and Catt, 1986a). Moffat's grid reference (TQ 147932) and elevation (133 m) place his section on the slope of the hill 200 m to the north of the GCR site, where *in situ* material would not be expected, although his description appears appropriate for the exposure under consideration here.

Interpretation

The Harrow Weald Common GCR site provides the best available section in the Stanmore Pebble Gravel (= 500 ft or High-level Pebble Gravel of earlier workers). The history of research on these high-level deposits is complex. Prestwich (1890c) regarded the Stanmore deposits as the residue of the Eocene Bagshot Beds, although he noted that materials foreign to the Palaeogene had been introduced into them. He assigned them to his 'Brentwood Group', associating them with similarly flint-dominated gravels in southern Essex. Prestwich

(1881, 1890b) correlated the more widespread lower-level Pebble Gravel (Northaw Pebble Gravel) with his 'Westleton Beds', so he clearly differentiated the Stanmore deposits from the latter. This differentiation was questioned by Monckton and Herries (1891, p. 112), who were inclined to regard the Stanmore deposit as 'a variety of the Westleton Group'. Reid (1900), like Prestwich, also considered the Stanmore gravels to be disturbed Bagshot Beds. This view received further support from Bromehead (in Woodward *et al.*, 1922) and Sherlock (1929), although Bromehead (1925) later adopted the views of A.E. Salter (see below). Conversely, other early workers made no distinction between the Stanmore deposits and the more widespread, lower (400 ft) Northaw Pebble Gravel of south Hertfordshire (Hughes, 1868; Wood, 1868; Whitaker, 1889).

Salter (1896) cited the locality at 'Harrow Weald' as an example of his 'Barnet Gate Type' of 'pebbly gravel', thus associating it with deposits capping high ground at Shooters Hill (south-east London) and the Brentwood area. Salter (1896, 1901) showed that material foreign to the London Basin was a small but important constituent of these deposits and attributed them to the activities of post-Miocene consequent streams, regarding them as 'the first deposits of the Glacial Series' (Salter, 1896, p. 404). In this statement, Salter intended to convey the meaning that the deposits were Early Pleistocene fluvial or glaciofluvial accumulations, rather than remnants of a Pliocene beach, as was the popular view.

Despite Salter's interpretation, most workers continued to favour a marine origin for the Stanmore deposits. Barrow (1919c) considered that the gravel at Harrow Weald Common resembled a beach deposit. He thought it similar to that at Little Heath (see above) and suggested that the two were contemporaneous. Barrow (1919a) interpreted the higher (Stanmore) Pebble Gravel as a beach deposit of equivalent age to the similar, more widespread deposits at the lower level of *c.* 400 ft (122 m) in south Hertfordshire (Northaw Pebble Gravel), regarding the latter as sea-floor sediments (see above, Introduction to Part 1).

Wooldridge (1927a, 1957, 1960; Wooldridge and Linton, 1939, 1955) considered the '400 ft' (Northaw Gravel) deposits to be the true Pebble Gravel, recognizing that they contain more non-flint material than the higher-level outliers of the Stanmore area; in particular, he noted that Greensand chert is locally abundant in the former type. This led Wooldridge (1927a) to conclude that the '400 ft' (Northaw) Pebble Gravel was of fluviatile origin, although he later changed his view, subsequently favouring deposition on 'the emergent (?Sicilian) sea-floor' (Wooldridge, 1957, 1960, p. 119). However, Wooldridge consistently regarded the higher-level gravels of the Stanmore/Bushey area as marine, agreeing with Barrow that they were the same age as the Little Heath deposits.

The first attempt to test the various hypotheses for the origin of the high-level gravel of the type exposed at Harrow Weald Common was by Hey *et al.* (1971), whose studies of the surface textures of sand grains from various types of Pebble Gravel (*sensu lato*) included samples from the Stanmore Formation at Bushey Heath and Arkley. These showed evidence (v-shaped breakage patterns and straight, or nearly straight, grooves or fractures) of having formed in a fairly low-energy beach environment. Samples from the Northaw Pebble Gravel, on the other hand, contained sand grains showing additional evidence of modification in a fluvial environment, as well as the aforementioned marine features. These results were cited by Hey *et al.* as evidence of a marine origin for the Stanmore Pebble Gravel. However, the possibility exists that sand grain surface textures indicative of beach conditions have been inherited from the Palaeogene deposits that presumably supplied most of the sand in these gravels.

Moffat (1980, 1986) divided the various high-level Neogene gravels of North London and the Chilterns, on the basis of petrology, into four groups, two of which were subdivided (with a further group for Palaeogene gravels). His work provided some support for Wooldridge's correlation of the Stanmore and Little Heath gravels: by comparison of the characteristics of their quartz components, both were classified in a group intermediate between the highly quartzose Westland Green-type Thames gravels and deposits containing little else but flint (the last are gravels of restricted local origins). However, the Stanmore Pebble Gravel, which he sampled at Harrow Weald Common, and the much higher gravel at Nettlebed, were separated by Moffat from other comparable deposits on the basis of particle size and mineralogy. Whereas other members of the group were interpreted as Pliocene/Early Pleistocene marine gravels, more or less *in situ* (see above, Little Heath), the Stanmore and Nettlebed deposits were attributed to Early Pleistocene fluvial activity (Moffat and Catt, 1986b).

Moffat (1986) cited variations in the distribution of different-sized quartz material ('quartz signatures') between types of high-level gravel as an indication that the quartz has come from three distinct sources. He found that the 'quartz signatures' from the supposed Pliocene/Early Pleistocene marine gravels at Little Heath and Lane End were similar to those from quartzose Reading Beds gravels and suggested that the latter had provided the quartz component of the former. Other quartzose gravels, including that at Harrow Weald Common, contain more quartz than could have been supplied by the Reading Beds, with a 'signature' suggesting a source rich in smaller clasts of quartz (4–8 mm), such as the Lower Greensand (see below). Gravels on the Chilterns dip slope, interpreted as remnants of the Stoke Row and Westland Green Formations of the Thames (Moffat, 1986; Moffat and Catt, 1986a), contain yet more quartz, with a third variety of 'quartz signature', indicating a supply of larger sized material. The association of this change with the appearance of 'Bunter quartzites' suggests that additional quartz in these gravels has come from sources outside the Thames catchment, adding support to the view that they were deposited by an early Thames flowing from the Midlands (see Part 2 of this chapter) and are clearly separable from the Pebble Gravel of North London (Stanmore and Northaw Gravels), which contains no material from the Midlands.

The plotting of the elevations of the high-level gravels discussed here against the projected downstream gradients of early Thames formations (see below) supports the interpretation of the Stanmore Gravel as fluviatile (Fig. 3.6). In particular, the distinct west to east gradient, recognized but dismissed as unimportant by Wooldridge (1927a), is readily apparent. A similar gradient is observed between outcrops of the lower Northaw Pebble Gravel between Barnet and Northaw Great Wood. In both cases the gradient is higher than projections of the early Thames gravels on the Chilterns dip slope (Fig. 3.6). Furthermore, gravels with larger quartz components and containing 'Bunter quartzites' have been recognized (Moffat, 1980; Moffat and Catt, 1986a) on the northern flanks of the Vale of St Albans at elevations well above the Westland Green Gravels of Hey (1965), which suggests an antiquity comparable to the North London Pebble Gravel. Moffat and Catt (1986a) suggested that gravels of this type at Bedmond (TL 104037) and Sherrards Park Wood (TL 225138) might be outliers of the Stoke Row Gravel of Gibbard (1983, 1985), a suggestion that is supported by their altitude (Fig. 3.6). These gravels must be at least as old as the Northaw Pebble Gravel, since they are at a comparable altitude.

The composition and steeper gradients of the two North London Pebble Gravel formations, the Stanmore and Northaw Gravels (Fig. 3.6), imply that they are the deposits of tributarics of the early Thames, not the Thames itself. This is supported by their location, well to the south of the earliest Thames route along the Chiltern dip slope. The distribution of these formations suggests north-eastward trending sediment bodies, now highly dissected, converging with the Thames route through the Vale of St Albans. Their composition is also consistent with deposition by a south-bank tributary of the Thames, a river draining the northern Weald and tapping sources of Greensand chert. The quartz component of these gravels, which is predominantly confined to the smaller size ranges, was probably secondarily derived from the Reading Beds and the Lower Greensand. Indeed, the 'quartz signature' determined by Moffat (1986) from the Harrow Weald Common gravel is suggestive of a Lower Greensand provenance. A similar quartz component has been encountered in (pre-Thames diversion) Mole-Wey deposits in the Finchley area, the Dollis Hill Gravel of Gibbard (1979) (see Fig. 3.1 and Tables 3.1 and 3.2). It seems likely, in fact, that several (as yet undifferentiated) left-bank terrace gravels of the erstwhile Mole-Wey river are represented in the area between the Lower Lea valley and the main Pebble Gravel outcrops of North London. The Northaw Pebble Gravel can be interpreted as a high-level element within this same terrace system; the term should be restricted to those outliers on the crest or the eastern side of the ridge, in which Greensand chert occurs (Fig. 3.1 and Table 3.2). Deposits at similar elevations further west, such as at Shenley, are unlikely to be *in situ* Mole-Wey deposits and may well be of colluvial origin, derived from the gravels capping the ridge. In a stone count of the Shenley gravel, Hey (1965) found no Greensand chert.

Small amounts of such chert are, however, recorded from the Stanmore Pebble Gravel (Wooldridge, 1927a; Moffat, 1980; Moffat and Catt, 1986a), suggesting that this deposit might represent the earliest evidence of the Mole-Wey drainage system. The altitudinal distribution and gradient of this deposit suggest correlation with the Stoke Row Gravel of the Thames, whereas the Northaw Gravel can probably be correlated with the Westland Green Gravels (see Fig. 3.6 and Table 3.1). Correlation of the deposit at Nettlebed with the Northaw Pebble Gravel, as suggested by Gibbard (1985) and Moffat and Catt (1986a), is not supported by this interpretation; the Nettlebed Gravel appears to be an older deposit than any part of the North London Pebble Gravel Subgroup (Table 3.1).

The exposure of the Stanmore Pebble Gravel at Harrow Weald Common is one of very few, in the Thames basin, in which Pliocene/Early Pleistocene deposits can be examined. The origin of this type of material has been a subject of considerable controversy but, until recently, most authors have favoured a shallow-marine origin. Recent reappraisal has considerably undermined the case for a Pliocene or Early Pleistocene marine episode in the London Basin, such that the Pebble Gravel Group is now generally thought to comprise fluviatile deposits, their marine characteristics having been inherited from earlier Cenozoic gravels. The suggestion, in this report, that the Stanmore Pebble Gravel represents a south-bank tributary of the early Thames is based on re-evaluation of published evidence. Further research on the deposit at Harrow Weald Common and other sites in the Stanmore and Northaw Gravels is required to test this theory and provide more information about the palaeogeography of the London Basin during the poorly understood Late Pliocene–Early Pleistocene period. The goal of such work would be a more complete understanding of the early history of the River Thames, prior to its acme, in the Early Pleistocene, as a huge river draining the West Midlands as well as the present catchment.

Conclusions

The gravel at Harrow Weald Common was once widely believed to have been deposited in a Pliocene sea, two million or more years ago. However, a reconsideration of the evidence from here and other sites suggests that these sediments are river deposits; their altitude and sedimentary characteristics are consistent with deposition by a south-bank tributary of the early Thames. This view stems from the application of more modern analytical methods, such as detailed studies of the constituent rock types and sand grains present in the gravels. The characteristic markings found on quartz grains when viewed under high magnification with a scanning electron microscope (SEM) provide a means of distinguishing between marine and fluviatile deposits. Such analyses suggest that the marine characteristics of the deposits were inherited from older sediments that have been incorporated in the fluviatile Stanmore Pebble Gravel. As a rare exposure of these early tributary deposits, the Harrow Weald Common site will be essential to any reinterpretation of the early history of the Thames catchment.

PRIEST'S HILL, NETTLEBED (SU 700872)

D.R. Bridgland

Highlights

A site high on the Chiltern escarpment, Priest's Hill is the location of what may be the oldest interglacial deposits to be preserved in the Thames catchment. Critical to the interpretation of the interglacial sediments and their implications for the evolution of the River Thames is the relation between the Priest's Hill deposits and the Nettlebed Gravel, which covers higher ground in the vicinity. The Nettlebed Gravel is widely believed to be the earliest true Thames gravel, derived through an early Goring Gap.

Introduction

This site lies on the western side of Windmill Hill, Nettlebed, which at 211 m O.D. is one of the highest points on the Chiltern escarpment in the area of the Goring Gap. The modern river passes through the gap in the escarpment *c.* 10 km to the south-west and 150 m below the level of the deposits at Nettlebed. These unbedded, clayey, sandy gravels, containing a high proportion of rounded flint pebbles, were classified as 'Westleton Beds' by Prestwich (1890b), a term also used by Monckton and Herries (1891) and White (1892, 1895). The deposits were subgrouped with the Pebble Gravel (Blake, 1891; Salter, 1896; Shrubsole, 1898; White, 1908b) and appear as such on the Geological Survey map (New Series, Sheet 254). On the basis of its elevation, Wooldridge (1927a) grouped this outlier with his high-level Pliocene (Diestian) marine deposits, which he later (Wooldridge, 1957, 1960) redefined as Early Pleistocene. The gravel remnants at Nettlebed have recently been reinterpreted as fluvial deposits (Horton, 1977, 1983; Gibbard, in Turner, 1983; Gibbard, 1985). Gibbard regarded them as early Thames deposits, classifying them as Nettlebed Gravel, the highest lithostratigraphical division within his system of Middle Thames terrace gravels and, therefore, probably the earliest Thames formation to have been preserved (if the interpretation of the Little Heath and Lane End gravels as marine is correct – see Little Heath).

Priest's Hill is of exceptional significance, since an Early Pleistocene interglacial organic deposit has been discovered there, in association with the Nettlebed Gravel (Horton, 1977, 1983; Turner, 1983; Gibbard, 1985). Although the precise relation between the organic sediments and the gravel is unclear at present, the former offers a potential means of determining the age of the earliest elements of the Middle Thames sequence.

Description

Several gravel remnants are preserved at Nettlebed, occurring slightly below the summit of Windmill Hill and on the subsidiary summit of Priest's Hill (203 m O.D.), in both cases overlying thin basal London Clay above Reading Beds. The summit of Windmill Hill, 211 m above O.D., is gravel-free (Blake, 1891; White, 1892; Horton 1977). White (1892) regarded the surviving drift remnants on Windmill Hill as the infills of 'pipes' and hollows in the Reading Beds and considered it likely that the hill-top was once covered with a similar deposit, since removed by denudation. The gravel consists primarily of flint (both reworked from Palaeogene beds and derived directly from the Chalk) and quartz, with subordinate Palaeozoic chert and quartzite (Table 3.2). The size and origin of the quartzite component is a subject of some doubt (see below).

Prestwich (1854), in the earliest record of the site, illustrated a section showing 1.5 m of crudely stratified gravel above *c.* 4 m of bedded white and yellow sand with flint and quartz pebbles, the latter having an erosive base cutting into Reading Beds sands and clays. The base of the gravel was also irregular and it is not clear whether Prestwich regarded the sand as part of the drift or of the solid geology. Monckton and Herries (1891), who described a comparable section at Priest's Hill, considered the sand to be associated with the gravel. It is possible that this sand is equivalent to that seen underlying the organic sediments in recent temporary sections (Horton, 1977, 1983; below).

The Nettlebed interglacial site has only been investigated from boreholes and temporary exposures. In July 1975 the following sequence was revealed, apparently filling a depression in the Reading Beds (Horton, 1977, 1983; Fig. 3.7):

		Thickness
6.	Soil	0.2 m
5.	Grey, poorly sorted clayey gravel (= Nettlebed Gravel?)	1.35 m
4.	Grey-white sand	0.04 m
3.	Humic, pebble-free clay	0.025 m
2.	Dark brown humic silt with scattered pebbles and plant remains	0.66 m
1.	White silty, pebbly sand	0.93 m

Turner (1983) published a preliminary pollen diagram from the Nettlebed organic deposits (beds 2 and 3), showing rather more than the first half of a typical Pleistocene interglacial cycle (see Fig. 3.8).

Interpretation

Prestwich (1890b) and White (1892, 1895) regarded the 'Westleton Beds' at Nettlebed as marine deposits of Pliocene or Early Pleistocene age. Salter (1896), who classified the deposits with his 'High Barnet Type' of pebbly gravel, was unusual amongst early workers in dissenting from this view, regarding all the high-level gravels as fluviatile. Wooldridge (1927a) closely associated the gravel at Nettlebed with that capping the Palaeocene outlier at Lane End (see above, Little Heath), *c.* 12 km to the north-east and 10 m lower in altitude. He correlated these and other high-level deposits on the Chilterns with the Lenham Beds of Kent, on the basis both of altitude and heavy-mineral content, attributing them to a Late Pliocene/ Early Pleistocene marine incursion into the London Basin (Wooldridge, 1927a, 1957, 1960; Wooldridge and Linton, 1939, 1955). Recent work by Moffat (1980; Moffat and Catt, 1986a) has confirmed the possibility that the Lane End gravel is a Pliocene/Early Pleistocene marine deposit, but they and most other recent workers have favoured Salter's fluviatile interpretation of the Nettlebed sediments.

Horton (1977) recorded channelling within the sequence at Nettlebed, including the channel filled with humic silts and clays (beds 2 and 3) containing terrestrial pollen. He therefore thought the deposits to be of fluvial origin. Further support for this interpretation came from a resistivity survey, which revealed a channel trending NNW–SSE, approximately parallel with the orientation of the Goring Gap (Turner, 1983). Turner (1983) and Gibbard (1985) also cited evidence from a study by scanning electron microscopy (SEM) of sand grains from the Nettlebed Gravel, the surfaces of which showed various stages in the eradication (presumably during fluvial transport) of characteristic features formed by attrition in a marine environment, presumed to have been inherited from Palaeogene deposits (see above, Harrow Weald Common). Both Turner (1983) and Gibbard (1985) attributed the Nettlebed Gravel to the Thames, but were uncertain of its relation to the organic channel deposits.

The palynological sequence from the Nettlebed interglacial sediments (beds 2 and 3) (Turner, 1983; Fig. 3.8) begins with a pre-temperate phase (biozone I), passes through an early temperate phase (biozone II) and includes the early part of a late-temperate phase (biozone III). Birch (dominant), pine and subordinate spruce pollen characterize biozone I. Oak (abundant) and elm appear in biozone II, whereas biozone III is heralded by the appearance in the pollen record of hornbeam and the marked expansion

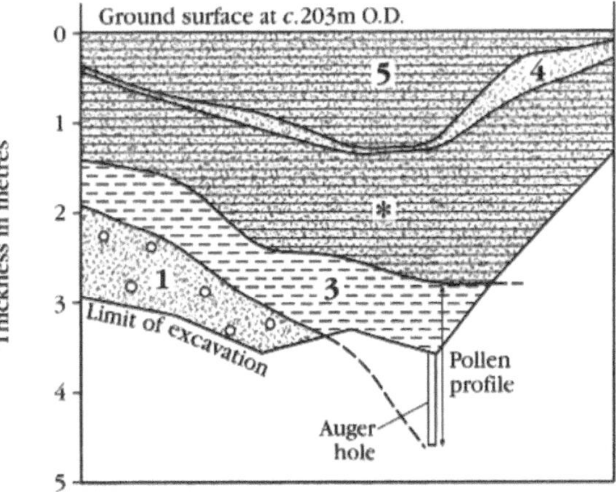

1 etc. Bed numbers used in description

* This unit appears in the illustration provided by Horton (1983), but not in his written description

Clayey sandy gravel

Sand

Humic clay with pebbles

Sand with scattered pebbles

Figure 3.7 The stratigraphy of the deposits at Priest's Hill, Nettlebed, as recorded from a temporary section. The position of the pollen profile is shown (after Horton, 1983).

of hazel. Despite the absence of exotic taxa from this record, spruce being the only non-native tree represented, Turner considered it quite different from interglacial pollen diagrams from the Middle and Upper Pleistocene. He pointed to the absence of temperate indicators such as ivy, holly and lime, to the late rise of hazel and to the early appearance of spruce as significant features, concluding that the deposit represents an interglacial episode prior to the Cromerian Stage, probably one hitherto unrecognized in Britain. Gibbard (1985) was in broad agreement with Turner's conclusions and proposed the provisional term 'Nettlebedian' for the stage represented. He suggested that the non-polleniferous sands (bed 1) underlying the organic sediments might have formed during a previous cold episode. Gibbard assessed the palaeobotanical evidence for correlation of the Nettlebed interglacial with the more complete continental Pleistocene

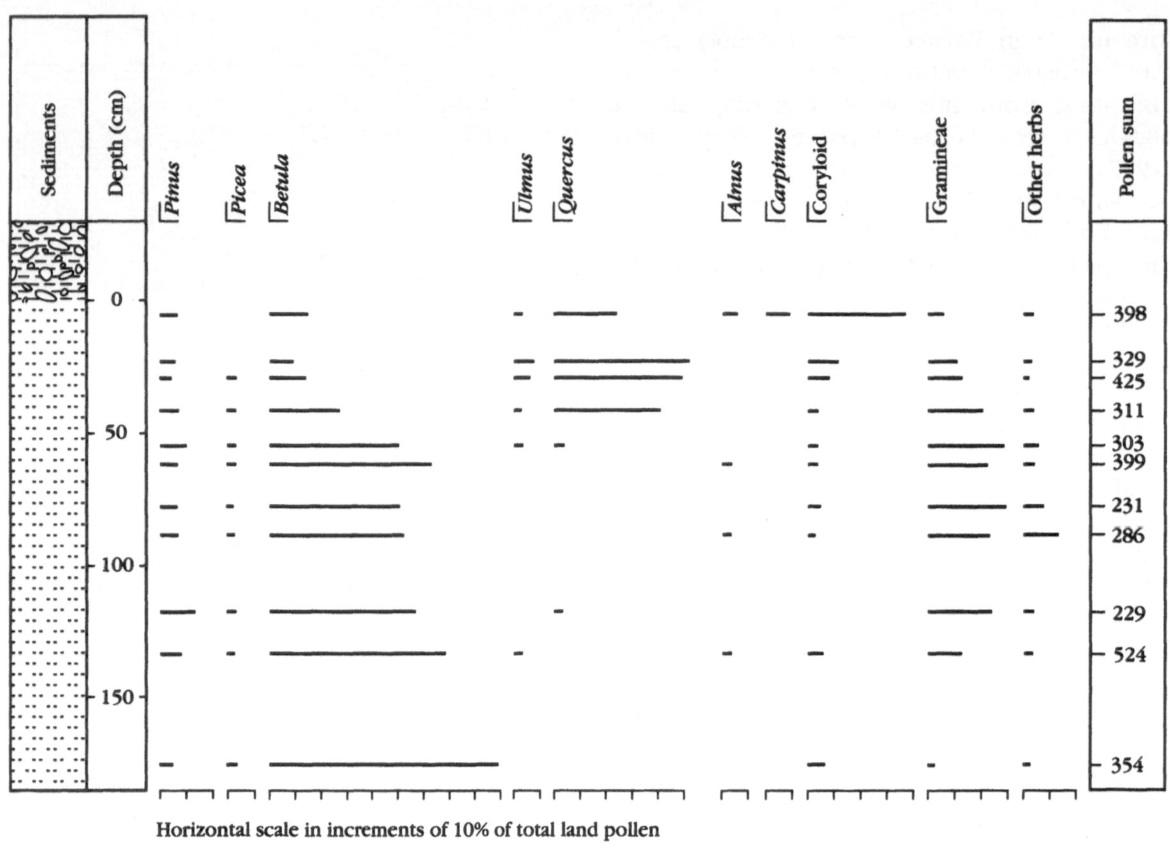

Figure 3.8 Pollen from the organic silts at Nettlebed (after Turner, 1983).

record and concluded that it probably belonged somewhere within the late Early Pleistocene/ early Middle Pleistocene 'Cromerian Complex' of The Netherlands. The stratigraphical position of the deposit, at a very high level within the Thames terrace system, implies a considerable age (Turner, 1983). Its relation to the Thames terrace system, as described in Chapter 1 (Figs 1.3 and 3.3), suggests that the Nettlebed deposit is very much older than that at Sugworth or any of the temperate-climate sediments in Essex that are ascribed to the 'Cromerian Complex' (see Chapter 5). It is considerably higher within the terrace sequence than the Gerrards Cross Gravel, for which an Early Pleistocene age now appears likely (see Chapter 1; below, Westwood Quarry). Further investigation of the Priest's Hill sediments is urgently required in order to address this question. In particular, evidence from palaeomagnetism, to establish whether the site pre-dates the Matuyama– Brunhes reversal (see Chapter 1), would be of considerable value.

The Nettlebed interglacial sediments have great potential as a means of dating the earliest Pleistocene deposits of the Thames system. Unfortunately, there is some doubt as to their precise relation to the Nettlebed Gravel. Gibbard (1985) observed that the clayey gravel (bed 5) overlying the organic channel-sediments closely resembles the Nettlebed Gravel of the Windmill Hill outlier in its clast-composition (Table 3.2), suggesting a direct correlation. However, the former has a very clayey matrix, in places reminiscent of the local Reading Beds, raising the possibility that the pollen-bearing sediments are capped by a soliflucted mixture of Nettlebed Gravel and Palaeogene clay and, therefore, that the channel deposits may post-date the deposition of the early Thames gravel at this altitude. The fact that Priest's Hill lies a few metres lower than the Windmill Hill outlier increases the feasibility of the latter interpretation.

It is therefore uncertain whether the Nettlebed Gravel was deposited before or after the interglacial sediments. However, the fact that both appear to be fluvial deposits belonging to the Thames terrace system suggests, given the small amount of vertical separation between them, that they are closely related in time. Even if the channel deposits are the product of a tributary, their position near the crest of the escarpment implies considerable antiquity. In terms of terrace stratigraphy, both the Nettlebed Gravel and the Priest's Hill channel deposits appear to pre-date the Stoke Row and Westland Green Gravels (Fig. 3.2), the earliest formations to provide unequivocal evidence for a Thames catchment that extended beyond the Cotswolds.

Further possible outliers of Nettlebed Gravel have been recognized at Russell's Water (SU 714889) and Kimble Farm (SU 749888), both downstream from the type locality (Turner, 1983; Gibbard, 1985; Fig. 3.6). This suggests a downstream gradient of *c.* 0.9 m per kilometre for this formation. Although it falls directly in line between Nettlebed and Kimble Farm, analysis of the Russell's Water deposit by Moffat (1980; Moffat and Catt, 1986a) shows it to be composed dominantly of angular flint, which suggests that it is not part of the Nettlebed Gravel.

Gibbard (1985) suggested a correlation of the Nettlebed Gravel with the 400 ft (Northaw) Pebble Gravel, but the Nettlebed deposit contains fewer rounded flints, reworked from Palaeogene deposits, than the latter (Table 3.2; see above, Harrow Weald Common). However, Nettlebed lies near the edge of the Tertiary sedimentary basin, beyond which Chalk would presumably have outcropped when the Nettlebed Gravel was deposited, as it does now, providing a supply of fresh flint. Rounded flint would have been progressively picked up from Palaeogene sources between Nettlebed and Hertfordshire. The compositional differences between the Nettlebed Gravel and the Northaw Pebble Gravel do not, therefore, preclude their correlation. However, Moffat and Catt (1986a) have identified gravels equivalent to Gibbard's Stoke Row Gravel, which post-dates the Nettlebed Gravel (see Fig. 3.2 and Table 1.1), on the Chiltern dip slope in Hertfordshire, at altitudes comparable to the Northaw Gravel (Fig. 3.6). Plotting of the altitudinal distribution of the Pebble Gravel remnants on the dip slope (Chiltern Pebble Gravel Subgroup) and in North London and south Hertfordshire (North London Pebble Gravel Subgroup) along a projection of the early Thames course across the area (Fig. 3.6), indicates that correlation of the Nettlebed Gravel with the Northaw Gravel is untenable. Instead, the latter is likely to represent a south-bank tributary stream and to correlate with the Westland Green Formation of the main river (Fig. 3.6; see above, Harrow Weald Common). Even the higher Stanmore Gravel, which is also interpreted as a tributary deposit, is too low to correlate with the Nettlebed Gravel; it probably equates with the Stoke Row Gravel (Fig. 3.6).

An outlier of gravel at Mardley Heath (TL 247185) is worthy, on the basis of its altitude and position on the Chiltern dip slope (Fig. 3.6), of consideration as a possible downstream correlative of the Nettlebed Gravel. Described by Moffat (1980; Moffat and Catt, 1986a), this deposit contains significant quartz and quartzite components and has been interpreted (Moffat and Catt, 1986a) as Westland Green/Stoke Row Gravel. However, interpretation of the Mardley Heath site is problematic. The deposit contains heavy minerals characteristic of Anglian glacial sediments and its gravel fraction is dominated by non-rounded flint (Moffat, 1980), both of which suggest that it is Anglian outwash and, therefore, of no relevance to the early evolution of the Thames.

Turner (1983) cited the presence of quartz and quartzites in the Nettlebed Gravel as an indication that a link between the Kennet-Thames and the Midlands already existed at the time of its deposition, rejecting White's (1906) theory that such material was reworked from the quartz-rich facies of the Reading Beds. However, the increased amounts of Midlands material that are present in later Thames formations indicate that the main input of 'exotic' material through the Goring Gap occurred after the deposition of the Nettlebed Gravel. Different estimates of the composition of the Nettlebed Gravel have been given by Moffat (1980, 1986; Moffat and Catt, 1986a) and Gibbard (in Turner, 1983, 1985). The most important discrepancy is that Moffat found less than 1% quartzite above the 8 mm sieve size, whereas Gibbard recorded *c.* 4% and concluded that the quartzose component of the gravel indicated a Midlands provenance (Table 3.2). Moffat (1986) interpreted the 'quartz signature' (see above, Harrow Weald Common) from the Nettlebed deposit as evidence for derivation from the Reading Beds and the Lower Greensand, but not from Triassic deposits. This would seem to imply that the Oxfordshire extension of the Thames was already in operation, but that no Triassic material had, up to that time, entered the catchment. Horton (1977) listed 18.5% quartzite from Nettlebed, but this figure clearly includes vein quartz and so must be disregarded. However, he did state that 'Northern Drift' pebbles were absent, suggesting support for Moffat's findings. Amongst the early workers, the views of Prestwich (1890b) are of some relevance to this problem, as he gave clear details of the composition of the gravel at Nettlebed. Prestwich (1890b, p. 140) considered that 'Pebbles of hornstone (?), veinstone, and Sarsen stone' accounted for 8% of the deposit. The record of sarsen suggests that he regarded the quartzitic material present as being of Palaeogene derivation. Hornstone (= schorl-rock?) is a highly durable lithology that is usually attributed to a Midlands Triassic source, but such material is also present in Mesozoic pebble beds within the present Thames catchment (Bridgland, 1986b). Thus it remains uncertain, at present, whether the Nettlebed deposits contain material from outside the present Thames basin. This question is of considerable importance, since it has major implications for Pleistocene palaeogeography and the evolution of Thames drainage.

The high-level gravel at Nettlebed is of major importance, therefore, as it probably represents the earliest evidence for Thames drainage that has survived. The GCR site at Priest's Hill is of even greater importance, however, on account of the very early interglacial sediments that have been preserved there in association with the Nettlebed Gravel. These are thought to represent an Early Pleistocene temperate episode hitherto unrecognized in Britain. The occurrence of fluvial sediments at this considerable altitude indicates that the present crest of the Chilterns was at valley-floor level at that time. This last fact demonstrates the enormous amount of erosion that has occurred during the Pleistocene.

Conclusions

At Priest's Hill, underlying a clayey gravel, are beds of silt and clay, containing pollen, that seem to fill an old river channel. Analysis of the pollen reveals a dominance of temperate-climate plant species and indicates that the silt and clay accumulated during a warm ('interglacial') phase of the Quaternary Ice Age. The high altitude of this site (around 200 m above sea level) suggests a very early age. It is similar to the level of the Nettlebed Gravel, the highest and oldest Thames gravel to be recognized. The Nettlebed Gravel is preserved in the vicinity of the Priest's Hill site and may be represented by the clayey gravel overlying the pollen-bearing sediments. This would suggest that the interglacial deposits were part of the Nettlebed Gravel Formation and would indicate that they are of very great antiquity. An alternative interpretation is possible, however. The clayey gravel may be a later mixture of material, largely derived from the Nettlebed Gravel, that has accumulated above the channel-fill sediments. In this case the interglacial beds could be considerably younger than the Nettlebed Gravel. However, their high altitude inevitably indicates that they are very early and of great significance; no major river can have flowed at this level for a very long time. The evidence from Nettlebed provides a clear indication of the vast amount of erosion that has taken place during the Pleistocene.

Part 2

PRE-DIVERSION DEPOSITS IN THE MIDDLE THAMES BASIN AND THE VALE OF ST ALBANS

D.R. Bridgland

Introduction

In this part of Chapter 3 the most extensive gravel formations to have been produced by the Thames are described. These were laid down after the river gained headwaters that drained a large part of the West Midlands, probably also extending into Wales (the 'Severn-Thames' of early writers), and before the diversion of the river by the Anglian Stage glaciation of the Vale of St Albans. Evidence for this diversion, which took the river into its modern valley through London, is derived from sites in the Vale of St Albans (see below, Moor Mill). The 'Severn-Thames', as envisaged by many Victorian authors and recently promoted by Hey (1986; see Chapter 2), was a huge river, probably the largest to have existed in Britain during the Quaternary. It appears to have come into existence in the Early Pleistocene, although its initiation probably post-dated the interglacial represented at Nettlebed (see above). The decline of the Thames appears to have begun before its diversion into the modern valley, in the Anglian. According to Whiteman (1990), the river had already lost its headwaters beyond the Cotswolds by the Anglian Stage, by which time it was more or less confined to the present catchment (see Chapter 1). Evidence from Essex suggests that the river's size was much diminished by 'Cromerian Complex' times, during which it formed a series of terrace deposits in the north-east of that county for which few equivalents have thus far been recognized upstream in the Middle Thames (see Chapters 1 and 5).

The deposits of the pre-diversion Thames are so extensive, and their far-travelled component so dominant, that they have often been interpreted as glacial outwash material and were mapped by the Geological Survey largely as 'Glacial Gravel'. The earliest 'Severn-Thames' gravels are, however, found within the high-level deposits of the Pebble Gravel Group, already described in Part 1 of this chapter. The oldest major fluvial formation to be recognized in the London Basin is the Nettlebed Gravel (Gibbard, 1985; see above), but the first appearance of a significant component of material from beyond the present Thames catchment is in lower-level formations within the Pebble Gravel, the Stoke Row and Westland Green Gravels (Table 3.1).

Salter (1896) had recognized that the lowest deposits of Pebble Gravel type could be separately defined on the basis of their distinctive clast composition, which included abundant quartz and quartzites. He classified such gravels as his 'Bell Bar type'. Hey's (1965) reinvestigation of these gravels confirmed that they represent a compositionally distinctive gravel type that can be traced across the London Basin from the Goring Gap to eastern Hertfordshire. He demonstrated that the quartz and quartzite content of these deposits, which he named the 'Westland Green Gravels', was consistently at least twice that of other deposits that had been classified as Pebble Gravel. He also noted that pebbles of orthoquartzite from the Triassic ('Bunter') pebble beds of the Midlands were present in the Westland Green Gravels, but not in other varieties of Pebble Gravel. He noticed that remnants of Westland Green Gravels occurred at progressively lower altitudes eastwards from the Goring Gap, falling from 174 m O.D. to 107 m O.D. near Bishops Stortford. He concluded that these deposits, which he later traced into Essex, Suffolk and Norfolk (Hey, 1980; Chapter 5) were of Thames origin and that they represented the first influx of material from the Midlands through the Goring Gap.

A large proportion of the high-level gravels in the Thames valley has been classified at various times as 'Glacial Gravel'. This terminology dates back to Wood (1867, 1870; Wood and Harmer, 1868, 1872), who assigned the till and associated gravel of East Anglia to his 'Glacial Series'. Wood's classification was adopted by the Geological Survey throughout the London Basin (for example, Whitaker, 1875, 1889; Sherlock and Noble, 1922; Bristow, 1985). Generally the gravels were classified as 'Middle Glacial' and the overlying till (later the Lowestoft Till) as 'Upper Glacial' (Wood and Harmer, 1868, 1872). The lowest tier of Wood's tripartite classification, the 'Lower Glacial', referred to a till of more localized distribution, later identified as part of the Scandinavian-derived North Sea Drift. Although Wood's original classification seems to have had predominantly stratigraphical (rather than genetic) implications, the term led inevitably to the misinterpretation of the gravel deposits as glaciofluvial outwash.

White (1895) was the first to advocate the idea that such gravels were the remains of early terrace deposits of the Thames, but this view did not gain widespread acceptance until Sherlock and Noble (1912) traced gravels rich in Midlands lithologies from the Goring Gap towards Watford and suggested that the Thames formerly flowed along this more northerly route. Shrubsole (1898) subdivided the various high-level gravels of Berkshire and Oxfordshire according to their clast content, which he listed in percentages. He defined four categories (in addition to 'local gravel'), as follows: 'Pebble Gravel', 'Goring Gap Gravel', 'Quartzose Gravel' and 'Quartzite Gravel'. The last two incorporated the lower pre-diversion Thames formations, classified as 'Glacial' by most of Shrubsole's contemporaries. The change from 'Quartzose' to 'Quartzite' type is reflected in the progressive decline in the quartz : quartzite + sandstone ratios subsequently recognized within the pre-diversion Thames sequence (Hey, 1980, 1986; Green and McGregor, 1986; Table 3.2). Shrubsole's work was therefore much ahead of its time, but he did not positively associate these gravels with the Thames.

The first attempt to define early terraces of the Thames within the so-called 'Glacial Gravels' of the lower Chiltern dip slope was by Saner and Wooldridge (1929). They recognized a 'Winter Hill Terrace' in addition to, and higher than, the previously identified terrace gravels. This and two older, much-dissected Thames aggradations, the 'Higher' and 'Lower Gravel Trains', were fully described by Wooldridge (1938). Hare (1947), on the basis of detailed geomorphological mapping of the area between Slough and Beaconsfield, redefined the Lower Gravel Train as the 'Harefield Terrace'. Later studies by Sealy and Sealy (1956), Thomas (1961) and Allen (1978) extended Hare's terrace mapping both westwards and eastwards. These workers added a number of new terraces to the Thames sequence, in some cases upgrading minor features that Hare had previously recognized. The separate existence of most of these has not been upheld in recent work by Gibbard (1985), although the Rassler Terrace of Sealy and Sealy (1956) has been interpreted in Chapter 1 as the product of a genuine and additional Thames aggradation, between the Harefield and Winter Hill Terraces as defined by Hare (1947).

Hey's (1965) work on the Westland Green Gravels pioneered an approach based on clast-lithological analysis (see Chapter 1). Using this technique, he distinguished the Westland Green Formation from the rest of the Pebble Gravel and recognized it as an early aggradation of the Thames, despite the fact that erosion has reduced this deposit to small patches of gravel capping hills. The Westland Green Gravels were the oldest Thames deposits to be recognized until Gibbard (1983, 1985) identified the earlier Nettlebed and Stoke Row Gravels (above, and Tables 3.1 and 3.2). Gibbard recognized another additional Thames formation, his Satwell Gravel, which is intermediate, within the terrace sequence, between the Higher Gravel Train and the Westland Green Gravels (Table 3.1). Gibbard (1977, 1978a, 1983, 1985) redefined the Thames terrace deposits on a lithostratigraphical basis, generally using the geographical names that had previously been applied to the terrace surfaces. He replaced the terms 'Higher Gravel Train' and 'Lower Gravel Train' (Harefield Terrace) with Beaconsfield Gravel and Gerrards Cross Gravel respectively, thus establishing the system of nomenclature presently in use.

It is interesting to note that the elevation of the deposit at Chorleywood (TQ 023953), claimed by Moffat and Catt (1986a) as a further representative of the Westland Green Gravels, appears from its altitudinal position (Fig. 3.6) more likely to result from a separate aggradation, hitherto unrecognized, between that formation and the later Satwell Gravel. According to analyses by Moffat (1986; Moffat and Catt, 1986a), this deposit, which is situated approximately 10 m below the expected level of the Westland Green Formation, contains higher levels of quartz and quartzites and less flint than other outliers of the Westland Green Gravels (Table 3.2). A further low-level Westland Green Gravels outlier with a high quartz and quartzite content appears to have been recorded by Green and McGregor (1978a), on the opposite side of the River Chess from Chorleywood, to the north-west of Croxley Green. Caution must be exercised in the interpretation of altitudinal evidence on the Chilterns dip slope, given the prevalence of Chalk solution phenomena. However, the apparent clast-lithological distinction is potentially significant, particularly as it is in keeping with the progressive trend, in successively lower pre-Gerrards Cross Gravel formations, of increasing quartzose rocks at the expense of flint (see Table 3.2). Further investigation of this area is called for, to assess whether an additional terrace formation is preserved there.

THE VALE OF ST ALBANS THAMES AND ITS DEMISE
D.R. Bridgland

Salter (1905) was apparently the first to recognize that fluvial drainage through the Goring Gap formerly continued north-eastwards into Hertfordshire by way of the Vale of St Albans. He observed that varieties of rock introduced into the area in glacial deposits are absent in the gravels associated with this former drainage route, but that they occur in all the gravels associated with the modern drainage system. He thus concluded that the initiation of this system must have coincided with the glaciation. Perhaps because these far-reaching conclusions were lost in a weighty description of deposits over the whole of south-eastern England, they received no immediate support. Thus Sherlock and Noble (1912), who claimed that the Thames formerly flowed from the Beaconsfield area towards Watford, through the Vale of St Albans, and that the glaciation of this district brought Thames drainage by this route to an end, are often given credit for the initiation of the theory. Later, Sherlock (1924), Sherlock and Pocock (1924) and Clayton and Brown (1958) described lacustrine deposits in the Vale of St Albans that, they suggested, were related to the ponding of the Thames by ice. Overspill from the proglacial lake so formed was considered to have effected the diversion of the river into its modern valley.

Wooldridge (1938, 1960) believed that the Thames had formerly occupied two different northerly courses, one running through the Vale of St Albans and a later route through Finchley (Fig. 3.4). He suggested that both of these routes were abandoned as a result of ice advances, the Vale of St Albans route because of an early 'Chiltern Drift' advance (Barrow, 1919a) and the Finchley route as a result of the main 'Chalky Boulder Clay' advance (now attributed to the Anglian Stage). Wooldridge considered that the Higher and Lower Gravel Trains and the Winter Hill Terrace all continued eastwards via the intermediate Finchley route.

Hey (1965) traced his Westland Green Gravels from the Goring Gap along the Chiltern dip slope to Hertfordshire, thus confirming that the river at this time flowed through the Vale of St Albans. Later work showed that the Beaconsfield, Gerrards Cross and Winter Hill Gravels also extended along this route (Gibbard, 1974, 1977, 1983, 1985; Green and McGregor, 1978a, 1978b; McGregor and Green, 1978), rather than through Finchley, as suggested by Wooldridge. Hare (1947) had recognized that the Winter Hill Terrace, in the area immediately west of the Colne confluence, flattens downstream until it has a very small eastward gradient. He attributed this marked decrease in downstream gradient to the ponding of the Thames at this time by ice. Gibbard (1977, 1985) confirmed this interpretation, although he attributed the flattened terrace surface in this area to an upper, deltaic division of the Winter Hill Formation (Chapter 1), his 'Winter Hill Upper Gravel'.

Gibbard traced the fluviatile Winter Hill Gravel downstream beneath the glacial deposits of the Vale of St Albans, where it is synonymous with the Westmill Lower Gravel (*sensu* Cheshire, 1986a; see below, Westmill). He showed that this gravel is overlain in the Watford area by proglacial lake deposits, which are themselves overlain by till. Gibbard concluded that this lake arose from the same glacial ponding event that produced the Winter Hill Upper Gravel, and that the overspill from the lake eventually brought about the diversion of the Thames from its early course into its modern valley, without there having been an intermediate route as envisaged by Wooldridge. The evidence for a 'Chiltern Drift' glaciation, which Wooldridge believed to have diverted the Thames from the Vale of St Albans into his intermediate route, has been challenged by recent workers (Avery and Catt, 1983; Green and McGregor, 1983; McGregor and Green, 1983a). Cheshire (1981, 1983a, 1983b, 1986a; in Allen *et al.*, 1991) has further refined the stratigraphical succession in the Vale of St Albans and has recognized greater complexity in the sequence of Anglian glacial events in the area (this chapter, Moor Mill, Westmill and Ugley). Cheshire's work was based on a detailed examination of particle size, carbonate content and small-clast lithology of till samples from sites throughout eastern

Hertfordshire and western Essex. Cheshire has recognized four separate ice advances within the Anglian glaciation of the Vale of St Albans, the first of which diverted the Thames into its modern valley.

CHALFONT ST GILES BRICK PIT (SU 977942) and FURNEUX PELHAM GRAVEL PIT (TL 442267)

D.R. Bridgland

Highlights

The Chalfont St Giles and Furneux Pelham pits are important sites that provide evidence for the early history of the River Thames. Such evidence has been used to reconstruct the river's former course across Hertfordshire and into East Anglia, prior to its southward deflection by Anglian Stage ice.

Introduction

Sites at Chalfont St Giles and Furneux Pelham are important in establishing the route taken by the Thames during the formation of the Early Pleistocene Westland Green Gravels, as defined by Hey (1965). These gravels have been traced from the Middle Thames through the Vale of St Albans towards East Anglia (Hey, 1965, 1980). The Chalfont St Giles pit lies within the Middle Thames region, at the northernmost and highest edge of the classic 'staircase' of terrace deposits preserved between the Thames and Colne valleys (Fig. 3.1). The pit at Furneux Pelham, on the other hand, lies at the eastern end of the Vale of St Albans, only 6 km from the type locality of the Westland Green Gravels (Hey, 1983). The deposits contain rocks derived from the West Midlands and, it is believed, North Wales, the latter suggesting that glaciation of upland areas may have occurred at this early stage of Thames evolution (Green *et al.*, 1980; Hey, 1980; Bowen *et al.*, 1986a).

The Westland Green Gravels were the first Thames deposits to be traced across the London Basin from the Reading area to East Anglia (Hey, 1980, 1982; Bowen *et al.*, 1986a). Allen (1983, 1984) and Gibbard (1983, 1985) later proposed subdivisions of this unit in Suffolk (Chapter 5, Part 1) and the Middle Thames respectively. Gibbard (1983, 1985) concluded that the westernmost outlier recognized by Hey (1965), at Stoke Row (SU 686834), represents a higher and even earlier Thames aggradation, which he termed the Stoke Row Gravel. Re-evaluation of the various outliers in Hertfordshire identified by Hey (1965, 1980) as Westland Green Gravels suggests that some, including that at Furneux Pelham, may represent the earlier Stoke Row Gravel (Fig. 3.6).

Recent work by Whiteman (1990) indicates that the deposits classified as Westland Green Gravels in the Middle Thames and East Anglia do not belong to the same formation, although the basic conclusion that the Thames followed the route envisaged by Hey is upheld (see Chapter 5).

Description

Chalfont St Giles Brick Pit, Buckinghamshire

This site is an intermittently worked brick pit exploiting the silty clays of the Reading Beds. Up to 2 m of poorly stratified Westland Green Gravels occur as overburden above the Reading Beds. The site lies near the north-eastern edge of a small plateau, occupied by Hodgemoor Wood, at about 140 m O.D. The Geological Survey showed 'Glacial Gravel' overlying Reading Beds in the area of the brick pit (Old Series, Sheet 7, 1871; New Series, Sheet 255, 1922). An early description of the site was provided by Barrow (1919a). According to Barrow (1919a, p. 38), the Chalfont St Giles pit showed 'quartz pebble gravel ... somewhat churned up by the passage of ice over it, and in consequence mixed with a considerable number of far-travelled stones'. Despite this supposed glacial contamination, Barrow included the Hodgemoor Wood outlier in the Pebble Gravel. Hey (1965) recognized that the Hodgemoor Wood plateau represents one of the largest remnants of the Westland Green Gravels. He pointed to the occurrence in these deposits of quartzites and Carboniferous chert from the Midlands as evidence of a Thames origin; such materials are common constituents of the Northern Drift of the Upper Thames catchment (see Chapter 2), leading Hey to conclude that they were introduced into the London Basin, through an early Goring Gap, by the ancestral Thames. Green and McGregor (1978a) published the results of a stone count of a sample that appears, from their map and profile diagram, to be from this site or nearby, their sample 86 (see Table 3.2). The Chalfont St Giles pit was reinvestigated by Moffat (1980, 1986; Moffat and Catt, 1986a), who confirmed that Westland Green Gravels occur there.

Furneux Pelham Gravel Pit (Hillcollins Pit), Hertfordshire

This site lies in the present catchment of the River Lea, 10 km north of Ware, where that river turns south from the Vale of St Albans into the lower part of its valley, towards its confluence with the Thames in east London (see Fig. 3.1). The pit is 6 km north of the Westland Green type locality, which is now overgrown. As at Chalfont St Giles, Geological Survey maps (Old Series, Sheet 18) indicate 'Glacial Gravel' at Furneux Pelham (showing that the early mappers separated these deposits from the Pebble Gravel). The pit exposes 3–4 m of well-bedded sandy gravel, aggraded to *c.* 107 m O.D. and showing the abundance of quartzites and flint pebbles reworked from the Palaeogene that characterizes the Westland Green Gravels. Matrix-supported gravel is interbedded with cross-stratified sand lenses. Over much of the pit, the deposits appear to dip sharply to the north-west, perhaps towards a solution hollow in the underlying Chalk. Hey (1983) described the following section:

		Thickness
4.	Wind-blown sand	up to 1 m
3.	Coarse, poorly sorted gravel, the uppermost beds clay-enriched and reddened	6 m
2.	Yellow sand, with one band of flint pebbles	2 m
1.	Dark purplish-brown clayey sand	1 m
	Base not seen	

A hitherto unpublished stone count of the gravel at this site (bed 3) shows *c.* 77% flint, over two-thirds of which consists of reworked Palaeogene pebbles, with 19% quartz and 6% quartzite and sandstone (R.W. Hey, pers. comm.; Table 3.2). This analysis falls within the range of counts from the Westland Green Gravels previously published by Hey (1965). Table 3.2 also includes clast-lithological data from this site supplied by Cheshire (1986a).

Interpretation

The recognition by Hey (1965) of an Early Pleistocene 'Westland Green Gravels' aggradation, extending across an area from the Goring Gap to Hertfordshire, was a landmark in Thames research. Salter (1896, 1905), who had attempted to trace various types of gravel from the Chiltern dip slope to East Anglia, had previously included many of the Westland Green Gravels outcrops in his 'Bell Bar type', although he was apparently not aware of the outlier at Hodgemoor Wood. Hey traced the Westland Green Gravels from Stoke Row, on the Chilterns, via Ashley and Bowsey Hills and across Hertfordshire, where it occurs at Hodgemoor Wood, Hatfield Park, Essendon, Little Berkhamsted, and several sites between the Mimram and Stort valleys. The last included the best exposure available at the time of his survey, at Westland Green (TL 422215), near Little Hadham. He renamed the deposits after this locality, since he had observed that the sediments at Bell Bar, taken as the type section for this gravel by Salter (1896), were not *in situ*. Warren (1945, 1957) had previously attributed the gravel at Westland Green to the early Thames.

Support for the fluvial origin of the Westland Green Gravels was derived from their regional and altitudinal distribution, shown by Hey (1965) to closely resemble the long-profile of a river. The steep upstream part of Hey's reconstructed long-profile was, however, removed when Gibbard (1983, 1985) redefined the highest outlier, at Stoke Row, as part of an earlier formation (see below). A study of sand grain surfaces, using scanning electron microscopy, provided confirmatory evidence that the Westland Green Gravels are of fluviatile origin (Hey *et al.*, 1971). The distribution and composition of the gravels, which contain a similar suite of pebbles to later Thames terrace deposits, led

Hey (1965) to claim that the Westland Green Gravels were the earliest product of the River Thames. The further suggestion that deposition of the Westland Green Gravels corresponded with glaciation in areas to the north-west of the Thames catchment (Hey, 1965) found favour with later authors, notably Bowen *et al.* (1986a). This was based on the occurrence in these deposits of volcanic rocks thought to be derived from Wales, although these appear to be more common in the lower pre-Anglian gravels (Green *et al.*, 1980).

Evans (1971, fig. 50), in his early attempt to relate the various Thames terrace aggradations to the cold and warm cycles of the deep-sea oxygen isotope record, allocated the Westland Green Gravels to his cycle 16W and suggested an age of about 620,000 years for this formation. Evans's cycle 16W would appear to equate with Stage 31 of the current oxygen isotope chronological nomenclature (see Chapter 1). According to the latest estimates (Ruddiman *et al.*, 1989), a correlation with Stage 31 would imply an age of just less than one million years. Evans, however, based his correlations on extrapolated interglacial sea levels, which he took to have fallen progressively during the Pleistocene. He considered the major terrace formations to have aggraded during interglacials, an idea largely superseded by the modern view that they were predominantly deposited during cold episodes. His model implied a sea level of 103 m above present ordnance datum during Westland Green times. This would require that the type area in Hertfordshire, where the gravels fall below 110 m O.D., lay close to the contemporary coastline. The modern association of gravel aggradation with cold episodes renders Evans's sea-level prediction obsolete (Chapter 1), but the broad correlation between climatic cycles and terrace sequences that he envisaged, essentially one of counting backwards, compares quite closely with more recent interpretations based on the oxygen isotope chronology (see Chapter 1). Only a very crude approximation of the age of the Westland Green Gravels would be claimed from this type of correlation today, however.

Hey (1976b, 1980) went on to address the problem of correlating the Westland Green Gravels of the Middle Thames with the Pleistocene sequence in East Anglia, attempting to trace the unit to north Norfolk. He recognized (1980) that deposits of Westland Green type occur within the higher levels of the Kesgrave Group (Rose *et al.*, 1976). He noted that a marine gravel on the foreshore at Beeston Regis, Norfolk (TG 260402), ascribed to the 'Pre-Pastonian a' Stage, contains a similar range of clast types to the Westland Green Gravels. Hey (1980) suggested that the Westland Green Gravels were the terrestrial equivalent of this marine gravel. Subsequent work has shown that the Kesgrave Group deposits in Norfolk that were correlated by Hey with the pre-Pastonian marine gravel at Beeston belong to a later Thames formation than the Westland Green Gravels, implying that the latter may be even older than the 'Pre-Pastonian a' Stage (Whiteman, 1990; Chapter 1).

Gibbard (1983, 1985) concluded that the deposit at Stoke Row on the Chilterns dip slope, the furthest upstream of Hey's Westland Green outliers, probably represents an earlier aggradational phase. He assigned this outlier to the newly defined Stoke Row Gravel. Hey (1965) had noticed that this remnant was, at 174 m O.D., some 6 m above the projected long-profile (thalweg) reconstructed from the other occurrences of Westland Green Gravels. He suggested an increased gradient in the upstream part of the river system as a possible explanation. Gibbard, however, found gravel at the expected elevation of the Westland Green Formation, lower down the dip slope in this area, around Crays Pond (SU 637805). Remnants of gravel, generally similar in composition to the Westland Green Formation but at a higher level, have subsequently been identified at Bedmond (TL 104037) and Sherrards Park Wood (TL 225138). These have been assigned to the Stoke Row Formation (Moffat, 1980; Cheshire, 1986a; Moffat and Catt, 1986a). Moffat and Catt (1986a) suggested that these outliers have larger quartz : quartzite + sandstone ratios than samples of the Westland Green Gravels; their clast-lithological data supports this view, but it is difficult to determine whether the same is true of the data published by Hey (1965) and Gibbard (1985), largely because of differences between the categories and size fractions used by these various workers (Table 3.2).

Plotting the remnants of Westland Green Gravels and Stoke Row Gravel along the generalized early-Thames route from the Goring Gap to eastern Hertfordshire (Fig. 3.6) suggests that a vertical separation of 12–15 m, as demonstrated by Gibbard in the Goring Gap area, is maintained as far downstream as Essendon. However, the various outliers in the Westland Green type area fall somewhat randomly between the projected Westland Green and Stoke Row levels, so that all are higher than would be expected for the former and lower than for the latter. In particular, the gravel at Furneux Pelham, with a surface height of 107 m O.D., plots significantly higher than the other remnants, given its position at the downstream end of Fig. 3.6; it is only a few metres below the projected Stoke Row level (based on the two sites at Bedmond and Sherrards Park Wood). It therefore seems likely that the fluvial deposit at the Furneux Pelham GCR site does not represent the Westland Green Gravels, as was asserted by Hey (1983), but is instead a degraded remnant of the Stoke Row Formation.

There is some support for this interpretation from clast-lithological data from the Furneux Pelham site. The ratios of quartz to quartzite + sandstone at Furneux Pelham, indicated by the available stone-count data, are amongst the highest recorded from gravels of Westland Green type in Hertfordshire (Table 3.2). Although the suggestion that higher ratios of this type might characterize the Stoke Row Gravel (Moffat and Catt, 1986a) has yet to be confirmed, the figures from Furneux Pelham appear to support the altitudinal evidence for correlation with that formation.

Further work is required to clarify the distribution of the earliest Thames gravels in and downstream from eastern Hertfordshire, in order to determine whether the division into Stoke Row and Westland Green Formations can be continued north-eastwards. There is some indication that the long-profiles of these formations converge downstream, as a result of a shallowing of the Westland Green gradient, a steepening of the Stoke Row gradient, or perhaps both (Fig. 3.6). Moffat and Catt's (1986a) observation that the Stoke Row Gravel may have larger quartz : quartzite + sandstone ratios than are typical of the Westland Green Gravels hints at a possible method, independent of altitude, for distinguishing the two aggradations.

The GCR sites at Chalfont St Giles and Furneux Pelham thus represent an extremely important period in the history of the River Thames, close to the time when the river was first initiated as a drainage route from the Midlands into the London Basin. Although both were originally ascribed to the Westland Green Gravels, reappraisal of the deposits in eastern Hertfordshire suggests that the Furneux Pelham site exposes the earlier Stoke Row Gravel. The latter appears at present to be the earliest Thames deposit containing abundant material from the Midlands, indicating that a link with the present Severn catchment may have existed across the Cotswolds (see above; Chapters 1 and 2). The two GCR sites provide rare opportunities for studying exposures in these formations and as such are of significant value to British Pleistocene stratigraphy.

Conclusions

The gravels exposed at these sites provide evidence for the early history of the River Thames. Their height, distribution and gravel content have all been used as evidence in tracing the river's early course. These sites show that the Thames once flowed along a more northerly course across Hertfordshire and eventually into East Anglia. Although both sites were thought to show gravels of the same type and age, differences in their topographical position and stone content suggest that deposits of somewhat different ages may be present at each site. The oldest of these (the Stoke Row Gravel), at Furneux Pelham, is possibly the earliest gravel to contain material carried by the early Thames from beyond the Cotswolds escarpment. Both gravels were deposited at a time when the river had a much larger catchment than at present, probably extending into the West Midlands and possibly as far as Wales.

WESTWOOD QUARRY (TQ 071993)

D.R. Bridgland

Highlights

This is an important locality for the study of terrace stratigraphy in the Middle Thames basin. The gravels at Westwood Quarry yield possible evidence, in the form of volcanic rocks transported from North Wales, for glaciation in the Early Pleistocene.

Introduction

Westwood Quarry exposes sand and gravel deposited by the Thames when it flowed north of London, through the Vale of St Albans and central Essex, towards East Anglia. These deposits belong to the Gerrards Cross Gravel (Gibbard, 1983, 1985), formerly known as the Lower Gravel Train (Wooldridge, 1938). The upper surface of this formation, where present, forms the Harefield Terrace of Hare (1947). The sections at Westwood Quarry reveal bedded gravel and sand overlain by probable soliflucted gravel (Green and McGregor, 1978c; McGregor and Green, 1983a, 1983b; Gibbard, 1985). The site is notable for the occurrence, in the upper levels of the bedded gravel, of a clay-enriched and iron-stained horizon of possible pedogenic origin.

Description

The sequence at Westwood Quarry can be summarized as follows (based on published reports and on the re-excavated GCR section):

		Thickness
3.	Coarse, very clayey gravel, heavily iron-stained	up to 4 m
2a.	Sandy gravel, clay-enriched and mottled (soil developed in top of bed 2)	
2.	Sandy, clayey, bedded gravel (Gerrards Cross Gravel)	up to 5 m
1.	Chalk, forming pinnacles	

The first description of sections at Westwood Quarry was by Green and McGregor (1978c), who recorded well-bedded sands and gravels (bed 2) overlying 'pinnacled Chalk' and in turn overlain by compact clayey gravel (bed 3). The clayey gravel, which contains a higher proportion of local material than the well-bedded gravel (Table 3.2), was interpreted by Green and McGregor (1978c; McGregor and Green, 1983a, 1983b) as a colluvial deposit. McGregor and Green (1983a) recorded a clast-fabric orientation from the upper gravel suggesting derivation by solifluction from the east, although no high ground now exists in that direction. They also presented palaeocurrent evidence (clast fabrics and foreset orientations) from the fluvial gravel (bed 2) compatible with emplacement by an eastward-flowing river. Surface elevations of 84 m and 88 m were recorded by Green and McGregor (1978a) and Gibbard (1985) respectively.

Similar deposits at Westwood Quarry were recorded by Gibbard (1985), who additionally referred to lenses of brown pebbly clay interbedded with the gravels. He attributed these poorly sorted lenses to mass-movement of material from the valley side. In all these descriptions the effects of solution of the underlying Chalk were noticed; this process has given rise to a pinnacled bedrock surface, to faulting in the overlying gravel and to the formation of pipes and associated collapse structures filled with later, usually fine-grained, sediments. The reddening, mottling and clay enrichment of the upper levels of the lower fluvial gravels (bed 2a) is interpreted as evidence for pedogenesis prior to the accumulation of the upper soliflucted gravel (P.L. Gibbard and J.A. Catt, pers. comms).

Interpretation

A principal part of the geological interest at Westwood Quarry arises from the interpretation of the gravels there as early Thames deposits, laid down at a time when the river flowed through the Vale of St Albans. The history of research leading to this interpretation is long and complex. It was formerly thought that high-level sand and gravel of this type was of glacial origin; early Geological Survey maps show the deposits of this area as 'Glacial Gravel' (Old Series, Sheet 7, 1871; New Series, Sheet 225, 1922; see above, Introduction to Part 2). However, some of the early authors recognized that these high-level deposits are disposed in terrace-like remnants similar to the well-established terraces of the valley gravels (White, 1895, 1897, 1907; Shrubsole, 1898).

It was also observed at an early stage (although not widely accepted until much later) that the far-travelled element of these gravels, mainly quartz, quartzite and chert pebbles from the West Midlands and the Welsh borderlands, are types that have been transported into the London Basin not by ice, but by an ancestral Thames flowing through the Goring Gap (White, 1895). The extension of these ancient gravel deposits from the Middle Thames valley into the Vale of St Albans, rather than along the modern valley through London, was recognized by Salter (1905).

Wooldridge (1938) provided the first systematic subdivision of the belt of old Thames drifts that extends from the Goring Gap to Ware, where it passes beneath true glacial deposits. He recognized three new terrace aggradations at higher levels than those mapped by the Geological Survey, although the two highest are so dissected that he declined to refer to them as terraces. Instead he called them the 'Higher Gravel Train' and the 'Lower Gravel Train'; these occur at 122 m and 113 m O.D., respectively, in the classic Middle Thames area north of Slough. Hare (1947) showed that the Lower Gravel Train is sufficiently well-preserved in the Beaconsfield district to be recognized as a terrace, which he called the 'Harefield Terrace'.

Wooldridge (1938, 1957, 1960; Wooldridge and Linton, 1939, 1955) did not believe that the Thames flowed through the Vale of St Albans after Pebble Gravel times. He suggested that the Lower Gravel Train river flowed in a course via the 'Middlesex Loopway' and 'Finchley Depression' (Fig. 3.4), rejoining the earlier northern route near Hertford. Hare (1947) traced this formation (in the form of his Harefield Terrace) to the Harefield area, where he supposed that it entered the 'Middlesex Loopway'. Wooldridge (1957, 1960) perpetuated this view, but no description of Thames gravels in either the 'Middlesex Loopway' or the 'Finchley Depression' was ever provided. Wooldridge (1938) applied the term 'Leavesden Gravel Train' to the deposits now exposed in Westwood Quarry, attributing them to a westward-flowing tributary of the Higher Gravel Train Thames, which he also believed to have flowed through the 'Middlesex Loopway' and the 'Finchley Depression'.

It was later demonstrated, from analyses of the distribution and composition of gravel deposits, that Thames drainage through the Vale of St Albans persisted until the river was diverted by ice during the Anglian Stage (Gibbard, 1974, 1977, 1978a, 1983; Green and McGregor, 1978a, 1978b; McGregor and Green, 1978). Gibbard (1979) also concluded that the 'Finchley Depression' was continuously occupied by the Mole-Wey tributary prior to the diversion of the Thames and that the latter river shifted directly from the Vale of St Albans route into its modern course. Gibbard (1974, 1977, 1978a) recognized that the maximum elevation of the fluvial part of the gravel at Westwood Quarry, and its clast composition, indicate that it is part of the Lower Gravel Train, a view supported by Green and McGregor (1978a, 1978b, 1978c; McGregor and Green, 1978). Gibbard proposed the name 'Leavesden Green Gravel' for this aggradation in the Vale of St Albans, but later (1983, 1985) dropped the term in favour of the name 'Gerrards Cross Gravel', which he applied to all the deposits formerly assigned to the Lower Gravel Train.

According to Gibbard, the Gerrards Cross Gravel represents the penultimate phase of terrace aggradation by the Thames in its Vale of St Albans route, the last being the Winter Hill/ Westmill Gravel (Gibbard, 1977, 1978a, 1983, 1985). The deposition of the latter immediately preceded the glaciation of the Vale of St Albans and the resultant diversion of the Thames (see below, Moor Mill and Westmill). Remnants in the Reading–Henley area of an intermediate Rassler Gravel Formation, previously recognized by Sealy and Sealy (1956), are here attributed to aggradation by the Thames in the interval between the deposition of the Gerrards Cross and Winter Hill Formations (see Chapter 1; Fig. 3.2).

The clast composition of the Gerrards Cross Gravel shows it to be the richest of all the Thames terrace formations in exotic material from beyond the present catchment (McGregor and Green, 1978, 1983b, 1983c, 1986; Green *et al.*, 1980; Gibbard, 1985; Table 3.2), particularly the familiar volcanic (including pyroclastic) pebbles that are believed to come from North Wales (Hey and Brenchley, 1977; Green *et al.*, 1980; Whiteman, 1983). McGregor and Green (1983b) noted that these igneous clasts are concentrated in certain areas, perhaps suggesting the localized fragmentation of larger ice-rafted blocks, and that the proportions of the various types differ from those in the Kesgrave Sands and Gravels of Essex and Suffolk, where they have also attracted attention. Despite his agreement that these exotic lithologies 'reach their acme' in the Gerrards Cross Gravel, Gibbard (1985, p. 16) presented data that shows twice as much igneous material (8–32 mm size range) in the Beaconsfield Gravel (average 0.46%) as in the Gerrards Cross Gravel (average 0.23%). Gibbard did, however, record high levels of igneous rocks at Westwood Quarry (average 0.44%). Data from 11.2–16 mm clast counts published by Green and McGregor (1978a, 1983), Green *et al.* (1980) and McGregor and Green (1986) clearly indicate that the highest frequencies of volcanic clasts occur in the Gerrards Cross Gravel; the record of 3.16% volcanics in a sample from the M1 motorway (McGregor and Green, 1986) appears to be the highest value from any of the various Thames formations.

The occurrence, in the Gerrards Cross Gravel, of the highest frequencies of igneous material within the Thames terrace succession has led to the suggestion that deposition of this formation coincided with a glacial advance from Wales and the north-west into the upper reaches of the catchment (Green and McGregor, 1978a). Since the Gerrards Cross Gravel is older than the Winter Hill Gravel, which immediately pre-dates the Lowestoft Till, this Welsh glaciation must have been earlier than any that occurred during the Anglian Stage. After comparison with the influx of far-travelled material into East Anglia in the Early Pleistocene (Hey, 1976b), Green and McGregor (1978a) suggested that a glaciation of Baventian age, or of post-Baventian/pre-Anglian age, had introduced exotic material into the Thames catchment during the deposition of the Gerrards Cross Gravel and, probably, on earlier occasions. Bowen *et al.* (1986a) interpreted the occurrence of such materials in the various early Thames formations as evidence for repeated glaciation in upland Britain. However, Gibbard (1983, 1985) favoured an early Anglian age for the Gerrards Cross Gravel and any corresponding glaciation of the upper reaches of the catchment. He reported that the study of sand-grain surfaces by scanning electron microscope had revealed little evidence for a proximal glacial source of the Gerrards Cross Gravel.

The recognition in this volume of an additional terrace formation in the Middle Thames between the Gerrards Cross and Winter Hill Gravels (above; see Chapter 1) makes an early Anglian age for the Gerrards Cross Formation unlikely. The conclusions of Whiteman (1990), and their implications for correlation between the Thames sequence and that in East Anglia (see Chapter 1), indicate that the Gerrards Cross Gravel is very much older than the Winter Hill/Westmill Formation. Early attempts to correlate the Gerrards Cross Gravel with parts of the Kesgrave Sands and Gravels (Green *et al.*, 1982; Green and McGregor, 1983; McGregor and Green, 1986; Bowen *et al.*, 1986a) were based largely on Hey's (1980) projection of the Westland Green Gravels into East Anglia, although Bridgland (1988a) also used deposits related to the Anglian glaciation and its diversion of the Thames within Essex as a stratigraphical marker. Whiteman's work suggests that the Gerrards Cross Gravel is the direct correlative of the deposits in Suffolk and Norfolk identified by Hey (1980) and Allen (1983, 1984) as Westland Green Gravels. The implication of this is that the sequence of lower pre-diversion gravels (Low-level Kesgrave Subgroup – see Chapter 5) recognized in southern East Anglia is poorly represented in the Middle Thames valley, there being a major gap in the terrace succession in the latter area between the Gerrards Cross and Winter Hill Formations. The suggested correlation of the early Thames gravels in Suffolk and Norfolk with the East Anglian sequence (Hey, 1980; see above and Chapter 1) points to an Early Pleistocene age for the Gerrards Cross Gravel, implying that deposition of the Westland Green and Winter Hill Gravels was separated by a considerable period of time, including the entire early Middle Pleistocene. The poorly preserved Rassler Formation, recognized in the Reading area (Chapter 1), appears to be the only deposit in the Middle Thames that represents part of this interval.

Whiteman's views represent a considerable challenge to workers in the Middle Thames valley, which has long been held to be the classic area for studies of the river's development. If his correlation scheme is correct, the pre-Anglian sequence in East Anglia is considerably more complete than that in the Middle Thames. Whiteman has interpreted the Gerrards Cross Gravel as the last aggradation by the 'Severn-Thames', subsequent formations reflecting a river, much reduced in size, that had lost its headwaters beyond the Cotswolds escarpment. The coincidence of this formation with the peak in exotic gravel lithologies suggests that a fairly major Early Pleistocene glaciation may have occurred at this time. This raises the possibility that this glaciation may have influenced fluvial evolution in the area beyond the Cotswolds and may have been responsible for diverting the drainage of this area away from the Thames, probably into the Trent system (which was represented at that time by a river flowing from southern Lincolnshire into northern East Anglia, the 'Ingham River' (Rose, 1987, 1989; Bridgland and Lewis, 1991)).

Little is known about the palaeosol that occurs in the upper levels of the Gerrards Cross Gravel at Westwood Quarry. This may be of similar origin and age to the Valley Farm Soil, which is developed on the Kesgrave Sands and Gravels in East Anglia (Rose *et al.*, 1976; Rose and Allen, 1977; Kemp, 1985a; see Chapter 5, Newney Green), and may therefore provide further evidence for correlating the Gerrards Cross Formation with divisions of the Kesgrave Group. Apart from the widely recognized Valley Farm Soil, the occurrence of a buried palaeosol is an extreme rarity within the Thames terrace sequence. Study of the soil horizon at Westwood Quarry may produce corroborative evidence for the great age of the Gerrards Cross Gravel implied by Whiteman's interpretation of Lower/lower Middle Pleistocene Thames stratigraphy.

Conclusion

The Gerrards Cross Gravel, exposed at Westwood Quarry, charts the course of the early River Thames, which once flowed through the Vale of St Albans towards eastern Hertfordshire and East Anglia. This river brought with it exotic rocks from as far afield as the West Midlands and North Wales. Such evidence has been used to suggest that, at this time (probably over three-quarters of a million years ago), the Thames drained a much larger area than at present. Some workers believe that glaciation in the upland area of North Wales brought some of the more exotic rock-types into the Thames catchment. If so, this is a rare indication of glaciation prior to the growth of the well-known Anglian ice sheets, about 450,000 years ago.

WESTMILL QUARRY (TL 344158)

D.R. Bridgland and D.A. Cheshire

Highlights

Exposures at this site have provided evidence for determining the Anglian history of the River Thames and its tributaries. They reveal a complex sequence of gravels and tills that records the replacement of the Thames in this area by a newly formed River Lea. This was a direct response to glacial inundation of the former Thames valley, which eventually deflected that river into its modern course through London.

Introduction

One of the thickest Pleistocene sequences in the Thames basin can be seen at Westmill Quarry, Hertfordshire, where over 20 m of sediments, all attributed to the Anglian Stage, are exposed. These deposits represent the drift sequence, made up of fluvial, glacial, glaciofluvial and glaciolacustrine sediments, that plugs the old valley of the Thames. This old valley separates the Chilterns and the hills of North London, occupying an area known as the Vale of St Albans. The fact that the Vale of St Albans was once glaciated was recognized by Walker (1871) and Whitaker (1889). Salter (1905) and Sherlock and Noble (1912) suggested that this glaciation brought about the cessation of north-eastward drainage, a view that was subsequently confirmed by the recognition of proglacial lake deposits in various parts of the Vale of St Albans (Sherlock, 1924; Sherlock and Pocock, 1924; Clayton and Brown, 1958; Gibbard, 1974, 1977; see below, Moor Mill).

The sequence of events during the Anglian glaciation of the Vale of St Albans was reconstructed by Gibbard (1974, 1977, 1978a), the Westmill Quarry section providing critical evidence for his interpretation. Gibbard recognized two glacial advances into the Vale of St Albans during the Anglian. He considered the Thames drainage to have survived the first of these advances, despite local ponding in the Ware area, but that the river was blocked and diverted by the second. Subsequently Cheshire (1981, 1983a, 1983b, 1986a), in a detailed study of till lithologies and stratigraphy, found evidence for four separate Anglian ice advances into the area. According to Cheshire, it was the first of these that effected the diversion of the Thames.

Description

Westmill Quarry is a modern excavation, first described by Gibbard (1974, 1977, 1978a, 1978b), who recorded a sequence of two gravels and two tills. A further till was subsequently recognized by Cheshire (1983a, 1983b, 1986a), who also suggested that the upper till at Westmill, the Eastend Green Till of Gibbard, should be renamed the 'Westmill Till', on the grounds that it is not stratigraphically equivalent to the tills exposed at Gibbard's type site at Eastend Green (see below). The sequence currently recognized at Westmill is therefore as follows (see Fig. 3.9):

			Thickness
5.	Chalky till, blue-grey to brown	(Westmill Till)	up to 4 m
	3b. Chalk-rich gravel	(Ugley Gravel)	0–3.5 m
4.	Very chalky till, pale brown (N.B. occurs as lenses within or below 3a)	(Stortford Till)	
	3a. Chalk-poor gravel	(Hoddesdon Gravel)	
3.	Sand and gravel, cross-bedded	(Westmill Upper Gravel)	7–11 m
2.	Chalky till, dark grey	(Ware Till)	0–3 m
1.	Sand and gravel, cross-bedded	(Westmill Lower Gravel)	up to 9 m

Chalk surface, 48–54 m O.D., rising towards the north-west.

Two separate fluviatile formations are represented in this sequence, which also includes glacigenic members. Gibbard (1974, 1977) regarded the fluvial deposits here as equivalent to the Winter Hill Formation of the Middle Thames, of which they appear to be a downstream continuation. However, further stratigraphical studies have shown that only the Westmill Lower Gravel is actually a continuation of the Winter Hill Formation; the Westmill Upper Gravel can be mapped in parts of the Lea catchment not formerly occupied by the Thames as a distinct body of sediment, probably equivalent to the Black Park Formation of the main river (Cheshire, 1986a; see below).

The Westmill Lower Gravel (unit 1) is generally more massive in its lower part, usually with a very coarse flinty 'lag' at the base, resting on an irregular Chalk surface that has suffered scouring and/or solution. Towards the top of the unit the proportion of cross-bedded sand and calcareous silt (the latter interpreted as channel-fill material) increases. Both tabular and trough cross-bedding occur and fining-upwards sequences are common. Imbrication in the gravel and foreset orientation in the sand reveal that palaeocurrent flow was towards the north-east. The gravel contains a mixture of flint (over 80%) and quartz and quartzite (accounting for most of the remainder), plus subordinate southern and far-travelled (exotic) material. This is a composition suggestive of a Thames origin (Table 3.2), an interpretation that is supported by the palaeocurrent evidence (Gibbard, 1974, 1977; Cheshire, 1983b, 1986a).

The Ware Till (unit 2) rests upon the eroded surface of the Westmill Lower Gravel, which varies from 55 m to 60.2 m above O.D. Occasionally present at the base of the till is a yellowish-brown laminated silty clay about 0.15 m thick, the laminations suggesting deposition in water; elsewhere the same till overlies the deposits of a local proglacial lake (Gibbard, 1974, 1977). The upper surface of the till is erosional, the deposit having been cut out completely by later channelling in some areas. Because its matrix consists largely of unoxidized clays (with occasional pyrite) derived from the Jurassic, the Ware Till is dark greyish-brown in colour and has a lower calcareous content than the other two tills in the Westmill area. Its particle-size distribution is remarkably uniform throughout, possessing a large and characteristic peak in the fine-sand fraction and a low proportion of pebble clasts in comparison with the later tills. Although the small-clast composition is more variable than the particle-size distribution, the till is characteristically richer in quartz and poorer in flint and *Rhaxella* chert than later tills in the Vale of St Albans sequence. The Ware Till in this area is brecciated; fabric studies show no preferred clast-orientation, indicating emplacement under low-stress glacial conditions or subsequent multi- or non-directional disturbance. *Rhaxella* chert is a rock made up largely of sponge spicules of very characteristic shape and size, occurring only in the Oxfordian and Portlandian (Middle Jurassic).

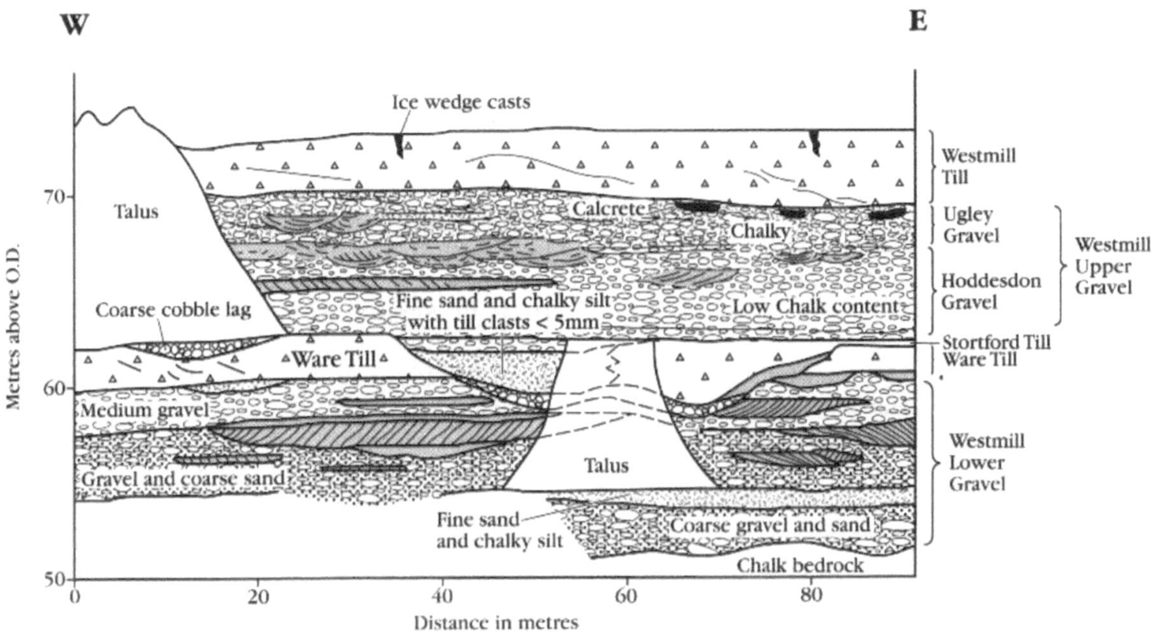

Figure 3.9 Section in the north face of Westmill Quarry, recorded in June 1981 (after Cheshire, 1983b).

It is believed to have been carried in quantity into the London Basin by Anglian ice, from sources in north Yorkshire (Bridgland, 1986a).

The Westmill Upper Gravel (unit 3) is variable in lithology and lateral extent. It contains lenses and larger bodies of cross-stratified sand and finer gravel, but is predominantly massive. The upper part of the deposit is also sandy and shows trough cross-stratification. There is a change in composition within these gravels that coincides with a major change in palaeocurrent direction. Immediately above the Ware Till the gravel contains relatively abundant Chalk (13%) and soft 'erratics'. Higher in the sequence these become less frequent, the Chalk component falling to 2–4%. Palaeocurrent measurements from these levels indicate flow towards the north-east. In the upper 3–4 m of the gravel, the Chalk content dramatically increases to 22%, again associated with an increase in soft far-travelled material, particularly from the Mesozoic of central and eastern England. Palaeocurrent measurements from this horizon show local flow to the south-west. Gibbard attributed this change in composition and palaeocurrent directions, which occurs at around 68 m O.D. in the Westmill sequence, to the initiation of a new stream flowing from the north-east. He considered this stream to represent outwash from an approaching ice sheet, that which deposited the overlying (Westmill) Till. Cheshire (1986a) has correlated this higher division of the Westmill Upper Gravel with his Ugley Gravel, as defined at Ugley Park Quarry (see below). The lower division of the Westmill Upper Gravel has been referred by Cheshire to his Hoddesdon Gravel, which contains less Chalk than the Ugley Gravel (Table 3.2); it contains no Chalk at all in the Lower Lea valley south of the Hoddesdon type locality (TL 354077).

The Stortford Till (unit 4) occurs as detached lenses or channel-fills above, within or cutting through the Ware Till (Fig. 3.9). In places a coarse gravel up to 0.8 m thick intervenes between the two tills (basal Hoddesdon Gravel). Although the occurrence of the Stortford Till is localized, at its maximum thickness of 3.5 m this deposit exceeds the laterally more persistent Ware Till. It may be easily distinguished from the latter in the field by its very much greater Chalk content and lighter colour. Unlike the Ware Till, the Stortford Till does not have a vertically uniform particle-size distribution. At its type locality (Stortford Green borehole, TL 479195; Cheshire, 1986a), the Stortford Till shows a progressive decline in the frequency of medium-fine sand from its base to its top. Only parts of this transition are seen in the Westmill area, some lenses appearing to have incorporated material from the subjacent Ware Till. The flint and *Rhaxella* chert (small-clast) content is greater than in the Ware Till, but the proportion of quartz is reduced (Cheshire, 1986a). Fabric studies indicate a regional ice-flow direction from the north-east.

The Westmill Till (unit 5) superficially resembles the Stortford Till, being light yellowish-brown and having a noticeably high Chalk content. It is the thickest of the three tills, increasing to more than 4 m where it overlies hollows in the surface of the Westmill Upper Gravel. The till crops out extensively in the Westmill area but has been eroded locally from the interfluve to the north and east. The particle-size distribution is vertically uniform and shows a low medium-sand content compared with the Ware and Stortford Tills (with well-defined modes in the 125 and 63 micron fractions). The proportion in the pebble fraction is also greater. The small-clast composition is remarkably uniform, being characterized by relatively high proportions of flint and *Rhaxella* chert, whereas there is less quartz than in other tills. Shear planes are present, generally dipping towards the north-east, indicating ice movement from that direction. Fabric studies show a very strong NE–SW preferred orientation (Gibbard, 1974; Cheshire, 1986a), confirming ice movement from the north-east.

Interpretation

Westmill Quarry, perhaps because of its considerable size and depth, has proved to be a most valuable site for examining the thick sequence of Anglian Stage deposits that fills the old Thames valley through the Vale of St Albans. Although this sequence has been dissected in post-Anglian times by streams of the Colne and Lea systems, a large proportion of the Anglian sediments remains intact. Detailed studies of the area in recent years (Gibbard, 1974, 1977, 1978a; Cheshire, 1981, 1983a, 1983b, 1986a) have increased knowledge of the characteristics and three-dimensional form of these deposits. The importance of the Westmill site in such studies is underlined by the fact that it is the type locality of the Westmill Lower and Upper Gravels, the Ware Till and the Westmill Till.

Gibbard (1974, 1977, 1978a, 1983, 1985) considered the Westmill Lower and Westmill Upper Gravels to be part of a single aggradation by the Thames, which he correlated with the Winter Hill Gravel of the main river. An ostracod fauna (Robinson, 1978, 1983), obtained from a bed of silt, 0.3–0.4 m thick, within the Westmill Lower Gravel, indicates that cold-climate conditions prevailed during the aggradation of this member, immediately prior to the arrival of Anglian ice in the area. Robinson pointed to the species *Paralimnocythere compressa* (Brady and Norman) as probably the most significant of seven taxa recognized. This species is believed to indicate a fluviatile setting within a cold steppe or tundra environment, as well as a broadly Middle Pleistocene age.

Gibbard did not recognize the Ware Till in the western part of the Vale of St Albans; he correlated the glacial deposits in that area with the upper till (Westmill Till) at Westmill, assigning both to his 'Eastend Green Till' (see below, Moor Mill). Gravel underlying the Ware Till in the latter area was termed Westmill Gravel by Gibbard, who considered it to be the lateral equivalent of both gravel members at Westmill Quarry. Gibbard's correlation of the Westmill Gravel with the Winter Hill Gravel of the Middle Thames, on the basis of elevation and composition, provided an important link between the Thames terrace sequence and the glacial stratigraphy of eastern England. Cheshire's (1981, 1983a, 1983b, 1986a) subsequent revision of this stratigraphical scheme, in which the Westmill Upper Gravel was recognized as a post-diversion River Lea deposit, demonstrated that only the Westmill Lower Gravel (Fig. 3.10A) is a correlative of the Winter Hill Gravel. The stratigraphical link established by Gibbard was upheld by Cheshire, however; on account of its close association with the glaciation of the Vale

Figure 3.10 Palaeodrainage during key phases of the Anglian evolution of the Vale of St Albans (from Cheshire, 1986a):

(A) During deposition of the Westmill Lower Gravel;
(B) During the existence of the Watton Road lake;
(C) During the existence of the Moor Mill lake;
(D) At the maximum extent of the Ware Till ice;
(E) At the maximum extent of the Stortford Till ice;
(F) At the maximum extent of the Ugley Till ice;
(G) During the deposition of the Westmill Upper Gravel and the Smug Oak Gravel;
(H) During the Westmill Till ice advance.

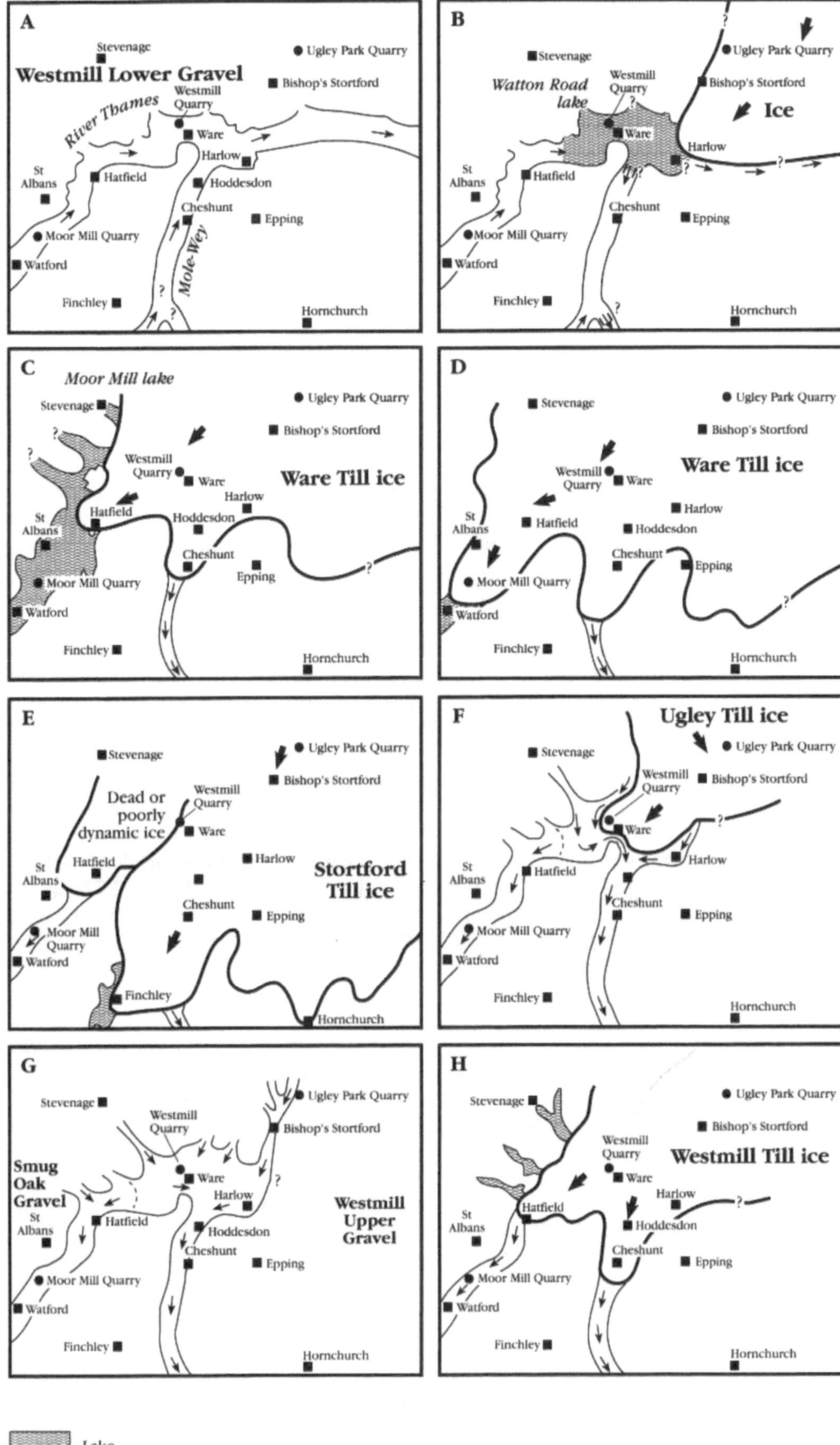

Lake

Fluvial/fluvioglacial flow direction

Interfluve

Ice flow direction

0 kilometres 10

N

of St Albans, the Winter Hill/Westmill Lower Gravel can be confidently dated to the Anglian glacial maximum, providing a clear chronostratigraphical marker within the Thames succession.

Gibbard recognized that the Ware Till ice advance had caused ponding of the Thames at the north-eastern end of the Vale of St Albans. Evidence for this ponding event is provided by laminated silts underlying the Ware Till in the Hertford area (Gibbard, 1974, 1977, 1978a). These 'Watton Road Laminated Silts' (type locality: Watton Road Quarry, TL 431149) were formed in a proglacial lake at the edge of the Ware Till ice, which later overrode the lake beds (Fig. 3.10B). Gibbard recorded a minimum of 485 varve-like couplets within this unit; if this rhythmic sequence was seasonally controlled, this would imply that the lake existed for at least 485 years. Gibbard considered that Thames drainage via the Vale of St Albans and the 'Mid-Essex Depression' (see Chapter 5) survived the Ware Till glacial advance and that the river was ponded again, at the western end of the vale, by the later advance that led to the deposition of his 'Eastend Green Till'.

Cheshire (1981) suggested that the Ware Till ice advance and the resultant formation of the Watton Road lake had caused an initial southward diversion of the Thames via the Lower Lea, which is essentially a reversed section of the early Mole-Wey-Wandle valley. This interpretation was based on the recognition of the continuation of the Westmill Upper Gravel at Hoddesdon, which is in the Lea valley and well to the south of the reconstructed Westmill Lower Gravel course of the Thames. Thames drainage via the Lower Lea had previously been suggested by Sherlock and Noble (1912), Hawkins (1922), Baker and Jones (1980) and Jones (1981), the last two of these on the grounds that the 'Mid-Essex Depression' was probably blocked by ice in Westmill Upper Gravel times. However, Gibbard (1983) suggested that the Thames at that time had been forced southwards to Hoddesdon around an ice lobe (that from which the Ware Till was deposited), but that it had curved north-eastwards to regain its original course, which remained ice-free at this time. However, the discovery of Westmill Upper Gravel at Bullscross Farm, Waltham Cross (TL 340007), demonstrated that this unit (which must now be regarded as having formation status) continues southwards down the Lea valley (Cheshire, 1983c, 1986a), contrary to Gibbard's suggestion. An excavation at Bullscross Farm by the GCR Unit (Anon., 1982a) revealed Mole-Wey gravel, rich in Greensand chert and with evidence of northward palaeocurrents, cut out by the edge of a channel containing the Westmill Upper Gravel, which has bedding structures indicative of southward palaeocurrents. These two gravels are separated by a thin 'diamicton' with a particle-size distribution almost identical to the Ware Till of the Vale of St Albans. Confirmation of this correlation on the basis of composition is precluded by the non-calcareous (?decalcified) condition of the 'diamicton' (Cheshire, 1986a). Macrofabric evidence shows this material to have a highly significant monopolar preferred orientation, which suggests that it represents remobilized (possibly soliflucted) Ware Till. The evidence from this site shows that the Westmill Upper Gravel stream flowed southwards down the Lower Lea valley after the Ware Till ice had extended almost as far south as the present location of the M25 motorway.

Later work by Cheshire (1983a, 1986a) showed that the Ware Till extends westwards beyond the area of the Watton Road lake and can be recognized at the western end of the Vale of St Albans, where it is equivalent to the deposit classified by Gibbard (1974, 1977) as 'Eastend Green Till'. According to Cheshire, it was the Ware Till ice that caused both the formation of a lake in the Watford area (the Moor Mill lake of Gibbard, 1974, 1977) and the ultimate diversion of the Thames into its modern course (Fig. 3.10C; see below, Moor Mill). The Westmill Upper Gravel is therefore not a Thames deposit, as Gibbard suggested. It is confined to the Lea catchment and was regarded by Cheshire (1986a) as the earliest Lea aggradation, albeit largely fed by outwash from Anglian ice. Cheshire recognized, from analyses of clast-lithological composition at sites throughout the catchment, lower (Chalk-poor) and upper (Chalk-rich) divisions of this formation, his Hoddesdon and Ugley Gravels (see below, Ugley).

The attribution of the complete sequence of tills and gravels at Westmill to the Anglian Stage was established by Gibbard (1974, 1977), who pointed to the occurrence of supposed kettle-hole infills in the surface of the uppermost till (Westmill Till; unit 5), including pollen-bearing sediments ascribed to the Hoxnian Stage. Organic deposits of this type have been found at Hatfield (first described by Sparks *et al.* (1969)) and Colney Heath (Gibbard, 1977, 1978a, 1978c). Successive pollen spectra obtained from these deposits reveal the change from cold conditions at the end of the Anglian glaciation to the ameliorating climate of the early Hoxnian Stage. The Hatfield sequence continues into the latter half of the Hoxnian (biozone HoIIIb) (Gibbard and Cheshire, 1983).

Recent reappraisal of the glacial stratigraphy in the Vale of St Albans by Cheshire (1983a, 1983b, 1986a) has raised doubts about the relations of the Hoxnian deposits to the till sequence. There is also a possibility that the till directly underlying the Hoxnian organic deposits at Hatfield is not *in situ*. It has been variously described as a slumped deposit, possibly deformed by the melting of buried ice (Rose, 1974), as flow till (Gibbard, 1978c) or as soliflucted till that has been derived from higher ground immediately to the east (Cheshire, 1986a). Below this disturbed unit is an *in situ* till that can be traced through a large number of quarries and boreholes south-westwards to Moor Mill and north-eastwards to Westmill; in both directions it appears to be continuous with the Ware Till (Cheshire, 1986a). The Ware Till has characteristic properties of particle size, carbonate and small-clast content that have allowed Cheshire (1986a) to recognize it throughout the Vale of St Albans and beyond. His work shows that only this, the earliest of the Vale of St Albans tills, is firmly dated as pre-Hoxnian by overlying polleniferous deposits. However, the various tills in the Vale of St Albans sequence are separated only by cold-climate glaciofluvial gravels, so there is no reason to suspect that any temperate episode(s) occurred between the deposition of any of them.

The Hatfield site has also yielded Mollusca, of species indicative of a temperate fluvial environment (Sparks *et al.*, 1969), examples of which have recently been subjected to amino acid analyses (Bowen *et al.*, 1989). Bowen *et al.* obtained D : L ratios from shells of *Lymnaea*, *Gyraulus* and *Valvata* (0.22 ± 0.02, 0.246 ± 0.002 and 0.247 ± 0.02 respectively) that are within the range of specimens believed to date from Oxygen Isotope Stage 9. In comparison with results from other sites attributed to the Hoxnian Stage, the ratios from Hatfield are similar to those from Hoxne itself, but significantly lower than those from Swanscombe and Clacton. The comparison with Swanscombe is important, since sediments there are also believed to have been deposited soon after the diversion of the Thames. Of the three genera from Hatfield that were analysed, the only one represented in the Swanscombe data is *Valvata*, from which Bowen *et al.* obtained a ratio of 0.30. This is significantly higher than the *Valvata* ratio from Hatfield (0.247) and was attributed to Oxygen Isotope Stage 11 by Bowen *et al.* These ratios suggest that the Hatfield organic sediments are significantly younger than the glacial deposits of the Vale of St Albans, since the Anglian glaciation is widely believed to have occurred during Stage 12 (see Chapter 1).

Cheshire (1986a) regarded the Westmill Upper Gravel as penecontemporaneous with the Smug Oak Gravel, an outwash-charged early Colne deposit that overlies the Ware Till in the western part of the Vale of St Albans (Fig. 3.10G; see below, Moor Mill). This is highly significant, since Gibbard (1977, 1978a) correlated the Smug Oak Gravel with the Black Park Gravel of the Thames, the first formation that can be traced into the modern valley through London. If Cheshire's interpretation is correct, the River Lea, during Westmill Upper Gravel times, was also feeding the Black Park Gravel Thames with outwash-derived material from the Anglian ice sheets that persisted in eastern Hertfordshire. This interpretation, which is supported by projections of the long-profiles of the Westmill Upper Gravel and the Black Park Gravel into the Lower Thames (see Fig. 4.7), further strengthens Gibbard's (1977, 1979, 1985) attribution of the Black Park Gravel to the Anglian Stage.

Cheshire (1978, 1981, 1983a, 1983b, 1986a, 1986b) has identified two tills in the Vale of St Albans that are additional to the sequence established by Gibbard (1974, 1977). The first of these additional tills was initially described in the Hertford area (Cheshire, 1978, 1981), where it is preserved only as thin lenticles within the Westmill Upper Gravel at Foxholes Quarry, Hertford (TL 340123). The stratigraphical position of this till is represented elsewhere in the area (including Westmill Quarry) by the change from low-Chalk (Hoddesdon Gravel) to Chalk-rich (Ugley Gravel) clast composition (Table 3.2). Cheshire originally called this unit the 'Foxholes Till', but the discovery that a more substantial remnant is preserved at Ugley (see below, Ugley Park Quarry) led him to adopt the name Ugley Till. The other additional till, the Stortford Till, is the middle of the three tills recognized at Westmill (unit 4). It represents a glacial advance into the Vale of St Albans intermediate between deposition of the Ware Till and the Ugley Till (Fig. 3.10E). The Stortford Till is therefore stratigraphically the second till in the regional sequence.

According to the revised stratigraphy established by Cheshire (1986a; in Allen *et al.*, 1991), the Ware Till, deposited by the earliest of the four ice advances, can be traced throughout the Vale of St Albans. It rests upon proglacial lake deposits in the Watford area (Gibbard, 1977; see Fig. 3.10C and D), indicating its importance in effecting the diversion of the Thames. In the area south-west of the northern suburbs of Hatfield, the Ware Till is not overlain, directly or indirectly, by any subsequent glacial deposit. Decay of the Ware Till ice initiated independent drainage in the Lea and Colne basins. The Stortford Till advance possibly occurred before the wasting of the Ware Till ice was complete. Dead ice from the Ware Till ice lobe would have formed a barrier to this second advance, which may explain why the Stortford Till does not extend across the low ground to the south-west of the Hertford area (Cheshire, 1986a). The main thrust of the Stortford Till advance reached further south than that of the Ware Till, with lobes extending to Finchley, in the former Mole-Wey valley (Fig. 3.10E) and Hornchurch (see Chapter 4). The Stortford Till ice then retreated to a position north of Bishop's Stortford before a third advance of ice, extending as far as Hertford, led to the deposition of the Ugley Till (Fig. 3.10F). The ice once more retreated to the north of Bishop's Stortford and the carbonate-rich Ugley Gravel, the upper division of the Westmill Upper Gravel, was added to the sequence (the drainage pattern before and after the Ugley Till advance was probably similar – see Fig. 3.10G). The final ice advance, leading to the deposition of the Westmill Till, extended further than the Ugley Till, sending lobes into the Vale of St Albans and the Lower Lea valley that reached terminal positions in north Hatfield and near Waltham Cross respectively (Cheshire, 1986a). All four tills record the advance of ice into the region from the north and/or north-east.

From the foregoing account it would seem that the southward-draining River Lea arose as a reversal of part of the pre-Anglian Mole-Wey-Wandle course, as a result of the advance of Ware Till ice into the Hertford area. Cheshire (1983c, 1986a) in fact suggested that the Mole-Wey may have already been captured by a tributary of the Darent-Medway prior to the glaciation, with a southward-flowing river occupying the Lower Lea valley by this time. He considered that southward through-drainage via the Lower Lea was initiated by overflow from the Watton Road lake, which was dammed by Ware Till ice during the early part of its advance into the Vale of St Albans (Fig. 3.10B). This event preceded the formation of the Moor Mill lake, which did not occur until the Ware Till ice had advanced beyond the Hertford area, blocking the Vale of St Albans drainage entirely and diverting the Thames southwards to the Windsor area (Fig. 3.10C and D). Thus for a brief interval, during the existence of the Watton Road lake and during the latter part of the interval represented by the Winter Hill/Westmill Lower Gravel, Thames waters presumably found their way from the Vale of St Albans into the Medway system via the Lower Lea valley, perhaps initiating deposition of the gravel identified in the Lea valley as Westmill Upper Gravel. The initial diversion of the Thames therefore seems to have resulted from the overflow of the Watton Road lake, not that at Moor Mill. This earlier lake-overflow produced a very short-lived Thames route via the Vale of St Albans and the Lower Lea, similar to that envisaged by earlier authors (Sherlock and Noble, 1912; Hawkins, 1922; Baker and Jones, 1980; Cheshire, 1981; Jones, 1981). If this earliest diverted route is represented at all within the sedimentary record of the Thames downstream from London, it would be expected to have contributed to the earliest post-diversion deposits within the channel-system recognized in eastern Essex, which are correlated with the Black Park Formation of the Middle Thames (Bridgland, 1988a; Chapter 5). Thus the latest parts of the Winter Hill/Westmill Lower Gravel aggradation would be contemporaneous with deposits at the Black Park level, east of London, although the latter have been buried by the subsequent (early Saalian) deposition of the Boyn Hill/Orsett Heath Gravel and have yet to be separately identified (see Chapter 4, Hornchurch and Wansunt Pit). The close association of the Winter Hill and Black Park Gravels is well-established; it has been suggested above that the downcutting from the Winter Hill to the Black Park level was simply a response to the diversion of the Thames (see above, Chapter 1).

The Westmill succession therefore represents a remarkably complete record of Anglian Stage deposition in the Vale of St Albans, including both fluvial and glacial deposits. The sequence in this district, established over several years from exposures in Westmill Quarry, is unique in the Vale of St Albans, in that tills related to three of the four Anglian glacial advances now recognized in Hertfordshire and western Essex are represented. The fluvial sediments in this sequence represent the last phase of pre-diversion Thames aggradation and the newly formed (post-Thames diversion) River Lea.

Conclusions

Westmill Quarry provides exposures through a complex series of gravels and tills (boulder clays), laid down during probably the most significant cold phase of the Quaternary ice age, around 450,000 years ago (the Anglian Stage). The lowest gravel (the Westmill Lower Gravel) was laid down by the Thames when it flowed through the Vale of St Albans and into Essex, along a more northerly route than the present river. This gravel was the last to be deposited by the Thames in this old northern course. The overlying deposits at Westmill consist of a complex series of tills (deposits laid down directly by ice sheets) and fluvial gravels deposited by the newly formed River Lea, the latter fed by meltwater streams flowing from ice sheets as they advanced and retreated across the area. It was these Anglian ice sheets that were responsible for the blocking and diversion of the Thames, eventually leading to the formation of the modern Thames valley through London.

In all, four separate advances of ice into this part of Hertfordshire have been recognized, three of which have left a direct record, in the form of till, at Westmill. Thus, collectively, the deposits exposed here are important for charting the course of the Thames before it was diverted southwards by the ice, as well as for establishing the number, extent and direction of movement of the various Anglian ice advances that are known to have affected this region.

MOOR MILL QUARRY (TL 145027)

D.R. Bridgland

Highlights

This is a classic locality that shows pre-diversion Thames deposits overlain by laminated, pro-glacial lake clays, Anglian till and a later suite of gravels deposited by the early River Colne, the sequence as a whole illustrating the glacial diversion of the Thames from this area.

Introduction

The sequence at Moor Mill Quarry, which is situated in the western part of the Vale of St Albans, is of major importance for demonstrating the glacial diversion of the Thames. The sequence includes glaciolacustrine sediments that are believed to have been deposited in a lake formed by the ponding of the Thames by Anglian Stage ice. Till overlies the lacustrine beds, indicating that the ice subsequently overrode the lake. Gravels below the lake beds are attributed to the Thames, whereas those above the till were deposited by the newly formed River Colne. The replacement of the Thames by the Colne in the western part of the Vale of St Albans can thus be shown to have occurred in the interval represented at Moor Mill by glacial deposits.

The overflow from the Moor Mill lake is considered to have brought about the diversion of the Thames into its modern valley through London (Gibbard, 1977, 1979), although overflow from an earlier lake in the Hertford area may have resulted in the diversion of the river into southern Essex via the Lower Lea (Cheshire, 1981, 1986a; see above, Westmill).

Description

The first description of exposures at Moor Mill was by Evans (1954), although Prestwich (1858b) had described a similar sequence (excluding the lacustrine sediments) at Bricket Wood, about 2 km to the south-west. The Moor Mill site was described in detail by Gibbard (1974, 1977, 1978d). The nomenclature applied to some elements of the sequence has been modified as a result of later work in the Vale of St Albans by Cheshire (1986a). The sediments exposed at Moor Mill are as follows:

		Thickness
4. Sand and gravel, cross-bedded	(Smug Oak Gravel)	5 m
3. Chalky till, blue-grey	(Ware Till)	6 m
2. Laminated clay	(Moor Mill Laminated Clay)	2.6 m
1. Sand and gravel, cross-bedded	(Westmill Lower Gravel)	6.5 m
Chalk		

Evans's (1954) sections, to the north and west of the present GCR site, differed from the above only in that he recorded slightly decreased thicknesses for all the units. He suggested that the laminated sediments were of glaciolacustrine origin, the banding being of seasonal origin and comparable to Swedish varves. Gibbard's (1974, 1977) descriptions, which are the basis of the sequence reproduced above, fully confirmed the stratigraphy reported by Evans.

The Westmill Lower Gravel (unit 1) rests on an uneven surface of brecciated Chalk. It comprises laterally persistent, massive, coarse gravels and current-bedded coarse sands and fine gravels, as well as localized silt. Gibbard (1974) reported an ice-wedge cast from the lower part of these deposits, indicating periglacial conditions during deposition (see Fig. 3.11). According to Gibbard, imbrication of the lowermost, coarsest gravel and cross-stratification in the sands both indicate a north-eastward palaeocurrent direction. Clast-lithological analysis revealed a flint-dominated composition (85–90%) with quartz and quartzite (9–12%), small amounts of Lower Greensand chert and various far-travelled minor components (Gibbard, 1974, 1977; Table 3.2). This is similar to the composition that characterizes the various terrace gravels of the Middle Thames, particularly those originally mapped as 'Glacial' (see above, Introduction to Part 2).

Moor Mill Quarry is the type locality for the Moor Mill Laminated Clay (unit 2). Evans (1954) noted that, in the sections he studied, the lower part of this deposit was of a brownish hue and that it resembled Reading Beds material, whereas the upper part comprised alternations of pale silt and dark grey clay, similar to the matrix of the local till. Only the upper part of the member appears to have been recorded in more recent exposures, which show laminated deposits directly overlying the Westmill Lower Gravel (Gibbard, 1974, 1978d; Cheshire and Gibbard, 1983). The upper part of the unit has been somewhat deformed, presumably when overriden by the ice that deposited the overlying till (Gibbard, 1974, 1978d). Trace fossils have been recognized on bedding planes within the laminated sediments (Gibbard and Stuart, 1974).

The Moor Mill Laminated Clay resembles a classic glacial lake deposit (Fig. 3.12). A minimum total of 342 laminar pairs was counted from a vertical section through this deposit at the nearby Harper Lane Quarry (TL 164019). If these laminar pairs can be interpreted as annual varves, their number points to the existence of a lake in the area for a period of at least that many years (Cheshire and Gibbard, 1983). Only 246 laminar couplets have been recorded at Moor Mill (Gibbard, 1974, 1978d).

The overlying Ware Till (unit 3) consists of a grey silty clay, passing upwards into a massive chalky till. Gibbard suggested that the lower part of the till was deposited from floating ice, this giving rise to its partial stratification. The superposition of this till over the lacustrine laminated clays indicates that the ice sheet finally advanced and overrode the proglacial lake. Fabric analysis of the till suggests that ice movement was from the north-east (Gibbard, 1974, 1977, 1978d).

The Smug Oak Gravel (unit 4), which also has its type site at Moor Mill, has an irregular base, filling channels in the upper surface of the underlying till. The gravel is weakly cross-bedded and somewhat disturbed. Palaeocurrent measurements, from imbrication and foreset orientation, indicate flow to the south-west (Gibbard, 1974, 1978d). Stone counts show, in comparison with the Westmill Lower Gravel, a larger proportion of quartz and quartzite, a higher chalk content and the disappearance of Greensand chert, confirming a significant change in provenance (Gibbard, 1974, 1978d; Table 3.2).

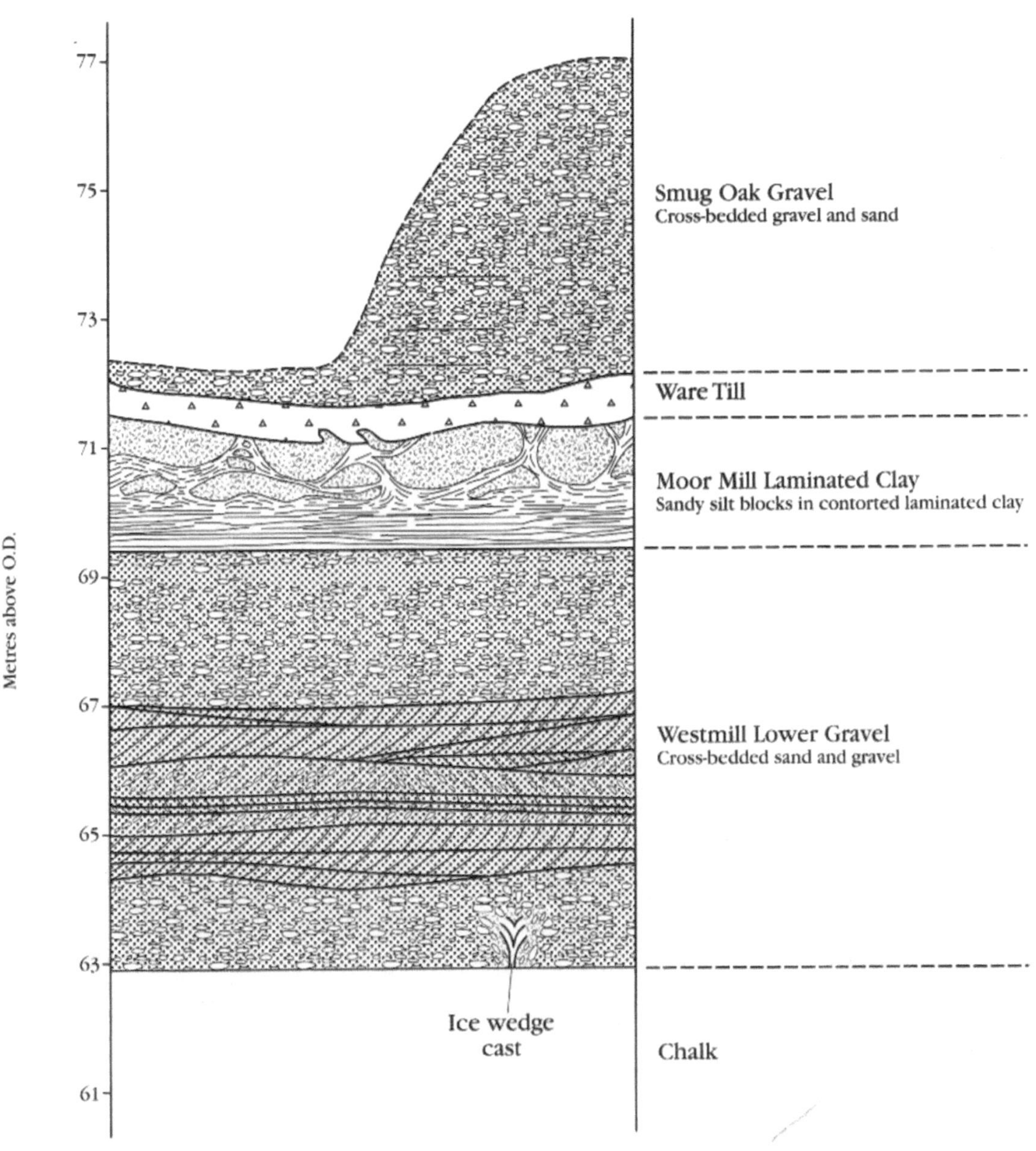

Figure 3.11 Section in the south face of Moor Mill Quarry, recorded in July 1972 (after Gibbard, 1978d).

Interpretation

The occurrence of glaciolacustrine deposits in the Vale of St Albans has been known since the 1920s. Sherlock (1924; Sherlock and Pocock, 1924) described possible lacustrine beds in the Watford area and suggested that these were related to the glacial ponding of the early Thames. Evans (1954) and Clayton and Brown (1958) provided further records of such lacustrine beds. Clayton and Brown described deposits of this type over a wider area, including the eastern Vale of St Albans and the 'Finchley Depression'. They postulated that an ice advance from the north-east had formed a large ice-dammed lake (Lake Hertford) in this area.

The involvement of the 'Chalky Boulder Clay' (Anglian) glaciation in the diversion of the Thames from the Vale of St Albans was not universally accepted, however. Wooldridge (1938, 1957, 1960) and Wooldridge and Linton (1939, 1955) believed that the river had abandoned the Vale of St Albans for an intermediate route through Finchley long before this

Figure 3.12 Detail of laminated lake beds at Moor Mill Quarry. (Photo: P.L. Gibbard.)

glaciation, perhaps in response to an earlier invasion by ice (Fig. 3.4; see above, Introduction). Zeuner (1945, 1959) preferred to attribute the diversion of the Thames to river capture. Rose *et al.* (1976) considered that the Thames had ceased to flow northwards to East Anglia by Cromerian times, since they recognized a Cromerian palaeosol developed in the upper surface of early Thames (Kesgrave Group) gravels in Suffolk (see Chapter 5, Part 1). Later work has, however, identified lower-level pre-diversion Thames gravels in north-east Essex that lack this fossil soil and are attributed to the Anglian Stage (Bridgland, 1988a), thus reconciling the disappearance of the Thames in Suffolk prior to the Anglian with the theory of Anglian glacial diversion from the Vale of St Albans.

The first systematic work in the Vale of St Albans was by Gibbard (1974, 1977, 1978a), who recognized the deposits of two separate proglacial lakes, both of Anglian age. One of these, his Watton Road lake (see above, Westmill), was in the Ware area (Fig. 3.10B), where its deposits were previously recognized by Clayton and Brown (1958). The second lake was at the western end of the Vale of St Albans, around Bricket Wood (Fig. 3.10C). Gibbard believed that these lakes were ponded by separate ice advances, but Cheshire (1983a, 1986a) demonstrated that both formed (albeit at different times) as a result of the first ice advance into the area, that which deposited the Ware Till (see above, Westmill). Lacustrine beds beneath the Ware Till at Moor Mill Quarry (Gibbard, 1974, 1978d) represent the second, larger lake (Fig. 3.10C). This lake was formed after the ice advanced upstream in the pre-diversion Thames valley, beyond the position of the Watton Road lake.

Gibbard (1974, 1977, 1978a) interpreted the lower gravel at Moor Mill as an upstream continuation of the Westmill Gravel of the Hertford area, on the basis of its composition, stratigraphical position and palaeocurrents. He attributed this deposit to the Thames. In its type area, the Westmill Gravel was divided by Gibbard into upper and lower members, separated by the Ware Till (see above, Westmill). On the grounds of distribution, composition and elevation, he correlated the Westmill Gravel with the Winter Hill Gravel of the Middle Thames, tracing a single formation, the Winter Hill/ Westmill Gravel, from the Reading area into the Vale of St Albans.

Hare (1947) had shown that the Winter Hill Terrace surface has a very low gradient as it approaches the Thames-Colne confluence. He attributed this to the effects of the aggradation of outwash from the 'Great Eastern Glaciation' (now assigned to the Anglian Stage) in this area, recognizing (following Wooldridge (1938)) that the Winter Hill Thames and the Vale of St Albans glaciation were broadly coeval. Gibbard found that the low-gradient part of the Winter Hill Terrace profile, around Stoke Common, Buckinghamshire (SU 983852), is underlain by gravel and sand with very large-scale cross-stratification, suggestive of a deltaic origin. He interpreted these deposits as a localized Winter Hill Upper Gravel Member, formed as a delta built out into the proglacial (Moor Mill) lake at the western end of the Vale of St Albans. This member is therefore of equivalent age to the Moor Mill Laminated Clay, which was deposited in the same lake, formed by the ponding of the Thames in front of the advancing Ware Till ice (Cheshire, 1986a; Fig. 3.10C). Although not disputing the deltaic origin of the Winter Hill Upper Gravel, Cheshire (1986a) pointed out that the flattening of the terrace gradient immediately upstream of the Vale of St Albans may be partly the result of isostatic uplift that followed the disappearance of the Anglian ice sheets from the eastern part of the area.

The till at Moor Mill was correlated by Gibbard (1974, 1977, 1978a, 1978b) with his 'Eastend Green Till' of the Hertford area, which he attributed to the second of two Anglian glacial advances into the Vale of St Albans. According to Gibbard, this same till capped the sequence at Westmill and thus post-dated both the Westmill Upper and Lower Gravels (see above, Westmill). However, subsequent detailed reappraisal of the tills of the Vale of St Albans by Cheshire (1983a, 1986a) has shown that the till at Moor Mill can be correlated with the lower till at Westmill, the Ware Till. The Westmill Upper Gravel, which overlies this till at Westmill, is therefore later than any part of the lower gravel at Moor Mill, which must as a result be redefined as Westmill Lower Gravel (Cheshire, 1983a, 1986a). This modification of the sequence of events established by Gibbard (1974, 1977) in no way undermines the importance of the Moor Mill sequence in illustrating the diversion of the Thames from the Vale of St Albans, although the river may have been initially diverted from its old route east of Hertford by the overflow of the earlier Watton Road lake, flowing for a brief interval by way of the Vale of St Albans and the Lower Lea valley (Cheshire, 1983c, 1986a; see above, Westmill).

The palaeocurrent data from the Smug Oak Gravel, which caps the Moor Mill sequence, contrasts with that from the Westmill Lower Gravel and indicates that the river responsible for its deposition drained in the opposite direction to the Thames. Gibbard (1977) traced this gravel to Uxbridge, where it forms the Black Park Terrace of the Colne (Hare, 1947). He therefore interpreted the Smug Oak Gravel as an early Colne deposit, equivalent in age to the Black Park Gravel, which has been shown to be the earliest terrace formation represented in the modern Thames valley through London (Wooldridge and Linton, 1955; Gibbard, 1979; Chapter 1). Gibbard (1977) considered the Smug Oak

Gravel to be of late Anglian age, deposited by meltwater from his 'Eastend Green Till' ice sheet. Cheshire (1986a) correlated the Smug Oak Gravel with the Westmill Upper Gravel of the Lea basin, which, he believed, was also fed by outwash from Anglian ice sheets as they fluctuated around the eastern end of the Vale of St Albans (see above, Westmill).

The change, in the western part of the Vale of St Albans, from a north-eastward to a south-westward flowing river is clearly seen from the sequence at Moor Mill Quarry to be associated with the advance of the Ware Till ice into the area. Since it could not have coexisted in the vale with the south-westward draining Colne, as represented by the Smug Oak Gravel, the diversion of the Thames must have taken place prior to the deposition of this gravel, during the interval represented at Moor Mill by the laminated clay and till. This strongly supports the hypothesis that the Thames was glacially diverted from the Vale of St Albans, presumably as a result of the blocking of its valley by the Ware Till ice sheet. This initially caused the ponding of a proglacial lake in the Moor Mill area, which was eventually overriden by a further advance of the ice, its waters escaping southwards through the Chertsey area into the modern valley (Gibbard, 1977). The valley through London is considered to have already been established, in pre-Anglian times, as a tributary of the River Medway, which flowed northwards across eastern Essex (Bridgland, 1980, 1988a).

The diversion of the Thames, as demonstrated by the sequence at Moor Mill, provides an important stratigraphical marker that can be used for correlation between parts of the Thames basin and particularly with the sequence downstream in Essex, where the diversion resulted in the Thames adopting the pre-existing valley of the Medway (Bridgland, 1980, 1983a, 1983b, 1988a; see Chapter 5, Part 2; St Osyth and Holland-on-Sea). This correlation provides a basis for wider comparison of the sequences in the Middle Thames/Vale of St Albans and East Anglia.

Conclusions

The sequence at Moor Mill Quarry is important for reconstructing Quaternary Ice Age events, at around 450,000 years ago, that had a major influence on the evolution of the Thames drainage system. It illustrates that the Thames formerly flowed through this area, that it was ponded by an ice sheet and that this resulted in a reversal of the local drainage, with the replacement of the Thames by the south-westward-flowing River Colne. The sequence begins with a Thames gravel, with evidence from bedding structures confirming deposition by this north-eastward-flowing river. This gravel is overlain by a distinctive laminated clay deposit, typical of sedimentation in a glacial lake. This lake is believed to have formed when the Thames was blocked by an ice sheet advancing south-westwards up its valley, during the Anglian glaciation. Direct evidence for the presence of this ice sheet is provided by the next element of the Moor Mill sequence, the Ware Till. This is a typical 'boulder clay' deposit laid down beneath the advancing ice sheet as it finally extended across the area of the lake. The final deposit in the Moor Mill sequence is a further gravel, this time deposited by the Colne, flowing towards the south-west. The change from a north-eastward-flowing river in the lower gravel to a south-westward-flowing river in the upper gravel thus records the diversion of the Thames from its old route through the Vale of St Albans. The newly formed River Colne, as represented by the upper gravel at Moor Mill, flowed into the diverted Thames at the western end of the vale.

UGLEY PARK QUARRY (TL 519280)

D.A. Cheshire and D.R. Bridgland

Highlights

Ugley Park Quarry is important for demonstrating the stratigraphical relations of the glacial and glaciofluvial deposits formed during the later part of the glaciation of the Vale of St Albans, and for correlating these with the fluvial sequences in the Thames catchment. It offers a rare opportunity for the study of ice-proximal glaciofluvial deposits.

Introduction

Ugley Park Quarry is the most north-easterly of the sites described in this chapter, being situated virtually on the interfluve between the catchments of the River Stort (a tributary of the Lea) and of the Cam or Granta (part of the Great Ouse system, draining to the Wash). The quarry is excavated in Pleistocene sediments that fill a valley cut through Reading and Thanet Beds (Palaeogene) into Chalk. This particular valley-fill lies to the west of the subglacially formed Stort-Cam tunnel valley (Woodland, 1970), which traverses the East Anglian chalklands from north to south. The valley at Ugley has been interpreted as a glaciofluvial spillway (Hopson, 1981). The fluvial sediments that now fill this valley, at elevations of 80–100 m O.D., were deposited by north-bank tributaries of the Thames shortly after its diversion from the Vale of St Albans, at a time when the ice sheets of the Anglian glaciation persisted in the vicinity. These ice sheets supplied outwash to the river system and, intermittently, advanced across the gravel floodplains, depositing the tills that are interbedded with the fluviatile sequence.

The significance of the site accrues partly from its geographical position on the modern watershed between the Thames system and the drainage of the Wash basin; it thus provides a link between the fluvial and glaciofluvial sediments of the Middle Thames/Vale of St Albans and the equivalent glacial deposits of central-southern East Anglia. The main value of the site is that it demonstrates more clearly than elsewhere the stratigraphical relations between the late Anglian sediments of both these areas.

Description

Ugley Park Quarry consists of separate east and west pits, of which the latter has been partly infilled. The floor of the west pit formerly exposed the junction between the Upper Chalk and the overlying Palaeogene, here represented by the Thanet Beds (with the characteristic Bullhead Bed at the base, at 80.5 m O.D.). The Thanet Beds and the succeeding Reading Beds formed the floor of the eastern part of the west pit, but have not been seen in the east pit, as the full Pleistocene sequence there has not been worked. London Clay is not seen, but occurs 300 m east of the east pit, below till. Silty sand and clay that have been interpreted as Palaeogene in age appear in the eastern face of the east pit but, as discussed later, are probably not *in situ*.

Attention was drawn to the deposits at Ugley Park Quarry by Hopson (1981) and they were discussed briefly by Wilson and Lake (1983). The latter authors recognized two Pleistocene sedimentary beds: up to 10 m of poorly sorted chalky flint-gravels succeeded by up to 6 m of chalky till. The chalky gravel has been attributed to a proximal glacial origin and is thought to be the infilling of a north-south aligned channel parallel to, but 1.5 km to the west of, the Stort-Cam tunnel valley (Hopson, 1981). Wilson and Lake (1983, p. 78) noted that, in the north-east part of the east pit, the chalky till contained 'intercalations of silts with laminated beds, and irregular silty sands with complex convoluted structures and lenticles of firm olive grey clay'. They asserted that the lenticles of clay were derived from soft Palaeogene bedrock.

Cheshire (1986a) recognized a further lithostratigraphical unit in the east pit. This unit, a stiff, dark grey chalky till, does not outcrop in the eastern face, the face that has received the most attention, and appears not to have been recognized earlier. Boreholes in the quarry floor have revealed that an additional gravel unit occurs beneath the till, although in places the latter rests directly upon Reading Beds. This basal gravel is poorly sorted and contains cobbles of flint and Chalk.

The full sequence is therefore as follows:

			Maximum thickness west pit/east pit
4.	Chalky till, brown/yellow-brown	(Westmill Till)	2.0 m/3.0 m
3.	Chalky gravel and sand, cross-bedded	(Ugley Gravel)	9.5 m/4.6 m
2.	Chalky till, dark grey	(Ugley Till)	—/6.9 m
1.	Clayey, silty sandy gravel		—/>3.7 m

The Ugley Till occurs only in the east pit, where the top 2 m has been exposed. It is stiff, massive, compact and apparently structureless. The particle size, acid solubility, and small-clast composition are similar to those of the Westmill Till, both at this site and at Westmill Quarry.

Fabric data from the Ugley Till reveals statistically significant preferred orientations suggestive of lodgement by ice moving from the north-west or north-north-west.

The Ugley Gravel lies in a channel with its base at about 85 m O.D. It overlaps the Ugley Till at elevations of between 88 m and 94 m O.D. in the east pit. The maximum thickness is seen in the west pit, where the gravel reaches 9.5 m; it thickens as the top of the Ugley Till descends eastwards across the east pit. The orientations of large-scale trough and tabular cross-bedding in the chalky gravel and sand in the west pit suggest southward palaeocurrents. These structures also suggest deposition in a braided river with fluctuating energy conditions. Similarly, southward palaeocurrents are indicated by structures in chalky sand in the east pit. The gravel in the east face of the east pit incorporates a till-derived debris flow and massive matrix-supported gravel in its lowest three metres. This is succeeded by a thin buff silty clay and up to 1.4 m of cross-bedded coarse chalky sand.

The Ugley Gravel has a clast composition that differs markedly from other gravels in the Thames system (Bridgland, 1980, 1983a, 1986b; Cheshire, 1986a; Bridgland *et al.*, 1990; Table 3.2). In particular, it has a very low proportion of rounded flint pebbles reworked from the Palaeogene; this can be as low as 1–2%, but is highly variable. The gravel also contains abundant Chalk and exotic limestones (mainly Jurassic), the calcareous fraction sometimes approaching half the total count. Another significant constituent of the gravel is *Rhaxella* chert, at up to 1.5% (including calcareous *Rhaxella*-bearing rocks). Clasts of this rock are believed to have been introduced in quantity into the London Basin for the first time by Anglian ice. Exotic rocks that are characteristic components of Thames deposits are relatively scarce in the Ugley Gravel, particularly in comparison with the pre-diversion Thames formations. The extremely calcareous Ugley Gravel contains the highest proportion of Chalk recorded in any fluvial deposit yet studied in Hertfordshire and Essex and has been taken as a standard for Anglian Stage ice-proximal outwash (Bridgland, 1980, 1986b; Bridgland *et al.*, 1990; Table 5.1).

Resting upon the eroded but generally even surface of the Ugley Gravel in both pits is the Westmill Till. This brown to yellowish-brown, very chalky till increases in thickness north-eastward, reaching 8.6 m at The Hall, Ugley (TL 521285). Its particle-size distribution is remarkably uniform both vertically and laterally and, in common with the same till at Westmill Quarry, it contains little medium-grade sand in comparison with the Ware and Stortford Tills. The small-clast composition includes relatively high proportions of flint and *Rhaxella* chert; quartz is less abundant and the acid-soluble content higher than in the Ware and Stortford Tills. Fabric analysis indicates ice-movement from due north.

The thickness of chalky Westmill Till is reduced to about 1 m in the north-eastern part of the east pit, where the lower part of this member contains a large body of silty medium-fine sand. This sand, which contains faint cross-bedding, appears similar to Palaeogene bedrock sediment, but lies 12 m above the maximum height at which Palaeogene strata are known to occur in this part of the quarry. However, in small pits at the base of the face, the silty sand was seen to overlie coarse chalky sand of the Ugley Gravel. This silty sand is probably the same material observed by Wilson and Lake (1983) in the north-eastern part of the site. Streaks and small lenses of chalky till occur in the upper 0.5 m of this sand body, which is thought to represent a large raft (at least 15 m across) of Palaeogene sediment transported, possibly as a frozen block, by the Westmill Till ice.

Interpretation

The sections at Ugley Park Quarry record the only known sequence that demonstrates conclusively the stratigraphical relations of the Ugley Gravel and the Ugley Till. The Chalk-rich Ugley Gravel can be traced, mainly through borehole records, down the Stort valley to the modern Lea valley, where it can be correlated with that part of the Westmill Upper Gravel occurring above 68 m O.D. at Westmill Quarry. Gibbard (1974, 1977) showed that the carbonate content of the Westmill Upper Gravel increases significantly at that elevation within the Westmill sequence, in conjunction with a change in palaeocurrent direction (see above, Westmill). This higher carbonate content (principally Chalk) has been recorded by Cheshire (1981, 1986a) in the upper part of the Westmill

Upper Gravel at other sites in eastern Hertfordshire and western Essex, the largest calcareous component occurring at Ugley Park Quarry. The latter site was therefore selected as the type locality for the upper division of the Westmill Upper Gravel, the Ugley Gravel. The thickness and Chalk-content of the Ugley Gravel is considerably reduced in the area between the type site and Westmill. The lower part of the Westmill Upper Gravel was redefined by Cheshire (1986a) as the Hoddesdon Gravel (type locality: Cock Lane Quarry, Hoddesdon, TL 354077).

The Ugley Till, unit 2 at Ugley Park Quarry, is absent at Westmill. However, at the former Foxholes Quarry (TL 340123 and TL 342125), south of Hertford, a till with particle-size, small-clast lithology and carbonate properties identical to the lower till at Ugley occurs below a Chalk-rich gravel that is correlated with the Ugley Gravel (see above, Westmill). At this site, the Ugley Till had a highly significant fabric orientation, suggesting lodgement from the north-east. Cheshire (1986a) considered that the Ugley Till at Foxholes Quarry was deposited near the ice margin, implying that it represents the least extensive of the four glacial advances that affected the Hertfordshire/western Essex region.

In previous schemes for Anglian glacial stratigraphy in this area, only two glacial advances were recognized. West and Donner (1956) and Clayton and Brown (1958) advocated the separation of upper and lower tills over a wide area of southern East Anglia on the basis of distinctive fabric orientations. According to these authors, fabrics from lower tills were indicative of ice-movement from the north-west and those from upper tills from the north. Gibbard (1977) agreed that tills in the wider region can be separated by their fabric properties and also recognized two advances in Hertfordshire, those associated with his Ware and Eastend Green Tills. At Quendon, 3 km north of Ugley Park Quarry, Baker (1977) identified a lower till, which he named the Quendon Till, in a similar sequence to that found at Ugley. The Quendon Till is 1.8–7.6 m thick, dark grey and has a preferred fabric orientation indicating ice-movement from the north-west, similar to the Ugley Till, to which it is quite possibly equivalent. However, Baker and Jones (1980) equated the Quendon Till with the Ware Till of Gibbard (1977) and the Maldon Till of Clayton (1957), on the basis that it is the lower of the two tills that were recognized over a wide area at that time. Baker and Jones cited evidence (after Baker, 1977), from proglacial varves in lacustrine sediments, implying that, between the deposition of their lower and upper tills, the ice front retreated to the Newport area in northern Essex and stabilized there for a minimum period of 5400 years.

Cheshire (1986a) has shown that tills in the south Hertfordshire/western Essex area possess distinctive petrographical properties, which, with the use of similarity indices, enable correlation from site to site. These methods show that the distinctive Ware Till signature may be traced from site to site towards Ugley, but cannot be recognized in the Ugley Till. Thus the Ugley Till cannot be equated with the Ware Till, despite having a fabric orientation that indicates ice advance from the north-west or north-north-west. Although the Ware Till, Stortford Till and Westmill Till may be differentiated from each other by their petrographical properties, the Westmill Till and Ugley Till are petrographically similar. They may, however, be differentiated by their stratigraphical positions respectively above or below the Ugley Gravel.

At Westmill Quarry, the Westmill Till shows a preferred fabric orientation that, coupled with structural discontinuities related to ice-movement, indicates lodgement from the north-east; this contrasts with the northerly origin of the same till at Ugley. This general pattern, which is reproduced to a greater or lesser extent in sediments deposited by each of the four ice advances into the Vale of St Albans, may be interpreted as the result of ice repeatedly approaching the Thames Basin from the north or the north-north-west, as proposed by Perrin *et al.* (1979). As the ice crossed the Chalk escarpment and entered the former valley of the Thames, it spread out to the south-west in Hertfordshire, to the south-east in Essex and Suffolk (Allen, 1983), and southwards in central Essex (Allen *et al.*, 1991).

Samples of the gravel detected in boreholes below the Ugley Till were not available for analysis of petrographical properties, but the material is probably equivalent to the Hoddesdon Gravel, the lower, less calcareous division of the Westmill Upper Gravel. This unit, like the Ugley Gravel, can be traced down the Lower Lea valley as far as Bullscross Farm (Cheshire, 1983c, 1986a; see above, Westmill). Thus all the sediments at Ugley Park Quarry were deposited after the Thames had been diverted from its Vale of St Albans course, in what had already become the catchment of the River Lea.

The sequence at Ugley Park Quarry shows the stratigraphical relations of the later Anglian glacial and glaciofluvial deposits in the northern part of the Thames Basin. It complements the site at Westmill, in which the Ugley Till is missing from the sequence. The Ugley Park Quarry site also provides the best exposure of the chalky Ugley Gravel outwash. Thus both Westmill and Ugley can be regarded as key Anglian sites, each exhibiting part of a complex sequence.

Conclusions

The complex sequence of sands, gravels and till (boulder clay) at Ugley Park Quarry is important for showing that, during the cold Anglian Stage of the Quaternary Ice Age (about 450,000 years ago), the Thames catchment was repeatedly invaded by ice moving from the north or north-north-west. The evidence from Ugley is critical, in conjunction with that from Westmill, in demonstrating that there were at least four of these Anglian glacial advances, each of which deposited characteristic tills. Gravels and sands, found between and underneath the tills at Ugley, were deposited by meltwater streams flowing from the ice sheets and feeding the newly formed River Lea system.

Part 3

THE SEQUENCE POST-DATING THE DIVERSION OF THE THAMES (SITES IN THE MIDDLE THAMES AND ITS TRIBUTARY, THE KENNET)

D.R. Bridgland

Introduction

This part of Chapter 3 is concerned with the terrace deposits laid down by the Thames since its diversion into its modern course. The Black Park Gravel Formation, aggradation of which occurred while Anglian ice persisted in the Vale of St Albans, post-dates the diversion and so is included at the beginning of this section, represented at Highlands Farm Pit and Hamstead Marshall Gravel Pit (the latter in the tributary Kennet valley). Subsequent to its diversion, the history of the river is recorded in the Middle Thames basin by an extensive sequence of terrace gravels, in marked contrast to the paucity of depositional evidence from the immediate pre-Anglian period (see Part 2 of this chapter).

The history and extent of research on this sequence is considerable. Prestwich (1855) divided the valley gravels into high-level and low-level terrace deposits, but the first systematic account of the drift geology of the Thames valley in which the importance of depositional river terraces was recognized was by Whitaker (1864, 1889). He identified three distinct terraces, which he illustrated in a map of the Maidenhead district (Whitaker, 1889, p. 391). This work, extended by Pocock (1903), formed the basis for the tripartite terrace system - Boyn Hill Terrace, Taplow Terrace and Floodplain Terrace - recognized by the Geological Survey from 1911 (Bromehead, 1912), the type localities for which occur in the Beaconsfield district (Sherlock and Noble, 1922). Upper and lower divisions of the Floodplain Terrace were recognized by Dewey and Bromehead (1921).

When the (higher) Winter Hill Terrace was added to the Middle Thames sequence (Saner and Wooldridge, 1929; Wooldridge, 1938; Wooldridge and Linton, 1939), it was thought to follow the modern valley through London. It was later discovered that the Winter Hill Terrace, as originally defined, was multiple. The lowest division of this multiple terrace, the Black Park Terrace of Hare (1947), was later recognized to be the first gravel formation in the modern Thames valley (Wooldridge and Linton, 1955; Gibbard, 1979). Hare (1947), who carried out detailed geomorphological mapping around Slough and Beaconsfield, recognized that there was a further important aggradation between the Boyn Hill and Taplow Terraces, forming his

Table 3.3 Post-Winter Hill terraces and gravel formations in the Middle Thames and Kennet valleys.

Thames		**Kennet**	
Terrace	**Gravel formation**	**Terrace**	**Gravel formation**
Lower Floodplain	Shepperton	Beenham Grange	*
Upper Floodplain	Kempton Park	within Thatcham?	*
Taplow+	Taplow+	Thatcham+	*
Lynch Hill	Lynch Hill	*	*
Boyn Hill	Boyn Hill	*	*
Black Park	Black Park	Hamstead Marshall	Silchester
Winter Hill	Winter Hill	Upper Winter Hill of Thomas (1961)	*

* not separately named.

+ lower and upper (geomorphological) divisions of the Taplow and Thatcham Terraces were recognized by Sealy and Sealy (1956) and Cheetham (1980) respectively. The validity of these is doubtful (see Fern House Pit).

Lynch Hill Terrace (Table 3.3). Both of these newly defined terraces had previously been subsumed within rather broader definitions of those already established. Hare's work was extended upstream by Sealy and Sealy (1956) and Thomas (1961), the latter taking the new scheme into the tributary Kennet and Blackwater-Loddon valleys, and downstream by Allen (1978).

Hare had recorded a minor, lower facet of the Taplow Terrace, which Sealy and Sealy (1956) subsequently redefined as a separate 'Lower Taplow Terrace', Hare's Taplow Terrace becoming their 'Upper Taplow Terrace'. Gibbard (1985) found the Upper Taplow Terrace to be formed by the fluviatile Taplow Gravel plus an overlying loessic silt (brickearth), whereas the Lower Taplow Terrace is formed by the same gravel, but with no overburden. Thus the lower of the two terraces recognized by the Sealys is the true fluviatile Taplow Terrace (see below, Fern House Pit).

Allen (1978) recognized an additional 'Stoke Park Terrace' in the Yiewsley area, between the Boyn Hill and Lynch Hill Terraces. He regarded this as a downstream extension of the minor erosional facet identified by Hare (1947) as the 'Stoke Park Cut'. Gibbard (1985) again rejected this as an additional aggradational terrace, pointing out that the gravel assigned by Allen to the Stoke Park Terrace probably represents the northern feather-edge of the Lynch Hill Formation, separated from the thicker part of the Lynch Hill sequence immediately to the south by diapirically uplifted London Clay.

Gibbard (1985) has provided a recent reappraisal of the Middle Thames terrace sequence, basing his work on studies of the deposits rather than surface morphology. He combined the reconstruction of the sediment bodies that form the various terraces with clast-lithological analysis of the gravels to establish a lithostratigraphical scheme. As Table 3.2 illustrates, there is a progressive decline in the exotic (far-travelled) component of the post-diversion Thames gravels, a trend that began following the deposition of the Gerrards Cross Gravel. Gibbard considered individual terrace aggradations to be of member status, emphasizing the role of clast composition in their differentiation. Bridgland (1988b, 1990a) argued instead that the principal basis for recognizing individual terrace aggradations is mapping; therefore each forms a mutually exclusive primary lithostratigraphical unit and should be regarded as a separate formation (see Chapter 1). Comparable lithostratigraphical schemes have been used to classify fluvial sequences in adjacent areas, providing a reliable mechanism for the description and correlation of terrace formations (Gibbard, 1977, 1978a, 1982, 1989; Allen, 1983, 1984; Bridgland, 1983a, 1983b, 1988a; Bridgland and Harding, 1985; Cheshire, 1986a; Gibbard *et al.*, 1988). The terms Boyn Hill Gravel and Taplow Gravel (Bromehead, 1912), already used on most New Series Geological Survey maps of the area, have been retained for the Middle Thames. Gibbard (1985) extended this nomenclature to include the additional terraces recognized by Hare (1947), establishing a parallel stratigraphical scheme but retaining Hare's place-name nomenclature (Table 3.3 and Fig. 3.1). He also provided names for the deposits underlying the Upper and Lower Floodplain Terraces, respectively his Kempton Park and Shepperton Gravels (Gibbard *et al.*, 1982; Gibbard, 1985; Table 3.3).

The stratigraphical scheme developed for the Middle Thames terraces has frequently been used to classify the gravel spreads of the Kennet valley, particularly since Thomas (1961) traced the terraces defined by Hare (1947) and Sealy and Sealy (1956) upstream to the Newbury area. White (1907) had previously provided local names for the higher gravels of the Kennet, equivalents of the Winter Hill, Gerrards Cross and older gravels of the Thames. Chartres *et al.* (1976), Cheetham (1980) and Chartres (1981) subsequently proposed a separate nomenclature for terraces in the Kennet valley. Gibbard (1982) used a modified version of White's term – 'Silchester Gravel' – to describe the terrace sediments in the Kennet valley that equate with the Black Park Gravel of the Middle Thames. Correlation with Middle Thames formations is not problematic, because the gravel-bodies can be traced and/or projected between the two valleys and the Thames sequence is well-preserved in the confluence area. The use of Thames nomenclature, except where correlation is problematic, would seem perfectly feasible therefore.

The Kennet sequence falls into two clear groupings. Firstly, the older gravels recognized by White (1907) are preserved as wide spreads that form a southward-declining sequence. This sequence culminates in the Silchester/Black Park Gravel, which caps extensive plateaux to the south of the modern river. Later deposits are much less extensive, perhaps because the Kennet became entrenched along the line of its present valley. Equivalents of the Boyn Hill and Lynch Hill Formations are poorly represented in the Kennet, being restricted to a few degraded patches of gravel either side of the valley near Brimpton (Fig. 3.1). Later formations are better preserved, remnants of Thatcham Terrace deposits (= Taplow Formation) occurring at Thatcham and Brimpton (see below, Brimpton), while Beenham Grange (Floodplain) Terrace (= Shepperton Gravel) deposits are well preserved at and downstream of Thatcham.

HIGHLANDS FARM PIT (SU 744813)

D.R. Bridgland

Highlights

This is the best-known site in the deposits forming the floor of the 'Ancient Channel' of the Thames (now correlated with the Black Park Gravel) in the Caversham–Henley area. Highlands Farm Pit provides important evidence for the Palaeolithic occupation of Britain prior to the Hoxnian Stage.

Introduction

Highlands Farm Pit, Oxfordshire, exposes gravel that is assigned to the Black Park Gravel Formation of the River Thames. In this district the Black Park Gravel forms the floor of an abandoned section of an early valley of the Thames, generally known as the 'Ancient Channel' (Treacher *et al.,* 1948; Wymer, 1956, 1961, 1968). This old valley, now much dissected by later erosion, runs from Caversham (Mapledurham) to Henley (Fig. 3.1). It lies to the north of the modern valley, which takes a more circuitous route between those towns (via Sonning). As the Black Park Gravel was the last formation to be deposited before the Thames abandoned the 'Ancient Channel', this gravel is preserved in this area as a dissected valley-floor rather than the more usual north-bank terrace. The particular importance of Highlands Farm Pit lies in the occurrence in the gravel there of abundant Palaeolithic artefacts, as first reported by White (1895). Since it is derived from the Anglian Stage Black Park Formation, this assemblage is one of the oldest from the Thames system. Extensive collections of Palaeolithic material from Highlands Farm have been described (Smith, 1917; Treacher *et al.*, 1948; Wymer, 1956, 1961, 1968). They have been central to the continuing controversy during recent decades over the timing of Palaeolithic man's earliest occupation of Britain.

The attribution of the gravel forming the floor of the 'Ancient Channel' to the Black Park Formation was established relatively recently (Clarke and Dixon, 1981; Gibbard, 1983, 1985). It had previously been included in the Winter Hill Terrace (Wooldridge, 1938; Treacher *et al.*, 1948; Wymer, 1961, 1968) or in the Lower Winter Hill terrace (Sealy and Sealy, 1956). The occurrence of well-made Palaeolithic artefacts in so high a terrace within the Thames sequence (Arkell and Oakley, 1948), with the implication of an Anglian age (Table 1.1), was until recently considered problematic, since many authorities believed that only crude implements, if any, were made in Britain prior to the Hoxnian Stage. This problem has been resolved following recent discoveries at Boxgrove, West Sussex (Roberts, 1986), which have demonstrated that flint tools of high quality were made in pre-Hoxnian times (see Chapter 1).

Description

All the published descriptions of exposures at Highlands Farm Pit post-date the reopening of the workings in the 1950s. This extension of the pit enabled a rich assemblage of palaeoliths to be obtained from sections specially cleared for the purpose (Wymer, 1956, 1961, 1968; Gibbard and Wymer, 1983; Gibbard, 1985). Wymer recorded *c.* 4 m of gravel at the site, aggraded to 76 m O.D. The gravel is much disturbed by solution of the underlying Chalk, which has caused the overlying deposits to collapse into pipes of varying size. The lowest 3 m comprise coarser gravel, but throughout the sequence lenses of current-bedded sand occur (Gibbard and Wymer, 1983). The upper levels are disturbed by cryoturbation and are overlain by 0.3 m of silt (brickearth). The gravel is dominated by flint, quartz and quartzites, with subordinate crystalline rocks, cherts and sandstones (Walder, 1967; Gibbard and Wymer, 1983; Table 3.2), a composition confirming deposition by the Thames.

Wymer's work at Highlands Farm Pit led to the discovery that Clactonian flakes and cores accompanied the (Acheulian) hand-axes that had already been found there in great numbers. The absence of Clactonian artefacts in the earlier collections from the site probably reflects both the inability of the gravel diggers to recognize the cores and flakes of this industry and the preference of the collectors for hand-axes. There are other reports of artefacts collected from Highlands Farm, by Case and Kirk (1952, 1955) and Wymer (1958, 1959, 1960, 1962, 1964a); detailed summaries of the material have been provided by Roe (1968a, 1981) and Wymer (1968). The only faunal remains to come from the pit, or from any site in the Black Park Gravel of the 'Ancient Channel', are a horse tooth from a few centimetres above the Chalk and a piece of elephant tooth from an unspecified location (Wymer, 1964a).

Interpretation

In order to demonstrate the significance of Highlands Farm Pit, it is necessary to outline the research history both of the site and of the 'Ancient Channel' in some detail. The various gravel deposits to the west of Henley, including that at Highlands Farm, appeared as 'Glacial Gravel' on the Old Series geological map of the area (Sheet 13, 1860: Hull and Whitaker, 1861), but were later designated 'Plateau Gravel' (New Series, Sheet 254, 1905: White, 1908b). Following their reinterpretation by Wooldridge (1938), however, the gravels of this district have been recognized as early Thames terrace deposits and have recently been shown as such on new Geological Survey maps (Squirrell, 1978; New Series, Sheet 254, 1980 revision).

The earliest reference to the GCR site was by White (1895, p. 20), who noted that no distinction could be made between the 'glacial' and river gravels of the district and that Palaeolithic artefacts had been found there by L. Treacher and G.W. Smith 'so far distant from the river as Highlands Farm' (outside the area mapped as river gravels). At that time the site consisted merely of one or more shallow pits, presumably for farm use, and the gravel had not been commercially exploited. The Treacher collection contains specimens from this site (then called Helen's Farm) labelled 1889 and 1892 (Arkell and Oakley, 1948). Highlands Farm was also noted by Smith (1917) as a locality where implements could be obtained from high-level deposits. Thus the site was established as unusual, in that it appeared to lie above the general level of the terrace gravels, yet it yielded palaeoliths, which are usually confined to the latter deposits.

White again referred to Highlands Farm in the Henley memoir (Jukes-Browne and White, 1908, p. 88), in which he described terraces within the plateau gravel of the area, including: 'south-west of Henley, a clearly defined terrace between 255 and 260 ft above sea level (160 ft above the Thames)'. He associated the Highlands Farm site with this feature and suggested a correlation with the 'Silchester stage' of the Kennet and Lodden (White, 1902, 1907, 1908b). However, as has already been noted, the Highlands Farm deposits are not part of a simple north-bank terrace of the Thames. They lie at the eastern end of an abandoned section of valley cut through the Chalk (Fig. 3.1). Treacher (1926) first recognized the significance of this feature, from the fact that higher-level gravel at Shiplake, on the interfluve between the 'Ancient Channel' and the present river, contains up to 60% of Lower Greensand chert, implying deposition by a south-bank Thames tributary on the north side of the present valley. The deposit at Shiplake is thought to represent the early Blackwater-Loddon River, which at that time drained a large area of Lower Greensand outcrop that now falls within the Wey catchment (Walder, 1967). Saner and Wooldridge (1929) and Ross (1932) attributed the gravels of the area south-west of Henley to a terrace that could be traced downstream to Winter Hill, Maidenhead and beyond, which they called the Winter Hill Terrace. Wooldridge (1938) ascribed the gravels of the 'Ancient Channel' (which he termed the 'Crowsley Park Trench') to this Winter Hill Terrace. He differentiated the Winter Hill Terrace deposits flooring the channel from older gravels to the north and south on the basis of composition and elevation and, from an exposure 2–3 km south-west of Highlands Farm, was able to demonstrate their fluvial origin.

The definitive description of the 'Ancient Channel', and of the Treacher collection of Palaeolithic implements, was by Treacher *et al.* (1948). Although Highlands Farm was mentioned, the main sites listed were commercial pits exploiting the 'Ancient Channel' gravel;

these provided the bulk of the palaeoliths discovered prior to the reopening of Highlands Farm Pit. Arkell and Oakley (1948) concluded that these collections represented an 'Abbevillian' or 'Early Acheulian' culture, which they regarded as typologically older than assemblages from the Boyn Hill Terrace of the Middle Thames. They suggested a correlation with similar industries from gravels at Fordwich (Kent) and Farnham (Surrey) that are thought to pre-date the Boyn Hill Formation. They noted that Hare (1947) had associated the Winter Hill Terrace with the glaciation of the London area, which suggested that the implements from the 'Ancient Channel' were made before this glaciation, thought to be of Mindel (Anglian) age.

Following the extension of Highlands Farm Pit in 1954, Wymer (1956, 1961, 1968) made the important discovery that Clactonian artefacts were also present. This material, mainly crude flakes and cores in a rolled condition, was in fact dominant in the new Highlands Farm assemblage, despite being barely represented in the Treacher collection. Wymer also found well-made hand-axes to be more common at Highlands Farm than the crude types that had been ascribed to the Early Acheulian. The archaeological record is, however, without stratification; all the artefacts are in an abraded condition, although the well-made hand-axes are generally less rolled (Wymer, 1961, 1968, 1977a). Wymer's conclusions were broadly confirmed by Roe (1964, 1968a), following statistical analyses in which he compared the Highlands Farm material with that from the Treacher collection. Roe (1968a, p. 59) concluded that the assemblage from Highlands Farm was 'heavily dominated by the only slightly disturbed output of a single nearby working site'.

The Clactonian Industry had been shown to be stratigraphically earlier than the Acheulian hand-axe industry at Swanscombe (see Chapter 4) where, as at the Clacton type site (Chapter 5), it is overlain by or incorporated in the lower part of a Hoxnian aggradation. Both the Clactonian and Acheulian industries were held to have appeared during the Hoxnian, but the Clactonian seemed, on the basis of the Swanscombe sequence, to have been present earlier and to represent the earliest Palaeolithic occupation of Britain (see Chapter 1). Thus it was thought that humans did not reach Britain until the late Anglian or early Hoxnian, certainly after the glaciation of the London district (Wymer, 1961). Because a correlation had been established between this terrace and the glaciation of the Vale of St Albans (see Part 2 of this chapter), the attribution of the 'Ancient Channel gravel' to the Winter Hill Terrace was difficult to reconcile with the occurrence, at sites such as Highlands Farm, of Palaeolithic artefacts. This was particularly the case since many of the artefacts from the 'Ancient Channel' show refined workmanship, whereas the only other assemblages from Britain that were widely believed to be pre-Hoxnian comprise crude stone-struck material (Wymer, 1961, 1977a). This conflict in views led to suggestions that the artefact-bearing gravels in the 'Ancient Channel' post-date the occupation by the Thames of this section of valley. One alternative interpretation of such deposits suggested emplacement by colluvial processes, at some time after the Anglian Stage, mixing reworked Thames gravels with later material, the latter containing the artefacts (Wymer, 1961, 1976, 1977a). A similar alternative interpretation attributed the deposits yielding palaeoliths to reworking by a later tributary stream flowing through this abandoned section of the valley formerly occupied by the main Thames (Gibbard and Wymer, 1983).

Because of the controversy that had arisen over the age and origin of the 'Ancient Channel gravel', the workers who sought to extend Hare's (1947) scheme to the Reading area and beyond (Sealy and Sealy, 1956; Thomas, 1961; Walder, 1967) paid particular attention to its correlation with the terrace sequence and to its abandonment by the Thames. White (in Arkell and Oakley (1948)) suggested that during aggradation of the '150 ft' (Winter Hill) gravel the Thames parted into two 'arms', one flowing through the 'Ancient Channel' and the other via the modern course further to the south, the latter receiving the (Greensand) chert-bearing waters of the Loddon. This view was further developed by Sealy and Sealy (1956), who considered that their 'Lower Winter Hill Terrace', to which they assigned the 'Ancient Channel gravel', converged with the (lower) Black Park Terrace in the Reading area. This resulted, according to the Sealys, from the unusually steep gradient of the Black Park Terrace and the shallow gradient of the 'Lower Winter Hill Terrace'. Sealy and Sealy postulated that in 'Lower Winter Hill' times the Thames flowed in two separate channels between Caversham and Henley (as suggested by White) and that rejuvenation prior to the formation of the Black Park Terrace incised the southern (modern) course more effectively, because of softer Tertiary bedrock in that area, causing the Thames to abandon the more northerly ('Ancient Channel') route through the Chalk. Thomas (1961) agreed with the terrace correlations proposed by the Sealys, but considered that the new route was likely to have been established before the pre-Black Park rejuvenation occurred in the Reading area, perhaps in response to the enhanced erosive capacity of the Kennet and Blackwater-Loddon tributaries in comparison with the main Thames.

Walder (1967), in a pioneering study based on the analysis of their clast-lithological content, showed that gravels forming the floor of the 'Ancient Channel' contain significantly more flint than earlier formations. She noted, however, that the 'Lower Winter Hill' and Black Park Terraces could not easily be separated in the area upstream from Henley. Clarke and Dixon (1981) went further, proposing, on the basis of altitude, that the 'Ancient Channel gravel' represents the Black Park and not the Winter Hill Terrace. This view was confirmed by Gibbard (in Gibbard and Wymer, 1983; Gibbard, 1985), who included the deposits between Caversham and Henley (previously designated as Lower Winter Hill Terrace) in his Black Park Gravel. Gibbard verified Walder's observation that this formation contains significantly more flint than the Winter Hill Gravel and other earlier Thames deposits (Table 3.2). The gravels previously ascribed to the Black Park Terrace in the modern Thames valley through Reading are attributed, according to this reinterpretation, to the tributary Kennet and Blackwater-Loddon streams, which, in Black Park times, united to join the 'Ancient Channel' Thames at Henley. Gibbard and Wymer (1983, fig. 36) reclassified these tributary deposits as Silchester Gravel (a name that implies a Kennet origin; see below, Hamstead Marshall). The above-mentioned tributaries appear to have established, in Black Park times, a substantial valley between Reading and Henley, subparallel to (and south of) that of the Thames. At some time between the aggradation of the Black Park and the later Boyn Hill gravels, the Thames itself adopted this more southerly route, possibly in response to its capture, in the Mapledurham area, by a tributary of the Kennet. The fact that the new Thames course was established in soft Palaeogene strata instead of Chalk may, as suggested by Sealy and Sealy (1956), have facilitated this minor diversion.

The correlation of the 'Ancient Channel gravel' with the Black Park Formation, instead of with the Winter Hill Formation, still implies an Anglian age, as ice sheets are believed to have continued to occupy parts of Hertfordshire and western Essex during Black Park times (see above, Westmill). Human occupation of the Middle Thames valley during or before this glaciation is therefore still indicated by the occurrence of artefacts at sites such as Highlands Farm Pit.

Reports of artefacts from the Black Park Gravel elsewhere in the Thames catchment are uncommon, but several instances are worthy of consideration. In the Thames they occur at Hillingdon (Town Pits, TQ 072824; Wymer, 1968). Palaeoliths from tributary-valley terraces thought to pre-date the Boyn Hill Gravel of the main river have been reported at Farnham in the Wey valley (Oakley, 1939; Arkell and Oakley, 1948), Fordwich in the valley of the Kentish Stour (Smith, 1933; Arkell and Oakley, 1948; Roe, 1977) and in the Silchester Gravel of the Kennet (Wymer, 1968, 1977b; see below, Hamstead Marshall). Only the last of these tributary sites has been firmly established as the equivalent of the Black Park Formation (Gibbard, 1983, 1985), but a similar age for the others seems likely. Few of these sites have produced well-made hand-axes of the type recovered by Wymer (1956, 1961) from Highlands Farm Pit. This typological evidence led Wymer (1961, p. 25) to state that: 'the final filling of the ancient channel of the Thames cannot be earlier than the Hoxnian interglacial', a view to which he still adhered over two decades later (Gibbard and Wymer, 1983).

Recently, evidence from sites such as Boxgrove (Roberts, 1986) and High Lodge has forced most workers to accept that well-made artefacts occur in pre-Hoxnian contexts, so the occurrence of such artefacts in gravels of Anglian age is no longer problematic. The Boxgrove site in particular, having yielded well-made implements in association with a 'late Cromerian' fauna, has undermined the use of hand-axe typology as a basis for the relative dating of Middle Pleistocene deposits (Chapter 1), which leaves no reason why the assemblage from Highlands Farm Pit cannot be derived from *in situ* Black Park Gravel, deposited during the Anglian Stage.

Conclusions

The gravels at Highlands Farm Pit do not lie in the main Thames valley, which lies further to the south, but in an abandoned section of an older Thames course between Caversham and Henley, which has been called the 'Ancient Channel'. This course was abandoned in favour of the modern route through Reading soon after the 'Ancient Channel' deposits were laid down. This was a minor rerouting in comparison with the earlier major diversion brought about by the Anglian ice sheets, when the Thames was deflected into its route through London.

The 'Ancient Channel' deposits (recently equated with the Black Park Gravel of the Slough area) are famous for the profusion of stone tools they have yielded. The occurrence of these man-made implements in such ancient river deposits has been important in showing that Man's earliest arrival in Britain was probably before the Anglian glaciation. The occurrence of well-made flint tools at 'early' sites such as Highlands Farm Pit has helped to dispel the belief that the quality of Palaeolithic workmanship is a reliable indication of age.

HAMSTEAD MARSHALL GRAVEL PIT (SU 414662)

D.R. Bridgland

Highlights

The Hamstead Marshall site provides a rare combination of sedimentological, archaeological and pedological evidence that can be used to interpret the Silchester Gravel of the River Kennet, one of the best-preserved terrace deposits in southern Britain.

Introduction

Hamstead Marshall Gravel Pit is of considerable importance for Pleistocene stratigraphy in the western London Basin, as it is the type locality of the Hamstead Marshall Terrace of the River Kennet (Chartres *et al.*, 1976; Chartres, 1981, 1984). This, the most extensive and best-preserved of all the Kennet terraces, was formerly referred to as the 'Silchester Stage' (White, 1907), a name revived by Gibbard (1982) for his Silchester Gravel. This aggradation was generally correlated with the Winter Hill Terrace of the main river (Ross, 1932; Wooldridge, 1938; Thomas, 1961; Wymer, 1968) until reappraisal of the terrace stratigraphy in the main Thames valley led Gibbard (1979, 1983, 1985) to equate it with the slightly later Black Park Gravel Formation. Either way, an Anglian age is implied, indicating aggradation during probably the most severe climatic interval during the Pleistocene.

The discovery of Palaeolithic artefacts in the gravel at Hamstead Marshall (Wymer, 1968; Roe, 1981) supports the view that humans occupied southern Britain before the Anglian Stage glaciation of the London Basin.

Description

Sections in the Hamstead Marshall pit reveal *c.* 3.0–3.5 m of gravels with occasional sandy and clayey lenses, overlying buff-coloured sands of the (Palaeogene) Reading Beds. The contact with the top of the latter is undulating, suggesting incision by a network of braided river channels prior to deposition of the basal gravels. The sequence can be divided as follows (Chartres *et al.*, 1976; Chartres, 1981, 1984; Fig. 3.13):

	Thickness
4. Topsoil	0.4 m
3. Silty gravel, unbedded	0.8 m
2. Clayey gravel, reddened matrix, especially in lower half; manganese stained near top; poorly bedded	1.0 m
1. Sandy gravel, well-bedded	up to 1.0 m
Reading Beds	

Chartres interpreted these divisions as of pedogenic rather than sedimentary origin. Micromorphological studies have shown that all but the basal part of the gravel has been affected by complex pedogenesis under various climatic conditions (Chartres, 1984).

The gravel, predominantly of coarse grade, is

composed of flints with occasional sarsens. Few sedimentary structures are evident and clear stratification is confined to the lowest division (unit 1). The most noticeable structures result from post-depositional disturbance by periglacial processes, particularly prevalent in the uppermost metre, where involutions and vertically orientated stones are in evidence (Fig. 3.13). Chartres (1975) interpreted several scoop- or bowl-shaped hollows as ice-wedge casts, but these are not laterally extensive and seem more likely to have resulted from solution of Chalk clasts within the gravels (D.T. Holyoak, pers. comm.).

Interpretation

Hamstead Marshall Gravel Pit is important as a representative section in the deposits that form the terrace for which it is the type locality, the Hamstead Marshall Terrace of Chartres *et al.* (1976). This aggradation was previously attributed to the 'Silchester Stage', the lowest of three early Kennet gravels identified by White (1907). The sequence described by White was as follows:

1. Cold Ash Stage (80 m above the floodplain)
2. Bucklebury Stage (70 m above the floodplain)
3. Silchester Stage (47 m above the floodplain)

Gibbard (1982) used the term Silchester Gravel to classify the lowest of White's 'stages'. This is one of the best-preserved terrace aggradations in southern Britain, forming a wide and nearly continuous plateau on the southern flank of the Kennet valley from Wash Common (SU 455648), near Newbury, to Burghfield Common, near Reading (Wymer, 1977b). The surface of this gravel falls from 120 m O.D. to 95 m O.D. between these points, while the width of the gravel sheet reaches a maximum 5 km at its eastern (downstream) end. The longitudinal gradient shows a flattening downstream, from 1.9 m to 1.3 m per km. This remarkable spread of gravel attracted the attention of numerous early geologists; notably, in addition to White, Buckland (1823), Monckton (1892), Blake (1900, 1903) and Hawkins (1928).

A possible reason for the exceptional preserv-

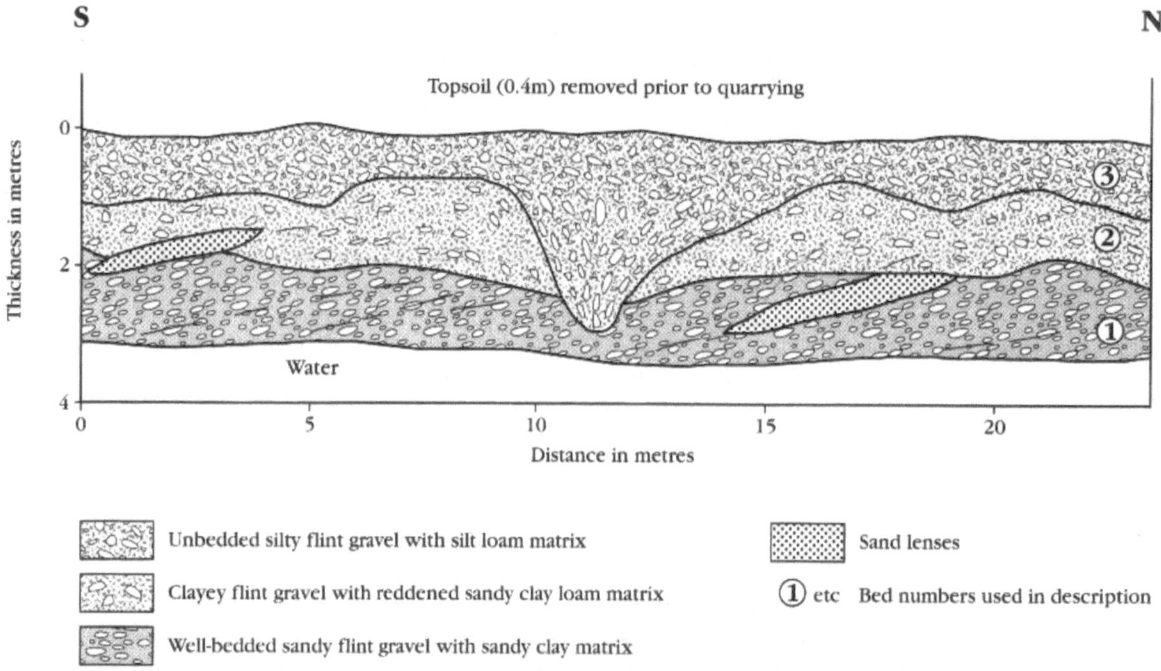

Figure 3.13 Section at Hamstead Marshall Gravel Pit, showing the three divisions of the Silchester Gravel at this site, as described in the text (numbers apply to description). Note that the boundary between divisions 1 and 2 cuts across the sedimentary bedding. Also note central ground-ice/solution structure (after Chartres *et al.*, 1976).

ation of this deposit is that subsequent phases of Kennet development were constrained by the river having cut through the Tertiary strata, which underlie the older gravels, to become entrenched in the Chalk. As noticed by Thomas (1961), the Silchester Gravel marked the end of a progressive southward migration by the Kennet, attributed by him to uniclinal shifting. Bridgland (1985b) has observed that neither migration nor the development of extensive terrace remnants are usual in valleys cut through Chalk.

Wooldridge (1938) correlated the Silchester Gravel with the then newly defined Winter Hill Terrace of the Thames (Saner and Wooldridge, 1929), an interpretation that was confirmed by Thomas (1961), who equated it with the Lower Winter Hill Terrace of Sealy and Sealy (1956). The spreads of gravel in the Thames valley with which correlations were made include the 'Ancient Channel gravel' between Mapledurham and Henley, originally included in the (Lower) Winter Hill aggradation but recently redefined by Gibbard (1983, 1985; Gibbard and Wymer, 1983) as Black Park Gravel (see above, Highlands Farm Pit). Thus the term Silchester Gravel is a synonym, applied in the Kennet valley, for the Black Park Gravel Formation.

Thomas (1961) summarized earlier observations on the gravels beneath the Hamstead Marshall (his 'Lower Winter Hill') Terrace surface, noting that their composition and appearance are remarkably consistent over a wide area. Significant variations were, however, observed in the depth of the gravels. On Wash Common, they attain a thickness of 6.1 m, but become progressively thinner downstream; they decline to 4.6 m on Greenham Common (SU 495648), 3.0 m on Brimpton Common (SU 570630) and further east to as little as 1.8–2.4 m, although they thicken towards their north-eastern margin, near Sulhamstead Abbots (SU 645677), to 3.0 m. Thicker remnants are also found on the north side of the valley, attaining a thickness of 3.7 m on May Ridge (SU 610707) and over 4.3 m at Dark Lane (SU 615745). Thomas (1961) argued that these variations, together with the character of the gravels and the great extent of the terrace, indicate a considerable period of planation followed by a short, rapid period of aggradation, during which the gravels were deposited by a fast-flowing braided river. He also suggested that the planation occurred under interglacial conditions, whereas the gravels were seen as possible products of solifluction from surrounding hillslopes under periglacial conditions.

In comparison with the deposits forming the later Thatcham and Beenham Grange Terraces of the Kennet (Chartres *et al.*, 1976), which are much less well developed, the Silchester Gravel is intensively cryoturbated and generally more weathered in appearance. It is possible that this extensive, relatively thin and uniform gravel-spread results from the frequent lateral shifting of a network of braided river channels. Alternatively, it could represent a long period of deposition in an essentially stable fluvial environment.

The occurrence of palaeoliths on and in the Silchester Gravel (Shrubsole, 1906; Smith, 1915; Wymer, 1968) has attracted considerable interest, on account of the correlation of the former with (originally) the Winter Hill Terrace and (since the early 1980s) the Black Park Gravel of the Thames. Both of these correlations imply contemporaneity with the (Anglian Stage) glaciation of the Vale of St Albans (Hare, 1947; see above, Highlands Farm Pit). There is no direct local evidence for the age of the Silchester Gravel, so correlation with the Thames sequence forms the basis for most attempts at relative dating (Thomas, 1961; Chartres *et al.*, 1976; Chartres, 1980).

The widely held view, prevalent until quite recently, that the earliest human occupation of Britain was during the Hoxnian Stage (see above, Highlands Farm Pit; Wymer, 1961), led Wymer (1968, 1977b) to suggest that most, if not all, of the finds from the Silchester Gravel were probably artefacts dropped on to the terrace surface and incorporated into the top of the deposits by cryoturbation. He noted, however, that the abraded condition of hand-axes from sites on this gravel at Sulhamstead (SU 645680) and, in particular, the pits at Hamstead Marshall indicate that they have been transported with the gravel and so are presumably contemporaneous with it or older. Palaeolithic artefacts have continued to be found in the gravel at the Hamstead Marshall GCR site by local collectors; Wymer (1968) listed five hand-axes (plus a broken fragment) but, by 1992, the number had increased to at least 23 hand-axes (J.J. Wymer, pers. comm.). The occurrence of hand-axes in the Black Park Formation is now well established (see above, Highlands Farm Pit), so the finds at this site need no longer be regarded as anomalous.

Roe (1981) suggested that the abraded hand-axes from sites on the Silchester Gravel, which can be confidently interpreted (because they are water-worn) as derived from the body of the Black Park Formation, are 'archaic-looking', thick and narrow in form. More 'advanced' ovate hand-axes from the same terrace are generally unabraded; Roe agreed with Wymer (1968) that the latter were probably surface finds or artefacts intruded into the top of the gravel by cryoturbation. Roe likened the small assemblage of 'crude' implements from the Silchester Gravel to collections of similar artefacts, from other high-level (pre-Boyn Hill) gravels, that he had previously interpreted as 'Early Acheulian' (Roe, 1964, 1968a). The view that a distinctive and stratigraphically significant 'Early Acheulian' culture existed in Britain during the Middle Pleistocene, promoted by Roe (1964, 1968a, 1981), is less convincing now that well-made hand-axes have been discovered in an apparently pre-Hoxnian context at Boxgrove (see Chapter 1).

The role of pedogenesis in the modification of the Silchester Gravel at Hamstead Marshall has been confirmed by studies of micromorphology and mineralogy (Chartres *et al.*, 1976; Chartres, 1981, 1984). Micromorphological studies showed that the clayey matrix in the middle division of the gravel sequence (unit 2) is of illuvial origin, indicating soil formation under temperate conditions. This clay is strongly orientated, but has been disrupted following illuviation, probably under permafrost conditions. Chartres considered the silt in the matrix of the upper division (unit 3) to be of aeolian (loessic) origin, an interpretation that was supported by the analysis of its mineralogy (Chartres, 1981). This revealed significant quantities of minerals such as epidote, which are rare in pre-Quaternary deposits in the Kennet valley but are common constituents of loess. Chartres concluded that a covering of loess had been deposited on the surface of the gravel and then mixed into the upper disturbed division (unit 3) by cryoturbation. He suggested that this intrusive silt had been prevented from penetrating to depths below 1.2 m from the surface by the clayey layer (unit 2), which had already been formed by pedogenesis prior to the accumulation of the loess.

Micromorphological studies have revealed significant differences, below the levels affected by recent pedogenic activity, between soils formed on the different terraces in the Kennet valley (Chartres, 1975, 1980, 1984; Chartres *et al.*, 1976). As might be expected, the older, higher-level gravels show more evidence for pre-Holocene pedogenesis than those of lower terrace levels; soils on the higher gravels also have a more complex history of modification. Chartres found that the high-level gravels, including those at Hamstead Marshall, were the only ones to contain concentrations of significantly rubified illuvial clay (ferriargillans and papules). These were considered by Chartres to be relict features, since they are generally embedded in yellow illuvial clay, thought to be the product of a later phase of illuviation. The soil has been thoroughly disturbed on at least two occasions; once between the two phases of illuviation, breaking up the reddened illuvial clay, and again after the second illuvial phase. Chartres (1980, p. 140) attributed these disturbance events to cryoturbation, considering that wetting and drying, the principal alternative mechanism, would not have been able to produce the 'general disturbed nature of the profiles'.

The pedogenic evidence from Hamstead Marshall implies that the Silchester Gravel has been subjected to soil-forming processes during at least two temperate intervals and two cold intervals. This may be taken as an indication of relative antiquity. The occurrence of reddened illuvial clay may provide further evidence for the relative antiquity of the gravel. Chartres (1980) cited work in the Paris Basin by Federoff (1971) that suggests that rubification to the degree observed in the earliest papules at Hamstead Marshall is only found there in pre-Saalian soils. The Kennet sequence, in which such reddened material is found in soils formed on the Silchester Gravel and higher formations, but not on the younger Taplow (Thatcham Terrace) Gravel, appears to follow a similar pattern, since the latter deposit is probably of Saalian or post-Saalian age (see below, Brimpton). It has been shown, however, that rubification has occurred under favourable soil conditions in Germany during the Holocene (Schwertmann *et al.*, 1982), raising doubts about the stratigraphical significance of rubified horizons in palaeosols (R.A. Kemp, pers. comm.).

Chartres's research was carried out before the greater complexity of climatic fluctuation during the Middle Pleistocene, as indicated by the oceanic oxygen isotope record (Chapter 1), was widely acknowledged in land-based studies. It is possible that detailed micromorphological analyses in the future will determine that more phases of illuviation and disruption have affected the gravel at Hamstead Marshall than were recognized by Chartres, although the resolution of complex sequences of pedogenic activity in relict soils is fraught with difficulty (Whiteman and Kemp, 1990).

Although the Silchester Gravel is now recognized as an upstream continuation of the Black Park Formation of the Middle Thames (see Fig. 3.1), it is possible that deposits equivalent to the (higher) Winter Hill Gravel, as redefined in this volume (see Chapter 1), are also represented within this aggradation in the upper reaches of the Kennet. This is because of the unique association between the Winter Hill and Black Park Gravels, both of which are at least partly coeval with the glaciation of the Vale of St Albans (see Part 2 of this chapter). The rejuvenation from the Winter Hill to the Black Park floodplain level is believed to have resulted directly from the diversion of the Thames. This rejuvenation is not thought to have extended upstream in the Upper Thames basin beyond the Abingdon area (see Chapter 2, Sugworth). The shallow and steep downstream gradients of the Winter Hill and Black Park Formations (respectively) provide important evidence in support of this interpretation (Fig. 1.3); these formations appear to converge upstream from the Goring Gap. In the Kennet valley, possible remnants of the Winter Hill Formation were recorded in the Beenham area by Thomas (1961), who classified them as Upper Winter Hill Terrace. He correlated these gravels with remnants in the Henley area that Sealy and Sealy (1956) also ascribed to the Upper Winter Hill Terrace. The Lower Winter Hill Terrace of both the Henley area and the Kennet valley is now assigned to the Black Park Formation and the remnants of Upper Winter Hill Terrace recognized by the Sealys have been reinterpreted in this volume as part of the Winter Hill Formation (Chapter 1). Thus the deposits at Beenham probably represent the continuation of the Winter Hill Formation upstream into the Kennet valley (Figs 3.1 and 3.3). As downstream gradients in the Kennet are generally steeper than in the main Thames, foreshortening the various terraces in comparison to their equivalents in the Upper Thames (Fig. 1.3), it might be expected that the convergence of the Winter Hill and Black Park terrace surfaces occurs a shorter distance upstream in the Kennet than in the Thames. This would mean that it occurs between Hamstead Marshall and the Beenham area (Fig. 1.3), so that the deposits at the Hamstead Marshall site include equivalents of both formations. This may explain why remnants of the Silchester Gravel are thicker in the higher reaches of the Kennet than those further down the valley; in the former area aggradation would have continued at the same floodplain level throughout the period of Winter Hill and Black Park Gravel deposition, whereas downstream these two formations are separately represented.

Conclusions

The Silchester Gravel at Hamstead Marshall was deposited during what was probably the most severe climatic phase of the Quaternary Ice Age – during the Anglian Stage. The Hamstead Marshall site shows deposits that form part of a huge spread, or terrace, formed by these ancient Kennet gravels. This terrace is one of the finest and most extensive examples in southern Britain, extending laterally some 5 km at its widest point. It probably accumulated at a time when the river flowed in many individual channels, separated by gravel bars, across a hostile, sparsely vegetated landscape. Detailed analyses of the gravels show that soils subsequently formed on the terrace surface on at least two separate occasions during the warmer 'interglacial' episodes of the Ice Age, with disruption of these soils by frost action during colder phases. The man-made flint implements recovered from the gravels at this site add a further controversial element to its interest, having a major bearing on the time when humans first appeared in Britain.

CANNONCOURT FARM PIT (SU 878831)

D.R. Bridgland

Highlights

A reference site for the Pleistocene Lynch Hill Gravel of the Middle Thames basin, Cannoncourt Farm Pit is also the source of one of the richest and most important assemblages of British Palaeolithic artefacts.

Introduction

Cannoncourt Farm Pit, at Furze Platt, Maidenhead, is one of the most celebrated Palaeolithic sites in Britain. It is a gravel pit, worked by hand during the early part of the century, in which large numbers of artefacts were found, many of elegant form and in excellent condition (Lacaille, 1940; Wymer, 1968). The deposits belong to the Lynch Hill Gravel of the Thames, not one of the earliest terrace formations to be recognized, but now acknowledged as one of the most important, particularly in view of its Palaeolithic content (Wymer, 1968, 1988). The Lynch Hill Terrace was originally defined by Hare (1947), although previously it had been observed locally, principally at Furze Platt, and given a number of different names, as follows: the Furze Platt Stage (Warren, 1926, 1933), based on observations by Treacher (1909); the Furze Platt Terrace (Wright, 1937); Taplow Terrace No. 1 (Burchell, 1934a), the upper of two Taplow Terraces recognized by Burchell (not to be confused with the two divisions of the Taplow Terrace recognized by Sealy and Sealy (1956); see above, Introduction to Part 3); the Iver Stage (King and Oakley, 1936) and the Lower Boyn Hill Terrace (Lacaille, 1940).

Reviews of the Furze Platt site and its Palaeolithic industry were provided by Wymer (1968, 1977c), Roe (1981) and Cranshaw (1983). A recent housing development in the area immediately to the north of the GCR site has provided new sections in the deposits underlying the Lynch Hill Terrace, prompting a reappraisal of the Pleistocene geology at Furze Platt (Harding *et al.*, 1991).

Description

Cannoncourt Farm Pit is part of a complex of old workings that once exploited the Lynch Hill Gravel. It lies in a part of the Thames valley where the river has flowed from north to south since the aggradation of the Black Park Gravel (Hare, 1947; Fig. 3.1). The earliest mention of a site at Furze Platt appears to be that by Treacher (1896), who reported that Palaeolithic implements and waste flakes were found close together at the bottom of the gravel, in a layer comprising large unrolled flints in a sandy matrix. This layer, which was overlain by 4 m of well-stratified gravel, was, according to Treacher, the nearest approximation to a Palaeolithic 'workshop' to have been seen in the area. Treacher (1904) later described only 2.5 m of gravel at the site, the lowest 0.5 m yielding artefacts, 500–600 of which (excluding waste flakes) had been found by that time. The implements described were hand-axes showing a minimum of secondary chipping, mainly of small size, although including a few large and massive specimens. In early descriptions of the Furze Platt artefacts, the economy of labour in their manufacture was noted, a minimum number of blows having been utilized to produce the finished implements (Treacher, 1896, 1904; Shrubsole, 1906). The various early records suggest that the present GCR site was not in existence before 1909 and that Treacher's earlier descriptions appertain to the former 'Cooper's Pit' immediately to the north (Fig. 3.14; Wymer, 1968; Cranshaw, 1983).

The first detailed description of the Furze Platt sections was by Lacaille (1940), who recorded *c.* 3 m of poorly stratified (but clean) sandy gravel overlying Chalk at 42 m O.D. The upper part of the gravel was channelled and overlain by solifluction deposits and sandy brickearth. In addition to palaeoliths, Lacaille found poorly preserved bone fragments at this site, amongst which only an antler of giant deer and a horse tooth were identifiable. He recorded Abbevillian (Early Acheulian), Acheulian and Clactonian artefacts from the site, illustrating many of these as well as a section through the deposits. Further examples of palaeoliths were illustrated by Lacaille (1960) and Wymer (1968). Wymer did not support Lacaille's (1940) claim that the site had also yielded artefacts made using the Levallois technique (see Chapter 1).

Wymer (1968) described the history of Cannoncourt Farm Pit in great detail and included photographs taken while the pit was operational; he had himself conducted excavations there in 1953–4, when he removed some of the last remaining patches of undisturbed gravel from the floor of the pit, finding a number of artefacts in the process. The face at this time showed 1.2 m of brickearth overlying dark-stained gravel, the lower, coarsest part of which yielded a hand-axe and a few flakes. There is a strong indication, from surviving records of the Furze

Platt workings, that implements, particularly the unabraded, well-made examples, were concentrated at the base of the gravel, in places resting directly on the Chalk (Treacher, 1896; Wymer, 1968, 1977c; Roe, 1981; Cranshaw, 1983). According to Roe (1981), the total number of artefacts from this context at the Furze Platt sites exceeds 2500.

A section re-excavated at the Cannoncourt Farm Pit GCR site in April 1987 confirmed the preservation there of nearly 4 m of bedded gravel (Fig. 3.15). This and recent investigations in the area to the north, where housing development has taken place, enabled comparison with exposures in a current working quarry at Switchback Road, less than 1 km to the north (Harding *et al.*, 1991; Fig. 3.14). These sections have allowed a more detailed analysis of the sediments than had been attempted previously. A variable thickness of sandy, silty clay, often with gravelly inclusions, was found to be prominent above the Lynch Hill Gravel throughout the area. This material is repeatedly let down through the gravel in a profusion of large pipes, resulting from solution of the Chalk bedrock. In a small area between Cannoncourt Farm Pit and Cooper's Pit, the clay is itself overlain by a later gravel containing more local material than the Lynch Hill Formation (Table 3.2). The clay beneath this upper gravel is mottled and reddened and contains evidence of a palaeosol (Harding *et al.*, 1991; see below).

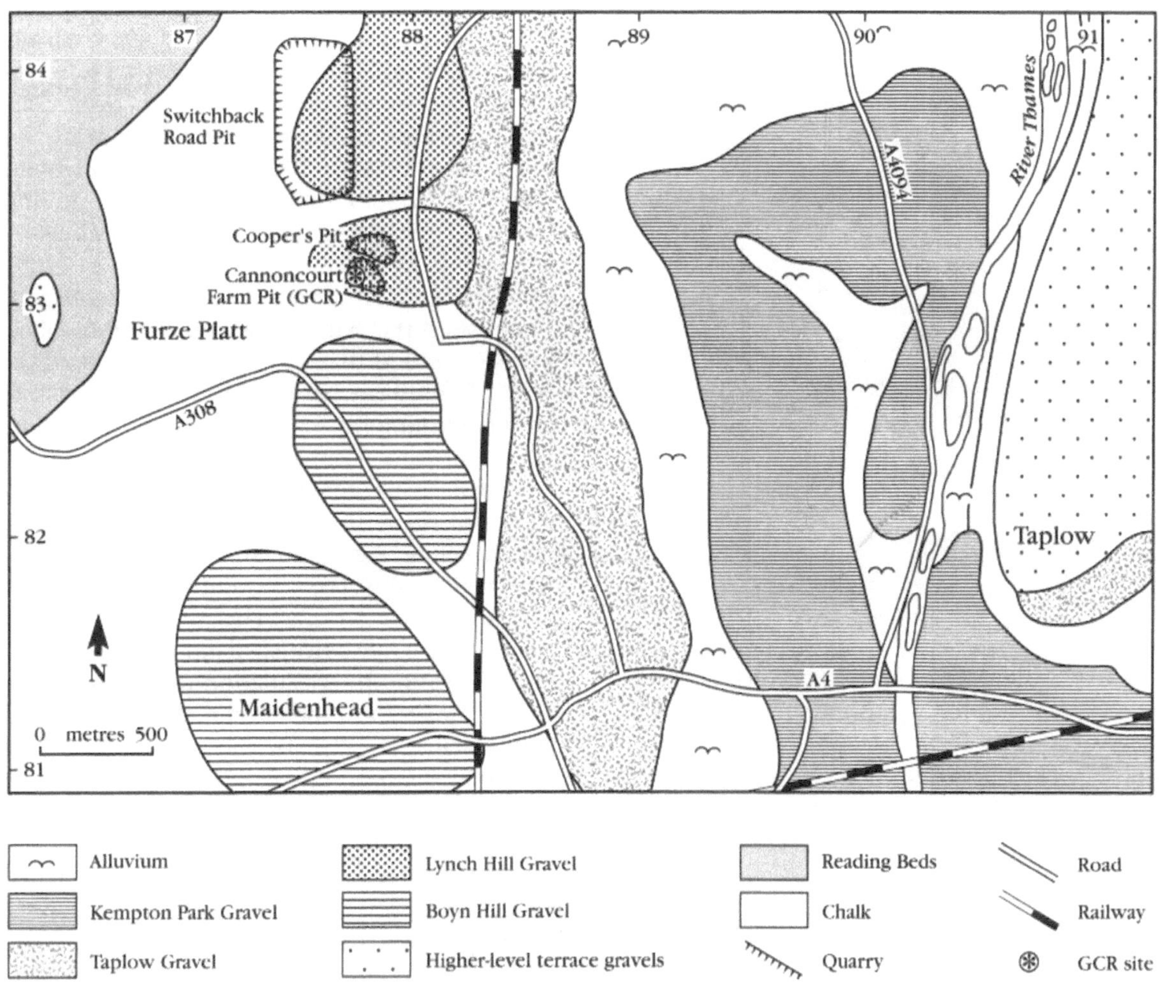

Figure 3.14 Map showing the various sites at Furze Platt and their relation to the Pleistocene geology.

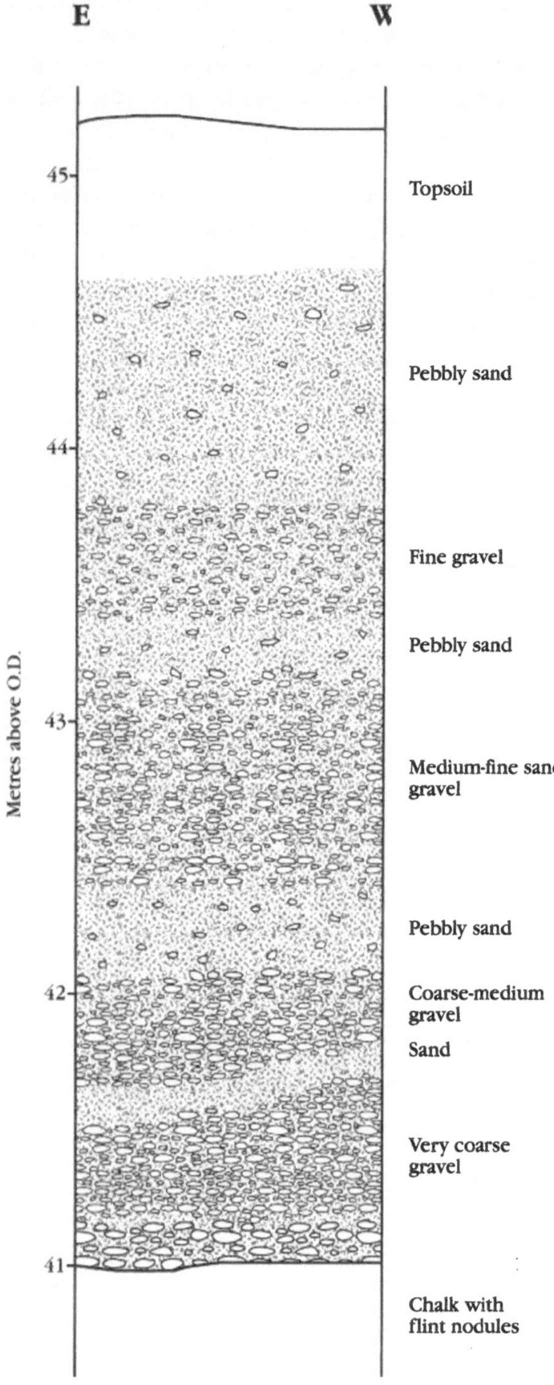

Figure 3.15 The GCR section excavated at Cannoncourt Farm Pit in April 1987. For location, see Fig. 3.14.

Interpretation

There is an extensive history of publication on the deposits at Furze Platt and their Palaeolithic content. Of particular significance was a Geologists' Association excursion visit to the locality in 1909, under the directorship of L. Treacher and H.J.O. White, who introduced on this occasion (Treacher, 1909) the idea that two distinct terraces existed in the area within the highest of the three originally recognized by Whitaker (1889). The higher of these two divisions of the 'High Terrace' (now recognized as the Boyn Hill Gravel) was exposed in a pit near Furze Platt church, in which several abraded artefacts were found, whereas the lower was represented at 'Cannoncourt (Furze Platt lower pit)', where the artefacts were more varied and less abraded (Treacher, 1909, p. 199). The latter working, almost certainly Cooper's Pit (Fig. 3.14), had at that time recently been abandoned, although a new pit opened nearby had already yielded a number of implements (Treacher, 1909). It seems likely that this newer working was the site later to be known as Cannoncourt Farm Pit and that records from before this date (Treacher, 1896, 1904; Shrubsole, 1906) all refer to Cooper's Pit.

The majority of the implements from Cannoncourt Farm Pit were discovered during the period from 1909 to 1931, mainly by the gravel diggers, who sold them to collectors, notably Llewelyn Treacher of Twyford and George Smith of Reading (Wymer, 1968; Cranshaw, 1983). Cranshaw (1983), after studying the numerous unpublished records that accompany the collections, concluded that 1919 probably marked the acme of hand-axe discovery. In this year a famous implement, 32 cm long, was unearthed in Cannoncourt Farm Pit by a gravel digger, Mr G. Carter, who achieved acclaim as the discoverer of many fine hand-axes (Lacaille, 1940; Wymer, 1968; Figs 3.16 and 3.17a). Many of these were of the slender pointed variety known as 'ficrons' (Fig. 3.17b). The 32 cm implement remains the largest British hand-axe, its enormous size leading to speculation that its intended purpose may have been ceremonial (Wymer, 1968; MacRae, 1987). By the end of the 1930s, over 1600 hand-axes had been recovered from Furze Platt, mostly from the newer pit (Roe, 1968b). No formal descriptions of the site or of the Palaeolithic assemblage were published during this period, however, although they were briefly mentioned in the Beaconsfield Geological Survey memoir (Sherlock and Noble, 1922).

Warren (1926, 1933) followed Treacher in recognizing the gravel at Furze Platt as distinct

Figure 3.16 Photograph taken in 1913 of Mr George (Deffy) Carter working at Cannoncourt Farm Pit, sieving gravel. The digging and sorting of gravel by hand, before mechanization, resulted in frequent discoveries of palaeoliths. Mr Carter found many of the artefacts in the collections from Cannoncourt Farm Pit, including the 32 cm specimen (Figure 3.17a) that remains the largest from Britain. (Photo: L. Treacher, reproduced by courtesy of J.J. Wymer.)

from the established (Boyn Hill and Taplow) terraces of the Middle Thames. He ascribed the artefacts from these deposits to his 'Grays Inn Lane Group', named after the site (in London) of the earliest recorded discovery in Britain of a Palaeolithic hand-axe, in 1690 (Evans, 1860). In a footnote, Warren (1926, p. 43) suggested that the gravels in which these artefacts occurred might be separated under the name 'Furze Platt Stage'. Warren (1933) further developed this theme by linking the Furze Platt industry with that from Swanscombe (Middle Gravels), both being assigned to his 'Grays Inn Lane Group'. Warren (1942) also correlated the Furze Platt aggradation with Palaeolithic gravels at Stoke Newington (Smith, 1894) and Leytonstone; he

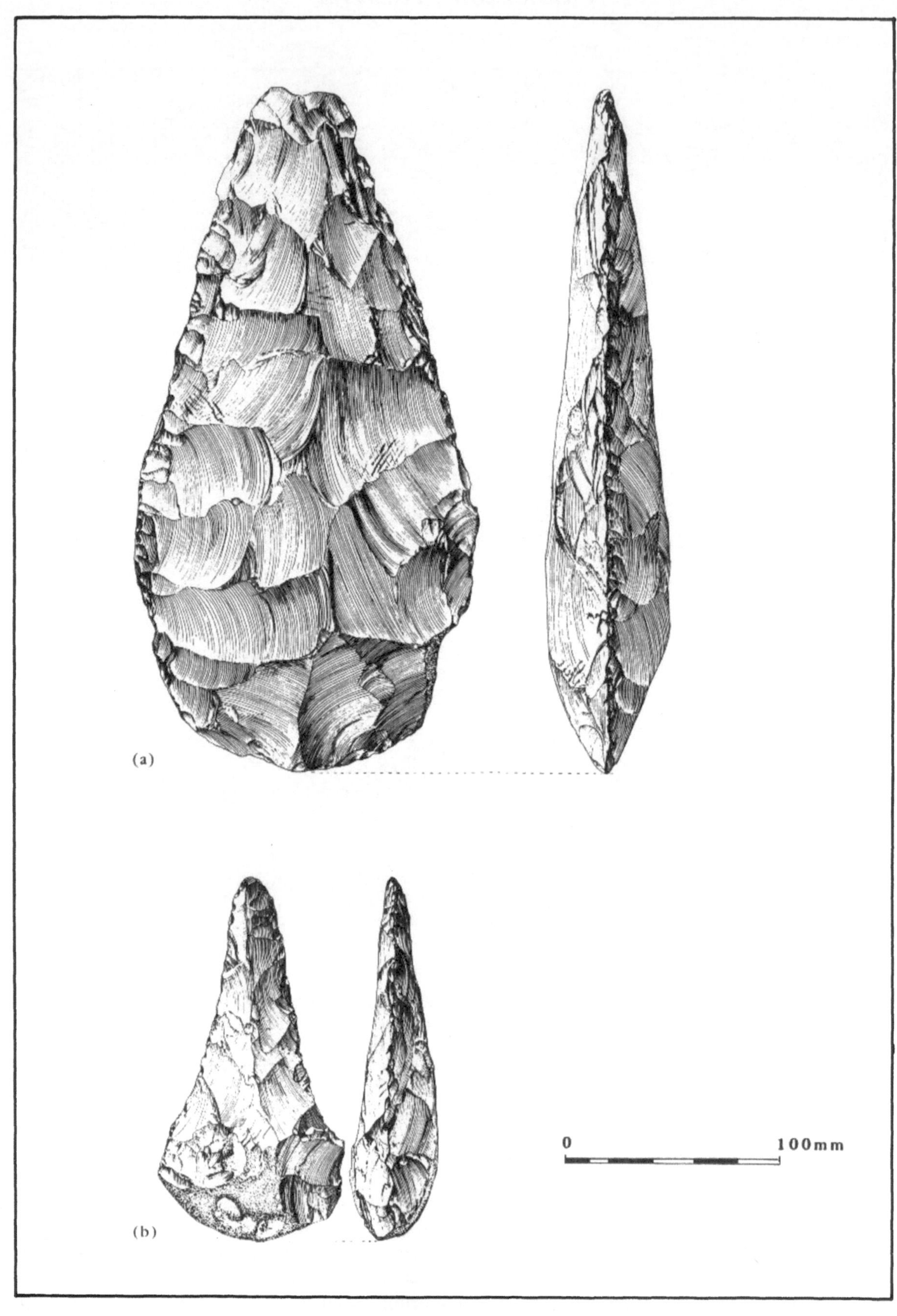

Figure 3.17 Hand-axes from Cannoncourt Farm Pit: drawings originally published by Lacaille (1940).
(a) The extraordinarily large pointed hand-axe found by Mr G. Carter in March 1919. This was acquired by L. Treacher and donated by him to the Natural History Museum. (b) A fine example of a 'ficron' hand-axe.

assigned the Clactonian gravels of Swanscombe (Lower Gravel), Grays and Clacton itself to the early part of the 'Furze Platt Stage'. In a brief review of the Thames sequence, Wright (1937), alluding to the deposits at a lower level than the Boyn Hill Gravel, adopted the name 'Furze Platt Terrace'.

Lacaille (1940, p. 247) also recognized three terraces above the floodplain in the Maidenhead area; he stated that: 'The Taplow Terrace, which is the most distinct, is banked against another terrace, the ground-surface of which, east of Cannoncourt Farm, stands between 154 ft and 140 ft O.D. A shelving and undulating rise of some 20 ft marks the step to another terrace, attaining a maximum surface altitude of 171.5 ft above O.D. and 93 ft above the Thames, which rests on Chalk and extends southward. This includes the classic locality of Boyn Hill'. Lacaille recognized his 'Lower Boyn Hill Terrace' (the middle of these three, now classified as the Lynch Hill Terrace) on both sides of the river. He described exposures in the deposits of this terrace at various sites, including Cannoncourt Farm Pit.

King and Oakley (1936) believed the Furze Platt (Lynch Hill Gravel) deposits and the higher Boyn Hill Gravel to be the product of a single aggradation, built up over two separate erosional 'benches' (Fig. 3.18). They regarded the abraded artefacts from the Boyn Hill Terrace *sensu stricto* at Maidenhead as having been reworked from industries in the lower gravel (that now attributed to the Lynch Hill Formation), such as at Furze Platt. They assigned this single aggradational phase to their 'Middle Barnfield Stage', supporting Warren's (1933) correlation of the industries of Furze Platt and the Swanscombe Middle Gravel. King and Oakley also recognized another aggradational phase between the Boyn Hill and Taplow Terraces, their 'Iver Stage' (based upon observations at Iver by Lacaille). Unlike that at Furze Platt, the deposit at Iver contains Levallois flakes, together with hand-axes in abraded condition. This suggested to King and Oakley that the Iver deposits were later than those at Furze Platt. However, Gibbard (1985) followed Hare (1947) in assigning the gravels at both Furze Platt and Iver to the Lynch Hill Formation. According to Gibbard, the Levallois industry at Iver was derived from an accumulation of

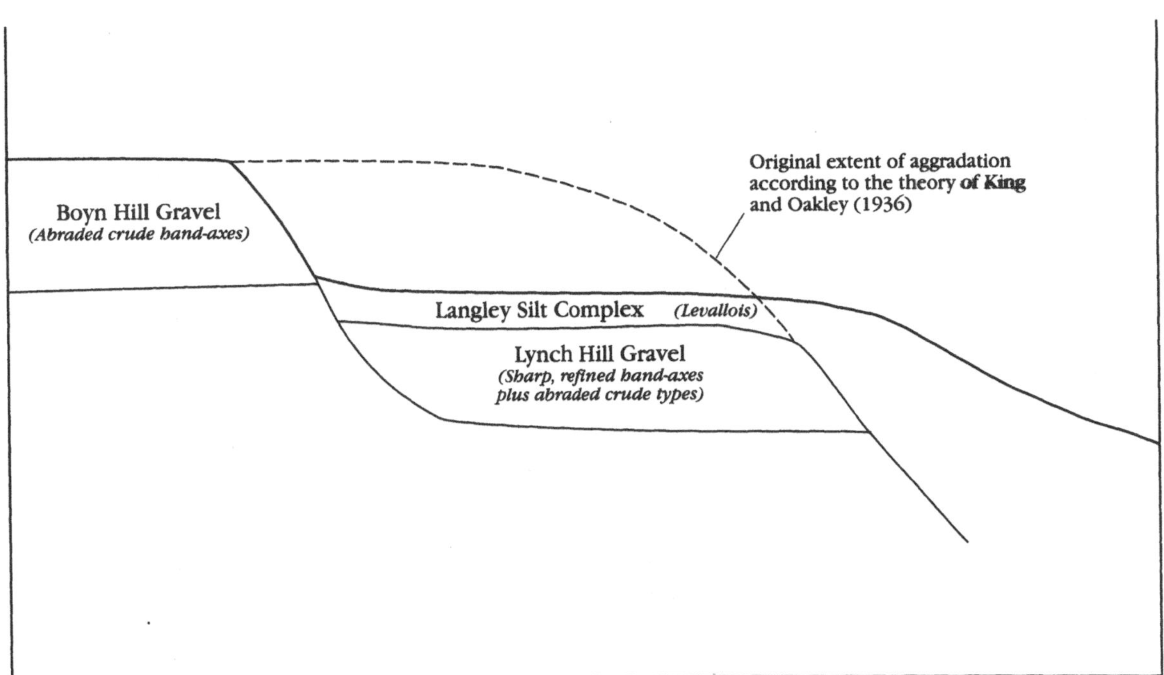

Figure 3.18 Diagrammatic representation of the relations between the Boyn Hill and Lynch Hill Gravels and the Langley Silt Complex. The types of Palaeolithic artefacts that characterize each deposit are shown. Also illustrated is the interpretation of these deposits by King and Oakley (1936).

fine-grained sediments above the Lynch Hill Gravel, dominated by wind-blown silt (loess). He assigned these fine-grained sediments to his 'Langley Silt Complex', which overlies (and therefore post-dates) the Lynch Hill Gravel, thus vindicating King and Oakley's interpretation of the archaeology.

The geomorphological and geological approaches of Hare (1947) and Gibbard (1985) produced very different interpretations of the fluvial sequence to those based on archaeology, however. Hare was unable to accept King and Oakley's single Furze Platt/Boyn Hill aggradation; he noted that the Furze Platt gravels underlie a separate terrace surface, lower than the true Boyn Hill level, which he named the Lynch Hill Terrace. Oakley (in Hare, 1947) suggested that later erosion could have produced this lower terrace feature within the single aggradation previously envisaged, a view that has continued to receive occasional support (see Wymer, 1977c; Roe, 1981). However, Cranshaw (1983) reversed King and Oakley's argument, suggesting that occasional abraded and undistinguished artefacts found at Furze Platt might have been reworked from the (older) Boyn Hill aggradation, which is characterized by similar implements. Conversely, it is clear from the various records that the well-preserved and skilfully made hand-axes found in Cannoncourt Farm Pit around 1919 have no parallel, abraded or otherwise, in the local Boyn Hill Gravel. Such implements would be expected to appear in the latter formation, in reworked condition, if the stratigraphy advocated by King and Oakley was correct. Detailed mapping by Gibbard (1985) of the two separate sediment bodies, the Boyn Hill and Lynch Hill Gravels, appears to have resolved this issue in favour of an interpretation of the Furze Platt deposits as later than those of the Boyn Hill Formation, in line with conventional terrace stratigraphy (Fig. 3.18).

Roe (1964), in his statistical analyses of Palaeolithic collections, used the Furze Platt assemblage as an example of a 'Middle Acheulian' industry. This study demonstrated similarities and differences in typology between the Furze Platt industry and those from other British Lower and Middle Palaeolithic sites. The site was also described by Roe (1968a, 1981) in his analysis of British Lower and Middle Palaeolithic hand-axe types. Roe's analyses confirmed the similarity of the Furze Platt artefacts to an assemblage from an identical stratigraphical situation, concentrated near the base of the Lynch Hill Gravel, on the opposite (eastern) side of the Thames at Baker's Farm (SU 958822). This was another pit from which collections were assembled by Treacher and Lacaille (Lacaille, 1940, 1960; Wymer, 1968). Roe (1968a, 1981) also grouped the Furze Platt industry with those from Stoke Newington and Cuxton and suggested a Mindel–Riss (Hoxnian) age for these assemblages. Wymer (1968) considered the industry from Cannoncourt Farm Pit to belong to a single phase of the Acheulian culture, with a few rolled specimens possibly derived from earlier assemblages. He denied the occurrence of Levallois flakes or cores, as had been claimed by Lacaille, although he was more convinced by Levallois material collected by Lacaille from Baker's Farm.

The past few years have seen renewed research activity at Furze Platt. The small-scale GCR excavation in Cannoncourt Farm Pit in 1987 closely followed an appraisal by the Trust for Wessex Archaeology of the nearby Switchback Road Pit (Fig. 3.14) prior to its extension (Harding *et al.*, 1991). In 1988 the same organization undertook a detailed study of those parts of the Cooper's Pit/Cannoncourt Farm Pit complex to the north of the GCR site, before and during construction of a new housing development.

These new studies included a detailed analysis of the sandy, silty clay overburden at the various sites. Particle-size and heavy-mineral analyses indicated that this material differs significantly from the 'Langley Silt Complex' of Gibbard (1985), as described elsewhere in the Middle Thames (Gibbard *et al.*, 1987). It seems instead to largely comprise sediment reworked, probably by solifluction, from the local Reading Beds. Of potential significance to the dating of the Lynch Hill Formation of the Middle Thames is the discovery of a palaeosol developed in this silty clay, beneath what appears to be a later tributary gravel (on the basis of clast-content – see Table 3.2). Micromorphological analysis of this buried soil horizon, which displays red-brown mottling not seen elsewhere in the silty clay, has confirmed that modification by pedogenesis occurred under temperate conditions prior to the deposition of the overlying gravel. If the latter deposit is attributed to a cold climatic interval, it being unlikely that a tributary stream could have carried gravel to this location during the Holocene interglacial (most such valleys in the area, being developed on Chalk, are dry at the present time), the palaeosol must date from a pre-Holocene temperate episode. This provides a minimum age for the Lynch Hill Gravel underlying the brickearth, which must pre-date the last interglacial (Harding *et al.*, 1991).

In fact, the correlation scheme for the Thames terrace sequence advocated in Chapter 1 suggests that the Lynch Hill Gravel is significantly older than this. It is ascribed, according to the climatic model for terrace formation also proposed in Chapter 1, to Oxygen Isotope Stages 10 (phase 2 gravel) and 8 (phase 4 gravel). Assuming the phase 2 part of the Lynch Hill Formation (see Chapter 1) to be present at Cannoncourt Farm Pit, this interpretation suggests manufacture of the artefacts that were found just above the Chalk surface during Stages 10 or 11. The latter age would indicate broad contemporaneity with the Swanscombe Middle Gravel industry, an interpretation that had been promoted, on typological grounds, by Warren (1933). However, it is also possible that the gravels overlying the Chalk at Furze Platt represent the phase 4 part of the Lynch Hill aggradation, in which case the artefacts could be later, dating perhaps from Stage 9 or early Stage 8. According to the correlations proposed in this volume, this would suggest a similar age to the assemblages from sites such as Stoke Newington, grouped with Furze Platt by Warren (1942) and Roe (1968a, 1981), and Purfleet (see Chapter 4). This alternative interpretation is perhaps the more likely, given the occurrence of artefacts showing Levallois technique at the related site at Baker's Farm.

Thus the Cannoncourt Farm Pit GCR site provides both a reference section in the Lynch Hill Gravel of the Middle Thames and a reserve of unexcavated ground to enable future research, including possible excavation to assess the Palaeolithic content of this formation. The site is one of the richest and most celebrated British Palaeolithic localities, a fact that is of considerable geological significance, given that the artefacts have all come from Thames gravel. Continuing research, both at Furze Platt and elsewhere, may soon provide an improved geochronological framework for this important Palaeolithic assemblage. Such studies will also provide more information about the geological history of the Lynch Hill Formation and related sediments, which form an important part of the Thames sequence.

Conclusions

The deposits at Cannoncourt Farm Pit are part of the Lynch Hill Gravel of the Thames, one of the lowest in the celebrated 'staircase' of terraces preserved in the Slough area. Here, and at adjacent sites, this gravel has long been famous for yielding a profusion of man-made flint tools and the waste-flakes from their manufacture – over 2500 such Stone-age artefacts have been found to date. Cannoncourt Farm Pit is particularly notable for the discovery there of the largest Palaeolithic hand-axe ever found in Britain.

The gravel at Cannoncourt Farm is overlain by a pebbly, silty clay deposit that, in places, has collapsed through the gravel into large solution hollows (pipes) that penetrate down into the underlying Chalk. Analysis of the silty clay shows that it is largely composed of redeposited material derived from the local Reading Beds (a much older set of sediments deposited around 60 million years ago), suggesting that it was formed not by the Thames, but through a process of slope movement, particularly during periods of intense cold within the Ice Age. A temperate-climate soil formed in this silty clay, preserved where it is buried by later gravel, indicates that there has been at least one temperate (interglacial) period since deposition of the Lynch Hill Gravel and the overlying cold-climate slope sediments.

FERN HOUSE GRAVEL PIT (SU 883885)

D.R. Bridgland

Highlights

This site provides an exposure in gravels of the Taplow Gravel Formation. The sediments here, and their contained fossils, may hold the key to elucidating a poorly understood part of the history of the Thames, that between the (glacial) Anglian Stage, when the river was diverted into its modern course, and the last (Ipswichian) interglacial.

Introduction

Fern House Gravel Pit, Buckinghamshire, also known as 'Ferns Pit' and 'Well End Pit', exposes sediments of the Taplow Gravel Formation overlain by several metres of colluvial gravels. This pit, which lies *c.* 10 km upstream from the Taplow type locality, was much visited earlier in this century, when it yielded a number of elephant remains, mostly teeth of mammoth.

The Taplow Terrace, the geomorphological feature formed by the Taplow Gravel, was one of the three Thames terraces originally defined by the Geological Survey (Bromehead, 1912). It had previously been referred to as the 'Middle' or '50 ft Terrace' (Hinton and Kennard, 1900, 1905, 1907; Pocock, 1903). The status of this formation as an important element of the Thames sequence was consolidated in later work (Dewey and Bromehead, 1915; Sherlock and Noble, 1922; King and Oakley, 1936; Oakley, 1937; Hare, 1947). Sealy and Sealy (1956) proposed a division into Upper and Lower Taplow Terraces, but these are not considered to represent separate fluvial aggradations (Gibbard, 1985).

According to Gibbard (1985), the Taplow Gravel is characterized by a mammalian fauna indicative of severe cold. However, the occurrence of occasional remains of temperate species suggests that deposition of this formation also spanned one or more temperate intervals.

Description

Fern House Pit appears first to have been operational in the early years of this century. Many years earlier, however, sections showing a 4 m thickness of (Taplow) gravel were recorded 10 km further downstream, at what was to become the type locality of both this terrace and the gravel that forms it, Taplow Station Pit (Owen, 1855; Prestwich, 1855; Sherlock and Noble, 1922; Oakley, 1937). This site yielded bones of mammoth, woolly rhinoceros and musk ox (Sherlock and Noble, 1922), the last being the first record of this species in Britain. Musk ox was later considered by Gibbard (1985) to be particularly characteristic of the Taplow Gravel.

The earliest description of Fern House Pit appears to be that by Treacher (1916) who, in a Geologists' Association excursion report, observed that two types of gravel were exposed: a lower sandy, current-bedded deposit, which he thought likely to belong to the Taplow Terrace, and an upper, less well-bedded, clayey spread, which he suggested was hillwash from Flackwell Heath to the north. During the visit, a fragment of mammoth's tooth was discovered in the lower gravel (Treacher, 1916). A pit west of Well End, probably the GCR site, was described by Sherlock and Noble (1922), who reported up to 8 m of gravel, the lower metre of chalky composition (this thickness presumably includes the upper colluvial gravel). The presence of Chalk clasts in fluvial gravels is exceptional in the Thames terrace sequence and is restricted to deposits immediately overlying the source (bedrock Chalk or coombe rock). The calcareous nature of the lower gravel may explain the preservation in it of mammalian remains in some quantity (see below).

The Geologists' Association again visited the pit in 1933 (Treacher, 1934), when the distinction between the lower fluviatile gravel and the upper 'slopewash' material was once again noted. Treacher suggested that the latter comprised material washed down from the neighbouring hills through a series of short steep valleys that converged to form a fan in the vicinity of the pit. According to Treacher, mammalian remains were commonly found in the gravel, especially molar teeth of mammoth (***Mammuthus primigenius***), although at least one tooth of straight-tusked elephant (***Palaeoloxodon antiquus***) had also been found. Treacher also noted on this occasion that the two types of gravel were separated by a band of 'loam' (silt?) averaging 1.2 m thick.

Wymer (1968) noted that the site had been overgrown since 1960, when he had observed 3 m of 'solifluction gravel' overlying sandy bedded gravel. According to Wymer, the upper gravel contains a high proportion of quartzite pebbles, indicating that it represents the redistribution of one of the early Thames drifts to the north, as suggested by Treacher. Although Treacher (1934) maintained that no artefacts had been found in the pit, Wymer believed that a very rolled primary flake in the Ashmolean Museum, found in 1941 (after Treacher's observation) and marked 'Bourne End Pit', might have originated there. Sherlock and Noble (1922) claimed the Taplow Terrace to be associated with Mousterian (Levallois) artefacts, but Gibbard (1985) considered such material to have come from wind-blown silt (loess) overlying the fluvial deposits, his 'Langley Silt Complex'. According to Gibbard, therefore, these artefacts post-date the deposition of the Taplow Formation (see, however, Chapter 4, Lion Pit and Northfleet). Gibbard provided the results of a stone count from this pit (Table 3.2), showing a typical Taplow Gravel composition, but lacking the Chalk reported by Sherlock and Noble. Gibbard's sample presumably came from the upper (Chalk-free) part of the bedded gravel at the site.

Interpretation

The Taplow Terrace and underlying Taplow Gravel have an important place in the later part of the Thames terrace succession. The middle of the three terraces originally mapped in the Maidenhead area (see above and Introduction to Part 3), this aggradation has been widely recognized throughout the Middle and Lower Thames basins (see Chapter 1). Correlation of Taplow remnants throughout this area is not without its problems, however. For example, recent work in the Lower Thames suggests that it has previously been misidentified downstream from London (see Chapters 4 and 5).

Following the introduction of the terms Taplow Terrace and Taplow Gravel by the Geological Survey (Bromehead, 1912; Dewey and Bromehead, 1915), King and Oakley (1936) proposed the name 'Taplow Stage' for the period represented by this aggradation. This was linked to the 18 m 'Main Monastirian Sea-Level' by Zeuner (1945), who believed the gradient of the terrace downstream from Ealing to be negligible (after Pocock, 1903). This now appears to be a mistaken impression that arose from miscorrelation across the London area; recent re-evaluation of the sequences in the Lower Thames and eastern Essex indicates that the Taplow Gravel declines continuously towards a low sea level (see Fig. 1.3; Chapters 4 and 5). This is consistent with aggradation of the Taplow Formation during a cold episode, as is indicated by the mammalian fauna. However, east of London the gravel is overlain by or interbedded with fine-grained sediments of probable estuarine origin that are ascribed to a high (interglacial) sea level, thought to be equivalent in age to Oxygen Isotope Stage 7 of the oceanic record (see Chapter 4).

Hare (1947), in his detailed geomorphological mapping of the Slough–Beaconsfield area, noticed a minor, lower sub-facet of the Taplow Terrace. Later authors used this as a basis for a division into Upper and Lower Taplow Terraces (Sealy and Sealy, 1956; Thomas, 1961; Evans, 1971). The importance of 'Ferns Pit', and the descriptions of it by Treacher, were acknowledged by Sealy and Sealy (1956). These authors attributed the upper gravels at this site to rapid deposition during a cold episode following the aggradation of the lower bedded gravel, which they assigned to their Upper Taplow Terrace. As this implies, they correlated the fluvial deposits at Taplow level in the Marlow district, including the lower gravel at Fern House Pit, with the Upper Taplow Terrace of the type area, around Slough, which they believed to be the major of the two divisions. Gibbard (1985), however, considered Sealy and Sealy's division of the Taplow Terrace to be of geomorphological significance only. He observed that the Upper Taplow Terrace of the type area is formed by the Taplow Gravel aggradation with an overlying layer of wind-blown silt (loess). The localized removal of this later silt has given rise to the minor, lower facet originally described by Hare. The upper surface of the Taplow Gravel, which is Sealy and Sealy's Lower Taplow Terrace, is therefore the true expression of the fluvial aggradation; the higher feature has no significance to the history of Thames drainage evolution. The layer of 'loam' at Fern House Pit, between the lower bedded gravel (presumed to represent the Taplow Formation) and the upper colluvial gravel, may well represent this wind-blown accumulation, the 'Langley Silt Complex' of Gibbard (1985).

Gibbard's reinterpretation of the Upper and Lower Taplow Terrace features applies only to the area downstream from Fern House Pit, however. The Taplow Formation has been entirely removed by erosion in the district immediately upstream from Marlow, but further west, in the Reading area, separate geomorphological terraces have again been recognized at both the Upper and Lower Taplow levels (Sealy and Sealy, 1956), despite the absence in this district of loessic overburden. Wymer (1968) recognized these as his Henley Road and Taplow Terraces, respectively. Gibbard (1985) considered the higher of these (Wymer's Henley Road

Terrace) to represent the Taplow Gravel *sensu stricto* and interpreted the lower as a unit peculiar to this area, his 'Reading Town Gravel'. He attributed this deposit to aggradation during the Early Devensian, although noting that possible temperate-climate deposits had been described from a site at this level known as Redlands Pit (SU 732727) (Poulton, 1880; Wymer, 1968). Gibbard considered that these were probably residual Ipswichian Stage deposits and that the Reading Town Gravel had been banked against them. The Redlands Pit site, some 25–30 km upstream from Fern House Pit, yielded reworked Jurassic, Cretaceous and Palaeogene Mollusca, together with (Pleistocene) bones of ox, horse, mammoth and rhinoceros (Poulton, 1880). Poulton also described tree trunks, which Wymer believed to have been reworked from the Palaeogene (leaf-beds occur in the immediate vicinity within the Reading Beds). Gibbard (1985) thought that these tree remains, tentatively identified as *Pinus* by Poulton, might equally well have been of Pleistocene age, although the original description indicates that they were partly mineralized, which would seem to support Wymer's interpretation. If they were of Palaeogene origin, the only evidence for a Pleistocene temperate episode from this pit would be the mammalian fossils, none of which can be regarded as indicating fully interglacial conditions.

Another pit in Reading on the Taplow Terrace of Wymer (1968), the 'Reading Town Gravel' of Gibbard (1985), provided stronger evidence for interglacial conditions in the form of teeth of hippopotamus (Shrubsole and Whitaker, 1902; Wymer, 1968). This site, on a separate remnant to Redlands Pit, was at Kensington Road (SU 698734). The records provide no information about the context of these remains, but there is no indication that they were associated with fine-grained or organic sediments. Gibbard (1985) cited the occurrence of these remains in his 'Reading Town Gravel' as important evidence for the Early Devensian age of the latter, since hippopotamus is believed to have been present in Britain, after the Anglian Stage, only during the last interglacial, the Ipswichian Stage (*sensu* Trafalgar Square) (see Chapter 1 and Chapter 2, Stanton Harcourt and Magdalen Grove). Attribution of the mammalian assemblage from Redlands Pit to the Ipswichian may be problematic, however, since horse is widely believed to be absent from that stage (Shotton, 1983), although Stuart (1982a) considered that it disappeared at the beginning of the Ipswichian, to reappear in pollen biozone IpIII.

Reassessment of the altitudinal distribution of gravel remnants in the Reading area attributed to the post-Lynch Hill aggradations of the Thames (Chapter 1) suggests a different interpretation to that of Gibbard (1985). If the Taplow Gravel of the type area is projected upstream, passing through the fluvial gravels at Fern House Pit (Fig. 3.3), the deposits in the Reading area that fall at the predicted elevation for this formation are those identified by Gibbard as his 'Reading Town Gravel'. It is thus concluded that this is the true upstream equivalent of the Taplow Formation at Reading (see Chapter 1), confirming the view of Wymer (1968). An alternative interpretation is therefore required for the higher gravel remnants, attributed by Gibbard to the Taplow, that lie between the true Taplow (Reading Town) Gravel and the Lynch Hill Gravel. Wymer tentatively associated these gravels, his 'Henley Road Terrace', with the deposits at Iver (see above, Cannoncourt Farm Pit), some 18 km downstream from Fern House Pit. Moreover, he regarded deposits in the Sonning area, which were also included by Gibbard in the Taplow Gravel, as part of the Lynch Hill Formation. It seems likely that all these problematic remnants represent degraded thicker parts of the Lynch Hill Gravel, separated from the downslope feather-edge of that formation by later erosion; the Henley Road Terrace may thus be an erosional terrace similar to the Stoke Park Cut, cut in Lynch Hill Gravel (see above, Introduction to Part 3 of this Chapter).

This re-evaluation of the terrace stratigraphy in the Reading area has major repercussions for correlation with deposits in the Upper Thames basin and for the relative dating of the Taplow Formation. Gibbard (1985) argued for correlation of his Reading Town Gravel with the Summertown-Radley Terrace of the Upper Thames, which he also attributed to the Early Devensian. Reappraisal of terrace correlations between the Upper and Middle Thames (see Chapter 1) supports Gibbard's view, thus implying that the downstream continuation of the Summertown-Radley Formation in the Middle Thames is the Taplow Gravel. The Summertown-Radley aggradation is seen by many workers to be a highly complex succession of both temperate- and cold-climate deposits, most if not all of which is pre-Devensian (see Chapter 2, Stanton Harcourt and Magdalen Grove). This formation is, according to the interpretation favoured in Chapter 2, dominated by cold-climate gravels that accumulated in Oxygen Isotope Stages 8 and 6, although it also incorporates interglacial sediments from Stage 7 and Substage 5e (= Ipswichian *sensu* Trafalgar Square) and may culminate in Lower Devensian gravels (Fig. 2.18).

The above correlation indicates that deposits of this wide range of ages might also be expected in the Taplow Formation downstream from the Goring Gap, certainly in the Reading area and perhaps at sites such as Fern House Pit. In the lower reaches of the river, however, the Taplow Formation is restricted in age to Oxygen Isotope Stages 8–6 (Chapters 1 and 4), incision to a new terrace level having occurred within Stage 6 (see Chapter 1). This incision (rejuvenation) was followed by the aggradation of the Kempton Park Formation, which has only been recognized in the Thames valley below Reading (see Chapters 1 and 2, Stanton Harcourt and Magdalen Grove). Within the Kempton Park Formation, each element of the aggradational sequence predicted by the climatic model for terrace formation promoted in Chapter 1 has been recognized (Fig. 2.18). After the erosional phase (phase 1), the first phase of cold-climate gravel deposition (phase 2) is represented by the Spring Gardens Gravel of Gibbard (1985), which underlies the Trafalgar Square (Ipswichian) sediments and can be attributed to Oxygen Isotope Stage 6. The sediments at Trafalgar Square (and equivalent deposits at Brentford) represent the interglacial phase of the model (phase 3), attributable in this case to the Ipswichian Stage (Oxygen Isotope Substage 5e). The major part of the Kempton Park Formation represents phase 4 of the model, the post-interglacial cold-climate aggradation, and is of well-established Devensian age (Gibbard, 1985).

The occurrence of hippopotamus in the Taplow Gravel at the Kensington Road pit suggests that Ipswichian deposits occur within that formation in the Reading area. This implies that no Stage 6 rejuvenation occurred in this area, so the terrace sequence there is equivalent to that in the Upper Thames and not to that in the London area (Fig. 2.18). It is interesting that Sandford (1932), reviewing the previously published evidence, observed that the Taplow Terrace had yielded fauna of both warm and cold affinities, pre-empting the above conclusions.

The above paragraphs provide a context for discussing the information from Fern House Pit. Evidence for the relative dating of the deposits of the Middle Thames sequence is extremely rare, there being no pre-Devensian sites within the lower terraces that have yielded molluscan remains or pollen. The only evidence to support hypotheses based on the reconstruction of terrace aggradations derives from mammals. The mammalian record from the Taplow Formation is relatively sparse and is restricted to large animals, but it appears to incorporate a cold-climate fauna, typified by musk ox (Gibbard, 1985), and temperate-climate species characteristic of both Oxygen Isotope Stage 7 and the Ipswichian Stage (*sensu* Trafalgar Square). The assemblage from the Redlands Pit at Reading includes horse and mammoth, two of the three important species of the 'Stage 7 fauna' described by Shotton (1983). The third species, straight-tusked elephant, was recorded at Fern House Pit by Treacher (1934), presumably from the main body of the Taplow Gravel. This is the most important of the three, since horse and mammoth are known from interstadial deposits and cannot alone be used as evidence for a major temperate-climatic event. Straight-tusked elephant occurs in Ipswichian deposits, however, so it could be attributed to the same origin as the hippopotamus from Kensington Road, Reading. It should be noted that most, if not all, of these large-mammal remains probably come from cold-climate gravels, into which they have been reworked; no *in situ* temperate-climate sediments have yet been recognized within the Taplow Formation in the Middle Thames area (except perhaps at Redlands Pit).

Fern House Pit is located at the extreme upstream limit of the area in which the Kempton Park Formation has been recognized. Gibbard (1985) regarded a low-level gravel at Marlow (SU 865874) as a potential outlier of the Kempton Park Gravel, the furthest upstream that he recognized. A beetle assemblage from organic sediments within this gravel is suggestive of the Upton Warren Interstadial (Coope, in Gibbard, 1985), supporting an early or mid-Devensian age. This deposit extends to *c.* 3 km upstream of Fern House Pit (Fig. 3.1). It thus seems likely that incision from the level of the Taplow Formation to that of the Kempton Park

Gravel did occur in the Marlow area during Oxygen Isotope Stage 6, following the aggradation, during the early part of this cold episode, of the main (phase 4) part of the Taplow Formation as recognized in the type area (see Fig. 2.18). This is an important concept to establish, because it implies that the straight-tusked elephant from Fern House Pit is pre-Ipswichian and that it was reworked from temperate-climate deposits laid down prior to Stage 6 (probably in Stage 7). It is possible, in the light of the recent discovery at Cannoncourt Farm Pit (see above), that the 'loam' separating the Taplow Gravel from the overlying colluvial gravel at Fern House Pit may show evidence of pedogenesis. This might provide a means for demonstrating the pre-Ipswichian age of the Taplow Formation at Little Marlow.

Fern House Pit therefore exposes an important and fossiliferous remnant of the Taplow Gravel Formation of the Middle Thames. Further work is clearly required on these sediments in order to test the revised stratigraphical interpretation of this formation outlined above. The reports of a basal chalky gravel suggest that the deposits at Little Marlow are more likely to have fossils preserved in them than the more typical non-calcareous gravel of the Middle Thames, a view supported by the early records of mammalian remains. It is hoped that re-excavation of sections in the fluviatile deposits will provide future opportunities for modern studies of the sediments and their faunal content.

Conclusions

Fern House Pit provides exposures of the Taplow Gravel, which forms the Taplow Terrace, one of the best-known and longest recognized of the Thames terraces. Well-bedded river gravel is here overlain by a thick accumulation of unbedded gravelly deposits that have moved down the valley side from the north, probably under intensely cold conditions. Teeth of woolly mammoth and the temperate-climate-dwelling straight-tusked elephant (an even larger extinct species) have been found in the bedded gravel. The dating of the deposits is controversial. It is generally agreed that they fall in an interval of geological time known as the Saalian Stage, which began around 400,000 years BP and finished 128,000 years BP. It used to be thought that this interval was a single cold (glacial) phase, intermediate between the Anglian, when the Thames was diverted by ice, and the 'last glacial', which ended only 10,000 years ago. However, it is now believed that a succession of climatic fluctuations occurred during the Saalian Stage, with three cold and two warm episodes. This view is far from being universally accepted, but the record of Thames terraces formed during the Saalian may prove critical in providing evidence for this complex series of climatic changes. To this end, the evidence from the Fern House Pit may hold the key to resolving some of the outstanding problems.

BRIMPTON GRAVEL PIT (SU 568651)

D.R. Bridgland

Highlights

Faunal and floral remains from silts and organic muds within the gravel sequence at this site provide unique but controversial evidence for reconstructing Middle and/or Late Pleistocene conditions in Britain. Although it has been argued that the biostratigraphical evidence here shows two interstadial and three stadial phases within the Devensian Stage, much of the sequence may prove to be pre-Devensian in age.

Introduction

Brimpton Gravel Pit, Berkshire, was operational during the late 1970s and early 1980s. It exploited the Thatcham Terrace deposits of the tributary Kennet valley (Chartres *et al.*, 1976). It is a site of major importance for Upper Pleistocene (possibly upper Middle Pleistocene) stratigraphy and palaeoclimatic reconstruction in Britain. Beds of silts and organic muds within a gravel sequence here provide a biostratigraphical record of two interstadial periods and at least three stadials (Bryant and Holyoak, 1980; Bryant *et al.*, 1983).

This is the type locality of the Brimpton Interstadial, as yet unknown elsewhere in Britain. Bryant *et al.* (1983) attributed the entire Brimpton sequence to the Devensian Stage, regarding the Brimpton Interstadial as a temperate interlude intermediate between the Chelford and Upton Warren interstadials within the British sequence. This view, based on biostratigraphical correlation between sediments at Brimpton and the Chelford type locality, is challenged here on the grounds that terrace correlation between the Thames and Kennet valleys implies that a pre-Devensian age for part of the Brimpton sequence is also possible.

Description

The Brimpton site is located on a low gravel-covered interfluve between the valleys of the Kennet and its tributary, the River Enborne (Fig. 3.1). The original land surface at the site was 6–7 m above the Kennet floodplain, forming part of a terrace feature that continues westwards (upstream) in the Kennet valley and southwards into the Enborne valley. This feature is the Thatcham Terrace of Chartres *et al.* (1976), which appears to be a continuation, in the Kennet valley, of the Taplow Terrace as recognized in the Reading area (see above, Fern House Pit; Fig. 3.3).

The sediments at Brimpton were progressively exposed as the working face moved westwards, between late 1979 and early 1982. At no time was the complete stratigraphical sequence visible. The deepest part of the pit coincided with a depression in the London Clay, *c.* 10 m deep and filled with bedded gravel. The available sections suggested this to be an enclosed basin, possibly representing a scour-hollow excavated at the confluence of the Kennet and Enborne rivers (Bryant *et al.*, 1983).

Lenses and more widespread bands of clay, silt and sand occur throughout the gravel sequence at Brimpton, with clasts of these lithologies occurring in the uppermost 3 m. The matrix of the upper 1–3 m of the gravels contains significant silt and clay, whereas these are generally scarcer in the matrix of the lower gravels. The division between upper (silty) gravels and lower (sandy) gravels is indistinct in some places but sharp in others, often cutting across bedding structures.

Persistent searching in the less oxidized of various clay units within the gravel sequence yielded pollen and Mollusca from numerous levels, as well as plant macrofossils and beetles from a few (Bryant *et al.*, 1983). A sequence of alternating woodland and open-country assemblages was recognized, providing the basis for a biostratigraphical subdivision of the Brimpton sediments. From the palynological succession Bryant *et al.* (1983) deduced that a complex sequence of climatic fluctuations, which they regarded as stadials and interstadials, was represented.

The sequence of fossiliferous beds or lenses, all separated one from another by gravel beds, is as shown in Table 3.4 (interpretations follow Bryant *et al.*, 1983; see also Fig. 3.19). The earliest stratigraphical level from which botanical evidence was obtained (A, Figs 3.19 and 3.20) yielded a pollen assemblage with birch

Table 3.4 Stratigraphy of the Thatcham Terrace deposits at Brimpton (after Bryant *et al.*, 1983).

Lithology	* – Local pollen zones	Interpretation/age
1. Upper silty gravels	F – Non-arboreal pollen	Cold (stadial) late mid-Devensian (^{14}C 27,400 ± 1250 BP (BM-1638))
2. Lower sandy gravels	E – Gramineae-*Betula-Pinus*	Brimpton Interstadial
	D – Cyperaceae-*Betula*	Cold (stadial)
	C – *Betula-Pinus-Picea*	Chelford Interstadial
	B – Non-arboreal pollen	Cold (stadial)
	A – Non-arboreal pollen-*Betula-Pinus*	Late Ipswichian or Wretton Interstadial

* See Fig. 3.20; A–F also coincide with labels in Fig. 3.19, which indicate sample points.

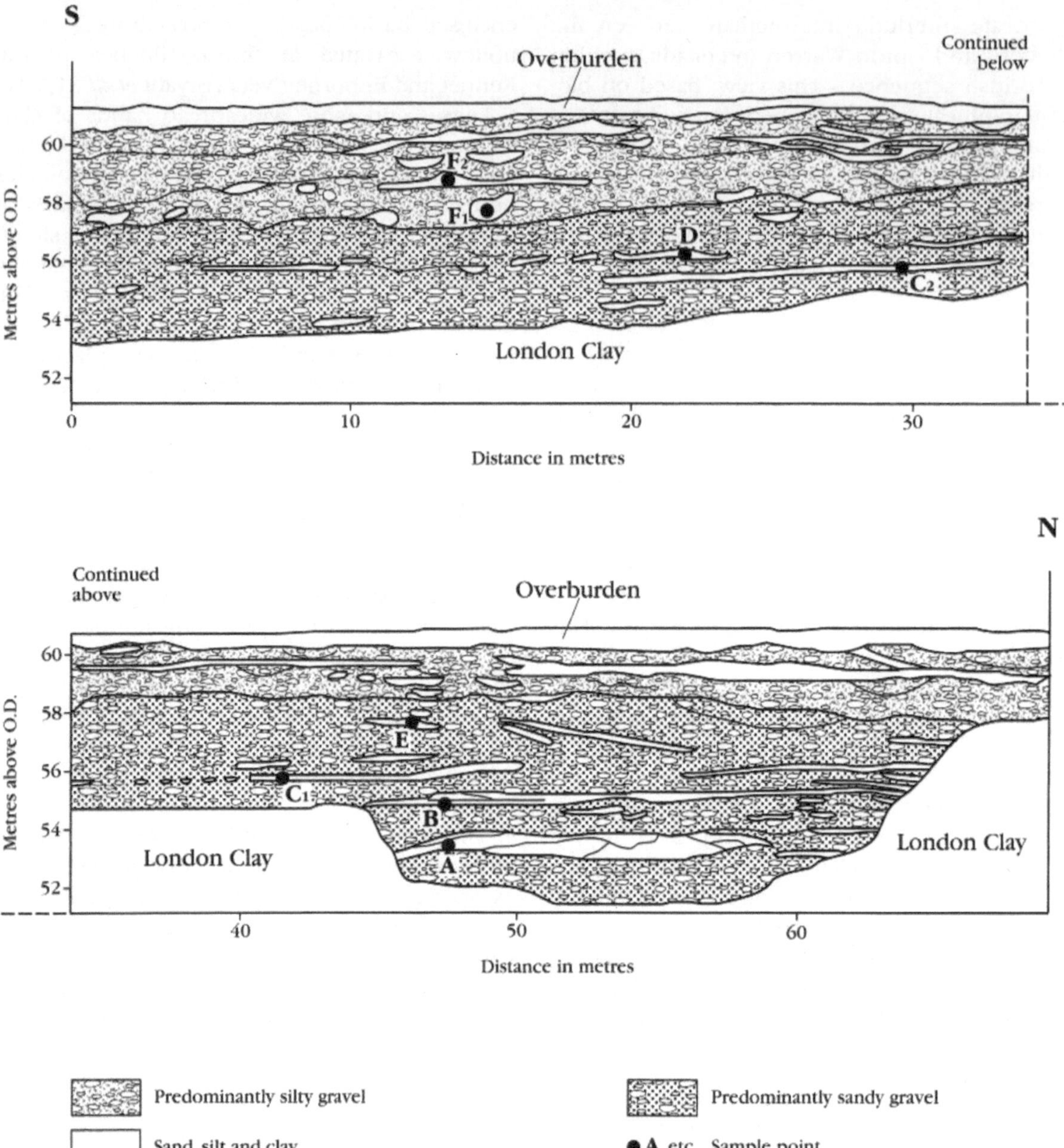

Figure 3.19 Composite section at Brimpton, showing the stratigraphical relations of the various deposits and the locations of samples of biostratigraphical significance (modified from Bryant *et al.*, 1983).

(11%), pine (16%), willow (9%) and occasional oak and spruce. Local grassland was suggested by abundant pollen of Gramineae (25%) and occasional grains from grassland herbs. Heathland, waterside and aquatic species were also represented in an assemblage suggestive of temperate (interstadial or late interglacial) conditions. There was support for this interpretation from a plant macrofossil assemblage that included pine and various herb remains. The occurrence of a single fruit of hornbeam (*Carpinus*), an indicator of fully interglacial conditions, was attributed to reworking, as this species was not represented amongst the pollen spectra (Bryant *et al.*, 1983). This level also produced 15 aquatic and four terrestrial molluscan species, including *Vallonia pulchella* (Müller), *Trichia hispida* and *Pisidium supinum*

(Schmidt), all of which are regarded as thermophilous taxa in the British Devensian. Their presence, and the overall richness of the fauna, lend support to the interpretation, based on the palaeobotany, of this level as temperate (Bryant *et al.*, 1983).

The second level from which evidence was obtained (B, Fig. 3.19) was a silty bar-top drape. This yielded pollen, comprising 55–62% Cyperaceae, 5–10% Gramineae and less than 3% tree pollen (Fig. 3.20). Bryant *et al.* (1983) regarded this assemblage as indicative of the typical open vegetation of full stadial conditions, with trees largely absent. They claimed support for this interpretation from an impoverished molluscan fauna, also obtained from this level, comprising three freshwater species (*Anisus leucostoma* (Millet), *Lymnaea peregra* (Müller) and *Pisidium nitidum*) and three terrestrial species, all of which today tolerate arctic conditions.

At a slightly higher stratigraphical level, unoxidized clay/silt bar-top drapes and/or small channel-fills were found to contain abundant well-preserved pollen (C_1 and C_2, Figs 3.19 and 3.20). Of these, C_2 also yielded plant macrofossils and molluscs, whereas C_1 yielded nine species of Coleoptera. These levels were dominated by tree pollen (62–76%), principally birch and pine, both also represented amongst the macrofossils. Spruce formed an abundant element of the latter, but only 1–4% of the pollen. As this tree is not a high pollen producer, these facts suggest that it was an important species in local forests at the time of deposition. Occasional grains of oak and alder were also encountered. The remainder of the plant community appears to have included both shade-tolerant (forest) and open-habitat taxa, the latter possibly growing close to the river. The freshwater molluscan fauna was similar to that from the previous level (B), but the land fauna was significantly different and highly distinctive. It included the boreal forest snail *Discus ruderatus* (Férussac), only previously recorded from interglacials in Britain, and a species found only in interglacials and in the Late Devensian, *Nesovitrea hammonis* (Ströns). Although these snails extend today into the Arctic region, their presence was seen by Bryant *et al.* (1983) to provide some support for the palaeobotanical evidence for climatic amelioration at this level. Further support for this view came from the beetles, which were considered indicative of an environment similar to central and southern Fennoscandia at the present time, with much colder winters and slightly cooler summers than are currently experienced in southern Britain.

The fourth pollen-bearing level (D, Figs 3.19 and 3.20), another set of bar-top drapes, yielded sparse pollen dominated by non-arboreal types. This material actually comes from seven samples at slightly different stratigraphical levels, but their similarity suggests that they represent the same biozone, one characterized by open-vegetation species (Bryant *et al.*, 1983). This interpretation is supported by the limited molluscan fauna from the same levels, comprising taxa tolerant of arctic conditions (such as *Columella columella* and *Pupilla muscorum*), all known from Devensian stadial sediments.

The fifth fossiliferous level to be recognized was represented in silt drapes and channel-fills within the remainder of the lower sandy gravel at Brimpton (E, Fig. 3.19). Tree pollen was markedly and consistently more abundant in these silts than in level D, although the overall assemblage is suggestive of a combination of open habitats and birch-pine woodland (Fig. 3.20). This evidence for a further temperate interval is supported by a diverse molluscan assemblage that includes taxa characteristic of shallow water bodies with plentiful vegetation (*Valvata cristata* (Müller) and *Anisus leucostoma*).

Finally, pollen and plant macrofossils occur at a number of different levels within the upper silty gravel, in lenses and clasts of fine-grained sediment. Some of these are richly fossiliferous and have yielded notably varied macrofossil floras as well as pollen, mollusc and beetle remains. The pollen and plant macrofossils point to open, treeless conditions with northern and arctic-alpine species well represented. Molluscs likewise suggest stadial conditions, with a preponderance of open-habitat terrestrial species such as *P. muscorum*. With the exception of a small sample from the highest levels (see below), all the molluscs present have modern ranges extending into the Arctic and all are known from stadial deposits in Britain. Certain samples within the upper silty gravel yielded insect remains, including a wide range of beetles. Amongst these, those species that have restricted climatic ranges suggest temperatures colder than at present (*Amara quenseli* (Schoenberr), *A. torrida* (Panzer), *Otiorhynchus nodosus* (Müller), *Notaris aethiops*

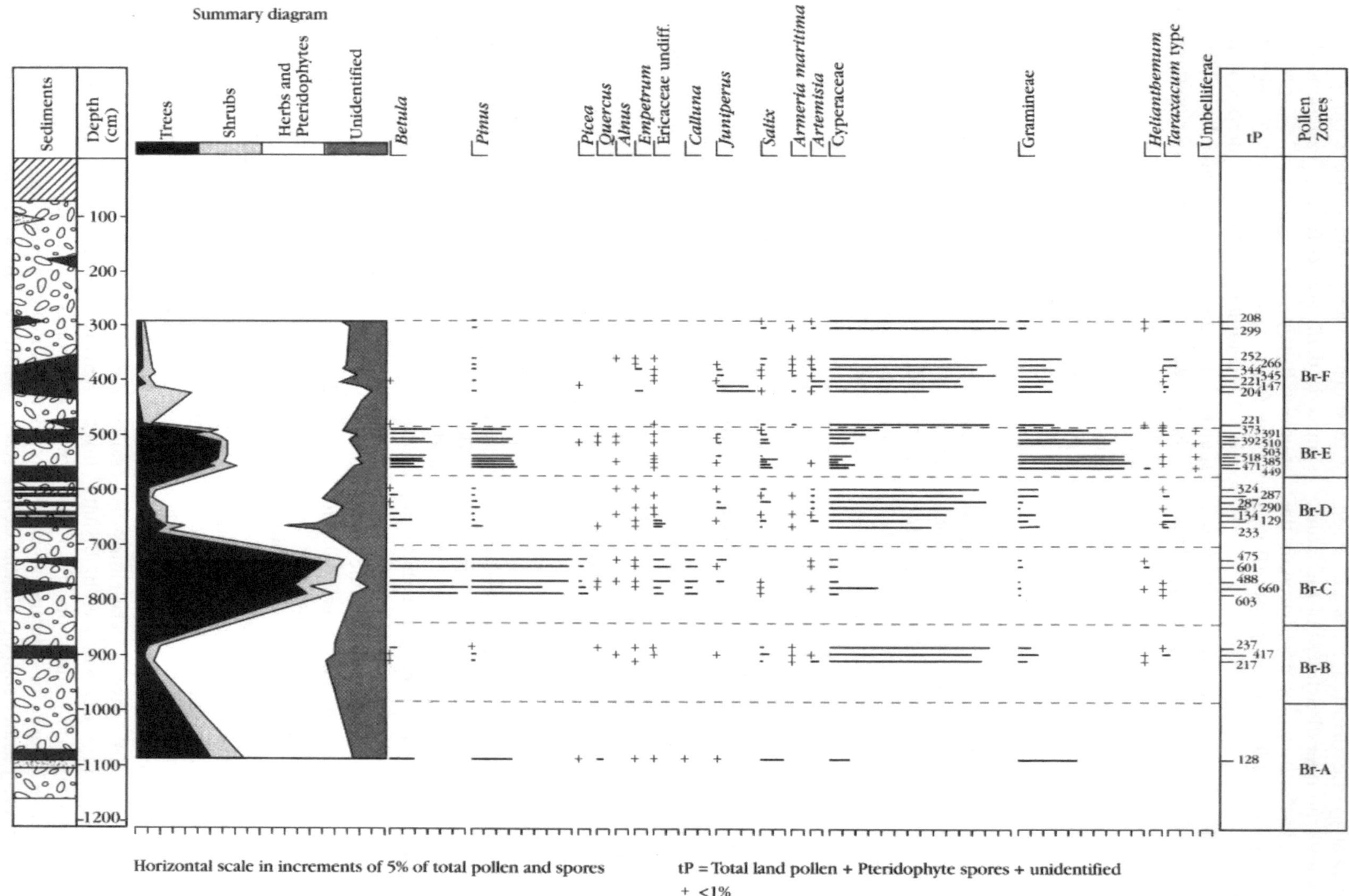

Figure 3.20 Pollen from the deposits at Brimpton (after Bryant *et al.*, 1983).

and *Tachinus jacuticus* Poppius), although with little indication of the degree of severity (Bryant *et al.*, 1983).

Radiocarbon dating of willow twigs from near the base of the upper silty gravel (F_1 and F_2, Fig. 3.19) has indicated ages of 29,500 ± 460 years BP and 26,340 ± 1210 years BP, respectively (mid-Late Devensian). This is the only geochronometrical dating evidence from the site, the attribution of the lower sandy gravel to the Devensian relying on the recognition, from faunal similarities, of the Chelford Interstadial within the Brimpton sequence.

Interpretation

The significance of the Brimpton site stems from the complex biostratigraphy that has been demonstrated, largely on the basis of palynology, from the sequence there. Six pollen assemblages have been recognized at different stratigraphical levels (see above; Figs 3.19 and 3.20). Bryant *et al.* (1983) regarded the *Betula-Pinus-Picea* pollen assemblage (C, Figs 3.19 and 3.20) as equivalent to the Early Devensian Chelford Interstadial; pollen spectra from this level at Brimpton resemble those from the Chelford type locality (Simpson and West, 1958). This correlation is supported by the presence of macrofossils of Norway spruce (*Picea abies* (L.) Karsten) at both sites. If this correlation is accepted, the implication is that the lower temperate levels at Brimpton (A, Figs 3.19 and 3.20) represent either the latter part of the Ipswichian Stage or an earlier, pre-Chelford, Devensian interstadial. The higher temperate horizon (E, Figs 3.19 and 3.20) does not resemble the mid-Devensian Upton Warren Interstadial of Coope (Coope *et al.*, 1961) in its floral and faunal content, particularly since trees appear to have been present, whereas they are unknown from deposits of Upton Warren age. For this reason, level E was attributed to a hitherto unrecognized interstadial within the Devensian, named after the site at Brimpton (Bryant *et al.*, 1983). Deposits of Chelford Interstadial age have been reported from only a few sites in England (Shotton, 1977), the nearest to Brimpton being at Wretton, in Norfolk (West *et al.*, 1974). None of the sites hitherto attributed to this interstadial have yielded Mollusca, so the assemblage from Brimpton is of especial interest. It includes two snail species (*Vertigo substriata* (Jeffreys) and *Discus ruderatus*) that in Britain are otherwise known only from interglacial deposits (Bryant *et al.*, 1983).

The pollen spectra attributed to the Brimpton Interstadial (level E, Figs 3.19 and 3.20) show appreciable frequencies of *Betula* and *Pinus* (together reaching over 30% of the total pollen), leaving little doubt that these trees were growing in the region. The land-snail fauna from this level supports this evidence for a relatively temperate climate. As well as the aquatic taxa noted above (see above, Description), the occurrence of snails *Myxas glutinosa* (Müller), *Cochlicopa lubrica* (Müller), *Vallonia pulchella* and *Trichia hispida*, as well as the general diversity of the molluscan assemblage, suggests interstadial conditions (Bryant *et al.*, 1983).

The radiocarbon dates from the upper silty gravel (*c.* 27,000 years BP – see above) suggest that it was deposited late in the Devensian Stage. Sediments of similar age have been studied at Thrapston (Bell, 1969), Beckford (Briggs *et al.*, 1975a), Brandon (Shotton, 1968) and Great Billing (Morgan, 1969). Bryant *et al.* (1983) provided a full discussion of possible correlations between the Brimpton sequence and that from the post-Eemian period on the continent, where more complete sedimentary records exist. They tentatively suggested a correlation of the Brimpton Interstadial with the continental Odderade Interstadial. However, Bowen (1989; Bowen *et al.*, 1989) has recently suggested that the Upton Warren Interstadial equates with the Odderade Interstadial, attributing both to Oxygen Isotope Substage 5a. Bowen (1989, p. 44) expressed misgivings about the evidence from Brimpton; he considered that a fluvial sequence of this type, dominated by coarse-grained sediments, would inevitably be punctuated by unconformities and that 'extensive reworking and redeposition should be expected'.

Consideration of regional stratigraphical evidence in the Middle Thames and Kennet valleys suggests that the attribution of the entire Brimpton sequence to the Devensian Stage is open to question. The distribution and elevation of the deposits at Brimpton suggest that, with a surface 7–8 m above the present-day river, they form part of the Thatcham Terrace of the Kennet and/or Enborne, as defined by Chartres *et al.* (1976). Thomas (1961) included these deposits in the Lower Taplow Terrace of the Kennet, which he correlated with the terrace of the same name in the Middle Thames valley at Reading, originally described by Sealy and Sealy (1956). The Lower Taplow Terrace at Reading was redefined by Gibbard (1985) as the Reading Town Gravel. Gibbard assigned this aggradation to the Early Devensian and suggested a correlation with the Summertown-Radley Terrace of the Upper Thames, the deposits of which (Stanton Harcourt Gravel) had also been ascribed to the Devensian (Seddon and Holyoak, 1985; Chapter 2, Stanton Harcourt and Magdalen Grove). These various terrace correlations, which would appear to be in close agreement with the supposed Devensian age of the Brimpton deposits, are supported by the reappraisal of terrace correlation within the Thames catchment, outlined in Chapter 1.

However, there are strong grounds for attributing the deposits at Reading, correlated with the Thatcham Terrace, with the Taplow Formation (see Chapter 1 and Fig. 1.3; Fig. 3.3 and above, Fern House Pit). The Summertown-Radley aggradation has also been correlated in this volume with the Taplow Formation and is considered to have aggraded over a lengthy period between the mid-Saalian (Oxygen Isotope Stage 8) and the mid-Devensian, with only its uppermost levels typically dating from the latter stage (see Chapter 2, Stanton Harcourt and Magdalen Grove; Fig. 2.18). The Taplow Formation downstream from Reading has been ascribed to the Saalian Stage (Chapter 1); its aggradation is thought to have spanned the period between Oxygen Isotope Stages 8 and 6 (inclusive). A later formation, the Kempton Park Gravel, occurs downstream from Reading, representing aggradation between Stage 6 and the mid-Devensian (Gibbard, 1985; see above, Fern House Pit; Fig. 2.18). The rejuvenation during Stage 6, which separated these two formations, appears not to have occurred in the Reading area or further upstream in the Thames valley. Therefore the Taplow Formation at Reading, like the Summertown-Radley Formation of the Upper Thames, is believed to represent a considerable period of floodplain stability, from Oxygen Isotope Stage 8 to the mid-Devensian (see above, Fern House Pit).

These interpretations have important implications for the age of the deposits forming the Thatcham Terrace of the Kennet valley. The Summertown-Radley Formation provides an analogue for the Thatcham Terrace deposits; both are the upstream equivalents of the Taplow Formation of the Reading area and both lie upstream of the limit of the Stage 6 rejuvenation, when the Thames further downstream was incised to the level of the Kempton Park Gravel (see above, Fern House Pit). This comparison suggests that the Thatcham Terrace surface may overlie sediments with a considerable range of ages, from Oxygen Isotope Stage 8 to the mid-Devensian. Sediments from the earlier parts of this time interval appear to dominate the Summertown-Radley sequence, the Ipswichian Eynsham Gravel and overlying ?Devensian gravels typically being restricted to the uppermost levels of the formation (Chapter 2). The bulk of the Brimpton sequence might therefore be expected to be pre-Ipswichian, despite the similarity between faunas from some levels within it to British Devensian interstadial assemblages. It is particularly pertinent to note, at this point, that there is very little available information about faunas and floras of Saalian Stage interstadials, which are essentially unknown in Britain and are rare in Europe as a whole. Since, by definition, such periods had relatively restricted biotas, it is likely that interstadials in different 'glacial stages' within the same part of the Pleistocene appear similar in terms of their palaeontology. The basis for the correlation of level C at Brimpton with the Chelford Interstadial may therefore require re-examination.

The only part of the Brimpton sequence for which geochronometric dates indicative of the Devensian Stage have been obtained is the upper silty gravel. The radiocarbon dates from this upper division indicate that the lower sandy gravel was laid down prior to the mid-Devensian. It is possible, however, that the upper division is significantly later than the lower deposits; it may even represent the addition of a mid-Devensian Enborne gravel to a sequence of older Kennet sediments. Bryant *et al.* (1983) suggested that the change from (lower) sandy to (upper) silty gravel resulted from an increase in the deposition of fine sediment as the river aggraded to the highest levels of the floodplain. Chartres (1981), however, showed that loessic silt had been incorporated in the upper levels of the terrace gravels in the Kennet valley by pedogenic activity; the matrix of the upper gravel at Brimpton might therefore be of a similar origin, particularly since its lower limit appears to be independent of the sedimentary bedding (see above, Description). Chartres described a mechanism whereby this loessic silt was restricted to a distinct upper layer: he thought that a clayey illuvial horizon, developed by soil-forming processes prior to loess accumulation, had acted as a barrier to downward movement of silt. An upper silty layer was thus recognized by Chartres (1980, 1984) on the Thatcham Terrace at Thatcham and attributed by him to post-depositional (pedogenic) modification of the original fluvial sediments. However, although the Brimpton site is also on the Thatcham Terrace and an upper silty division has been recognized there, no illuvial clay layer has been observed at the top of the lower gravel. The question of whether the upper silty gravel at Brimpton is of sedimentary or pedogenic origin therefore remains open.

Study of soil profiles on the Kennet terrace gravels may nevertheless provide information of significance to the dating of the Brimpton sediments. Work of this type has revealed differences in levels of complexity between subsoil horizons on different terraces within the Kennet system, with the highest and oldest terraces having the most complex soils (Chartres *et al.*, 1976; Chartres, 1980, 1984; see above, Hamstead Marshall). Chartres did not work at the Brimpton site, but studied the upper levels in the same terrace formation at Thatcham, some 6 km upstream. Micromorphological analyses of samples from Thatcham revealed evidence for complex soil formation over at least one full climatic cycle (Chartres, 1980, 1984). Within the lower part of the soil profile, Chartres recognized concentrations of yellow illuvial clay (argillans), which he attributed to soil formation under temperate conditions. These clays had been disrupted and incorporated into the matrix as a result of disturbance of the soil structure by cryoturbation, under periglacial conditions (in a similar way to the disruption of presumably older red argillans in soils developed on the higher Kennet gravels, such as those at Hamstead Marshall – see above). This interpretation implies that the illuviation phase that produced the yellow argillans was followed by at least one periglacial episode and therefore must have occurred prior to the Holocene. Thus, the Thatcham Terrace surface must have been in existence during a pre-Holocene temperate episode. Chartres duly ascribed the phase of yellow-clay illuviation at Thatcham to the Ipswichian interglacial. In contrast to the evidence from the Thatcham Terrace, no disrupted argillans were observed in soils developed on the lowest terrace in the Kennet sequence, the Beenham Grange Terrace. The implication of this is that no pre-Holocene temperate-climate soil formation has affected this terrace, which fully accords with its correlation with the Late Devensian Shepperton Gravel of the Middle Thames. Chartres's observations led him to propose that pedological evidence from the various Kennet terraces provided a stratigraphical framework, with larger numbers of climatic cycles represented in the soils on higher terraces. His view that soils on the Thatcham Terrace incorporate relict Ipswichian features lends support for the proposal (above) that this aggradation is largely pre-Devensian in age.

Thus two opposing interpretations of the deposits at Brimpton have emerged. The first, that proposed in the original report on the site by Bryant *et al.* (1983), considers them to be Devensian (with possible Ipswichian remnants at the base of the sequence). The second, influenced by regional stratigraphy and by arguments for a more complex Pleistocene record, holds that the deposits may span a longer period and may be dominantly, if not wholly, of pre-Devensian age. The argument between these opposing views hinges largely on the correlation between the Chelford Interstadial and level C at Brimpton. This correlation is based on a sequence of five pollen counts from level C, which indicated a similar floral assemblage to that at Chelford. However, as has been noted above, interstadial assemblages are relatively impoverished and non-distinctive, and little is known about the characteristics of pre-Devensian interstadials. Given the contrary evidence from terrace stratigraphy, supported by pedological studies, the validity of this correlation must be questioned. The radiocarbon dates from Brimpton only indicate a Devensian age for the upper gravel; they have no bearing on the age of the largest part of the sequence. There may in any case be grounds for doubting the accuracy of these dates; the soil evidence, already described, appears to indicate a pre-Ipswichian age for the entire Thatcham Terrace sequence at Thatcham. An assessment of the soil at the top of the Brimpton sequence is clearly desirable, as are further attempts to determine the ages of the various deposits. Amino acid analyses of mollusc shells from the sequence might provide important geochronological evidence of this kind.

The Brimpton pit, with its succession of alternating cold- and temperate-climate deposits, is clearly a stratigraphical site of considerable importance. If its attribution to the Devensian Stage is correct, the site must represent one of the most complete Lower and Middle Devensian sequences in Britain. Comparable sites at Wretton (West *et al.*, 1974) and Four Ashes (Morgan, 1973; Shotton, 1977) have deposits of presumed Chelford Interstadial age, but fossil molluscan faunas are absent. The Brimpton Interstadial temperate episode has been recognized at no other site. The occurrence of molluscan and insect faunas at Brimpton, together with considerable palaeobotanical evidence, also makes this a site of considerable interest.

If the alternative interpretation of the Brimpton sequence as being largely of pre-Devensian age were to be proved correct, however, the importance of the site would not in any way be diminished; pre-Devensian interstadial sediments are extremely rare in Britain and no lengthy biostratigraphical record from any pre-Devensian glacial episode has so far been recognized at a single site.

Conclusions

Brimpton Gravel Pit reveals unique evidence for conditions in southern England during the latter part of the Quaternary Ice Age. Here, beds of gravel deposited by the River Kennet, a tributary of the Thames, alternate with silty and organic layers containing the remains of plants (including pollen) and snails. These fossil remains allow the prevailing climatic conditions at the time of deposition to be determined. Differences between the assemblages from different levels record a sequence of three warm and three cold episodes at Brimpton. It has been suggested that the sequence of deposits here was formed during the last major cold phase of the Quaternary, between about 80,000 and 10,000 years ago, when glaciers covered much of northern Britain, but failed to reach this part of southern England.

However, the Brimpton deposits form part of a terrace of the River Kennet that is correlated with the Taplow Terrace of the main Thames. This terrace, in the Thames valley, is thought to have been formed between 250,000 and 130,000 years ago. There is some reason to suspect that the Brimpton sequence may be of a comparable antiquity, for at Thatcham there is a soil developed in the same terrace that seems to indicate deposition of the gravel prior to the last interglacial (which occurred between 130,000 and 120,000 years ago). Whatever the outcome of this controversy, it is clear that the Brimpton site exposes an important sequence of sediments laid down during a period of fluctuating climate. Very few sites in Britain reveal so complex a history of climatic change, either in the last glacial or earlier; the resolution of the question of dating the Brimpton sequence is of acute urgency. Regardless of whether the Brimpton sediments are of Devensian or Saalian age (they may even include both), the site is of major significance.'

Chapter 4

The Lower Thames

Introduction

The previous two chapters have covered the upper part of the Thames catchment, upstream from the London Basin (Chapter 2), and then the middle part of its course, the area in which the longest record of the river's depositional history is preserved and where detailed evidence for its glacial diversion can be observed (Chapter 3). This chapter is concerned with the valley of the Lower Thames, which contains a complex sequence of Middle and Upper Pleistocene deposits laid down under a variety of climatic conditions in both fluvial and estuarine environments.

The division between the Middle and Lower Thames, a purely arbitrary boundary, is conventionally placed in central London. Application of the term Lower Thames to the valley downstream from the Colne confluence at Staines, which has only existed as a Thames course since the glacial diversion, would perhaps provide a more meaningful geographical and geological division. Any problems of definition are largely academic, however, since most of the scientific interest and all of the GCR sites in the post-diversion valley fall either within or downstream from Greater London.

Despite an abundance of important palaeontological and archaeological localities and a long history of research, the succession in the Lower Thames has been poorly understood and a satisfactory correlation with the terraces of the classic Middle Thames region has been demonstrated only recently (Gibbard, 1985; Bridgland, 1988a; Gibbard *et al.*, 1988).

Research history

The first detailed appraisal of the Lower Thames sequence was by Hinton and Kennard (1900, 1905, 1907), who identified (1905) four gravel terraces numbered in declining sequence. The mapping of the three named 'valley gravel' terraces (Boyn Hill, Taplow and Floodplain), established in the Middle Thames (Chapter 3), was extended into the Lower Thames by the Geological Survey on its 'New Series' maps (Sheets 257 and 271; Dewey *et al.*, 1924; Dines and Edmunds, 1925). This mapping did not distinguish between Hinton and Kennard's 1st and 2nd terraces, both of which were classified as 'Boyn Hill Gravel'. King and Oakley (1936) concluded that a simple altitudinal sequence was not applicable to the Lower Thames because, in that part of the valley, repeated rejuvenations and aggradations had resulted in deposits of various ages being laid down at similar elevations. They described a complex sequence of downcutting and depositional phases, the evidence for which was largely archaeological or palaeontological. For example, sediments at different levels were considered to have been laid down sequentially, on account of their having similar assemblages of Palaeolithic artefacts; conversely, deposits beneath single terrace surfaces were sometimes attributed to depositional events widely separated in time, on the grounds of archaeological differences. King and Oakley's sequence offered little hope for the classification of the more widespread unfossiliferous gravels that make up the bulk of the Pleistocene record in the Lower Thames and contain only mixed assemblages of abraded artefacts. Their scheme did, however, gain widespread acceptance, a possible reason for the paucity of subsequent work on terrace stratigraphy in the area, although the rich Palaeolithic and palaeontological sites continued to receive much attention.

The post-war years saw a number of additions made to the original tripartite sequence in the Middle Thames, notably the Lynch Hill Terrace (between the Boyn Hill and the Taplow) and various pre-Boyn Hill aggradations, from the Black Park Terrace upwards (see Chapter 1; Chapter 3 and Fig. 3.2). Few attempts were made to trace these into the Lower Thames, although Wooldridge and Linton (1955) suggested that the Lynch Hill Terrace was probably represented within the belts of gravel to the east of London mapped as Boyn Hill and Taplow. Wooldridge and Linton provided a description of the terraces in the eastern part of the Lower Thames area in which they adhered closely to the Geological Survey's sequence. However, Evans (1971) returned to a complex scheme in his pioneering attempt to correlate the Thames terrace sequence with the deep-sea record (see Chapter 1). There were considerable similarities between his model and that of King and Oakley, but Evans took the radical step of allocating the various aggradations in the Lower Thames to seven separate warm periods. With the exception of Evans's model, the interpretation of the Lower Thames terrace sequence has, over the past two to three decades, generally followed the conventional Middle and Upper Pleistocene chronological scheme, as set out by Mitchell *et al.* (1973; see Chapter 1). Various interglacial deposits were described at a number of important sites and were correlated with the Hoxnian or Ipswichian Stages of the East Anglian sequence: examples include the Swanscombe deposits, correlated with the Hoxnian (Kerney, 1971), and the biogenic sediments at Aveley, Ilford, Purfleet and Trafalgar Square, which were assigned to the Ipswichian (Franks, 1960; West *et al.*, 1964; West, 1969; Hollin, 1977). Doubts about some of these correlations were first expressed by Sutcliffe (1960, 1964, 1975, 1976) and later by Allen (1977). As these reservations were expressed during detailed consideration of the most important sites in the area, they will be discussed in more detail in appropriate site reports in this chapter.

In the most recent work in the Lower Thames, local lithostratigraphical nomenclature has been applied to the terrace deposits (Table 4.1 and Figs 4.1, 4.2 and 4.3; Bridgland, 1983a, 1983b, 1988a; Gibbard *et al.*, 1988). The New Series Geological Survey mapping of the Lower Thames was found to be broadly accurate and was used as the basis for the reclassification of the three gravel formations originally recognized (see Figs 4.1, 4.2 and 4.3 and Table 4.1). However, Gibbard (1979) correlated high-level deposits at Dartford Heath, mapped as Boyn Hill/ Orsett Heath Gravel, with the Black Park Terrace, thus adding a possible fourth formation to the Lower Thames sequence (see Wansunt Pit). Bridgland (1983a, 1988a; Gibbard *et al.*, 1988) showed that the Lower Thames sequence below the Boyn Hill level had been miscorrelated with the Middle Thames in the original Geological Survey mapping. He concluded (in collaboration with Gibbard) that the Corbets Tey Gravel of the Lower Thames correlates with the Lynch Hill Gravel of the Middle Thames (Table 4.1), not recognized in the original tripartite classification of the terraces in the latter area. This formation appears on the geological maps of the Lower Thames as Taplow, but the true correlative of the Taplow Gravel of the Middle Thames type area is the Mucking Gravel, which was mapped east of London as Floodplain Gravel.

Beneath the alluvium of the modern floodplain, Bridgland (1983a, 1988a) recognized a further gravel formation, the East Tilbury Marshes Gravel, which is interpreted as a downstream continuation of the Kempton Park (Upper Floodplain Terrace) Gravel of the Middle Thames (Bridgland, 1988a; Gibbard *et al.*, 1988; Table 4.1). The gravel sequence in the Lower Thames may be summarized as follows:

Hinton and Kennard (1905)	**Geological Survey**	**Bridgland (1988a)** ***Gibbard et al. (1988)***
		East Tilbury Marshes Gravel
Terrace 4	Floodplain	*West Thurrock Gravel*[2] Mucking Gravel
Terrace 3	Taplow	Corbets Tey Gravel
Terrace 2 Terrace 1	Boyn Hill	Orsett Heath Gravel *Dartford Heath Gravel*[1]

Notes: (1) The separation of the Dartford Heath from the Orsett Heath Gravel is not advocated in this volume (see Hornchurch and Wansunt Pit). (2) The separation of the Mucking and West Thurrock Gravels is dependent on the interpretation of palynological evidence from the West Thurrock brickearth and is not supported in this volume (see Lion Pit).

Differences in detail between the stratigraphical schemes of Bridgland (1988a) and Gibbard *et al.* (1988) reflect a different emphasis in interpreting and ranking various types of evidence. Bridgland gave priority to terrace stratigraphy, his scheme closely following the altitudinal sequence of gravel formations, whereas Gibbard *et al.* favoured biostratigraphical (particularly palynological) evidence, closely adhering to the post-Anglian chronology of Mitchell *et al.* (1973). This has led to important differences in the interpretation of sites such as Purfleet, Globe Pit (Little Thurrock) and the Lion Pit tramway cutting (West Thurrock); these differences are discussed below (see appropriate reports).

An alternative dating model for the Lower Thames terrace succession is proposed in this chapter, based on the stratigraphical relations between the bedded, largely unfossiliferous gravels, ascribed to periglacial episodes, and the interglacial sediments that occur at various sites. This model, which adheres closely to that

Table 4.1 The Pleistocene fluvial sequence in the Lower Thames (first published usage of lithostratigraphical terms in reference given in parentheses), with proposed correlations with the Middle Thames sequence, Pleistocene stages and oxygen isotope stages.

Formation *etc* (First publication)	Type locality (National Grid Ref.)	Middle Thames equivalent	Stage	^{18}O
East Tilbury Marshes Gravel (Bridgland, 1983b)	East Tilbury Marshes (TQ 688784)	Kempton Park Gravel	mid to late Devensian	6-2[7]
(West Thurrock Gravel) (Gibbard *et al.*, 1988)[6]	Lion Pit tramway cutting (TQ 597779)	(Reading Town Gravel)[6]	(early Devensian)	(?5d)
Interglacial Beds at Trafalgar Square		Brentford deposits[5]	Ipswichian	5e
Mucking Gravel (Bridgland, 1983b)	Mucking (TQ 689815)	Taplow Gravel	late Saalian	8-6[4]
Interglacial beds at West Thurrock, Aveley etc.			Intra-Saalian	7
Corbets Tey Gravel (Gibbard, 1985)	Corbets Tey (TQ 570844)	Lynch Hill Gravel	mid-Saalian	10-8[3]
Interglacial beds at Purfleet and Grays			Intra-Saalian	9
Orsett Heath Gravel (Bridgland, 1983b)	Orsett Heath (TQ 668803)	Boyn Hill Gravel	early Saalian	12-10[2]
Interglacial beds at Swanscombe			Hoxnian *sensu* Swanscombe	11[2]
(Dartford Heath Gravel) (Gibbard, 1979)[1]	Wansunt Pit (TQ 5147360)	(?Black Park Gravel)	(late Anglian)	(12)

1 The separate existence of the Dartford Heath Gravel, the subject of a lengthy controversy, is doubtful (see Wansunt Pit). This is thought to be part of the late Anglian to early Saalian Orsett Heath Formation.

2 The Boyn Hill/Orsett Heath Formation includes the interglacial sediments at Swanscombe, here attributed to ^{18}O Stage 11 (referred to as Hoxnian *sensu* Swanscombe in this volume).

3 Aggradation of the terrace deposits included within the Corbets Tey Formation began prior to the interglacial represented at Purfleet and Grays.

4 Aggradation of the terrace deposits included within the Mucking Formation began prior to the interglacial represented at West Thurrock, Aveley etc.

5 Described by Trimmer (1813) and Zeuner (1959).

6 The separate existence of the West Thurrock and Reading Town Gravels is disputed in this volume. These are believed to be part of the late Saalian Taplow/Mucking Formation (see West Thurrock and Fern House Pit).

7 The Ipswichian sediments at Trafalgar Square and Brentford are regarded here as part of the Kempton Park Formation (see Chapter 3, Fern House Pit). This formation is considered to represent aggradation from the end of Stage 6 (gravel underlying the Trafalgar Square sediments, the Spring Gardens Gravel of Gibbard, 1985) to the mid-Devensian.

outlined by Bridgland (1988a), recognizes two additional fully temperate episodes within the sequence, between the conventional Hoxnian (*sensu* Swanscombe) and Ipswichian (*sensu* Trafalgar Square) Stages (Fig. 4.3; Table 4.1). It forms the principal basis for the stratigraphical scheme for the Thames catchment as a whole and for the climatic model for terrace formation, both of which were put forward in Chapter 1. If the Anglian Stage, during which modern Lower Thames drainage was initiated, is correlated with Stage 12 of the oxygen isotope chronology (Bowen *et al.*, 1986b) and the last interglacial (Ipswichian *sensu* Trafalgar Square) with Sub-stage 5e (Gascoyne *et al.*, 1981; Shotton, 1983), three earlier post-Anglian interglacial episodes remain to be identified on land. Bowen *et al.* (1989) have recently published

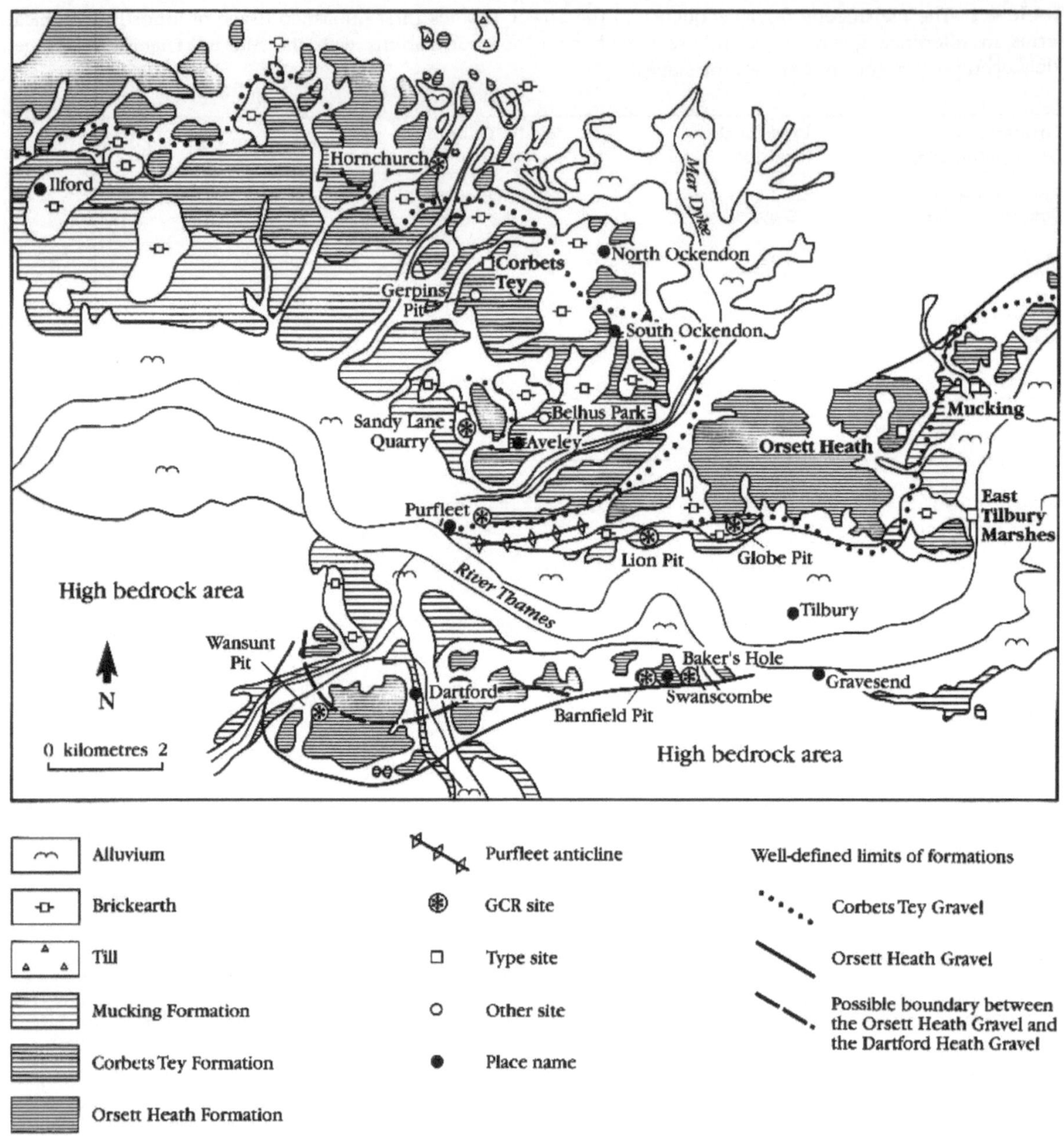

Figure 4.1 The Pleistocene deposits of the Lower Thames (after Bridgland, 1988a).

amino acid ratios from various post-Anglian sediments and suggested that these can be divided into four groupings, equating with the four major post-Anglian temperate episodes, Oxygen Isotope stages 11, 9, 7 and 5. The same number of temperate episodes is now recognized in the Lower Thames sequence; on this basis correlations with the deep-sea record are suggested in Table 4.1. Details of the evidence from which this model has been assembled are given in the various site descriptions in this chapter.

HORNCHURCH RAILWAY CUTTING (TQ 547874)

D.R. Bridgland

Highlights

This locality demonstrates the maximum southern limit of the Anglian ice sheets. At Hornchurch a remnant of Anglian till is overlain by a post-diversion Thames gravel. This gravel is part

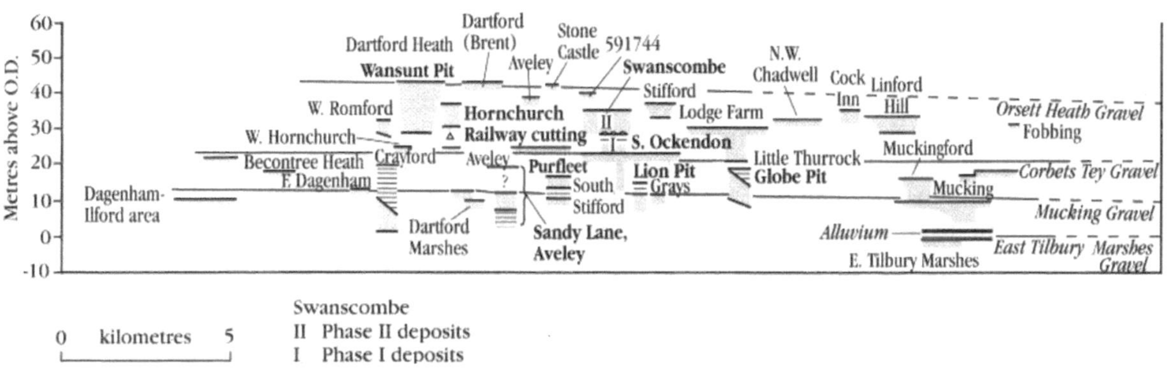

Stipple indicates gravels and sands, horizontal ornament indicates fine-grained deposits.

Figure 4.2 Longitudinal profiles of terrace deposits in the Lower Thames.

of the highest terrace of the Lower Thames, a fact that implies that the terrace sequence in this part of the valley is entirely of post-diversion (late Anglian/post-Anglian) age.

Introduction

Sections at Hornchurch showing chalky till overlain by Thames terrace gravel have been famous since the last century, when the original descriptions were written (Holmes, 1892a, 1892b, 1892c). The stratigraphical importance of these sections, created during construction of the Romford to Upminster railway line, has long been recognized. The sequence in the Hornchurch–Romford district is unique, in the Lower Thames valley, in that terrace deposits there are in contact with Anglian glacigenic sediments. The superposition of the Boyn Hill/Orsett Heath Gravel above till at Hornchurch has formed a principal basis for considering the entire Lower

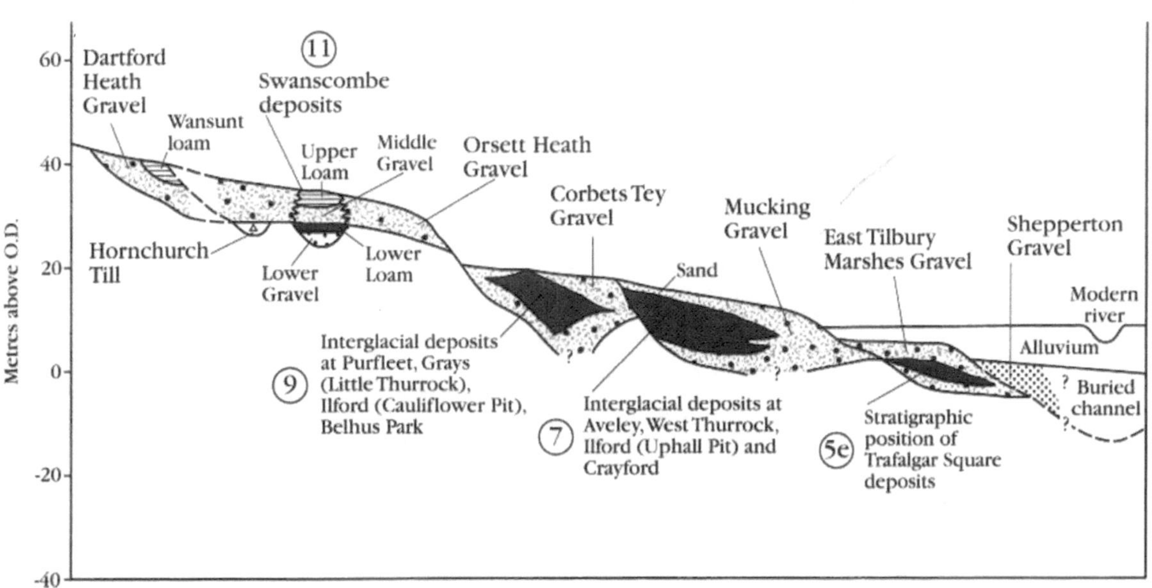

Figure 4.3 Idealized transverse section through the terraces of the Lower Thames. The odd-numbered (warm) oxygen isotope stages to which the various interglacial deposits are attributed are indicated (numbers in circles). The stratigraphical position of the Trafalgar Square deposits is shown.

Thames terrace system to be later than the main glaciation of eastern England (Whitaker, 1889; Holmes, 1892a, 1893; Wooldridge, 1957), although this was once a controversial interpretation (Hinton, 1910, 1926a; Kennard, 1916; Woodward *et al.*, 1922). Hornchurch is the southernmost locality at which the 'Chalky Till' of East Anglia has been recognized.

Description

The GCR site is part of a railway cutting excavated in the 1890s through a ridge of gravel-capped land running north-eastwards from the parish church at Hornchurch. When newly excavated, a section here (TQ 547874), up to 8 m deep and 600 m long, showed *c.* 5 m of till overlain by sand and gravel (Fig. 4.4). The till was observed over a distance of *c.* 300 m in the central part of the cutting; it appeared to occupy a depression in the London Clay, as the sand and gravel directly overlay London Clay to the north-west and south-east of the till outcrop (Holmes, 1892a, 1892b, 1892c, 1893, 1894; Fig. 4.4). At the Romford end of the same railway line a second cutting (TQ 525887) also showed till between the London Clay and Thames gravel, in this case associated with 'dark silt' deposits, interbedded with sand and pebbles (Holmes, 1894).

The gravel overlying the till at the original railway cutting site was included by Dines and Edmunds (1925) in the Boyn Hill Terrace of the Thames. On their Geological Survey map of the Romford area (New Series, Sheet 257), a tongue of Boyn Hill Gravel is shown running north-eastwards from the church (TQ 544870), across the railway cutting site, and terminating *c.* 200 m to the north-east, where a strip of boulder clay continues the northward trend of the gravel (Fig. 4.1). This strip of dissected 'drift' seems to form an erosional terrace on the western side of the Ingrebourne valley, whereas the 'boulder clay' and brickearth of the Upminster district form a complementary feature on the eastern side of the valley (Pocock, 1903; New Series Sheet 257; Fig. 4.1).

Attempts to locate till beneath the thickest part of the Orsett Heath Gravel at Hornchurch, exposed in a pit near the church (TQ 544868), revealed only London Clay bedrock, although chalky till was reported when an electricity substation was built in part of this pit (Anon., 1982b). A section was therefore cleared in the railway cutting in 1983 (Anon., 1984a), revealing the sequence illustrated in Figs 4.5 and 4.6. The land surface at this, the GCR site (*c.* 33 m above O.D.), is clearly erosional, so that the full thickness of the Orsett Heath Gravel is not preserved. However, 4 m of bedded gravel, considerably disturbed by an ice-wedge cast, was exposed above 3 m of till (Figs 4.5 and 4.6). Clast-lithological analysis of the gravel shows it to be

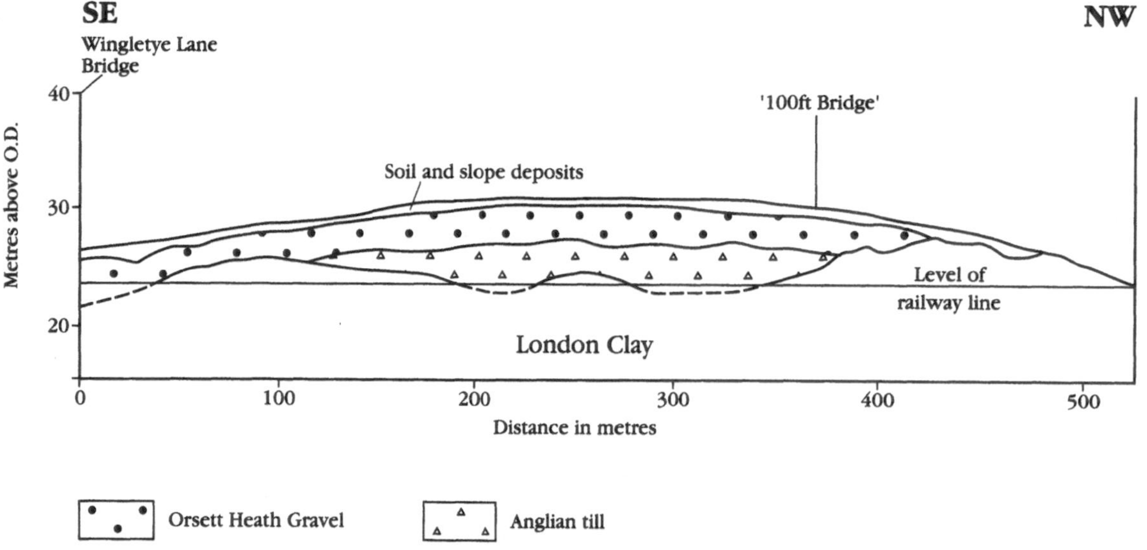

Figure 4.4 Section in Hornchurch Railway Cutting, recorded during its original construction (after Holmes, 1892a).

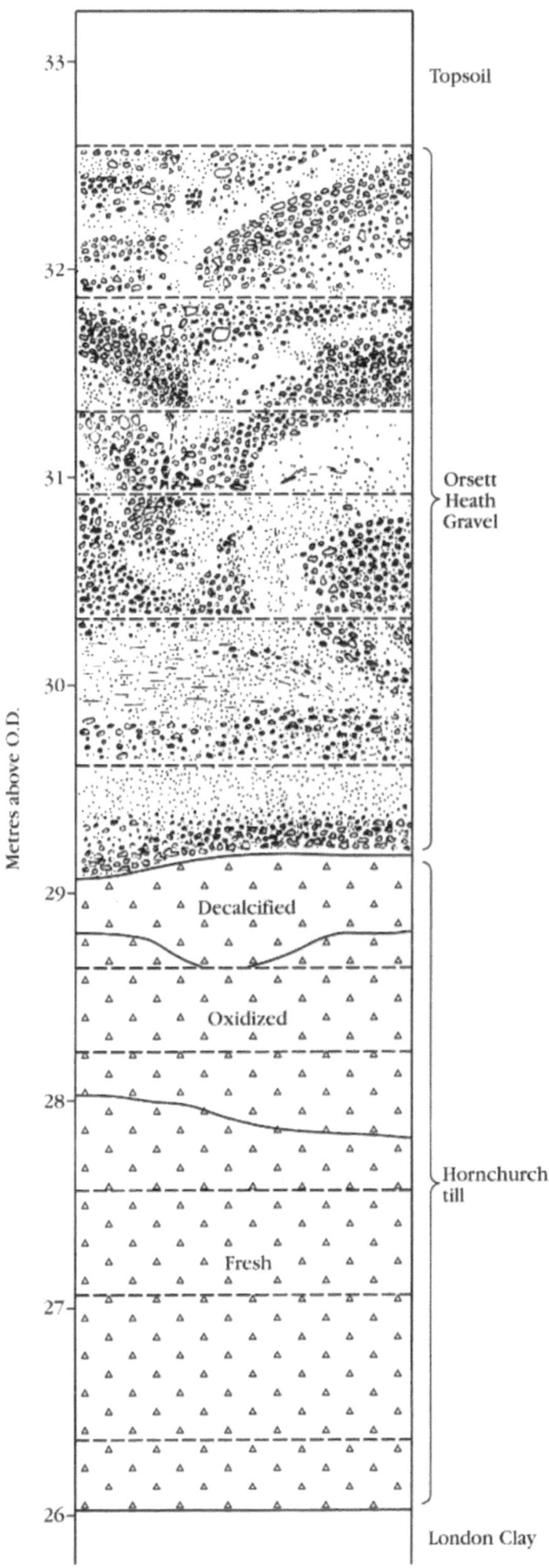

Figure 4.5 The GCR section, Hornchurch Railway Cutting, 1983. Pecked horizontal rulings denote steps in the section. The cutting side slopes at approximately 45° (see Figure 4.6).

typical of Lower Thames deposits upstream from the Darent confluence (Table 4.2), thus supporting its attribution to the Orsett Heath Formation. The lowest 1.5 m of the till in this exposure was unweathered; above this it was oxidized, with the top 0.1–0.3 m also decalcified.

Interpretation

Holmes (1892a, 1892b, 1893, 1894) believed that the gravel overlying the till at Hornchurch belonged to the oldest terrace of the Lower Thames valley. He therefore concluded that the fluvial drifts of the area were entirely 'post-glacial'. This interpretation was generally accepted, although some workers preferred to place the 'Chalky Boulder Clay' of south-eastern Britain later than the 'High Terrace' of the Thames (Hinton, 1910, 1926a; Kennard, 1916; Woodward *et al.*, 1922). Kennard (1916) was perhaps the staunchest opponent of Holmes's interpretation. He believed the Lower Thames and its tributaries to be of 'pre-glacial' age and considered the gravel above the Hornchurch till to be the product of 'a tributary stream, possibly the Ingrebourne, or ... not a river gravel at all' (Kennard, 1916, p. 264). According to Preece (1990a), Kennard's view reflected his strong monoglacialist convictions; the evidence from Hornchurch was of fundamental importance in the replacement, during the early decades of this century, of a monoglacial interpretation of the Pleistocene by one involving multiple glacials and interglacials.

Holmes also noted that the till at Hornchurch, only *c.* 80 ft (25 m) above O.D., is considerably lower than the general level of similar deposits to the north (see Fig. 5.1). He considered that the deposit occupied a valley or hollow and, following a suggestion by Monckton (in discussion of Holmes, 1892a), concluded that a valley system had existed in the area prior to the arrival of the ice (Holmes, 1893). The later discovery of till at Romford (above), at a similar elevation to the Hornchurch remnant, indicated to Holmes (1894) that the glacial deposits were laid down over a valley floor of considerable width. This led him to suggest that there existed a major valley running north of the present Thames and passing out to sea via the Blackwater estuary. He later developed this idea further, attributing this hypothetical

Figure 4.6 The GCR section, Hornchurch Railway Cutting, 1983. The Boyn Hill/Orsett Heath Gravel occupies the upper part of the section, its base occurring on the step beneath the tree root. The remainder of the visible section is in till, although London Clay was reached in the base of the excavation. (Photo: P. Harding.)

valley to a 'Romford River' (Holmes, 1896).

This hypothetical phase of fluvial development, presumed to pre-date the deposition of any of the terrace gravels of the Lower Thames, became widely accepted by later writers as an early course of the river (Saner and Wooldridge, 1929; Baker and Jones, 1980; Baker, 1983), but recent work in the area through which it was thought to have passed has yielded no evidence for drainage by the Thames or any other river (Bridgland, 1986c). On the contrary, studies of the fluvial deposits of eastern and southern Essex suggest that any precursor of the modern Lower Thames was a minor tributary of the Medway, occupying much the same geographical area as the present river between Dartford and Southend (Bridgland, 1980, 1983a, 1988a). The 'Romford River' lowland can be interpreted as a classic example of 'inverted relief', since it represents an interfluve area between the pre- and post-diversion courses of the Thames. In contrast to the areas to the north and south, there were no early gravels on this interfluve, so the non-resistant London Clay was therefore afforded no protection from erosion during the latter part of the Pleistocene (Bridgland, 1986c).

Table 4.2 Clast-lithological data from the Lower Thames. All counts by the author, at 16–32 mm size range, except those in italics, which are 11.2–16 mm counts. Note that non-durables (including Chalk) are excluded from the calculations, but Chalk is shown in this table as a relative % of the total durables.

			Flint			Chalk	Southern		Exotics									
Gravel	**Site**	**Sample**	**Tertiary**	**Nodular**	**Total**	**(1)Chalk**	**Gnsd chert**	**Total**	**Quartz**	**Quartzite**	**Carb chert**	**Rhax chert**	**Igneous**	**Total**	**Ratio (qtz:qtzt)**	**Total count**	**National Grid Reference**
East Tilbury	E.Tilbury Mshs	1 *D*	58.9	9.9	96.2		0.9	1.1	0.9	0.7	0.5	0.3	0.3	2.7	1.40	745	TQ 6880 7843
Marshes Gr.	11.2-16	1 *D*	49.5	6.6	92.2		1.5	1.6	3.2	1.4	0.6	0.2	0.1	6.1	2.21	979	
Mucking	Lion Pit - lwr gravel	1 *D*	47.8	35.9	97.5	(1.1)	0.7	0.7	0.7	1.1				1.8	0.67	276	TQ 5978 7821
Gravel	('floor') *11.2-16*	*1 D*	*50.2*	*19.6*	*95.7*	*(0.3)*	*0.6*	*0.6*	*1.8*	*0.9*	*0.6*		*0.3*	*3.7*	*2.00*	*327*	
	upper gravel (2)	2 *D*	67.1	5.9	95.3		0.8	0.8		3.5				3.9		255	TQ 5978 7809
	11.2-16	*2 D*	*59.4*	*3.2*	*94.2*		*1.1*	*1.1*	*1.9*	*1.5*	*0.4*	*0.4*		*4.7*	*1.29*	*465*	
	Mucking	1A *D*	64.0	9.3	97.0		1.1	1.1	0.9	0.6		0.1		1.8	1.50	708	TQ 6892 8154
	11.2-16	*1A D*	*57.7*	*4.9*	*92.1*		*1.9*	*1.9*	*3.1*	*1.2*	*1.1*	*0.2*	*0.1*	*6.0*	*2.55*	*901*	
		1B *D*	37.4	13.3	92.5		4.9	4.9	1.2	0.6	0.6	0.3		2.6	2.00	345	
Corbets Tey	Stifford	1A	51.6	8.4	94.0		0.4	0.4	2.9	1.2	0.6	0.1	0.4	5.5	2.33	730	TQ 5900 7908
Gravel		1B	52.5	•	92.9		0.9	1.0	3.5	1.4	0.5	0.1		5.9	2.46	918	
	11.2-16	*1B*	*39,2*	*8.3*	*88.3*		*1.1*	*1.4*	*6.0*	*2.6*	*1.1*	*0.2*	*0.1*	*10.3*	*2.30*	*1277*	
	Belhus Park, organic bed (3)	1	47.5	9.8	90.2	(0.3)	0.7	0.7	2.0	4.4	2.0	0.7		9.1	0.46	297	TQ 575811
	Belhus Park, upper gravel (3)	1	49.0	9.7	93.8				3.5	1.4	0.7		0.7	6.2	2.50	145	
	Purfleet, Esso Pit	1A	44.8	16.9	91.8		0.5	0.5	2.5	3.0	1.6			7.4	0.82	366	TQ 5607 7837
	11.2-16	*1A*	*36.3*	*7.6*	*86.6*		*1.0*	*1.1*	*3.9*	*3.7*	*3.1*	*0.5*	*0.2*	*11.7*	*1.04*	*618*	
		1B	47.7	18.1	95.0	(37.3)	1.5	1.5	0.8	1.5	0.8	0.4		3.5	0.50	260	
	Globe Pit	1 *D*	57.9	11.2	93.1		3.2	3.5	0.8	1.1	1.1	0.2		3.4	0.71	653	TQ 6251 7830
		2 *D*	50.2	10.5	93.2		3.1	3.1	1.3	0.7	0.7	0.8		3.7	2.00	617	TQ 6251 7828
	11.2-16	*2 D*	*40.7*	*5.4*	*90.5*		*4.4*	*4.7*	*2.1*	*0.8*	*1.2*	*0.2*	*0.1*	*4.5*	*2.73*	*1456*	
		3 *D*	64.6	8.9	94.4		2.4	2.4	1.5	1.0	0.4			3.2	1.40	463	TQ 6251 7827
	Barvills Fm Pit	1 *D*	67.9	11.8	92.9		3.3	3.3	1.7	1.1	0.4	0.1		3.6	1.50	722	TQ 6811 7774
	11.2-16	*1 D*	*55.6*	*5.6*	*91.8*		*2.7*	*2.9*	*2.2*	*1.1*	*1.1*	*0.3*	*0.3*	*5.3*	*2.08*	*1138*	
Orsett Heath	Hornchurch	1	41.8	0.7	92.6		2.3	2.3	2.0	1.4	0.6	0.6		5.1	1.17	352	TQ 5464 8739
Gravel	railway cutting	2	28.9	11.7	90.2		1.6	1.9	1.9	2.3	1.6	0.9	0.9	7.9	0.80	429	TQ 5464 8739
	Hornchurch Dell	1	54.0	7.7	91.7		1.5	1.5	2.1	2.8	1.2	0.4		6.7	0.78	676	TQ 5440 8675
	Globe Pit North (4)	1A *D*	41.4	9.0	90.4		4.1	4.4	0.6	1.4	1.6	0.3		5.2	0.40	365	TQ 6245 7855
	Linford	1 *D*	64.6	11.6	96.0		2.2	2.4	0.7		0.2		0.2	1.7		424	TQ 6681 8028
		2 *D*	84.2	4.0	95.7		1.4	1.6		0.5		0.2	1.2	2.7		625	TQ 6681 8028
	11.2-16	*2 D*	*28.0*	*3.6*	*91.3*		*1.1*	*1.2*	*3.9*	*2.3*	*0.6*	*0.2*	*0.5*	*7.4*	*1.73*	*665*	
Swanscombe	Barnfield Pit 1	1 *D*	58.2	9.8	93.9		0.9	1.2	2.4	1.8	0.5			4.8	1.37	1081	TQ 5973 7430
Lower Middle	*11.2-16*	*1 D*	*50.9*	*5.3*	*89.9*		*2.1*	*2.3*	*4.4*	*2.0*	*0.8*		*0.1*	*7.7*	*2.21*	*1703*	
Gravel		2 *D*	48.5	12.7	92.7		1.9	2.0	1.9	1.8	0.5	0.1	0.2	5.0	1.05	992	TQ 5973 7430
	11.2-16	*2 D*	*41.6*	*5.5*	*89.7*		*3.0*	*3.1*	*3.5*	*1.5*	*0.5*	*0.2*	*0.2*	*6.8*	*2.42*	*1785*	
Swanscombe	Barnfield Pit	3 *D*	55.5	8.3	94.3		1.0	1.0	2.3	1.3	0.5	0.2	0.1	4.5	1.75	931	TQ 5974 7430
Lower	*11.2-16*	*3 D*	*36.5*	*5.9*	*89.0*	*(0.1)*	*2.5*	*2.7*	*4.0*	*2.9*	*0.5*	*0.1*	*0.1*	*8.3*	*1.40*	*1391*	
Gravel		4 *D*	30.5	11.8	94.1	(0.4)	2.7	2.8	1.1	0.8	0.4	0.1		2.7	1.29	857	TQ 5974 7430
	11.2-16	*4 D*	*28.1*	*8.8*	*90.6*	*(0.3)*	*3.5*	*3.8*	*2.7*	*1.5*	*0.9*	*0.2*		*5.6•*	*1.74*	*1494*	

• Not separately recorded

D (after sample number) indicates that the sample concerned came from downstream of the contemporary Darent confluence.

(1) -Chalk, a non-durable, is only present locally and was therefore excluded from calculations, but shown instead as a % of the total durable material.

(2) Lion Pit tramway cutting sample 2 is from the upper gravel in section 2;

(3) The Belhus Park samples are from the organic sediments within the Corbets Tey Formation and from the gravel overlying the organic sediments;

(4) The Globe Pit North sample is from the Orsett Heath Gravel outcrop in the northern part of the old workings, outside the GCR site.

Calculations in this table (and in Tables 5.2 and 5.5) are based on the durable content only. Non-durables such as London Clay pebbles, clay ironstones, fragments of septaria and even Chalk, are highly localized in their occurrence and are excluded because they inhibit comparison of widespread gravel characteristics. Space does not permit the inclusion of all available data. Where the addition of the flint, southern and exotic totals falls short of 100%, the occurrence of other local material is indicated, predominantly sarsen. Extra material in the southern category comprises Greensand sandstones and (where not shown separately) Hastings Beds sandstones, siltstones and ironstones. Extra material in the exotic category comprises arkosic sandstones, unidentified cherts and (where not shown separately) igneous rocks. Note that: the Tertiary flint category comprises rounded pebbles (sometimes subsequently broken) reworked from the Palaeogene; Gnsd = Greensand; Carb chert = Carboniferous/Palaeozoic chert; Rhax = *Rhaxella*; the igneous category includes metamorphic rocks (very rarely encountered); the quartzite category includes durable sandstones.

By the beginning of the present century the Hornchurch cuttings were no longer available for study, but attention moved to the opposite side of the Ingrebourne valley when till was discovered beneath brickearth deposits near Upminster, *c.* 2 km to the north-east of the Hornchurch site (Pocock, 1903). The Upminster brickyards exposed over 7 m of horizontally bedded brickearth with occasional seams of gravel (Dalton, 1890; Pocock, 1903; Woodward, 1904). This deposit was laminated in its lower part, the laminations showing contortions, leading Pocock (1903, p. 200) to suggest that the sediment had been disturbed 'by ice-floes'. These contortions, and the association with till, suggested to Pocock that the brickearth was a glacial-lake deposit. Woodward went further and, believing that the ice had reached the Grays area and caused disturbances there in the Chalk, suggested that 'the waters of the Thames Valley were pounded up by an icy dam' (Woodward, 1904, pp. 483–484).

Warren (1912) supported theories of lake development in association with the glaciation of the Hornchurch area. He claimed that the Hornchurch-Romford till probably rested on an overdeepened lake bottom, since it was unlikely that the whole Thames valley had been excavated at that time to the depth of the Hornchurch deposits. The same author later recorded new sections near Hornchurch, opened up on either side of the Ingrebourne valley during the construction of the Southend Arterial Road (A127), showing till overlain by laminated silts that he interpreted as lacustrine (Warren, 1924a). In the eastern section (TQ 565890), Palaeolithic artefacts were found in gravel and sand interbedded with the silts (Dines and Edmunds, 1925; Dewey, 1930, 1932; Warren, 1942; Wymer, 1968, 1985b), interpreted by Warren (1942) as a product of the Ingrebourne. According to Dewey (1932), some of these artefacts suggested the use of the Levallois technique. There is little indication that artefacts have been recovered from the gravel overlying the till at Hornchurch (see, however, Wymer (1985b, p. 297)), although Palaeolithic material is widespread in the Boyn Hill/Orsett Heath Gravel. Dines and Edmunds (1925) described both of the Arterial Road sections in some detail, recording several steep-sided channels about 2 m deep filled with chalky till. They suggested (1925, p. 32) that these occurrences of till represented 'remnants of a spread which filled a valley now occupied by the Ingrebourne river', as previously envisaged by Woodward (1909; Woodward *et al.*, 1922).

Zeuner (1945, p. 155) claimed that the Hornchurch till lies on what he termed the 'Boyn Hill bench', meaning the erosion surface beneath the Boyn Hill/Orsett Heath Gravel (formed by the downcutting phase separating this from the preceding Black Park Formation). Zeuner recognized that the earliest occupation by the river of the Lower Thames valley coincided with his 'Kingston Leaf', later redefined as the Black Park Terrace (Wooldridge and Linton, 1955; Gibbard, 1979), and not with the Boyn Hill Terrace. His interpretation of the Hornchurch site therefore implied that the diversion of the Thames into its modern valley and the formation of the Kingston Leaf (Black Park) aggradation both pre-dated the Hornchurch glaciation. Zeuner, in fact, favoured river capture as the mechanism for the diversion of the Thames, not glacial intervention.

Wooldridge (1957) remapped the deposits of the Hornchurch area and concluded that the till was part of a dissected lobe that descends into the Ingrebourne valley from the plateau to the north, reflecting the former presence of a tongue of ice 'of glacier-like dimensions' (1957, p. 13). Wooldridge suggested a correlation between the till at Hornchurch and the Maldon Till of Clayton (1957), on the basis that both are confined to valley floors (see Chapter 5, Maldon). He suggested that these low-level tills could be equated with the Lowestoft Till of Suffolk, implying an Anglian age. Wooldridge considered that the major part of the excavation of the Lower Thames valley had occurred since the emplacement of the till, thus removing any evidence for the latter having extended further southwards. Wooldridge also suggested a correlation between the excavation of the valley in which the Hornchurch till was deposited and the erosion of the Clacton Channel. He followed King and Oakley (1936), however, in referring the erosion of the Clacton Channel to their 'Inter-Boyn Hill Erosion Stage'; he thus implied, indirectly, that the lowest 'Boyn Hill' deposits at Swanscombe were older than the till at Hornchurch, an interpretation that would seem to assign the Hornchurch glaciation to the mid-Hoxnian (interglacial) Stage – clearly an untenable view.

Ideas that a major valley system existed in the Lower Thames region prior to the Hornchurch glaciation, as implied by Zeuner (1945) and Wooldridge (1957), cast doubt on the theory that the diversion of the Thames was brought about by the glaciation of its former, more northerly route. In recent years, however, work in the Vale of St Albans has confirmed the role of Anglian Stage ice sheets in this diversion event (Gibbard, 1974, 1977, 1979; Chapter 3). Gibbard (1979) also demonstrated that the Black Park Gravel, which he assigned to the late Anglian, is the oldest Thames formation in the new valley through London. Later work has confirmed the Anglian age of the Black Park Formation and has suggested that it was emplaced, at least in part, while ice still occupied parts of the London Basin (Gibbard, 1983, 1985; Cheshire, 1986a; Chapter 3, Part 2).

The reopening of the section in the Hornchurch railway cutting, as part of the GCR programme (Anon., 1984a), allowed confirmation of the sequence described by Holmes and the application of modern analytical techniques to the deposits exposed. Sedimentological and chemical analyses of the till (C.A. Whiteman, pers. comm.) have shown that it comprises 65–77% clay (with subordinate silt), up to 25% sand and up to 10% gravel, the latter dominated by Chalk (40–65%) and flint (12–35%). Limestones, calcareous fossils and non-durable igneous rocks are also present. Fabric data show preferred east-west clast orientations, interpreted by Whiteman as transverse to ice flow, which he considered to be from the north. He interpreted similar fabrics in the lower till of the Chelmsford area in the same way. This fabric data, together with the colour and chemical composition of the deposit, led Whiteman to correlate the till at Hornchurch with his Newney Green Member of the Lowestoft Formation (Whiteman, 1990; Allen *et al.*, 1991; Chapter 5, Newney Green). Cheshire (1986a), however, correlated the till at Hornchurch with his Stortford Till, which was formed by the second of four separate ice advances into Hertfordshire and south Essex, all part of the Anglian (Lowestoft) glaciation (see Chapter 3 and Fig. 3.10E).

Analysis of the clast composition of the overlying gravel (Bridgland, 1988a) supports its interpretation as a main-stream Thames deposit (Table 4.2); it was included by Bridgland (1988a) and Gibbard *et al.* (1988) in the Orsett Heath Gravel. As the latter is correlated with the Boyn Hill Formation of the Middle Thames, this confirms the attribution of the gravel at Hornchurch to this terrace (the former 'High Terrace' of the Lower Thames). This is not the oldest terrace in the post-diversion Thames valley; as stated above, it has been demonstrated that the Black Park Gravel is the earliest post-diversion formation. The recognition of the Black Park Formation in the Lower Thames is, however, controversial (see below, Wansunt Pit).

Many previous authors have noted that the Hornchurch till lies significantly lower than the general level of the Anglian glacial sediments of southern East Anglia and, as stated, this led to suggestions that a valley system existed in the Hornchurch area prior to the glaciation (Holmes, 1892b, 1893; Woodward, 1909; Woodward *et al.*, 1922; Dines and Edmunds, 1925; Warren, 1942; Wooldridge, 1957). The low altitude of the Hornchurch till, overlying London Clay at 25 m O.D., poses problems for Lower Thames stratigraphy and for the reconstruction of chronological events following the diversion of the river. According to Gibbard (1979), the downstream correlative of the earliest post-diversion formation, the Black Park Gravel, has a base level of *c.* 38 m O.D. in the Dartford area (his Dartford Heath Gravel), which was at least 10 km downstream from Hornchurch along the route taken by the pre-Mucking Gravel Thames (see Fig. 4.1). The aggradation of this formation is considered to have been coeval, at least in part, with the continued glaciation of the Vale of St Albans (see Chapter 3, Part 2). However, the altitude of the till at Hornchurch suggests that the valley system there had already been excavated to over 12 m below the supposed Black Park base level (at Dartford) at the time of the ice advance (the Stortford Till advance of Cheshire, 1986a). This is difficult to reconcile with the correlation of the Black Park and Dartford Heath gravels, which has long been a subject of debate (see below, Wansunt Pit). It must be stated at this juncture that the elevation of glacial sediments is of no stratigraphical significance unless they are interbedded with fluviatile or marine deposits; glaciers can erode and infill closed hollows of very large dimensions, which can be well below the level of other contemporaneous sediments. It is therefore possible that the till at Hornchurch was deposited in an overdeepened hollow (as suggested by Warren (1917, 1924a)) or a 'tunnel-valley'. However, neither of these explanations seems likely given the location so close to the limit of what appears to have been a narrow lobe of ice. Moreover, there is further evidence that a valley system was deeply excavated in the Lower Thames by Late Anglian times, from projections of the Westmill Upper Gravel from the Vale of St Albans southwards down the Lea valley (Cheshire, 1983c, 1986a). This gravel, contemporaneous with the later part of the glaciation of the Hertford area and charged with its outwash (Chapter 3, Part 2), has a steep downstream gradient, indicating that its base level would have been as low as 35 m O.D. when it joined the Lower Thames valley and around 30 m O.D. by the Hornchurch area – 8 m below the base of the Dartford Heath Gravel some 10 km further downstream (see Fig. 4.7).

Bridgland (1980, 1983a) suggested a revised Lower Thames stratigraphy, in which the surface overlain by the Hornchurch till was correlated with the Black Park 'bench', the erosion surface beneath the Black Park Gravel. This surface, equivalent to that underlying Zeuner's Kingston Leaf, appears to fall below the level of the Boyn Hill Gravel east of London (Evans, 1971; Bridgland, 1980). It was probably formed as a result of rapid downcutting by the newly diverted Thames along its adopted course. This interpretation allows the reconciliation of an Anglian age for the till at Hornchurch with the evidence for diversion of the Thames from the Vale of St Albans during the same glacial period. According to Cheshire (1986a), the Thames was diverted by the first of the four ice advances into the Vale of St Albans, whereas the till at Hornchurch represents the second advance. Thus by the time the ice reached Hornchurch, the Thames had already excavated its newly adopted valley to the base level of the Hornchurch till.

There remains a significant difficulty, however. This arises from the fact that Anglian glacial deposits and Hoxnian fluvial sediments apparently overlie the same erosion surface, 26 m above O.D. at Hornchurch and *c.* 23 m O.D. at Swanscombe; yet the accepted terrace stratigraphy of the Lower Thames requires the (Late Anglian) Black Park/Dartford Heath Gravel to have been aggraded to over 40 m O.D. (Fig. 4.3) between the deposition of these two sets of sediments. This problem (discussed below – see Wansunt Pit) raises serious doubts about the correlation of the gravel at Dartford Heath with the Black Park Formation.

The Hornchurch railway cutting GCR site contains an important stratigraphical reference section, illustrating the relations between Anglian till and one of the oldest gravels of the Lower Thames terrace sequence. The details of the initiation and evolution of the Thames valley through London, during and following the glaciation, are at present uncertain. Further studies are required of deposits in the critical areas on the northern side of the present valley, where glacial sediments are preserved beneath fluvial (Thames) deposits. In particular, knowledge of the precise geometry of the till remnants would be useful, as would further information about the possible lacustrine sediments at Upminster. The reference section at Hornchurch will be a starting point for further work on the Anglian glacial deposits and palaeogeography of the Lower Thames region.

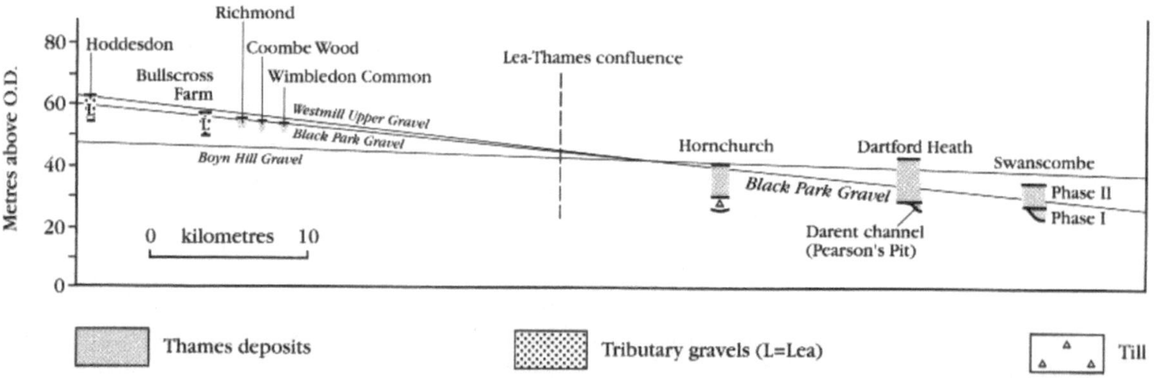

Figure 4.7 Long profile projections of the Black Park and Boyn Hill Formations between the Middle and Lower Thames. The correlation with the Westmill Upper Gravel of the Lea basin is also shown.

Conclusions

This locality is a unique reference site, providing important evidence that the glaciation of the North London area by East Anglian ice (during the Anglian Stage, around 450,000 years BP) occurred before the deposition of the highest terrace gravel of the Lower Thames. During this glaciation, ice sheets repeatedly invaded the old Thames valley across Hertfordshire and central Essex. Hornchurch is the most southerly point known to have been reached by these ice sheets. A narrow lobe of till (boulder clay), directly deposited by the ice, now occupies the Ingrebourne valley to the north of Hornchurch, as was first discovered when the railway was constructed. The till at Hornchurch is overlain by gravel of the Boyn Hill Terrace. This juxtaposition underlines the fact that the Lower Thames came into existence only after the Anglian ice sheets blocked the old route of the river, diverting it into its modern valley through London.

WANSUNT PIT, DARTFORD HEATH (TQ 515738)

D.R. Bridgland

Highlights

Controversial Thames gravels, variously correlated with the Black Park or Boyn Hill Formations, are overlain at this site by palaeolith-bearing fine-grained sediments.

Introduction

Wansunt Pit is one of many old workings that once exploited the Pleistocene gravels and the underlying Thanet Sand of Dartford Heath. Although most have been infilled, the floor of Wansunt Pit has become a factory site and steep faces on its south-eastern side are still available for study. There is a long history of interest in the Dartford Heath deposits, which have been correlated by some workers with the famous Palaeolithic fossiliferous sediments at Swanscombe, a few kilometres to the east (Chandler and Leach, 1911, 1912; Leach, 1913; Smith and Dewey, 1914; Dewey *et al.*, 1924). A considerable controversy has existed since the turn of the century over the relations between the gravel at Dartford Heath and the various deposits at Swanscombe (Hinton and Kennard, 1905; Chandler and Leach, 1907; Zeuner, 1945), a problem that is by no means resolved (Bridgland, 1988a). The former was named the Dartford Heath Gravel by Gibbard (1979, 1985), who considered it to be earlier than the Swanscombe deposits and to be a downstream continuation of the late Anglian Black Park Gravel of the Middle Thames (see Chapter 3). It is difficult, however, to reconcile this interpretation with evidence from other sites in the region and with the projected downstream gradients of the Black Park and Boyn Hill Gravels of the Middle Thames.

Description

Wansunt Pit lies on the western side of Dartford Heath, excavated into the edge of the largely gravel-covered 'plateau'. The thickest depth of gravel recorded from any Thames valley site was once exposed here, almost 20 m in all (Smith and Dewey, 1914). Few detailed descriptions of exposures in the gravel exist. There has been speculation, however, that two separate aggradations were represented, one banked against the other (Cornwall, 1950; Fig. 4.8A). No observations of two gravel bodies with such a relation have ever been made, however; published descriptions of pits on Dartford Heath all suggest that a single aggradation was represented (Chandler and Leach, 1911, 1912; Smith and Dewey, 1914; Dewey *et al.*, 1924; Dewey, 1959).

Chandler and Leach (1912) reported that up to 13 m of gravel was exposed in Wansunt Pit, the upper 4 m loamy and the remainder cross-bedded and sandy (Fig. 4.8B). They believed that most of the non-local gravel clasts occurred in the lower sandy gravel, which suggests that the upper 4 m may have represented locally-derived, possibly colluvial material. The lower sandy gravel also yielded mammalian remains, the most exhaustive list being provided by Leach (1913). He recorded ***Palaeoloxodon antiquus, Cervus elaphus, Cervus*** sp., ***Bos*** or ***Bison*** sp., ***Equus caballus*** and indeterminate rhinoceros.

Over part of the site the gravel was cut out by

(A)

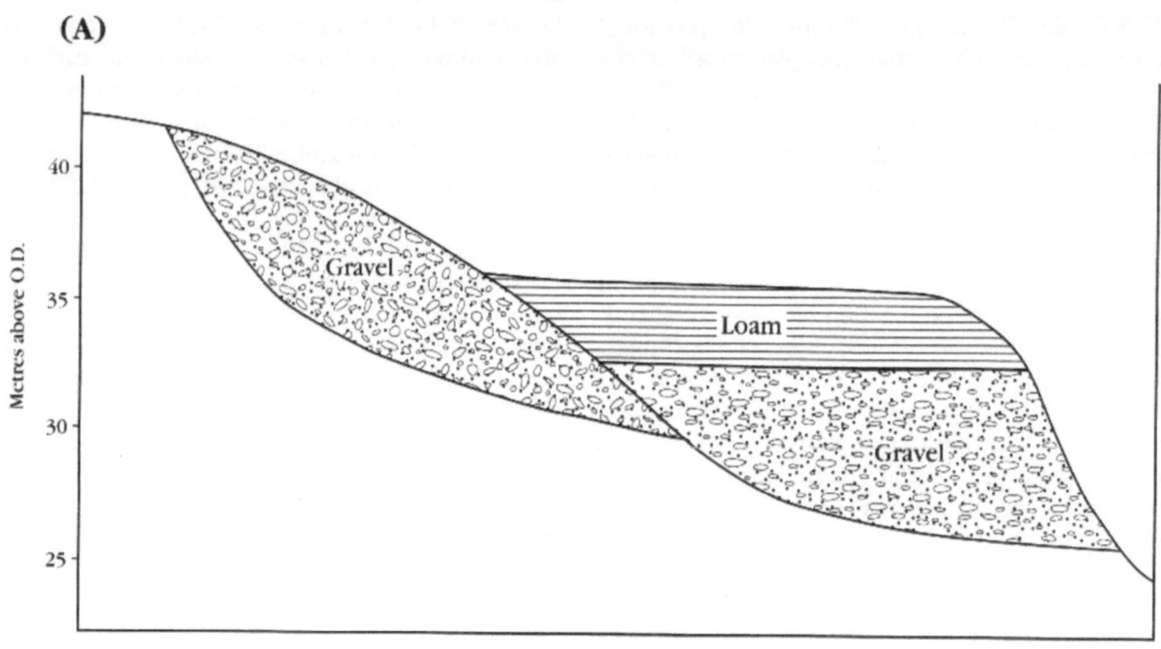

(B)

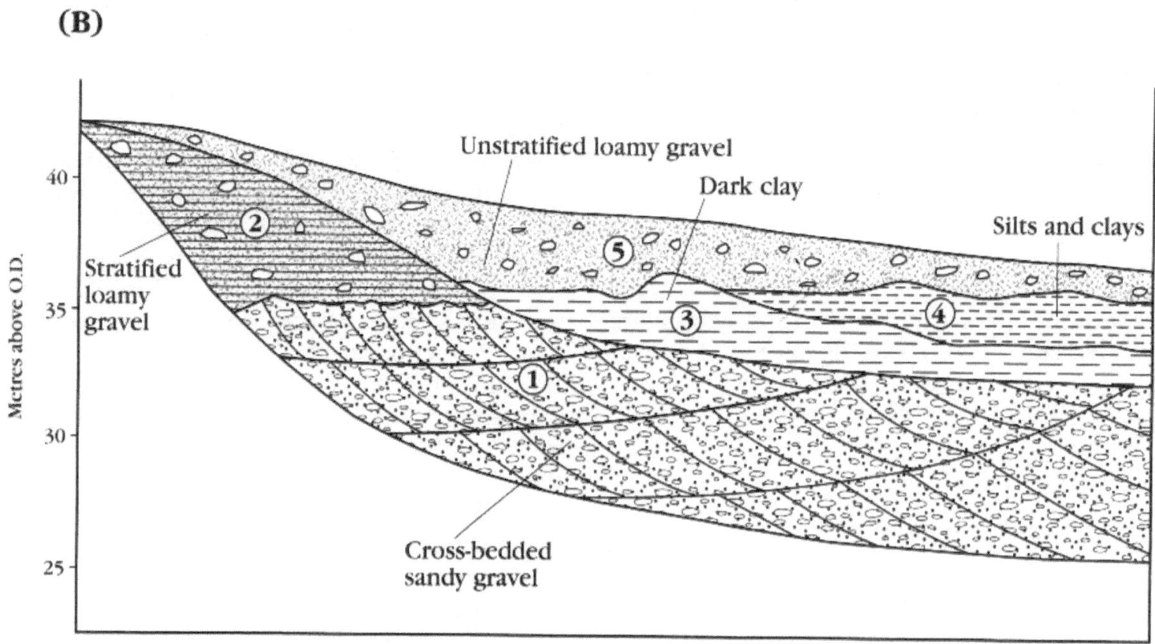

Figure 4.8 Contrasting interpretations of the sediments at Dartford Heath: (A) hypothetical and idealized section, after Cornwall (1950); (B) a composite section based on observations of exposures in Wansunt Pit (after Chandler and Leach, 1912). Bed numbers used in the description section are indicated.

overlying fine-grained deposits ('Wansunt loam') from which Palaeolithic artefacts were obtained (Chandler and Leach, 1911, 1912; Smith and Dewey, 1914). The sequence therefore comprised (see also Fig. 4.8B):

		Thickness
5. Unstratified, loamy gravel	(?colluvium)	uncertain
4. Stratified silts and clays	'Wansunt loam'	0–3 m
3. Dark clay		
2. Loamy gravel, planar-bedded	Dartford Heath Gravel	4 m
1. Sandy gravel, cross-bedded		up to 11 m

Smith and Dewey (1914) recorded a greater thickness for bed 2 (over 6 m) and reported that another 3 m of clayey gravel occurred beneath the floor of the pit, bringing the total thickness of Pleistocene deposits to nearly 20 m.

The stratified silts and clays (bed 4) yielded most of the flakes and implements, as well as fragments of mammal teeth (*Equus caballus* and *Bos* sp.), although the dark clay (3) contained a few flakes. Beds 3 and 4 were interpreted by Chandler and Leach as the fill of a small stream channel ('Wansunt loam' – see below). The uppermost unstratified gravel (5) was interpreted as slope wash from the higher part of the heath to the south (see Fig. 4.8B; Chandler and Leach, 1912).

A number of reviews and catalogues of the archaeology of Wansunt Pit have appeared since the initial discoveries, notably those of Clinch (1908), Smith (1926), Anon. (1931) and Wymer (1968).

Interpretation

Wansunt Pit provides valuable exposures in what many have regarded as the oldest gravel deposited by the Thames in its post-diversion valley. It is the type locality of the Dartford Heath Gravel (Gibbard, 1979), the name given to this high-level deposit in the Lower Thames. Part of the scientific importance of this site arises from the long-standing controversy over the age of the gravel and its relation to other deposits in the Middle and Lower Thames. For this reason, a detailed examination of past references to this and other Dartford Heath localities will be included here, to illustrate the development of the controversy.

Probably the earliest description of the Dartford Heath Gravel was by Trimmer (1853), who noted its continuation on the eastern side of the Darent valley as far as Greenhithe. Trimmer also traced his 'Dartford Gravel' southwards up the Darent valley to Sutton-at-Hone, but later mapping by the Geological Survey has shown the latter to be a more recent deposit (New Series, Sheet 271). Trimmer recorded 15 ft (4.5 m) of gravel exposed in a pit at Wilmington, on the south-eastern side of the heath. He described at length the composition of the gravel in the Dartford area, recording the presence of flints, southern lithologies from the Weald and quartzite pebbles from the Midlands. He noted that the spreads of gravel at Dartford Heath and Dartford Brent (on the opposite side of the Darent) coincide with the former confluence of the rivers Darent and Thames.

Evans (1872) figured a hand-axe found by F.C.J. Spurrell some 8 ft (2.5 m) beneath the surface of the Dartford Heath Gravel. He considered this gravel to be 'that of the upper level of Dartford Heath' (Evans, 1872, p. 532) and attributed it to the Thames rather than the Darent or the Cray. The above quotation from Evans is of interest, since it suggests that he may have observed more than one gravel unit at Dartford Heath, many years before the controversy over the terrace sequence in the Dartford area was initiated. The comment may simply mean, however, that Evans attributed the Dartford Heath deposits to a higher rather than a lower terrace. A second implement was found nearby in 1879 (Spurrell, 1880; Evans, 1897). Spurrell (1880) remarked that the 'Dartford Gravel' extended for many miles on either side of the present River Thames and that it contained Palaeolithic implements. He later illustrated a section at Dartford Heath, showing brickearth channelled into the edge of the gravel (Spurrell, 1886, plate 1). As no details of location or orientation were given, it is impossible to judge whether this might be an early reference to the 'Wansunt loam', although Spurrell (1886, p. 102) attributed it to the 'uppermost layers at Crayford' (for summary of Crayford deposits see below, Lion Pit).

Prestwich (1891) correlated the Dartford Heath aggradation with his 'Upper Valley Gravels' of the Darent, which he traced from the modern Darent–Eden watershed at Limpsfield, where palaeoliths were recovered. No artefacts or fossils were recorded at Dartford by Prestwich in this paper; a reference to this author, without date or source, by Chandler and Leach (1912), attributing rhinoceros, mammoth and *Corbicula fluminalis* to the Dartford Heath Gravel, remains a mystery. According to Newton (1895), however, Spurrell had found bones of mammoth, rhinoceros, ox, horse and deer in pits at Northfleet and Dartford Brent and, at the latter (probable location TQ 555743), had also found the shells of *C. fluminalis, Bithynia tentaculata* (L.), *Valvata piscinalis* and *Pisidium fontinale*. This fauna indicates that temperate-climate sediments are (or were) included within the deposits covering Dartford Heath and Dartford Brent. Spurrell himself recorded finds of mammoth and rhinoceros from the Dartford Heath gravels, in a brief report of a Geologists' Association excursion that visited Wansunt Pit (Spurrell, 1893). In another excursion report, Salter (1903) noted that sections 54 ft (16.5 m) deep were observed.

Spurrell (1886) and Salter (1903) both associated the Dartford Heath deposits with the high-level gravels of Kingston Hill and Wimbledon Common, which later appeared as 'Glacial Gravel' rather than terrace gravel on the New Series Geological Survey map (1921, Sheet 270). Spurrell's and Salter's views thus pre-empted the controversy that followed. This began, however, with the first real dissension from the view that the Dartford Heath and Swanscombe gravels were part of the same 'High Terrace' sheet, by Hinton and Kennard (1905). They believed that the Dartford Heath Gravel represented the earliest Lower Thames terrace (their 1st Terrace), aggraded to 136 ft (41.5 m) O.D., whereas the Swanscombe deposits, which reach only 100 ft (30.5 m) O.D., were assigned to their 2nd Terrace.

Chandler and Leach (1907, 1911, 1912) described more than 12 m of cross-bedded sands and gravels at Wansunt Pit, resting on Thanet Sand at 90 ft (27.5 m) O.D., and took these to be 'fluviatile drifts lying at the same level as the Swanscombe and Galley Hill gravel spreads; contemporaneous and, arguably, continuous with them, but now separated from them by the subsequent erosion of the Darent valley' (1907, p. 122). In this series of papers they consistently refuted the idea that the gravel at Dartford Heath was the result of an earlier, higher-level aggradation. They noted, with the collaboration of E.T. Newton, the admixture in the gravels of the characteristic suite of 'exotic' lithologies, which occurs throughout the Thames Basin, and southern rocks from the Darent catchment. Their observation of *Rhaxella* chert was the earliest record in the Thames valley of this important rock-type, attributed at that time to the Arngrove Stone of Oxfordshire and Buckinghamshire (Newton, 1907). Recent studies have shown that most, if not all, of the *Rhaxella* chert in the Pleistocene of the London Basin was derived from North Yorkshire during the Anglian glaciation (Bridgland, 1980, 1986b). Its presence in Pleistocene deposits in south-east England is therefore important evidence for an age contemporary with, or later than, this glacial event.

The interpretation of the Dartford Heath deposits promoted by Chandler and Leach was supported by Burchell (1933), King and Oakley (1936) and Marston (1937). The fluctuation of opinion concerning these deposits is reflected in the various Geological Survey memoirs describing the area. In the earliest of these volumes, Whitaker (1884, 1889) listed exposures at Dartford Heath and Dartford Brent and grouped the deposits there with gravels extending further eastwards to Swanscombe, considering them all to represent a dissected 'high terrace'. In the first edition of *The Geology of the London District*, Woodward (1909) reproduced Hinton and Kennard's classification of the Thames terraces, distinguishing the gravels of Dartford and Swanscombe as 1st and 2nd Terraces respectively. In the second edition of this memoir (Woodward *et al.*, 1922), the distinction was dropped and both Dartford Heath and Swanscombe were again included in the 'High Terrace'. Following a resurvey of the London district, undertaken before the First World War, the geomorphological feature and associated deposits formerly referred to as the 'High' or '100 ft' terrace were reclassified as 'Boyn Hill Terrace' (Bromehead, 1912; Dewey and Bromehead, 1915; Sherlock and Noble, 1922). On the New Series geological map of Dartford (Sheet 271), the Dartford Heath deposits, in common with patches of gravel east of the Darent from Dartford to Northfleet, duly appeared as Boyn Hill Gravel. In the accompanying memoir, Dewey *et al.* (1924) remarked that a number of exposures between Dartford and Northfleet indicated that the Boyn Hill Gravel occupies and overlies a channel running west to east; the base of the deposits thus rises to the south and to the north, where their lower levels are separated from the modern river by a buried ridge of bedrock. Rapid rises in the bedrock surface towards the edges of a large trough-feature of this kind may explain the variations in bedrock surface level beneath different parts of the Dartford Heath deposits that are apparent from published descriptions over the years.

The controversy initiated by Hinton and Kennard (1905) was rekindled by Zeuner (1945), who returned to a model in which the Dartford Heath Gravel was classified with the high-level deposits of Kingston and Wimbledon. Zeuner believed that all these deposits were related to the recently described Winter Hill Terrace of the Middle Thames (see Chapter 3). He considered that the 'bench' of the Winter Hill Terrace (the base of the terrace deposits) divided near the Thames–Colne confluence and that the higher of the two 'benches' so produced passed through the Finchley Depression, following Wooldridge's (1938) intermediate course to Essex (Chapter 3 and Fig. 3.4), whereas the lower 'bench' followed the modern course of the river. Zeuner named the deposits overlying these two erosional features the 'Finchley Leaf' and 'Kingston Leaf' respectively. The Kingston and Wimbledon gravels had previously been assigned to the Winter Hill Terrace by Saner and Wooldridge (1929). Zeuner (1945) traced his Kingston Leaf 'bench' from Burnham Beeches (67 m O.D.) downstream to Hillingdon (55 m O.D.), Kingston and Richmond Hills (46 m O.D.) and Dartford Heath (27.5–30.5 m O.D.), therefore correlating the Dartford Heath Gravel with the lower division of the Winter Hill Terrace. He believed the Dartford Heath deposits to be older than those at Swanscombe, which overlie his 'Boyn Hill bench' at 75 ft (23 m) O.D., and that the formation of the two 'benches' was separated by the glaciation of the London area (see above, Hornchurch). In reintroducing the idea that the Dartford Heath deposits were distinct from the Boyn Hill Formation, as represented elsewhere in the Lower Thames, Zeuner perpetuated a controversy that has considerable significance for the classification and dating of the Lower Thames succession and its correlation with sequences elsewhere. Later contributions to this debate will be considered below, in the section on correlation.

Other important Dartford Heath sites

Following the early descriptions of Wansunt Pit, most work on the Dartford Heath deposits has taken place at other sites. Newton (1930) described gravels penetrated by curious cones of sandy clay (loam) in Pearson's Pit (TQ 533737), on the eastern side of the Heath. These features may have been similar to the 'pipes' that disrupt the Lynch Hill Gravel at Furze Platt, which are thought to result from Chalk solution (see Chapter 3, Cannoncourt Farm Pit). Pearson's Pit also produced implements from the body of the Dartford Heath Gravel (Marston, 1937; Dewey, 1959; Wymer, 1968). Marston concluded from these discoveries that the gravel represented 'a later stage of the Barnfield later Middle Gravels and Sands' (Marston, 1937, p. 351) and was of later age than the Swanscombe Lower and Lower Middle Gravels. This was in general agreement with King and Oakley (1936), who placed the Dartford Heath Gravel in their 'Middle Barnfield', or 'Late Boyn Hill Stage'. Wymer (1968) suggested that the implements may have come from the overlying 'loam', but Dewey (1959) clearly reported that some of Mr Pearson's personal collection of hand-axes had been derived from the lowest part of the gravel, the base of which was about 33.5 m above O.D.

A third important quarry that exploited the Dartford Heath Gravel was Bowman's Lodge Pit (TQ 518738), a short distance to the east of the Wansunt Pit. Opened after the Second World War, the excavation of this pit was closely monitored by the archaeologist, P.J. Tester, who described Palaeolithic artefacts obtained from the surface of the gravel, beneath an overlying brickearth that he interpreted as a continuation of the 'Wansunt loam' (Tester, 1951, 1975). Cornwall (1950) described Bowman's Lodge Pit, noting that the base of the gravel rose eastwards from 26 m O.D. to 29 m O.D. in only a few hundred metres along the northern side of the pit, the gravel correspondingly thinning from 7.5 to 6 m. This appears to represent the eastern edge of gravel deposition on Dartford Heath.

Excavations in 1952 on the eastern fringe of Dartford Heath yielded Acheulian implements from fluviatile sands and loams overlying Chalk at 26.5 m O.D. (Tester, 1953). Tester concluded that these were typologically similar to the Middle Barnfield industry of Swanscombe; strong evidence, he considered, for the Boyn Hill age of at least the eastern portion of the Dartford Heath Gravel.

The 'Wansunt loam' and its Palaeolithic industry

Chandler and Leach (1911) were the first to observe the clays and silts ('Wansunt loam'), cut into the Dartford Heath Gravel, that yielded Palaeolithic artefacts and mammal teeth (see above, Description). They later described in detail how the gravel was truncated by a steep slope, in the north-eastern corner of Wansunt Pit, against which the silts and clays were banked (Chandler and Leach, 1912; Fig. 4.8B). They interpreted this feature as the edge of a channel cut into the Dartford Heath Gravel, probably formed by a tributary stream and of significantly younger age (Chandler and Leach, 1911, 1912). Leach (1913) described a further, lower channel in Wansunt Pit, filled with hillwash and loam and containing charcoal and occasional palaeoliths. Chandler and Leach (1912) also recorded occasional finds of artefacts in the Dartford Heath Gravel itself, from Wansunt Pit and other sites, mainly in the upper loamy gravel, but including a rolled example from the cross-bedded gravel.

In 1913 Smith and Dewey (1914) conducted excavations for the British Museum at Wansunt Pit. Various cuttings were made in and to the north-east of the pit, in order to test the interpretation by Chandler and Leach of the clayey and silty deposits as the channel-fill of a minor tributary, a hypothesis put forward without the northern side of the inferred channel having been observed. The results of this work revealed that no northern side is preserved and that the clays and silts extend to the front edge of the terrace remnant at Dartford Heath, where they are truncated by the slope to the modern Thames floodplain. It was concluded, therefore, that the fine-grained sediments were emplaced above and against the gravel by the Thames itself, possibly near the end of 100 ft (Boyn Hill) Terrace times (Smith and Dewey, 1914; Dewey *et al.*, 1924; Smith, 1926). These authors recorded many Acheulian implements from the clays and loams and a few from the Dartford Heath Gravel itself. Following Smith's and Dewey's work, the deposits formerly attributed to the 'Wansunt Channel' have been generally referred to as the 'Wansunt loam'.

The artefacts collected from the 'Wansunt loam' were in an excellent state of preservation and included several sets of conjoinable flakes, suggesting the close proximity of a working floor. Indeed, Chandler and Leach (1912, p. 108) reported that 'At least one 'nest' of flakes was found such as would result from flaking on the spot'. Finished tools were also found, dominantly cordate and ovate hand-axes, although side- and end-scrapers were also identified (Wymer, 1968). Occasional flakes have been interpreted as showing evidence of Levallois technique (Chandler and Leach, 1912; Wymer, 1968), although Roe (1968b) listed no Levallois material from the site.

The Bowmans Lodge industry, from a supposed continuation of the 'Wansunt loam', is similar to that from the original site, mainly comprising typical Acheulian material, but with the important addition of a collection of small 'chopper-cores' (Tester, 1951, 1975; Wymer, 1968). The occurrence together of 'chopper-cores' and hand-axes in similar states of preservation, as they are in the Bowmans Lodge collection, is extremely rare in the Thames valley (Wymer, 1968). As at Wansunt Pit, occasional evidence of Levallois technique has been claimed (Tester, 1951, 1975; Wymer, 1968; Roe, 1981, p. 228). The Bowmans Lodge assemblage was used by Roe (1964, 1968a, 1981) in his statistical analysis of the typology of British Lower and Middle Palaeolithic hand-axe assemblages. Roe was impressed by the undisturbed nature of this material, which fell within his Group VI (ovate tradition, more pointed types), as did the assemblage from the Swanscombe Upper Loam, whereas the industry in the Middle Gravels at Swanscombe fell within his Group II (pointed tradition with some ovates). Roe's analyses, based on measurements of breadth and thickness, add support to the suggestion that the 'Wansunt loam' post-dates much of the Swanscombe sequence, although this is based largely on the superposition of Roe's Group VI over Group II at Swanscombe itself. It is clear, however, that the Swanscombe situation, with 'pointed' industries preceding 'ovate' ones, does not hold good for Britain as a whole (Roe, 1981, pp. 302–3). What remains uncertain is whether this model is stratigraphically valid for the immediate Swanscombe-Dartford area.

A hand-annotated 1:10,560 map of the

Wansunt Pit site formerly belonging to A.L. Leach, a copy of which was supplied to the Nature Conservancy Council by J.N. Carreck, clarifies a number of questions about site locations in the vicinity of Wansunt Pit. Wymer (1968) suggested that the 'Wansunt loam' was first discovered in Martin's Pit; the abovementioned map, however, shows Martin's Pit to have been to the south of Wansunt Pit, quite remote from the edge of the loam. The annotated map suggests that the southern edge of the 'Wansunt loam' passes through the northern part of the present GCR site.

Correlation with the Lower Thames sequence

The continuing controversy over the correlation between the Dartford Heath deposits and the Swanscombe sediments has hampered the integration of these terrace remnants in north Kent with the established Thames sequence.

Cornwall (1950) noted that there was no evidence for a protracted period of subaerial weathering of the Dartford Heath Gravel surface prior to the deposition of the 'Wansunt loam', a fact that suggested to him that the two deposits were not separated by any lengthy interval of time. He claimed, however, that the industry from the surface of the gravel appears, on typological grounds, to be slightly later than the 'Middle Barnfield Stage' at Swanscombe. These conclusions argue against the Hinton and Kennard model of the Dartford Heath Gravel as a separate, pre-Boyn Hill aggradation, especially if Smith and Dewey (1914) were correct in attributing the 'Wansunt loam' to the Thames. It is difficult to envisage the Thames depositing the latter sediments above the Dartford Heath Gravel after rejuvenation had taken the river on to a lower terrace level, yet Cornwall considered the 'Wansunt loam' to be of 'late Boyn Hill' age (therefore post-dating a supposed rejuvenation from the Dartford Heath to the Boyn Hill level). This would appear to favour the interpretation promoted by Chandler and Leach, in which the Dartford Heath deposits are seen as the culmination of the Boyn Hill Gravel aggradation. In an attempt to find a compromise between the two theories, Cornwall suggested that the highest parts of the heath might represent an earlier terrace, against which the Boyn Hill Gravel has been banked (Fig. 4.8A), although he presented no field evidence in support of this idea; a similar suggestion had been made by Zeuner (1945, p. 120), in a footnote. If this was the case, the higher of the two units would be the Dartford Heath Gravel and the lower would be attributable to the Orsett Heath Gravel (Bridgland, 1988a; see Introduction to this chapter).

Dewey (1959) reviewed this continuing controversy, as well as making a number of important observations and describing sites about which little or nothing has been published. He noted that at Pearson's Pit (TQ 530733) the Dartford Heath Gravel overlay a buried channel, some 2.5 m deep and 30 m wide, cut into the Thanet Sand. He attributed this feature to a tributary flowing across the area to join the main valley and correlated it with the Lower Gravel and Lower Loam channel at Barnfield Pit, with which it is altitudinally comparable, concluding that the excavation of the latter pre-dated the deposition of the Dartford Heath Gravel. This interpretation implies that the Dartford Heath Gravel is part of the same terrace aggradation as the upper part of the sequence at Swanscombe, reaffirming the views of Chandler and Leach (1907, 1911, 1912).

The controversy was further intensified by Evans (1971). According to Evans, the first three terraces to be formed in the modern Thames valley had progressively shallower downstream gradients, indicating that base level rose between each, rather than falling (as would be expected in a terrace sequence – see Chapter 1). These three terraces were the Kingston Leaf, Black Park and Boyn Hill aggradations, for which Evans estimated contemporary sea levels of 27 m, 29 m and 32 m respectively. Wooldridge and Linton (1955) had attributed Zeuner's 'Kingston Leaf' at Wimbledon and Kingston to the Black Park Terrace (as defined by Hare, 1947), but Evans did not accept this correlation, regarding the Kingston deposits as the result of a separate, higher-level aggradation, somewhat earlier than the Black Park Terrace. Evans's hypothesis implied that the Kingston Leaf and Black Park gravels, because of their steeper downstream gradients, should fall below the Boyn Hill Terrace surface somewhere in the London area, so that the products of the three aggradations are superimposed in the Lower Thames. Therefore (according to Evans), the deposits at Dartford, which are aggraded to the highest level of any Lower Thames gravels, would represent the latest of these three aggradations, that of the Boyn Hill Terrace. The Black Park Gravel, if represented at all at Dartford, would fall within the lower part of the sedimentary sequence there, beneath the Boyn Hill deposits.

However, P.L. Gibbard (pers. comm.) has refuted Evans's idea of a separate Kingston Leaf; he has pointed out that the Kingston Hill deposits are higher than the main Black Park Gravel because they represent the equivalent terrace of the tributary River Wey. Gibbard (1979) reaffirmed the conclusions of Hinton and Kennard (1905) and Zeuner (1945) by correlating the Dartford Heath Gravel with the Black Park Gravel of the Middle Thames, citing the similarity in gravel composition between Dartford Heath and Richmond in support of this argument. He later reiterated this view and suggested that a downstream continuation of the Dartford Heath Gravel could be recognized in Essex, at Orsett (TQ 655811; Gibbard *et al.*, 1988). However, Bridgland (1980) followed Evans (1971) in concluding that the Black Park Gravel falls below the Boyn Hill aggradation in the London area. He suggested that the Swanscombe Lower Gravel and Loam might represent part of the Black Park aggradation, but that the Swanscombe Lower Middle and Upper Middle Gravels and the Dartford Heath deposits represent the Boyn Hill Gravel. Projection of the Black Park and Boyn Hill Gravels downstream from the Middle Thames, using data presented by Gibbard (1985) in the latter area, indicates that they should intersect between London and Dartford (Fig. 4.7). The Black Park surface would therefore be expected to be lower, not higher, than the Boyn Hill surface in the Dartford area, as Evans predicted. However, the elevation of the upper surface of the Dartford Heath Gravel, 41.5 m O.D., is *c.* 7 m higher than the general level of the Boyn Hill Terrace surface in this area.

Other explanations of the anomalously high level of part of the Dartford Heath spread must be considered, therefore. It is difficult to envisage a mechanism that could result in the localized accumulation of fluviatile deposits so high above the normal floodplain level. The location of the Dartford Heath deposits at the back edge of the terrace, at the apex of a major bend in the floodplain at that time and in the region of the multiple Darent/Cray/Thames confluence (Fig. 4.1), may be of relevance. Dewey *et al.* (1924, p. 90) opened their account of the Boyn Hill (100 ft) Terrace of the Dartford area by noting that: 'the gravel spreads are, in part, of the nature of deltas at the confluence of the tributaries with the main river'. The Orsett remnant, claimed by Gibbard *et al.* (1988) as a downstream correlative of the Dartford Heath Gravel, is only 2–3 m higher than the Boyn Hill/Orsett Heath Gravel at the Orsett Heath type locality, nearby. Its extra elevation may well reflect its position at the back-edge of an Orsett Heath Gravel terrace remnant.

Records of buried channels beneath the terrace sediments of the Dartford area (Dewey *et al.*, 1924; Dewey, 1959) may be of some significance to this discussion. Bridgland (1980, 1983a, 1983b, 1988a) described a major buried channel system underlying the downstream continuation of the Boyn Hill/Orsett Heath Gravel across eastern Essex, which he correlated with the Lower Gravel Channel at Swanscombe and the Clacton Channel (see Chapter 5). The terrace gravels overlying this channel system, the highest post-diversion Thames deposits recognized in eastern Essex by Bridgland, were correlated with the Boyn Hill/Orsett Heath Gravel, no continuation of the Black Park Gravel apparently being represented. This supports the view, first suggested by Evans (1971), that the Black Park Terrace passes beneath the later Boyn Hill aggradation east of London; the former aggradation is probably represented within the channel deposits that underlie the latter in this area (Bridgland, 1980, 1988a; Fig. 1.3). This interpretation gains further support from the low height of the till at Hornchurch (see above, Hornchurch) and the gradient of the earliest River Lea deposit, the Westmill Upper Gravel (Cheshire, 1983c, 1986a; Chapter 3, Part 2; Fig. 4.7), both of which imply that the Lower Thames valley was excavated to below 30 m O.D. by the late Anglian, the recognized age of the Black Park Gravel.

The balance of evidence, therefore, appears to favour the interpretation of the Dartford Heath Gravel as part of the Boyn Hill/Orsett Heath Formation. The most significant points are: (1) the recorded thickness of the gravel at Dartford Heath, which implies a single aggradation up to 20 m thick; (2) the record, by Dewey *et al.* (1924), of a major channel beneath the mapped spreads of gravel east of Dartford Heath, apparently equivalent to that beneath the Boyn Hill/Orsett Heath Gravel of the Swanscombe area; (3) Dewey's (1959) observation of a tributary channel, cut to the base level of the Swanscombe Lower Gravel, beneath the Dartford Heath Gravel at Pearson's Pit (which reinforces 2); (4) the fact that the Black Park Formation, with its steep long-profile, appears likely to fall below the Boyn Hill level upstream of Dartford, therefore precluding correlation with the Dartford Heath Gravel (Fig. 4.7); and (5) evidence from the glacial history of the area to the north-west, which indicates that the Lower Thames valley system was excavated to below the level of the Dartford Heath Gravel by late Anglian times. These points combine to provide a strong case for assigning the entire Dartford Heath sequence to the Boyn Hill/Orsett Heath Formation.

The interpretation of the Dartford Heath Gravel and associated deposits by Gibbard (1979; Gibbard *et al.*, 1988), and that favoured in this report, are clearly incompatible. These different views represent the latest episode in a long-standing controversy regarding the interpretation and wider correlation of the Lower Thames sequence. The resolution of this controversy is a task requiring urgent attention. The last remaining exposures of this deposit, at Wansunt Pit, will be of prime importance for future work on the gravel. The possibility that part of the 'Wansunt loam' remains *in situ* at the site heightens its scientific importance. Future work may determine whether a surviving remnant of the loam is indeed present and provide new information about the origin and age of both the loam and the underlying gravel.

Conclusions

Wansunt Pit provides exposures in the Dartford Heath Gravel, the highest-level terrace deposit in the Lower Thames. Controversy has continued throughout this century over the relation of this deposit to others in the terrace 'staircase' in this part of the valley, particularly the famous Swanscombe skull site. Some workers have regarded the gravel at Wansunt Pit as a remnant of a very early terrace, one that is not preserved downstream. Others have regarded it as merely an unusually high 'feather-edge' remnant of a high-level terrace that is widely recognized elsewhere in the valley, including Swanscombe – the view favoured in this volume. The resolution of this dispute is of considerable urgency, as it has implications for the dating of the Pleistocene terrace sequence in the Lower Thames and for relating it to the original formation of this part of the valley, following the river's diversion by ice around 450,000 years ago. An important aspect of the geological interest at Wansunt Pit is the occurrence there, in a loam deposit overlying the gravel, of a rich assemblage of well-preserved Palaeolithic artefacts.

SWANSCOMBE (BARNFIELD PIT – SKULL SITE NNR AND ALKERDEN LANE ALLOTMENTS SSSI)

D.R. Bridgland

Highlights

A sequence of gravels, sands and loams at Swanscombe has yielded important assemblages of interglacial mammals and molluscs. The sediments also contain a wealth of Palaeolithic artefacts, demonstrating the rare superposition of different industries – Acheulian above Clactonian. The site is famous for the discovery, in association with Acheulian artefacts, of a human skull, intermediate in form between *Homo erectus* and Neanderthal Man. The site has long been regarded as Hoxnian, but this has recently become the subject of controversy, although it is clear that the first post-Anglian interglacial is represented.

Introduction

Barnfield Pit, now the 'Swanscombe skull site' NNR, is probably the most famous Pleistocene locality in Britain and is certainly the best known in the Thames valley. The site became the first geological NNR (National Nature Reserve) when it was donated to the nation by the former owners, the Associated Portland Cement Company Ltd, in 1954. In fact two old pits, the original Barnfield Pit and Colyer's Pit, were amalgamated during the early decades of this century to form the site later known as Barnfield Pit or, sometimes, as Milton Street Pit (Wymer, 1968). The Swanscombe site is of international renown as a result of the discovery there, *in situ* in an old

Thames gravel, of the fragmentary skull of an early human. Remarkably, three pieces of the same skull were found on separate occasions, in 1935, 1936 and 1955, from the same sedimentary unit (Marston, 1937; Wymer, 1955, 1964b). Together they form the back half of the skull of a young adult, possibly a woman (Stringer, 1985).

Quite apart from the discovery of 'Swanscombe Man', Barnfield Pit is an extremely important Pleistocene stratigraphical site. Products of two fluvial aggradations occur in superposition, separated by a soil horizon and overlain by overbank ('floodloam') and slope deposits (Wymer, 1968; Conway and Waechter, 1977; Bridgland *et al.*, 1985). All these sediments contain Palaeolithic artefacts, three distinct assemblages being represented in stratigraphical sequence (Wymer, 1968; Roe, 1981). In fact, the Swanscombe district has produced more palaeoliths than any other in Britain (Wymer, 1968). Many of the sediments are also fossiliferous, with mammals (Sutcliffe, 1964) and molluscs (Kerney, 1971) providing the most significant environmental and biostratigraphical evidence. In recent years there has been general agreement that these indicate a Hoxnian age, but current uncertainty about the correlation between terrestrial stratigraphies and the marine oxygen isotope sequence (Bowen *et al.*, 1989; Chapter 1), together with a controversial pollen record from Swanscombe suggesting a complex succession of climatic fluctuations (Hubbard, 1982; Turner, 1985), have raised new doubts about this interpretation.

Sections in the south-western part of the skull site were opened in 1977 for a visit by INQUA (International Union for Quaternary Research; Conway and Waechter, 1977) and reopened in 1985 for a celebration of the 50th anniversary of the discovery of 'Swanscombe Man' (Duff, 1985). Further research excavations have taken place at Swanscombe under the auspices of the Geological Conservation Review. Firstly, in 1982, the full sequence was uncovered in the north-west corner of the site (Figs 4.9, 4.10, 4.11 and 4.12) for geological appraisal and sampling, including the collection of material for thermoluminescence dating (Bridgland *et al.*, 1985) and, in particular, for the detailed examination of the weathering horizon at the top of the Lower Loam (Kemp, 1985b). Secondly, in 1986, the Lower Gravel and Lower Loam were re-exposed to allow sampling of the former for sieving in search of small vertebrates, and to collect calcite coatings from pebbles in the Lower Gravel in order to obtain a uranium-series date (results of the latter are not yet available). The upper part of the Swanscombe sequence has been quarried away from the area of Barnfield Pit, now the NNR. However, sections in the full sequence, which is preserved *in situ* in the Alkerden Lane Allotments SSSI, are available at the western margin of the old pit (Figs 4.9 and 4.11).

Within the confines of the present volume, a comprehensive review of the literature and research on the Swanscombe site is not possible. In addition to over a hundred specialist reports dealing with various different aspects of the scientific interest, two major symposium volumes exist that are devoted to Swanscombe (Swanscombe Committee, 1938; Ovey, 1964). Swanscombe is also the most frequently cited Thames (Pleistocene) site, there being innumerable passing references to it in the geological, palaeontological and archaeological literature. An archaeological excavation had already taken place at Barnfield Pit prior to the discovery of the Swanscombe skull (Smith and Dewey, 1914), underlining the importance of the site as a Palaeolithic locality, irrespective of its anthropological significance. Several further excavations have been carried out since the discovery, of which two were major undertakings (Wymer, 1964b; Waechter, 1969, 1970, 1971, 1972). Comprehensive, multi-disciplinary reviews of the site and its importance (with extensive bibliographies) were produced by Oakley (1952) and Wymer (1968). The most recent archaeological excavations, which were confined to the lower part of the sequence (Waechter, 1969, 1970, 1971, 1972), have yet to be published in detail. Meanwhile Roe (1981) has provided the most recent detailed review of the archaeological significance of the Swanscombe skull site and associated localities.

Important reviews of the palaeontology have appeared from time to time, the most recent being by Sutcliffe (1964) and Stuart (1982a) on the mammals and by Kerney (1971) on the molluscs. The present report will concentrate on developments over the past two decades and on the significance of Swanscombe in respect to the current interpretation of the Thames sequence and the British Pleistocene as a whole.

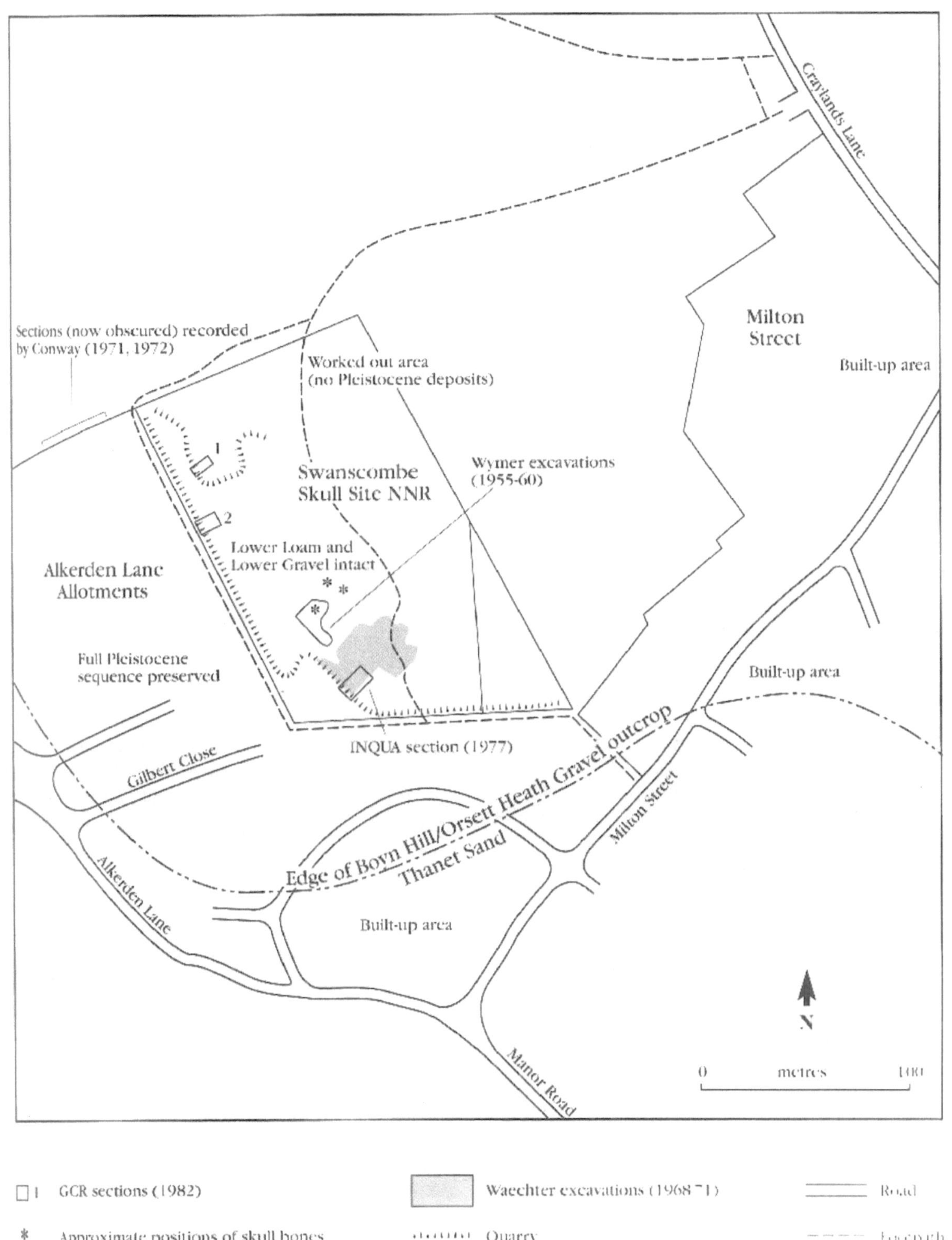

Figure 4.9 Map of the Swanscombe skull site and adjacent areas.

Description

Detailed first-hand descriptions of the Swanscombe (Barnfield Pit) deposits have been provided by Smith and Dewey (1913, 1914; Dewey and Smith, 1914), Dewey (1932), Dines *et al.* (1938), Wymer (1964b, 1968), Conway (1969, 1970a, 1971, 1972, 1985) and Bridgland *et al.* (1985). The summary that follows is derived from a combination of the more recent descriptions. It is necessary to describe the stratigraphy of the Swanscombe deposits in some detail, since part of the unique importance of the site lies in the occurrence there of a complex succession of sediments with different palaeontological and archaeological contents.

The complete Swanscombe sequence is as follows (after Conway and Waechter, 1977) (see also Figs 4.12 and 4.13):

Members (beds)	**Thickness**	**Distribution**
Phase III		
IIIe Higher loams	up to 1 m	South-west only
IIId **Upper Gravel**	2 m	Most of site
IIIc **Upper Loam**	1 m	Most of site
IIIb Channel deposits	0–2 m	Localized channel-fill
IIIa Soliflucted clay	0–1 m	South-west only
Phase II		
IIb **Upper Middle Gravel**	1.5–3 m	Most of site
IIa **Lower Middle Gravel**	0.5–2 m	Most of site
Phase I		
Id **Lower Loam**	2–2.5 m	Wide channel-fill
Ic 'Midden' complex	0–0.75 m	Localized
Ib **Lower Gravel**	up to 5 m	Wide channel-fill
Ia Basal gravel	0–0.5 m	Localized

Conway and Waechter (1977) recognized three phases of deposition (they called them stages) at Swanscombe, reflected in their numbering system, which is adopted here (see above). They believed the junctions between these to represent significant breaks in the succession. All the principal fluvial sediments were laid down during Phases I and II; these deposits are classified here as members and beds within the Boyn Hill/Orsett Heath Formation (Members IIIb and IIIc are probably also part of the fluviatile sequence).

Phase I: Basal Gravel (Ia)

An interpretation of the basal part of the Lower Gravel as a solifluction deposit was first suggested by Marston (1937, p. 31) on the grounds of its 'rough and tumble appearance ... and the presence of scratched flakes ...'. This view was reiterated by Paterson (1940) and Conway (1969, 1970a), the latter citing poor sorting and the occurrence of '... 'nests' of thermally fractured pebbles ...' (Conway, 1970a, p. 90) in the basal part of the unit as evidence of a periglacial origin. The idea was, however, refuted by Kerney (1971), who recorded mollusc remains from near the base of the Lower Gravel that indicated a climate no less temperate than those from the remainder of the deposit or those from the Lower Loam. In contrast to both Conway's and Kerney's observations, Bridgland *et al.* (1985) considered that the basal part of the Lower Gravel, as exposed in the 1982 GCR section, differed from the bulk of the unit in that it was clast-supported and lacked shells, but was clearly fluvially deposited. They concluded that this basal layer might date from the final part of the cold-climatic interval that preceded the interglacial represented by the overlying deposits. The principal evidence in support of this claim is the absence of fossil material in this basal layer; this may, however, be a result of lithological differences, such as the paucity of matrix in this clast-supported gravel.

There are, amongst the voluminous literature on Swanscombe, a number of similarly conflicting reports of this and other parts of the sequence, arising from observations of sections in different parts of the site. This is probably a reflection of lateral variation within the sediments, a source of some confusion when attempting an appraisal of past research at Swanscombe.

Figure 4.10 Exposure in the Swanscombe Lower Gravel and Lower Loam, GCR Section 1 (see Fig. 4.9), opened in October 1982 but photographed the following winter. The section has been sampled for clast-lithological analysis and palaeontological studies. Articulated bivalves are visible in the upper part of the Lower Gravel. (Photo: D.R. Bridgland.)

Lower Gravel (Ib)

The main body of this member is a coarse, sandy, horizontally-bedded gravel up to 5 m thick, occupying a wide channel excavated in the Thanet Sand (see Fig. 4.13; Wymer, 1968; Conway and Waechter, 1977). Molluscan and mammalian faunas from this deposit are well documented (Sutcliffe, 1964; Kerney, 1971; Stuart, 1982a) and point to fully temperate conditions. More equivocal is a palynological interpretation, based on analyses following pollen concentration by heavy liquid flotation, of open grassland with mixed oak forest environments (Hubbard, 1982). The Lower Gravel contains abundant flint artefacts of Clactonian type, comprising cores and flakes but no well-finished tools (Smith and Dewey, 1913; Chandler, 1930, 1931, 1932a, 1932b; Waechter, 1969, 1970, 1971, 1972).

Figure 4.11 Exposure in the Swanscombe Lower Loam, Lower Middle Gravel, Upper Middle Gravel and Upper Loam, GCR Section 2 (see Fig. 4.9), October 1982. The Lower Loam, exposed in a pit (see Fig. 4.15) at the base of the main section, has been sampled for studies of micromorphology. The shovel is standing on the top of the Lower Middle Gravel (see also Fig. 4.12). (Photo: P. Harding.)

'Midden' complex (Ic)

This localized bed is channelled into the upper part of the Lower Gravel to a depth of 0.75 m. It comprises thin alternations of silt, sand and fine gravel and contains important concentrations of mammalian remains and Clactonian artefacts in very fresh condition, possibly indicating a primary context (Conway, 1970a, 1971, 1972; Waechter, 1970, 1971, 1972; Conway and Waechter, 1977). It was originally described as part of the Lower Loam (Conway, 1970a; Waechter, 1970), but Conway (1971, p. 60) later recognized a 'distinct stratigraphic break' between the 'midden' deposits and the Lower Loam. The 'Midden' complex has been associated with the Lower Gravel in subsequent publications. The level of concentration of

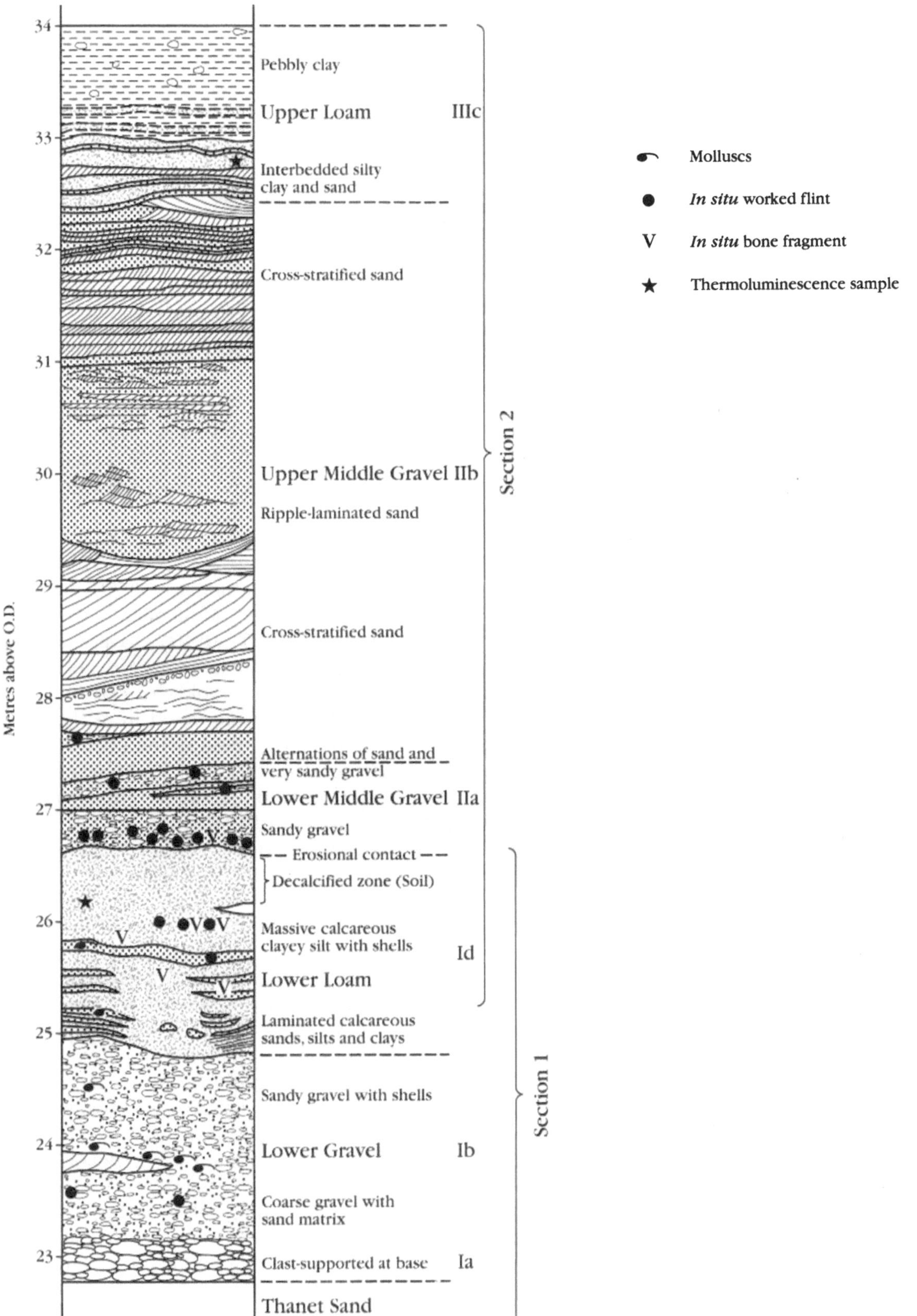

Figure 4.12 The sequence at Swanscombe, based on the exposures excavated by the GCR Unit in 1982 (after Bridgland *et al.*, 1985).

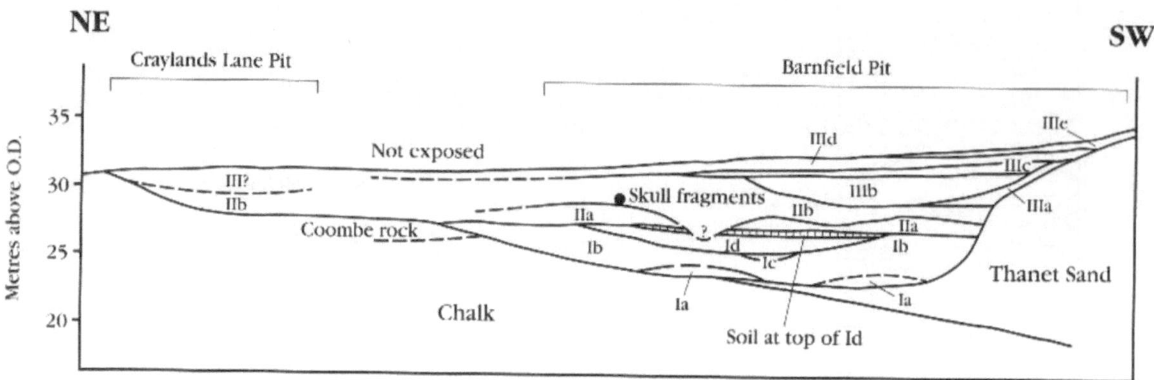

Figure 4.13 Section through the terrace deposits at Swanscombe. The notation follows the description section in the text.

mammalian remains in this bed is suggestive of an artificial accumulation, perhaps the work of Palaeolithic people (the association with artefacts is also suggestive in this connection), hence the use of the term 'midden' (J. McNabb, pers. comm.)

Lower Loam (Id)

The Lower Loam occupies a channel, *c.* 200 m wide and aligned SW–NE, excavated into the Lower Gravel to a depth of 2.5 m (Conway and Waechter, 1977). This member is made up of alternations of sand and/or silt, with significant clay and calcium carbonate components. It is partly laminated, with frequent channelling and lenticular bedding. The upper 0.5 m of the deposit, which is decalcified, has been interpreted as a buried soil (Zeuner, 1959; Conway, 1969, 1972; Kemp, 1985b), thus providing important evidence for a non-sequence following the deposition of the Lower Loam. The excavations from 1968 to 1971, in which attention was concentrated on this member, showed that it contains a wealth of mammalian remains, as well as Mollusca and a microfauna. Unabraded and slightly rolled Clactonian artefacts are also scattered throughout the deposit. Desiccation levels have been recognized, associated with concentrations of fauna and artefacts (Conway, 1969; Conway and Waechter, 1977). The latter include a number of small collections of conjoined flakes, important evidence for contemporary (Palaeolithic) occupation of the site (Newcomer, 1971; Waechter, 1971). Animal footprints (fallow deer and ox) have been exhumed at these levels and on the eroded surface of the loam (Waechter, 1970; Conway and Waechter, 1977; Conway, 1985; Fig. 4.14).

Pollen was extracted from sections created in the Lower Loam during the 1968–1971 archaeological excavations (Hubbard, 1972). Palynological analyses by W. Mullenders (in Wymer, 1974) and Hubbard (1982) have indicated deposition during the late-temperate zone of an interglacial. Pollen assemblages dominated by pine and alder were recovered from the lower part of the member, these giving way to grasses and herbs towards the top. Both palynologists suggested correlation with the Hoxnian Stage (pollen biozone HoII), Hubbard (1982) on the basis of the occurrence of the palynomorph known as 'Type X'. A significant aberration in the Mullenders pollen diagram is an abrupt change some 0.6 m from the top of the Lower Loam, interpreted by Hubbard (1982) as a disconformity, above which alder is replaced as the dominant arboreal taxon by pine, with grasses and herbs also much increased in relative importance. This assemblage, which was believed to represent cooler, more open conditions, perhaps came from a channel-fill of the type described by Conway (1970a; Hubbard, 1982). It was attributed to a post-Hoxnian temperate interval, possibly an interstadial (Mullenders, in Wymer, 1974; Hubbard, 1982). This interpretation, which was strongly challenged by Turner (1985), has important implications for the higher parts of the Swanscombe succession, in which there is further evidence for fully interglacial conditions (see below).

Turner (1985) considered the palynological

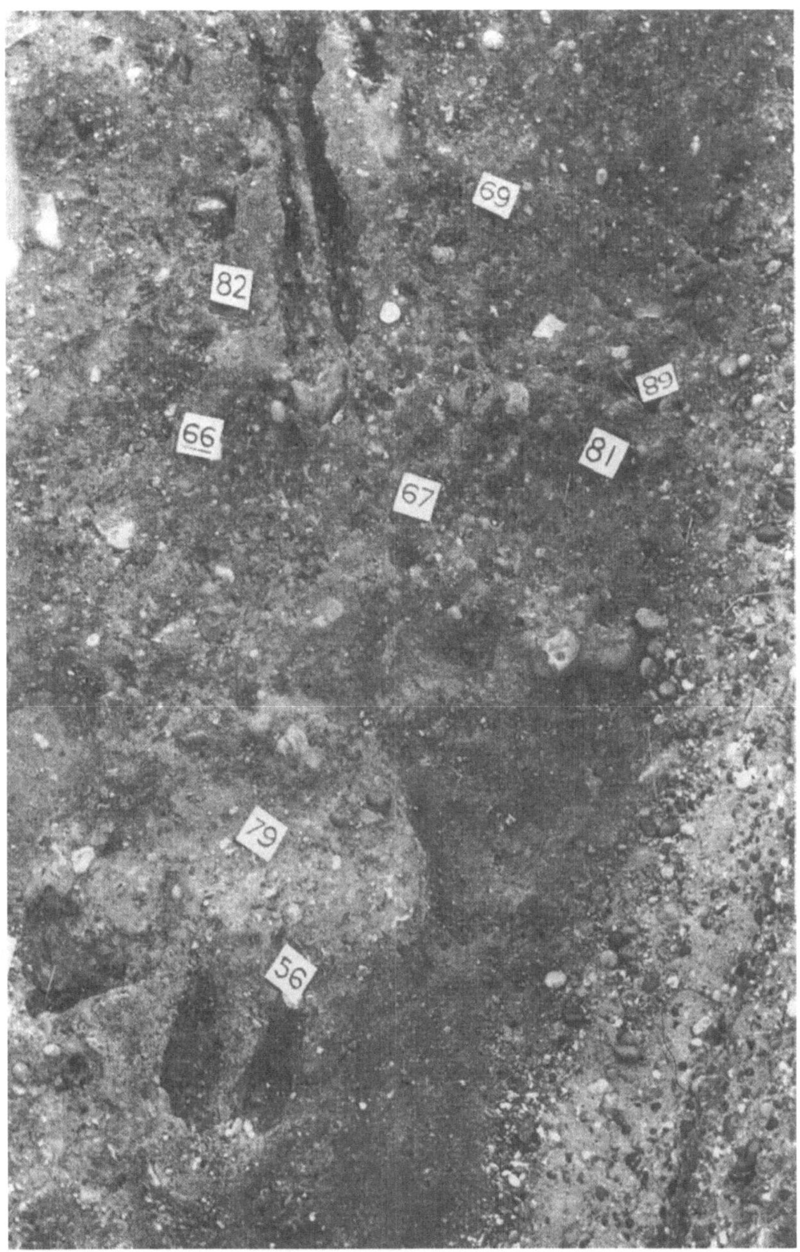

Figure 4.14 Animal footprints in the top of the Swanscombe Lower Loam (1972). (Photo: A.J. Sutcliffe.)

differences at the top of the Mullenders pollen diagram likely to result from the oxidation of the upper levels of the Lower Loam, rather than from any significant vegetational change; he pointed out that the pollen types encountered were those most resistant to destruction under such conditions. This view gains support from the long-established observation that the upper metre or so of the loam is weathered and decalcified, and from the recent confirmation that a buried soil is present at this level (Kemp, 1985b; Fig. 4.15). Following micromorphological analysis of various levels within the upper part of the Lower Loam, Kemp pointed to features such as root pseudomorphs and channels formed by soil animals, the antiquity of which is confirmed by coatings of iron, manganese and calcite, as evidence for pedogenic modification prior to the deposition of the overlying Lower Middle Gravel. The survival of these features indicated to Kemp that they were formed after the decalcification of the upper levels of the loam, since this produced a thorough reorganization of the microscopic structure of the material, with redeposition of calcium carbonate in lower parts of the profile. He was therefore able to argue strongly that the decalcification occurred as a result of pedogenesis while the loam was at the land surface, not as a result of leaching by groundwater percolating through the later sediments, as was suggested by Kerney (1971). Kemp thought it possible that this soil formed during part of a single temperate episode, presumably that represented by the interglacial fauna from the Phase I and II deposits at Swanscombe.

Phase II: Lower Middle Gravel (IIa)

The second phase of fluvial deposition at Swanscombe commenced with the aggradation of the Lower Middle Gravel, which took place over a wider area than is covered by any of the earlier sediments (see Fig. 4.13). A 'lag' deposit at the base of this member comprises coarse flints with occasional bone fragments (Conway and Waechter, 1977; Bridgland *et al.*, 1985). This member, which consists of loose sandy gravel interbedded with minor layers of sand, represents the only true gravel amongst the Phase II deposits. It varies from 0.5 to 2 m in thickness, largely as a result of erosion prior to deposition of the Upper Middle Gravel. According to some descriptions, the Lower Middle Gravel is in some areas completely cut out as a result of this erosion (Marston, 1937; Wymer, 1968), but recent reinterpretation has favoured collapse in response to the solution of the underlying Chalk as the reason for its absence from parts of the site (Conway, 1972). The Lower Middle Gravel is the oldest of the Swanscombe deposits to contain (Acheulian) hand-axes, which are extremely abundant, the member probably having yielded several thousand (Wymer, 1968). These finished tools are vastly outnumbered by flakes; section cleaning and sampling in 1982 yielded 25 flakes from the Lower Middle Gravel (from a section 3 m wide, in which the member was just over 1 m thick), although no hand-axes were found (Bridgland *et al.*, 1985). Sandy horizons within the Lower Middle Gravel have yielded molluscan remains, a number of species appearing at Swanscombe for the first time at this level, including taxa once believed to be indicative of a connection with the Rhine system (Kennard, 1938; below). The land snails present in the Lower Middle Gravel indicate that woodland was more firmly established during its deposition than at any other time during the aggradation of the Swanscombe sequence (Kerney, 1971).

Mammalian remains are common, but poorly preserved. Many come from the basal 'lag' horizon. Poor recording and labelling makes it difficult to separate material collected from the two members representing Phase II of the Swanscombe sequence (Sutcliffe, 1964). Hence Wymer (1974) was prepared to list only *Palaeoloxodon antiquus* and *Bos primigenius* from the Lower Middle Gravel (both of which also occur in the overlying Upper Middle Gravel). It seems likely that the bulk of the collections from the undifferentiated Middle Gravel came from the Upper Middle Gravel, especially as they suggest a change to more open conditions (Sutcliffe, 1964), which is consistent with other environmental evidence from the Upper Middle Gravel.

Upper Middle Gravel (IIb)

The term Upper Middle Gravel has been applied to a series of cross-bedded and ripple-laminated beds, predominantly of sand, but with subordinate silt and gravel horizons. It was from one such bed that three pieces of human skull were found (Marston, 1937; Wymer, 1955, 1964b; Stringer, 1985). In the only extensive excavation of the Phase II deposits, Wymer (1964b) found that certain beds within the Upper Middle Gravel contained concentrations of mammalian remains and Palaeolithic material. In this same excavation, silt horizons in this part of the sequence were found to contain specks of organic matter that may represent burnt vegetable material, possibly resulting from man-made fires (Oakley, 1964; Wymer, 1968). Mollusca are rare in this member, the assemblage generally resembling that from the Lower Middle Gravel. Mammalian remains are abundant, the commonest being horse and giant ox, but with important records of wolf, lion, the Clacton fallow deer, giant deer and a number of rodents (Sutcliffe, 1964). The latter include Norway lemming, which, coupled with a terrestrial molluscan fauna indicative of damp, open conditions, implies a climatic cooling (Kerney, 1971).

In the 1982 sections it was observed that the

Figure 4.15 Photograph of buried soil in the upper part of the Swanscombe Lower Loam. (Photo: D.R. Bridgland.)

cross-stratified sands comprising the highest part of the Upper Middle Gravel were interbedded with brown silty clay laminae. The latter became progressively thicker until they predominated, the sands finally dying out at *c.* 33 m O.D. This was interpreted as a transition between Upper Middle Gravel and Upper Loam facies (Bridgland *et al.*, 1985). A similar description of this part of the sequence was given by Dines *et al.* (1938). This observation is in conflict, however, with several other previous records, which suggest that an unconformity separates these two members, citing evidence for erosion, for the periglacial disturbance of the surface of the Upper Middle Gravel and for the emplacement of a lobe of colluvial gravel at this level (Marston, 1937; Paterson, 1940; Wymer, 1968; Conway and Waechter, 1977; Hubbard, 1982).

Deposits above the Upper Middle Gravel have therefore been assigned to a separate phase, Phase III of the Swanscombe succession (Conway and Waechter, 1977). There is, however, some support for interpretation of the Upper Middle Gravel/Upper Loam boundary as gradational, in the description by Conway (1971) of his Section G, cut in the northern edge of the Alkerden Lane Allotments. Given the location of the 1982 sections (Fig. 4.9), it seems that there may be no significant (visible) break in the sequence at this level in the northern part of the site.

Phase III: Soliflucted clay (IIIa)

A wedge of soliflucted material (clayey diamicton) was recognized between the Upper Middle Gravel and the Upper Loam near the back edge of the Swanscombe terrace remnant by Dines *et al.* (1938). This has recently been interpreted as the first of a series of cold-climate deposits marking the beginning of Phase III of the Swanscombe succession (Conway and Waechter, 1977). As stated above, other observations suggest little evidence for a major break at this level. The significance of such wedges of colluvial material adjacent to a former valley side, in terms of the length of time or severity of climate represented, is uncertain. A sequence of sands and silts at West Thurrock, thought to represent an estuarine accumulation, is punctuated by several wedges of unbedded chalky material that are not considered to represent major breaks in the succession (see below, Lion Pit); the interpretation of the West Thurrock sediments as estuarine implies a high sea level and, therefore, a temperate climate.

Channel deposits (IIIb)

More convincing evidence for cold-climatic conditions is provided by a channel *c.* 2 m deep, cut into the Upper Middle Gravel and infilled with horizontally-bedded, fine loamy sands with thin seams of silty clay. The occurrence in these sediments of ice-wedge casts, cryoturbation structures and microfaulting, together with an absence of pollen, has led to their attribution to a period of severely cold climate (Conway and Waechter, 1977; Hubbard, 1982; Conway, 1985). No fauna or archaeological material has been found in these sediments, which occur only beneath the north-western part of the Alkerden Lane Allotments site (Conway, 1972). However, they are of importance in providing evidence for a cold interval separating the deposition of the Upper Middle Gravel and the Upper Loam.

Upper Loam (IIIc)

According to Conway and Waechter (1977), the channel-fill sands (IIIb) pass upwards conformably into the Upper Loam. As noted above, in other parts of the site there appears to be little evidence for a major break between the Upper Middle Gravel and the Upper Loam. The latter is a poorly bedded to massive silty clay, brown or red-brown, with scattered flints. It has usually been interpreted as an overbank (floodplain) deposit, an interpretation supported by occasional reports of current bedding (Chandler, 1930). Dines *et al.* (1938) believed that the deposit may have been decalcified. They considered it to mark the final phase of Boyn Hill Terrace aggradation.

A further possible indication of a hiatus between the Upper Middle Gravel and the Upper Loam is the occurrence of patinated artefacts from the extreme base of the latter (Dewey, 1919, 1930; Burchell, 1931), comprising an assemblage quite distinct from that in the Middle Gravels (Wymer, 1968; Conway and Waechter, 1977). A further industry has been recovered from the upper part of the Upper Loam, made up of unabraded flint-knapping debris and white patinated hand-axes of twisted ovate type (Conway and Waechter, 1977). Ovates are unknown from Barnfield Pit below the level of the Upper Loam, although they make up a significant part of the assemblage from the gravels of the Craylands Lane Pit (TQ 604746; Dewey and Smith, 1914; Smith and Dewey, 1914; Dewey, 1932; Wymer, 1968). Wymer (1968) considered the implements from that site to be derived from the same stratigraphical horizon as the base of the Upper Loam in Barnfield Pit.

The Upper Loam has yielded no faunal remains, but sparse pollen spectra obtained by sieving large samples of the deposit have been considered indicative of temperate woodland conditions, resembling the middle part of an interglacial (Conway and Waechter, 1977; Hubbard, 1982). Hubbard (1982) tentatively correlated the deposit with the Ipswichian Stage on the basis of palynology. However, the validity of the palynological record from Swanscombe has been severely questioned (Turner, 1985; below).

Upper Gravel (IIId)

This member is a poorly sorted mixture of clay, sand and coarse, angular gravel, up to 2 m thick, with little sign of bedding. It has been universally interpreted as a solifluction deposit, an interpretation supported by evidence that periglaciation affected the surface of the Upper Loam before the gravel was deposited (principally from ice-wedge casts originating from the junction between the two members). The Upper Gravel has yielded several patinated twisted-ovate hand-axes, thought to have been reworked from the Upper Loam (Wymer, 1968; Conway and Waechter, 1977). The only faunal remains from the unit are *Ovibos* (musk ox) from near the base (Conway and Waechter, 1977), a further indication of intensely cold conditions.

Higher loams (IIIe)

Minor accumulations of loamy sand are recorded from above the Upper Gravel in the south-western part of the site (Conway and Waechter, 1977). These deposits have probably been washed or soliflucted over the bluff at the edge of the former floodplain. They extend the sequence up to nearly 35 m O.D., but appear to be of little significance to the fluvial record at the site.

Interpretation

The unique importance of the Swanscombe (Barnfield Pit) site is readily apparent. It has yielded the oldest fossil hominid remains in Britain and the only unequivocal Lower Palaeolithic human bones in England (there is a second British site with Lower Palaeolithic human fossils, of more recent age than the Swanscombe skull, at Pontnewydd, in North Wales (Green *et al.*, 1984)). The site is also unique in that three separate Palaeolithic industries occur there in stratigraphical superposition. The fact that all these discoveries are associated with a complex succession of gravels, sands and silts that contain abundant faunal remains and form part of the terrace record of Britain's major river, the Thames, is extremely fortuitous. This allows the anthropological and archaeological information from Swanscombe to be assimilated into the context of Pleistocene chronology and palaeogeography. Unfortunately, the Pleistocene evolution of the Thames system and its relation to the sequence of Middle Pleistocene climatic fluctuations are the subjects of controversy. The information from Swanscombe plays an important part in the investigation of these problems.

The Swanscombe skull

The interpretation of the fragmentary human skull from Swanscombe has involved a good deal of controversy, the absence of frontal bones preventing clear comparison with other, more complete finds from elsewhere. Thus the Swanscombe skull has been attributed by different authors to an early form of modern man (*Homo sapiens*) (Morant, 1938; Keith, 1939; Vallois, 1954, 1958; Montagu, 1960; Leakey, 1972), to Neanderthal Man (*Homo neanderthalensis*), again an early form (Weidenreich, 1940, 1943; Breitinger, 1952, 1955, 1964; Howell, 1960), to a late form of *Homo erectus* (Wolpoff, 1971) or to a Neanderthaloid 'early *sapiens*' form (Clark, 1955). Even the often quoted female sex of the individual is open to question; although most authors have adhered to this interpretation, Keith (1939) considered the individual to have been a male and Weiner and Campbell (1964) cited statistical analyses of detailed measurements of this and other human skulls in support of this latter view. There have even been attempts to erect new species on the basis of the Swanscombe specimen, namely *Homo marstoni* (Paterson, 1940) and *Homo swanscombensis* (Kennard, 1942).

Most authorities have considered the nearest match to the Swanscombe specimen to be a more complete skull from Steinheim, near Stuttgart, Germany (Berckhemer, 1933), recovered from gravels believed to date from the Holsteinian Stage (widely regarded as equivalent to the Hoxnian Interglacial – see Chapter 1). This specimen retains its frontal parts, which display prominent brow-ridges, a characteristic archaic feature. The consensus view at present holds that the Swanscombe and Steinheim skulls both belong in the early part of the Neanderthal lineage, ancestral to classic Neanderthal Man, who flourished during the last glaciation, but not necessarily an ancestor of modern man (Howell, 1960; Stringer, 1974, 1978, 1983, 1985, 1986; Stringer *et al.*, 1984; for summaries see Day, 1977; Cook *et al.*, 1982).

The Palaeolithic record at Swanscombe

Swanscombe is one of very few British localities to yield distinct Palaeolithic industries from different stratigraphical levels (in superposition). At Barnfield Pit the association of the Phase I sediments with Clactonian material and of Phases II and III with Acheulian artefacts was established at an early date (Smith and Dewey, 1913; Chandler, 1930, 1931; Marston, 1937; Wymer, 1968; Roe, 1981). There is also a typological distinction between the hand-axe assemblages from the Phase II and Phase III deposits, ovate forms being confined to the latter (see above; Roe, 1968a, 1981; Wymer, 1968). In the Lower Thames a stratified sequence of different artefact types has also been claimed by Wymer (1985b) at Bluelands Quarry, Purfleet (see below, Purfleet). Whether the stratigraphical relations between the particular Palaeolithic industries demonstrated at Swanscombe are of more than local significance is, at present, uncertain. Further information is required about British Palaeolithic development before this question can be answered.

Swanscombe is also of major importance as one of only three British Clactonian sites with material preserved in primary context, the others being Barnham, Suffolk, and Clacton (Wymer, 1985b). During 1970, excavations in the Swanscombe Lower Loam uncovered a thin horizon within the deposit, thought to represent a former subaerial surface, in which lay scatters of conjoinable flint flakes (Newcomer, 1971; Waechter, 1971). These have been interpreted as the debris of knapping activity that took place on site during breaks in the deposition of the Lower Loam. This interpretation is strongly supported by the discovery of small secondary broken pieces, formed as accidents of flaking, that can be fitted back on to the larger flakes (Newcomer, 1971). Newcomer was able to partly reconstruct the shapes of the cores from which some of these flakes were struck, even though the cores themselves were not present amongst the debris. These collections comprise the best-preserved archaeological material to have been discovered to date from any part of the Swanscombe sequence. All the other levels, many of which produce artefacts in much greater abundance, contain only allochthonous material washed together by the river and included in its deposits. This is true of the skull level in the Upper Middle Gravel, the skull itself representing part of the fluvial sediment-load. The discovery of these occupation horizons in the Lower Loam is ironic, since most earlier workers had considered this member to be practically sterile, with only Marston (1937) claiming to have found material in it, which he attributed to a Clactonian industry. Wymer (1968) concurred with Marston's interpretation, on the basis of a number of mint or near mint chopper-cores and flakes in the British Museum (Kennard Collection). However, Waechter (in Ohel, 1979) was not prepared to accept the material collected in 1970 from the Lower Loam as unequivocally Clactonian. He considered the size of the assemblage to be too small to allow 'firm "cultural" designation' (Waechter in Roe, 1981, p. 72). Newcomer (in Ohel, 1979) shared these doubts. Despite this caution, there are two lines of evidence suggesting that the material from the Lower Loam represents a continuation of the Clactonian industry from the underlying Lower Gravel: firstly, no hand-axes or hand-axe finishing flakes have been recovered from the Lower Loam; secondly, geological and palaeontological evidence indicates that the Lower Gravel and Lower Loam represent a separate phase of aggradation (Phase I), earlier than the hand-axe-bearing Phase II and III deposits at Swanscombe. In addition, recent reappraisal of the 1970 collections from the Lower Loam (J. McNabb, pers. comm.) indicates that the material can be considered technologically identical to the industry from the Lower Gravel.

The Acheulian assemblage from the Lower and Upper Middle gravels (IIa and IIb) is dominated by pointed hand-axes (Wymer, 1964b, 1968; Roe, 1968a, 1981), which, according to Roe (1981), account for almost 80% of the total. Roe (1968b) reported that at least 625 hand-axes were known to have come from these deposits. Although both these Phase II members contain artefacts, most authors have considered the two together and much of the material in the various collections is simply marked 'Middle Gravel' (Roe, 1968a; Wymer, 1968). There is little or no apparent difference between the material that can be precisely attributed to each of the two individual members (Roe, 1968a; Wymer, 1968).

This assemblage of predominantly pointed implements contrasts with the collections from the Upper Loam and Upper Gravel, which contain much higher proportions of cordate or ovate forms and are dominated by well-made flat ovates (Wymer, 1964b, 1968; Roe, 1968a, 1981; Waechter, 1973). Some of the latter material came from the base of the Upper Loam, lying on the surface of the Upper Middle Gravel. Here, and at various horizons within the Upper Loam, working floors were reported by Marston (1937, 1942), but there are no well-documented records of excavations demonstrating these. Some of the surviving material from these Phase III deposits, despite a white patina, is in extremely fresh condition (Wymer, 1968), which lends some support to Marston's claims. However, many such finds were recovered from the Upper Gravel, having presumably been reworked from the subjacent loam.

The possible stratigraphical significance of the superposition of these three separate industries at Swanscombe is considered below.

Palaeontological evidence from Swanscombe

The mammalian fauna from Swanscombe (from the Phase I and II deposits) forms an important part of the British Middle Pleistocene record (Sutcliffe, 1964; Stuart, 1982a). In addition to the more common species, which generally assist environmental interpretation and correlation with other sites, there are a number of records of very rare taxa. In particular, the cave bear *Ursus spelaeus* (Rosenmüller and Heintoth), discovered in the Lower Gravel (Kurtén, 1959; Sutcliffe, 1964), is possibly the earliest from western Europe. Sutcliffe (1964, p. 91) suggested that this species was present in Britain only during the early Hoxnian, all later British bears being assigned to the brown bear species (*Ursus arctos* (L.)). He listed the following features of the Swanscombe faunas that he considered to be significantly similar to other Hoxnian assemblages (such as that from Clacton – see Chapter 5) and significantly different to fossiliferous deposits from later temperate phases:

1. The rhinoceroses *Dicerorhinus kirchbergensis* (Jäger) and *D. hemitoechus* are both present (the former in units Ib-d and IIa and IIb, the latter in Ib).
2. The fallow deer is of a large race, *Dama dama clactoniana* (Falconer) (units Ib-Id, IIa and IIb).
3. Horse is present (units Ib-Id, IIa and IIb – becoming increasingly abundant in higher units).
4. Hippopotamus is absent.
5. Hyaena is absent.

Lister (1986) has recently discussed the characteristics and significance of fossil deer from Swanscombe. He emphasized that the occurrence of fallow deer in each of the main fluviatile levels (see 2, above) implies that these all reflect deposition under temperate conditions. The attribution of fallow deer remains from Swanscombe to *Dama dama clactoniana*, a large form restricted in its occurrence to post-Cromerian/pre-Ipswichian (*sensu* Trafalgar Square) sediments, could be unequivocally demonstrated only for specimens from the Phase I deposits, since only those had yielded antlers sufficiently well-preserved for determination. Lister also observed that antler fragments of *Megaloceros giganteus* (Blumenbach) from Swanscombe (admittedly only three in number) share morphological characteristics with specimens from Steinheim, the German site that has also yielded hominid remains closely comparable with those from Swanscombe (see above). The antlers from these sites have widths at the extreme upper limit of the variation seen in Devensian examples from Ireland, suggesting that this may be a feature of early *M. giganteus* populations (the occurrence of this species extends from the Anglian to the Late Devensian).

Relatively little has been published on small mammals from Swanscombe, although the occurrence of lemming in the Upper Middle Gravel (IIb) has been cited frequently as evidence for climatic cooling (Kerney, 1971; Sutcliffe and Kowalski, 1976; see above). Rodent remains have been recorded from the Phase I deposits at Barnfield Pit (Carreck in Kerney, 1959a; Sutcliffe and Kowalski, 1976) and from a silt bed within the Upper Middle Gravel (IIb), slightly higher than the level from which the skull fragments were obtained (Schreuder, 1950). Sutcliffe and Kowalski (1976) noted similarities between the rodent assemblage from Swanscombe and that from the Cromerian stratotype (the Upper Freshwater Bed of West Runton), particularly the occurrence in both of *Microtus arvalinus, M. ratticepoides* Hinton and *Pitymys arvaloides* (Hinton). Conversely, the vole *Mimomys savini*, which occurs at West Runton, is replaced by *Arvicola cantiana* at Swanscombe; at one time this was considered to provide strong support for a Hoxnian age for the Swanscombe deposits (Sutcliffe, 1964), but the recent recognition that a number of 'late Cromerian' faunas contain *A. cantiana* (Bishop, 1982; Stuart, 1982a, 1988; Currant in Roberts, 1986; Chapter 1) appears to negate this evidence.

The molluscan fauna from Swanscombe has provided the strongest evidence for correlating the sequence with the established Pleistocene chronostratigraphy (Castell, 1964; Kerney, 1971). The malacological collections from the various pits are amongst the richest from the British Middle Pleistocene, with several species unique to Swanscombe. Of particular importance is the so-called 'Rhenish suite', recognized by Kennard (1938), which first appears in the Phase II deposits. This comprises an assemblage of predominantly southern species, including *Corbicula fluminalis, Viviparus diluvianus* (Kunth) and *Theodoxus serratiliniformis* (Geyer), several of which have been claimed to be characteristic of Rhine deposits. According to Castell (1964), *T. serratiliniformis* is characteristic of the 'Great Interglacial' (Hoxnian Stage), when it was restricted to the Rhine-Thames basin. The record of this species from Swanscombe (Fig. 4.16) is the only one from Britain. The only other British occurrence of *Viviparus diluvianus* (Fig. 4.16) is in the Clacton Estuarine Beds. Kennard (1938) interpreted the appearance of these 'Rhenish' Mollusca as evidence that the Thames and Rhine systems became linked after deposition of the Lower Loam (Id).

By comparing the molluscan faunas from Swanscombe and Clacton, Kerney (1971) was able to suggest correlations between the various fossiliferous parts of the Swanscombe sequence and the Hoxnian pollen biozones established at Hoxne (West, 1956) and recognized at Clacton (Turner and Kerney, 1971). He attributed the Phase I deposits at Swanscombe to biozone HoII and the basal Phase II deposits to subzone HoIIIb, considering the transition between biozones II and III to be missing from the sedimentary record at Swanscombe, lost in the hiatus (marked by the period of soil formation) at the top of the Lower Loam.

Thus the sediments at Swanscombe were correlated with the Hoxnian Stage interglacial, which was originally defined palynologically, despite the absence of a pollen sequence from this site. Pollen assemblages have since been obtained from much of the Swanscombe sequence (Mullenders in Wymer, 1974; Hubbard, 1982; see above, Description). Interpretation of these assemblages has led to suggestions of considerable stratigraphical complexity, with two or even three climatic cycles being recognized (Hubbard, 1982). However, this interpretation has failed to gain wide acceptance; the validity of palynological studies of deposits with such low pollen concentrations has been questioned (Bridgland *et al.*, 1985; Turner, 1985). Turner considered it likely that various types of contamination have an important influence on palynological assemblages from sediments with a pollen-yield as poor as those at Swanscombe. Contamination could occur by the reworking of older pollen, by later pollen being introduced from percolating groundwater or by airborne modern grains being introduced during sampling and laboratory preparation. Turner also pointed out that sediments in which pollen preservation has been poor frequently contain only the more robust grains, leading to concentrations of certain taxa that result from diagenetic rather than vegetational factors; assemblages from such deposits need very careful interpretation, therefore. These points appear to seriously undermine the palaeoenvironmental and stratigraphical determinations based on pollen from this site, where none of the sediments yield abundant well-preserved pollen. Thus the pollen assemblages from Swanscombe should be interpreted with caution; where the palynological interpretation of the sediments contradicts evidence from other sources, the former should probably be disregarded.

Palaeoenvironmental and palaeogeographical significance of the Swanscombe sequence

A wealth of palaeoenvironmental information has been obtained from studies of the fossil

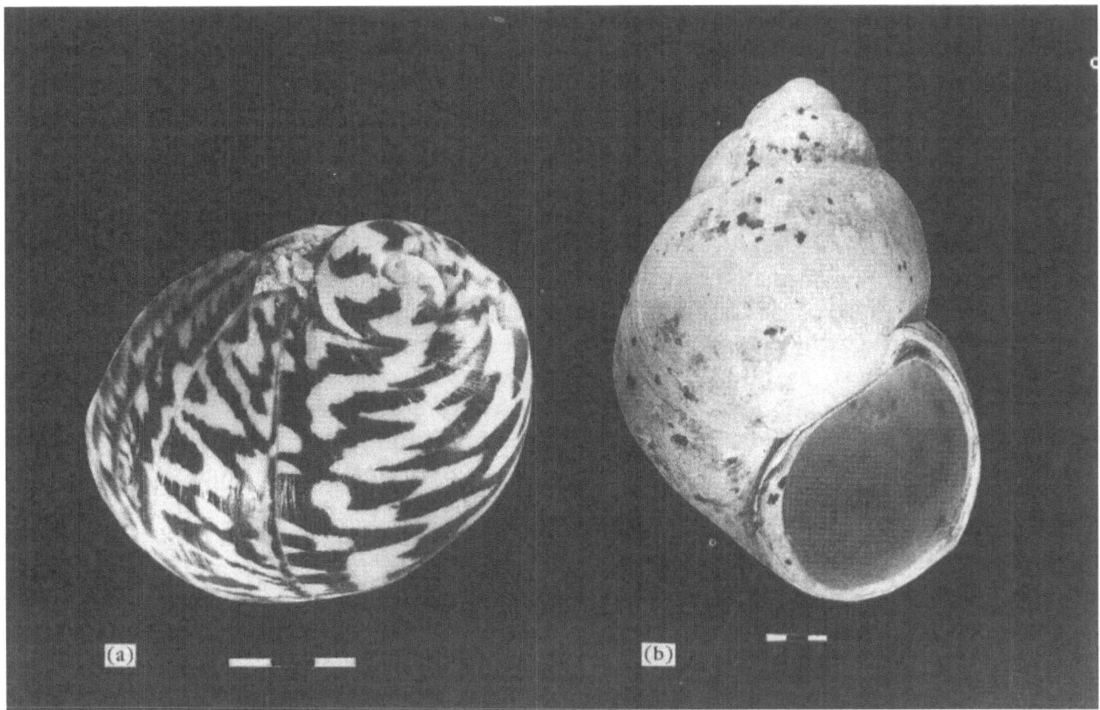

Figure 4.16 The two most characteristic molluscan species of the so-called 'Rhenish' fauna from Swanscombe: (A) *Theodoxus serratiliniformis* (Geyer); (B) *Viviparus diluvianus* (Kunth). Scale bars graduated in mm. (Photos: Department of Zoology, University of Cambridge).

content of the Swanscombe sediments. In particular, the prevailing climate and vegetation at the time of deposition of the individual units has been reconstructed, principally from the molluscan faunas, but also from mammal remains and, controversially, from pollen (see description of individual members and beds, above).

The environment of deposition represented by the various units can be determined from their sedimentological characteristics, although little appears to have been published on this subject. Bridgland *et al.* (1985) discussed the various depositional environments represented by the sequence exposed in 1982. They noted the progressive decline in fluvial energy in the Phase I sequence, from an actively aggrading gravel-bed river in the basal gravel (Ia), possibly representing the end of a cold period, through the increasingly sandy Lower Gravel (Ib), into the sandy basal parts of the Lower Loam (Id) and culminating in the siltier upper horizons of this member. The upward decrease in grain size in the Lower Loam was considered to represent a transition from channel-fill to overbank (floodplain) deposits, an interpretation supported by the molluscan fauna. The Phase II sequence again commences with high-energy deposits in the form of the Lower Middle Gravel (IIb), alternations of sand and gravel reflecting considerable current variability. The overlying Upper Middle Gravel (IIb) reflects continued fluctuations of current strength, but with a general decrease in energy indicated by the progression from planar cross-bedded to ripple-laminated sands. The interdigitation of silty horizons near the top of the 1982 section appears to indicate the further reduction in flow energy. This was thought to reflect a transition into the Upper Loam (IIIc), which has usually been interpreted as an overbank deposit. Palaeocurrent measurements have seldom been recorded from Swanscombe, although Bridgland *et al.* (1985) presented evidence, from foreset orientations in the Upper Middle Gravel (IIb), for flow towards the south-south-east. Interestingly, Wymer (1964b) estimated the same flow direction, based on gross sediment geometry, from his studies of the Upper Middle Gravel in connection with the skull discovery.

There has recently been a suggestion by Wymer (1985b, p. 321) that the Swanscombe deposits might be the product of the River Darent rather than the mainstream Thames. The composition of the various gravels at Swanscombe (Baden-Powell, 1951; Bridgland *et al.*, 1985) points strongly, however, to a Thames origin. The suite of 'exotic' (far-travelled) rock-types (Bridgland, 1986b), which can be traced downstream from the Evenlode valley and which characterizes all Thames formations from the Stoke Row Gravel onwards, is well represented. To this material (principally quartz, quartzites and Carboniferous chert) is added a further important 'exotic' lithology, *Rhaxella* chert, reworked from Anglian glacial deposits (above; Bridgland, 1986b). The predominance of these 'exotic' rocks over southern lithologies such as Greensand chert, which characterize south-bank tributaries such as the Darent, provides seemingly unequivocal evidence for the Thames origin of the Swanscombe deposits.

Correlation of the sequence at Barnfield Pit with other nearby sites in the Thames system

The initial task in correlating the sequence at Barnfield Pit with the Pleistocene record elsewhere is to determine its relation to the various other sites in the immediate neighbourhood of north Kent. These range from the other fossiliferous and Palaeolithic localities at Swanscombe to other sites in the terrace deposits on the south side of the Lower Thames, some of which (such as those at Dartford Heath) are also within the 'Boyn Hill Gravel' as mapped by the Geological Survey (see above, Wansunt Pit).

The most important of the other Swanscombe localities is probably the Ingress Vale site, or Dierden's Pit (TQ 595748; Fig. 4.17), separated from Barnfield Pit by a dry valley (the Ingress Vale). The former pit (now defunct) produced a wealth of faunal material and a confusing archaeological assemblage, but lacked the stratigraphical complexity of the skull site. It was so rich in molluscan remains that it was frequently referred to as the 'shell pit'. These came from a sandy 'shell bed' that also yielded a large collection of mammalian remains (Sutcliffe, 1964, Appendix; Wymer, 1968). There has, unfortunately, been considerable uncertainty about the correlation of these deposits with the sequence at Barnfield Pit. The earliest discoveries of palaeoliths in Dierden's Pit were well-made, sharp, patinated hand-axes found around the turn of the century, apparently in the shell bed (Stopes, 1900, 1903; Newton, 1901; Kerney, 1959b; Wymer, 1968). These discoveries led to detailed investigations, which produced collections of faunal remains and artefacts from the shell bed that were considered comparable with material from the Lower Gravel at Barnfield Pit (Dewey and Smith, 1914; Smith and Dewey, 1914), pointing to a correlation that is also supported by the altitudinal relations of the two sites (Wymer, 1968). In particular, large collections of Clactonian flakes and several chopper-cores were obtained from the deposit by Dewey and Smith and, more recently, by P.J. Tester and J.N. Carreck (Wymer, 1968). Wymer (1968) suggested that the earlier finds of Acheulian implements may have been from loamy deposits overlying the shell bed and was inclined to support the correlation of the latter with the Lower Gravel. However, Kennard (1916) was adamant that Acheulian material occurred in the Ingress Vale shell bed and that the fauna was contemporary with this industry. This controversy appears to have been resolved by Kerney (1959b, 1971, in Sutcliffe, 1964), who recorded the discovery of an Acheulian industry in the shelly deposits at Dierden's Pit, from what was apparently the reopening of the sections studied by Dewey and Smith. Kerney's material was entirely made up of flakes, including many of Clactonian type. Some, however, were distinctive hand-axe finishing flakes (Kerney, 1959b). The assemblage was studied by Marston (in Kerney, 1959b), in whose opinion the objects were unequivocally Acheulian and closely comparable to a series in his possession from the Upper Middle Gravel (IIb) of Barnfield Pit.

Kerney also reassessed the molluscan and vertebrate faunas from Dierden's Pit. He noted the presence of the characteristic 'Rhenish' molluscan suite and concluded that the Ingress Vale deposits belonged to the 'late temperate substage' of the interglacial, placing them above the Lower Loam of Barnfield Pit but earlier than the bulk of the Middle Gravel. The faunal assemblage from Dierden's Pit is indicative of temperate woodland, thus contrasting with the open-habitat fauna from the Upper Middle Gravel (Sutcliffe, 1964; Kerney, 1971). Kerney considered the latter assemblage to post-date that from Dierden's Pit and to imply climatic deterioration later in the interglacial. However, the presence of ovate hand-axes in the collections from the Dierden's Pit shell bed

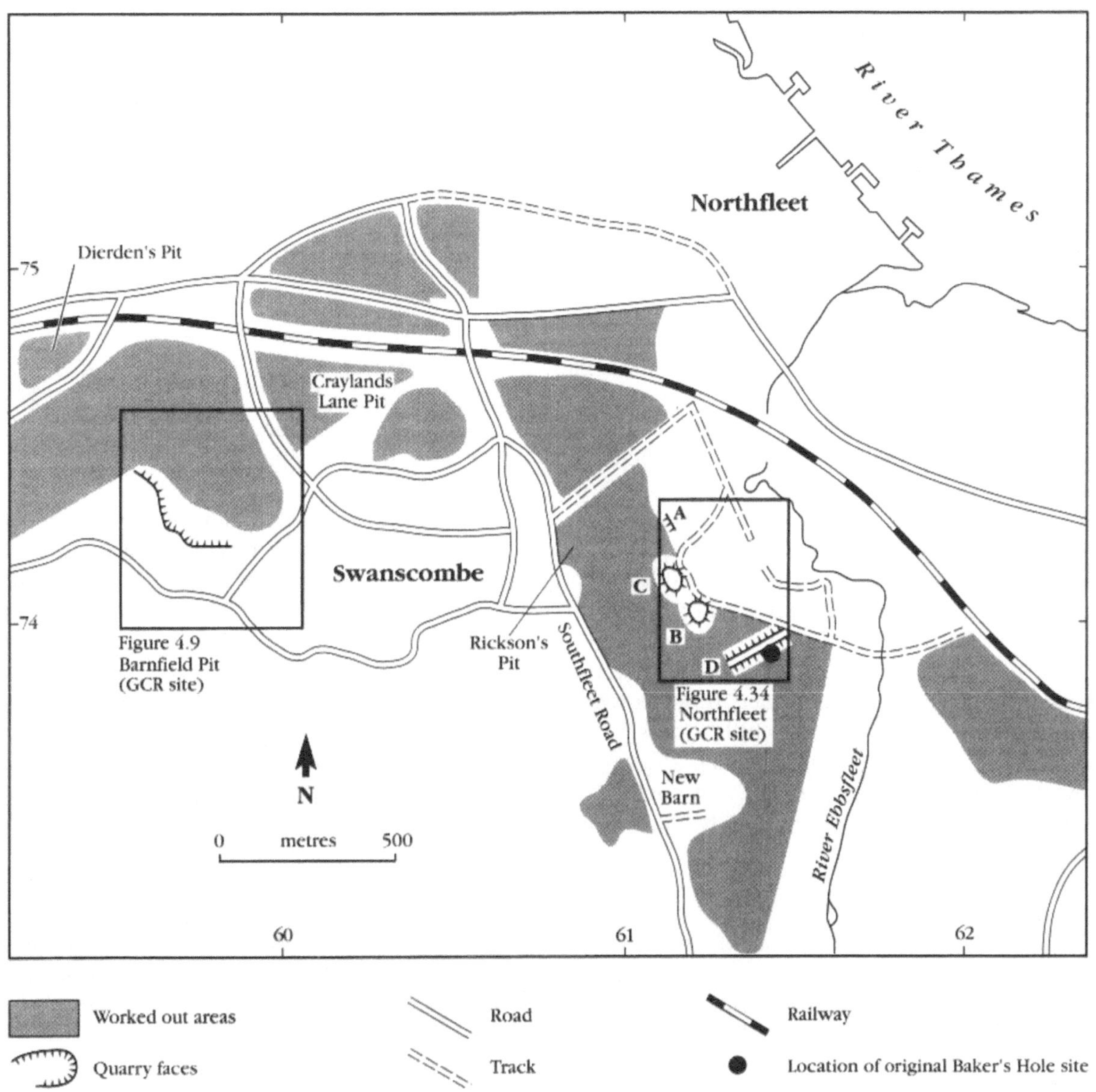

Figure 4.17 Map of the Swanscombe–Northfleet area, showing the locations of the various Pleistocene localities.

(Stopes, 1903) may indicate an age slightly younger than the Upper Middle Gravel, comparable perhaps with the higher deposits at Rickson's Pit (Kerney, 1959b; below). A unique find from Dierden's Pit is a vertebra of bottle-nosed dolphin (*Tursiops truncatus* (Montague)) in the Hinton collection (Sutcliffe, 1964), the presence of which has sometimes been claimed as evidence of proximity to the contemporary coast (Kerney, 1971). However, Stuart's (1982a) suggestion, that this represents an individual that swam up the river and became stranded, seems more plausible, considering the paucity of other indications of a marine influence at Swanscombe.

Another important site at Swanscombe was Rickson's Pit (TQ 608743; Fig. 4.17), now entirely quarried away (Tester, 1955), which shared with Barnfield Pit the important stratigraphical superposition of Acheulian above Clactonian industries. The succession at Rickson's Pit was less complex than that at Barnfield Pit, however, comprising a sequence of gravels and sands, and generally lacking the well-defined loams of the latter site (Dewey, 1932, 1934; Wymer, 1968; Waechter, 1973; Roe, 1981). The lowest gravel, 1 m thick and conspicuously coarser than any above, contained an uncontaminated Clactonian industry, indicating broad correlation with the Phase I deposits at Barnfield Pit. Above this was *c.* 3 m of sandy gravel rich in shells, likened to the Ingress Vale shell bed by Kerney (1971). This was in turn overlain by a further 2–3 m of even- and current-bedded sand and gravel. The upper part of the sequence contained hand-axes, confirming correlation with post-Phase I deposits at Barnfield Pit. However, the uppermost layers have yielded finely made ovate hand-axes of a type unknown at Barnfield Pit before the Upper Loam (IIIc), as well as at least one Levallois tortoise core (Burchell, 1931; Wymer, 1968). Burchell (1933) classified the industry from the Swanscombe Upper Loam as 'Levalloisian A'. He also noted remnants of a much denuded loam capping the sequence at Rickson's Pit, beneath which patinated artefacts occurred in identical fashion to those beneath the Upper Loam at Barnfield Pit (Burchell, 1931, 1934b). These observations suggest that the sequence at Rickson's Pit once represented a condensed version of the full Swanscombe succession. Levallois material from the upper levels of this and other sites at Swanscombe has probably been incorporated from above, perhaps by cryoturbation.

A further important site formerly existed at Swanscombe, the Craylands Lane Pit (TQ 604746), which lay to the east of the GCR site (Fig. 4.17). It exposed gravels and sands overlying Chalk at a level comparable to the base of the Lower Middle Gravel at Barnfield Pit. Correlation with post-Phase I deposits is confirmed by the occurrence of hand-axes in the gravels at Craylands Lane. As with Rickson's Pit, ovate hand-axes and Levalloisian material were recovered (Dewey and Smith, 1914; Smith and Dewey, 1914), indicating that sediments equivalent to the Phase III deposits of Barnfield Pit were present (Wymer, 1968; Waechter, 1973). Other nearby sites, of less significance, were documented by Wymer (1968).

The relations between the Swanscombe sequence and the gravels and 'loams' covering Dartford Heath, 7 km to the west, has been the subject of much debate (see above, Wansunt Pit). The deposits at the latter locality are aggraded to *c.* 42 m O.D., at least 8 m higher than the fluvial part of the Swanscombe sequence (Phases I and II). However, the deposits at both sites were mapped by the Geological Survey as part of their 'Boyn Hill Gravel' (Sheet 271) and, according to Dewey *et al.* (1924), form part of a single valley-fill sweeping across this part of north Kent. The Dartford Heath Gravel was attributed by King and Oakley (1936) to their 'Middle Barnfield Stage', implying a correlation with the Phase II deposits at Swanscombe. This correlation accords with the views of many other authors (Chandler and Leach, 1907, 1912; Marston, 1937; Dewey, 1959). Others have taken the view that at least part of the Dartford Heath deposits represent an earlier terrace aggradation (Hinton and Kennard, 1905; Zeuner, 1945; Cornwall, 1950; Gibbard, 1979). This controversy is fully discussed below and in the Wansunt Pit report.

Correlation of the Swanscombe sequence with the Lower Thames terrace deposits on the Essex side of the river is not entirely straightforward. The only sites north of the Thames for which a Hoxnian age has been suggested are at Grays and Little Thurrock, which are at a lower altitude than the Swanscombe sequence and are regarded here as part of a separate, later formation (Fig. 4.3). This difference in elevation led to the suggestion that the Grays deposits represent the time interval between the deposition of the Phase I and II sequences at Swanscombe (King and Oakley, 1936; below). The deposits on the Essex side of the Lower Thames valley mapped as Boyn Hill Gravel, reclassified recently as Orsett Heath Gravel (Bridgland, 1988a; Gibbard *et al.*, 1988), comprise unfossiliferous sands and gravels suggestive of a cold climate. The identification of these sediments as a downstream continuation of the Boyn Hill Gravel of the Middle Thames has recently been upheld (Bridgland, 1988a; Gibbard *et al.*, 1988). The Boyn Hill/Orsett Heath Gravel is interpreted as a periglacial braided-river deposit and has been attributed by Gibbard (1985) to the early Saalian Stage (Wolstonian). This dating is, in fact, largely based on the interpretation of the stratigraphical relations between the Boyn Hill/Orsett Heath Gravel and the Swanscombe sequence (Gibbard, 1985, pp. 136–137). Gibbard considered the palaeontological evidence for climatic cooling in the higher levels of the Swanscombe sequence (above the Lower Middle Gravel – IIa) to reflect a transition from the Hoxnian Stage interglacial to the glacial conditions of the early Saalian (Wolstonian) Stage. He pointed to the similarity in elevation between the top of the sequence at

Swanscombe and the local Orsett Heath Gravel (both may be considered to underlie the Boyn Hill Terrace surface) as evidence supporting their proximity in age. However, no record exists of cold-climate sediments of Orsett Heath Gravel type overlying any part of the Swanscombe sequence. According to the climatic model for terrace generation presented in Chapter 1 of this volume, interglacial sediments represent the middle of three phases of aggradation that can typically be recognized within individual terrace formations. The accumulation of such temperate-climate deposits is both preceded and followed by aggradation in a cold climate. The Basal Gravel at Swanscombe (Ia) has sometimes been attributed to a cold climate and so may represent the pre-interglacial aggradational phase of the Boyn Hill/Orsett Heath Formation. The subsequent post-interglacial aggradational phase, which is usually the best-represented in surviving terrace deposits (Chapter 1), appears not to have contributed to the sequence at Swanscombe; the river may have migrated to another part of the floodplain during this phase, thus fortuitously allowing the survival of the Swanscombe interglacial sediments.

Biostratigraphical correlation and geochronometric dating

The Swanscombe deposits have for many years been correlated with the Hoxnian Stage (formerly the 'Great Interglacial' or the 'Penultimate Interglacial'), and have long been regarded as a stratigraphical marker for that interval in the Thames valley (King and Oakley, 1936; Castell, 1964; Sutcliffe, 1964; Kerney, 1971). However, this may be in need of re-appraisal in the light of evidence that the record of climatic fluctuation during the Middle and Late Pleistocene is more complex than was thought hitherto (see Chapter 1).

To date, the most convincing biostratigraphical argument for the Hoxnian age of the Swanscombe sediments has being derived from the molluscan record and its comparison with that from Clacton (Kerney, 1971). The record of the palynomorph 'Type X' in the Lower Gravel and Lower Loam (Hubbard, 1982) appears to add weight to the argument, since this grain is considered to be characteristic of Hoxnian pollen assemblages (Turner, 1970; Phillips, 1976). Although the validity of pollen assemblages from Swanscombe has been questioned, and published interpretations of these (Mullenders, *in* Wymer, 1974; Hubbard, 1982) have been strongly criticized (Turner, 1985; above), the palynological record from the Lower Loam may be significant. Turner (1985, p. 366) accepted that this deposit was formerly more richly organic and regarded the pollen spectra obtained by Mullenders (*in* Wymer, 1974) and Hubbard (1982) as 'residual temperate pollen assemblages dominated by *Alnus* and *Pinus* which are on the one hand resistant to decay and on the other easy to recognize in a mutilated condition'. This implies that the record of 'Type X' from this deposit by Hubbard (1982) may indeed be significant, even if his elaborate interpretation of the Swanscombe sequence cannot be upheld.

Geochronometric dating techniques have been applied in recent years to the Swanscombe sequence. Szabo and Collins (1975) used the uranium-series method to date a bone from the basal Lower Middle Gravel (IIa). This indicated a minimum age of 272,000 years for the specimen, supporting claims for a Middle Pleistocene (Hoxnian *sensu lato*) age for the Upper Middle Gravel. Thermoluminescence dates were obtained from samples of Lower and Upper Loam (members Id and IIIc) collected from the 1982 GCR sections. These suggested ages of 228,800 years (±23,300) and 202,000 years (±15,200) respectively (Bridgland *et al.*, 1985). As these authors pointed out, these dates would place the full Swanscombe sequence within Oxygen Isotope Stage 7. Given the stratigraphical evidence presented in this volume for correlation of the Swanscombe deposits with an earlier temperate episode (probably Stage 11), together with uncertainties about the validity of early (pre-late Devensian) thermoluminescence dates (see, for example, Parks and Rendell, 1988), it seems likely that the dates from the Swanscombe loams are underestimates.

Recently, Bowen *et al.* (1989) have attempted to relate elements of the British terrestrial sequence to the oceanic (oxygen isotope) record, on the basis of amino acid analyses of mollusc shells. Amongst results from interglacial molluscan species from post-Anglian sediments, these authors recognized four groups of ratios that, they suggested, represent four separate temperate intervals. They correlated these with Oxygen Isotope Stages 11, 9, 7 and 5(e) (see Chapter 1). This technique has been applied to shells from all four main fluviatile members at Barnfield Pit (Ib, Id, IIa and IIb), as well as from the Ingress Vale shell bed (Bowen *et al.*, 1989). Results have been obtained from four different species. The D-alloisoleucine : L-isoleucine (D : L) ratios from Swanscombe are closely clustered around 0.3, the small range providing support for the view that the entire aggradation occurred during a single temperate episode. These results confirm the findings of early amino acid analyses of shells from Dierden's Pit (Ingress Vale) by Miller *et al.* (1979). However, ratios obtained from the Hoxnian stratotype are lower, indicating to Bowen *et al.* (1989) that the type-Hoxnian sequence relates to a later episode, which they correlated with Oxygen Isotope Stage 9. For this reason Bowen *et al.* suggested that Swanscombe should be regarded as the 'stratotype' for a previously undefined post-Cromerian/pre-Hoxnian temperate episode, equivalent to Oxygen Isotope Stage 11. This interpretation implies that there were two separate temperate intervals during the late Middle Pleistocene during which vegetation developed in Britain in similar ways; sediments that accumulated during both episodes have yielded pollen spectra that have been attributed to the Hoxnian interglacial. The strongest evidence for assigning the Swanscombe sediments to the Hoxnian Stage is probably the correlation with the palynologically dated Hoxnian sequence at Clacton (Kerney, 1971). The revised chronostratigraphical scheme of Bowen *et al.* (1989) upholds this correlation, since shells from the Clacton deposits also yield amino acid ratios suggestive of Stage 11 (Chapter 5), implying that they too may be pre-Hoxnian (*sensu* Hoxne). Because of the problems that exist in reconciling biostratigraphy with geochronology in this part of the British Pleistocene, the term Hoxnian Stage *sensu* Swanscombe is used to refer to the Stage 11 interglacial episode in this volume (Chapter 1).

The significance of the Swanscombe deposits within the Thames terrace sequence

For many years the association of the Swanscombe deposits with the Hoxnian Stage (formerly the 'Great Interglacial') was a key factor in dating the Boyn Hill Terrace and, therefore, the Thames terrace system as a whole. The evidence for a Hoxnian age for the Swanscombe sequence, chiefly from mammalian and molluscan faunas (see above), was reinforced by Baden-Powell (1951). He claimed to have recognized clast-types in the gravels at Swanscombe that were characteristic of the Lowestoft (Anglian) glaciation, but none that were characteristic of the later (post-Hoxnian) 'Gipping glaciation'. Since no post-Hoxnian glaciation is now recognized in southern East Anglia (Chapter 5), this evidence can no longer be accepted. However, the occurrence in the Swanscombe gravels of ***Rhaxella*** chert, a lithology first introduced into the London Basin in quantity by the Anglian (Lowestoft) glaciation (Bridgland, 1986b), was confirmed by Bridgland *et al.* (1985; see Table 4.2). These authors also recorded non-durable Jurassic clasts, including fragments of ***Gryphaea*** sp., in the Lower Gravel. They suggested that this material was secondarily derived from Anglian glacial deposits, the nearest representatives of which are at Hornchurch (see Fig. 4.1 and above, Hornchurch).

The established view of the Swanscombe sediments as evidence for a Hoxnian age for the Boyn Hill Terrace throughout the Thames system has been superseded in recent years, although it is still to be found in many texts. Gibbard (1985) demonstrated that the Boyn Hill Gravel (the deposit normally underlying this terrace) is, in common with most Thames terrace deposits, a cold-climate accumulation formed in a periglacial braided-river environment (see Chapter 3). Although the Swanscombe deposits do not therefore represent the Boyn Hill Gravel in its typical form, they may be considered as part of the Boyn Hill Formation, which, according to the stratigraphical scheme proposed in Chapter 1 of this volume, includes cold-climate deposits both pre-dating and post-dating the Swanscombe sediments (see Table 1.1).

There is, furthermore, a significant body of opinion that holds that the sequence at Swanscombe may represent sedimentation during more than a single temperate interval. The view that the sequence is of considerable complexity was first stated by King and Oakley (1936). These authors outlined a stratigraphical scheme for the Lower Thames terraces in which the deposition of the Swanscombe Phase I deposits (their 'Lower Barnfield Stage') was followed by major incision to a much lower base level. Aggradation at this lower level followed, represented by the Clacton Channel sediments and deposits containing Clactonian artefacts and faunal remains at Little Thurrock and Grays (their 'Clacton-on-Sea Stage'; see below, Globe Pit; Chapter 5, Clacton). This aggradation continued to the full height of the Boyn Hill Terrace surface as mapped by the Geological Survey (over 40 m O.D. at Dartford Heath), the Phase II sediments at Swanscombe representing part of this process (their 'Middle Barnfield Stage'). The bases for King and Oakley's reconstruction were (1) the archaeological record from the three Clactonian sites (Swanscombe, Little Thurrock and Clacton) and (2) the occurrence at Grays of a mammalian fauna that was believed to be of similar age to that from the Swanscombe deposits. This scheme was widely accepted for many years, although Marston (in Bull, 1942) made clear his scepticism about the downcutting phase between the Phase I and II deposits of the Swanscombe sequence.

Despite the complexity implied by their interpretation, King and Oakley attributed the entire fluvial succession at Swanscombe to the 'Great Interglacial' (Hoxnian). It is difficult to reconcile the implied chronostratigraphical position of the hiatus between the Lower Loam and the Lower Middle Gravel, in the mid-Hoxnian, with the incision event envisaged by King and Oakley at this stratigraphical level; the former would seem to indicate a high interglacial sea level, whereas the latter would suggest a marked lowering of sea level. However, there appears to be supporting evidence for a low sea-level event at this time from the molluscan faunas. Kennard (1938) considered the occurrence of the characteristic suite of 'Rhenish' molluscs in the Phase II deposits at Swanscombe to indicate that the Thames and Rhine became joined after the deposition of the Lower Loam. Such a connection would appear to require a low sea-level phase at precisely the time postulated by King and Oakley. However, Kerney (1971) was not convinced by the evidence for a major hiatus above the Lower Loam, coinciding with the linking of the Thames and Rhine. He pointed out that the first traces of the 'Rhenish' faunas appeared near the top of the Lower Loam (after Davis, 1953) and that the molluscan record showed no indication of a significant hiatus at the base of the Lower Middle Gravel. He also suggested a diagenetic origin for the weathered zone at the top of the loam, which had been widely regarded hitherto as evidence for a period of prolonged subaerial exposure. This horizon has now been confirmed as a buried soil (Kemp, 1985b; see above), but its characteristics suggest formation under temperate-climate conditions during a relatively brief period, so again no indication of a major break in the interglacial succession is provided.

Evans (1971) presented another complex interpretation of the Swanscombe sequence in his pioneering attempt to relate the Thames terraces to the global marine (oxygen isotope) record. He correlated the Lower Gravel and Lower Loam at Swanscombe with the Kingston Leaf aggradation of Zeuner (1945), attributing both to his half-cycle 8w (equivalent to Oxygen Isotope Stage 15 of Shackleton and Opdyke, 1973, 1976). Evans correlated the Swanscombe Lower Middle Gravel with the Black Park Gravel of the Middle Thames, which he attributed to his half-cycle 7w (Stage 13), and saw only the Upper Middle Gravel as a true Boyn Hill deposit, dating from the next warm half-cycle (6w = Stage 11). Evans believed that the two earlier aggradations were represented within the lower parts of the sequence at Swanscombe, rather than as higher terrace remnants on the valley side, because their steeper downstream gradients had taken them below the Boyn Hill Terrace level in the London area (see above, Wansunt Pit). This interpretation argues for two major breaks in the succession at Swanscombe, each equivalent to a cold event in the oxygen isotope record (Stages 14 and 12). The boundary between the Lower Loam and Lower Middle Gravel was already well-established as a non-sequence, with evidence for subaerial weathering (see above), and had been correlated by King and Oakley (1936) with erosion elsewhere in the valley. The later hiatus envisaged by Evans, between the Lower and Upper Middle Gravels, was less well-established, although Marston (1937) had reported evidence for major erosion at this level in the sequence. As stated above, Conway (1972) questioned the evidence for this erosive phase, suggesting that features previously interpreted as channels may have been produced by solution of the underlying Chalk. In addition to Conway's suggestion, the continuity of the mammalian, molluscan and Palaeolithic (Acheulian) assemblages between the Lower and Upper Middle Gravels argues against Evans's interpretation.

Reappraisal, in the area west of London, of

the late Anglian Black Park Gravel has indicated that Zeuner's (1945) Kingston Leaf is a correlative of this formation (Gibbard, 1979; see above, Wansunt Pit), rather than representing a separate, older aggradation as suggested by Evans (1971). As a result, Bridgland (1980) proposed a modified version of Evans's interpretation of the Swanscombe sequence, in which Boyn Hill deposits were considered to overlie Black Park deposits, rather than the tripartite sequence favoured by Evans. Bridgland suggested that the Lower Gravel channel was a product of pre-Black Park Formation downcutting, thus agreeing with Zeuner (1945) that the Clacton Channel was excavated at this time, rather than after deposition of the Lower Loam. In fact, Bridgland (1980, 1983a, 1983b, 1988a) traced an equivalent channel across eastern Essex between Southend and Mersea Island, suggesting a direct link between the Lower Gravel channel at Swanscombe and the Clacton Channel (see Chapter 5). He recognized that much of the Lower Gravel and Lower Loam and the Clacton sequences are of interglacial origin and therefore post-date the late Anglian Black Park Gravel, the aggradation of which began while ice still occupied parts of the old Thames valley in Hertfordshire (see Chapter 3). He suggested, however, that the basal deposits at the two sites, already claimed as late Anglian by previous authors (above; Chapter 5, Clacton), may be equivalent to the Black Park Gravel of the Middle Thames.

A problem exists in reconciling the different interpretations of this part of the Lower Thames sequence proposed by Gibbard (1979) and by Bridgland (1980, 1988a). This problem, which has already been discussed above (see Hornchurch and Wansunt Pit), hinges on whether the Lower Thames valley was deeply excavated in Black Park times (late Anglian), as the elevation of the Hornchurch till remnant suggests, or whether the valley floor was much higher; *c.* 37 m O.D. was suggested by Gibbard (1979) as the base of his Dartford Heath Gravel, which he correlated with the Black Park Gravel. However, the correlation of the Dartford Heath and Black Park Gravels has been rejected earlier in this volume (see above, Hornchurch and Wansunt Pit), on the grounds that the downstream projection of the Black Park Formation confirms Evans's suggestion that this deposit falls below the level of the Boyn Hill Terrace east of London. It therefore passes well below the level of the Dartford Heath outlier and supports the evidence from the till at Hornchurch for deep excavation of the valley by the late Anglian (Fig. 4.7). The Black Park Formation of the Middle Thames can, according to this reinterpretation, be correlated with the basal deposits of the Swanscombe Lower Gravel channel. At Swanscombe, however, little of this late Anglian gravel is preserved; it has been replaced by a Hoxnian temperate sequence. Aggradation continued to the Boyn Hill Terrace level, the river occupying this, its highest floodplain level in the Lower Thames, in the early part of the subsequent cold interval (the early Saalian).

The rejuvenation that separated the Boyn Hill floodplain from that on which the Black Park Gravel was deposited cannot, therefore, be recognized downstream from London. This event, which presumably occurred before the end of the Anglian Stage, may have had little effect in this area, where sediments equivalent to the Black Park Formation of the Middle Thames are thought to be directly overlain by gravels of the Boyn Hill Formation. It is difficult to determine whether any significant downcutting occurred in the valley downstream from London at this time, as any evidence would have been buried by the subsequent aggradation of the Boyn Hill Gravel to over 40 m O.D. Further upstream this rejuvenation is of considerable importance; for example, it coincided with the abandonment of the 'Ancient Channel' between Caversham and Henley (Chapter 3, Highlands Farm Pit). It has already been noted (Chapters 1 and 3) that the initiation of Black Park Gravel deposition was intimately related to the glaciation of the London Basin and the resultant diversion of the Thames; special circumstances therefore controlled the formation of this terrace, which was not generated by climatic fluctuation in the way that most other Thames terraces have been (see Chapter 1 – climatic model for terrace generation). It may be that the diversion of the Thames pre-empted the climatically induced rejuvenation event that was due to occur towards the end of the Anglian cold episode, causing the incision from the Winter Hill to the Black Park level. The time interval represented by the Black Park Gravel was evidently short; it was insufficient for a separate terrace feature to become established throughout the catchment, since the Black Park

Formation can be shown to converge upstream with the earlier Winter Hill Formation (see Chapter 1 and Fig. 1.3). The subsequent incision to the Boyn Hill level may have occurred in response to isostatic uplift following the disappearance of the Anglian ice sheets from Hertfordshire. Certainly the evolution of the terrace system at this time reflects the disruption and instability brought about by the Anglian glaciation and cannot be explained solely by the climatic model applied to other parts of the sequence.

In the revised stratigraphical scheme for the Thames terraces presented in Chapter 1 of this volume, the Swanscombe sediments provide important evidence for allocating the Hanborough/ Boyn Hill/Orsett Heath Formation to Oxygen Isotope Stages 12–10, by indicating a Stage 11 age for the interglacial phase (phase 3 of the climatic model) that separates the two cold-climate phases of aggradation recognized within this formation (Table 1.1). The post-interglacial aggradational phase (phase 4) appears to totally dominate the formation, possibly because of the disruption to the normal cycle caused by the glaciation of the London Basin, discussed above. Thus palaeontological evidence for the Stage 11 temperate episode has rarely been preserved; it is recognized in reworked mammal bones in the (phase 4) Hanborough Gravel of the Upper Thames (Chapter 2), but Swanscombe is the only locality in the present valley where *in situ* Stage 11 sediments are well-documented. The recorded occurrence of temperate-climate Mollusca at Dartford suggests that a further, smaller remnant may have existed there (see above, Wansunt Pit), but other surviving sediments of this age all lie downstream in Essex, in the former continuation of the Thames (Thames-Medway) course (see below, Chapter 5).

Final commentary

The Swanscombe (Barnfield Pit) site is one of the most important British Pleistocene localities. The wealth of information that has been gleaned from this site is apparent from the vast literature that exists. There is, however, considerable potential for gaining new information from the Swanscombe sequence. This is particularly the case at a time when the established terrestrial Pleistocene record is receiving such a critical re-examination. Swanscombe provides one of the foundations of this record and any re-appraisal will have to take account of the evidence from this site.

The Swanscombe record has a major shortcoming: despite the efforts of Mullenders (in Wymer, 1974) and Hubbard (1982), it lacks a convincing palynological sequence to compare with the established pollen-based biostratigraphy of the major British interglacials. Palynological studies have provided the basis for the identification of temperate intervals in the Pleistocene sequence both in Britain and western Europe (West, 1963; Mitchell *et al.*, 1973; de Jong, 1988), based on differences in vegetational development between different stages. Further work may enhance what has been achieved in this field at Swanscombe already, but the nature of the sediments probably precludes the preservation of a continuous record of vegetational change through the interglacial. However, other methods may supersede palynology as the principal technique for stratigraphical comparison in the British Pleistocene, particularly if mounting evidence that different temperate intervals can have nearly identical pollen records is substantiated.

There is some indication that faunal remains in the remaining Phase I deposits within the Swanscombe nature reserve have deteriorated since the higher parts of the sequence were removed, particularly the mammal bones. The third skull fragment was itself considerably weathered in comparison to the two previous finds (Wymer, 1964b), perhaps reflecting weathering during the intervening 19 years. If this is the case it is doubtful whether any similar material in the remaining sediments will have survived the further four decades of weathering that have followed the 1955 discovery. Further evidence that weathering is damaging the faunal remains in these deposits is provided by the poor state of the bones recovered from the Lower Loam in the 1982 investigation and the near absence of small vertebrate remains in the sieved samples from 1986 (A.P. Currant, pers. comm.). This emphasizes the critical importance of the Alkerden Lane Allotments SSSI, where the full Swanscombe sequence is preserved. In this area the survival of the higher layers, particularly the clayey slope and overbank deposits that cap the sequence, will hopefully have protected the faunal content of the lower material from weathering.

Conclusions

Swanscombe is probably the most famous British Pleistocene site; it is of international renown because of the discovery there of the fragmentary skull of an early Stone Age woman. Even without this find, the site would be of enormous importance. It combines a complex record of deposition by the Thames during a temperate (interglacial) episode, when the river laid down a series of gravels and loams containing abundant mammalian and molluscan fossils, with a rare occurrence of different Stone Age industries one above the other. In the first of these, confined to the lower part of the sequence, only flint flakes and 'cores' (blocks from which flakes have been removed) are found. The later industry, from the higher part of the sequence (including the level at which the skull was found), includes deliberately shaped tools called hand-axes, many hundreds of which have been found at Swanscombe. At the level within the sequence at which this change in the artefacts occurs, there are also important changes in the molluscan content of the sediments. In particular, a number of new species appear for the first time, including extremely rare forms known only from this locality in Britain. There is important evidence for a break in the sedimentary sequence of this same level, in the form of a soil that is developed in the top of the Swanscombe Lower Loam.

This combination of complex sedimentary, archaeological and faunal records makes Swanscombe a key reference site for other localities in Britain and western Europe. It is clear, from the position of the site within the 'staircase' of Lower Thames terraces, that the interglacial represented at Swanscombe is that which followed the Anglian Stage, when the Lower Thames drainage was initiated by the glacial diversion of the river. This would indicate deposition of the Swanscombe sequence at around 400,000 years before present. Hitherto, the Swanscombe sediments have generally been attributed to the Hoxnian Stage, named after a series of lake beds at Hoxne in Suffolk. However, it has recently been suggested that these lake beds relate to a more recent warm episode and that the Swanscombe section should be regarded as the type locality for the first post-Anglian interglacial, which is correlated with Stage 11 of the deep-sea record. This view contradicts the well-established intercorrelation between the Swanscombe deposits and the Hoxne lake beds by way of the important Thames site at Clacton. The Swanscombe and Clacton sequences are closely correlated on the basis of their molluscan faunas, whereas Clacton and Hoxne have been correlated using pollen analysis. The Swanscombe sequence is certain to be central to future research aimed at resolving this controversy. It is fortunate that the full sequence is preserved intact beneath the Alkerden Lane Allotments SSSI, adjacent to the partly worked-out Swanscombe Skull Site NNR.

PURFLEET – BLUELANDS, GREENLANDS, ESSO AND BOTANY PITS

D.R. Bridgland

Highlights

Adjacent quarries at Purfleet afford an unrivalled opportunity for the study of lateral variations in the deposits of a single Thames terrace. This complex site reveals interglacial sediments sandwiched between cold-climate gravels. The interglacial sediments have yielded an important molluscan fauna, while the gravels have been found to contain artefacts from several different Palaeolithic industries. The interglacial represented at Purfleet is believed to fall within the complex Saalian Stage, probably equating with Stage 9 of the oxygen isotope record.

Introduction

Four adjacent quarries are included in the Purfleet GCR site (Fig. 4.18), three of which are chalk pits revealing important sections in overlying Pleistocene deposits. These are, from north-east to south-west, Bluelands Quarry, Greenlands Quarry and Botany Pit. Between Botany Pit and Greenlands Quarry is a smaller gravel working, now within the confines of an oil storage depot, known as the Esso Pit. The Pleistocene sediments exposed in these pits are part of a spread of gravel, 'brickearth' and 'coombe deposits' that abuts against the northern side of the Purfleet Anticline, an

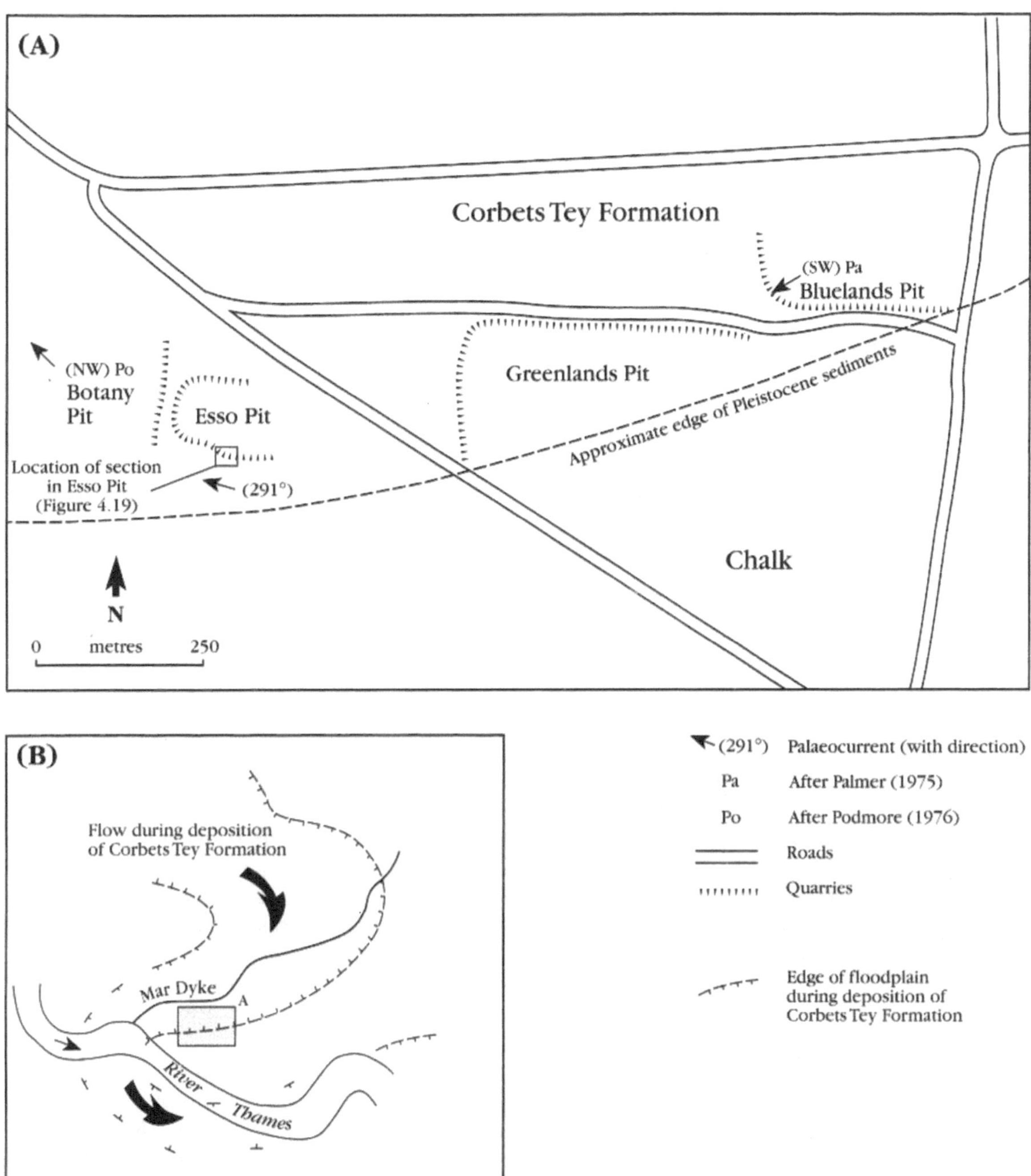

Figure 4.18 (A) Map showing the location of the various exposures at Purfleet; (B) the extent of the Thames floodplain in the Purfleet area during the deposition of the Corbets Tey Gravel. Note that where the floodplain passed through the Chalk outcrop of the Purfleet Anticline it was considerably restricted.

east-west trending structure that causes the Upper Chalk to outcrop in a ridge between Purfleet and Grays (Geological Survey, New Series, Sheet 271; Fig. 4.1). The deposits contain important assemblages of Palaeolithic artefacts and Mollusca (Wymer, 1968, 1985b; Palmer, 1975; Snelling, 1975; Allen, 1977; Hollin, 1977) and have also yielded pollen (Hollin, 1977).

Thames terrace deposits have been widely recognized to the south of the Purfleet Anticline, similarly banked against the Chalk. The area to the north of the Chalk ridge is now drained by the Mar Dyke, a westward-flowing tributary stream that joins the Thames at Purfleet, its valley dissecting the Pleistocene deposits on the northern side of the anticline (see Fig. 4.1). This situation has led to speculation that the latter, which include the deposits exposed in the Purfleet quarries, were laid down by the Mar Dyke (Dewey *et al.*, 1924; Wymer, 1968, 1985b; Palmer, 1975). Recent reconstructions of Lower Thames palaeodrainage do not support this view (Bridgland, 1988a). The Purfleet sediments are now considered to occupy a sinuous abandoned section of the main Lower Thames valley and to be part of the Lynch Hill/Corbets Tey Formation, of mid-Saalian age (Bridgland, 1988a; Gibbard *et al.*, 1988; Chapter 1).

Description

The Pleistocene deposits to the north and north-east of Purfleet seem to have been largely overlooked in early descriptions of the area, which tended to concentrate on the gravels and brickearths on the southern side of the Purfleet Anticline (for example, Whitaker, 1889; Hinton and Kennard, 1900; see below, Globe Pit). The Purfleet deposits, preserved on either side of the Mar Dyke valley (Fig. 4.1), are aggraded to *c.* 15 m O.D. and therefore form part of the 'Middle Terrace', as recognized in early work in the Lower Thames (see above, Introduction to this chapter). They were mapped by the Geological Survey (New Series, Sheet 271) as 'Taplow Gravel', but were interpreted by Dewey *et al.* (1924) as Mar Dyke sediments.

It is uncertain when these deposits were first quarried, but detailed research in the area began in the early 1960s, when Botany Chalk Pit was extended. The discovery of large numbers of flint artefacts in the overlying sand and gravel led to the publication of a description by Wymer (1968). He noted that the gravel was current-bedded and that it passed laterally into 'coombe rock' in the southern part of the pit. In the eastern face, now graded, the gravel could be seen banked against Chalk containing bands of flint nodules. The gravel immediately adjacent to this rising bank of Chalk contains one of the richest concentrations of worked flint in the British Palaeolithic (Wymer, 1968; A.J.R. Snelling, pers. comm.). The assemblage from Botany Pit includes hand-axes, chopper-cores of Clactonian type and 'proto-tortoise cores', the latter suggesting an early use of Levallois technique (Wymer, 1968, 1985b). Wymer (1985b, p. 313) suggested that the natural outcrop of flint nodules in the ancient river bank at Botany Pit had been exploited as 'a Palaeolithic flint quarry where the flint was knapped on the spot and selected pieces taken away'. He thought that the wealth of available raw material may have given rise to the early use of the extravagant Levallois technique (described in Chapter 1). Roe (1968b) classified only a small proportion (0.8%) of the flakes from Botany Pit as Levalloisian. Sections visible in this pit during the 1970s were recorded by Podmore (1976), Shephard (1976) and Lonsdale (1978).

It is uncertain when the smaller gravel working known as the Esso Pit was excavated, as no published descriptions have been found. A small-scale exploratory excavation here in 1986 by the author and N.D.W. Davey revealed deposits comparable to those in nearby Botany Pit, although Chalk-bearing gravel was encountered (Fig. 4.19 and Table 4.2), similar to the deposits associated with the shelly sediments in Greenlands and Bluelands Quarries (Fig. 4.20). The section proved to be rich in Palaeolithic artefacts; several flakes and a core were recovered from an exposure 5 m high and 1–2 m wide, with minimal disturbance of *in situ* material. These were largely undiagnostic, although a probable hand-axe finishing flake was included. Amongst a number of unstratified finds was a flake indicative of Levallois technology, with a 'faceted butt' (archaeological determinations by P. Harding).

Greenlands Quarry was also opened in the early 1960s, the resultant exposures revealing the existence of rich shell beds. The section at Greenlands and the fauna from these beds were described by Snelling (1975), who recorded over 25 ft (7.5 m) of Pleistocene deposits overlying Chalk and Chalk rubble in the north face (Fig. 4.20). The lower 4 m of these sediments are dominated, above a basal gravel (0.5 m), by shelly clays and sands, the best-preserved molluscan remains coming from the upper, more sandy half of these lower beds. Above the shelly beds, alternating gravels and clays form the remainder of the sequence. Snelling listed 27 molluscan species from the Greenlands shell beds (all but seven of them aquatic), together with mammal bones and teeth, including straight-tusked elephant (*Palaeoloxodon*

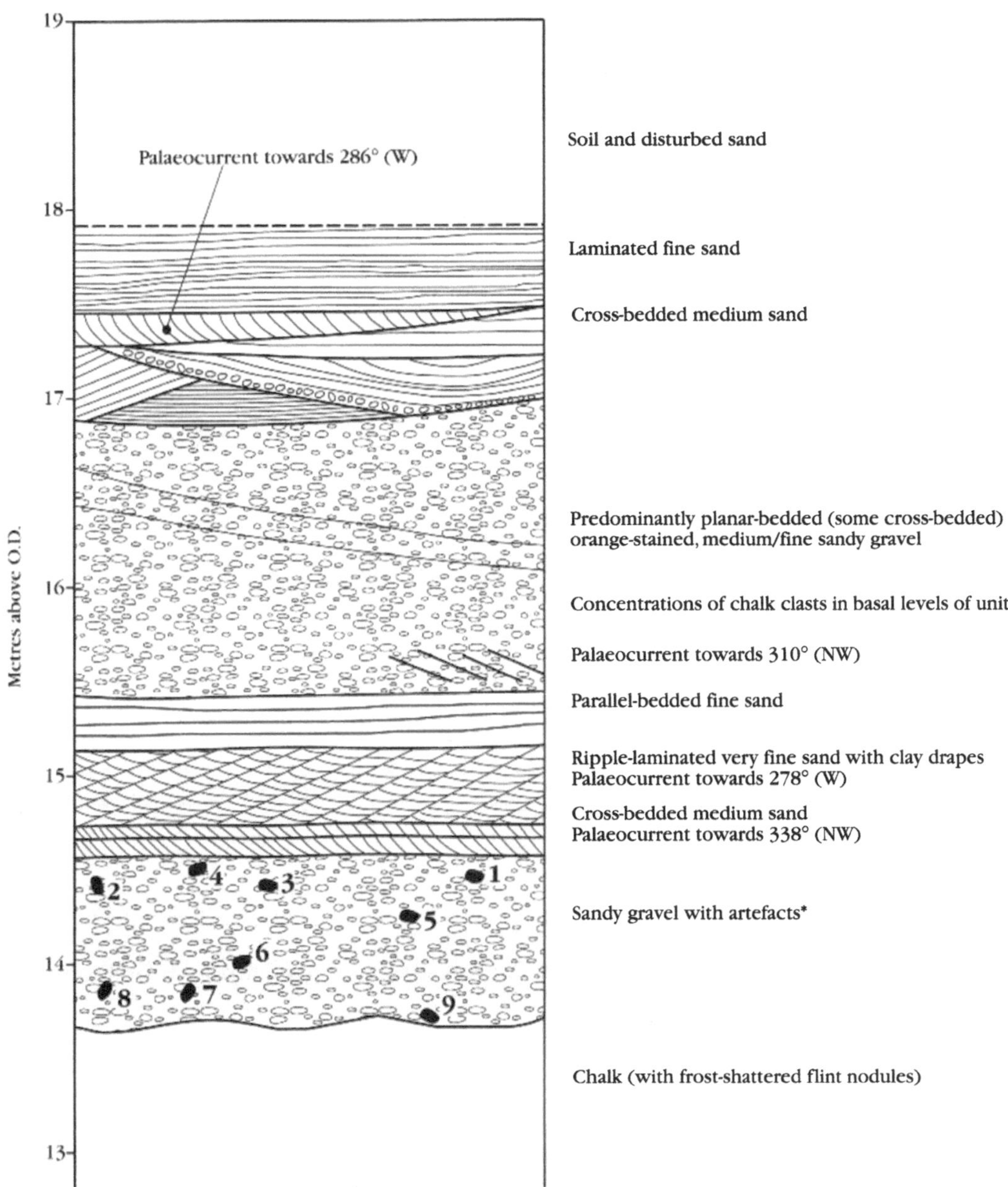

Figure 4.19 Section excavated in the Esso Pit by the GCR Unit in 1986. * Artefacts from the sandy gravel immediately overlying the Chalk are as follows (numbered on the figure): (1) undiagnostic sharp flake; (2) large preparation flake in sharp condition; (3) small cortical flake that may have formed naturally in the river's bed load; (4) undiagnostic hard-hammer flake in slightly rolled condition; (5) broken flake (the break probably occurred at the time of knapping); (6) broken unstained flake that may have formed naturally in the river's bed load; (7) core, utilizing a broken nodule – approximately four flakes have been removed by alternate flaking; (8) small sharp flake that may have formed naturally in the river's bed load; (9) a sharp flake, thick in section (particularly towards the distal end), with semi-converging scars on the dorsal surface – flakes of this type are produced during the shaping of hand-axes. Other material was found elsewhere at the site. The collection has been lodged with the British Museum. Archaeological determinations by P. Harding.

antiquus), bison and indeterminate small rodents. The molluscan assemblage includes the aquatic snail *Belgrandia marginata* (Michard) in great abundance; *Bithynia tentaculata* is also common, as are the bivalves *Pisidium amnicum* (Müller), *P. supinum*, *P. henslowanum* (Sheppard) and *Corbicula fluminalis*. Later descriptions of the sediments

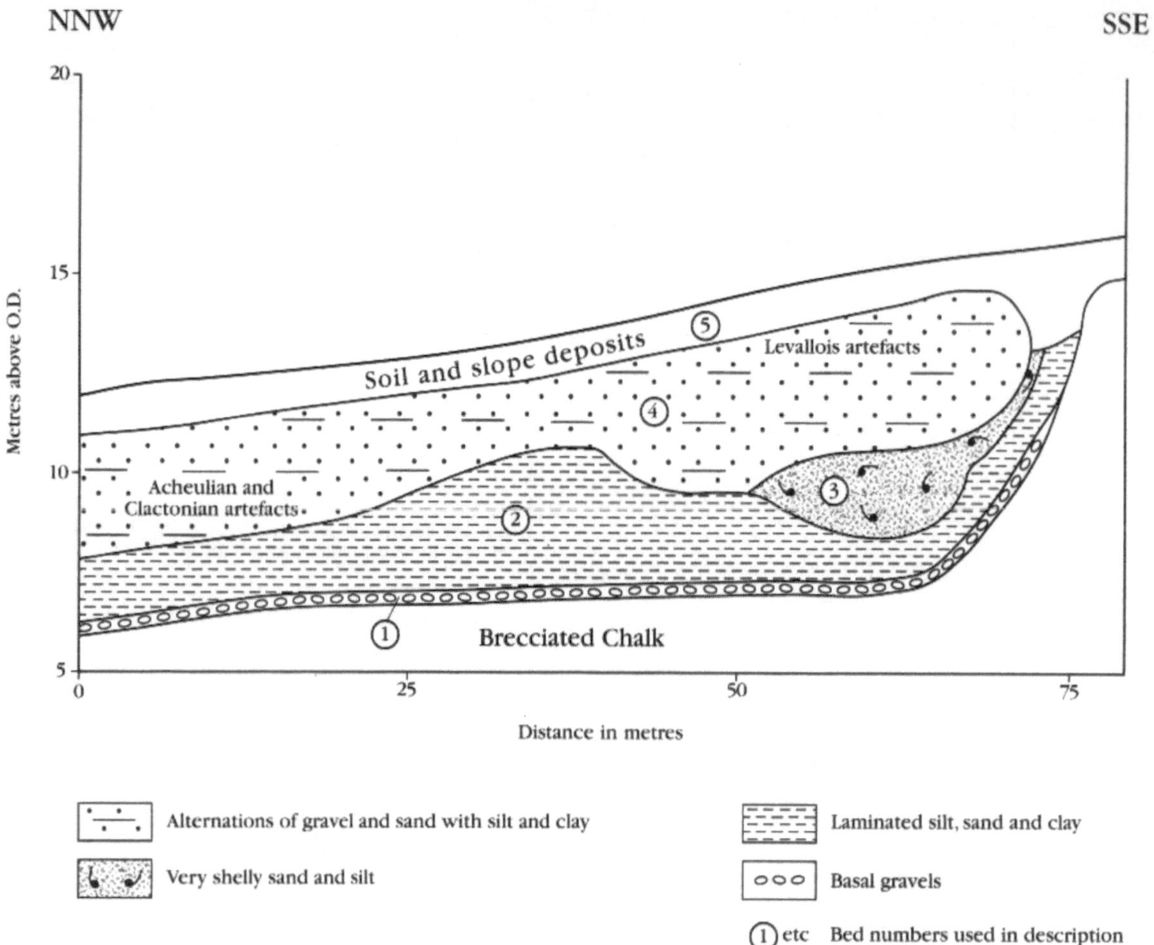

Figure 4.20 Idealized section through the terrace deposits at Purfleet (modified from Hollin, 1977). Bed 1 contains Clactonian artefacts.

here, and in the adjacent Bluelands Quarry, were provided by Palmer (1975), Hollin (1977), Lonsdale (1978) and Wymer (1985b).

Palmer (1975) discovered Palaeolithic material in both Greenlands and Bluelands Quarries; she assembled a mixed collection of scrapers, flakes and cores, principally from Bluelands. Palmer interpreted this assemblage as 'Middle Acheulian', with a considerable Clactonian element, much of which was rolled and therefore possibly reworked from earlier deposits. Some Levallois influence was observed, but Palmer considered the assemblage to differ significantly from the proto-Levallois material from Botany Pit. Wymer (1985b) suggested that three different industries were included in the collections assembled by Palmer, possibly reflecting a stratigraphical sequence of Clactonian, Acheulian and Levallois occupations (see below).

Observations by Hollin (1971, 1977) broadly confirmed earlier descriptions, although he noted that Snelling's main shell-bearing bed occupies a channel cut into more argillaceous, laminated deposits. He attributed the latter to deposition in an intertidal environment. Hollin also described an upper sand body at Purfleet, directly overlying a Chalk platform at 14 m O.D. The existence of this deposit, which has the appearance, in Hollin's section, of a separate higher terrace remnant, has not been confirmed by recent investigations (P. Allen, pers. comm.).

From the various descriptions of the Purfleet sections, the following generalized sequence can be reconstructed (see also Fig. 4.20, which indicates approximate thicknesses):

5. Colluvium and possible loessic material (disturbed), not part of the fluvial sequence.

4. A sequence of gravels, sands and silts with subordinate planar-bedded clay. The gravel and sand were attributed to braided-river deposition by Hollin (1977). Wymer (1985b) has suggested that Clactonian and Acheulian artefacts occur in the lower part of this sequence, with Levallois artefacts near the top. This bed overlies the interglacial beds (1–3) at Bluelands and Greenlands. Snelling (1975) suggested that the gravel containing Levallois material at Botany Pit is the equivalent of this unit. This view is in keeping with Wymer's (1985b) observation that the basal gravel at Botany Pit contains (Acheulian) hand-axes.

3. A channel cut into (2), filled with sand and silt, with a very large comminuted shell component (interglacial species). It also contains articulated bivalve specimens. It is preserved only in the Bluelands and Greenlands sections.

2. Laminated sand, silt and clay (brickearth), containing Mollusca, ostracods and pollen (interglacial species). The fossiliferous bed is restricted to Bluelands and Greenlands, but similar deposits, although without fossils, have been recorded in the Esso Pit and in Botany Pit (now destroyed). Without biostratigraphical control, it would be unwise to correlate these with bed 2, however.

1. Sandy gravel with Chalk and calcareous concretions, containing Mollusca (interglacial species). This bed contains Clactonian artefacts, some in sharp condition (Wymer, 1985b). Palmer (1975) distinguished a very sandy and calcareous upper division and a coarse lower division within this unit, the Mollusca apparently occurring in the former.

   Chalk rubble (above solid Chalk).

The fossiliferous beds 1–3 represent interglacial conditions. They are, however, of limited lateral extent; in their absence, the basal (unfossiliferous) part of bed 1 would be indistinguishable from bed 4 (except, perhaps, from their Palaeolithic contents).

Interpretation

As noted above, Dewey *et al.* (1924) and Wymer (1968) attributed the Purfleet deposits to the Mar Dyke, because they occupy the valley of that tributary rather than the main river. Palmer (1975) recorded the elevation of the Chalk surface in the area of Bluelands and Greenlands Quarries in some detail and concluded that the Pleistocene deposits occupied a channel with an approximate north-east to south-west alignment. She carried out a fabric study, measuring gravel clast imbrication, which indicated deposition by water flowing towards the south-west (Fig. 4.18). On the basis of this analysis, she too suggested that the gravels were deposited by the Mar Dyke. Palaeocurrent evidence from the Esso Pit, obtained during the 1986 investigation by measuring foreset orientations, confirms a broadly westward flow direction (Fig. 4.18). The results of clast-lithological analysis show the deposits here to be typical of Lower Thames gravels upstream from the Darent confluence (Table 4.2). This is not entirely incompatible with a Mar Dyke origin, since that stream would largely have been reworking Thames sediments, but some dilution by rounded flint pebbles from the Palaeogene outcrop to the north (and, secondarily, from earlier Pleistocene deposits derived from this) might be expected in gravel laid down by the tributary (see below).

The distribution of the Corbets Tey Gravel (mapped by the Geological Survey as 'Taplow Gravel'), in which the Purfleet sediments are included, suggests that the Lower Thames followed a markedly sinuous course in the Ockendon–Grays area at the time that this formation was deposited (Figs 4.1 and 4.18B; Bridgland, 1988a). The relations between the Corbets Tey Formation and higher ground to the north, east and south, which constrains the reconstruction of the Corbets Tey floodplain, reveal the extent of this sinuosity. When this deposit was aggrading, the river, flowing eastwards from London, turned sharply to the south-west in the Ockendon area, flowed in that direction as far as Purfleet, then swung southwards through a gap in the Chalk of the Purfleet Anticline, ultimately to resume its eastward heading (Fig. 4.18B). This interpretation explains why the Corbets Tey Gravel between Ockendon and Purfleet occupies the Mar Dyke valley and why it is banked against the northern side of the Chalk ridge at Purfleet: it is apparent that the Mar Dyke tributary has adopted this abandoned part of the early, sinuous Thames valley. An explanation is also provided for the palaeocurrent data from Purfleet, which records the westward flow between the two 180° bends in the valley (Fig. 4.18). In support of this interpretation is the fact that the wide and almost continuous belt of Corbets Tey Gravel between Ilford and Ockendon, clearly forming a terrace of the Thames, cannot be traced eastwards beyond Ockendon, as a higher area of bedrock stands in the way. Furthermore, the Mar Dyke valley upstream from the Corbets Tey Gravel outcrop is entirely devoid of gravel; there is nothing to indicate that this stream is of sufficient antiquity to have deposited any part of the Purfleet sequence and no evidence that it ever laid down substantial pre-Holocene sediments.

Deposits ascribed to a north-bank tributary of the Thames, perhaps ancestral to the Mar Dyke, were described by Hinton and Kennard (1900); this description probably pertains to the deposits located at around TQ 610790 (Fig. 4.21). These deposits occupy a 'col' to the north of Grays that separates two spreads of Orsett Heath Gravel capping the ridge formed by the Purfleet Anticline. Brickearth has been mapped here (Geological Survey, Sheet 271; Fig. 4.21), presumably corresponding to loams and clays described in the upper part of the Hinton

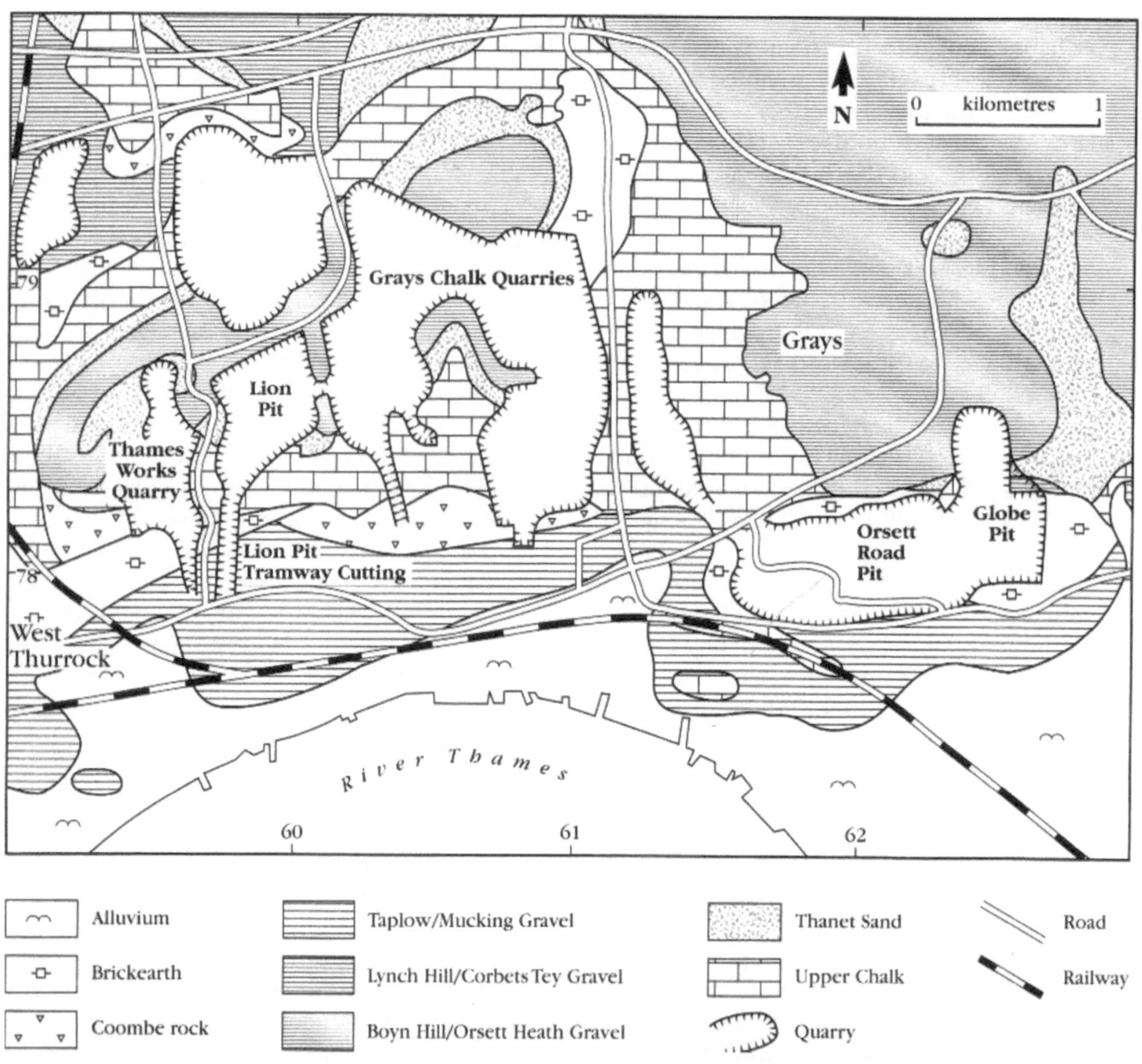

Figure 4.21 Map showing the various sites in the Thurrock-Grays area.

and Kennard section. Beneath this material was up to 2 m of gravel, largely composed of rounded flint pebbles (from the Palaeogene), but with some subangular flint. Rocks foreign to the area, of the type that characterize the Lower Thames gravels, were conspicuously scarce. The sequence fills a north–south-trending channel cut into the Thanet Sand (Hinton and Kennard, 1900). These deposits, of uncertain age, are clearly very different to the gravels (including those at Purfleet) that occupy the sinuous Thames course described above; this fact lends support to the argument that the Purfleet deposits are products of the main river.

Hollin (1971, 1977) ascribed laminated silts in the lower part of the Greenlands succession (bed 2) to an intertidal environment, suggesting that this part of the Lower Thames was within the estuarine zone at the time of their deposition. However, Robinson (in Hollin, 1977) recorded various freshwater ostracods from these laminated deposits, although specimens of *Cyprideis torosa* (Jones) from the later shelly channel-fill (bed 3) show the tuberculate ornament that this species develops in brackish conditions. Allen (1977) recorded further evidence for deposition in a slightly brackish environment, in the form of ostracods and foraminifera from a shelly seam within the laminated beds at Greenlands Quarry (bed 2). This seam also yielded the remains of insectivores, rodents, amphibians and fish (Palmer, 1975; Allen, 1977). A herpetofauna from the site has been recorded recently by Holman and Clayden (1988).

Stratigraphy and correlation

Very different stratigraphical interpretations of the Purfleet locality have been put forward, depending on whether terrace stratigraphy, archaeology or palaeontology have been given priority as evidence for relative dating. From the molluscan evidence, Snelling (1975) concluded that the interglacial deposits at Greenlands were of Hoxnian age. This view was based principally on the occurrence of the freshwater snail *Valvata antiqua* (Morris), not known from post-Hoxnian sediments (Castell, in Snelling, 1975). However, he observed that the deposits appeared to belong to the same terrace as the gravel at Botany Pit, which had yielded an abundant Proto-Levallois industry that he was reluctant to accept as of Hoxnian age. He therefore suggested that the upper (post-interglacial) part of the Greenlands sequence and the Palaeolithic gravel at Botany Pit were lateral equivalents and that they were of post-Hoxnian age.

Palmer (1975) likened the Palaeolithic assemblage from Bluelands Quarry to the Acheulian industry in the Middle Gravel at Swanscombe, thus supporting correlation with the Hoxnian. In a reappraisal of Palmer's findings, Wymer (1985b) concluded that three separate Palaeolithic industries were represented amongst her collection. He recognized Clactonian elements, generally somewhat abraded, from within and immediately above the shelly deposits (in beds 1–3 and in the lower part of 4). In the latter context (lower bed 4), he also recognized abraded Acheulian material. A Levallois flake was attributed to the highest gravel layer (upper bed 4) (Fig. 4.20). In a similar reappraisal of the records of finds in Botany Pit, Wymer (1985b) noted that sharp hand-axes occurred at the base of the gravel there, immediately above the Chalk. He suggested that these pre-date the Levalloisian material from the site. Thus there is archaeological support for the correlation of the gravel at Botany Pit with bed 4 of the general Purfleet sequence, in which Levallois artefacts also appear in the upper levels. This concurs with Snelling's (1975) suggestion that the Levallois material post-dates the interglacial beds. The implication that Palaeolithic assemblages may occur in a meaningful stratigraphical sequence in the Corbets Tey Gravel is potentially of great significance (see below, Globe Pit).

Hollin (in Palmer, 1975; Hollin, 1977) collected pollen samples from the laminated deposits (bed 2) in Greenlands Quarry, but obtained well-preserved material at only four levels. Analysis of these pointed to pollen biozone II of an interglacial, but the assemblage was insufficiently distinctive to allow correlation with either of the two post-Anglian temperate stages recognized in Britain at that time, the Hoxnian and the Ipswichian. However, Allen (in Hollin, 1977) considered that the occurrence, at Purfleet, of the bivalves *Unio* sp. and *Corbicula fluminalis* so early in an interglacial (biozone II) was suggestive of the Ipswichian rather than the Hoxnian. Hollin was inclined to accept this view rather than Castell's interpretation of the assemblage as Hoxnian (see above). Hollin suggested that the laminated deposits represented an estuarine phase resulting from a rapid rise in sea level in response to an Antarctic ice surge during the Ipswichian Stage.

Allen (1977) recognized that problems exist in ascribing the Purfleet sediments to either the Hoxnian or the Ipswichian Stage. He observed that if the deposits were Hoxnian, the Mollusca would indicate a late part of the interglacial (largely on the basis of a comparison with Swanscombe, where *Corbicula fluminalis* is absent from the early part of the sequence – see above, Swanscombe); this is contrary to the palynological evidence, which implies that biozone II, the early temperate phase, is represented. Conversely, Allen regarded the elevation of the deposits as too high, in comparison with Ipswichian sites elsewhere in the Lower Thames, for correlation with that stage. He suggested that 'the deposits aggraded during another interglacial between the Hoxnian and the Ipswichian' (Allen, 1977, p. 3).

Preece (1988) has contributed recently to this debate. He noted the occurrence at Greenlands of the large freshwater mussel *Margaritifera auricularia* (Spengler), which is only recorded from one other British interglacial site, namely the well-known Ipswichian deposits of Trafalgar Square. Despite this and other similarities, he pointed out that the molluscan faunas of Trafalgar Square and Purfleet show 'some striking differences' (Preece, 1988, p. 51). Preece cited, as an example, the absence of *C. fluminalis* at Trafalgar Square, in sediments that appear to derive from a similar depositional environment to that represented at Purfleet, where the species is abundant. The potential stratigraphical significance of this species, and in particular the theory that it was absent from Britain during the Ipswichian Stage (*sensu* Trafalgar Square), has been discussed in Chapter 2 (see Stanton Harcourt and Magdalen Grove).

Bridgland (1983a, 1988a) offered a re-appraisal of Lower Thames terrace stratigraphy based on the traditional method of recognizing the products of the various individual terrace aggradations from their distribution and elevation. He correlated the deposits at Purfleet with the Lynch Hill/Corbets Tey Formation of the Thames and demonstrated that they were laid down by the main river, not the Mar Dyke tributary (see above). This correlation implies a mid-Saalian age (Tables 4.1 and 1.1). Despite this, Gibbard *et al.* (1988) preferred to assign the Purfleet sediments to the Mar Dyke, believing them to relate to the infilling of this tributary valley during an Ipswichian high sea-level phase. In this difference of opinions, palaeogeographical reconstructions and age determinations are closely interrelated. Bridgland (1983a, 1988a) showed that the Thames, after depositing the Taplow/Corbets Tey Gravel in what is now the Mar Dyke valley, had abandoned this course by the late Saalian Stage, when the Mucking Gravel was laid down. Therefore, if the Purfleet interglacial deposits are Ipswichian, they must be the product of the Mar Dyke and not the Thames; conversely, if they were deposited by the Thames, they must be pre-Ipswichian. Comparison with the north-bank tributary gravels that were described by Hinton and Kennard (1900; see above), strongly suggests that the Purfleet deposits are of Thames origin, which, as pointed out above, carries the implication of a pre-Ipswichian age for the interglacial beds at Bluelands and Greenlands Quarries.

A further indication of the age of the Purfleet deposits has been obtained by the amino acid analysis of shells from the locality. The results of preliminary work of this type suggested that specimens of *C. fluminalis* from Purfleet (which gave D-alloisoleucine : L-isoleucine ratios of between 0.33 and 0.39) were the oldest of any from the Lower Thames, including examples from the Swanscombe Middle Gravel and from Stoke Newington (Miller *et al.*, 1979). These findings have been broadly confirmed by Bowen *et al.* (1989), who published a ratio of 0.34 (± 0.24) based on two *Bithynia* shells from Purfleet. The large standard deviation of the latter ratio, however, suggests that it could be imprecise. Bowen *et al.* also published an updated *Corbicula* ratio of 0.38 (± 0.07) from the shelly gravel at Greenlands and Bluelands. These ratios are comparable with results from sites attributed to the Cromerian Stage (Bowen *et al.*, 1989). These data must be misleading, since the deposits cannot pre-date the diversion of the Thames into the valley through London (and Purfleet) during the Anglian Stage. These high amino acid ratios may, however, lend support to the view that the Purfleet deposits are of pre-Ipswichian age, since shells from sites attributed to that stage (Trafalgar Square and Bobbitshole, for example) have given uniformly low ratios, in the range 0.09 to 0.13 (Bowen *et al.*, 1989).

From an examination of the various published descriptions and discussions of the Purfleet sediments, it is apparent that they have been correlated with practically every other interglacial site in the Lower Thames at one time or another. Those who favoured an older, Hoxnian age have suggested correlation with the Swanscombe Middle Gravel (Palmer, 1975), but those favouring an Ipswichian age have broadly linked the site with Ilford, Aveley, West Thurrock, Crayford and even Trafalgar Square (Hollin, 1977; Gibbard *et al.*, 1988), despite the fact that the Purfleet deposits form part of a different, higher terrace to all of these (Fig. 4.3).

There are, however, other interglacial sediments associated with the Corbets Tey Gravel, at Grays and Little Thurrock (see below, Globe Pit), at Belhus Park (Wymer, 1985b) and near the type locality at Corbets Tey (Ward, 1984). It is also likely that the complex of sites at Stoke Newington, discovered in the last century (Smith, 1883, 1894; Wymer, 1968; Kerney, 1971; Harding and Gibbard, 1984), belong to the Lynch Hill/Corbets Tey Formation. Amino acid ratios from specimens of *Corbicula* from Stoke Newington and Hackney support a correlation with Grays (Miller *et al.*, 1979), although similar ratios were also obtained from Swanscombe, considered here to be older. All the above sites, perhaps significantly, have yielded *Corbicula fluminalis*.

A borehole at Corbets Tey (TQ 550850) produced a very similar molluscan fauna to that from Greenlands, as well as plant, ostracod, small vertebrate and insect remains (Ward, 1984). The presence of the gastropod *Hydrobia* cf. *ventrosa* (Montagu) and the ostracod *Cyprideis torosa* provides evidence for estuarine or brackish conditions (Ward, 1984). Old records (in Wymer, 1985b) of a gravel pit at Gerpins (TQ 555840), 1 km to the south-west, appear to describe organic deposits containing wood beneath 8 m of sand belonging to the Corbets Tey Formation. A correlation of these deposits with that at Belhus Park, 3.5 km further to the south-east, was suggested by Wymer (1985b). Eight hand-axes are recorded from Gerpins Pit (Wymer, 1985b), possibly representing a lateral equivalent of the assemblage from Purfleet. The Corbets Tey, Gerpins Pit and Belhus Park sites are associated with a major channel feature within the Corbets Tey Gravel outcrop, revealed by borehole data (P.L. Gibbard, pers. comm.). According to Gibbard, this channel is excavated through the Corbets Tey Gravel and was filled during the Ipswichian Stage by Mar Dyke deposits. However, at Belhus Park the interglacial sediments were seen to be sandwiched between gravels, both apparently of Corbets Tey type. The upper gravel, in addition to containing a full suite of Thames far-travelled lithologies (Table 4.2), also yielded a number of Palaeolithic artefacts, some in very sharp condition (Wymer, 1985b). According to Wymer, these artefacts, Acheulian hand-axes and cleavers, 'cannot have been derived from any earlier deposit and must be contemporary with the aggradation that produced the gravel overlying the organic deposit' (1985b, p. 314). Artefacts of this type would be out of place in a post-Ipswichian context (see Wymer, 1988). It is therefore considered likely that the interglacial sediments at Belhus Park are pre-Ipswichian Thames deposits. Further evidence in support of this view has recently come to light, from amino acid analyses of shells from the Belhus Park interglacial deposits (Bowen, 1991; see below).

In the revised stratigraphical scheme for the Thames terrace system presented in Chapter 1, the interglacial represented within the Lynch Hill/Corbets Tey Formation is correlated with Oxygen Isotope Stage 9 (Table 1.1). The Purfleet sediments provide important evidence with regard to this interpretation. Despite many suggestions that they are Hoxnian, they have been differentiated from that stage, as represented at Swanscombe, on the basis of differences in the molluscan faunas from the two sites and their comparison with the record for vegetational history during the Hoxnian, derived from the Clacton sequence (Allen, 1977; see above, Swanscombe). An Ipswichian (*sensu* Trafalgar Square) age, still favoured by some authors (Gibbard *et al.*, 1988), may be ruled out by the presence of *Corbicula fluminalis*, which appears to have been absent from the Thames valley during that stage (Chapter 2, Stanton Harcourt and Magdalen Grove). Notwithstanding these biostratigraphical arguments, sediments that are attributed to these two established stages form part of different terrace formations to that (the Wolvercote/Lynch Hill/ Corbets Tey Formation) recognized at Purfleet (Table 1.1). Terrace stratigraphy, taken in isolation, suggests correlation of the Purfleet interglacial sediments with those at Wolvercote, Stoke Newington, and Grays and Little Thurrock (Table 1.1). The same line of evidence suggests that all these were laid down during the second of four post-Anglian interglacials (Fig. 4.3); the correlation of the Anglian with Oxygen Isotope Stage 12 therefore points to a Stage 9 age for the Purfleet deposits and their correlatives (see Chapter 1 and Table 1.1). Recently, amino acid ratios have been obtained from shells from the interglacial deposits at Belhus Park, described above. These ratios (0.26) support the correlation with Stage 9 (Bowen, 1991). The corroboration (or otherwise) of the above correlations, on biostratigraphical grounds or from additional amino acid analyses, must await further research on these various deposits.

Summary

The complex of pits at Purfleet together constitute one of the most important localities for the reconstruction of Thames drainage development. They provide evidence from every discipline of relevance to Pleistocene stratigraphy. In particular, important Palaeolithic and molluscan assemblages from different parts of the composite site provide significant stratigraphical and palaeoenvironmental evidence. The site has long been recognized as possibly representative of an undefined and unnamed temperate interval between the Hoxnian and Ipswichian. There is still considerable controversy about whether the sediments at Purfleet, which belong to the Lynch Hill/Corbets Tey Formation, represent a post-Hoxnian/pre-Ipswichian temperate episode or a partly estuarine valley-fill of Ipswichian age. There is also a closely related dispute as to whether the deposits are the product of the Thames or of the tributary Mar Dyke stream. In fact, several lines of evidence combine to suggest that the temperate deposits at Purfleet correlate with Oxygen Isotope Stage 9. Phases 3 (interglacial) and 4 (post-interglacial) of the model for terrace formation (proposed in Chapter 1) are clearly well-represented at Purfleet, by the various fossiliferous sediments and the overlying gravels of bed 4 (respectively). It seems likely that the unfossiliferous lower division of bed 1, recorded by Palmer (1975), represents phase 2 of the model, therefore dating from the end of Stage 10 (the same age as the gravel at the Globe Pit site – see below). The importance of the fossiliferous beds at Purfleet is heightened by the destruction of the last remaining exposures of the Grays brickearth (see below, Globe Pit), which is also considered to have accumulated during Oxygen Isotope Stage 9.

Conclusions

The various sections at Purfleet can be pieced together to provide a detailed record of the sediments that make up the Corbets Tey Gravel, which forms the second of the three terraces preserved on the Essex side of the Lower Thames valley. The picture that emerges is one of widespread cold-climate gravel aggradation, following the more localized deposition of fossiliferous sediments during an interglacial. The gravels contain Lower Palaeolithic (early Stone Age) flint artefacts – these are particularly abundant at the Botany and Esso Pits. The lowest gravels, beneath and/or part of the interglacial beds, contain artefacts belonging to the early and rather primitive Clactonian industry, whereas the later gravels contain the important addition of later types, including material flaked using the 'Levallois technique' (a distinctive method of flint working that produced artefacts of characteristic types). This appearance of Levallois material, implying the first use of this technique after the interglacial represented at Purfleet, is the earliest in the Lower Thames sequence and is of considerable stratigraphical importance.

The age of the interglacial at Purfleet is controversial. On the basis of rather meagre pollen evidence it has been correlated with the last interglacial, only 120,000 years BP. This view is contradicted by the rich molluscan assemblage from Bluelands and Greenlands Quarries, which indicates that the sediments are older than the accepted last interglacial site at Trafalgar Square. The relations of this site to the terrace sequence in the Lower Thames suggests that the second of four post-Anglian interglacials is represented at Purfleet, implying correlation with Stage 9 of the deep-sea record and an age of around 300,000 years BP.

GLOBE PIT, LITTLE THURROCK (TQ 625783)

D.R. Bridgland

Highlights

Globe Pit provides important evidence that contributes to the stratigraphical record of the Lower Thames terrace sequence and, in particular, to the parallel record of Palaeolithic occupation in southern Britain. A gravel here, the feather-edge of the Lynch Hill/Corbets Tey Formation, yields a prolific Clactonian industry. The occurrence of this industry, which has been regarded as 'early' within the Lower Palaeolithic, within deposits formerly ascribed to the 'Middle Terrace', has long been regarded as anomalous. The fact that hand-axes occur at higher-level sites such as Swanscombe has given rise to interpretations of the Lower Thames sequence involving complex fluctuations of base level. Consideration of the relations between these artefact-bearing deposits and fossiliferous sediments elsewhere in this and other terrace aggradations suggests that the gravel at Globe Pit is indeed slightly later than the Swanscombe sequence, probably dating from the latter part of Oxygen Isotope Stage 10 (early Saalian).

Introduction

Globe Pit is the first of two sites in the Grays area, a district famous for the fossiliferous Pleistocene brickearth (a mixture of silts, sands and clays) that was exploited there until early in the present century. Two main spreads of brickearth appear on the Geological Survey map (Sheet 271) in the vicinity of Grays, one at West Thurrock and the other at Little Thurrock (Fig. 4.1). A third spread to the north of Grays (around TQ 611793; Fig. 4.21) is not part of the fossiliferous Thames brickearth, but overlies north-bank tributary gravels (Hinton and Kennard, 1900; see above, Purfleet) and has itself been attributed to an ancestral Mar Dyke (Dewey *et al.*, 1924). The Chalk quarry at Globe Pit was an extension of early workings in the Little Thurrock brickearth spread.

The brickearth of the Grays area has generally been associated with the 'Middle Terrace' of the Lower Thames, mapped by the Geological Survey as 'Taplow Gravel', but now recognized as Corbets Tey Gravel and correlated with the Lynch Hill Formation of the Middle Thames (Bridgland, 1988a; Gibbard *et al.*, 1988). The term brickearth is in this instance applied to well-bedded, often laminated, fluviatile silts and fine sands with subordinate clay (West, 1969; Hollin, 1977).

The brickearth and associated gravels at Grays (Little Thurrock) have been studied by numerous workers over more than one and a half centuries. Early contributions included those of Morris (1836), Wood (1848), Jones (1850), Wood Jun. (1866a, 1867, 1868, 1872), Dawkins (1867), Tylor (1869), Hughes (in Whitaker, 1889, p. 420), Woodward (1890), Reid (1897), Hinton and Kennard (1900) and Dewey *et al.* (1924). A full list of previous literature describing the area was given by Hinton and Kennard (1900), much of which refers to faunal remains found in the brickearth. Of particular importance were sections in a pit south of Orsett Road, in which a highly fossiliferous lenticular bed of fine sand yielded abundant remains of molluscs and small vertebrates (Hinton and Kennard, 1900; Hinton, 1901). According to Sutcliffe and Kowalski (1976), this pit was *c.* 650 m to the west of the GCR site. It was certainly within the same terrace remnant (Fig. 4.21), suggesting that records from Grays and Little Thurrock can be considered together.

Elsewhere the Grays and Little Thurrock brickearth has yielded vertebrate remains, molluscs, ostracods, plant remains, pollen and, less frequently, Palaeolithic artefacts (Hinton and Kennard, 1900; Wymer, 1968, 1985b; West, 1969; Hollin, 1977). The principal source of palaeoliths has been Globe Pit, situated at the extreme eastern end of the brickearth spread, where a rich Clactonian industry has been recognized (King and Oakley, 1936; Wymer, 1957). The artefacts occur principally in gravels underlying the brickearth, but have also been reworked into the latter deposit (Wymer, 1957, 1968, 1985b; Snelling, 1964). The artefact-bearing gravel is believed to be part of the Lynch Hill/Corbets Tey Formation, of mid-Saalian age (Bridgland, 1988a; see above, Introduction and Purfleet). The recovery of a large Clactonian assemblage from the mid-Saalian Corbets Tey Gravel is somewhat problematic, as this industry has normally been associated with late Anglian or Hoxnian sediments, such as the Lower Gravel at Swanscombe and the Clacton Channel Deposits (Wymer, 1974; see above, Swanscombe; Chapter 5, Clacton). This has led to suggestions that the deposits at Little Thurrock are older than their position within the terrace sequence suggests (King and Oakley, 1936; Wymer, 1968).

Description

The surviving Pleistocene sediments in Globe Pit are concentrated in two areas. At the northern

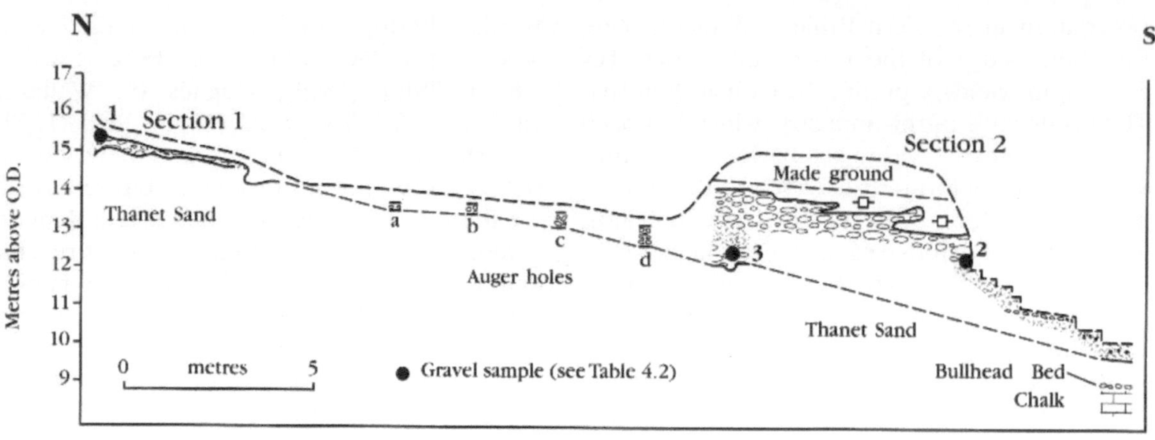

Figure 4.22 The GCR sections at Globe Pit.

edge of the site, in a small area occupied by allotments, a deposit mapped as Boyn Hill Gravel overlies Thanet Sand at 20.5 m O.D. (see Fig. 4.22). There are no clear records of artefacts from this deposit (Wymer, 1968, 1985b), which has been confirmed recently as part of the Boyn Hill/Orsett Heath Gravel Formation (Bridgland, 1988a). The Globe Pit GCR site is limited to an elevated area on the north-eastern side of a large Chalk quarry, behind the gardens on the south side of Overcliff Road, where a surviving remnant of Corbets Tey Gravel is located. In this area the Pleistocene deposits, which thin rapidly northwards, are banked against Palaeogene Thanet Sand (Fig. 4.22). The area has been partly excavated for gravel and it is difficult to ascertain how much of the original land surface remains. The gravel is overlain in the southern part of the site by a wedge of unbedded clayey sand, probably the feather-edge of the Grays brickearth (Fig. 4.22). Published descriptions of the site by West (1969) and Hollin (1977) indicate that *in situ* deposits formerly extended further south, where brickearth containing Mollusca and pollen of temperate-climate affinities occurred. Lamentably, despite the site having SSSI status since the 1950s, this fossiliferous material was entirely quarried away by 1980.

The earliest detailed description of exposures in the Pleistocene deposits at Little Thurrock was by Morris (1836), who recorded laminated beds with comminuted shell debris amongst various sediments occupying a 'valley' between the higher ground to the north (formed by the Chalk of the Purfleet Anticline) and a much lower 'ridge' of Chalk to the south. The latter was clarified by Dewey *et al.* (1924), who noted that the channel filled with brickearth is separated on its southern side from the alluvium of the modern valley by a low gravel-covered ridge of Chalk. This gravel, which was described by Tylor (1869), may form part of the Taplow/Mucking Formation (see Introduction to this chapter; Fig. 4.23).

The first published illustration of a section in Globe Pit was by Hinton and Kennard (1900, p. 364). This showed gravel and brickearth of their 'Middle Terrace Series' overlying material that, in the caption to the illustration, they termed 'Gravel and Sand washed down from [the] valley to the north (High Terrace Series derived)'. This last bed, which formed the northern edge of the Pleistocene deposits in Hinton and Kennard's section, was probably the equivalent of the gravel at the GCR site. Descriptions of the site have been provided in recent years by Wymer (1957, 1968, 1985b), Snelling (1964), West (1969) and Hollin (1977). The most detailed section through the sequence, most of which is now quarried away, was illustrated by Wymer (1985b). This showed brickearth above a lower gravel resting on Chalk at 6 m O.D. and overlain by a later gravel (Fig. 4.23). The lower gravel is shown to extend higher up the valley-side than the later sediments, where it overlies a narrow, higher 'bench' cut in Thanet Sand at 15 m O.D. Wymer (1957, 1968, 1985b) considered the gravel on this higher bench to be older than that at 6 m O.D. and regarded the deposits covering the slope between the two as of colluvial origin.

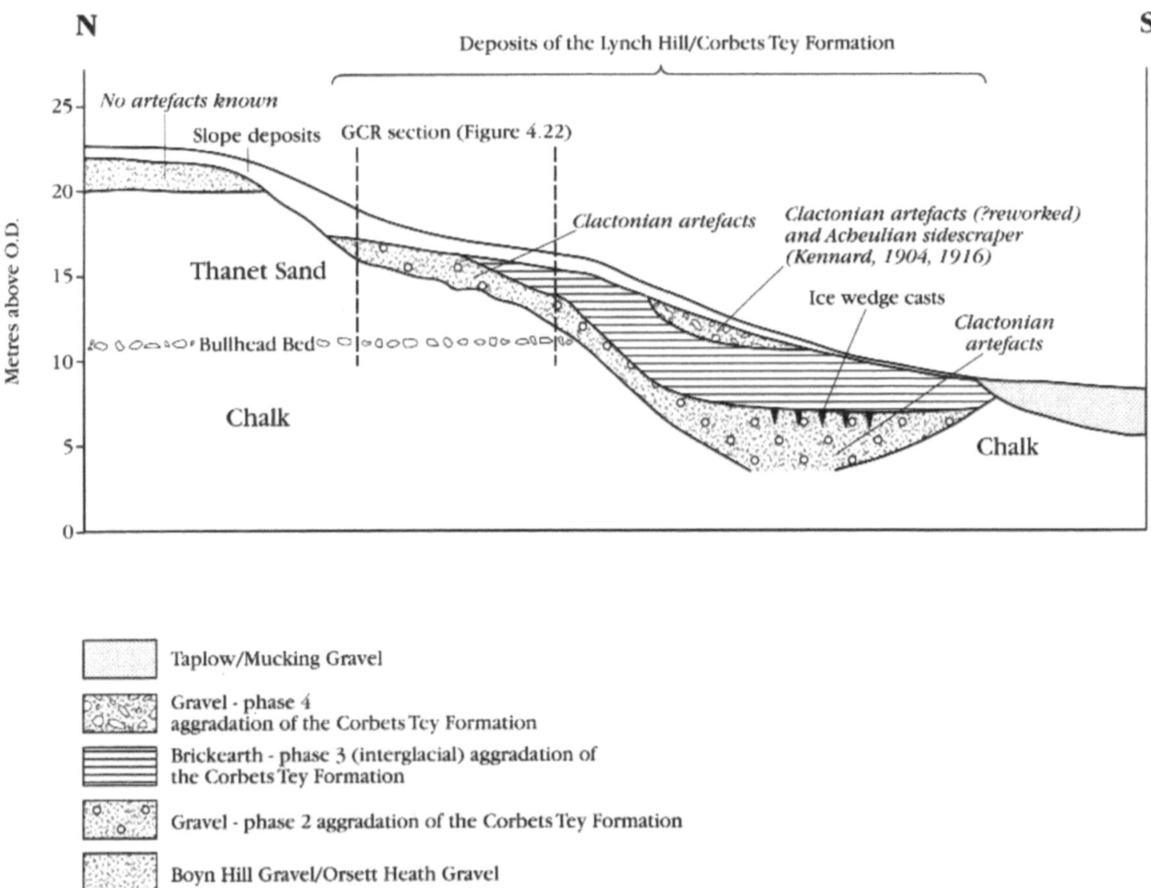

Figure 4.23 Section through the Pleistocene deposits in the area of Globe Pit (after Wymer, 1985b).

This interpretation, which was supported by West (1969) and Evans (1971), would appear to link back to the above quotation from Hinton and Kennard, who were presumably describing the same sloping gravel body.

Reappraisal of the site in 1983, as part of the Geological Conservation Review (Anon., 1984b), has raised serious doubts about this interpretation. Sections cut in the remaining deposits indicated that fluvially bedded sand and gravel, albeit with penecontemporaneous deformation structures, can be traced to below 10 m O.D. (Figs 4.22 and 4.24). Although the deposits further south, which extended down to 6 m O.D. (and, according to Tylor (1869), to well below ordnance datum), have now been quarried away, the GCR section extended well into the material interpreted by Wymer as a slope deposit. The recognition that *in situ* fluvial sediments extend to below 10 m O.D. leads to the suggestion that the 15 m 'bench' described by Wymer is simply the feather-edge of a much thicker aggradational sequence, that represented by the Corbets Tey Formation as a whole (see Fig. 4.23).

Wherever the Thanet Sand surface was uncovered at Globe Pit in the 1983 excavations, it proved to be extremely uneven, with what appeared to be 'potholes' in the old river bed at the base of the gravel. However, in a larger area of Thanet Sand surface that was uncovered in 1984 (section 1; Fig. 4.22), there were indications of a linear trend to the undulations. The largest of these features, at the southern end of section 1, appears to be coincident with the 'step' in the bedrock surface observed by Wymer (1957). This feature has been undercut on its northern side. Few previous descriptions of the form of bedrock surfaces beneath Pleistocene gravels have been published (see, however, Chapter 2, Wolvercote). Harding and Gibbard (1984) provided a record of similar features at

Figure 4.24 GCR Section 2 at Globe Pit, showing bedded gravel extending to below 10 m O.D. (looking north). The surveying staff rests on the exhumed Thanet Sand surface. (Photo: P. Harding.)

Stoke Newington; as at Globe Pit, these features, which otherwise resembled potholes, had a linear trend and showed undercutting. They were attributed by Harding and Gibbard to fluvial erosion of the London Clay. A similar explanation can therefore be offered for the features at Globe Pit. A summary of the tripartite sequence formerly exposed in this part of Globe Pit can therefore be given, as follows:

3. Upper gravel (now removed)
2. Fossiliferous brickearth (now removed)
1. Lower Gravel, containing Clactonian artefacts (feather-edge survives).

This sequence is assigned here to the Lynch Hill/Corbets Tey Formation. The clast-lithological contents of the basal Corbets Tey Gravel at Globe Pit, as well as those of the Orsett Heath Gravel in the northern part of the workings (outside the GCR site), are recorded in Table 4.2. Both deposits have clast compositions typical of Lower Thames gravels downstream from the Darent confluence.

Interpretation

The principal scientific interest in the remaining sediments at Globe Pit, in addition to the important evidence they provide for the depositional history and stratigraphy of the Thames terrace system, arises from the Clactonian artefacts they contain. Considerable emphasis has been placed on the stratigraphical significance of this assemblage by past workers. King and Oakley (1936) regarded the occurrence at Little Thurrock of an uncontaminated Clactonian industry as evidence that the deposits there were older than the Lower Middle Gravel at Swanscombe, which is at the 'High Terrace' level. They concluded that the Little Thurrock sediments, although at a 'Middle Terrace' elevation, filled a channel excavated by the Thames during the interval between the deposition of the Swanscombe Lower Loam and Lower Middle Gravel, during which time the upper surface of the Lower Loam was subjected to subaerial weathering (see above, Swanscombe). The well-known channel deposits at Clacton, containing the type-Clactonian industry, were attributed to the same time interval, their 'Clacton-on-Sea Stage'. This model, although rejected by Marston (in Bull, 1942), was supported by Oakley and Leakey (1937) and Warren (1955). It implied (1) that much of the 'High Terrace' (now recognized as Boyn Hill/ Orsett Heath Gravel), which contains Acheulian hand-axes, was younger than the 'Middle Terrace' deposits at Little Thurrock; (2) that aggradation had been continuous, following deposition of the Little Thurrock deposits, until reaching the highest level of the 'High Terrace', at *c.* 42 m O.D. at Dartford Heath; and (3) that the geomorphological 'Middle Terrace' feature had resulted from a combination of erosion and deposition, as the river subsequently incised its valley for a second time to the 'Middle Terrace' level. The Clactonian industry was, until recently, considered to appear earlier in Britain than the Acheulian (see Wymer, 1974), so its presence at Little Thurrock, uncontaminated by Acheulian material, provided an important basis for the above interpretation of the aggradational sequence.

Early records of archaeological material from Little Thurrock, although rarely containing details of location, probably refer to the Globe Pit or sites nearby. Spurrell (1892, p. 194) described 'numerous "waster" flint flakes' from the easternmost of the Grays pits, 'that at Little Thurrock', and Smith (1894, p. 271) illustrated a 'worked' red deer antler that, along with numerous other fragments of antlers, bones, tusks, and 'keen flakes and implements', he claimed to have found ***in situ*** on 'the Palaeolithic floor at Little Thurrock'.

The only record of a non-Clactonian artefact from the site is of a 'side scraper', probably of Acheulian affinities, reported to have been found ***in situ*** in the 'Middle Terrace' by Kennard (1904). He recognized that this differed from 'the true Middle Terrace implements' (1904, p. 112) and suggested that it had been reworked from the 'High Terrace' (Boyn Hill/Orsett Heath Gravel). Kennard (1916) later referred to 'gravel overlying the brickearth at the Globe Pit', which he claimed to have yielded a number of implements, including the one he had himself described in 1904. This presumably refers to the upper gravel figured by Wymer (1985b; Fig. 4.23), a record based largely on unpublished observations in the mid-1960s by B.W. Conway. This is an important record, as it indicates that the non-Clactonian artefact was from a different gravel to that which has yielded the extensive Clactonian assemblage (see below).

King and Oakley's (1936) correlation of the Little Thurrock sediments, and the artefacts they contain, with those at Clacton was reaffirmed by Warren (1942, 1955), who further suggested (1947) that the incision event represented by the channels at Grays and Clacton also preceded the aggradation of his 'Furze Platt Stage' deposits in the Maidenhead area. This is an interesting suggestion, since both the Furze Platt deposits (see Chapter 3, Cannoncourt Farm Pit) and the artefact-bearing gravel at Globe Pit are now believed to be part of the Lynch Hill/ Corbets Tey Formation (Table 1.1).

Wymer (1957) pinpointed the source of Clactonian artefacts at Little Thurrock to a small remnant of gravel overlying a 'bench' at 49 ft (15 m O.D.) in the north-eastern corner of

Globe Pit (approximately coinciding with the GCR site). He collected 289 flakes and five 'chopper-cores' from this gravel. Many of the former show secondary working and over half are in mint or fairly sharp condition, indicating minimal transport prior to incorporation in the gravel (Wymer, 1957, 1968). No Acheulian implements or finishing flakes were encountered, leading Wymer to conclude that the collection represents a single industry, with no admixture of material from any other. Wymer's sections were reopened and extended in 1961 by Snelling (1964), when some 280 worked flints were obtained, including two hammerstones, two waste cores and very occasional core-tools similar to those described by Wymer.

Hart (1960), who also described the deposits at Globe Pit, recorded a sequence of brickearth with gravelly partings, with Clactonian artefacts occurring in the latter. In an undated, unpublished report on file with English Nature, Hart recorded gravel containing Clactonian artefacts and mammalian remains occupying a channel with a base level of 6.5 m O.D., well below the level of the 'bench' described by Wymer (1957). The former existence at this level of deposits yielding (mainly sharp) Clactonian material has been confirmed recently by Wymer (1985b; Fig. 4.23).

Biostratigraphy and correlation

Even before the discovery of a Clactonian industry at Little Thurrock, a number of authors had concluded from its mammalian fauna that the Grays brickearth was of a greater age than had usually been attributed to deposits of the 'Middle Terrace' (Hinton, 1910, 1926a, 1926b; Kennard, 1916; Warren, 1923a). Hinton (1910, 1926a, 1926b) placed considerable emphasis on evidence from small mammals in support of this conclusion. This view was perpetuated in the stratigraphical model of the Lower Thames terraces proposed by King and Oakley (1936), who correlated the Little Thurrock deposits with the hiatus between the Lower Loam and Lower Middle Gravel at Swanscombe (see above). Since the Swanscombe sequence has been almost universally ascribed to the Hoxnian Stage (Sutcliffe, 1964; Kerney, 1971), a Hoxnian age for the Little Thurrock deposits is implied by this interpretation. Kerney (1959b) also considered a Hoxnian age to be likely on the basis of similarities between the molluscan fauna at Grays and that in the Swanscombe Middle Gravels. Many elements of the characteristic assemblage from the Middle Gravels at Swanscombe (see above) also occur in the Grays collections; an example is the woodland snail *Macrogastra ventricosa* (Draparnaud), which is restricted, within the Thames system, to these two localities.

The western tract of brickearth in the Grays district was considered by some early workers to be of a later age than that at Little Thurrock, as it yielded different mammalian species (particularly voles) and was associated, at West Thurrock, with a Levallois industry (Kennard, 1916; Warren, 1923a, 1923b; see below, Lion Pit). Both sets of deposits have, however, been attributed in recent years to the Ipswichian Stage, principally on the basis of palynology (West, 1969; Hollin, 1977; Gibbard *et al.*, 1988). West (1969) obtained pollen from the brickearth at Globe Pit (now destroyed), which was banked against the gravel that yields Clactonian artefacts (Fig. 4.23). He recorded a section similar to that described by Hart, with a base level of 9 m O.D., comprising up to 0.5 m of gravel, overlain by 3 m of brown silt and sand (brickearth) containing *Corbicula fluminalis.* The pollen showed the brickearth to have accumulated under interglacial conditions, but was insufficiently diagnostic to distinguish between the Hoxnian and Ipswichian Stages, both possible interpretations on the basis of the molluscan evidence. However, since brickearth at comparable elevations at Aveley and Ilford, respectively 8 km and 19 km upstream, had yielded Ipswichian pollen sequences (West *et al.*, 1964; West, 1969), West favoured a similar age for the Little Thurrock deposit. This suggestion was disputed by Conway (1970b), who regarded the Clactonian artefacts, found throughout the lower gravel and in the brickearth (Fig. 4.23), as evidence for a Hoxnian age, as originally implied by King and Oakley.

The rich mammalian fauna from the Grays and Little Thurrock brickearth, although much celebrated by early collectors (lists were supplied by Whitaker (1889), Hinton and Kennard (1900) and Hinton (1926b)), has failed to provide clear biostratigraphical evidence for the age of the deposits. It has therefore been possible to reconcile the assemblage with attributions to both the Hoxnian and Ipswichian stages. The view that this eastern spread of brickearth is older than that to the west of Grays (Kennard, 1916; see below, Lion Pit) has, however, persisted (see Wymer, 1985b). The record of hippopotamus from Little Thurrock is of considerable significance, as it would seem to indicate correlation with the Ipswichian Stage (*sensu* Trafalgar Square); however, the record of horse from the same deposit appears to be contradictory, since that animal is believed to have been absent during the Ipswichian – as was the bivalve *Corbicula fluminalis*, also present at Little Thurrock (Chapter 1; Chapter 2, Stanton Harcourt and Magdalen Grove). Wymer (1985b) noted that there are many difficulties in assessing the early faunal collections from the Grays district, as the precise provenance of many specimens is unknown. He was therefore inclined to dismiss the record of hippopotamus from the Little Thurrock brickearth. There is considerable justification for this revision: Abbott (1890) had recorded hippopotamus remains from West Thurrock (although this record must also be regarded as doubtful – see below, Lion Pit) and Hinton and Kennard's (1900) faunal assemblage for the Grays deposits, the basis of most later lists, was an amalgamation of records from West Thurrock and Grays and Little Thurrock. Only later did these authors realize that the brickearths to the west and east of Grays were of different ages (Kennard, 1916; above). In fact, hippopotamus appears in faunal lists from most fossiliferous sites in the Lower Thames, including Swanscombe, where the record was based on a single fragment (subsequently discredited by the analysis of its fluorine content (Oakley and Gardiner, 1964; Sutcliffe, 1964)). Thus Hinton (1926a, p. 339), referring to two of the rare exceptions, was able to state that 'Hippopotamus ..., so characteristic of the earlier Thames horizons, [has] not been found at Crayford or Erith'. The opposite view would now be taken; hippopotamus is currently regarded as present only in the latest of the Thames interglacial deposits (Chapter 1).

Hollin (1971, 1977), who obtained further supplementary pollen samples from Globe Pit and elsewhere in the Grays brickearth (*sensu lato*), considered their high pine and low birch frequencies to support West's correlation of the deposits with the Ipswichian Stage. He interpreted the brickearth as a tidal deposit and suggested that, as at Purfleet (see above), it recorded estuarine aggradation to 14 m O.D. in response to a rise in sea level brought about by an Antarctic ice surge during the Ipswichian. The interpretation of this and other sites as representing the infilling of the Lower Thames valley during an Ipswichian sea level rise has recently been supported by Gibbard *et al.* (1988).

The correlation of the Grays and Little Thurrock sediments with the Ipswichian Stage is not supported by amino acid ratios from shells from the early collections (Miller *et al.*, 1979; Bowen *et al.*, 1989). Both sets of authors obtained D : L ratios of around 0.29 (using the genera *Corbicula* and *Bithynia*) and both grouped the site with Swanscombe, which, according to the interpretation of Bowen *et al.* would imply correlation with Oxygen Isotope Stage 11 (see above, Swanscombe; Chapter 1).

Terrace stratigraphy

Bridgland (1983a) considered the Little Thurrock sequence to fall within his Barvills Gravel, which was later reclassified as Corbetts Tey Gravel (Bridgland, 1988a; Gibbard *et al.*, 1988). This formation was correlated by Bridgland (1988a) with the Rochford Gravel of the Southend area, implying correlation of the basal gravel and interglacial sediments (brickearth) at Little Thurrock with the Rochford Channel Gravel and Rochford Channel interglacial deposits respectively. This correlation has recently been retracted, the Corbets Tey Gravel now being regarded as equivalent to the Barling Gravel of the Southend area (Bridgland *et al.*, 1993; see Chapter 5, Part 2). The channel at Grays, recognized by early workers such as Morris (1836), Tylor (1869) and Dewey *et al.* (1924), is thus regarded as an upstream equivalent of the Shoeburyness Channel. Since the Corbets Tey Gravel is correlated with the mid-Saalian Lynch Hill Formation of the Middle Thames (Bridgland, 1988a), this interpretation provides a stratigraphical argument for an intra-Saalian age for the Grays interglacial (Table 1.1).

The above correlations provide the basis for part of the revised stratigraphical scheme for the Thames terrace sequence presented in Chapter 1. Besides the Shoeburyness Channel Deposits, which have been investigated in detail only very recently (H.M. Roe, pers. comm.), the Little

Thurrock interglacial sediments are considered, according to this scheme, to correlate with temperate-climate deposits at Stoke Newington, Corbets Tey, Belhus Park and Purfleet, all of which have yielded comparable molluscan faunas with *C. fluminalis* (see above, Purfleet; Table 1.1).

The Globe Pit GCR site, however, retains little if any evidence to bear on the age of the Grays interglacial. Its main importance, as stated above, stems from the Clactonian artefacts that occur in the gravel underlying the interglacial beds. This gravel (bed 1, above) is presumed to represent the pre-interglacial aggradational phase (phase 2) of the Lynch Hill/Corbets Tey Formation, according to the climatic model for terrace formation presented in Chapter 1. The equivalence of the gravel underlying the brick-earth and that from which Wymer (1957) and Snelling (1964) obtained Clactonian artefacts was demonstrated by the GCR excavations. The gravel that was observed above the brickearth (Kennard, 1916; Wymer, 1985b; Fig. 4.23) is assumed to represent the post-interglacial (phase 4) part of the formation (and is therefore equivalent to the upper gravel at Belhus Park – see above, Purfleet). According to the stratigraphical scheme favoured here, this later gravel dates from Oxygen Isotope Stage 8, whereas the gravel with Clactonian artefacts dates from Stage 10; the interglacial beds are attributed to Stage 9 (Fig. 4.3).

The interpretation outlined above clearly implies that the Clactonian artefact-bearing gravel (bed 1) at the GCR site post-dates the higher-level Orsett Heath Gravel, which contains hand-axes, thus refuting the stratigraphical significance of the former industry. Of probable relevance to this argument is the recovery of two possible hand-axe finishing flakes in abraded condition from the GCR excavations (Bridgland and Harding, in press), which appears to confirm that the Little Thurrock gravel post-dates hand-axe manufacture in the area. However, the question remains as to whether the Clactonian industry at Globe Pit is contemporaneous with the Corbets Tey Gravel or whether it is derived from an older deposit, of similar age to the Swanscombe Phase I sediments. The balance of evidence suggests that the Globe Pit assemblage results from the accumulation, in the feather-edge of the Corbets Tey Gravel, of a vast amount of material from a nearby Clactonian working site. The Corbets Tey Formation would be expected to contain reworked Palaeolithic material from the higher Orsett Heath Gravel, which may be represented by the hand-axe finishing flakes. The relation between the age of the Clactonian working site and the age of the gravel cannot be determined. The archaeological material may have been preserved for a considerable time in an earlier deposit in the vicinity and then reworked into the Corbets Tey Gravel. The largely unabraded condition of the artefacts implies minimal transport by the Thames, which is in tune with their apparent concentration at the edge of the channel (whether they are contemporaneous or reworked).

Abraded Clactonian material has also been recognized within a mixed assemblage of artefacts from interglacial and post-interglacial deposits of the Corbets Tey Formation (phases 3 and 4) at Purfleet (Palmer, 1975; Wymer, 1985b; see above, Purfleet) and from the upstream correlative of this aggradation, the Lynch Hill Gravel of the Middle Thames (Wymer, 1988). Wymer (1988, p. 89) suggested that 'a Clactonian Industry of crude chopper-cores and flakes, without hand-axes, is present in the Lynch Hill Gravels, as several such artefacts came from this gravel at Deep Lane, Burnham'. There is mounting evidence, therefore, that the use of the Clactonian knapping technique persisted after deposition of the Swanscombe Phase I deposits. If fresh Clactonian material is restricted, in the Lynch Hill/Corbets Tey Formation, to the pre-interglacial (phase 2) deposits, the implication is that the industry persisted into the next oxygen isotope stage after that represented at Swanscombe, Oxygen Isotope Stage 10 (Table 1.1), to be replaced by Levalloisian knapping practices in Stage 8 (see above, Purfleet). The possibility that this is the case requires further stratigraphical investigation, however.

Summary

This review of previous descriptions and interpretations of the Little Thurrock site reveals that, despite considerable attention from geologists and archaeologists, the age of the deposits and their position within the Lower Thames terrace succession remains controversial. The role of Palaeolithic archaeology as a potential source of stratigraphical and relative dating evidence is currently under review. The Globe Pit industry provided an important foundation for past stratigraphical reconstructions of the Lower Thames sequence. The deposits were held to be older than their position in the 'Middle Terrace' (Corbets Tey Gravel) suggests and were frequently attributed to the Hoxnian Stage. Other workers, influenced by the relatively low elevation of the site, favoured an Ipswichian age. Interpretations of the archaeological and palaeontological evidence from these deposits have been revised in recent years and it now appears that some, perhaps all, of the early views were erroneous.

In current attempts at correlation between the Thames sequence and the deep-sea (oxygen isotope) record (Chapter 1), evidence from Globe Pit continues to be of considerable importance. In particular, the site complements the palaeontological evidence from Purfleet, which is considered to be its broad correlative. The two provide a picture of Lower Thames development during the aggradation of the Lynch Hill/Corbets Tey Gravel Formation, which appears to have occurred between Oxygen Isotope Stages 10 and 8 (inclusive).

Conclusions

The remaining deposits at Globe Pit provide an important reserve of gravel rich in Lower Palaeolithic (early Stone Age) flint artefacts of Clactonian type (named after another GCR Thames site, at Clacton in Essex). These comprise flakes and 'cores' (pieces of flint from which flakes have been removed), rather than the crafted tools that appeared in more advanced industries. The occurrence of these primitive artefacts at this site has been the subject of considerable interest for many years. The gravel here represents the middle of three terraces recognized on the north side of the Lower Thames valley, which would normally suggest that it was intermediate in age between the higher (older) terrace and the lower terrace. However, the high terrace at Swanscombe yields numerous hand-axes, part of an industry that is traditionally regarded as advanced and later than that found at Globe Pit. This led many workers to devise complex explanations of how the deposits here, at Little Thurrock, could be older than their position in the terrace sequence would seem to suggest. Reconsideration of the Lower Thames terrace sequence and the fossiliferous sediments within it, as well as the archaeological evidence from the gravels in the Lower and Middle Thames, leads to the conclusion that the gravel and the Palaeolithic industry at Globe Pit are indeed younger than the entire Swanscombe sequence. It is suggested here that they date from early in the Saalian, at around 350,000 years BP.

LION PIT TRAMWAY CUTTING (WEST THURROCK; TQ 598783)

D.R. Bridgland and P. Harding

Highlights

A spectacular Pleistocene section occurs at this site, showing interglacial deposits sandwiched between Thames terrace gravels, all banked against a fossil Chalk cliff. The entire sequence belongs to the Taplow/Mucking Gravel Formation, dating from the mid to late Saalian. The lower gravel, which is believed to date from Oxygen Isotope Stage 8, contains the slightly disturbed debris of a Levallois (Palaeolithic) working floor. The interglacial beds are attributed to Stage 7, but this is controversial, some workers regarding them as Ipswichian (Substage 5e) in age.

Introduction

This site, the second associated with the brickearth deposits of the Grays district (Fig. 4.21), is part of an old tramway cutting leading into one of the many Chalk quarries in the area. This cutting provides sections through Pleistocene deposits banked against a fossil Chalk cliff at the northern end of the site. These deposits comprise a sequence of silts and sands, possibly of intertidal origin, with an intervening bed of silty clay, thought to represent the fossiliferous West Thurrock brickearth described at the end of the last century (Whitaker, 1889; Abbott, 1890). The latter has yielded mammalian, molluscan and pollen assemblages indicative of an interglacial environment (Abbott, 1890; Hinton and Kennard, 1900; Carreck, 1976; Hollin, 1977; Gibbard *et al.*, 1988). Beneath this sequence is a basal gravel that contains, near the fossil river cliff, abundant Levallois knapping debris in a near-primary context (Warren, 1923a, 1923b, 1942). At the southern end of the cutting the above sequence is overlain by a later gravel, the stratigraphical interpretation of which is disputed (Bridgland, 1988a; Gibbard *et al.*, 1988), although it is here assigned to the Taplow/Mucking Formation.

A number of similar and, presumably, broadly equivalent sections have been described, mainly before the First World War (Whitaker, 1889; Abbott, 1890; Hinton and Kennard, 1900), although in the early 1980s a new road cutting, 0.9 km to the west of the GCR site, provided a further opportunity to study the West Thurrock sequence. Important differences between the faunal and archaeological evidence from West Thurrock and that from Grays and Little Thurrock were recognized at an early date, leading to the conclusion that the deposits at West Thurrock are the younger (Kennard, 1916; Warren, 1923a). However, following the establishment of the pollen-based stratigraphical scheme for the British Pleistocene (West, 1963, 1968; Mitchell *et al.*, 1973), many recent authors have interpreted the deposits both to the west and to the east of Grays as part of a 'Middle Terrace' Ipswichian aggradation (Carreck, 1972; Hollin, 1977; Gibbard *et al.*, 1988). Others have continued to regard the Grays and Little Thurrock deposits as older, probably of Hoxnian age (see above, Globe Pit), but the dating of the West Thurrock brickearth as Ipswichian has been widely accepted. Only with the realization that there were more temperate intervals in the Middle and Late Pleistocene than had been hitherto recognized (Chapter 1) has it been suggested that sediments such as those at West Thurrock might represent additional, unidentified interglacial episodes.

Description

The West Thurrock sequence has been exposed at a number of different locations, mainly as a result of Chalk quarrying activities (the extent of which is readily apparent on the Geological Survey map – New Series, Sheet 271), but also in railway (Holmes, 1890) and road cuttings. As with the deposits at Grays and Little Thurrock (see above), there is an extensive history of research on the West Thurrock brickearth sections. Whitaker (1889) described Pleistocene beds overlying and banked against the Chalk at several sites, including the tramway cutting at the Lion Cement Works Pit (this appears to be the earliest reference to the GCR site).

Abbott (1890) published a description of a section at the West Thurrock Tunnel Cement Works, further to the west than the sites described by Whitaker. His section showed a lower gravel overlain by thick, partly cross-bedded sands with clay seams, overlain in turn by coarse gravel. The sand was reclassified as brickearth by Dines (in Dewey *et al.*, 1924). Abbott claimed to have noticed flint artefacts (flakes) in the lower gravel and also recorded mammalian fossils from the section. These were ***Palaeoloxodon antiquus***, ***Mammuthus primigenius***, ***Bos primigenius***, ***Cervus elaphus***, ***Bison?*** sp., ***Hippopotamus amphibius***, ***Dicerorhinus kirchbergensis***, ***D. hemitoechus*** and ***Coelodonta antiquitatis***. Abbott also described several species of mollusc from this section (see below). In recent years Abbott's faunal collection has been located and reassessed (Carreck, 1972; below). This record is of particular importance, as it is one of very few from the heyday of fossil mammal collection in the Grays area in which the West Thurrock and Little Thurrock brickearths are clearly distinguished. As noted above (see Globe Pit), most early faunal lists are amalgamations of collections from these two deposits, which were only later discovered to be of different ages. To Abbott's mammalian assemblage from the Tunnel Cement Works can be added horse, identified from the brickearth at the Thames Works Quarry (Hinton, 1901).

Hinton and Kennard (1900) described a number of sections in the western development of the Grays 'brickearths'. These were: (1) the Tunnel Cement Works (first described by Abbott, 1890), at the extreme western end of the brickearth outcrop (TQ 575777); (2) a pit west of Milwood (now Mill) Lane, first described by Whitaker (1889) and probably the pit known later as the Thames Works Quarry (Fig. 4.21); (3) the tramway cutting leading to the Lion Cement Works (the GCR site); (4) another similar section at the Grays Portland Cement Works (now within the large area of the Grays Chalk Quarries – Fig. 4.21); and lastly (5) a new excavation at the Lion Works, south of the road (and therefore, south of the GCR site), where 4 m of coarse, poorly stratified gravel was observed. This gravel was presumed to overlie the brickearth exposed in the Lion tramway cutting, a relationship that was confirmed in the recent road cutting (see Fig. 4.25; below).

The accuracy of these early descriptions has been largely confirmed by recent studies. Hollin (1977) excavated three profiles in the overgrown sides of the Lion Pit tramway cutting. He illustrated a section showing a Chalk cliff rising from 6 to 16 m O.D., the Chalk apparently shelving to below ordnance datum beneath the terrace deposits to the south. According to Hollin, the latter comprise (near the cliff) 9 m of sand, which is overlain further south by 2 m of clayey brickearth, in turn overlain by an upper sand, which reaches 15 m O.D. These horizons were seen to rise gently towards the cliff, with the entire section capped by 'trail' (colluvial deposits). Hollin recorded freshwater molluscs and ostracods from the base of the brickearth, although higher layers proved sterile. Attempts at obtaining pollen spectra met with little success, although a single countable sample yielded an assemblage rich in *Carpinus*, which Hollin attributed to biozone III of the Ipswichian interglacial.

In 1983/4 a new road cutting, 0.9 km to the west of the GCR site (TQ 590780), provided sections through a very similar sequence, again banked against Chalk on the southern side of the Purfleet Anticline (Fig. 4.25). This enabled the observation of large-scale sediment geometry and stratigraphical relations over the full north-south extent of the surviving interglacial sequence. The deposits here comprise *c.* 2.5 m of grey silty clay (brickearth), its upper two-thirds oxidized to brown. This overlies a massive bed of sand that was proved from 6.5 m O.D. (road level) to 7.9 m O.D. (base of the brickearth). Above the brickearth is a further sand bed, which is in turn overlain by unbedded gravelly, clayey sand of probable colluvial origin. The unoxidized part of the brickearth yields pollen spectra of Ipswichian affinities (P.L. Gibbard, pers. comm.). These deposits are unconformably overlain to the south by a well-bedded, medium to coarse sandy gravel (Fig. 4.25). The similarity of these deposits to those in the Lion Pit tramway cutting was confirmed in April 1984 by excavations at the latter site by the GCR Unit.

The main section excavated in the tramway cutting in 1984 (section 1, Figs 4.26, 4.27 and 4.28) was located at the northern limit of the

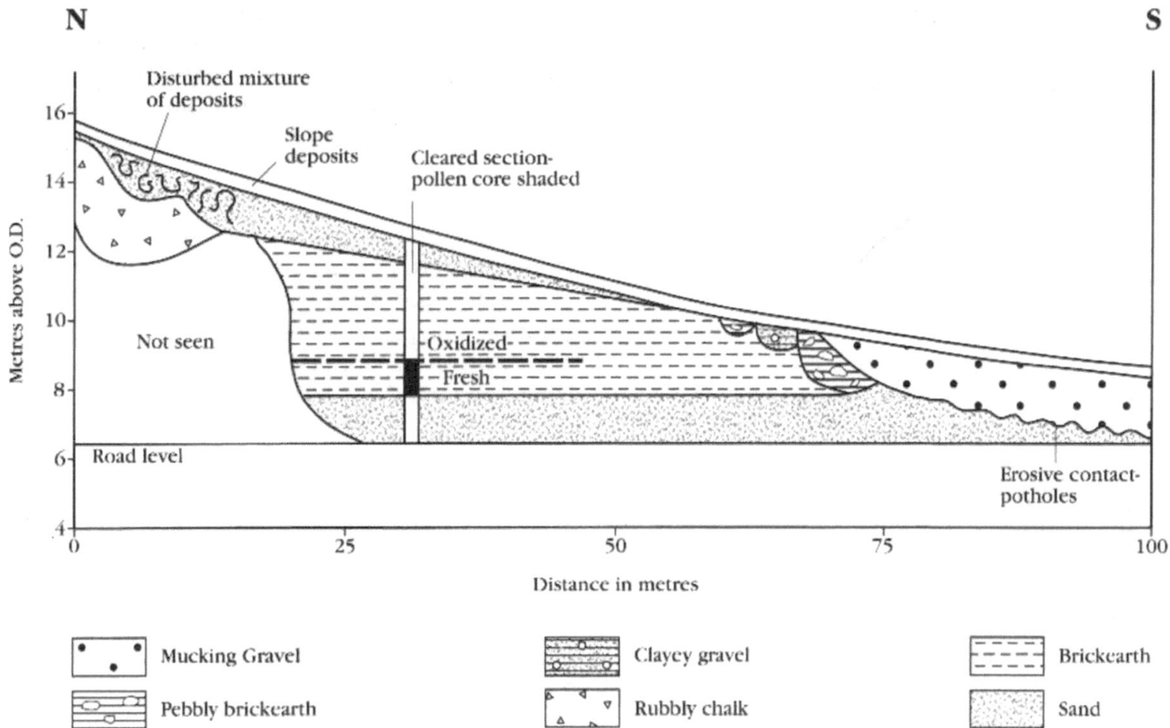

Figure 4.25 Section through the deposits of the Mucking Formation revealed in a road cutting in 1983–4.

Pleistocene deposits, where they abut against the old river cliff. It was in this area that Warren had reported a Levallois working floor. The excavation (Figs 4.27 and 4.28) revealed up to 12 m of well-bedded Pleistocene sediments overlying a surface, eroded in coombe rock, that slopes progressively southwards, although it appears to level off somewhat near the southern end of the section, where it is broken up by scour features (potholes) and/or solution hollows (Figs 4.27 and 4.29). Four separate lobes of coombe rock, the upper two disrupted by solution, project from this sloping surface and are interbedded with the waterlain sediments (Fig. 4.27). Nowhere was the junction between the coombe rock and the solid Chalk observed.

The sequence overlying the coombe rock can be summarized as follows (see also Figs 4.26 and 4.27):

	Thickness
6. Overburden: unbedded gravelly, clayey sand	
5. Upper gravel. Present in sections 2 and 3 only (see below and, for composition, Table 4.2)	up to 3 m
4. Upper sand. Interbedded fine sands and silts, including cross-stratified and ripple-laminated horizons	2.0 m
3. Silty clay (brickearth), unbedded and oxidized	0.5 m
2. Lower sand. Coarse at the base, becoming silty and clayey in higher levels (possibly matrix introduced from bed 3 above), where there are also 'stringers' of small pebbles. The unit is horizontally bedded throughout. Upper 1 m (approximately) forms a distinctive clay-enriched unit, capped with a pebbly layer 0.2 m thick	8.5 m
1. Basal gravel. Contains large, scarcely abraded flint nodules together with smaller gravel clasts in a matrix of sand (see Table 4.2).	up to 1 m
Coombe rock	

The basal gravel (bed 1) is divided into two by a thin seam of horizontally bedded sand. This shows deformation structures, including small-scale normal and reverse faults, possibly the result of settling in response to post-depositional solution of the underlying coombe rock. This basal gravel (Fig. 4.29) yields a large amount of worked flint, including characteristic Levallois artefacts of the type described by Warren (see below; Fig. 4.30). Several of the pieces collected in 1984 have been found to refit together (Fig. 4.30), supporting Warren's claim that a working site existed at West Thurrock.

The basal gravel is overlain by a thick sequence of sands, silts and clays, extending from 2 m O.D. to just over 13 m O.D. (Fig. 4.27). Within this sequence, Hollin's three divisions (lower sand, brickearth and upper sand) can be distinguished (beds 2–4). The lower sand (bed 2) is horizontally bedded throughout, but this is superficially masked by striking post-depositional ferruginous staining, which parallels the coombe rock surface (Fig. 4.27). The upper half of this bed is interbedded with the four lobes of coombe rock. Post-depositional solution of parts of the upper two of these has led to compensatory collapse and associated faulting of the overlying sediments (see Fig. 4.27).

Immediately below the brickearth, which is represented in this section by only 0.5 m of oxidized silty clay, is a distinctive layer of horizontally bedded sand, with clay-enriched seams that transcend the bedding, capped by 0.2 m of coarse pebbly sand. The base of the brickearth slopes markedly towards the Chalk cliff, as do all the higher units. This is probably the result of collapse of the upper beds into voids created by the solution of the coombe rock near the Chalk cliff (Fig. 4.27). This solution was probably also the cause of faulting observed in the upper sand (bed 4). The overlying unbedded gravelly, clayey sand, which is presumed to be of colluvial origin, has been cryoturbated into the top of the upper sand (see Fig. 4.27).

Two smaller sections (sections 2 and 3) further south revealed a sequence similar to that observed in the road cutting (Fig. 4.25). The brickearth (bed 3) was much thicker in both of these sections and was seen to comprise up to 3 m of grey clay, similar in appearance to London Clay. In section 3, opercula of ***Bithynia*** sp. were abundant in the lower part of this bed, but no other faunal or floral remains were re-

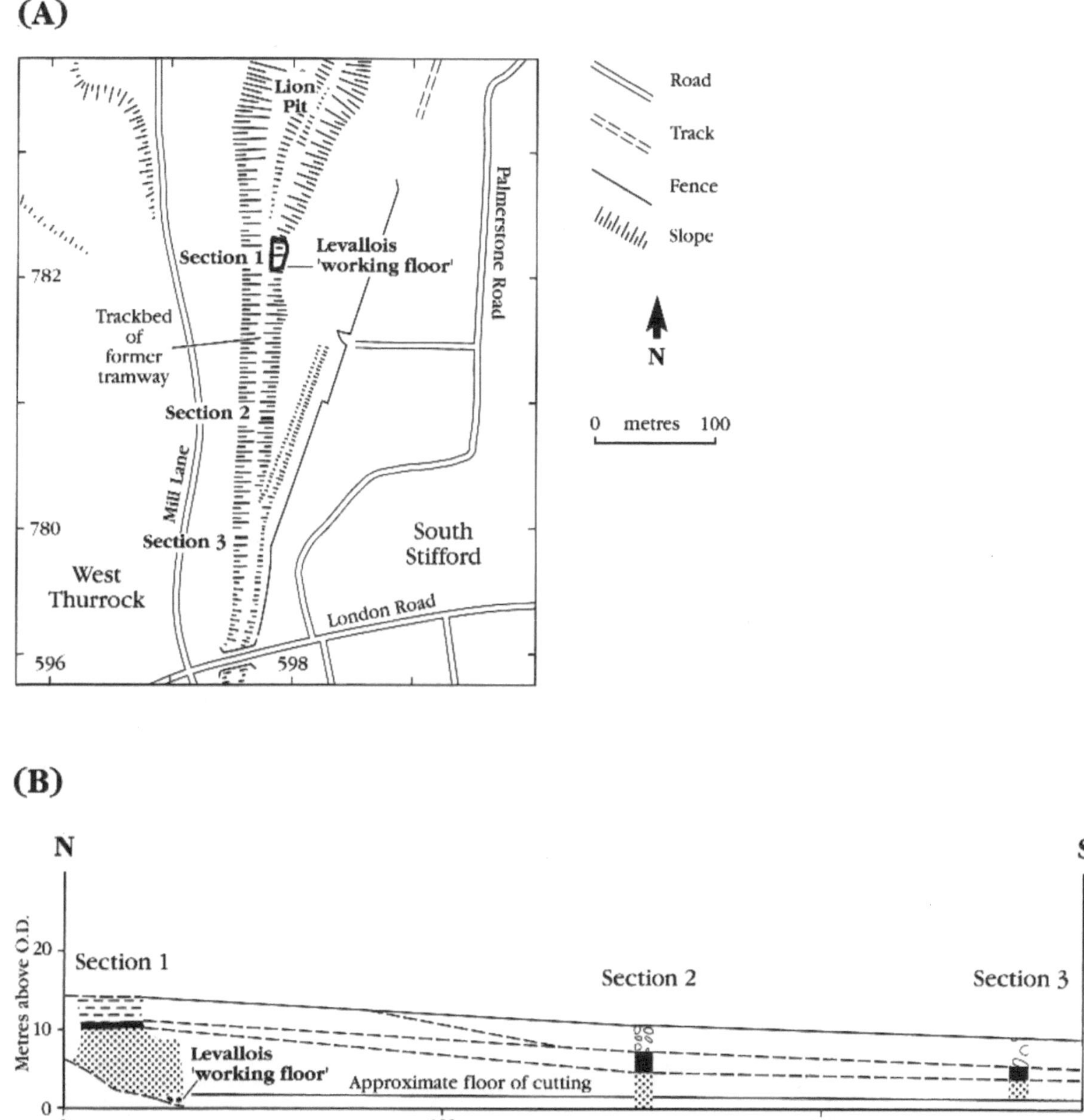

Figure 4.26 (A) Plan and (B) section, Lion Pit tramway cutting, showing the relative positions of the GCR sections and the relations of the various deposits.

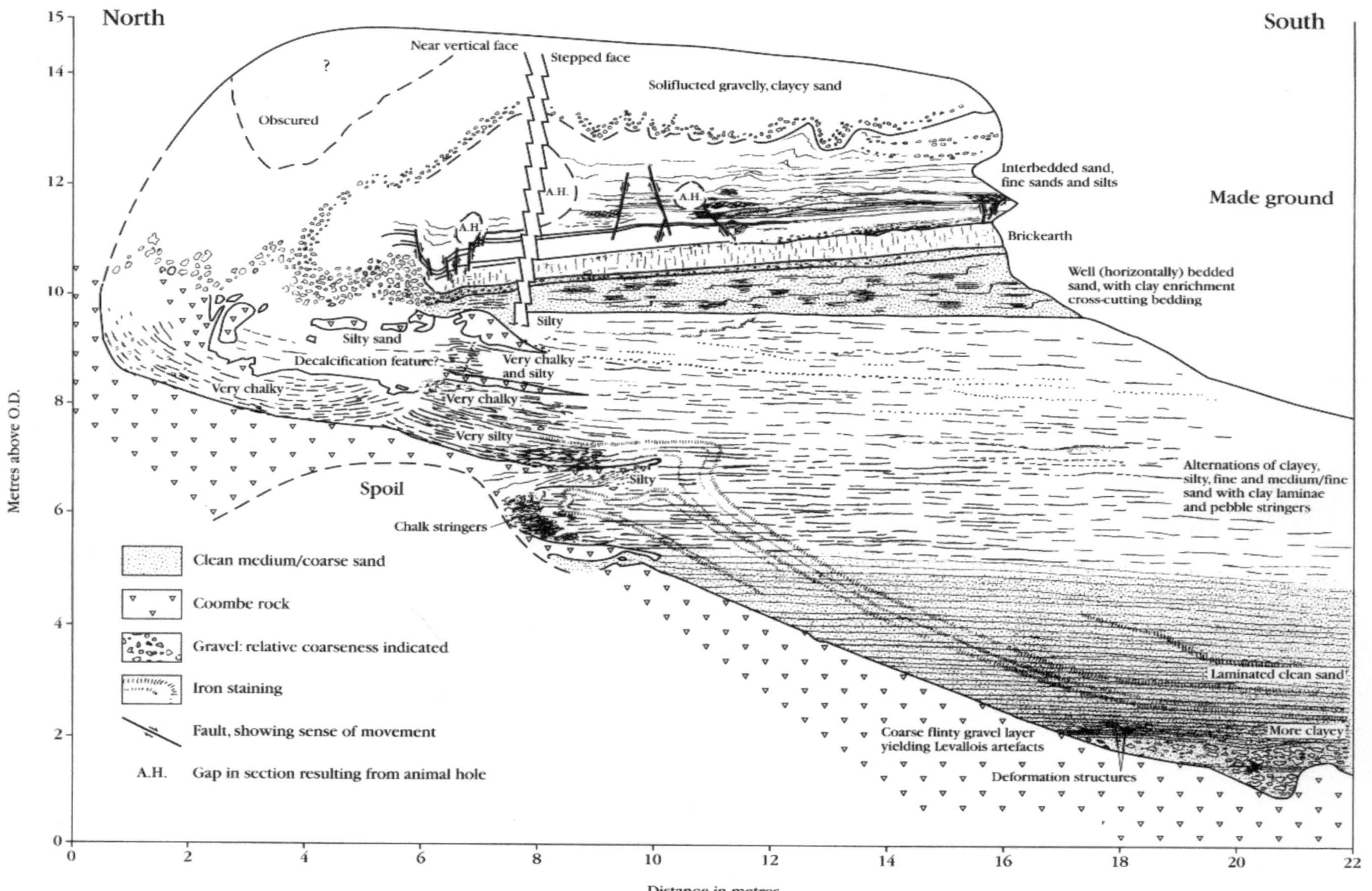

Figure 4.27 GCR Section 1 at the Lion Pit tramway cutting (1984).

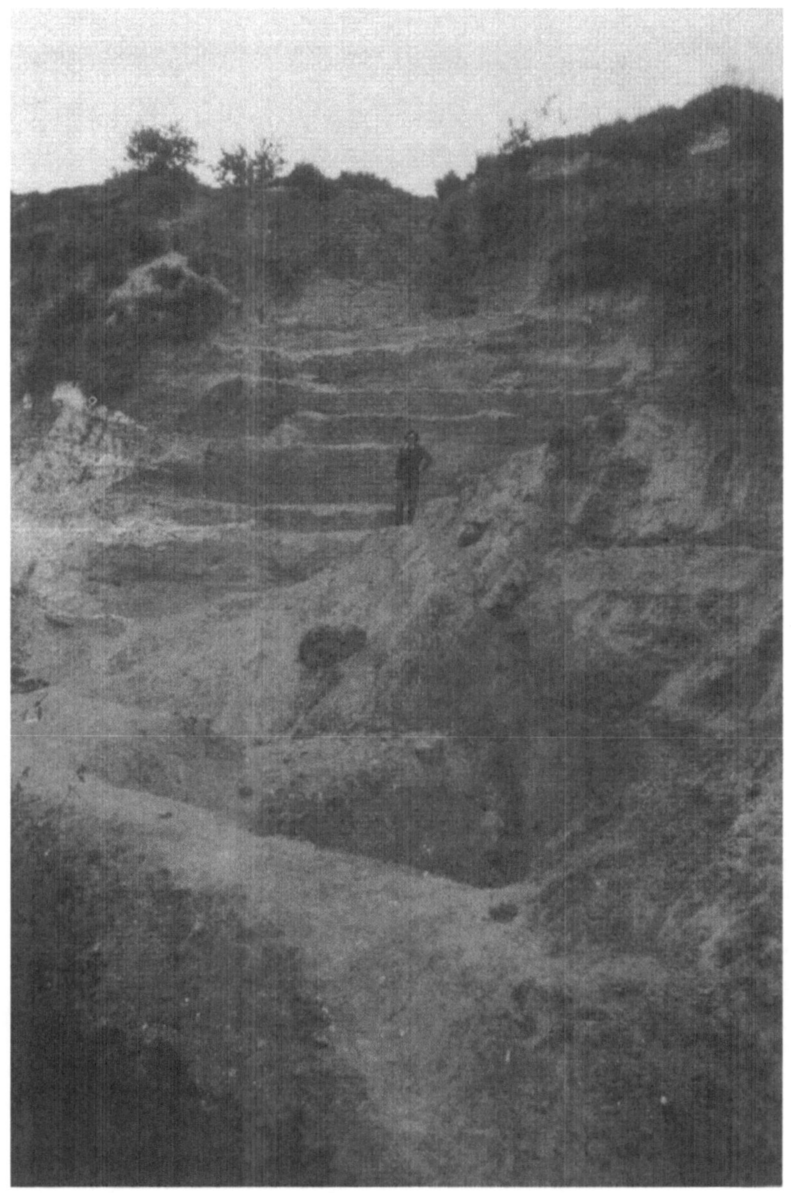

Figure 4.28 Photograph of Section 1 at the Lion Pit tramway cutting. (Photo: P. Harding.)

covered. However, Hollin's section WT1, which was close to (or possibly coincident with) GCR section 3, also yielded *Anodonta* sp. and *Sphaerium* sp. or *Pisidium* sp. (Evans, in Hollin, 1977), as well as the freshwater ostracod *Candona* (Robinson, in Hollin, 1977). The gastropods *Bithynia tentaculata*, *Valvata piscinalis*, *Planorbarius corneus* (L.), *Gyraulus albus* (Müller), *Hippeutis complanatus* (L.), *Lymnaea truncatula* (Müller), *L. peregra* (Müller) and *Pupilla muscorum* are also listed amongst the early collections from the West Thurrock brickearth (Woodward, 1890; Carreck, 1972).

In sections 2 and 3 the lower sand was again observed beneath the brickearth, but the latter was overlain by a medium-coarse bedded gravel rather than the upper sand. This is considered to be the same gravel that was observed cutting out the sand and brickearth sequence in the southern part of the 1983/4 road cutting exposure (Fig. 4.25). In section 2 of the tramway cutting the gravel, less than 2 m thick, was cryoturbated into the top of the brickearth (Fig. 4.26). Although not present at section 1, this gravel represents part of the waterlain succession at West Thurrock and has

Figure 4.29 Excavations at the base of Section 1, Lion Pit tramway cutting. The two layers of coarse flint near the base of the fluviatile deposits are clearly seen, above a potholed surface cut in coombe rock. The Lion Pit Palaeolithic industry occurs in these coarse layers. The material in the top right of the view is made ground. (Photo: P. Harding.)

therefore been added as bed 5 to the sequence described above.

Interpretation

The Lion Pit tramway cutting GCR site combines extensive sections in the stratigraphically significant West Thurrock Pleistocene beds with an important Palaeolithic locality. The archaeological interest here is poorly documented, but the recent reinvestigation by the GCR Unit has shown that it is of considerable importance, probably representing an inundated 'working floor'.

Hinton and Kennard (1900) claimed that several Palaeolithic flakes had been found in this pit, but Kennard (in Dibley and Kennard, 1916) was the first to record Levallois material. According to Kennard, a number of Levallois flakes were found near the base of the old buried cliff. Although he referred to it as the 'Wouldham Cement Company's quarry at Grays', the site visited by Kennard was probably the Lion Pit, which has also been called the 'Wouldham Cement Works' (Hollin, 1977).

During a Geologists' Association excursion visit to the cutting, Warren (1923b) demonstrated the results of subsoil pressure on flint nodules at the base of the Pleistocene gravel, which had produced eolith-like objects by natural flaking. He pointed out that natural flakes, the production of which provide important evidence against the human origin of eoliths, were mingled here with the genuine artefacts of the proto-Mousterian (Levalloisian) industry discovered some years earlier by Kennard. In the same year, Warren (1923a, p. 607) made another passing reference to the 'proto-Mousterian' industry at West Thurrock, stating that it was 'characterized by the familiar 'tortoise-cores' and Levallois flakes' and overlain by beds containing ***Mammuthus primigenius***. Warren (1942) subsequently referred the West

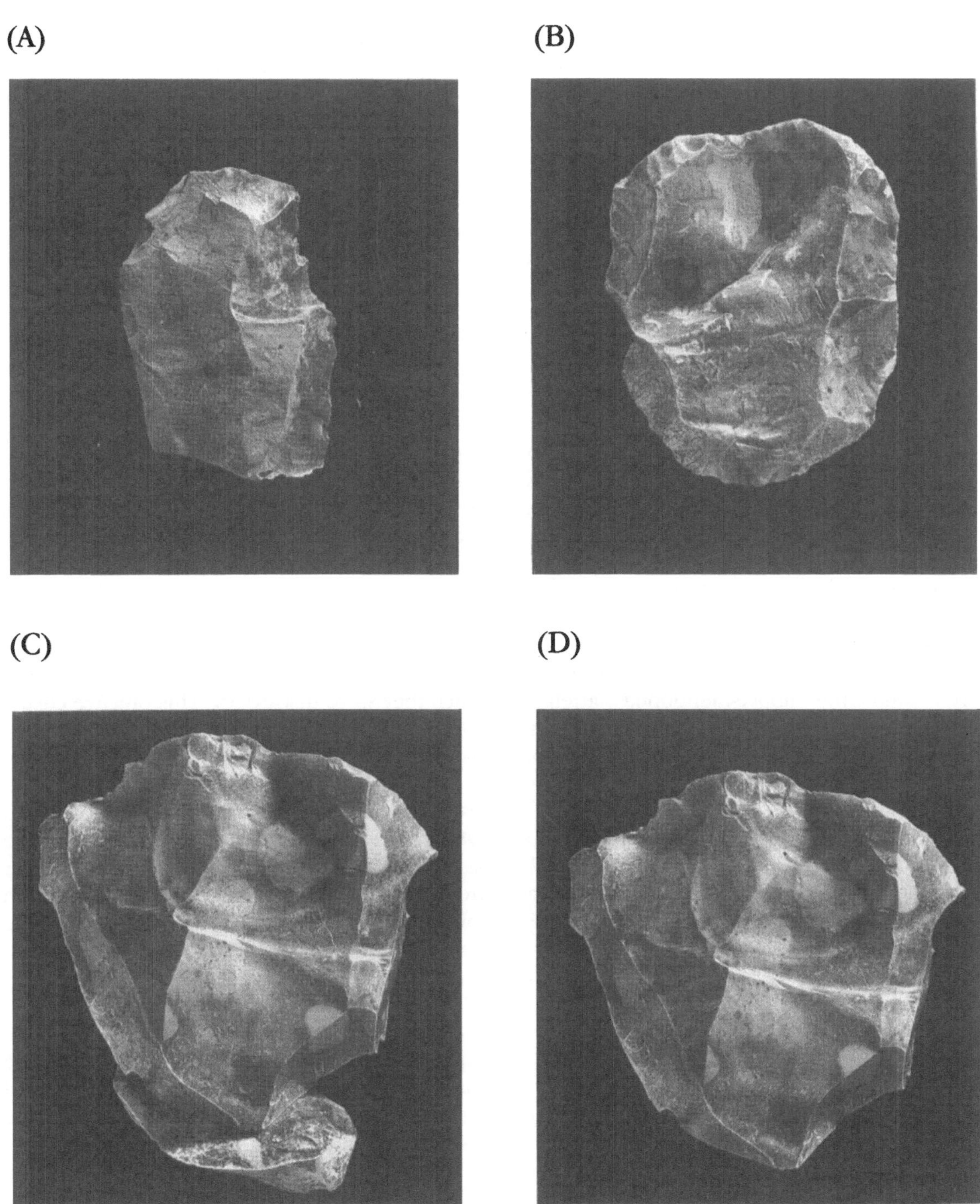

Figure 4.30 Flint artefacts from the 1984 GCR excavation at the Lion Pit tramway cutting. (A) Large broken flake (length 9 cm). The butt (at the top of the view) is faceted, showing evidence of the preparation of the striking platform (this cannot be seen in this dorsal view). At least five scars from previous flaking can be seen. The sharp condition is typical of material from the site. (B) Tortoise core (height 12 cm). This is a classic Levallois core from which a large flake has been detached, thus removing the central part of the 'tortoise', which was formed by radiating flake scars. By preparing a core of this type, a flake of predetermined shape and size has been removed (Levallois technique). (C) Single platform core with refitted flake (height of core plus flake 15.5 cm). The flake, found separately in the gravel, was produced during the shaping of the core. The presence of refitting material at the site is important evidence for the occurrence of knapping debris in a primary context. The striking platform of the core is at the top of the view. (D) The same core as in C, without the refitted flake (height 14 cm). (Photos by Elaine A. Wakefield.)

Thurrock industry to the 'Crayford Stage' of King and Oakley (1936). In this last paper, which included his most detailed description of the locality, Warren interpreted the Palaeolithic artefacts as representing a mid-Levallois working site on an old foreshore, inclined from the foot of the buried cliff towards the river, and covered by fluvial deposits to a height of 50 ft (15 m) O.D. In correspondence with the Nature Conservancy (pers. comm. to W.A. Macfadyen), he described the exact location of the working floor in the tramway cutting, information that enabled the re-exposure of the Levallois level in 1984 (Bridgland, 1985a).

Burchell (1933) recorded the discovery of an 'Upper Mousterian/Aurignacian' flake from loam banked against coombe rock in a cutting south of Belmont Castle, Grays. He noted that these deposits underlay the gravel of the '25 ft Terrace', which would now be referred to as the Mucking Formation. From Burchell's description it appears that this section was probably in the Lion Pit tramway cutting or a comparable exposure nearby. On the basis of the stratigraphy and the above-mentioned artefact, Burchell suggested that the deposits at West Thurrock and the sediments overlying coombe rock at his Northfleet site (see below, Northfleet) were closely related.

In 1984, at the GCR site, a narrow strip (1 m x 4 m) of basal gravel (bed 1, section 1; Fig. 4.27) was excavated under controlled archaeological conditions and yielded 87 flakes and four cores (Fig. 4.29). This more than doubled the known assemblage from the site, which previously comprised 73 flakes, seven cores and one hand-axe located in the Warren Collection at the British Museum. Both the material from the 1984 excavation and the Warren Collection include a number of flakes that are considered to result from natural processes, as described by Warren (1923b, 1923c). The Warren Collection generally comprises larger material, with a higher proportion of cores, than the 1984 assemblage; this difference is suggestive of selectivity in the older collection. The 1984 assemblage therefore appears to come from the only controlled excavation to have taken place at the site.

The 1984 assemblage was predominantly obtained from the upper half of the basal gravel (bed 1), above the interbedded sand (Fig. 4.27). The material includes a number of flakes that can be refitted on to one another or on to cores (Fig. 4.30). The distribution of such conjoinable material indicates some small degree of movement since its manufacture, but the assemblage clearly represents knapping debris that suffered little disturbance during incorporation into the gravel. The condition of the material – 91% was mint or sharp, according to the definition of Wymer (1968) – is consistent with abrasion by suspended particles washing over it, but there is nothing to suggest that the artefacts themselves have been transported. The assemblage is therefore interpreted as a collection of knapping debris that has been preserved in a near primary context following burial by later sediments. Warren's 'working floor' is thus confirmed.

The Palaeolithic material from the Lion Pit tramway cutting is principally a flake and blade industry, some of which has been produced using the Levallois technique. Levallois points and retouched flake tools are also present. The use of this extravagant technique must have been severely constrained by the availability of flint resources. The cores were carefully prepared in order to produce a small number of flakes or blades of a desired size and shape, after which they were discarded. This practice clearly failed to make optimum use of the available raw material. It is therefore likely that the technique was only employed when good quality flint was available in abundance (see, for example, Roe, 1981, p. 81). In the Lion Pit tramway cutting the basal gravel includes a considerably larger proportion of nodular flint than is typical in Lower Thames deposits (Table 4.2), confirming that the supply of fresh flint was indeed abundant at this locality.

The Lion Pit industry may be of considerable stratigraphical significance. The Levallois technique appears to have been first used in the Thames valley at some time during the interval between the Swanscombe interglacial and the Ipswichian Stage (*sensu* Trafalgar Square), within the period referred in this volume to the Saalian Stage (see above, Purfleet; Chapter 1). The recognition of artefacts made using the Levallois technique in the basal gravel of the Lion Pit tramway cutting suggests comparison with Palaeolithic sites at Crayford and Northfleet (see below and Northfleet).

Stratigraphy and correlation

The earliest suggestion that, within the 'Middle Terrace' of the Lower Thames, there occurred deposits of different ages was by Hinton (1910).

Hinton considered that the mammalian faunas from Grays and Little Thurrock did not match those from elsewhere in the 'Middle Terrace', particularly those from Crayford and Erith. The deposits at Grays (at sites such as the Orsett Road and Globe Pits – see above and Fig. 4.21) yielded an assemblage that includes the early vole species, *Arvicola cantiana*. Hinton noted that all the species recognized at Grays were absent from the deposits at Crayford and Erith, believing them to have been replaced in these sediments by a 'later' assemblage. Sutcliffe and Kowalski (1976), however, have pointed out that the Crayford and Erith rodent fauna has cold affinities and so is not directly comparable with that from Grays.

Hinton concluded that the Grays and Little Thurrock deposits were amongst the earliest within the 'Middle Terrace'. This view was supported by Kennard (1916), who established a sequence of four biostratigraphical stages within the 'Middle Terrace', as follows:

4. Crayford and Erith
3. Ilford (Cauliflower Pit)
2. Ilford (Uphall Pit)
1. Grays–Thurrock

Discussing the Grays area, Kennard (1916, p.255) assigned what he called 'the new brickearth to be seen in the various tramway cuttings' to stage 4. This description clearly appertains to the West Thurrock deposits, placing them, along with Crayford, in Kennard's most recent division of the 'Middle Terrace'. Later work, particularly the discovery of the Clactonian industry at Little Thurrock and of the more advanced Levallois industry at West Thurrock, served to consolidate this view (Warren, 1923a; Oakley and Leakey, 1937), although there has been disagreement about the difference in ages between the two Grays aggradations. Some authors have placed the earlier Grays and Little Thurrock deposits, directly or by implication, in the Hoxnian Stage (King and Oakley, 1936; Oakley and Leakey, 1937; Warren, 1942; Zeuner, 1945, 1959; Conway, 1970b; Wymer, 1985b), whereas others have favoured an Ipswichian age for all the Grays deposits (West, 1969; Carreck, 1972, 1976; Hollin, 1971, 1977; see above, Globe Pit).

Carreck (1972, 1976) conducted an extensive review of the available information concerning the Pleistocene of the Grays area, aided in his later work by the rediscovery in 1974 of Abbott's collection of mammalian fossils from West Thurrock. Carreck concluded that the West Thurrock deposits were of late Ipswichian age. Hollin (1977) cited the Levallois industry at the base of the section and the high incidence of *Carpinus* pollen in samples collected from the brickearth in support of an Ipswichian age (biozone III) for the West Thurrock brickearth. He suggested that the transition from the coarse lower sand (bed 2) to the finer-grained laminated sediments (bed 4) might represent a change from fluvial to estuarine conditions. He envisaged that this change resulted from a sudden rise in sea level, caused by an Antarctic ice surge during the mid-Ipswichian, that he had recognized at a number of Lower Thames sites (Hollin, 1971, 1977; see above, Purfleet and Globe Pit). Pollen spectra obtained from the brickearth in the 1983/4 road cutting (probably equivalent to bed 3 at the GCR site), which proved to be a considerable improvement on the material collected by Hollin, also point to an Ipswichian age (P.L. Gibbard, pers. comm.); the publication of a pollen diagram is anticipated in the near future.

Hollin's interpretation of the horizontally bedded sands as intertidal or estuarine deposits is supported by observations of the sequence exposed in GCR section 1 (Fig. 4.27). The thick (>7 m) horizontally bedded sand (bed 2) beneath the brickearth and the thinner sequence of interbedded sands and silts above the brickearth (bed 4) are both sedimentologically consistent with an intertidal depositional environment. In particular, the laminae of bed 4 resemble the closely interlayered bedding that is common in the sediments of intertidal flats (Reineck and Singh, 1975). Neither bed 4 nor bed 2, however, has yielded faunal evidence for such an environment, whereas the sparse molluscan assemblage from the brickearth itself clearly indicates freshwater conditions. If the unfossiliferous beds are of estuarine origin, they reflect a high relative sea level, implying an interglacial episode. They may, therefore, have formed during the same temperate interval as the brickearth, which is interpreted as an interglacial deposit on the basis of its molluscan and pollen content. The interdigitation of lobes of coombe rock into the edge of the lower sand appears to refute this suggestion, as such material (poorly sorted chalky debris) is generally attributed to periglacial slope processes, such as solifluction. The formation of minor lobes of material like those observed at Lion Pit may not require periglacial conditions, however; mudflows and slope failures, such as occur at the present time in Britain, are capable of producing similar deposits.

Given the doubts that have been expressed about the distinction of different late Middle and Late Pleistocene temperate episodes using palynology (Sutcliffe, 1960, 1964, 1975, 1976; Sutcliffe and Kowalski, 1976), the record of hippopotamus from West Thurrock (Abbott, 1890) would appear to be important in establishing an Ipswichian (*sensu* Trafalgar Square) age, since this animal is widely regarded as a reliable indicator of that stage in Britain (see Chapter 1). However, hippopotamus was not amongst the material described by Carreck (1976) in the rediscovered Abbott collection and it is possible that the record of this species at West Thurrock was based on a misidentification (A.P. Currant, pers. comm.).

The position of the interglacial beds at West Thurrock in relation to the Lower Thames terrace sequence is of some stratigraphical significance. The gravel that overlies these beds in both the road and tramway cuttings, mapped as 'Floodplain Gravel' by the Geological Survey (Sheet 271), was reclassified by Bridgland (1983a, 1988a) as the Mucking Gravel and equated with the late Saalian Taplow Formation of the Middle Thames (Bridgland, 1988a). This interpretation, which precludes an Ipswichian age for the West Thurrock brickearth, has proved controversial. Gibbard *et al.* (1988) considered the palynological evidence for an Ipswichian age for the West Thurrock brickearth to be sufficiently persuasive to indicate that the overlying gravel represents an early Devensian aggradation, hitherto unrecognized in the Lower Thames. They named this deposit, which has a composition and altitudinal distribution indistinguishable from the Mucking Gravel, the 'West Thurrock Gravel', citing the Lion Pit tramway cutting as the type section. Gibbard *et al.* (1988) suggested that this deposit could be correlated with the Reading Town Gravel of the Middle Thames, also attributed to the early Devensian (Gibbard, 1985). No equivalent unit has been recognized elsewhere in the Thames valley, however, and the separate existence of the Reading Town Gravel has been challenged earlier in this volume, where it has been reinterpreted as part of the late Saalian Taplow Formation (see Chapter 1; Chapter 2, Fern House Pit).

The pollen-based interpretation of the West Thurrock sequence can be challenged on a number of counts. Without the palynological evidence, the upper gravel (bed 5) would be assigned to the Taplow/Mucking Formation. Nearby sites within that formation, at Aveley and Ilford, have yielded similar pollen spectra to the West Thurrock brickearth, yet they have been assigned on other evidence (biostratigraphy, amino acid geochronology and terrace mapping) to a post-Hoxnian/pre-Ipswichian temperate episode (see below, Aveley). A straightforward interpretation of the regional stratigraphy in the Grays area is therefore favoured here and the West Thurrock sequence is attributed, in its entirety, to the late Saalian Mucking Formation. According to the climatic model for terrace formation (Chapter 1), the basal gravel (bed 1) would appear to represent the 'phase 2' (pre-interglacial) aggradation, whereas the interglacial beds (beds 2–4) and the upper gravel (bed 5) correspond to aggradational phases 3 and 4 respectively.

The above interpretation rejects the dating of the West Thurrock interglacial beds as Ipswichian. These sediments appear, however, to represent a later temperate episode than the Grays and Little Thurrock brickearth, since the two deposits are separated by both the post-interglacial (phase 4) gravel aggradation of the Lynch Hill/Corbets Tey Formation (see above, Globe Pit) and the subsequent downcutting that preceded the formation of the Lion Pit coombe rock and overlying basal gravel. As the interglacial sediments at Grays (and elsewhere within the Corbets Tey Formation) are attributed to Oxygen Isotope Stage 9 (see above, Globe Pit and Purfleet), and the Ipswichian Stage (*sensu* Trafalgar Square) is generally considered to correlate with Oxygen Isotope Substage 5e (Chapter 1), a correlation of the temperate episode represented at West Thurrock with Oxygen Isotope Stage 7 is suggested. The post-interglacial (phase 4) part of the Corbets Tey Gravel and the basal gravel at Lion Pit are both attributed, according to the above interpretation, to Oxygen Isotope Stage 8 (Table 1.1). There is archaeological support for the close association of these two gravel aggradations, despite their occupation of different terrace levels; both have yielded artefacts that show evidence of the use of the Levallois technique (see above, Purfleet and Globe Pit).

The altitude and stratigraphical position of the Ipswichian deposits at Trafalgar Square (Franks *et al.*, 1958; Gibbard, 1985) lend further support to the above interpretation. These deposits lie 30 km upstream from the Grays area, yet they are close to ordnance datum, nearly 10 m below the level of the West Thurrock brickearth. Furthermore, they represent the phase 3 (interglacial) part of the Kempton Park Formation, since they are underlain by a periglacial gravel sequence (the Spring Gardens Gravel Member of Gibbard (1985)), which represents the phase 2 aggradation of the Kempton Park Formation (see Table 1.1), and overlain by the Kempton Park Gravel as defined by Gibbard (1985). The Ipswichian (*sensu* Trafalgar Square) sediments therefore fall within a lower terrace than those at West Thurrock (Table 1.1). The altitude of the former deposits indicates that the Lower Thames valley was excavated to well below the level of the West Thurrock gravel by the beginning of the Ipswichian Stage. Indeed, it is clear that the Kempton Park Formation is aggraded to a much lower base level; it is correlated with the East Tilbury Marshes Gravel, which underlies the floodplain of the Lower Thames in the Tilbury area, where its upper surface is close to ordnance datum (Bridgland, 1983a, 1988a). Thus, if any sediments dating from the Ipswichian Stage (*sensu* Trafalgar Square) are preserved in the Grays area, they would be expected to be close to, or below, ordnance datum and to be represented within the 'Buried Channel' (Bridgland, 1988a; Table 1.1).

Relation to other sites in the Mucking Formation

The attribution of the interglacial sediments at West Thurrock to Oxygen Isotope Stage 7 implies correlation with other sites within the Thames system that have been attributed to that stage. In particular, sediments at Aveley (see below) and Stanton Harcourt (Chapter 2) have been important in the recognition and identification of this intra-Saalian temperate interval in Britain (Shotton, 1983; Bowen *et al.*, 1989; Chapter 1). The stratigraphical scheme for the Thames terraces established in this volume suggests that interglacial sediments at Ilford, Crayford and Northfleet also correlate with those at West Thurrock. There is some support for these correlations from both mammalian and molluscan faunas. Where temperate mammalian assemblages are recorded from the above sites, they resemble the fauna, which includes mammoth, straight-tusked elephant and horse, that is believed to characterize Oxygen Isotope Stage 7 (Shotton, 1983; see below, Aveley). In addition, *Corbicula fluminalis*, a bivalve that is characteristic of pre-Ipswichian temperate episodes but was probably absent from Britain during the Ipswichian Stage (*sensu* Trafalgar Square), occurs at all these sites. It has not, however, been recorded from the West Thurrock site, where its absence may be a reflection of the fact that molluscs in general (and bivalves in particular) do not appear to be well-preserved in the sediments there.

Where pollen has been obtained from the above sites, it has proved indistinguishable from assemblages from the Ipswichian Stage (*sensu* Trafalgar Square) and many palynologists are still of the opinion that all these localities represent the last interglacial (Ipswichian). Thus Ipswichian pollen sequences have been described from Aveley (West, 1969) and Ilford (West *et al.*, 1964), as well as West Thurrock (Gibbard *et al.*, 1988). Correlation of other sites in the Lower Thames with the Ipswichian Stage has been based on their stratigraphical relations to these polleniferous sites (Hollin, 1977; Stuart, 1982a).

The sites at Aveley, Ilford and Northfleet will be described below (see Aveley and Northfleet). At Crayford, areas of *in situ* deposits have been identified and it is hoped that temporary exposures will be investigated in the near future. No conservable remnants of these sediments remain, however. Nevertheless, a brief summary of the evidence from this important locality will be given here, as it seems likely to be a close correlative of the West Thurrock site. The Pleistocene deposits at Crayford (and Erith) are better known than their supposed correlatives at West Thurrock. Various pits in the Crayford area have yielded rich assemblages of molluscs, mammals and Palaeolithic artefacts, some of the latter in primary context (for summaries, see Kennard, 1944; Wymer, 1968; Roe, 1981). The Crayford sequence comprises brickearth above gravel, the latter descending to well below ordnance datum. The brickearth can be divided into lower fluviatile and upper colluvial elements, the former, which yielded the important faunal assemblages, extending to *c.* 11 m O.D. Palaeolithic artefacts occur in the gravel, between the gravel and the brickearth, and at several levels within the brickearth. Levalloisian working floors were reported from the base of the brickearth in two of the Crayford pits, Stoneham's Pit (TQ 517758) and Rutter's Pit (TQ 514765), both yielding conjoinable material (Spurrell, 1880; Chandler, 1916). The fluvial brickearth has frequently been ascribed to the Ipswichian Stage in recent years (Stuart, 1974, 1976, 1982a; Hollin, 1977; Roe, 1981; Gibbard *et al.*, 1988), but it has yielded both horse and *Corbicula fluminalis*, species believed by some workers (Chapter 1) to have been absent in Britain during that stage. Hippopotamus, regarded widely as indicative of the Ipswichian Stage (*sensu* Trafalgar Square) in Britain, has not been found at Crayford; indeed, the site has been interpreted, on the basis of its mammalian assemblage (which incorporates both temperate and cold elements), as intra-Saalian (Sutcliffe and Kowalski, 1976). Unfortunately, only the upper colluvial brickearth has been exposed in recent years. A large spread of 'Floodplain Gravel' appears on the Geological Survey map (Sheet 271) to the east of the brickearth outcrop at Crayford. In all records of the Crayford sites that refer to this gravel, it has been regarded as continuous with that underlying the brickearth (see Chandler, 1914; Kennard, 1944). This would therefore appear to represent the phase 2 aggradation of the Mucking Formation, with the Crayford interglacial sediments representing phase 3. The post-interglacial phase (phase 4) of Mucking Gravel aggradation may be represented within the outcrop to the east, but has not been recorded from above the brickearth. The biostratigraphy appears to indicate that the interglacial sediments accumulated during Oxygen Isotope Stage 7, as with other temperate-climate deposits within the Mucking Formation. Strong parallels can be recognized between the Crayford sequence and that at the Lion Pit tramway cutting; both have fine-grained fluviatile sediments containing interglacial faunas suggestive of an intra-Saalian age and in both cases these overlie Levallois 'floors'. Both sequences can be assigned to the Mucking Formation.

As has been noted above, evidence for the use of the Levallois technique first appears within the Corbets Tey Gravel, probably in the post-interglacial (phase 4) part of that formation. This suggests that the technique was first employed during Oxygen Isotope Stage 8. The Levallois artefacts at Crayford, Northfleet and West Thurrock all occur within or above gravels that underlie the temperate (phase 3) part of the Taplow/Mucking Formation, implying that they too date from Oxygen Isotope Stage 8. At Crayford, further finds from within the fluvial brickearth suggest that humans using the Levallois technique continued to occupy the area into Stage 7, although the occurrence of cold as well as temperate faunas in this deposit raises some doubts about the stratigraphical level of the Stage 8–Stage 7 transition; this may be near the top of the fluviatile sequence, at the base of the '*Corbicula* bed'. There is, nevertheless, important evidence from these three sites that the Levallois technique may be of chronostratigraphical significance.

Recently, the application of amino acid analysis to mollusc shells from sites in the Mucking Formation has proved to be a valuable source of evidence for correlation. Amino acid ratios from West Thurrock are not yet available, but results from sites that may be correlatives, such as Aveley, Ilford and Crayford (see below, Aveley), strongly support the claim that a Stage 7 interglacial event is represented at this level within the Lower Thames sequence (Bowen *et al.*, 1989).

Conclusions

The spectacular section in the Lion Pit tramway cutting combines important evidence from sediments, fossils and archaeology. A thick sequence of deposits occurs here at the northern edge of the terrace formed by Taplow/Mucking Gravel of the Lower Thames – the lowest of the three terraces recognized above the modern floodplain in this area. The basal deposit at West Thurrock is a mass of redeposited Chalk ('coombe rock') that has accumulated beneath the cliff, probably under cold conditions. Several further lenses of similar material interdigitate with the edge of the later sediments, where they are banked against the cliff. The lowest Thames deposit is a coarse gravel, containing the debris of a flint-working site almost as it was left by ancient man on the river's shoreline. Several of the pieces of flaked flint, which has been worked using the distinctive 'Levallois technique', can be fitted back together, a phenomenon that only occurs when such debris is little disturbed. This gravel is overlain by a thick sequence of laminated sands and silts, within which is a thicker bed of silty clay or 'brickearth'. The brickearth is better represented in the southern part of the site, where it has produced pollen and molluscs and, when the cutting was originally excavated, an extensive collection of mammal bones. The sands and silts are of intertidal (estuarine) origin, implying a high sea level, which again indicates deposition during an interglacial (sea levels were low during glacials, because much sea-water was locked up in larger polar ice caps). A later gravel, assumed to be a cold-climate deposit, overlies the interglacial sediments at the southern end of the cutting.

The age of the West Thurrock sequence is controversial. A more extensive pollen record has been obtained from the same brickearth in a road cutting a short distance to the west, where it has been attributed to the last interglacial (120,000 years BP). The position of the site within the 'staircase' of Lower Thames terraces, on the other hand, leads to the conclusion that the sequence represents part of the Saalian Stage, with the interglacial equivalent to the third of four interglacials recognized in the post-diversion Thames valley. This would indicate deposition at around 200,000 years BP, in Oxygen Isotope Stage 7 of the oceanic record.

AVELEY, SANDY LANE QUARRY (TQ 551808)

D.R. Bridgland

Highlights

This site, famous for the discovery of two elephant skeletons in the mid-1960s, preserves an important Pleistocene sequence thought to represent an interglacial between the Hoxnian and Ipswichian Stages. This temperate event is believed to equate with Oxygen Isotope Stage 7 of the deep-sea record. This interpretation continues to be a matter of controversy, however, as the Aveley sediments have also been assigned to the Ipswichian Stage (Substage 5e) on the basis of pollen analysis.

Introduction

This GCR site is adjacent to the former Sandy Lane Quarry, a large gravel and clay pit that occupied almost a square kilometre of land to the north-west of the village of Aveley. In this area the Thames floodplain runs approximately NNW–SSE, by-passing the loop through South Ockendon and Stifford that it followed in Corbets Tey Gravel times (see above, Purfleet and Fig. 4.1). Of the Lower Thames terrace deposits, only the Mucking Formation (mapped by the Geological Survey as 'Floodplain Gravel') appears to follow this shorter route, although later editions of the Geological Survey map (Sheet 257) show patches of brickearth and 'Taplow Gravel' between the Mucking Gravel and an outlier of Boyn Hill/Orsett Heath Gravel to the east of the GCR site (Fig. 4.1). The Orsett Heath Gravel outlier represents a 'meander core', as recognized by Wooldridge and Linton (1955), formed when the sinuous Corbets Tey Gravel course was abandoned. The addition of the 'Taplow Gravel' and brickearth to the later maps was a direct result of important exposures created in Sandy Lane Quarry, in an area hitherto mapped as Palaeogene Thanet Sand.

The eastern end of Sandy Lane Quarry exploited the above-mentioned outlier of Orsett Heath Gravel. The deposits mapped (later editions only of Sheet 257) as 'Taplow Gravel' were seen in sections at the western end of the quarry. They comprise a number of highly fossiliferous beds, which have yielded important mollusc, mammal, insect and pollen assemblages, indicative of interglacial conditions (Blezard, 1966, 1973; West, 1969; Cooper, 1972; Stuart, 1976; Sutcliffe, 1976; Sutcliffe and Kowalski, 1976; Hollin, 1977; Holyoak, 1983). Entirely separate from other mapped occurrences of 'Taplow Gravel', which have been reclassified as Lynch Hill/Corbets Tey Gravel (see Introduction to this chapter), these sediments have also been attributed to the 'Upper Floodplain Terrace' (West, 1977). Although the Upper Floodplain Terrace of the Middle Thames is correlated with the Kempton Park Formation (Chapter 1), the deposit mapped as 'Floodplain Gravel' in the Lower Thames is now referred to the Taplow/Mucking Formation (Bridgland, 1988a; Gibbard *et al.*, 1988; Introduction to this chapter). 'Floodplain' (Mucking) Gravel is mapped immediately to the west of the Aveley pit (Sheet 257), but its relation to the sediments there has yet to be determined.

Three alternative stratigraphical positions can thus be envisaged for the Aveley deposits within the Thames terrace sequence: they may belong (1) within the Lynch Hill/Corbets Tey Formation, as the later maps suggest, (2) within the Taplow/Mucking Formation or (3) within the Kempton Park Formation. The sediments at Aveley are, in fact, mainly sands, silts and clays, some with a high organic content, channelled into the London Clay. They were originally assigned to the Ipswichian (West, 1969; Hollin, 1977), but have also been ascribed to a hitherto unrecognized temperate episode between the Hoxnian and Ipswichian Stages (Sutcliffe and Bowen, 1973; Sutcliffe, 1975, 1976; Sutcliffe and Kowalski, 1976; Shotton, 1983; Wymer, 1985b; Bowen *et al.*, 1989).

The site is likely, on the basis of its faunal content and altitude, to correlate with Pleistocene deposits formerly exposed in south-west Ilford (Uphall Pit), which produced a rich molluscan fauna and considerable numbers of mammal bones when exploited for brick-making (Cotton, 1847; Dawkins, 1867; Phillips, 1871; Woodward and Davies, 1874; Hinton, 1900a, 1900b; Johnson, 1901; Rolfe, 1958). The Ilford deposits, like those at Aveley, have been attributed both to the Ipswichian (West *et al.*, 1964; Stuart, 1976, 1982a; Gibbard *et al.*, 1988) and to a post-Hoxnian/pre-Ipswichian temperate episode (Sutcliffe and Bowen, 1973; Sutcliffe, 1975, 1976; Sutcliffe and Kowalski, 1976; Shotton, 1983; Wymer, 1985b). The name 'Ilfordian' has been applied to this undefined 'stage' (Bowen, 1978; Wymer, 1985b), which is thought to correlate with Oxygen Isotope Stage 7 (Shotton, 1983; Bowen *et al.*, 1989). However, localities in north Ilford were on higher level, older terrace deposits, a fact that has led to much biostratigraphical confusion (see below) and makes Ilford, where no exposures have been available in recent years, unsuitable as a type locality.

Description

Unlike the Ilford pits, with which it is frequently correlated, there is no large body of early literature on the Aveley locality. A pit near the road junction (TQ 560808) north-west of Aveley was recorded by Whitaker (1889). He reported that 6 m of gravel was exposed here, part of the outlier of the deposit now classified as Boyn Hill/Orsett Heath Gravel, but it is unclear whether this working was within the area of the modern Sandy Lane Quarry. No early descriptions of the fossiliferous beds exist; it seems these were only discovered as the pit was extended westwards during London Clay extraction in the 1960s.

The sequence at Aveley is as follows (after West (1969) and Hollin (1977); see also Fig. 4.31):

	Thickness
6. Silt, pale yellow, with sand and gravel at the base	*c.* 1.0 m
5. Sand, grey and silty in its lowest 1 m, yellow and with scattered gravel above	up to 5.0 m
4. Silty clay, orange-brown, massive (brickearth). Contains Mollusca in its basal (calcareous) part only	2.0–2.5 m
3. Peaty layer with compressed wood (detritus mud)	up to 0.6 m
2. Silts and clays, grey (yellow-brown near base), containing freshwater Mollusca and fish, small vertebrates, wood and pollen	up to 7.0 m
1. Basal gravel below dominant sand, the latter with brown clay layers	*c.* 3.0 m
London Clay	

The first detailed description of the sediments exposed in the western part of Sandy Lane Quarry was by West (1969), reporting on a programme of pollen analyses, a preliminary account of which had been given by Blezard (1966). West described a sequence of Pleistocene deposits occupying a channel cut into the London Clay to a base level of 1.8 m O.D. (Fig. 4.31). The later description by Hollin (1977) indicates that the deposits thicken westwards, as revealed by later quarrying, reaching -4.3 m O.D. (it is possible that further extension of the quarry in this direction, towards the mapped outcrop of Mucking Gravel, might have revealed

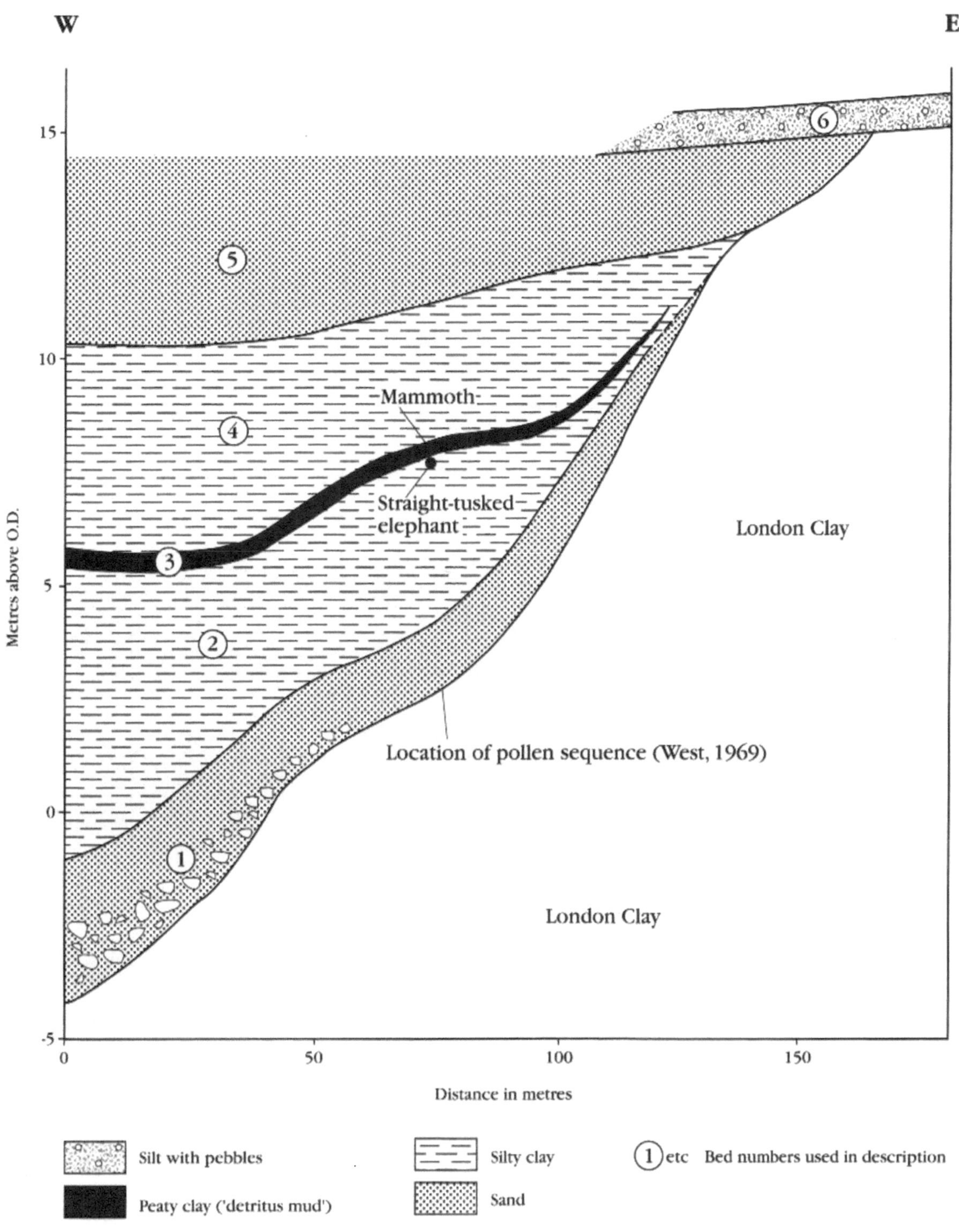

Figure 4.31 Section through the deposits of the Mucking Formation at Sandy Lane Quarry, Aveley (modified from Hollin, 1977).

(A)

Figure 4.32 The discovery of elephant skeletons at Aveley in 1964. (A) View of the working face in the Sandy Lane pit in 1964, during the excavation of the skeletons. The site was worked from west to east, so this view was taken looking towards the north-east. The excavation was located at approximately TQ 552808. (B) Close-up view, showing the mammoth bones at the higher level in the background and the straight-tusked elephant bones, at a lower stratigraphical level, in the foreground. (Photos: A.J. Sutcliffe.)

an even thicker sequence). Many of the beds are, therefore, of variable thickness, thinning against the London Clay 'cliff' to the east (Fig. 4.31). The lowest part of the channel infill (bed 1) comprises unfossiliferous silts, clays and thin sands above a basal gravel. These are overlain by organic silts and sands (bed 2), yielding freshwater shells and wood, that in turn are overlain by up to 0.6 m of compressed 'detritus mud' with wood (bed 3). The highest bed recorded by West was his brickearth, a stiff silty clay (bed 4).

Cooper (1972) provided detailed descriptions of the stratigraphy of the deposits based on a section drawn by G.R. Ward during the excavation of the elephant skeletons (Blezard, 1966;

(B)

see below). This included an additional 4 m of sand, overlying the brickearth described by West, that according to Ward had been removed prior to the excavation of the clay pit. This sand (bed 5), preserved in the north-west corner of the site, brought the surface height of the deposits up to nearly 15 m O.D., similar to that of the highest deposits at Ilford (Seven Kings; see below). Hollin (1971, 1977) noted that this upper sand extends eastwards beyond the limits of the fossiliferous channel-fill, where it directly overlies a steep London Clay surface. He also recorded a further bed above the sand, a pale yellow silt (bed 6) that he interpreted as aeolian or colluvial overburden, probably dating from the Devensian Stage.

The majority of the palaeontological evidence from Aveley comes from beds 2 and 3. This includes mammalian remains, amongst which are the elephant skeletons for which the site is famous (Anon., 1966; Fig. 4.32). Although found in close proximity, the skeletons are from different species; one is a straight-tusked elephant (*Palaeoloxodon antiquus*) and the other a woolly mammoth (*Mammuthus primigenius*). The former was in a silty clay at the top of bed 2 and the mammoth, only 0.3 m higher, was in the peaty 'detritus mud' of bed 3 (Anon., 1966;

West, 1969). The latter bed also yielded horse and the only British Pleistocene record of lesser white-toothed shrew (*Crocidura* cf. *suaveolens* (Pallas)) (Stuart, 1974, 1976, 1982a). Insect remains, primarily beetles, have also been recorded from these two beds (Coope, in Blezard, 1966; in Hollin, 1977; in Shotton, 1983), but they and their precise provenance have not been described in detail. Occasional ostracods were recovered from the same beds, while the brickearth (bed 4) produced an antler of red deer (Stuart, 1976). A second skeleton of straight-tusked elephant was excavated from the site, *c.* 25 m to the east of the first, by J.N. Carreck. This specimen, now in the Natural History Museum, was at the same stratigraphical level as the earlier *Palaeoloxodon* skeleton, beneath the detritus mud (A.J. Sutcliffe, pers. comm.).

Wiseman (1978) illustrated the exposures in the north face of Sandy Lane Quarry, from the Orsett Heath Gravel in the east to the fossiliferous deposits first described by West. His section traces the upper sand (bed 5) of the latter sequence to only 11 m O.D. and also shows a considerable amount of gravel within this bed. A separate development of sands and gravels occurred between 15 m and 20 m O.D. and was shown to be related to a minor geomorphological terrace feature visible in neighbouring unquarried land and on aerial photographs of the area taken prior to quarrying (Wiseman, 1978). These deposits, preserved primarily in pockets in the top of the London Clay, coincide with brickearth on later editions of the Geological Survey map (Sheet 257).

A temporary section in the eastern edge of the upper sand (bed 5) was observed by the present author in 1983. This showed cross-bedded sand banked against a steeply dipping London Clay surface (confirming Hollin's observation), presumed to be a channel edge or river cliff. The clay surface showed slickenside striations, however, suggesting relative movement between it and the sand. This might result from diapiric upwelling of the clay at the edge of the Pleistocene terrace, a phenomenon that appears commonly in the British Pleistocene (Allen, 1991), which would explain the extreme steepness of the observed London Clay surface.

Sandy Lane Quarry has now been infilled, but an unexcavated area immediately to the west has been identified as an alternative GCR site and it is hoped that the Aveley sequence will be exposed there in the near future. This may have the advantage of revealing the relation between the fossiliferous sequence and the spread of Mucking Gravel mapped immediately to the west of the pit.

Interpretation

The scientific importance of the Aveley site arises from its contribution to the palaeoenvironmental reconstruction and correlation of the Thames terrace sequence. The site is at the centre of the controversy over the stratigraphy and dating of post-Boyn Hill/Orsett Heath Gravel (= post-Oxygen Isotope Stage 10) interglacial sediments in the Lower Thames (see above, Purfleet, Globe Pit and Lion Pit).

Sandy Lane Quarry became well known as a result of the discovery there in 1964 of the two elephant skeletons, as described above (Anon., 1966; Blezard, 1966; Fig. 4.32). Straight-tusked elephant, represented by the lower of the two Aveley skeletons, is generally regarded as an interglacial animal, whereas mammoth (the upper specimen) is regarded as a cooler-climate species. The juxtaposition of these finds therefore presented a taphonomic problem. However, Blezard and Sutcliffe both pointed out that a considerable time gap could be represented by the vertical separation of the two skeletons, a view supported by the results of pollen analysis. A pollen diagram was produced by West (1969) from the fossiliferous sediments (beds 2 and 3). These yielded palynological spectra comparable to those previously obtained from Seven Kings, Ilford (West *et al.*, 1964), and were similarly attributed to the Ipswichian Stage (West, 1969). This analysis showed that the straight-tusked elephant lay in sediments dating from biozone IIb of the interglacial, whereas the mammoth lay in deposits of biozone III age. West (1969) discussed the possibility that the mammoth might date from after biozone III, considering that it may have been entombed in the older deposits while they were still soft. Sutcliffe (in West, 1969) reported that there was little sedimentological evidence to support such an interpretation. Correlation between the sites at Aveley and Ilford on the basis of their mammalian faunas has also been established (Sutcliffe and Bowen, 1973; Stuart, 1976; Sutcliffe, 1976), although there has been little recognition of the fact that two separate sets of deposits are present at Ilford (see below and Fig. 4.33).

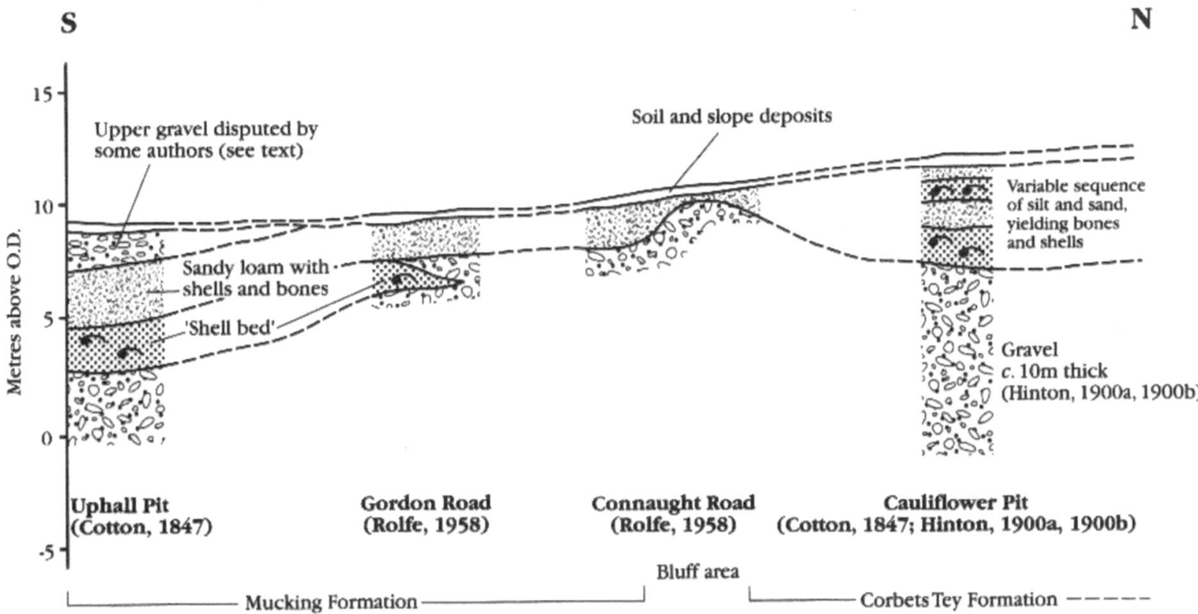

Figure 4.33 North–south section through the terrace deposits at Ilford. Compiled from published records, as shown. Note that information on the base levels of the Pleistocene deposits is generally lacking.

Correlation of the deposits at Ilford with the Ipswichian Stage (*sensu* Trafalgar Square) was the subject of controversy, however, even before the discovery of the elephant skeletons drew attention to the site at Aveley. The Aveley and Ilford sites both occur in association with deposits mapped as 'Taplow Gravel' by the Geological Survey. Sutcliffe (1960, 1964) considered the Ilford deposits to belong to a separate 'Ilford Terrace', intermediate in age between the Boyn Hill and 'Upper Floodplain' terraces. This view was based on differences in the interglacial mammalian assemblages from the deposits of the Boyn Hill/Orsett Heath Formation at Swanscombe, from the 'Ilford Terrace' and from the 'Upper Floodplain Terrace' at Trafalgar Square; it was also supported by the separate recognition of the terraces themselves. Sutcliffe made no distinction between the northern and southern pits at Ilford, which are here regarded as representing separate formations (the Lynch Hill/Corbets Tey Formation and the Taplow/Mucking Formation, respectively; Fig. 4.33, and see below). He noted that hippopotamus occurs only in the 'Upper Floodplain Terrace', as at Trafalgar Square, whereas horse and mammoth occur in the 'Ilford Terrace' but not the 'Upper Floodplain Terrace' (Sutcliffe, 1964, 1976; Sutcliffe and Bowen, 1973; Sutcliffe and Kowalski, 1976). Sutcliffe considered these assemblages to be so different that they could not be contemporaneous; they must either represent different parts of the same interglacial or two different interglacials. A considerable stratigraphical problem was therefore posed by the description of Ipswichian IIb pollen spectra in deposits of both the 'Ilford Terrace', at Ilford (Seven Kings) and Aveley (West *et al.*, 1964; West, 1969), and the 'Upper Floodplain Terrace' at Trafalgar Square (Franks, 1960). At Trafalgar Square the interglacial sediments lie slightly below ordnance datum, although the terrace surface is at 9 m O.D. The 'Ilford Terrace' (south Ilford deposits), on the other hand, is aggraded to above 10 m O.D. (the original terrace surface(s) in the Ilford area are degraded – see Fig. 4.33), some 15 km downstream from Trafalgar Square, with the biozone IIb deposits at 7 m O.D. (West *et al.*, 1964; Sutcliffe, 1976; Sutcliffe and Kowalski, 1976; see below). Sutcliffe attributed the Trafalgar Square deposits to the Ipswichian Stage, as represented at the Bobbitshole type locality in Suffolk, but considered the Aveley and Ilford deposits to have accumulated during an undefined post-Hoxnian/pre-Ipswichian temperate interval. The name 'Ilfordian' has subsequently been suggested for this episode (Bowen, 1978; Wymer, 1985b), but has not gained widespread recognition.

In his earlier papers, West (in West *et al.*,

1964; West, 1969) concurred with the Geological Survey's mapping of the 'Ilford Terrace' deposits as 'Taplow Gravel'. He later suggested that the Ilford–Aveley area might have been subjected to local uplift, associated perhaps with continuing movement of the Purfleet Anticline (West, 1972), and followed Evans (1971) in including the fossiliferous sediments at both sites in the 'Upper Floodplain Terrace' (West, 1977). This interpretation was influenced by his conclusion that they could be correlated with the deposits at Trafalgar Square on the basis of palynology. Cooper (1972) found that there were no criteria on which the molluscan assemblages from Aveley, Ilford and Trafalgar Square could be separated; he regarded the differences between these faunas, such as the occurrence of *Corbicula fluminalis* at Aveley and Ilford but not at Trafalgar Square, as of minor significance. Holyoak (1983) observed that the molluscan assemblage from Aveley consisted entirely of species known from Ipswichian sites; however, current opinion holds that many sites that have been attributed to that stage are, in fact, of intra-Saalian age, a view that originally derived from studies of mammalian faunas (Sutcliffe, 1960, 1964, 1975, 1976; see Chapter 1). It is suggested above that *C. fluminalis*, which Holyoak clearly regarded as present in British Ipswichian faunas, was in fact absent from this country during that stage (see Chapter 2, Stanton Harcourt and Magdalen Grove). Therefore molluscan evidence may support that from mammals in arguing for the distinction between the sediments at Aveley (and Ilford) and those at Trafalgar Square.

Stuart (1976) confirmed the distinction between the Ilford–Aveley and Trafalgar Square mammalian faunas, finding probable equivalents of both in deposits outside the Thames valley. However, he considered that the characteristics differentiating these assemblages could be related to pollen biozones within the Ipswichian. He noted that mammoth and horse were always absent from biozone II, even at Ilford and Aveley, whereas hippopotamus was recorded only from biozone II and the beginning of biozone III. He inferred that the Ilford fauna might in fact be later, not earlier, than the Trafalgar Square fauna and suggested that the Ilford and Aveley deposits were laid down in tributary valleys, therefore explaining their greater elevation than the sediments at Trafalgar Square.

Hollin (1977) tentatively interpreted the brickearth at Aveley (bed 4) as another product of the rapid submergence of the Lower Thames during the Late Ipswichian, evidence for which he had observed at other nearby sites (see above, Purfleet, Globe Pit and Lion Pit). He envisaged this submergence to have resulted from an Antarctic ice surge at the end of Ipswichian biozone III, which caused a sudden rise in sea level to around 14 m O.D. The overlying upper sand was interpreted by Hollin as a beach deposit formed as a result of the same submergence. He recognized that the lack of palaeontological evidence in these higher sediments, which he attributed to leaching, made it difficult to substantiate these interpretations. However, he argued that sedimentary features and the lack of any downstream gradient provided some support for the view. There appears to be little faunal evidence for a marine influence in the underlying fossiliferous deposits; ostracods from bed 2 were described as showing no sign of a marine influence (Robinson, in Hollin, 1977), although Cooper (1972) and Holyoak (1983) reported molluscan species that might indicate the proximity of the contemporary estuary. The particle-size distribution of the deposits at Aveley, evaluated by the plotting of standard deviation against the coarsest first percentile (after Friedman, 1967), is suggestive of a fluvial origin (Wiseman, 1978).

Correlation within the Lower Thames

The recent stratigraphical reappraisal of the Lower Thames terraces by Bridgland (1983a, 1988a) and Gibbard *et al.* (1988) allows further consideration of the relations of the interglacial deposits of the so-called 'Upper Floodplain' and 'Ilford' Terraces to one another and to the terrace sequence as a whole. The deposits mapped as 'Taplow Gravel' in the Lower Thames have now been reclassified as Corbets Tey Gravel and are correlated with the Lynch Hill Gravel of the Middle Thames (Bridgland, 1988a; Gibbard *et al.*, 1988). The deposits at Aveley, however, are not related to the Corbets Tey Formation, which follows a separate course to the east of the Orsett Heath Gravel outlier at Aveley (see above, Purfleet). They are more likely, on the grounds of their location and elevation, to represent part of the Mucking Formation, which is the true equivalent of the Taplow Formation of the Middle Thames

(Table 4.1). Their relation to the Mucking Gravel, mapped immediately to the west of the GCR site, is probably similar to that of the comparable sequence at West Thurrock. At the latter site the Mucking Gravel unconformably overlies a succession of gravel, lower sand, fossiliferous brickearth and upper sand, the last containing interbedded silts and clays and reaching *c.* 15 m O.D. (see above, Lion Pit). Hollin (1977) correlated the upper sand at West Thurrock with that at Aveley, an interpretation that is entirely plausible on altitudinal grounds. If the upper parts of the Aveley and West Thurrock sequences are equivalents, correlation of the basal gravels and interglacial sediments is implied.

The available records of the exposures at Ilford strongly indicate that deposits of more than one age are represented there. Geological Survey Sheet 257 shows the 'Taplow' (Corbets Tey Gravel) outcrop, beneath brickearth, projecting southwards to cover the whole of Ilford. However, Rolfe (1958) illustrated a north-south section through this area (from TQ 446868 to TQ 447865) that suggests that two separate terrace formations are represented at Ilford; the higher one relates to the Lynch Hill/Corbets Tey Formation and the lower one to the Taplow/ Mucking Formation (Fig. 4.33). His section showed a distinct rise in the gravel surface (beneath overlying brickearth) at about TQ 446867. This suggests that the deposits to the south of this point are part of the Mucking Formation, the boundary between this and the Corbets Tey Gravel occurring further north than shown on the Geological Survey map. The southward extension of the Corbets Tey Gravel outcrop is therefore erroneous; the boundary between the two formations continues the trend followed both west and east of Ilford (Fig. 4.1).

The Uphall Pit, from which a large proportion of the mammalian and molluscan faunas was obtained, exploited the lower-level deposits of the Mucking Formation, whereas the other main Ilford site, the Cauliflower or High Road Pit, was on the higher Corbets Tey Formation. The latter site also produced molluscan and mammalian remains, notably the collection of Hinton (1900a, 1900b). Records from this and other sites from the higher terrace level at Ilford are not always easily distinguished from those from the Uphall Pit, but should not be included in consideration of the 'Ilford Terrace'; if the latter is taken to include the Aveley deposits, it is clearly synonymous with the Mucking Formation, to which the Uphall deposits (alone amongst the Ilford sediments) belong. Unqualified references to Ilford will therefore be confined to the southern, Uphall Pit deposits, whereas the higher level sediments will be referred to under the name Seven Kings. Since the Uphall Pit sediments represent the back edge of the lower (Mucking Formation) terrace and the Seven Kings deposits occupy the leading edge of the higher terrace, the difference in their elevations is less than that separating the two formations as a whole (Fig. 4.33). This also explains why the higher fossiliferous deposits at Seven Kings are at the same height as the extreme feather-edge of the Mucking Formation at Aveley and West Thurrock. The deposits that reach this height at the last two sites are possibly of estuarine origin and thus influenced by a high interglacial sea level; they therefore reach considerably higher elevations than the general level of the Mucking Formation. It should be noted that Kennard and Woodward (1900) suggested that deposits of different ages were represented by the two main Ilford sites, an opinion that seems to have been based purely on their elevation. Kennard (1916), however, considered that the lower (Uphall Pit) deposits were the younger of the two, the opposite to the opinion expressed here.

Records from the Uphall Pit indicate that a rich molluscan fauna, including the bivalve *Corbicula fluminalis*, was obtained there from a sand interbedded with the gravel of the Mucking Formation. Most of the mammalian fauna, however, was from brickearth overlying the shelly gravel (see Fig. 4.33). A section published by Wood (1866a) showed a further gravel overlying the fossiliferous brickearth in the Uphall Pit; this was presumably the deposit later mapped as 'Floodplain Gravel', which would appear to confirm that the sequence seen at the Uphall Pit is part of the Mucking Formation. According to West *et al.* (1964), Wood (in Woodward and Davies, 1874) subsequently retracted this observation, but this later statement by Wood merely expresses the opinion that the upper gravel at Ilford was not the same deposit that underlies the fossiliferous beds at Grays and Crayford. That Wood continued to recognize an upper gravel at Ilford is indicated by the reproduction of his Uphall Pit section by Woodward and Davies (1874, p. 394), in which the gravel bed above the brickearth is reclassified as 'newer gravel'. Phillips (1871, p. 470) also illustrated sand and gravel above the fossiliferous deposits in the Uphall Pit. Dines and Edmunds (1925) suggested that this upper gravel corresponds with that mapped as 'Floodplain Gravel' by the Geological Survey, thus pre-empting the conclusions outlined above.

Only part of the early fossil collections from Ilford appears, therefore, to come from the Taplow/Mucking Formation – that part that was collected from the Uphall Pit. Nevertheless, the mammalian and molluscan collections from that site are consistent with correlation with the Aveley deposits. The higher-level fossiliferous deposits at the Cauliflower Pit and Seven Kings appear to belong to the Lynch Hill/Corbets Tey Formation. They may represent a further occurrence of temperate-climate sediments similar to those recorded at Grays, Purfleet and Belhus Park, also from the Corbets Tey Formation (see above, Purfleet and Globe Pit). The pollen-bearing deposits described by West *et al.* (1964) at Seven Kings, a series of silts, clays and 'detritus muds', were from a site (TQ 453872) only 0.3 km to the east of the Cauliflower Pit. These sediments overlie the Corbets Tey Gravel, but were attributed by West *et al.* to a later tributary stream. They yielded a different molluscan fauna to that from the Mucking Formation at the Uphall Pit. The Seven Kings assemblage was dominated by *Bithynia tentaculata* and lacked *Corbicula fluminalis*. The relation of the fossiliferous sediments at the Seven Kings pollen site to the Corbets Tey Formation remains uncertain, since they are not covered by a further aggradation of Thames gravel. There is nothing to suggest, however, that they are in any way related to the fossiliferous sediments within the Mucking Formation at Ilford and Aveley.

The sequences at Aveley and West Thurrock, and at the Uphall Pit in south Ilford, can thus be assigned to the Mucking Formation. Their component beds can be interpreted according to the climatic model for terrace formation (Chapter 1): the basal gravels at each site represent the pre-interglacial aggradational phase (phase 2) of the Mucking Formation, whereas the interglacial deposits represent the mid-sequence temperate phase (phase 3). The main Mucking Gravel aggradation, that which appears on the Geological Survey maps as 'Floodplain Gravel' and which overlies the interglacial beds, represents the post-interglacial phase (phase 4) of the climatic model. These three elements of an essentially tripartite sequence are considered here to respectively date from Oxygen Isotope Stages 8, 7 and 6 (Fig. 4.3; Table 1.1).

Correlation with sediments outside the Lower Thames

The conclusion that the mammalian assemblages from Aveley and Ilford represent a distinctive fauna, which relates to an undefined post-Hoxnian/pre-Ipswichian temperate interval, has received support in recent years from the recognition of similar assemblages elsewhere (see Chapter 1). One site that has produced critical evidence is at Marsworth, Buckinghamshire, where a fossiliferous channelfill, producing a mammalian assemblage similar to that at Ilford and Aveley, occurs stratigraphically below another deposit containing hippopotamus. Periglacial colluvium separates the two fossiliferous deposits, indicating that different temperate episodes are represented (Currant and Wymer, in Shotton, 1983; Green *et al.*, 1984; Wymer, 1985b).

Further evidence for an intra-Saalian temperate episode comes from the Upper Thames, where sediments yielding a mammalian assemblage of the Ilford-Aveley type occur within the lower part of the Summertown-Radley Formation, whereas hippopotamus-bearing gravels, attributed to the Ipswichian Stage (*sensu* Trafalgar Square), occur in the upper part of the formation. There is abundant evidence from the intervening gravels that periglacial conditions prevailed between the accumulation of the two sets of temperate-climate sediments (see Chapter 2, Stanton Harcourt and Magdalen Grove). The post-interglacial (phase 4) aggradation of the Mucking Gravel, which overlies the interglacial sediments at West Thurrock, Ilford and (probably) Aveley, provides similar stratigraphical evidence for a cold-climate interval separating these from the Trafalgar Square deposits, although in the Lower Thames the stratigraphical evidence is complicated by the incision event between the Mucking and Kempton Park Formations. The Thames sites at Aveley, Ilford and Stanton Harcourt therefore provide fundamental evidence for the recognition of an additional temperate cycle separating the Hoxnian and Ipswichian interglacials as defined by Mitchell *et al.* (1973). This additional episode has been correlated with Oxygen Isotope Stage 7 of the deep-sea record

(Shotton, 1983; Wymer, 1985b; Bowen *et al.*, 1989; Chapter 1).

Recent analyses of beetle faunas from Aveley, Stanton Harcourt, Marsworth (upper channel) and a further site with similar affinities, at Stoke Goldington, Buckinghamshire, have provided supporting evidence for the correlation of these localities and their distinction from sediments ascribed to the Ipswichian Stage (*sensu* Trafalgar Square) (Coope, in Shotton, 1983). In particular, Coope cited the presence of a species now resident in the Caucasus, *Anotylus gibbulus*, as a dominant feature in the above four faunas. This species had not been recorded from Ipswichian sediments, although it is a minor component of the Devensian interstadial faunas at sites such as Upton Warren and Chelford. Coope suggested that the presence of *A. gibbulus* in abundance may be characteristic of deposits dating from Oxygen Isotope Stage 7. However, the recent discovery of this beetle in sediments at Coston, Norfolk, which also yield abundant hippopotamus and are therefore presumed to be of Ipswichian age (R.C. Preece, pers. comm.), appears to undermine its biostratigraphical value.

Geochronology

The chronostratigraphical interpretation of the Aveley site has been assisted by amino acid analyses of shells from the interglacial beds (see Chapter 1). Early work of this type by Miller *et al.* (1979) yielded amino acid ratios from *C. fluminalis* shells from various Lower Thames sites, including both Ilford and Aveley. From the latter site an average ratio of 0.19 ± 0.023 was obtained. Amino acid ratios from *Bithynia tentaculata* from Aveley, published recently by Bowen *et al.* (1989), suggest correlation with Stanton Harcourt and Crayford, both sites variously interpreted as Ipswichian or post-Hoxnian/pre-Ipswichian (see above, Lion Pit; Chapter 2, Stanton Harcourt). This species produced ratios of 0.170 ± 0.02 (Crayford), 0.154 ± 0.007 (Stanton Harcourt) and 0.148 ± 0.016 (Aveley). In contrast, the same species from the Bobbitshole Ipswichian type locality gave a ratio of 0.09 ± 0.015 and from Trafalgar Square gave 0.11 ± 0.005. These results strongly support the view, already argued on the basis of mammalian faunas (see above; Chapter 1), that the deposits at Aveley, Crayford and Stanton Harcourt are pre-Ipswichian. Bowen *et al.* (1989) regarded the amino acid ratios from these sites as indicative of correlation with Oxygen Isotope Stage 7.

However, an amino acid ratio of 0.23 ± 0.02 from *B. tentaculata* from Ilford (Bowen *et al.*, 1989) indicates a greater antiquity than the Aveley sediments, raising doubts about the correlation of these two sites. Bowen *et al.* ascribed these specimens to the 'shelly bed' at Ilford, which suggests that they came from collections from one of the early sites. *Bithynia tentaculata* was recorded from both the Cauliflower and Uphall pits and was also abundant at the Seven Kings pollen site (West *et al.*, 1964). The above ratio falls within the range interpreted by Bowen *et al.* as indicative of Oxygen Isotope Stage 9. This suggests that the specimens may have come from the Seven Kings deposits (Cauliflower Pit), as the interglacial sediments elsewhere within the Corbets Tey Formation have been attributed to Stage 9 (see Chapter 1 and Table 1.1; Fig. 4.3). The *Corbicula* shells analysed by Miller *et al.* (1979) were claimed to have come from the Uphall Pit. They yielded a mean ratio of 0.23 ± 0.038, identical to that obtained by Bowen *et al.* (1989) from *Bithynia*. Ratios from individual specimens were also published by Miller *et al.*, as follows: 0.19, 0.21, 0.21, 0.26, 0.28. It may be that two separate groupings can be recognized amongst these results, one at around 0.20 and the other from 0.26 to 0.28. The first of these would represent Stage 7 and the second group of higher ratios would represent Stage 9. This might indicate that the shells obtained by Miller *et al.* (from the Natural History Museum) were a mixture of specimens from the Uphall and Cauliflower pits. It is also possible that reworked shells from the older deposits were mixed with indigenous specimens in the sediments in the Uphall Pit. It must, however, be noted here that *Bithynia* shells from Little Thurrock and Purfleet, both of which are broadly correlated in this volume with the Cauliflower Pit sediments and attributed to the same interglacial episode, have given higher ratios than would be expected for Stage 9 (see above, Purfleet and Globe Pit).

Conclusions

The fossiliferous Pleistocene deposits at Aveley are of considerable importance both to the history of the Lower Thames and to British

Pleistocene stratigraphy as a whole. Their interpretation remains a subject of controversy, but most workers now regard this as a key site for the recognition of an additional interglacial between what used to be regarded as the next to last and the last interglacials - the Hoxnian and Ipswichian (respectively) of the established chronology of two decades ago. The controversy hinges on contradictory evidence from mammals and molluscs on the one hand and pollen on the other. It also involves consideration of an inaccessible site that is universally attributed to the last interglacial, at Trafalgar Square. Both sites have produced similar pollen sequences, leading to both being attributed on this basis to the last interglacial (Ipswichian Stage). However, there are marked differences between the mammal and mollusc faunas at the two sites: mammoth, for example, is present only at Aveley, whereas hippopotamus is present only at Trafalgar Square. This led to early suggestions that an older interglacial was represented at Aveley, intermediate in age between the Hoxnian and Ipswichian. In support of this view are two further pieces of evidence. Firstly, the Trafalgar Square sediments form part of a lower terrace, the downstream slope of which takes it below the level of the modern floodplain by the time the Aveley area is reached. Secondly, the analysis of amino acids in shells from the two sites confirms that Aveley is older than Trafalgar Square. Such evidence has led to the widespread recognition that the interglacial at Aveley represents the true penultimate interglacial, equivalent to Stage 7 of the deep-sea record, which occurred at around 200,000 years BP.

NORTHFLEET (EBBSFLEET VALLEY): BAKER'S HOLE COMPLEX

D.R. Bridgland

Highlights

This locality exposes a complex sequence of predominantly fine-grained sediments, probably representing a mixture of fluvial, colluvial and aeolian deposition in a tributary valley of the Lower Thames. The deposits have yielded sporadic mammalian and molluscan remains, mostly of species with cold affinities, but one level in particular suggests interglacial conditions. The rich 'Levallois industry' that occurs in the lower part of the sequence makes Northfleet the most significant Levallois site in Britain.

Introduction

Pleistocene deposits in the Northfleet district constitute the most famous and most prolific source of Levallois artefacts in Britain. These sediments are not of mainstream Thames origin, however, but belong to the tributary Ebbsfleet valley. For this reason, their relation to the Thames terrace system has been difficult to

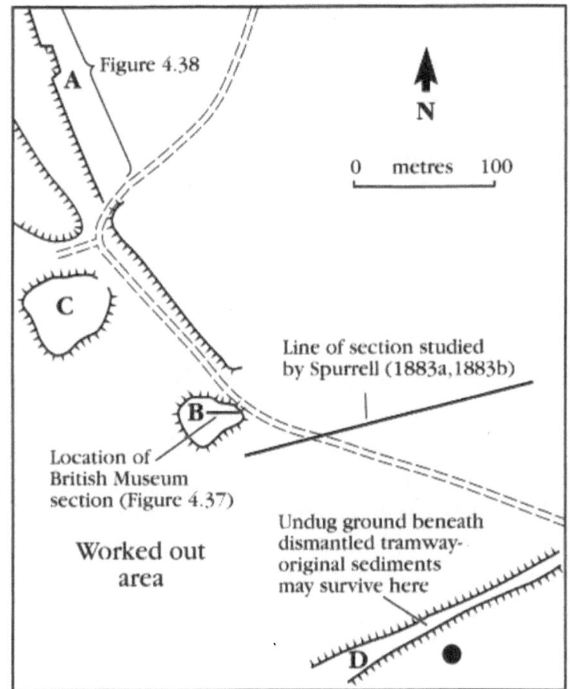

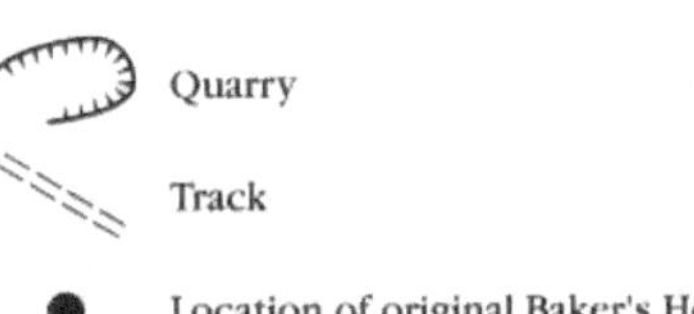

Figure 4.34 Plan of the surviving remnants of the Baker's Hole Complex, Northfleet. A–C are the three parts of the GCR site. Updated information regarding the location of the original Baker's Hole site (and the possibility that sediments related to this survive at D) has been supplied by F.F. Wenban-Smith (pers. comm.).

determine, despite the abundance of archaeological evidence they have yielded. The sediments, which comprise a series of gravels, sands and silts overlying a substantial sheet of soliflucted Chalk (coombe rock), have generally been correlated with the '50 ft' or 'Middle Terrace' of the main river (Smith, 1923; Burchell, 1933, 1934a, 1936a). The sediments above the coombe rock are often referred to as the 'Ebbsfleet Channel' deposits, a name first used by Burchell (1936a, 1936b). The fact that many of the huge numbers of Palaeolithic flakes from Northfleet were made using the Levallois technique (see Chapter 1) has long been recognized (Spurrell, 1883a; Smith, 1911; Dewey, 1932).

Three separate remnants of the Ebbsfleet deposits are included in the GCR site (Fig. 4.34). The stratigraphical relations of these to one another and to the earliest sections at Baker's Hole are imperfectly known. The Baker's Hole site, described in the early years of this century (Abbott, 1911; Smith, 1911), has only been relocated in recent years, having previously been regarded by many authors as entirely quarried away. Although never used by the owners of the site, the name Baker's Hole has consistently been cited in the archaeological and geological literature since 1911. A local legend attributes the name to a drunkard called Baker who perished by falling into the pit (Abbott, 1911). This name was applied by King and Oakley (1936) to the interval during which the periglacial deposits at the base of the Northfleet sequence were supposedly deposited, their 'Baker's Hole or Main Coombe Rock Stage'. The site has also been described under the names 'Southfleet Pit' and 'New Barn Pit'. Chalk extraction at Northfleet has now ceased and the surviving remnants of the Ebbsfleet sediments within the GCR site are to be incorporated in a large-scale restoration scheme in such a way that useful exposures and a reserve of deposits for future investigation both remain.

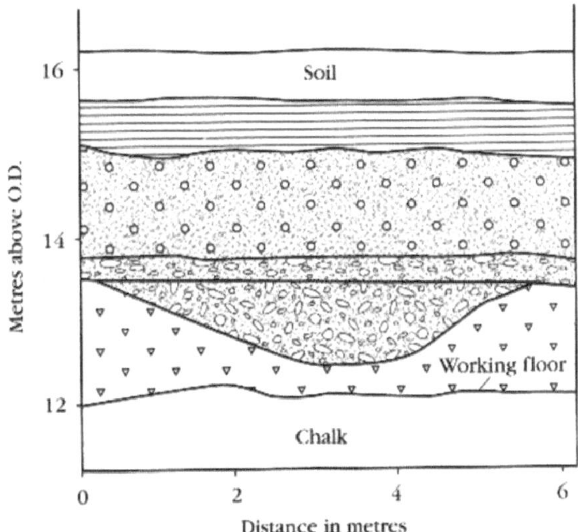

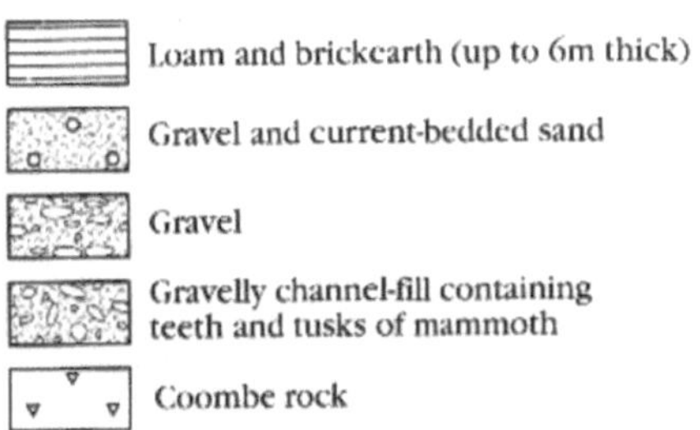

Figure 4.35 Section at the original Baker's Hole site (after Dewey, 1932).

Description

The occurrence of Palaeolithic artefacts in the Ebbsfleet valley at Northfleet was first reported by Spurrell (1883a, 1883b), although it is uncertain precisely where his observations were made. In the early years of this century new quarrying led to the discovery nearby of the celebrated Baker's Hole site, detailed descriptions being provided by Abbott (1911), Smith (1911) and Dewey (1932), the last including an illustration of the stratigraphy (see Figs 4.35 and 4.36). Smith (1911) described Palaeolithic implements and fossils from an accumulation of coombe rock likened, by Reid (in Smith, 1911), to that of the South Downs. The fossils included teeth and bones of elephant, horse, rhinoceros and deer. Few could be identified to species level because of their fragmentary nature and weathered state, but teeth of *Mammuthus primigenius* and *Dicerorhinus hemitoechus* were recognized, together with antler fragments of *Cervus elaphus*.

Dewey (1930, p. 148) described the Baker's Hole site as 'a working floor ... lying under masses of unassorted chalk and flint rubble'. He also observed that there were gravel and sand-filled channels cut into the upper surface of the coombe rock, in which mammoth tusks occurred (Dewey, 1930, 1932). These basal

(A)

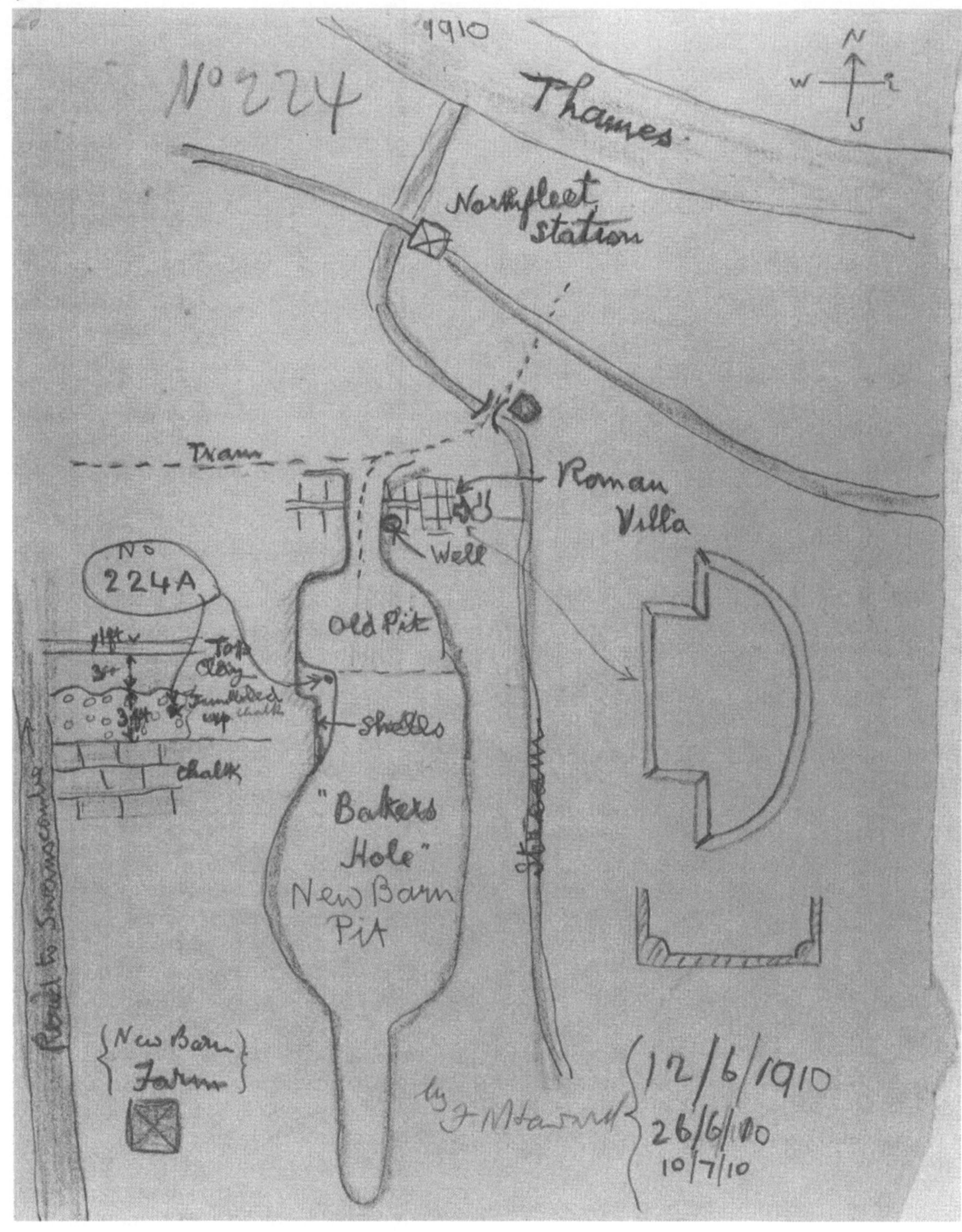

Figure 4.36 Early records from Northfleet, sketches of the original Baker's Hole site. (A) Map showing the location of the quarry known as 'Baker's Hole' and of the section there that yielded Palaeolithic artefacts, drawn by F.N. Haward in 1910. Reproduced by courtesy of the British Museum, London. (B) Measured drawing by F.N. Haward of the section on the west side of 'Newbarn Pit (Baker's Hole)' as seen in 1906, although the drawing is dated November 1920. The locations of artefact discoveries are indicated. (Reproduced by courtesy of the Natural History Museum, London). Thanks are due to F.F. Wenban-Smith, who drew the author's attention to the existence of these archival records, and S. Parfitt, who discovered the section drawing.

(B)

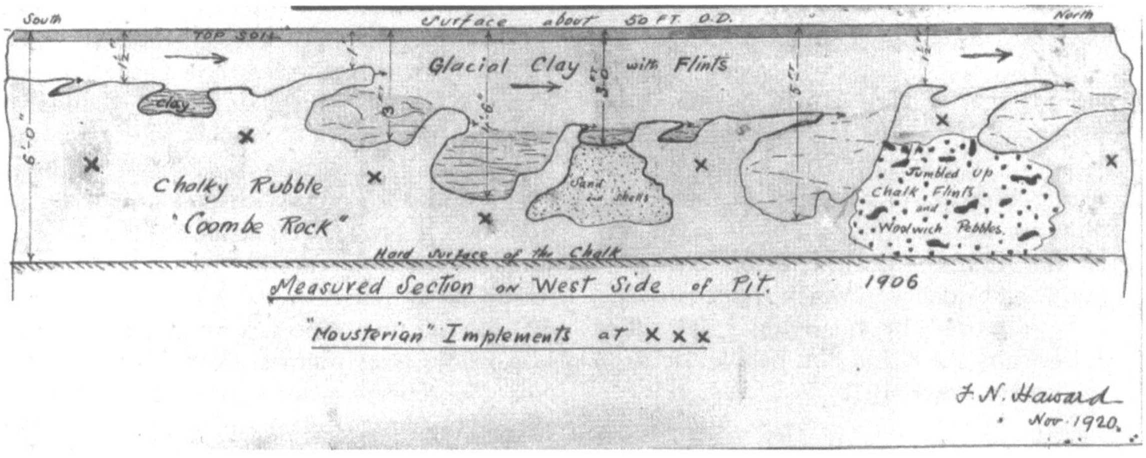

deposits were overlain by gravel and sand, the latter cross-bedded, then 'loam' and brickearth (Fig. 4.35). Spurrell (1883b), however, had referred to the artefacts lying on a 'kind of beach', which led Roe (1981) to question whether the industry might be associated with a fluvial deposit that pre-dates the coombe rock. It seems likely, however, that Spurrell's observations were made in the vicinity of the site later studied by Burchell (see below), so that the 'beach' to which he referred was probably a gravel forming part of the infill of the Ebbsfleet Channel.

Burchell (1933) described a series of gravels and 'brickearths' in the Ebbsfleet valley, overlying a 'bench' cut into the coombe rock and underlying Chalk at 7.5 m O.D. (Burchell, 1933; Fig. 4.37). The stratigraphical sequence here, equivalent to that represented in part of the GCR site, was pieced together over a lengthy period of observation (Burchell, 1933, 1935a, 1935b, 1936a, 1936b, 1936c, 1954, 1957; Boswell, 1940; Zeuner, 1945, 1946, 1954; Kerney and Sieveking, 1977). It can be summarized as follows (few indications of thickness have been recorded, perhaps because of variability in different parts of the channel; see Fig. 4.37):

12. 'Trail' – as bed 9 (formerly undifferentiated from 9).

11. Sandy 'fluvial brickearth'.

10. '*Cailloutis*' (thin gravel bed), yielding Levallois artefacts.

9. 'Trail': gravelly 'loam' with rafts of coombe rock. Published descriptions and illustrations suggest that this bed (and possibly bed 7) were cryoturbated.

8. Silt, aeolian/colluvial (brickearth). This yielded ***Pupilla muscorum***. Its upper part was decalcified and devoid of shells. Bands of ferruginous staining were observed near the top.

7. Upper coombe rock, with derived artefacts and land snails.

6. Freshwater silt, fossiliferous (temperate-climate). This bed contains ***Corbicula fluminalis*** (see, however, below), amongst remains of 40 species of terrestrial, marsh and freshwater Mollusca (see below), as well as mammoth, giant deer, horse and indeterminate rhinoceros.

5a. Buried soil developed in the top of bed 5.

5. Silt (brickearth). This is interbedded with numerous minor lobes of 'coombe rock' and/or gravel. It contains ***Pupilla muscorum***, ***Vallonia costata*** (Müller) and ***Limax*** sp., the last two from the lower part of the bed only. Descriptions and photographic records of sections excavated by the British Museum in 1969 reveal clear indications of aqueous bedding. These records also suggest interdigitation with (or incision through) beds 2 and 4. A total thickness of over 6 m is indicated (see Fig. 4.37). An

assemblage of small vertebrates is also recorded from this bed (Carreck, 1972; see below).

4. Gravel, with remains of woolly rhinoceros, mammoth and horse, together with artefacts (concentrated in 4a?). This is probably the higher-level gravel recorded in the British Museum sections (Fig. 4.37). If so, its separation from 2 is unclear in the absence of bed 3, all evidence of which appears to have been removed by quarrying. Reworked Palaeogene shells and flint pebbles occur in this bed (Carreck, 1972).

4a. Palaeolithic horizon at the base of bed 4, with a mixture of hand-axes, cores and flakes, including Clactonian and Levallois types (unabraded and unpatinated). These are accompanied by remains of mammoth, woolly rhinoceros and horse (Burchell, 1936a).

3. Sand, fossiliferous, yielding *Bithynia tentaculata.* Not seen in British Museum excavations? Carreck (1972), citing unpublished sources, also listed shells of *Arion* sp. and *Limax marginatus* (Müller), as well as giant deer and a number of small mammals. The latter were listed as *Arvicola abbotti*, *Microtus* sp. and *Clethrionomys* sp.

2. Coarse gravel, cryoturbated into or filling scour/solution hollows in the top of bed 1. Built up to >2 m at the edge of the channel, where it is interbedded with lenses of coombe rock and appears to interdigitate with the lower part of 5 (see Fig. 4.37).

1. Main Coombe Rock, thought to be equivalent to that at the Baker's Hole site. The working floor at Baker's Hole was at the base of this deposit (Fig. 4.35).

Frost-shattered Chalk.

In his early reports, Burchell took the gravel that appears above as bed 4 to mark the base of the Ebbsfleet sequence. Later excavations (Burchell, 1936a) revealed additional beds below this (2 and 3) and showed that the channel had been eroded prior to the deposition of the basal coombe rock (bed 1). The latter formerly occupied the channel, but was largely removed by erosion before its denuded remnant was covered with coarse gravel (bed 2). This gravel was in turn overlain by a fossiliferous sand (bed 3), above which another gravel (bed 4) was observed, from which unabraded hand-axes, amongst other types of artefact, were recovered (Burchell, 1936a, 1936b; see above). Burchell (1935a) also recorded small-mammal remains from bed 4 (4a?). The latter record has been supplemented in recent years from material placed by Burchell in the British Museum, so that the full microtine assemblage from the site as a whole is as follows: *Clethrionomys glareolus* (Schreber); a form transitional between *Microtus arvalis* (Pallas) and *M. agrestis* (L.); *M. anglicus* (Hinton); *M. nivalis* (Martins); *Arvicola cantiana* (Carreck, 1972; Sutcliffe and Kowalski, 1976). Whether all these taxa were present in bed 4/4a is uncertain; Burchell (1936a) also recorded a 'microtine fauna' from bed 3, but gave no details. Carreck (1972) recorded different elements from the above assemblage in beds 3, 4 and 5. Bed 4 also yielded a large Palaeolithic assemblage, listed by Burchell (1933) as: (1) much-rolled Clactonian and Acheulian artefacts, derived from earlier deposits, (2) less-abraded Levallois specimens, washed from the adjacent coombe rock, and (3) unrolled Levallois artefacts of a later type, some with marked Aurignacian (Upper Palaeolithic) characteristics.

Both Burchell and Zeuner interpreted the basal gravels and sands (beds 2–4) as fluviatile, but Kerney and Sieveking (1977) attributed them to solifluction. Detailed records and photographs of the British Museum sections show bedding structures in bed 5 that suggest the supply of silt from the valley side. Once in the channel this silt, of apparent loessic origin (see below), probably joined the sediment load of the Ebbsfleet, so that fluvial silt was deposited away from the margins of the channel. These records also raise questions about the stratigraphical significance of beds 1–4 and of Burchell's interpretation of the sequence of erosional and depositional events. It appears that the sections revealed the edge of a channel cut into frost shattered and soliflucted Chalk, filled by a gravel lag followed by fluviatile silts, the latter interdigitating with further lobes of gravel and coombe rock near the sides of the channel. A widening of the channel was followed by the formation of a later, more widespread gravel lag, which continues over the lower sequence in the deeper part of the channel. This may be the later gravel described by Burchell (bed 4). However, without the fossiliferous bed 3, it is impossible to distinguish beds 2 and 4 amongst a succession of gravels that interdigitate with the silt of bed 5 (Fig. 4.37).

The upper part of the sequence (beds 5 and above), predominantly silts (brickearth) with land snails, was deposited over a wider area than the lower channel deposits. From these later sediments Burchell (1933) obtained artefacts that he classified as Upper Mousterian, Aurignacian and Solutrian (it is difficult to relate these finds to the beds described above, which were not recognized at that time). The surviving brickearth (bed 5) has a particle-size distribution and mineralogical characteristics that suggest that it is predominantly of loessic origin, although with additional sand and gravel material (J.A. Catt and A.H. Weir, pers. comm.).

Continued excavations at Northfleet allowed Burchell (1935a) to observe that a previously unrecognized higher spread of coombe rock (bed 7), containing derived artefacts, occurred within this predominantly loessic (fluvially redeposited?) sequence. The recognition of this and other subordinate beds of coombe rock within the higher part of this succession led to the widespread use of the term 'Main Coombe Rock' for the deposit first described (bed 1). Burchell (1936c) recognized an additional fluviatile cycle at the top of the sequence, within what was formerly recognized as 'trail', thus adding beds 10–12. These additional beds were not recognized by Zeuner (1945, 1946, 1954, 1959).

Burchell originally recorded only the snail *Pupilla muscorum* from the brickearth sequence, which led him to conclude that a cold climate was represented (Burchell, 1935a).

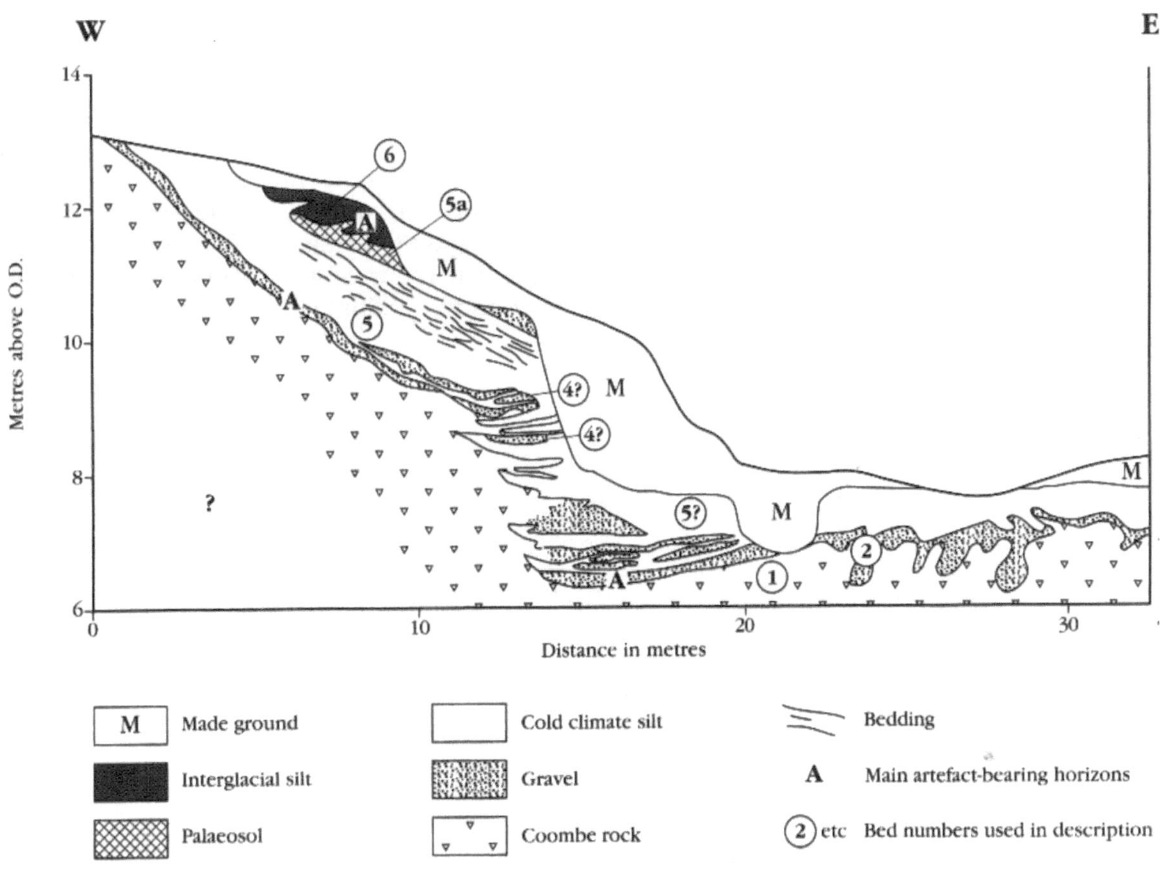

Figure 4.37 Section excavated at Northfleet (B, Fig. 4.34) by the British Museum (after Kerney and Sieveking, 1977). This is believed to coincide with Burchell's main 'Ebbsfleet Channel' section. Numbers refer to the description in the text.

However, later collecting revealed the presence of additional species within the silt (see above, bed 5) and, more importantly, led to the recognition of a temperate-climate bed (bed 6) (Zeuner, 1945, 1946, 1954, 1959; Burchell, 1954, 1957). This produced a rich molluscan fauna numbering some nine freshwater and 16 terrestrial species (Burchell, 1957); amongst the former *Corbicula fluminalis* and amongst the latter *Discus rotundatus* (Müller) were regarded by Burchell as particularly indicative of a climate at least as warm as at present. Carreck (1972) considered the record of *C. fluminalis* to be dubious. Another species listed by Burchell that is important as an interglacial indicator is *Azeca goodalli* (Férussac) (R.C. Preece, pers. comm.). Bed 6 also yielded bones and/or teeth of *Megaloceros giganteus* (giant deer), *Equus ferus* (horse), *Mammuthus primigenius* (mammoth) and rhinoceros, together with an assemblage of pointed hand-axes, classified by Burchell (1957) as Micoquian (see, however, Roe, 1981).

Burchell described a change in the molluscan fauna in the upper part of Bed 6 to a much more restricted assemblage, which, he believed, heralded the return of cold conditions prior to the formation of the immediately overlying (upper) coombe rock (bed 7). Beneath the temperate-climate silt a zone of weathering has been recognized at the top of bed 5 (Zeuner, 1945, 1946, 1954, 1955, 1958, 1959; Dalrymple, 1958; Catt, 1979; Kemp, 1984, 1991; Fig. 4.37).

The precise locations of the Baker's Hole section and of exposures studied at later dates have been subjects of considerable debate, largely because of the imperfect records made at the time. Wymer (1968) placed the original locality approximately 0.8 km to the south-east of the GCR site. Roe (1968b), on the other hand, cited a location 0.3 km to the west of the latter. Carreck (1972) considered Smith's Baker's Hole site to have been *c.* 200 m to the south-east of the surviving sections, at around TQ 614738. A recent detailed study of early publications and of notes and maps preserved with the Palaeolithic collections in the British Museum has led to the confirmation of Carreck's location (Wenban-Smith, 1990, and pers. comm.). It is possible that remnants of this important sequence survive beneath the bed of a disused tramway (Fig. 4.34), thought to be located within 20 m of the faces originally studied by Smith (1911), records of which have recently been discovered in the Natural History Museum (F.F. Wenban-Smith, pers. comm.; Fig. 4.36). The sections studied by Burchell in the 1930s are thought to have been close to the present GCR Section B (Carreck, 1972; Fig. 4.34).

The remnants of Burchell's Ebbsfleet channel site were reinvestigated by the British Museum in 1969. This work remains largely unpublished, but a short note appeared in 1977, coinciding with a visit to the site by INQUA (Kerney and Sieveking, 1977). A section included in that report largely confirmed the earlier descriptions, although showing numerous

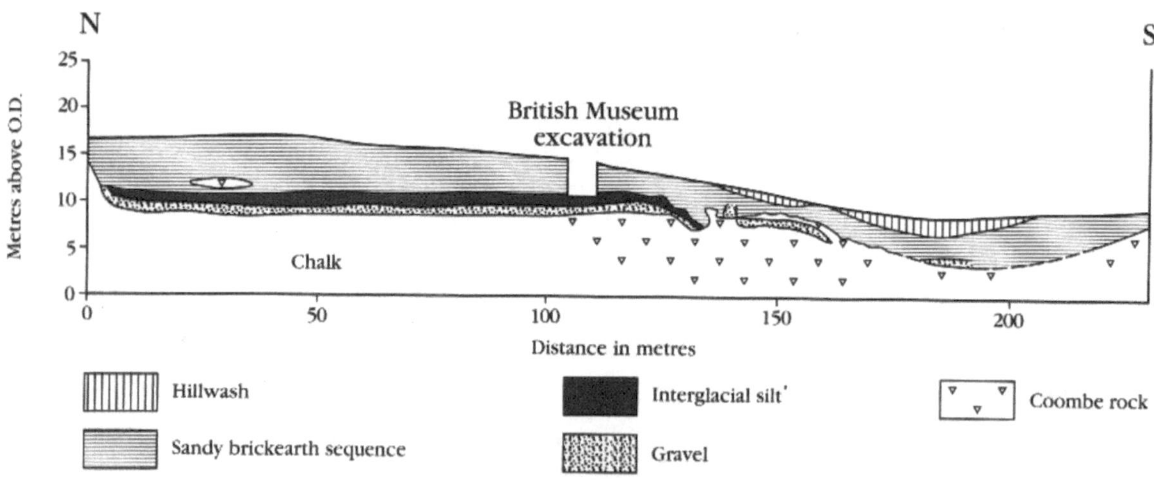

Figure 4.38 Section excavated at Northfleet (A, Fig. 4.34) by the British Museum (after Kerney and Sieveking, 1977).

interdigitations between the coombe rock and the Ebbsfleet loams (Fig. 4.37). The site had unfortunately been damaged by this time by the removal of the highest deposits (the top of bed 6 and all higher beds had been entirely quarried away by that time), and by tipping against the old faces. The occurrence of Levallois artefacts and a cold mammalian fauna in the basal gravels (bed 4?) was confirmed (Kerney and Sieveking, 1977), the palaeoliths corresponding typologically with Smith's (1911) Baker's Hole industry.

All that now remains of Burchell's Ebbsfleet valley locality are two residual islands and a linear face trending NNW–SSE (see Fig. 4.34). These show Chalk overlain by coombe rock, into which the Ebbsfleet deposits are channelled. One of the islands (B, Fig. 4.34) was excavated by the British Museum in the late 1960s (Kerney and Sieveking, 1977) and found to show the edge of the fluvial deposits, banked against coombe rock and Chalk (Fig. 4.37). This may possibly be close to the area studied by Spurrell (1883a, 1883b), before the Baker's Hole site was discovered (Carreck, 1972; Fig. 4.34). The other island (C, Fig. 4.34), shows a predominantly colluvial sequence, but without the channel edge being visible (Carreck, 1972; F.F. Wenban-Smith, pers. comm.). The linear face further to the north (A, Fig. 4.34) has also been investigated by Carreck (1972) and the British Museum (Kerney and Sieveking, 1977). The sequence in this section, overlying Chalk and coombe rock at *c.* 9 m O.D., comprises waterlain gravels and silts capped by sandy loessic or colluvial/loessic deposits (Fig. 4.38). These deposits have yet to be directly related to Burchell's section, but the waterlain silts have yielded freshwater molluscs, land snails and mammals indicative of a temperate climate and an open habitat. The molluscan assemblage is dominated by aquatic gastropods, particularly *Lymnaea truncatula*, *L. peregra* and *Anisus leucostoma* (M.P. Kerney, pers. comm.). This assemblage differs markedly from that in the temperate silt (bed 6) in section B, but contains no stratigraphically diagnostic species. It appears to represent stagnant swampy conditions rather than a typical fluviatile environment, as is signified by the fauna from bed 6. The contrast between the temperate-climate shelly beds in sections A and B may indicate a difference in facies rather than in age. The altitude of the two deposits is closely comparable (*c.* 11 m, section A; *c.* 12 m, section B), which suggests, if both can be confirmed as part of fluviatile sequences, that they may be of similar ages or even lateral equivalents. It is important to establish whether this is the case; if so, the deposits overlying the temperate-climate shelly silt in section A might equate with the higher part of Burchell's sequence, now removed by quarrying from the area of section B.

According to Carreck (1972), a record by Burchell (1935b, p. 330) of 25 molluscan taxa, from 'between Swanscombe and Northfleet', is an early reference to bed 6 in section B. Dominated by ***Trichia hispida***, this assemblage was interpreted by Burchell as indicating a climate at least as warm as that at present. However, he provided no details of the location from which this fauna came and did not refer back to this earlier report in 1957. Furthermore, not all the species in the earlier list are present in the later one (Burchell, 1957). There are also some similarities between the unprovenanced assemblage and that obtained from section A; it is therefore possible that the 1935 assemblage is transitional between the two recorded later. It remains to be demonstrated whether all these molluscan records are from a single, variable bed within the Northfleet sequence.

Polished facets on bone fragments and flints from the temperate-climate silts in section A were interpreted by Kerney and Sieveking (1977) as possible human artefacts. Carreck (1972) had previously recorded polished flints and bone fragments from the surviving sediments at Northfleet (not just from section A). He had concluded, however, that these resulted from natural processes. Carreck did record a right ilium of horse from section A, from a stratigraphical level that places it later than the temperate-climate shelly bed, that showed signs of having been cut. He regarded this as the best candidate for a bone artefact from his own collection, but stressed the need for an assessment of the material obtained from the British Museum excavations. This assessment is still awaited.

The deposits overlying the temperate-climate sediments in section A have particle-size and mineralogical characteristics conforming with a loessic origin, but differ in their mineralogy from bed 5 in section B (J.A. Catt and A.H. Weir, pers. comm.; see below).

Interpretation

Abbott (1911) observed the association at Baker's Hole of a depression scoured into the Chalk with brecciation of the bedrock, festooning of the Pleistocene beds, deformation of these strata with slickensides along highly inclined slip-surfaces, together with a general mixing of fossils, archaeological relics and Pleistocene deposits. This led him to suggest that the movement of a heavy frozen mass passing from higher to lower ground had occurred. Of the various early explanations of coombe rock formation, this is one of the closest to the modern interpretation of such deposits as the result of mass-movement (solifluction).

Reid (in Smith, 1911) regarded the coombe rock at Baker's Hole as the product of rainwash derived from chalk slopes, the surfaces of which were partly frozen, during a period of intense cold. Bromehead (in Dewey *et al.*, 1924) suggested that the deposit was laid down by torrents of water produced by summer thaw, likening this to modern processes in Siberia. Similar ideas of torrential deposition of the Baker's Hole deposits were proposed by Jessop (1930). Dewey (1930) was the first to attribute the Main Coombe Rock to mass-movement associated with permafrost conditions, although he regarded the inundation of the working floor as a sudden and catastrophic event.

Interpretation of Palaeolithic assemblages

The importance of the Ebbsfleet valley as a source of unusual and distinctive Palaeolithic material was recognized at an early date. Spurrell (1883a, 1883b) noted the great variety of size, shape and freshness of artefacts from Northfleet and was clearly aware of the method of their manufacture (he gave (1883a) an accurate description of what was later to be termed the Levallois technique). Smith (1911) likened the industry at Baker's Hole to that from Le Moustier in the Dordogne but, although he mentioned that Levallois flakes occurred, the latter term was first applied to the assemblage as a whole by Dewey (1932).

Abbott (1911) repeatedly visited the site from 1892 and recovered implements and 'debitage' of the typical Northfleet type. Following an expansion of quarrying in 1907, large numbers of worked flints were collected by J. Cross; these formed the majority of the implements described by Abbott, who suggested the names 'Prestwichian' and 'Ebbsfleetian' to describe the industries represented at Baker's Hole, terms that did not receive widespread acceptance.

Smith (1911) considered that a small proportion of the Palaeolithic assemblage from Baker's Hole was derived from the earlier '100 ft Terrace' (Boyn Hill/Orsett Heath) deposits, through which the Ebbsfleet valley was eroded prior to the formation of the coombe rock. At least 99% of the palaeoliths were of a distinctive type, however, and were taken by Smith to represent the indigenous Baker's Hole industry. These consisted of flakes and cores, unrolled and usually unpatinated, which were classified by Smith as of 'Le Moustier' type. He concluded that this material represented the debris of a working floor that had been inundated by deposition of the coombe rock.

Burchell (1931) described the industry from the Baker's Hole working floor as Early Mousterian, with Levallois in parentheses. He later considered the unabraded Palaeolithic material collected from the gravels and silts overlying the coombe rock to represent a distinctive later industry, which he classified as 'Levallois D', recognizing certain Aurignacian (Upper Palaeolithic) characteristics in this assemblage (Burchell, 1933). In this paper (1933), he classified the industry from the Main Coombe Rock (bed 1) as 'Levallois B', believing it to post-date the industry from the Upper Loam at Swanscombe, which he classified as 'Levallois A' (see above, Swanscombe). Burchell believed there to be a typological succession of Levallois industries that could be related to terrace stratigraphy in the Lower Thames. He later described unrolled and unpatinated hand-axes from the junction between beds 3 and 4, regarding these as contemporaneous with the Levallois industry, a similar association having been recorded in the Somme valley (Burchell, 1936a, 1936b).

Correlation

The early workers were impressed by the similarity between the coombe rock at Baker's Hole and that on the Sussex coast (see Reid, in Smith, 1911). The occurrence of narrow-nosed rhinoceros *Dicerorhinus hemitoechus* at Baker's Hole was somewhat problematic, since the Sussex coombe rock, the palaeontology of which is otherwise similar, contains the woolly rhinoceros *Coelodonta antiquitatis*. Smith (1911) argued that the solitary tooth of *D. hemitoechus* on which the identification was based might have been derived from earlier deposits, along with a small proportion of the palaeoliths. This is certainly a possibility, since *D. hemitoechus* occurs in the Swanscombe sediments, which cap higher ground to the west of the Ebbsfleet valley (see above, Swanscombe).

Newton (in Smith, 1911) likened the restricted mammalian fauna at Baker's Hole to those from the 'Middle Terrace' of the Thames at Grays and Ilford (see above, Globe Pit and Aveley). Bromehead (in Dewey *et al.*, 1924) considered that the coombe rock (bed 1) had been deposited during the period of erosion that followed the deposition of the 'Middle (Taplow) Terrace' deposits of the Lower Thames, thus supporting Newton's correlation. Dewey (1932) correlated the cold period during which the coombe rock was deposited with the second of the two East Anglian glaciations recognized at that time, which would seem to imply a Saalian age. A similar correlation was proposed by Burchell (1931). Breuil (1932a, 1934) also assigned the Main Coombe Rock to the Riss (glacial) Stage (= Saalian) and interpreted the artefacts as 'early Levallois'. He regarded the deposits as earlier than, or contemporaneous with, the nearby Crayford sediments, now ascribed to the late Saalian Mucking Formation (see above, Lion Pit), which also yield Levallois artefacts.

Burchell (1933) disagreed with Bromehead's interpretation of the Main Coombe Rock as a post-'Taplow Terrace' accumulation, pointing out that coombe rock had never been found overlying this terrace; instead he suggested that the gravels and 'brickearths' that he had observed in the Ebbsfleet valley, cut into the Main Coombe Rock (Fig. 4.37), were probably of 'Taplow' ('50 ft Series') affinities. Burchell likened his 'Levallois D' artefact assemblage from the Ebbsfleet deposits to the industry at Crayford (see above, Lion Pit), thus concurring with Breuil, and claimed support for this view from the fact that both sets of deposits yield *Coelodonta antiquitatis*. He also considered that a comparable sequence of brickearth overlying coombe rock could be observed to the north of the Thames in the Grays area, an apparent reference to the sediments at West Thurrock (see above, Lion Pit).

In their well-known synthesis of Pleistocene deposits in the Thames valley, King and Oakley (1936) proposed two new stage names that relied heavily on evidence from the Northfleet area. The first was the ill-defined, possibly multiple, 'pre-Coombe Rock Erosion Stage(s)'. To this King and Oakley attributed the cutting of the various 'benches' recognized beneath Taplow terrace deposits in the Thames valley, the erosion and infilling of the Wansunt 'channel' (Dartford Heath – see above, Wansunt Pit) and the occupation of the Levallois working site at Baker's Hole. They also proposed a 'Baker's Hole or Main Coombe Rock Stage', which they correlated with the 'Little Eastern glaciation' (Saalian Stage). They believed that the coombe rock covering the Baker's Hole working floor was emplaced and subsequently dissected by the Ebbsfleet during this 'stage' and attributed the Ebbsfleet Channel deposits to the succeeding 'Taplow Stage', thus agreeing with Burchell rather than Bromehead. They also followed Burchell (1936c) in placing the uppermost part of the Ebbsfleet Channel sequence (beds 10 and 11) in their 'Ponders End or Upper Floodplain Terrace No 1 Stage', suggesting that the uppermost silt bed (11) had more the nature of hillwash than fluviatile alluvium. This last statement was disputed by Burchell (1936a), who claimed that, despite its steeply sloping base (from +15 m to -3 m O.D.), this bed contained sedimentary evidence for a fluvial origin.

Zeuner (1945), summarizing the Ebbsfleet sequence, interpreted the predominant silts as loessic deposits, citing the results of mechanical analyses. This view has received recent support from Catt (1977, 1978, pers. comm.; see below). Zeuner (1945, 1955) also described a weathering horizon at the top of bed 5, immediately beneath the temperate-climate silt (bed 6), that showed intense rubification, indicative of an interglacial climate. Catt (1979), however, observed that the reddened horizon contains less illuvial clay than would be expected in a typical interglacial soil. Kemp (1984, 1991) has recently confirmed Catt's observation; he noted that a number of pedogenic features can be recognized, not only in the reddened zone, but also more extensively within bed 5. Kemp considered the surviving material to represent only the basal part of a truncated soil profile, perhaps below the main levels of clay illuviation. He refuted Zeuner's claim that the Northfleet buried soil is strongly developed.

Zeuner (1945) correlated the aggradation (to *c.* 12 m O.D.) of the fluviatile temperate-climate deposits in Burchell's section with his 'Late Monastirian' sea level. He considered that this high-sea-level phase occurred around 125,000 years BP, which would imply correlation with the Ipswichian Stage (*sensu* Trafalgar Square). However, he noted that the sediments at the Baker's Hole site were aggraded to *c.* 16 m O.D., which he thought sufficient to suggest a correlation with the earlier 'Main Monastirian' sea level, which he dated at *c.* 150,000 years BP. He attributed both these high-sea-level phases to the 'last interglacial'; however, the last-mentioned date would now be considered to fall late within the Saalian Stage and to equate with the Oxygen Isotope Stage 6 cold episode (Table 1.1).

Examination by Oakley of the Palaeolithic assemblage collected by Spurrell from Northfleet led Oakley and King (1945) to express misgivings about their earlier correlation of the Baker's Hole coombe rock with the Saalian (King and Oakley, 1936). They decided, in collaboration with A.D. Lacaille, that the typology of the Baker's Hole assemblage placed it later within the 'Levalloisian' than had previously been thought. Oakley and King (1945) concluded that if the Baker's Hole industry, buried by the Main Coombe Rock, was 'late Levalloisian', the aforementioned bed must be post-Saalian. However, Breuil (1947) suggested that local variation in cultural development was responsible for this apparent anomaly and that the Baker's Hole industry, although similar to the 'Early Upper Levallois' in France, was in fact of 'Lower Levallois' (Saalian) age. He thought that a possible reason for such diachronism in the archaeological record was the southward migration of Palaeolithic Man from England in response to climatic deterioration, introducing more advanced techniques, developed in England, into the French sequence at higher stratigraphical levels.

Tester (1958) concurred with Breuil, pointing out that the Main Coombe Rock, if not Rissian (Saalian), would have to be considered of Würm (Devensian) age, which would imply a 'last glacial' age for the Taplow Terrace and leave a large hiatus between this and the (Hoxnian) Boyn Hill Terrace. Tester suggested that the character of the Baker's Hole industry resulted in part from the unusual abundance of excellent flint that was available following the pre-Main Coombe Rock downcutting. This allowed the extravagant Levallois technique to be used, there being no need to conserve raw material. This suggestion by Tester pre-empted modern interpretations, which recognize the availability of raw material as a major influence on the use of the Levallois technique and no longer consider it possible to recognize an evolutionary sequence within the Levallois assemblages from the Lower Thames (Roe, 1981).

In his later work, Burchell (1957) suggested a direct correlation between the temperate-climate silt (bed 6) in the Ebbsfleet succession and the similar '*Corbicula* bed' at Crayford (see above, Lion Pit). The temperate-climate sediments at Northfleet were attributed by Kerney and Sieveking (1977) to the Ipswichian Stage, mainly on the basis of their elevation, since they contain no pollen or stratigraphically significant fauna. However, Stuart (in Sutcliffe and Kowalski, 1976) noted that teeth of the vole *Arvicola cantiana* from the Burchell collection were of early type, implying (according to Sutcliffe and Kowalski) a pre-Ipswichian (*sensu* Trafalgar Square) age. The occurrence of *C. fluminalis* may provide support for this view, since this species is now regarded by some authorities as having been absent from Britain during the Ipswichian Stage (*sensu* Trafalgar Square) (see Chapter 2, Stanton Harcourt and Magdalen Grove). Note, however, that the occurrence of *C. fluminalis* at Northfleet requires confirmation. Carreck (1972) noted that reworked fragments of a Palaeogene *Corbicula* species are common in the Ebbsfleet sediments. Unfortunately, amino acid analyses of shells from Northfleet, which might provide a further indication of their age, have yet to be carried out (see, however, note at end of Interpretation section).

Care must be exercised in applying the new stratigraphical scheme for the Lower Thames terraces, presented in this volume, to the Ebbsfleet deposits, as they occur in a tributary valley, in which steeper gradients might be expected. However, the elevation of the water-lain sediments, between 7.5 m and 12 m O.D., suggests an association with the Taplow/Mucking Formation of the main river. This would confirm a correlation of the Northfleet sediments with the Taplow aggradation, as advocated by Burchell and others, but not with the deposits mapped as 'Taplow' in the Lower Thames (these are older; see above, Purfleet and Globe Pit), a fact that may have led to the dispute between Burchell and Bromehead (see above). This interpretation implies a correlation of the temperate-climate bed at Northfleet with the various interglacial deposits within the Mucking Formation of the main valley, such as those at Aveley and West Thurrock. Although regarded by many previous authors as being of Ipswichian age, these have been interpreted in this volume as representative of an undefined post-Hoxnian/pre-Ipswichian temperate interval, the equivalent of Oxygen Isotope Stage 7 (see above, Lion Pit and Aveley). The underlying fluviatile and aeolian beds at Northfleet can therefore be ascribed to Oxygen Isotope Stage 8, whereas the sediments above the temperate-climate silt probably date from Stage 6. This interpretation receives support from the association, at other sites in the Lower Thames, of sediments ascribed to Stage 8 with Levallois artefacts (see above, Purfleet and Lion Pit). The evidence from Northfleet may therefore help to reinforce the view that the recognition of this technique within the British Palaeolithic record may be of considerable stratigraphical significance.

The thermoluminescence dating technique has recently been applied to the Northfleet deposits (Parks and Rendell, 1988), sediments having been sampled from section A (H.M. Rendell, pers. comm.). These range between 149,200 and 115,600 years BP, results that would seem to point to a younger age than that suggested above. However, Parks and Rendell emphasized that, because of problems encountered in using this technique to date pre-Devensian sediments, these should be regarded as 'minimum age estimates' (the implications of these dates cannot in any case be determined until full stratigraphical details are published).

Mineralogical analyses of the loess-derived sediments exposed in both sections A and B (Figs 4.37 and 4.38) by J.A. Catt and A.H. Weir (pers. comm.) provide further clues to the possible age of the Northfleet sequence. By determining the relative proportions of non-opaque heavy-mineral species in the silt and fine-sand size ranges, it has been shown that Devensian loesses, which are common throughout southern and eastern Britain, have a characteristic suite of such minerals, strongly dominated by epidote (Catt *et al.*, 1971, 1974; Eden, 1980; Bridgland, 1983a). The fluvially bedded silt (bed 5) below the temperate-climate horizon in section B was found to have significantly different mineralogical characteristics; for example, it contains less epidote, amphiboles and rutile than typical Devensian loess, but more zircon, tourmaline and kyanite. Bed 5 also contains a rare brown and green spinel, which has not been reported in Devensian loessic deposits. Whether these differences result from genuine distinctions between loesses of different ages (which would be of considerable stratigraphical value), or whether they reflect the addition of silt-grade material from other sources into the Ebbsfleet deposits, remains uncertain. However, J.A. Catt and A.H. Weir (pers. comm.) have observed similar heavy-mineral distributions in pre-Eemian loess in Belgium. They have found that the Ebbsfleet silt (bed 5) also resembles this pre-Eemian loess in its clay mineralogy; both are composed mainly of smectite, illite and kaolinite. Devensian loess at Pegwell Bay, Kent (Weir *et al.*, 1971), has a quite different clay mineralogy in that it also contains significant proportions of vermiculite.

Catt and Weir have also found that the suite of heavy minerals in the temperate-climate sediments in section A at Northfleet resembles that in the pre-'*Corbicula* bed' silt (bed 5) in section B and in the Belgian pre-Eemian loess. The clay mineralogy confirms this affinity, which suggests that reworking of sediment from the underlying silt into the temperate-climate deposit has occurred. The brickearth that overlies the temperate-climate silts in section A has a similar clay mineralogy to the other sediments at Northfleet, but its heavy-mineral assemblage has considerably more epidote and less zircon. Although these differences mean that this upper brickearth has a mineral content closer to Devensian loess than any other bed at Northfleet, its composition remains transitional; in fact it resembles the pre-temperate silt (bed 5) more closely than it resembles Devensian loess. These results indicate that, if this upper brickearth is primarily of loessic origin (as its particle-size distribution suggests), either it too is of pre-Devensian age or it has largely been reworked from the older silt (bed 5).

Summary

The complex site at Northfleet is of considerable importance to British Quaternary stratigraphy. It represents the most significant occurrence of Palaeolithic material in Britain showing the use of the Levallois technique. Past researchers have probably read more into the Palaeolithic sequence at this site than can be supported by more rigorous assessment, but knowledge and understanding of the Levallois technique as represented in Britain is heavily dependent on evidence from this locality.

Important evidence for dating the industry and for reconstructing the palaeoenvironment that prevailed when the deposits were laid down is provided by the fossiliferous parts of the Northfleet sequence. The succession represents an important phase in the fluvial history of the Lower Thames region, one in which the presence of Palaeolithic Man was significant. It is probable that a previously unrecognized temperate episode, intermediate between the Hoxnian and Ipswichian Stages of the traditional chronology, is recorded by part of the Ebbsfleet sediments. These temperate-climate sediments have been widely attributed to the Ipswichian, but according to the new Lower Thames stratigraphical scheme, adopted in this volume, they are believed to correlate with Oxygen Isotope Stage 7 (intra-Saalian). This implies broad correlation with sediments of the main Lower Thames at Ilford (Uphall Pit), Aveley, West Thurrock and Crayford (Table 1.1).

Note: since writing this report, amino acid ratios have been obtained from shells of *Lymnaea peregra* from section A. Three sets of analyses yielded similar results, giving the following mean ratios: 0.177 ± 0.020 (n = 5); 0.188 ± 0.022 (n = 5); 0.169 ± 0.038 (n = 8) (D.Q. Bowen and F.F. Wenban-Smith, pers. comm.). These results strongly support the arguments given above for a Stage 7 age for the temperate-climate deposits at Northfleet.

Conclusions

The Northfleet site records the deposition of sediments in an old channel of the Ebbsfleet, a tributary of the main River Thames. The sequence here is largely of cold-climate origin, commencing with a basal 'coombe rock', a deposit formed by the accumulation of Chalk debris as a result of slope movement under periglacial conditions. At the bottom of this deposit, at the original Baker's Hole site, a Stone Age working area was discovered. The flint debris of this working site, some of which could be fitted back together, showed the use of a technique known as 'Levallois'. Indeed, Baker's Hole is the most important Levallois site in Britain. A complex sequence of gravel, sand and silt overlies the basal coombe rock and has produced further Levallois material, as well as occasional mammalian bones and teeth and snail shells, all suggestive of cold conditions. Other types of flint artefacts have also been recovered. Silty material at two locations (at least) within the Baker's Hole complex has yielded shells of temperate-climate molluscs, implying that a warm episode is represented within the predominantly cold Northfleet sequence. Traces of soil formation in the earlier sediments beneath this bed seem to confirm that temperate conditions prevailed and that there was a brief break in sedimentation at this time. This deposit was overlain by further cold-climate sediments, now mostly quarried away, which showed evidence of frost action during periods of intense cold.

There is little information from which to date either the temperate episode represented at Northfleet or the important artefact-bearing sediments that make up the lower part of the sequence. Comparison with the terrace sequence in the adjacent Lower Thames valley, however, suggests that the deposits are of mid- to late Saalian age and that the temperate episode probably equates with Stage 7 of the deep-sea record – about 200,000 years BP.

Chapter 5

Essex

Introduction

Several sites in the extreme south-west of Essex have been described already, in Chapter 4, as they fall within the Lower Thames valley. Pleistocene fluvial deposits are widespread in the remainder of the county and their study has been of great importance in reconstructing the evolution of the Thames drainage system. The succession in Essex comprises not only the deposits of the River Thames (both pre- and post-diversion), but also of its major right-bank tributary, the Medway. A number of more recently formed left-bank Thames tributaries, which now drain the northern part of the county (Fig. 5.1), are also represented. Pleistocene sites in this area included in the Geological Conservation Review invariably reflect aspects of drainage development. At a few sites, interbedded glacial sediments are included, from which can be demonstrated the advance of the Anglian ice sheets into the area. Buried soil layers form an important part of the interest at some sites. There are also a number of important Palaeolithic localities, the most notable of which is at Clacton, the type locality for the Clactonian Industry and a site of great importance to Pleistocene geology.

A large part of Essex is covered by Pleistocene deposits. In addition to fluvial sediments, a widespread covering of till dominates the higher land in the north-western half of the county (Fig. 5.1; Whiteman, 1987). Considerable spreads of mostly fluviatile gravel both underlie and overlie much of the till and extend over large parts of the remaining, unglaciated districts. In addition to the Lower Thames deposits, described in Chapter 4, these gravels can be separated into three main divisions (Fig. 5.1): (1) the Kesgrave Sands and Gravels of central and northern Essex, which are pre-diversion Thames deposits laid down prior to the Anglian glaciation, (2) the East Essex Gravels, Medway and Thames-Medway deposits that form terraces running parallel to the Essex coast, and (3) the deposits of local rivers that have developed as tributaries of the Thames-Medway system following the Anglian glaciation. This chapter is divided into three parts, corresponding to these three categories of fluvial deposits.

Both the Kesgrave and East Essex Gravel Groups are made up of several component (terrace) gravel formations; High-level and Low-level Subgroups can be recognized within both groups (Table 5.1).

Research in this area has been less extensive than in the present Thames valley, although the fluvial record in Essex is now acknowledged as critical for reconstructing the development of Thames drainage during the Pleistocene. Prior to the definition of the Kesgrave Sands and Gravels by Rose and Allen (1977; Rose *et al.*, 1976), few authors regarded any of the gravels in Essex outside the Lower Thames valley as products of that river. Because the pre-diversion (Kesgrave) Thames gravels are overlain by Lowestoft Till over a wide area (Fig. 5.1), they were commonly attributed to glaciofluvial processes, an interpretation that has continued to receive support (Wilson and Lake, 1983). Recent work in north-eastern Essex has revealed remnants of temperate-climate deposits interbedded with the various formations of the Low-level Kesgrave Subgroup (Bridgland, 1988a;

Table 5.1 Lithostratigraphy of fluvial gravels in Essex.

Central and northern Essex		Eastern Essex	
Deposits of local rivers		Deposits of local rivers Low-level East Essex Gravel Subgroup	East Essex Gravel Group (Low-level)
Anglian glacial deposits		High-level East Essex Gravel Subgroup	East Essex Gravel Group (High-level)
Low-level Kesgrave Subgroup	Kesgrave Sands and Gravels Group		
High-level Kesgrave Subgroup	Kesgrave Sands and Gravels Group		

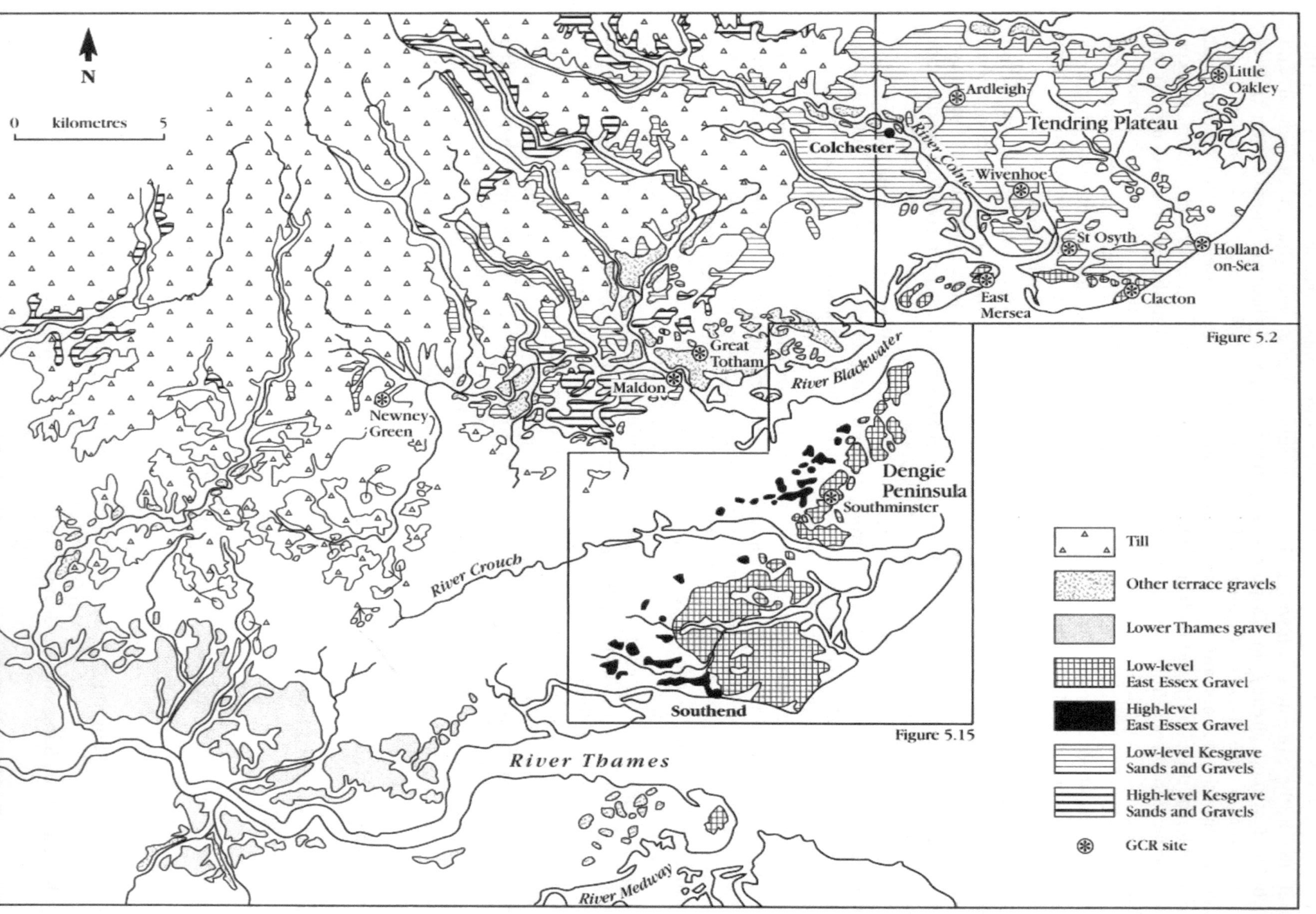

Figure 5.1 Pleistocene geology of Essex, showing the various types of gravel described in this chapter, the extent of the Anglian till sheet and the relation of these to the existing drainage systems (modified from Bridgland, 1988a).

Bridgland *et al.*, 1988). These promise to facilitate both the dating of the pre-diversion Thames formations and correlation of the Thames sequence as a whole with the established Pleistocene stratigraphy of East Anglia and that of north-west Europe (Gibbard *et al.*, 1991).

The East Essex Gravels have been variously attributed to marine processes, to the River Medway, to the Thames-Medway, or to glaciofluvial sedimentation (see Part 2 of this chapter). Recent work (Bridgland, 1980, 1983a, 1983b, 1988a) has shown that Medway and Thames-Medway deposits are represented by the High-level East Essex Gravel and Low-level East Essex Gravel Subgroups, respectively. The change from the former to the latter coincided with the diversion of the Thames (by Anglian ice) into its modern valley and, thereby, into the old Medway valley, which was already in existence across eastern Essex. The Low-level East Essex Gravel is represented in GCR sites at Clacton, East Mersea and Southminster.

Relatively little information has come from the deposits of the more recently formed tributary rivers, those which now have separate estuaries on the North Sea coast. Their deposits have yielded occasional artefacts and fossils, but interglacial sediments have rarely been recorded. However, these rivers provide the only fluvial record of the Late Pleistocene in Essex, as a combination of subsidence of the North Sea Basin and the low base level of the main river during the Late Pleistocene have resulted in Thames deposits of this age being confined, east of London, to the buried channel beneath the modern floodplain (see Chapter 4). The complex GCR site at East Mersea includes the only sediments within the GCR Thames coverage that can be ascribed with any confidence to the Ipswichian Stage (*sensu* Trafalgar Square); these are the Blackwater deposits described in Part 3 of this chapter, not the complex Thames-Medway sequence at Cudmore Grove, which appears in Part 2. Another Blackwater site, at Great Totham, provides probably the only example of Devensian fossiliferous deposits in the present volume (see, however, Chapter 3, Brimpton).

Part 1:

THE KESGRAVE SANDS AND GRAVELS
D.R. Bridgland

Introduction

The first of the three divisions of this chapter is devoted to sites associated with the pre-diversion Thames deposits of the Kesgrave Group. These deposits have only been accepted as the direct products of Thames drainage since the definition of the Kesgrave Sands and Gravels (Rose *et al.*, 1976; Rose and Allen, 1977). However, the typical sequence of Pleistocene sediments in which they occur, which is widespread in East Anglia, has been the subject of research extending back over a century. The main components of this sequence are the Kesgrave Sands and Gravels, mapped until very recent years as 'Glacial' by the Geological Survey, and the (Anglian Stage) Lowestoft Till, which overlies the early Thames gravels over much of their distribution. Localized occurrences of outwash gravel and aeolian deposits (coversand and loess) have been observed between the Thames gravels and the till. Other stratigraphically important parts of the sequence are superimposed warm- and cold-climate fossil soils, which occur at the top of the Kesgrave Group deposits throughout the area.

Early research

The occurrence of gravels beneath the East Anglian till sheet was first noted by Wood (1867, 1870; Wood and Harmer, 1868), who devised a tripartite stratigraphical scheme, based on the till-sand-till sequence of north-eastern Norfolk (now classified as Cromer Till, overlain by Corton Sands, overlain in turn by Lowestoft Till). In this scheme, the three divisions were respectively classified as 'Lower Glacial', 'Middle Glacial' and 'Upper Glacial'. Wood's classification was subsequently used by the Geological Survey when mapping Essex (Whitaker, 1877, 1889; Dalton, 1880). The upper (Lowestoft) till was widely recognized in this area and gravels underlying it were assumed to equate with those classified as 'Middle Glacial' in Norfolk. This view was strengthened by the recognition, at a few sites (such as Maldon – see below), of a till below gravel of similar type (Whitaker, 1889).

Prestwich (1881) observed that the majority of gravels underlying the till in eastern Suffolk were composed of large quantities of rounded flint and quartz, with significant amounts of subangular flint and Lower Greensand material, suggesting a southern derivation, in contrast to the northern provenance associated with the 'Glacial' drifts (including the 'Middle Glacial Gravel' of north-east Norfolk). He noted that these gravels, which he termed 'Westleton Beds', ranged as far afield as Essex and the Thames valley. He later traced possible equivalents of these beds, which he regarded as pre-'Glacial' marine deposits, over much of southern England (Prestwich, 1890b, 1890c). The distinction between the 'Glacial Beds' and 'Westleton Beds' of the Braintree area was outlined in some detail by French (1891). Despite these early attempts at clarification, the deposits later to be recognized as early Thames gravels (belonging to the Kesgrave Group) were variously classified by early workers as both 'Glacial Gravel' and 'Westleton Beds'. Furthermore, the early Geological Survey mapping recognized only 'Glacial Gravel', a category that included both the early Thames deposits and true glaciofluvial deposits. Whitaker (1889, p. 299) emphasized that the term 'Glacial' (with a capital G), as applied by the Geological Survey, was 'a proper name for a geologic period' (this can be correlated broadly with the post-Cromerian Pleistocene of modern usage) and not to be confused with the adjective meaning associated with or produced by ice. Unfortunately, such confusion became commonplace, the deposits mapped as 'Glacial Gravel' being widely interpreted as glaciofluvial outwash until the 1970s (see below).

Salter (1896) disputed the evidence for the marine origin of the deposits in Essex classified as 'Westleton Beds' by Prestwich. He pointed out that the rounded character of many of the flints in these gravels was inherited from the Palaeogene pebble beds that were their immediate source. Salter (1905) later included these beds in a series of gravels, traceable along the Chiltern dip slope from the Goring Gap towards the North Sea (Chapter 3), which he regarded as the product of an early north-eastward flowing drainage system that existed prior to the formation of the modern Lower Thames valley. Important evidence for the former existence of this drainage system, according to Salter, was the concentration of southern rocks in high-level deposits between the Lower Thames valley and the Chilterns. He realized that such gravels, containing abundant clasts of southern origin, were the products of right-bank tributaries of the old north-eastward flowing river.

Gregory (1894, 1922) regarded the bulk of the gravels of central Essex as 'pre-glacial' and divided them into 'High Level Quartzite Gravels' (or 'Danbury Gravels') and 'Low Level Quartzite Gravels' (or 'Brain Valley Gravels'), the latter coinciding largely with Prestwich's 'Westleton Beds'. Gregory believed that all these gravels had been deposited by streams flowing from the west, on a regional slope resulting from late Oligocene to early Miocene uplift. However, Solomon (1935) considered the two levels of 'Quartzite Gravels' recognized by Gregory to belong to a single series disturbed by tectonic activity, incorporating the 'Westleton Beds' (*sensu* Prestwich) and Wood's 'Middle Glacial Gravel'. Solomon classified these deposits as his 'Westleton Series', which he regarded as marine, although with a glacial intercalation in Norfolk. He ascribed lateral differences in composition to local variations in provenance within a single depositional basin. Solomon (in Clayton, 1957) claimed a distinction, based on heavy-mineral analysis, between two types of gravels in Essex: one, previously classified as 'Middle Glacial Gravel' or 'Westleton Gravel', pre-dated the Essex till sheet, whereas the other he regarded as a true outwash deposit, intimately associated with the till.

Warren (1955, 1957) reported that gravels deposited by the Thames in pre-Chalky/Jurassic Till (Lowestoft Till) times had been traced by Baden-Powell and himself from Oxford, across central Essex to the Clacton area. This interpretation, which confirmed the views of Salter (1905), was the first to indicate that Thames deposits could be recognized within the Pleistocene record of southern East Anglia. Based on clast lithology, but without systematic analyses, it anticipated the results of later work by Rose *et al.* (1976; Rose and Allen, 1977; see below).

Tills in Essex

Whilst progress was being made in the interpretation of the early gravel aggradations of southern East Anglia, research was being carried out in parallel on the glacial deposits that commonly overlie them. The 'monoglacial' model favoured by the 19th century workers was replaced by one that recognized four separate glaciations, as first envisaged in the Alpine region by Penck and Brückner (1901–1909). However, deposits of only the last three of these were identified in East Anglia, with the last confined to the extreme north-western corner of Norfolk (West, 1963). The two previous glaciations were considered to be represented by superimposed tills in East Anglia, differentiated on the basis of composition by Baden-Powell (1948) into a lower 'Lowestoft Till' and an upper 'Gipping Till'. Confirmation of this distinction was provided by studies of till fabrics by West and Donner (1956), who found significant differences in this respect between the Lowestoft and Gipping Tills and suggested respective correlations with the continental Elsterian and Saalian glaciations.

Clayton (1957) proposed a complex subdivision of the glacial deposits of the Chelmsford area, in which he recognized an older, weathered and dissected till (Hanningfield Till), largely covering plateaux, and a younger 'sandwich' of deposits filling valleys, comprising lower and upper tills separated by gravel. To this later tripartite sequence he applied the names Maldon Till, Chelmsford Gravels and Springfield Till, in ascending stratigraphical sequence. The Maldon Till was thought to be highly localized, whereas the Springfield Till was considerably more widespread. The Chelmsford Gravels, generally equivalent to the 'Middle Glacial Gravel' of Wood and Harmer (1868), were interpreted as outwash deposits. Clayton (1957) suggested a correlation between the Hanningfield Till and the continental Elsterian (= Anglian) glaciation and between the later tripartite sequence (Maldon Till, Chelmsford Gravels and Springfield Till) and the Saalian Stage. However, in an appendix to a later paper he proposed correlations with glacial drifts in Norfolk and Suffolk that implied an Anglian age for the Hanningfield and Maldon Tills (seen as equivalent to the Cromer and Lowestoft Tills, respectively) and a Saalian (Gipping Till) age for the Springfield Till (Clayton, 1960). In this later paper he suggested a Hoxnian age for the Chelmsford Gravels, which would seem to preclude their interpretation as glaciofluvial sediments, although no explanation of the change of view was provided. Clayton's earlier interpretation of the gravels as glaciofluvial was largely reiterated by Baker (1971), who suggested that the area may have been occupied by a preglacial Thames and that the deposits of this river may have been reworked and incorporated in the outwash.

The distinction of the products of two separate glaciations in southern East Anglia was seriously questioned by Baker (1971), who found that the Hanningfield and Springfield tills could not be differentiated either by lithological, stratigraphical or morphological evidence in north-west Essex. Baker also suggested that the Maldon Till resulted from a minor advance of the main 'Chalky Boulder Clay' glaciation of East Anglia (see below, Maldon), implying that all the glacial deposits of Essex are the product of a single glaciation. Subsequent mapping and lithological analyses of tills throughout East Anglia and the East Midlands led to confirmation that only a single glacial episode is represented amongst the Chalk-rich glacial deposits of these areas, equivalent to the Lowestoft Till of Norfolk (Bristow and Cox, 1973; Perrin *et al.*, 1973; Perrin *et al.*, 1979). Separate recognition of the partly Scandinavian Cromer Tills/ North Sea Drift glaciation was maintained, but the Gipping Till in its type area (the Gipping valley, near Ipswich) was found to be nothing more than a weathered profile in the upper part of the Lowestoft Till. Evidence for a glaciation equivalent to the continental Saalian was, however, accepted in the West Midlands and, with a good deal of caution, in the Breckland (Turner, 1973); a new name 'Wolstonian' was given to this post-Hoxnian and pre-Ipswichian glaciation (Shotton, 1973b; Chapter 1). Straw (1979, 1983) has continued to argue for an earlier 'Wolstonian' glaciation in northern East Anglia, principally on geomorphological grounds, and Wymer (1985b) has described sites in Suffolk and Norfolk where glacial deposits might overlie Hoxnian sediments. However, the view that only a single (Anglian Stage) glaciation, albeit with multiple ice advances, can be recognized in Essex has been consolidated in recent reviews (Baker and Jones, 1980; Whiteman, 1987; Allen *et al.*, 1991).

Even in the Wolstonian type area around Coventry there is growing uncertainty over the distinction between the deposits attributed to this glaciation and the Anglian deposits to the east (Perrin *et al.*, 1979; Sumbler, 1983; Rose, 1987). There are clear stratigraphical reasons for considering the whole of the 'Chalky till' to have been deposited during a single glaciation, during the Anglian Stage. Doubts have, however, been expressed elsewhere in this volume about the tenability of any model that places all these glacial deposits within a single glacial episode. These doubts stem from stratigraphical correlations of Chalk-rich tills on the Cotswolds and in the Vale of St Albans with the Thames terrace sequence, which suggest that the glaciations recognized in these two areas may have been separated by a temperate episode (see Chapter 1; Chapter 2, Long Hanborough and Wolvercote; Chapter 3, Part 2). It may be feasible, in the light of the more complex chronology indicated by the record from deep-sea cores (Chapter 1), for two separate glacial episodes to be represented within the 'Chalky Till' of eastern England, both of them post-Cromerian (*sensu* West Runton) and pre-Hoxnian (*sensu* Hoxne).

The recognition of early Thames deposits in southern East Anglia

Re-evaluation of the gravels underlying the till sheet of southern East Anglia has also taken place over the past quarter of a century. Hey (1967), following considerable reinvestigation of the deposits, concluded that the 'Westleton Beds' of the type area in north-east Suffolk were truly marine. However, he found elsewhere that gravels described by Prestwich (1881, 1890b, 1890c) under that name, and attributed by Solomon (1935) to his 'Westleton Series', differed from the type Westleton Beds and could frequently be shown to be younger. These gravels, generally mapped by the Geological Survey as part of the glacial sequence, were defined as the Kesgrave Sands and Gravels by Rose *et al.* (1976) and Rose and Allen (1977), who interpreted them as periglacial fluvial deposits of probable Thames origin. These authors demonstrated a distinction, on the basis of clast-composition, between the early Thames gravels and less extensive glacial outwash deposits, which they named Barham Sands and Gravels. They confirmed, therefore, the distinction made by Solomon (in Clayton, 1957) using heavy-mineral analysis.

Rose *et al.* (1976) proposed an important new stratigraphical scheme for southern East Anglia. From a study of sites throughout the area between Ongar (in the south-west), Thetford (in the north-west) and the coast, they found that the Lowestoft Till overlies widespread fluvial deposits of the pre-diversion Thames, their Kesgrave Sands and Gravels. The upper part of these deposits was frequently found to be rubified and clay-enriched, features indicative of pedogenic activity in a warm climate (Rose and Allen, 1977; Kemp, 1985a; Rose *et al.*, 1985b). This horizon occurs immediately beneath the Lowestoft Till or its associated outwash (Barham Sands and Gravels), indicating that soil development occurred prior to the Anglian glaciation. It was therefore concluded that the uppermost parts of the Kesgrave Group gravels incorporate the illuvial horizon of an interglacial soil, named the Valley Farm Soil (Rose *et al.*, 1976; Rose and Allen, 1977; Kemp, 1985a). As the Kesgrave Sands and Gravels overlie the Chillesford Beds at Chillesford Church Pit, which at that time were ascribed to the Pastonian Stage (Turner, 1973), the early Thames gravels were placed by Rose *et al.* (1976) and Rose and Allen (1977) in the subsequent cold stage, the Beestonian, and the palaeosol, its age constrained by the overlying Anglian glacial sediments, in the Cromerian Stage.

It is clear that intensely cold conditions prevailed following the formation of the Valley Farm Soil, even before the deposition of the Lowestoft Till; at many localities, including the GCR site at Newney Green, a periglacial soil is superimposed on the earlier warm-climate one. This periglacial soil, termed the Barham Soil, was recognized from both large- and small-scale structures related to frost activity; the larger structures include involutions and ice-wedge casts, whereas the smaller ones include disrupted clay skins and fractured gravel clasts (Rose *et al.*, 1976, 1985a; Rose and Allen, 1977). As well as being developed in the upper parts of the Kesgrave Group gravels (incorporating the earlier Valley Farm Soil), the Barham Soil has been shown to have developed in Cromerian interglacial sediments in north Norfolk. In such cases, where overlain by Anglian Stage tills, this soil has been established as a 'soil stratigraphic unit' of early Anglian age (Rose *et al.*, 1985a). Without these overlying deposits, in areas where Kesgrave Group gravels form the present land surface, it is possible to identify relict features of both the Barham and Valley Farm Soils (Rose *et al.*, 1976, 1985a, 1985b; Rose and Allen, 1977; Kemp, 1985a). However, later temperate-climate soils have been developed in the top of the Lowestoft Till (Rose *et al.*, 1978; Sturdy *et al.*, 1978) and cryoturbation has probably occurred during several cold episodes since the Anglian glaciation, so the recognition of pedogenic features characteristic of the Valley Farm and Barham Soils is of equivocal stratigraphical value beyond the glaciated area. Loess and coversand have also been recorded between the Kesgrave Sands and Gravels and the Lowestoft Till (Rose *et al.*, 1976, 1985a; Rose and Allen, 1977). These sediments, ascribed to the Anglian Stage, are frequently incorporated in the large-scale cryoturbation structures of the Barham Soil.

The recognition of a widespread Cromerian soil developed on the early Thames deposits of central and northern Essex led Rose *et al.* (1976) to suggest that the river had migrated to a more southerly route by that stage. However, subsequent work in the Vale of St Albans (Gibbard, 1974, 1977, 1978a; Green and McGregor, 1978a, 1978b; Chapter 3) and in eastern Essex (Bridgland, 1980, 1983a, 1983b, 1988a; Part 2 of this chapter) has indicated that the Thames was not diverted from its early course through Hertfordshire and central Essex until the Anglian glacial maximum. A possible explanation for this apparent contradiction is that the palaeosol seen beneath the Lowestoft Till in many gravel pits in southern East Anglia was formed on the terraces of the pre-glacial Thames valley, whereas the valley floor and channel occupied by the river immediately prior to its diversion, in the centre of the Mid-Essex Depression of Wooldridge and Henderson (1955), is deeply buried by Lowestoft Till and not exposed. In the unglaciated area east of Colchester, the lowest formation within the Low-level Kesgrave Subgroup has been ascribed to the Anglian Stage (Bridgland, 1980, 1983a, 1988a; Bridgland *et al.*, 1988, 1990). The Valley Farm Soil is not developed on this formation (see below, St Osyth and Holland-on-Sea).

Study of the clast composition of the Kesgrave Sands and Gravels reveals important minor components of value as provenance indicators. The gravels generally contain 0.5–2% sponge-spicular chert derived from the Lower Greensand of Kent and Surrey (Bridgland, 1980, 1986b; Green *et al.*, 1982). Carboniferous chert, for which there are a number of potential sources to the north and west, all of them outside the London Basin (Bridgland, 1986b), typically accounts for 1–2% of the total gravel content (Bridgland *et al.*, 1990). Also important is a restricted suite of volcanic rocks, some of which have been tentatively traced to sources in North Wales (Hey and Brenchley, 1977; Whiteman, 1983). The abundant quartz and quartzite (up to 35%) must also have been derived from outside the London Basin, the Midlands and Welsh borderlands representing the most likely source areas (Bridgland, 1986b). As was recognized by Rose *et al.* (1976), these various non-flint components are, when found in gravels in the London Basin, indicative of a Thames origin. A similar assemblage of component rocks has been recognized in the early terrace gravels of the Middle Thames (Green and McGregor, 1978a; McGregor and Green, 1978; Gibbard, 1985) and in the Northern Drift of Oxfordshire, also believed to represent early Thames aggradations (Hey, 1986; Chapter 2). Whether these rock-types were introduced into the Thames system by glaciation(s), by fluvial transport in a more extensive catchment or by a combination of the two remains a subject of controversy (Bowen *et al.*, 1986a; Bridgland, 1986b, 1988c; Hey, 1986; Whiteman, 1990; see Chapter 1).

Rose *et al.* (1976) and Rose and Allen (1977) recognized that their Kesgrave Sands and Gravels might represent a series of terrace aggradations, but the first formal subdivision was proposed by Hey (1980), who recognized the downstream continuation of his Westland Green Gravels of the Middle Thames (Hey, 1965; Chapter 3) within the older, higher-level part of the group (Fig. 5.1). Hey's distinction of the Westland Green Gravels from other Kesgrave Group deposits was based on both altitudinal and compositional considerations, which enabled him to trace that unit to Suffolk and Norfolk. Allen (1983, 1984) has further subdivided the Kesgrave Group in Suffolk, recognizing new terrace formations immediately above and below Hey's 'Westland Green Gravels', the Baylham Common Gravel and Waldringfield Gravel respectively. The Waldringfield Gravel has also been identified in northern Essex (Bridgland, 1988a; Fig. 5.2). Later studies of the Valley Farm Soil have shown that it is more complex than previously recognized, particularly on the older and higher formations within the Kesgrave Group, where evidence for several alternating phases of temperate-climate pedogenesis and periglacial disturbance have been determined from micromorphological studies (Rose, 1983a; Kemp, 1985a; Rose *et al.*, 1985b; see below, Newney Green). A distinction, based partly on clast lithologies, between high-level and low-level divisions of the Kesgrave Sands and Gravels, was first recognized by Hey (1980) and was adopted in the Chelmsford area by Bristow (1985). These divisions are each made up of a number of component formations and are regarded here as subgroups. Similar categories to these subgroups were proposed by Gregory (1922), his High Level and Low Level Quartzite Gravels.

Recent work by Whiteman (1990) has shown that the Westland Green Gravels as recognized in Suffolk do not, in fact, correlate with the unit of the same name originally defined by Hey (1965) in Hertfordshire and the Middle Thames. By determining (largely from borehole records) the distribution of the various formations within the Kesgrave Group in Essex and Suffolk, Whiteman has demonstrated that the formation in Suffolk that has been termed Westland Green Gravels is, in fact, the downstream continuation of the Gerrards Cross Gravel of the Middle Thames (see Chapters 1 and 3). It is therefore necessary to make considerable changes to the schemes for correlating the pre-diversion gravels of the Thames valley with the formations of the Kesgrave Group that have been published in recent years (Hey, 1980; Green *et al.*, 1982; Gibbard, 1983; Green and McGregor, 1983; Bowen *et al.*, 1986a; Bridgland, 1988a).

Whiteman's work has wider repercussions than are outlined above. He was able to suggest correlations, based on altitude and variations in clast composition, between central Essex and the higher parts of the terrace sequences upstream in both the Middle and Upper Thames regions (see Chapters 3 and 2 respectively). He concluded that the earliest true Thames deposits, from the Stoke Row Gravel up to and including the Gerrards Cross Gravel, are the product of a very large river, with a catchment extending far beyond the present Upper Thames (as already envisaged by Hey (1986)). In East Anglia these formations have all been included within the High-level Kesgrave Subgroup.

The formations of the Low-level Kesgrave Subgroup, thought until recently to represent the downstream continuation of Middle Thames terrace formations between the Satwell and Winter Hill Gravels (inclusive), have been shown by Whiteman to fall entirely between the deposition of the Gerrards Cross Gravel (last of the High-level Kesgrave Subgroup formations) and

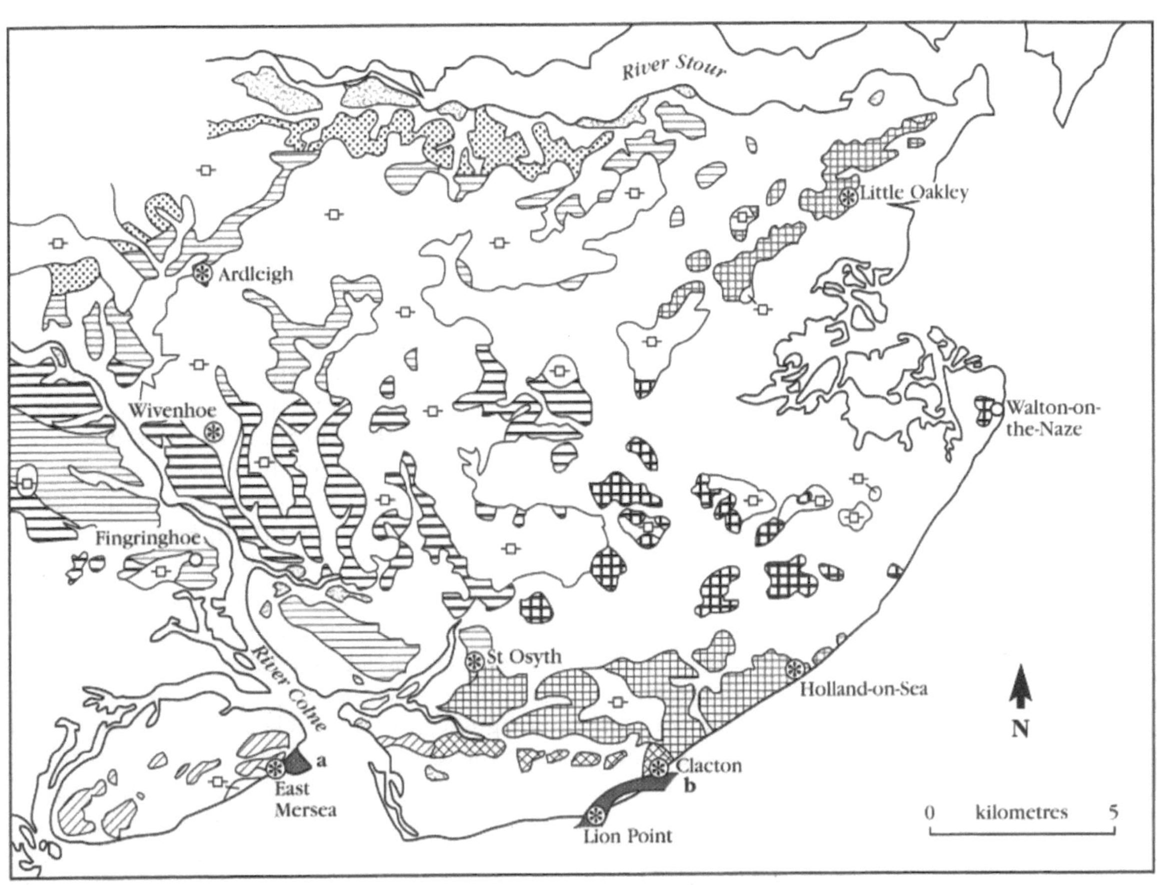

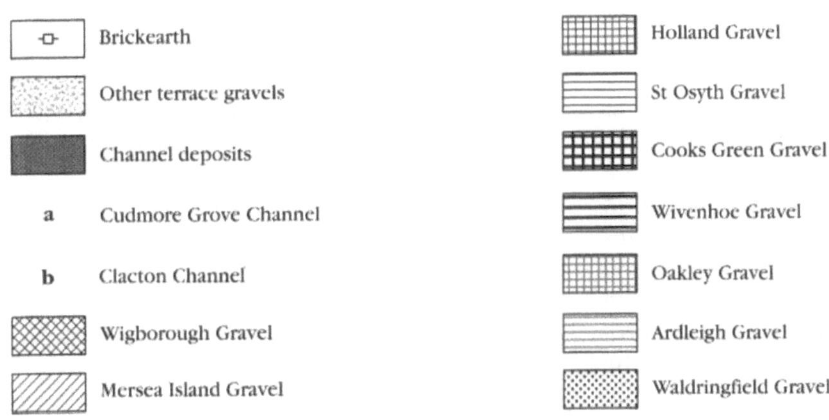

Figure 5.2 Pleistocene gravels of the Tendring Plateau (after Bridgland, 1988a).

the diversion of the river. Therefore an interval that in the Middle Thames is represented only by the Winter Hill and Rassler Gravels (see Chapters 1 and 3) is more fully represented by the Low-level Kesgrave Subgroup in Essex, which comprises four separate formations. Despite this revision, the correlation of the lowest of these with the Winter Hill Formation of the Middle Thames (Bridgland, 1980, 1983a, 1988a), based on stratigraphical evidence resulting from the diversion of the river, is upheld (see below, St Osyth and Holland-on-Sea).

Bridgland (1988a) has suggested that the deposits in northern Essex that are now classified as Low-level Kesgrave Sands and Gravels occupy a separate valley from that in which the formations of the High-level Kesgrave Subgroup were deposited. The latter valley lies further to the north and west and is separated from the former, in south-east Suffolk, by an interfluve of residual Red Crag (Allen, 1983). Further upstream the Low-level Kesgrave Subgroup formations are preserved in central Essex as the lowest tiers in a terrace 'staircase' of Kesgrave gravels that is largely obscured by later (Anglian) glacial deposits. Formations within the High-level Kesgrave Subgroup form the higher part of this 'staircase', which declines from north-west to south-east, towards the modern Chelmer and Blackwater valleys. To the south and east of these valleys, possible outliers of High-level Kesgrave Sands and Gravels have been identified on Danbury Hill and on the Tiptree Ridge (Hey, 1980; Bristow, 1985; Fig. 5.1). All these deposits, unfortunately, appear on the Geological Survey map (Sheet 241) as 'Glacial Sand and Gravel', a category that, in the Maldon area, appears also to include true glaciofluvial sediments as well as early post-glaciation Blackwater terrace deposits (see Part 3 of this Chapter, Maldon). Furthermore, the high-level outliers at Danbury and Tiptree have been thought to be affected by glaciotectonic processes at the edge of the ice sheet that filled the old Thames valley (Clayton, 1957; Hey, 1980; Bristow, 1985), which raises the possibility that they might have been substantially disturbed, elevated or even transported for some distance by the ice.

Important information about the age and stratigraphy of the Low-level Kesgrave Sands and Gravels has been gained in recent years from the discovery, at a number of sites in Essex, of interglacial sediments interbedded with these early Thames gravels (Bridgland, 1988a; Bridgland *et al.*, 1988, 1990; see below, Ardleigh, Little Oakley and Wivenhoe). It is hoped that the detailed interpretation of these sites will provide a framework for the relative dating of the lower Middle Pleistocene Low-level Kesgrave Subgroup, which can be extended to assist with the dating of the sequences in other parts of the Thames Basin and permit correlation with the sequences further north in East Anglia and on the Continent.

In addition to the recognition of individual (terrace) formations within the Low-level Kesgrave Subgroup, differences in gravel composition enable the distinction of western and eastern subdivisions of the lowest three of these formations on the Tendring Plateau (Bridgland, 1988a; Bridgland *et al.*, 1990; Figs 5.2 and 5.3), a name given by Warren (1957) to the peninsula between the Colne and Stour estuaries. The western subdivisions are lithologically typical of the Low-level Kesgrave Subgroup, but these pass eastwards (downstream) into deposits containing significantly more clasts of southern origin, predominantly Lower Greensand chert, with a complementary decrease in 'exotic' material such as quartz, quartzite and Palaeozoic chert (Tables 5.2 and 5.3). This change in gravel composition has been shown to result from the contemporary confluence between the Kesgrave Thames and the early Medway, which flowed from the Weald across eastern Essex to join the Thames in the area of the present Tendring Plateau (Bridgland, 1980, 1983a, 1988a; Fig. 5.4). The highest Low-level Kesgrave Subgroup formation, the Waldringfield Gravel, cannot be divided in this way, perhaps because the contemporary Thames-Medway confluence lay to the east, beyond the present coast (Bridgland, 1988a). The sequence of Low-level Kesgrave Subgroup formations on the Tendring Plateau is therefore as follows (Figs 5.2 and 5.3 and Table 5.3:

West (Thames)	**East (Thames-Medway)**
Waldringfield Gravel	
Ardleigh Gravel	Oakley Gravel
Wivenhoe Gravel	Cooks Green Gravel
Lower St Osyth Gravel	Lower Holland Gravel

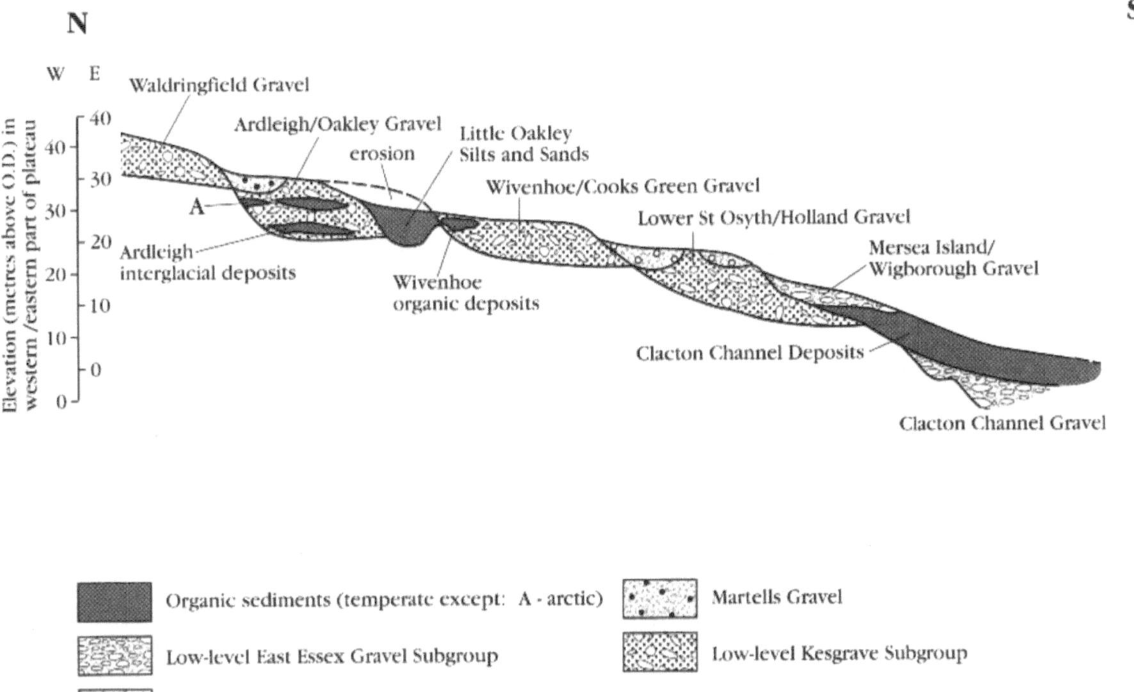

Figure 5.3 Idealized N–S transverse section through the Pleistocene deposits of the Tendring Plateau (after Bridgland, 1988a).

The Lower St Osyth Gravel and its Thames-Medway equivalent, the Lower Holland Gravel, are so called because they are overlain by later deposits, the Upper St Osyth and Upper Holland Gravels, that are not typical of the Kesgrave Group. These deposits can be closely correlated with sediments and events further upstream in the old Thames valley. They indicate the cessation of normal Thames drainage when the river was blocked by the Lowestoft Till ice sheet. The Upper St Osyth and Upper Holland Gravels both contain distal outwash material from the ice sheet, but as the Medway valley was unglaciated, the latter river made a significant and uninterrupted contribution to the Upper Holland Gravel (Bridgland, 1983a, 1988a; Fig. 5.4F; see St Osyth and Holland-on-Sea). The newly diverted Thames adopted its modern valley through London, from which it joined the former Medway valley across eastern Essex (Bridgland, 1980, 1983a, 1983b, 1988a). The deposits of this newly diverted Thames are represented on the Tendring Plateau as part of the Low-level East Essex Gravel Subgroup (Fig. 5.5A and 5.5B; see Part 2 of this chapter, especially Clacton-on-Sea).

NEWNEY GREEN QUARRY (TL 648065)

D.R. Bridgland

Highlights

This site provides evidence for the route of the pre-Anglian Thames through East Anglia and for the major glaciation that led to the diversion of the river into its modern course. Superimposed warm-climate and cold-climate soils separate Thames gravels and Anglian till at Newney Green and represent an important element of the regional stratigraphy.

Introduction

Newney Green Quarry is a key site that has been used to establish the lower Middle Pleistocene succession of the Chelmsford area (Rose *et al.*, 1976, 1978). This sequence comprises fluvial deposits, part of the Kesgrave Group, overlain by various products of glaciation during the Anglian Stage, including outwash gravel,

Table 5.2 Clast-lithological composition of the gravels described in Chapter 5, Part 1.

Gravel	Site	Sample	Flint			Southern		Exotics									
			Tertiary	Nodular	Total	Gnsd chert	Total	Quartz	Quartzite	Carb chert	Rhax chert	Igneous	Total	Ratio (sthrn: q/qtzt)	Ratio (qtz:qtzt)	Total count	National Grid Reference
Anglian glacial gravels (1)	Ugley	1	41.9	23.7	87.9			3.5	0.8	1.5	0.4	1.9	11.9		4.55	520	TL 516278
		2	3.6	37.6	87.1			2.6	1.7	2.1	1.7	1.9	12.6		1.56	420	
Upper St Osyth Gravel	Fingringhoe	1A	15.4	13.0	80.8	2.4	2.4	4.1	4.3	4.1	1.4	0.8	16.8	0.29	0.95	369	TM 0419 2017
		1B	15.9	13.7	81.7	0.7	0.7	5.7	6.8	0.9	0.9	0.9	17.7	0.05	0.84	453	
	St Osyth	2	8.7	19.1	89.8	2.1	2.1	4.2	0.9	2.1		0.2	8.1	0.41	4.35	530	TM 1196 1704
	11.2-16	*2(b)*	*14.9*	*9.4*	*78.4*	*2.4*	*2.4*	*11.9*	*1.0*	*2.7*	*0.4*	*0.3*	*18.8*	*0.18*	*12.50*	*714*	
Upper Holland Gravel	Bypass Rd.	1A	9.9	21.3	82.1	8.8	9.5	3.8	0.4	2.3		0.4	8.4	2.27	2.50	263	TM 1161 1703
		1B	12.6	16.1	74.8	18.6	19.6	2.5	0.6	1.3		0.6	5.7	6.20	4.00	317	
	Earls Hall 1	1	11.5	11.3	75.8	21.7	21.7	0.3	0.8	0.5			2.2	19.75	0.33	364	TM 1432 1625
	11.2-16	*1(b)*	*16.0*	*8.8*	*80.3*	*10.2*	*10.5*	*4.2*	*1.5*	*2.7*	*0.3*	*0.1*	*9.1*	*1.86*	*2.78*	*932*	
		2	13.6	•	77.6	15.0	15.0	3.0	0.3	2.8		0.8	7.2	4.50	11.11	361	TM 1429 1625
	Burrs Road	1	10.8	•	64.8	30.0	31.4	0.7		1.4	1.4		3.8	45.00		287	TM 1926 1735
		2	11.6	•	66.0	28.4	28.8	3.1		1.3	0.6		5.3	9.20		320	TM 1927 1734
	Holland-on-Sea	1	15.5	9.7	70.7	24.5	24.7	2.2	0.2	1.0	0.5		4.6	10.20	11.11	413	TM 2108 1662
		2A	15.7	9.0	68.9	25.1	25.1	3.0	0.7	0.7		0.3	6.0	6.70	4.35	267	TM 2109 1663
	(transitional?)	2B	23.7	13.5	71.3	15.6	16.1	4.7	5.2	1.4	0.2		12.6	1.62	0.90	422	
Lower St Osyth Gravel	Fingringhoe	1C	31.4	12.5	85.1			4.8	8.0	1.9			14.9		0.60	376	TM 0419 2017
	Moverons	1	29.0	16.0	80.8	0.6	0.6	8.4	7.0	1.4		1.0	18.3	0.04	1.20	929	TM 0711 1818
		2	30.8	16.5	79.6	1.1	1.1	11.2	5.3	0.7		0.5	19.3	0.07	2.13	1031	TM 0712 1819
	11.2-16	*2*	*32.3*	*5.9*	*73.5*	*1.6*	*1.7*	*14.2*	*7.6*	*1.7*		*0.8*	*24.7*	*0.08*	*1.89*	*1330*	
		3	31.8	13.1	77.5	0.6	0.7	11.4	7.8	1.3		0.7	21.7	0.04	1.47	994	TM 0699 1825
	St Osyth	1A	35.4	•	77.1	0.5	0.5	11.1	7.7	1.8		0.2	22.4	0.03	1.45	559	TM 1161 1704
		1B	30.6	•	79.8	1.5	1.6	10.4	4.9	1.3		0.7	18.6	0.10	2.13	748	
	11.2-16	*1B*	*30.1*	*7.7*	*78.0*	*1.7*	*1.7*	*12.5*	*4.8*	*2.0*		*0.5*	*20.2*	*0.10*	*2.63*	*1325*	
Lower Holland Gravel	St Osyth	3	31.6	16.8	83.1	1.4	1.4	10.3	2.7	1.6		0.5	15.3	0.11	3.29	561	TM 1201 1703
		5	21.8	10.8	80.0	4.6	4.9	5.8	6.8	1.8		0.3	15.1	0.39	0.85	325	TM 1213 1665
		6	29.5	16.0	81.2	2.2	2.5	8.8	5.3	1.3			16.0	0.18	1.67	319	TM 1225 1688
	Bush Paddock	1	43.3	10.5	83.9	4.8	5.1	5.9	3.7	0.8		0.3	11.0	0.53	1.59	647	TM 1357 1611
	11.2-16	*1*	*40.8*	*5.7*	*75.6*	*10.5*	*10.8*	*9.2*	*2.4*	*0.8*	*0.1*	*0.2*	*13.6*	*0.79*	*3.85*	*1215*	
	Holland-on-Sea	2C	32.8	11.9	80.6	2.2	2.2	8.0	7.5	1.0		0.2	17.2	0.14	1.06	412	TM 2109 1663
		2D	26.7	13.9	81.5	1.8	1.8	9.2	5.6	1.1		0.3	16.5	0.12	1.64	655	
	Holland Haven	1A	24.9	•	84.0	2.4	2.9	7.3	3.7	1.6	0.3	0.3	13.1	0.26	1.96	382	TM 2208 1744
		1B	34.6	14.6	83.1	2.3	3.1	9.6	4.6	0.4			13.9	0.24	2.08	260	
		2	25.3	•	82.2	2.8	3.0	8.4	3.9	1.7	0.2	0.2	14.8	0.24	2.17	534	TM 2205 1743
	11.2-16	*2*	*31.4*	*7.2*	*76.8*	*5.2*	*5.2*	*12.4*	*2.2*	*1.4*	*0.2*	*0.9*	*18.0*	*0.36*	*5.64*	*939*	
	Clacton cliffs	4C	33.9	15.2	81.3	8.5	9.0	3.2	4.4	1.2	0.7		9.7	1.18	0.72	433	TM 1739 1433
		4D	38.7	12.6	81.5	5.3	5.6	6.4	2.8	2.0		0.3	12.9	0.61	2.27	357	
	11.2-16	*4D*	*39.9*	*6.6*	*78.7*	*10.2*	*10.4*	*5.5*	*3.6*	*1.2*	*0.4*		*10.8*	*1.15*	*1.52*	*804*	
Wivenhoe Gravel	Wivenhoe	1B	25.1	17.8	80.1	0.8	0.8	5.4	9.7	2.7		0.3	18.3	0.05	0.56	371	TM 0494 2330
	(Wiv.U.Gr.)	2A	30.4	14.7	74.6	0.4	0.7	12.4	9.5	1.4		0.4	24.7	0.03	1.31	283	TM 0495 2358
	Arlesford	1	36.0	8.1	73.6	0.4	0.4	14.2	8.1	2.6		0.7	26.0	0.02	1.75	458	TM 0711 2192
	11.2-16	*1*	*31.1*	*4.7*	*66.1*	*1.4*	*1.4*	*17.7*	*10.6*	*2.1*		*1.1*	*32.4*	*0.05*	*1.67*	*716*	
		2	21.5	16.9	82.6	0.9	0.9	7.3	7.0	1.5		0.6	16.6	0.06	1.04	344	TM 0711 2192
Cooks Green Gravel	Cooks Grn	1A	21.3	•	83.8	3.2	3.2	7.2	3.5	1.0		0.5	13.0	0.30	2.04	625	TM 1889 1856
		1B	27.2	14.4	84.2	2.0	2.0	8.3	2.2	2.8			13.8	0.19	3.70	492	
	11.2-16	*1B*	*26.9*	*7.1*	*72.6*	*3.7*	*3.7*	*16.4*	*4.7*	*1.2*		*0.3*	*23.7*	*0.17*	*3.45*	*1205*	
		2	29.4	12.7	83.0	3.3	3.3	8.1	4.1	0.3		0.3	13.5	0.27	1.96	394	TM 1898 1840
	Gt Holland	1	25.5	19.1	84.0	1.7	1.7	8.4	6.0	0.7			16.0	0.12	1.41	419	TM 2112 1892
	11.2-16	*1*	*25.9*	*8.6*	*80.3*	*3.1*	*3.2*	*8.9*	*5.5*	*1.8*	*0.1*	*0.2*	*16.5*	*0.22*	*1.61*	*1289*	
Little Oakley Silts and Sands	L Oakley	AB	33.6	12.6	87.4	0.8	0.8	4.2	5.9	1.7			11.8	0.08	0.71	119	TM 2233 2952
	11.2-16	*AB*	*26.4*	*7.2*	*72.9*	*2.0*	*2.0*	*11.2*	*9.8*	*2.0*			*24.8*	*0.09*	*1.14*	*295*	
	AC	33.7	16.7	83.7	0.4	0.4	9.5	4.8	0.4			0.4	15.9	0.03	1.96	252	TM 2223 2951
	11.2-16	*AC*	*29.7*	*8.5*	*73.1*	*1.6*	*1.6*	*13.2*	*9.1*	*1.2*	*0.1*	*0.6*	*25.2*	*0.07*	*1.45*	*674*	
		AF	26.0	14.3	80.3	1.8	1.8	9.4	6.3	1.3			17.9	0.11	1.49	223	TM 2233 2946
	11.2-16	*AF*	*28.8*	*6.4*	*70.8*	*2.2*	*2.3*	*15.3*	*6.9*	*3.8*		*0.6*	*26.9*	*0.11*	*2.23*	*640*	
Martells Gravel	Ardleigh	3	20.5	14.1	76.6	1.6	1.6	10.4	8.2	1.4	0.2	0.8	21.9	0.08	1.26	512	TM 0515 2800
		4C	19.3	13.2	76.7	0.7	0.7	11.7	7.3	2.5	0.2	0.5	22.6	0.03	1.61	605	TM 0519 2807
	11.2-16	*4C*	*26.2*	*6.5*	*72.5*	*1.0*	*1.2*	*14.3*	*7.6*	*2.7*	*0.3*	*1.0*	*21.8*	*0.05*	*1.89*	*596*	
Ardleigh Gravel	Ardleigh	1	26.8	15.4	75.6	0.7	0.7	11.9	7.5	0.8		1.5	23.6	0.04	1.58	590	TM 0536 2802
	11.2-16	*1*	*27.1*	*7.7*	*72.3*	*1.3*	*1.7*	*15.4*	*7.3*	*2.2*		*0.2*	*25.9*	*0.07*	*2.13*	*1008*	
		2	23.7	19.2	80.0	1.3	1.5	9.3	5.4	2.0			17.7	0.10	1.72	615	TM 0533 2805
	11.2-16	*2*	*29.0*	*6.4*	*69.9*	*0.7*	*1.0*	*14.9*	*9.5*	*3.0*		*0.7*	*29.1*	*0.04*	*1.57*	*1219*	
		4B	29.3	13.0	75.4	1.3	1.3	9.8	9.4	1.1		1.1	23.0	0.07	1.04	447	TM 0519 2807
	(Ardleigh L.Gr.)	4A	33.3	12.7	72.0	0.4	0.4	11.4	13.9	1.1		0.9	27.5	0.01	0.82	553	

Gravel	Site	Sample	Flint: Tertiary	Flint: Nodular	Flint: Total	Southern: Gnsd chert	Southern: Total	Exotics: Quartz	Exotics: Quartzite	Exotics: Carb chert	Exotics: Rhax chert	Exotics: Igneous	Exotics: Total	Ratio (sthrn: q/qtzt)	Ratio (qtz:qtzt)	Total count	National Grid Reference
Oakley Gravel	Dovercourt	DA	30.3	12.1	79.2	2.1	2.4	9.8	4.7	1.6		1.1	18.5	0.17	2.08	379	TM 2328 3027
	11.2-16	*DA*	*25.3*	*8.3*	*75.0*	*4.7*	*4.9*	*11.7*	*5.5*	*1.7*			*20.1*	*0.28*	*2.13*	*783*	
	Little Oakley	KA	30.3	15.5	80.2	2.0	2.0	8.6	7.2	0.6	0.2	0.3	17.8	0.13	1.19	653	TM 2191 2947
	11.2-16	*KA*	*25.7*	*9.1*	*76.4*	*2.1*	*2.2*	*11.9*	*7.0*	*1.9*	*0.2*	*0.3*	*21.3*	*0.12*	*1.70*	*673*	
Waldring-field Gravel	Newney	1	52.9	*	72.5	0.5	0.5	10.5	11.7	0.8		1.2	26.5	0.02	0.90	599	TL 645064
	Green	2	55.3	4.1	70.6	0.6	0.6	14.7	10.2	0.8		0.6	24.9	0.02	1.45	490	
	Mistley Hth	1	26.3	13.7	82.7	1.6	1.6	9.6	9.0	1.6			15.6	0.09	1.06	365	TM 1282 3125

* Not separately recorded
(1) (for comparison. N.B. non-durables excluded - see, however, Table 3.1, and notes appended to Table 4.2., page 181)

coversand and till. The site has also provided exceptional opportunities to study a buried soil horizon that occurs at the top of the Kesgrave Sands and Gravels. This horizon, of considerable stratigraphical significance, comprises superimposed warm- and cold-climate fossil soils, the Valley Farm and Barham Soils respectively (Rose *et al.*, 1976, 1985a, 1985b; Rose and Allen, 1977; Kemp, 1985a).

The site at Newney Green is also of importance for studies of the glacial (Lowestoft Formation) deposits. The site is the type locality of the Newney Green (Till) Member (see below), a subdivision of the Lowestoft Formation proposed by Whiteman (1987, 1990; in Allen *et al.*, 1991). Although there has been considerable work on the palaeosols and till at Newney Green, the regional stratigraphical relations of

Table 5.3 Correlation of gravel formations in Essex within the Kesgrave Group with deposits in other areas.

See Chapter 3, Part 2: Middle Thames	See Chapter 3, Part 2: Vale of St Albans	Tendring Plateau: Low-level Kesgrave Thames	Tendring Plateau: Low-level Kesgrave Thames - Medway	South of Blackwater: High-level East Essex Gravel Medway	Climate	Stage
Winter Hill U.Gr.	Moor Mill Clay	Upper St Osyth Gr.[1]	Upper Holland Gr.[1]	Chalkwell/Caidge Gr.	Glacial	Anglian[2]
Winter Hill L.Gr.	Westmill L.Gravel	Lower St Osyth Gr.	Lower Holland Gr.	Chalkwell/Caidge Gr.	Periglacial	early Anglian
No equivalent formations recognized in the area upstream from Essex, with the possible exception of the Rassler Gravel of the Reading area (see Chapter 1 and Fig. 1.3)		– – – –Rejuvenation event– – – –				
		Wivenhoe Formation { Wiv.U.Gr. intgl.seds Wiv.L.Gr. }	Cooks Green Gravel	Canewdon/St Lawr.Gr.	Periglacial Temperate Periglacial	
		– – – –Rejuvenation event– – – –				
		Ardleigh Formation { Ard.U.Gr. intgl.seds Ard.L.Gr. }	? L.Oakley Slts & Sds[3] Oakley Gravel	Belfairs/Mayland Gr.	Periglacial Temperate Periglacial Temperate Periglacial	'Cromerian Complex'
		– – – –Rejuvenation event– – – –				
		Waldringfield Gr.	None recognized	Ashingdon Gravel?	Periglacial	

1 Not part of the Kesgrave Group (deposited while the Thames was blocked).
2 Anglian glacial maximum.
3 The Little Oakley Silts and Sands may date from the same temperate episode as the Ardleigh interglacial deposits.

the underlying Kesgrave Group gravels have been largely overlooked. This same situation exists throughout central Essex, where the various formations of the Low-level Kesgrave Subgroup are largely buried by Anglian Stage glacial deposits, precluding a regional synthesis using normal mapping techniques. A recent evaluation of borehole data (Whiteman, 1990) promises, however, to resolve this problem.

Description

Sections at Newney Green have varied considerably over the 15 years of the quarry's life. In particular, the till sequence is missing in lower areas, where it has been removed by post-Anglian erosion. The stratigraphical sequence is set out in Table 5.4 (see also Fig. 5.6).

The Kesgrave Sands and Gravels (1) are represented by the lowest and most extensive Pleistocene unit present at Newney Green. This comprises predominantly cross-stratified sands and gravels with foreset orientations indicating an eastward palaeocurrent direction (Rose *et al.*, 1978). The sedimentary characteristics of this deposit point to deposition by a braided river (Rose *et al.*, 1976). It has a gravel composition typical of the Kesgrave Group, with conspicuous cobble-sized clasts of volcanic rocks of the type attributed by Hey and Brenchley (1977) to sources in North Wales.

The Valley Farm Soil is well-developed at the top of the fluvial deposits, except in areas where the till has a strongly erosive base and directly overlies the lower parts of the Kesgrave gravels. This soil is apparent as a clay-rich, reddened horizon, with patches of grey mottling. This represents the illuvial horizon of an argillic soil, the formation of which required a temperate climate. The higher layers of this soil were presumably removed by erosion prior to the deposition of the overlying sediments. Other indications of clay illuviation in this temperate

Table 5.4 The Pleistocene sequence at Newney Green (after Rose *et al.*, 1978; Whiteman, 1990; see also Fig. 5.6).

			Thickness
6. Glaciofluvial gravel with laminated sand and silts			0–2.5 m
5. Lowestoft Till Formation	Great Waltham Member	5d Decalcified till	up to 3.75 m
		5c Chalk-rich till	
	Newney Green Member	5b Brown sandy till, rich in material reworked from the underlying deposits	
		5a Deformation till: comprises remobilized palaeosol and coversand, attenuated into laminae	
4. Glaciofluvial gravels (Barham Sands and Gravels)			0–1.5 m
3. Coversand and, rarely, loess. Both in localized lenses or, more commonly, in involutions within 2			0–1.0 m
2. Complex buried soil horizon (Valley Farm and Barham Soils)			0–1.0 m
1. Kesgrave Sands and Gravels (Waldringfield Formation?)			up to 3.4 m
London Clay			

soil are the presence of clay skins around gravel clasts and clay infillings of voids. The red coloration comes from haematite, the formation of which is evidence of a warm temperate environment with high seasonality (J. Rose, pers. comm.). Whiteman (1990) recognized five separate horizons within this soil in part of the Newney Green Pit, but these appeared to be primarily of sedimentary origin, reflecting original differences in parent lithologies.

The reddened Valley Farm Soil is usually deformed by the superimposed periglacial Barham Soil, which manifests itself as a complex of regularly spaced involutions, with occasional ice wedge pseudomorphs (Rose *et al.*, 1976, 1985a; Rose and Allen, 1977). Other characteristic features are frost-cracks, sand-wedges, fractured and vertically orientated stones, silty-clay cappings of sand grains and features associated with ground-ice development, such as platy aggregates and banded fabrics (for summary, see Rose *et al.*, 1985a). The cores of the involutions and wedges are often filled with coversand, in which wind-faceted stones (ventifacts) may be found (Rose *et al.*, 1978). During the working of the quarry, removal of the Lowestoft Till revealed, from time to time, the polygonal nature of involutions and wedges within the periglacial soil (Rose *et al.*, 1978, 1985a; Fig. 5.7). Possible remnants of higher, humic horizons have been observed on rare occasions, preserved within the cryoturbation structures of the periglacial soil. These were found to have a very low organic content and probably represent the diffuse humic horizons of an Arctic Brown Soil (J. Rose, pers. comm.). No pollen or microfauna has been found in them.

The glaciofluvial gravel (bed 4) occurs as localized lenses between the buried soil horizon and the till. It can be distinguished from the underlying Kesgrave gravels on the basis of its clast and heavy-mineral content and, within individual beds, its poorer sorting (Rose *et al.*, 1978). The clast component of this gravel includes Chalk and other (exotic) non-durable material, of types characteristic of the overlying till complex (Whiteman, 1990). The heavy minerals are particularly diagnostic; a rich assemblage has been recognized (Catt, in Rose *et al.*, 1978), including poorly durable varieties such as apatite and collophane. This assemblage is comparable with that from the coversands, but contrasts markedly with the mineral suite from the Kesgrave Sands and Gravels, which is restricted to residual grains such as tourmaline and zircon. This change from durable, residual minerals to a more varied assemblage, including easily weathered types, suggests an input of fresh material into the region. It therefore provides important evidence for the association of the upper gravel with the glaciation that deposited the overlying till (Rose *et al.*, 1978).

The Lowestoft Till uniformly covers the highest ground throughout the area of the Newney Green workings. In early descriptions it was reported to be generally homogeneous and structureless, with local banding; other recorded variations include lenses of flow till, identified at the base of the main (lodgement) till unit and differences in the frequency of calcareous material, particularly Chalk clasts (Rose *et al.*, 1978). Rose *et al.* also noticed that the lower part of the till was less calcareous, its gravel-sized component being dominated by flint and quartzites reworked from the underlying Kesgrave Group deposits. This distinction was later used by Whiteman (1987; in Allen *et al.*, 1991) to define separate lower (Chalk-poor) and upper (chalky) divisions, his Newney Green and Great Waltham Members (respectively). These members are themselves subdivided (Whiteman, 1987, 1990; in Allen *et al.*, 1991). The Newney Green Member is made up of a lower, laminated deformation till, comprising sheared and remobilized material from the palaeosols and coversand (beds 2 and 3), and an upper homogenized till of similar composition (also including much reworked material from the underlying sands and gravels). Subdivision of the Great Waltham Member reflects the decalcification of the upper part of this very chalky member.

Coarse, chalky gravel together with laminated sands, silts and clays have recently been observed occupying an irregular and undercut depression in the surface of the till, possibly pointing to a final phase of Anglian deposition in the area (Whiteman, 1990). The clast composition of the gravel resembles that of the underlying glacial beds. Whiteman considered that these deposits had been let down into the upper part of the till after deposition, perhaps in response to the melting of a block of ice incorporated in the underlying sequence. They may thus be considered to represent a kettle hole infill.

A conservation section was identified at

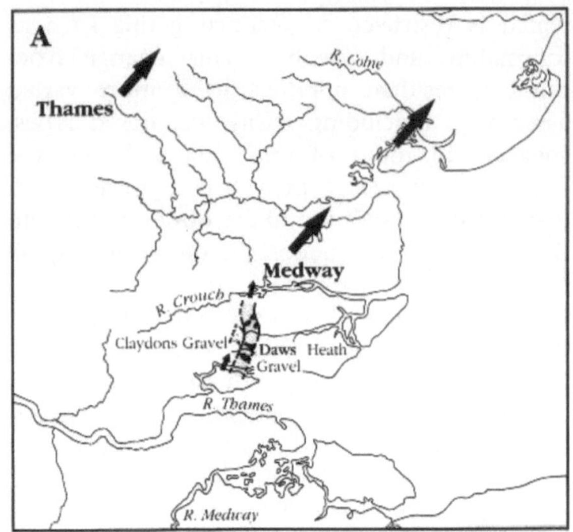

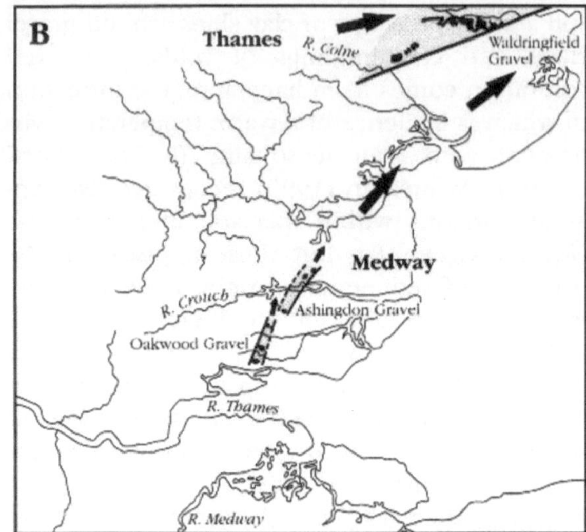

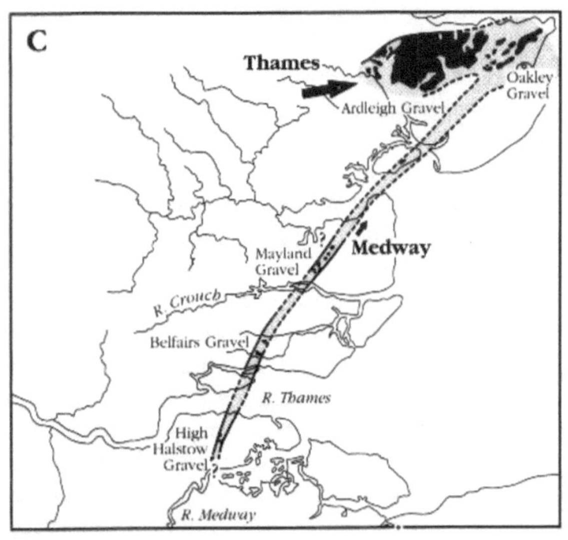

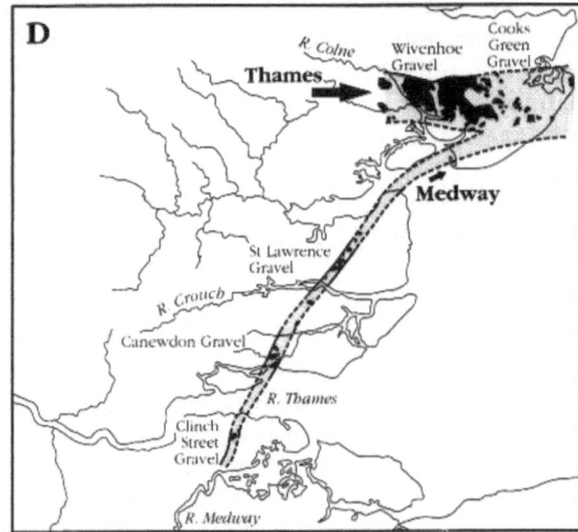

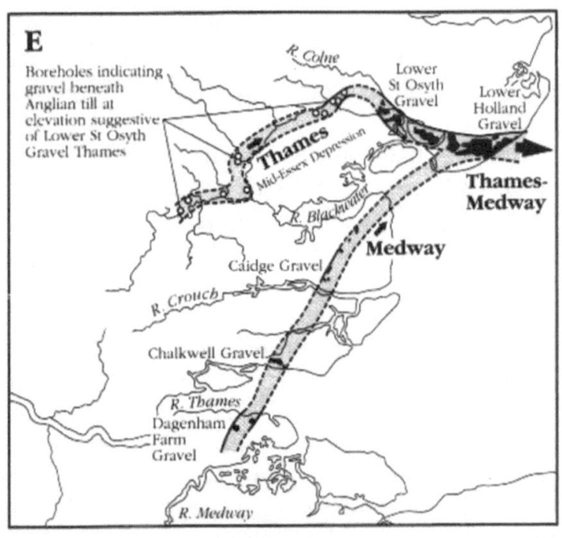

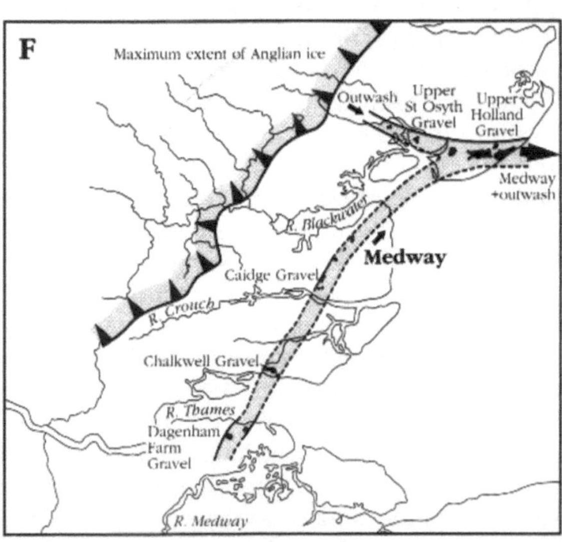

0 kilometres 20

N

Newney Green in 1988, as commercial operations were drawing to a close. A preliminary excavation revealed that all the essential elements of the stratigraphy for which the site is important are present (Fig. 5.8). The complex palaeosol is well-developed in the northern part of the section, but is cut out by the erosive base of the glacial deposits to the south. Incorporated in the cryoturbation features of the Barham Soil are unweathered gravels and sands of presumed Kesgrave affinities, as well as reddened material from the Valley Farm Soil. A bed of grey clay is also incorporated in these structures, mottled with the red coloration of the warm-climate palaeosol. This material appears to be derived from the London Clay bedrock; it may have been redeposited in standing water, but a more likely explanation is that it was brought to its present level, prior to the cryoturbation episode, by diapirism, a phenomenon that has been observed at this site previously (Rose *et al.*, 1978; Whiteman, 1990). Whiteman (1990) noted that both London Clay and till have been injected into the lower sands and gravels by this process. An upper orange sand, probably coversand, is also present in the cores of involutions and, unusually, as a small undisturbed remnant above the cryoturbated palaeosol (Fig. 5.8).

Strong lateral deformation of the cryoturbation structures of the Barham Soil can be recognized in the GCR section (Fig. 5.8). This may be related to similar lateral deformation of such structures previously observed at this and other sites (Rose *et al.*, 1978; Allen, 1983, 1984), attributed to a glaciotectonic effect caused by the pressure of the overriding ice sheet. The tilting of the upper parts of the structures to the south (Fig. 5.8), in keeping with the notion that the ice was moving generally southward, lends support to this interpretation.

Figure 5.4 Palaeodrainage of eastern Essex up to the Anglian glaciation (after Bridgland, 1988a): (A) Palaeodrainage at the time of deposition by the Medway of the Claydons and Daws Heath Gravels, part of the Rayleigh Hills gravels. The Thames and Medway are thought to have had separate routes to the North Sea at this time. (B) Palaeodrainage at the time of deposition by the Medway of the Oakwood and Ashingdon Gravels. The Waldringfield Gravel, which might be a correlative of the Ashingdon Gravel, is also shown. It is believed that the Thames and Medway joined during Waldringfield Gravel times, but this confluence is believed to have been situated to the east of the present coastline. (C) Palaeodrainage at the time of deposition by the Thames of the Ardleigh Gravel. (D) Palaeodrainage at the time of deposition by the Thames of the Wivenhoe Gravel. (E) Palaeodrainage during the early Anglian Stage, prior to the inundation of the Thames valley by the Lowestoft Till ice sheet. (F) Palaeodrainage during the Anglian glaciation, prior to the diversion of the Thames but after its valley became blocked by ice. The highly distinctive Upper St Osyth and Upper Holland Gravels were laid down at this time.

Interpretation

There is a considerable history of debate on the stratigraphy and origin of the gravels underlying the Lowestoft Till of southern East Anglia, dating back to the original definition of the 'Glacial Beds' by Wood (1867; Wood and Harmer, 1868). Much of the published material dealing with these deposits incorrectly assumes that they are of glaciofluvial origin (Clayton, 1957; Baker, 1971; Wilson and Lake, 1983). The idea that most of the gravels underlying the till sheet of Essex and Suffolk are the product of the early Thames, rather than the Anglian glaciation, dates back to the beginning of the century, to the views of Salter (1896, 1905). It is only recently, however, that this interpretation has found widespread acceptance, through the work of Rose *et al.* (1976; Rose and Allen, 1977), who gave them the name Kesgrave Sands and Gravels. A recent regional synthesis of these early Thames deposits (Whiteman, 1990, pers. comm.) has suggested that the gravel at Newney Green represents the Waldringfield Gravel, previously recognized in south-east Suffolk and north-east Essex (Allen, 1983, 1984; Bridgland, 1988a; Fig. 5.2).

The early Thames terrace gravels of the Kesgrave Group in fact extend well beyond the limits of the Anglian glaciation and can be demonstrated at many sites to be pre-Anglian. The principal evidence for this is the preservation of the warm-climate Valley Farm Soil in the upper levels of the Kesgrave deposits, beneath the Anglian till (see above). The degree of reddening and clay enrichment recorded from various sites suggests that fully interglacial conditions would have been necessary for its formation (Rose and Allen, 1977; Kemp, 1985a). Kemp (1983, 1985a) discovered, from a study of the micromorphological properties of the Valley Farm Soil at several sites, that it has a complex

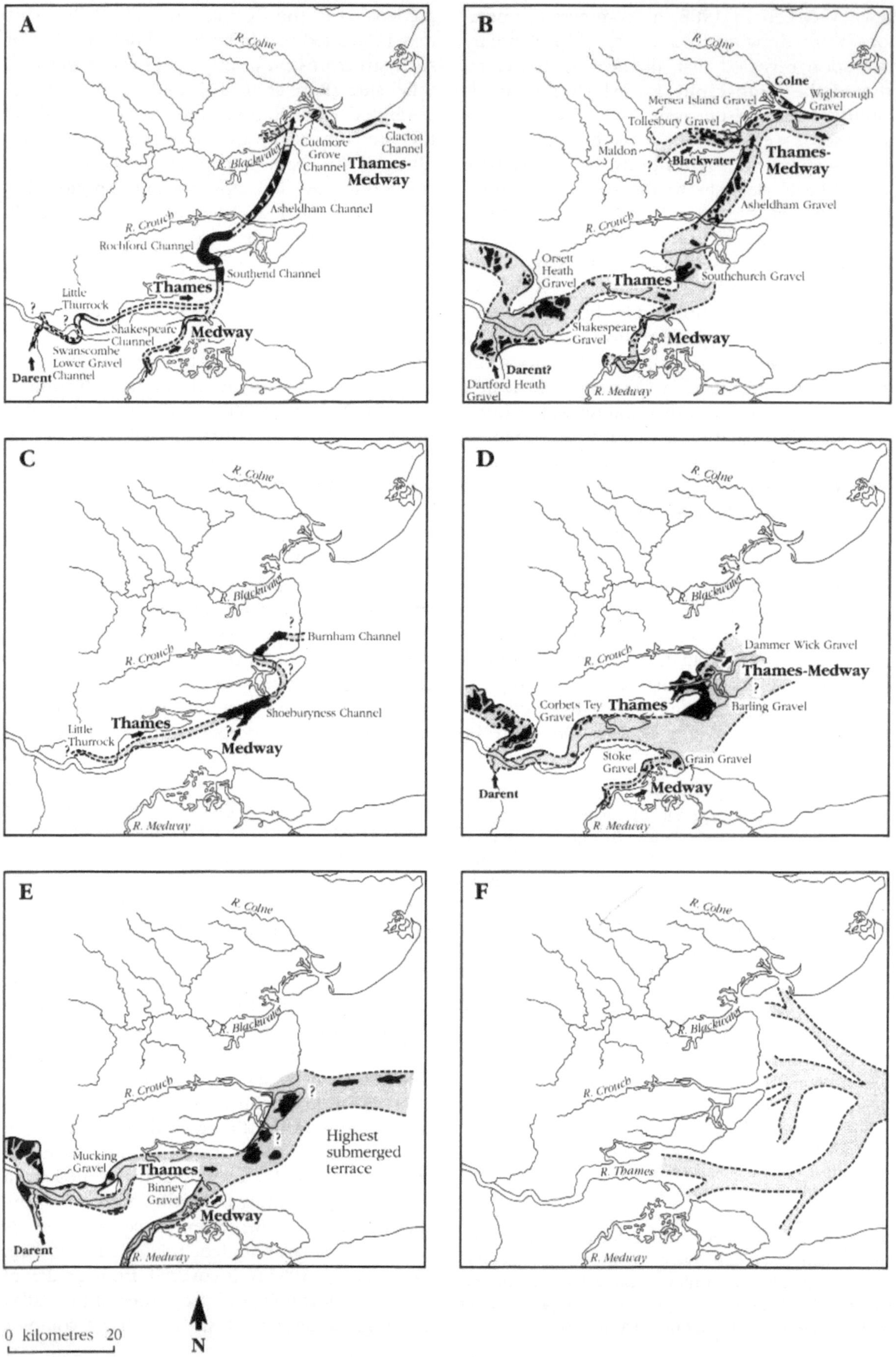

A
R. Colne
R. Blackwater
Cudmore Grove Channel
Clacton Channel
Thames-Medway
Asheldham Channel
R. Crouch
Rochford Channel
Southend Channel
Little Thurrock
Thames
Shakespeare Channel
Medway
Swanscombe Lower Gravel Channel
Darent

B
R. Colne
Colne
Mersea Island Gravel
Wigborough Gravel
Tollesbury Gravel
Maldon
Blackwater
Thames-Medway
Asheldham Gravel
R. Crouch
Orsett Heath Gravel
Thames
Southchurch Gravel
Shakespeare Gravel
Medway
Darent?
Dartford Heath Gravel
R. Medway

C
R. Colne
R. Blackwater
Burnham Channel
R. Crouch
Shoeburyness Channel
Little Thurrock
Thames
Medway
R. Medway

D
R. Colne
R. Blackwater
Dammer Wick Gravel
R. Crouch
Thames-Medway
Corbets Tey Gravel
Thames
Barling Gravel
Stoke Gravel
Grain Gravel
Medway
Darent
R. Medway

E
R. Colne
R. Blackwater
R. Crouch
Highest submerged terrace
Mucking Gravel
Thames
Binney Gravel
Medway
Darent
R. Medway

F
R. Colne
R. Blackwater
R. Crouch
R. Thames
R. Medway

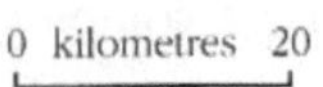

history of formation, with alternating periods of rubification and disruption of its fabric by periglacial processes. Different degrees of complexity are found at various sites, with the greatest number of climatic cycles indicated where the soil is developed on formations of the High-level Kesgrave Subgroup. Unfortunately, it is difficult to determine the precise number of climatic fluctuations responsible for a particular soil type, but Kemp's work provided important new evidence that the stratigraphical sequence first outlined by Rose *et al.* (1976) was oversimplified. It is now accepted that the Valley Farm Soil, as recognized on different formations within the Kesgrave Group, represents different numbers of climatic cycles, reflecting the deposition of the Kesgrave Sands and Gravels over a considerable period of Pleistocene time (Rose, 1983a; Kemp, 1985a; Rose *et al.*, 1985b).

Whiteman (1990) found evidence from micromorphological analysis of the palaeosol (unit 2) at Newney Green for only a single period of illuviation, uninterrupted by any significant disturbance prior to the formation of the superimposed Barham Soil. Subsequent to the formation of the various disruption features associated with the Barham Soil, illuviation of calcium carbonate together with further clay had apparently occurred. However, Whiteman concluded that this later illuviation episode postdated the emplacement of the till, since he could envisage no potential source for the calcium carbonate other than Chalk clasts in the glacial deposits. Since illuviation appears, according to this interpretation, to have affected the palaeosol horizon after its burial beneath the till, Whiteman (1990) was forced to conclude that the thickness of till at Newney Green (maximum 3.75 m) was insufficient to isolate the fossil soil from post-Anglian pedogenic activity. His failure to recognize pre-Anglian disturbance of the soil fabric at Newney Green is difficult to reconcile with his attribution of the gravel there to the Waldringfield Formation. Stratigraphical evidence from north-eastern Essex suggests that several climatic cycles intervened between the deposition of the Waldringfield Gravel and the Anglian Stage glaciation (Bridgland, 1988a; see above, Introduction to Part 1). A possible explanation is that the upper layers of the gravel may have been stripped during pre-Anglian periglacial episodes, a process that is likely to have affected the upper surfaces of Kesgrave Group terrace formations throughout the area. Only on the most stable land surfaces would the full potential complexity of the Valley Farm Soil have been realized.

The superimposed Barham Soil represents the final modification, by periglacial processes during the early Anglian Stage, of the complex Valley Farm Soil. It was initially termed an Arctic structure soil (Rose *et al.*, 1976; Rose and Allen, 1977), but this term was subsequently found to be too restrictive, as a wider range of pedogenic properties was observed at different sites, and the Barham Soil was recognized as a complex and variable periglacial soil (Rose *et al.*, 1985a, 1985b). When it is developed on the complex Valley Farm Soil, it may be difficult or impossible to distinguish certain of the features of the Barham Soil, such as the fracturing of clasts and the break up of clay skins (Rose *et al.*, 1985a),

Figure 5.5 Palaeodrainage of Essex following the Anglian glaciation (modified from Bridgland, 1988a). (A) Palaeodrainage during the filling of the Southend/Asheldham/Clacton Channel. The Swanscombe Lower Gravel Channel and the Cudmore Grove Channel are both thought to be lateral equivalents. The Rochford Channel is now thought to represent an overdeepened section of the same feature (see text). This channel was excavated in the late Anglian by the newly diverted Thames and filled during the Hoxnian Stage (*sensu* Swanscombe). (B) Palaeodrainage during the deposition of the Southchurch/Asheldham Gravel. This aggradational phase is believed to have culminated during the earliest part of the Saalian Stage, early in Oxygen Isotope Stage 10. (C) Palaeodrainage during the filling of the Shoeburyness Channel. The channel beneath the Corbets Tey Gravel of the Lower Thames is believed to be an upstream equivalent of this feature. It is thought that both the excavation and filling of the channel were intra-Saalian events, dating from Oxygen Isotope Stages 10 and 9 respectively. (D) Palaeodrainage during the deposition of the Barling Gravel. This is regarded as an intra-Saalian deposit, aggraded during Oxygen Isotope Stage 8. (E) Palaeodrainage during the deposition of the Mucking Gravel of the Lower Thames. The Thames–Medway equivalent of this formation is buried beneath the coastal alluvium east of Southend and can be traced offshore (Bridgland *et al.*, 1993). This aggradational phase occurred towards the end of the complex Saalian Stage, culminating early in Oxygen Isotope Stage 6. (F) Palaeodrainage during the last glacial. The submerged valley of the Thames–Medway has been recognized beneath Flandrian marine sediments in the area offshore from eastern Essex (after D'Olier, 1975).

from pre-existing features of the Valley Farm Soil, formed by earlier periglacial disruption (Kemp, 1985a). The characteristic features of the Barham Soil are particularly well-developed at Newney Green (Fig. 5.8). For example, the polygonal pattern of the larger-scale structures, reminiscent of landscapes in modern tundra regions, is superbly illustrated on the exhumed upper surface of the Barham Soil (Fig. 5.7).

The Barham Soil is closely associated throughout the area of its occurrence with various glacially-derived wind-blown sediments, namely coversands and loess (Rose *et al.*, 1976, 1985a; Rose and Allen, 1977). These are usually preserved in wedges and in the cores of involutions within the Barham Soil (Fig. 5.6), but occasional remnants are found, apparently in situ, between the Barham Soil and the overlying glacial sediments (see Fig. 5.8). Further evidence of the importance of aeolian activity during the early Anglian is the occurrence of wind-polished or faceted pebbles in the coversand.

The overriding of the Barham Soil, later in the Anglian, by the ice sheet that deposited the Lowestoft Till, resulted in glaciotectonic deformation of the pre-existing features, sometimes producing overfolds, shears and nappe-like structures (Rose *et al.*, 1985a). Whiteman (1987; in Allen *et al.*, 1991) recognized a basal 'deformation till', immediately overlying the palaeosol horizon, at Newney Green (Table 5.4) and at other nearby sites. He noted that it is often difficult to determine the junction between the deformed palaeosol horizon and the overlying deformation till. Ideally, however, there is a plane of *décollement* separating this lowest division of the till from the underlying, glacio-tectonically deformed sediments. This plane of *décollement* serves as the most effective lower boundary of the till sequence. In the GCR section, glaciotectonic effects can be observed in the form of lateral deformation of the structures within the Barham Soil (Fig. 5.7), broadly reflecting ice movement towards the south or south-east.

Whiteman (1987, 1990; in Allen *et al.*, 1991) has subdivided the Anglian till of central Essex into a number of facies-related beds, the gross lithological properties of which suggest that the sequence as a whole represents two distinct members of the Lowestoft Till Formation. The sections at Newney Green have been of considerable importance in the definition of these members, the lower of which is named after

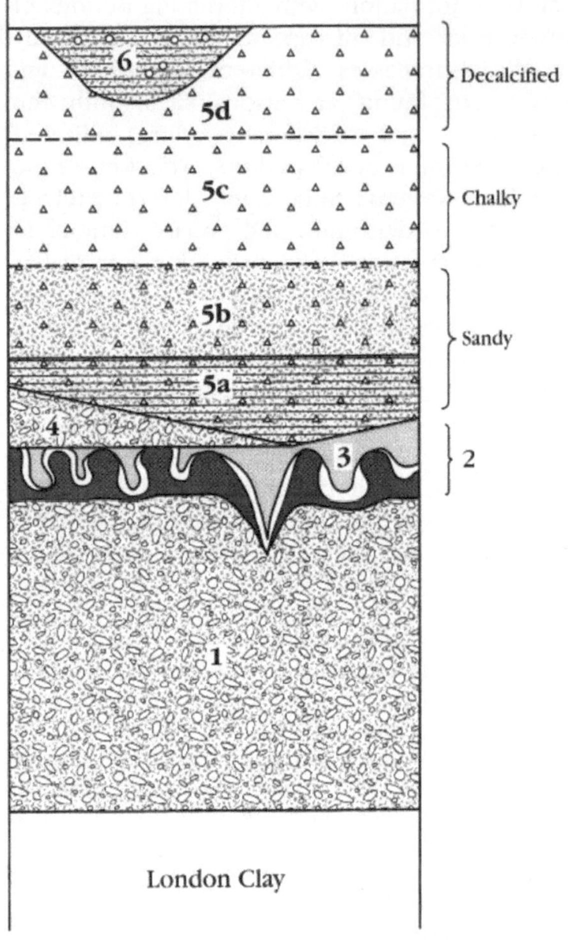

6 ? Kettle hole infill (see text)

5 Lowestoft Till: d Decalcified till, c Chalky till (Great Waltham Member); b Sandy till, a Deformation till (Newney Green Member)

4 Barham Sand and Gravel (glaciofluvial)

3 Coversand (in wedges, cores of involutions and occasionally in discrete lenses)

2 Complex buried soils horizon (superimposed palaeosols) comprising:

involutions and wedges of intensely cold-climate soil (Barham Soil) and :-

[dark shading] Rubified and mottled temperate soil (Valley Farm Soil)

1 Kesgrave Group Gravel (Waldringfield Formation?)

Figure 5.6 Idealized stratigraphical sequence through the Pleistocene deposits at Newney Green (after Whiteman, 1990).

Figure 5.7 'Patterned ground' at the top of the Kesgrave Group gravel at Newney Green, as exhumed from beneath the Lowestoft Till by quarrying. The pattern results from the polygonal distribution of sand wedges in the Barham Soil (see text). (Photo: P. Allen.)

the site. The higher, stratigraphically younger 'Great Waltham Member' is named after another site in the Chelmsford area, 8 km to the north of Newney Green.

The Newney Green Member, largely derived from the underlying gravels, palaeosol and coversand, is commonly banded throughout, not just in its lower 'deformation till' division, in which the banding is attributed to shearing and attenuation of layers within the original sediment (see above). Many different mechanisms to explain banding in tills have been proposed and the origin of this phenomenon in the upper part of the Newney Green Member is uncertain at present (see Allen *et al.*, 1991).

The overlying Great Waltham Member is a compact grey, chalky till of classic 'Chalky Boulder Clay' type, although Whiteman subdivided it into lower and upper units, the latter representing its weathered and decalcified form. Decalcified Great Waltham Till may account for the full sequence in areas where the till cover is thin. In some instances differences between the lower and upper divisions of the Great Waltham Member cannot be explained entirely as the result of weathering. This led Whiteman (in Allen *et al.*, 1991) to suggest that differences in the direction of ice-movement and, therefore, in provenance are implicated.

The junction between the Newney Green and Great Waltham members is generally sharp, but Whiteman (1987, 1990; in Allen *et al.*, 1991) found no indication that they represent more than a single ice advance. There is no evidence as yet to relate this advance to the more complex glacial sequence in south-west Essex and Hertfordshire, where four separate ice advances occurred during the Anglian Stage (Cheshire, 1986a; Chapter 3, Part 2). One important difference between the two till members at Newney Green, however, is that their fabrics are generally at 90° to one another, with that in the Great Waltham Till most closely in agreement with other evidence for ice flow direction, such as the configuration of subglacial

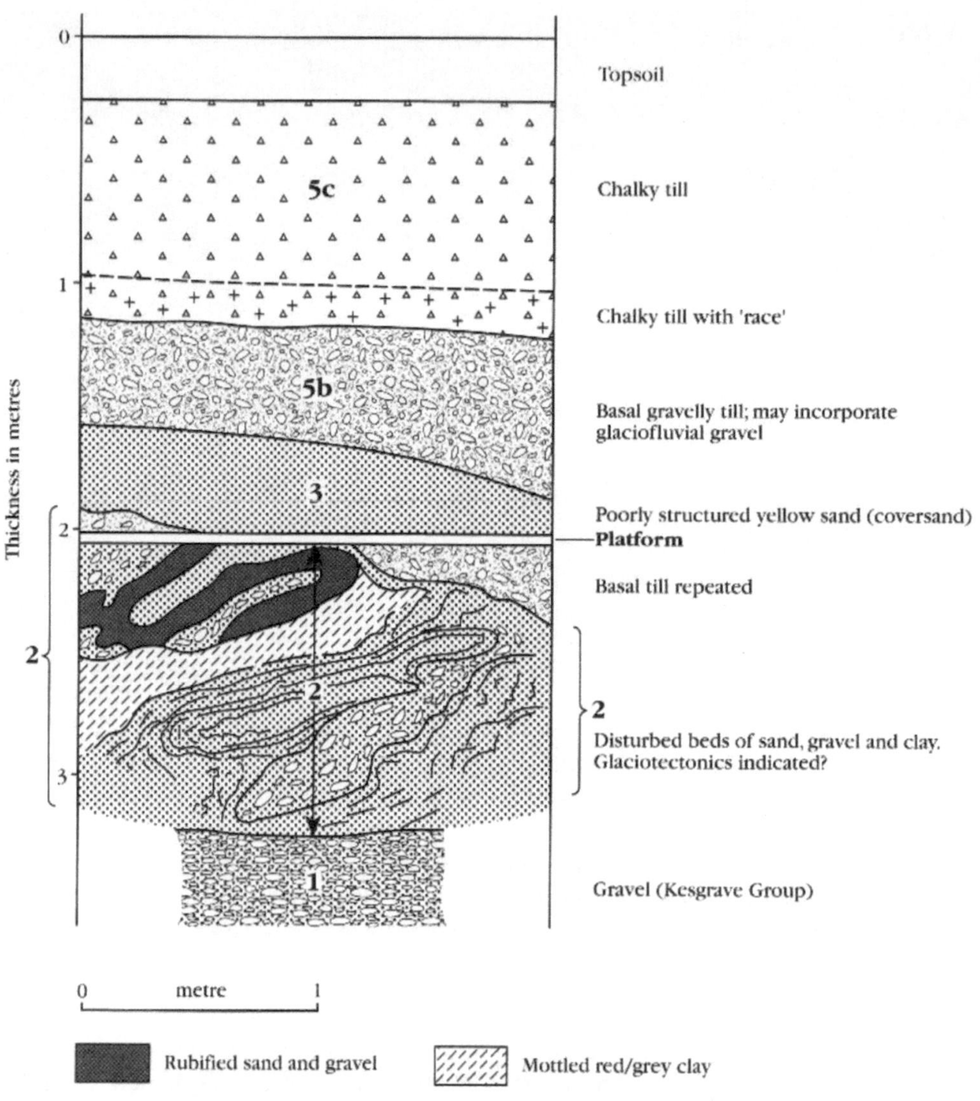

Figure 5.8 Section excavated in 1988 at the Newney Green GCR site.

topography and the directional trend of glaciotectonic deformation. Whiteman concluded that the fabric of the Newney Green Member was orientated transverse to ice flow directions. He regarded this as suggestive of strong compressional forces within a highly deformed till, formed above a rough glacier bed.

Conclusions

There are four important elements to the geological interest at Newney Green. Firstly, the site shows gravels deposited by the early (pre-diversion) River Thames at a time when it flowed across Hertfordshire and East Anglia. Secondly, a temperate-climate soil was developed at the top of this gravel. Following this temperate phase, the climate deteriorated until permafrost conditions prevailed, leading to the formation of a periglacial soil with characteristic structures (large-scale structures include patterned ground and ice wedges, such as are formed at the present day in tundra regions through the growth and subsequent melting of ground-ice). This period of extreme cold culminated in the covering of the area by (Anglian) ice, which moved southwards, depositing till (boulder clay), the fourth element in the Newney Green sequence. It was at this time that the course of the Thames was blocked by the ice and the river adopted a new route through what is now the London area. The temperate-climate soil is of considerable importance, as it shows that the gravels were not deposited in the same cold episode as the overlying glacial deposits.

ARDLEIGH (MARTELLS QUARRY; TM 053280)

D.R. Bridgland

Highlights

At this site cold-climate river gravels, assigned to the pre-diversion Thames, are interbedded with various organic sediments, some of which are indicative of temperate (interglacial) conditions. The latter have yielded the remains of deciduous trees. The site therefore reveals an early Middle Pleistocene cold-warm-cold climatic cycle expressed in Thames sediments. The complex sequence at Ardleigh also includes cold-climate organic deposits, a later (?tributary) gravel and soils formed under both warm and cold climatic conditions.

Introduction

Martells Quarry, Ardleigh, is the type locality for the Ardleigh Gravel, second highest of the four Low-level Kesgrave Group formations recognized on the Tendring Plateau (Bridgland, 1988a; see above, Introduction to Part 1; Figs 5.2 and 5.3 and Table 5.3). There has been a pit at Ardleigh for many years, attention first being drawn to the site by Spencer (1966), following the discovery of bones, including a skull fragment of a ziphiid whale. Later work has revealed the presence at Ardleigh of lenses and beds of organic sediments within the Ardleigh Formation, two distinct types being recognized, one indicative of a temperate climate and the other, higher in the sequence, of intensely cold conditions (Bridgland, 1988a; Bridgland *et al.*, 1988; Bridgland and Gibbard, 1990).

A complex succession of Pleistocene deposits is now recognized at Ardleigh, with evidence for periglacial conditions at more than one level and a complex palaeosol at the top of the sequence (Bridgland *et al.*, 1988). The interglacial represented by the lower set of organic sediments at Ardleigh is thought to correlate broadly with the 'Cromerian Complex' of The Netherlands (see Chapter 1; Zagwijn *et al.*, 1971; Zagwijn, 1986; de Jong, 1988).

Description

Spencer (1966) produced a diagrammatic illustration of a section at Ardleigh (without a scale), showing a partly submerged sequence of sands and gravels with a cryoturbated zone just above the water level. He considered the deposits above this zone of cryoturbation to be glacial outwash and those below to be of fluvial origin.

Recent work at this site has revealed a much greater stratigraphical complexity than was envisaged by Spencer (Bridgland *et al.*, 1988). The sequence now recognized is as follows:

			Thickness
4a.	Complex rubified and cryoturbated relict soil	(Valley Farm and Barham Soils)	up to 1.5 m
4.	Brown/orange gravel and sand. This has a low frequency of rounded flint pebbles; *Rhaxella* chert is present (Table 5.2). Palaeocurrents are towards the south-west	(Martells Gravel)	up to 3 m
3a.	Silty clay, dark grey and organic, with plant macrofossils. This occurs as variable beds and lenses *c.* 1.5 m below the top of member 3	(cold-climate deposits)	0.1–1 m
3.	Pale buff gravel and sand. This has a relatively high frequency of rounded flint pebbles; *Rhaxella* chert is absent (Table 5.2); palaeocurrents are towards the north-east	(Ardleigh Upper Gravel)	up to 5 m

2. Sand, dark grey/black and organic. This has high silt, clay and organic contents. It contains plant macrofossils and pollen. It occurs as variable beds and lenses.	(Ardleigh interglacial deposits)	up to 0.5 m
1. Pale, buff gravel and sand. This has a relatively high frequency of rounded flint pebbles; *Rhaxella* chert is absent; palaoecurrents are towards the north-east	(Ardleigh Lower Gravel)	up to 2 m
London Clay		

Two separate gravel deposits occur here in superposition. They are distinguished on the bases of differences in clast lithology and sedimentological criteria, particularly palaeocurrent evidence. These are the Ardleigh Gravel (members 1 and 3) and the Martells Gravel (member 4). With member 2, the Ardleigh interglacial deposits, members 1 and 3 combine to form the Ardleigh Gravel Formation, which belongs to the Low-level Kesgrave Subgroup. The GCR site is the type locality for all these units.

The Ardleigh Gravel has a composition typical of the Low-level Kesgrave Subgroup in this area (Table 5.2), with evidence for north-eastward palaeocurrents adding support to its interpretation as a Thames deposit (Bridgland, 1988a; Bridgland *et al.*, 1988). It is divided into two members, (1) the Ardleigh Upper Gravel and (3) the Ardleigh Lower Gravel, by the interglacial deposits (member 2), which represent a temperate interval (Bridgland *et al.*, 1988; Fig. 5.9). The Ardleigh Lower and Upper Gravels clearly represent different cold-climate episodes, but they cannot be distinguished in the absence of the organic interglacial deposits.

The temperate-climate organic sediments comprise variable lenses and beds of predominantly sandy deposits, frequently showing deformation structures suggestive of internal collapse while waterlogged (Fig. 5.10). These beds contain pollen, plant macrofossils and beetle remains. Poorly preserved wood remains and occasional indeterminate, abraded mammal bones that have been found in the lower parts of the Ardleigh Upper Gravel may have been reworked from the interglacial level. The Ardleigh interglacial deposits have been revealed intermittently by continued quarrying over a wide area, occurrences probably representing fills of isolated shallow channels (Bridgland and Gibbard, 1990). Pollen analyses have indicated that these various remnants are not all contemporaneous (Bridgland and Gibbard, 1990). These analyses have allowed the compilation of a preliminary and fragmentary pollen diagram, combining evidence from three of these isolated channel-fills (Bridgland and Gibbard, 1990). All three parts of this diagram record the occurrence of deciduous trees, namely birch (*Betula*), oak (*Quercus*), elm (*Ulmus*), alder (*Alnus*), willow (*Salix*) and hazel (*Corylus*). Pine and spruce were also present throughout (with the exception of the basal part of the oldest channel-fill). In two of the channel-fills, regarded as earlier than the third, the pollen was dominated by herb taxa, indicating cooler conditions and suggesting an earlier part of the interglacial. The basal layers of one of the channel-fills was practically devoid of tree pollen, suggesting that the very onset of interglacial conditions was represented. It was presumed that this was the oldest of the sampled sediment bodies (Bridgland and Gibbard, 1990). In the third part of the diagram, tree pollen constitutes over 60% of the total and a greater diversity of forest trees is indicated, suggesting the middle part of an interglacial.

At a higher stratigraphical level within the Ardleigh (Upper) Gravel there occur further organic sediments. These contain macrofossils of cold-climate plants, predominantly mosses, grasses and sedges, and clearly represent the vegetation of a periglacial episode. It is apparent, from the occurrence of ice-wedge casts originating from the upper surface of the Ardleigh (Upper) Gravel (Fig. 5.9), that permafrost conditions prevailed prior to the deposition of the overlying Martells Gravel.

In contrast to the Ardleigh Gravel, the Martells Gravel (member 4) contains a lower proportion of rounded flint pebbles (of the type reworked from the Palaeogene) than is usual in the Kesgrave Group. It contains significant amounts of *Rhaxella* chert, a rock that is extremely scarce in the Kesgrave Sands and Gravels upstream (to the south and west) of the Crag basin (Table 5.2). Foreset orientations indicate palaeocurrents towards the west-south-west, essentially a reversal of the flow direction indicated by palaeocurrent measurements from the Ardleigh

Gravel (Fig. 5.9). The combination of these various lines of evidence implies that the Martells Gravel was deposited by a river flowing from the north-east, which leads to the conclusion that it is not a Thames deposit and not part of the Kesgrave Group.

At the top of the sequence at Ardleigh, a rubified and cryoturbated palaeosol is developed in the upper part of the Martells Gravel, immediately beneath the modern topsoil (Fig. 5.9). Ice-wedge casts originating from the top of the fluvial sequence have also been recognized, superimposed on the earlier system of wedges developed from the surface of the Ardleigh (Upper) Gravel (Fig. 5.9).

Interpretation

The principal significance of the Ardleigh GCR site lies in the occurrence there, within a gravel formation ascribed to the Low-level Kesgrave Subgroup, of lower Middle Pleistocene temperate-climate sediments. As yet the

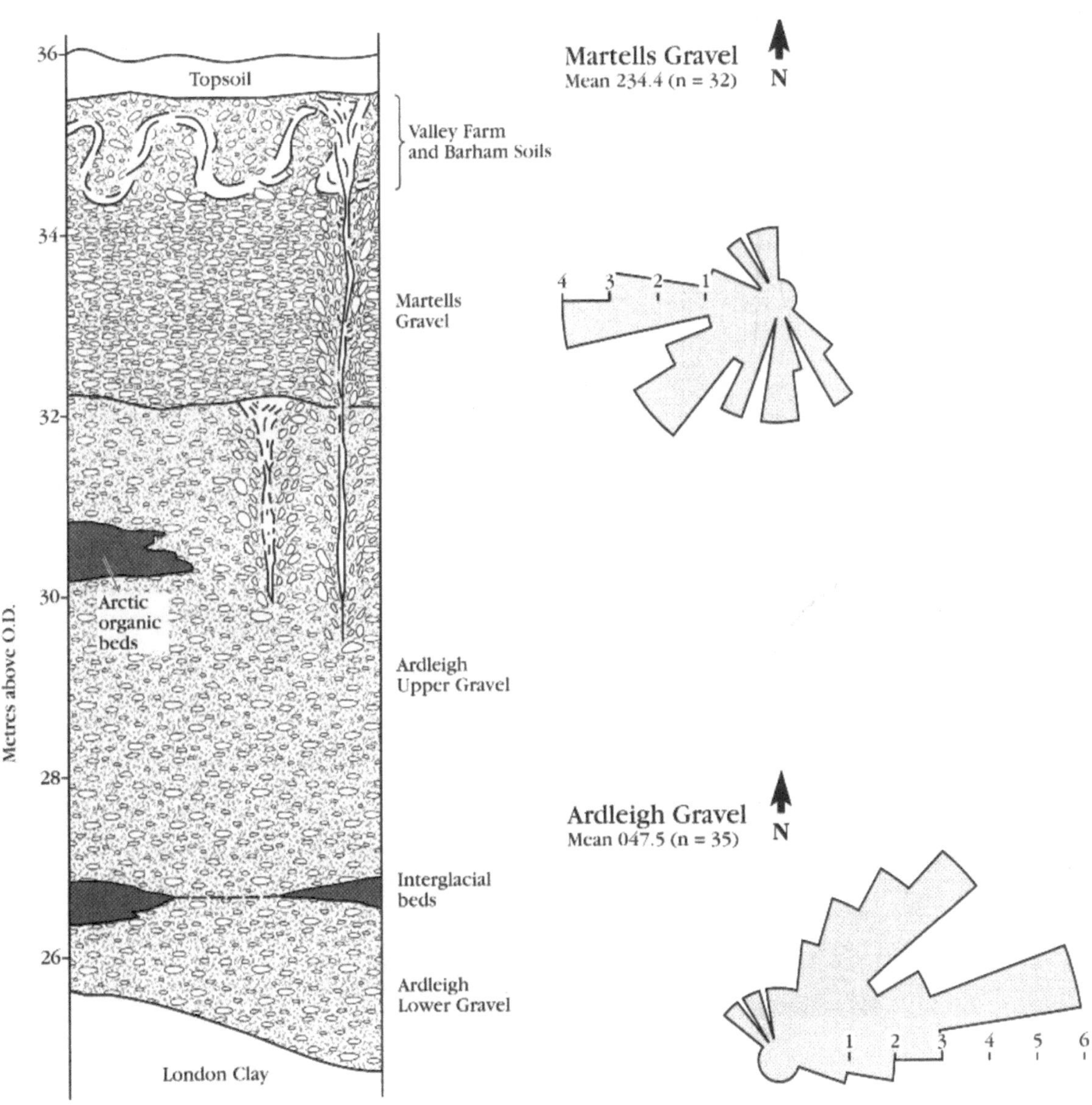

Figure 5.9 Idealized Pleistocene sequence at Ardleigh (after Bridgland *et al.*, 1988).

palaeontological evidence from these sediments is insufficiently distinctive to allow the temperate episode they represent to be identified. The vegetational sequence revealed in the fragmentary pollen diagram (Bridgland and Gibbard, 1990) is similar to that found in the early parts of other lower Middle Pleistocene interglacial deposits. The presence and persistence of spruce, the early expansion of elm and the low frequency of hazel were all identified as significant by Bridgland and Gibbard (1990), but there are no features in the pollen record at Ardleigh that are of biostratigraphical or chronostratigraphical significance.

Palaeomagnetic measurements from the Ardleigh interglacial deposits (T. Austin, in Bridgland and Gibbard, 1990) have indicated a normal polarity, implying that they post-date the Matuyama–Brunhes magnetic reversal, which is taken as the base of the Middle Pleistocene (Chapter 1). The oldest known interglacial episode with which they could be correlated is therefore 'Cromerian Complex Interglacial II' of the Dutch sequence (see Chapter 1).

The relation of these sediments, which lie near the base of the Ardleigh Gravel sequence, to the interglacial channel-fill at Little Oakley (Bridgland *et al.*, 1988, 1990; see below) is of great importance. The latter deposits, which contain molluscan and vertebrate remains as well as pollen, are interpreted as of broadly Cromerian age and thought to belong within the 'Cromerian Complex' as identified in The Netherlands. They occur in the same terrace formation as the Ardleigh sediments, but are 17 km to the east-north-east (downstream). This places them within the area of the confluence between the Kesgrave Thames and the early Medway, within the Oakley Gravel Formation, the lateral Thames-Medway equivalent of the Ardleigh Gravel (Figs 5.2 and 5.4C). The channel at Little Oakley cuts through

Figure 5.10 The Ardleigh interglacial deposits exposed above Ardleigh Lower Gravel in a drainage channel in the floor of Martells Quarry, Ardleigh (1987). (Photo: D.R. Bridgland.)

the local Oakley Gravel, implying that the interglacial represented there may be later than that at Ardleigh. This would suggest that the Ardleigh Upper Gravel was laid down in the interval between the Ardleigh and Little Oakley interglacials (see, however, Little Oakley). Gibbard (1988b) has attributed the Ardleigh interglacial deposits to a temperate interval between the Pastonian and Cromerian and has suggested correlation with either Interglacial II or Interglacial III of the 'Cromerian Complex' (see Chapter 1).

With cold-climate gravels both above and below the interglacial sediments, it is apparent that the sequence within the Ardleigh Formation at the GCR site represents all three aggradational phases of the climatic model for terrace formation promoted in Chapter 1, as follows:

Phase 4	(post-interglacial aggradation)	Ardleigh Upper Gravel	(member 3)
Phase 3	(interglacial aggradation)	Ardleigh interglacial sediments	(member 2)
Phase 2	(pre-interglacial aggradation)	Ardleigh Lower Gravel	(member 1)

The relation of the Little Oakley channel deposits to this model will be discussed below (see Little Oakley).

Assuming the Ardleigh temperate episode to be earlier than that represented at Little Oakley, only one other site is currently known in Britain where a possible correlative of the Ardleigh interglacial sediments occurs, this being at Broomfield, near Chelmsford. As at Ardleigh, both temperate- and cold-climate organic deposits occur at Broomfield (Gibbard, 1988b; Whiteman, 1990). Whiteman (pers. comm.) considers that the gravel at Broomfield, which is buried beneath Lowestoft Till, may be an upstream continuation of the Ardleigh Formation.

The interpretation of the Martells Gravel, which overlies the Ardleigh Gravel at the GCR site, is somewhat problematic. As described above, the Martells Gravel contains significantly less rounded flint (reworked from the Palaeogene) than the underlying Ardleigh Gravel and it also differs in that it contains *Rhaxella* chert. Both differences suggest an apparent affinity to outwash gravels from the Anglian glaciation, which also have these particular ('northern') characteristics (Bridgland, 1980, 1986b; Bridgland *et al.*, 1988). However, Anglian outwash deposits also differ from Kesgrave Group gravels in that about half of their exotic component (derived from the north and west) is made up of non-quartzose lithologies (an example is the ice-proximal Ugley Gravel at Ugley Park Quarry (see Chapter 3, Part 2). The exotic suite found in Kesgrave Group gravels typically includes over 80% quartz and quartzites. The Martells Gravel shares this high proportion of quartzose exotics with the Kesgrave Group, thus differing markedly from outwash gravel. The palaeocurrent evidence from the Martells Gravel, which indicates flow from the north-east, would appear to preclude the interpretation of the unit as either outwash or a Kesgrave Thames deposit. In the latter case flow towards the north-east would be expected, the exact opposite of what is indicated, whereas the location of the Lowestoft Till margin well to the west of Ardleigh (Fig. 5.4F) suggests that any outwash streams crossing the Tendring Plateau would have flowed in a broadly eastward or southeastward direction.

The most plausible interpretation of the Martells Gravel, based on these facts, is as the product of a river flowing from the north-east into the pre-diversion Thames valley or, if of post-Anglian age, into the Colne valley (Bridgland *et al.*, 1988). Such a river would have drained an area covered by earlier Kesgrave Group Thames gravels and by Red Crag. The former would have supplied much of the material in the Martells Gravel, ensuring a high quartz and quartzite content, but the inclusion of pebbles reworked from the Crag would have diluted the rounded flint, which is not found in quantity in the Crag, and added the *Rhaxella* chert, which is an important component of Crag pebble beds. The river presumably had an insufficiently large catchment to tap the Westleton Beds of northern Suffolk, as these would have yielded abundant rounded flint, largely indistinguishable from that reworked from the Palaeogene of the London Basin. No other remnants of gravels that may have been aggraded by this hypothetical river have been identified.

There is little evidence for the age of the Martells Gravel. It is clearly later than the Ardleigh/Oakley Gravel and presumably post-dates Thames occupation of the Ardleigh/Oakley

Formation floodplain. No soil development has been observed at the top of the Ardleigh Gravel, beneath the Martells Gravel, in recent studies. However, Spencer (1966) recorded a zone of cryoturbation at Ardleigh beneath an upper gravel, although no details of thickness were given. He quoted the grid reference of his section, which leaves no doubt that he referred to earlier sections in the present Martells Pit. If the upper gravel identified by Spencer was the Martells Gravel, his description may record a zone of cryoturbation at the top of the Ardleigh Gravel, where only ice-wedge casts have been seen in recent exposures. Spencer's record was cited by Rose *et al.* (1985a), who claimed that his cryoturbated zone represented the early Anglian Barham Soil. The only cryoturbated horizon recorded recently is that developed in the top of the Martells Gravel (Fig. 5.9); this would appear to lie too close to the modern land surface to be the horizon identified by Spencer, unless his section revealed a higher, later gravel that is missing from more recent exposures. Confirmation of a cryoturbated and/ or weathered horizon at the top of the Ardleigh Gravel might provide additional information about the difference in age between it and the overlying Martells Gravel. C. Turner (pers. comm.) has suggested, however, that Spencer's cryoturbation layer was in fact the deformed lower (temperate) organic member, the penecontemporaneous deformation structures that disrupt this deposit (see Description and Fig. 5.10) having been mistaken for involutions.

The rubified and cryoturbated zone at the top of the Martells Gravel is reminiscent of the superimposed (temperate) Valley Farm and (cold) Barham Soils, which have been identified in the upper levels of the various Kesgrave Group formations (Rose *et al.*, 1976, 1985a, 1985b; Rose and Allen, 1977; Kemp, 1985a; see above, Introduction to Part 1). The Valley Farm Soil was recorded at the top of a sequence at Ardleigh by Rose *et al.* (1976), although the later Barham Soil was not recognized. The combination of this record and the reinterpretation by Rose *et al.* (1985a) of Spencer's section suggests that palaeosols may be represented at two different stratigraphical levels at Ardleigh. The Valley Farm Soil has recently been tentatively identified over much of the Tendring Plateau 'in relict form' (Kemp, 1985a). However, in this area, where it is at or near the modern surface and not overlain by Anglian glacial deposits, it is impossible to demonstrate the pre-Anglian origin of the reddened material. The correlation of this soil layer with the pre-Anglian Valley Farm Soil would be of considerable significance, since the age and origin of the Martells Gravel cannot be established by other means. However, a post-Anglian rubified soil has been recorded from the Chelmsford area (Rose *et al.*, 1978), so the presence of a reddened zone in the Martells Gravel may be of little stratigraphical value.

Spencer (1966) suggested that the lower part of the sequence at Ardleigh was of Hoxnian age. He appears to have based this suggestion on the occurrence of the whale bone. However, as Spencer noted, ziphiid whale bones of this type are common in the basement bed of the Red Crag at Walton-on-the-Naze, some 20 km to the east of Ardleigh. It seems likely that this bone had been reworked from such a source and that its provenance at Ardleigh was the Martells Gravel, which has already been claimed to contain certain gravel material reworked from the Crag, and has been shown to be the product of a river flowing from the direction of the Crag outcrop (see above). Further investigation of sections in the area may throw more light on the relation of Spencer's section to the sequence described at the GCR site and on the relations of the relict soils at Ardleigh to the Valley Farm and Barham Soils of East Anglia.

Conclusions

Martells Quarry, Ardleigh, is an important site for Pleistocene stratigraphy and palaeoenvironmental reconstruction. The sequence here includes sediments deposited by an ancestral River Thames at a time when it flowed across East Anglia, long before its diversion, by ice, into its modern valley. Deposits from two cold-climate episodes separated by a temperate (interglacial) interval are represented here. The interglacial deposits are of considerable significance in that they may be unique in Britain. The interglacial represented probably belongs within a complex of cold and warm episodes recognized in The Netherlands (referred to as the 'Cromerian Complex', after the most famous site of this general age, near Cromer in Norfolk), implying a date somewhere between 750,000 and 450,000 years BP. The site has been identified as the type locality for an important formation within the Kesgrave Group of early Thames deposits, the Ardleigh Gravel, which is itself subdivided by the interglacial beds where these are present. The Ardleigh Lower and Ardleigh Upper gravels therefore represent two different cold periods. Further fossiliferous beds occur within the Ardleigh Upper Gravel. These are most unusual in that they contain plant remains representative of a tundra environment. A later, enigmatic upper deposit, the Martells Gravel, appears to be the product of a later river draining from the north-east, but its age is indeterminate at present.

LITTLE OAKLEY (TM 223294)

D.R. Bridgland

Highlights

Extremely rare Cromerian deposits, with important molluscan and mammalian faunas, occur here in a channel cut through pre-diversion Thames-Medway gravels. This association is important for the correlation of the Thames terrace sequence with the type Cromerian of Norfolk and the more complete 'Cromerian Complex' succession of The Netherlands.

Introduction

At Little Oakley, in the north-eastern corner of the Tendring Plateau (Fig. 5.2), fossiliferous interglacial sediments occupy a large river channel cut through the local Oakley Gravel Formation (this is the oldest Thames-Medway gravel within the Low-level Kesgrave Subgroup, no Thames-Medway equivalent of the older Waldringfield Gravel having yet been identified – see above, Introduction to Part 1). The channel is believed, on the basis of the clast content of its infill, to have been formed by the pre-diversion Thames at a point immediately upstream from its confluence with the Medway (Bridgland *et al.*, 1988, 1990). The palaeontological evidence, which includes rich assemblages of mammals, molluscs and ostracods as well as a detailed pollen record, suggests correlation with an early Middle Pleistocene interglacial, probably within the 'Cromerian Complex' as defined in The Netherlands (Zagwijn *et al.*, 1971) and possibly the Cromerian Stage *sensu* West Runton (Bridgland, 1990b; Bridgland *et al.*, 1990; Gibbard and Peglar, 1990; Lister *et al.*, 1990; Preece, 1990b; Robinson, 1990). This correlation is supported by results of amino acid analyses of shells from this and other sites (Bowen *et al.*, 1989; Bridgland *et al.*, 1990), as well as by palaeomagnetic measurements, which indicate a normal geomagnetic polarity (Bridgland *et al.*, 1988, 1990).

The interglacial deposits at Little Oakley were first discovered in 1939 by Warren (1940; Sutcliffe *et al.*, 1979), who recognized that they were of 'Forest Beds' (Cromerian) age. The site has been frequently cited in subsequent literature as an important Cromerian locality, a considerable rarity outside the type area of north Norfolk (Oakley, 1943; Kerney, 1959a; Turner, 1973; Sutcliffe *et al.*, 1979). However, prior to the recent investigations (Bridgland *et al.*, 1988, 1990), which included re-excavation of the deposits as part of the GCR programme (Bridgland, 1985a), no detailed study of the sediments or *in situ* sampling had been attempted.

Description

Little Oakley lies near the eastern end of a ridge of London Clay, capped with Pleistocene gravels and small remnants of Red Crag, lying between the Stour estuary to the north, Hamford Water to the south and the North Sea to the east. On the only available Geological Survey map (Old Series, Sheet 48), the deposits capping this ridge are classified as 'Glacial Gravel' (Whitaker, 1877). However, gravels of this type throughout Suffolk and Essex have been shown to pre-date the Anglian Stage, during which the principal glacial deposits in this region were laid down, from the fact that their upper layers show evidence of warm-climate soil formation (the Valley Farm Soil) prior to burial by Anglian till. Such gravels are now interpreted as deposits of the pre-diversion Thames and classified as the Kesgrave Sands and Gravels Group (Rose *et al.*, 1976; see above, Introduction to Part 1).

The Kesgrave Sands and Gravels have been progressively subdivided during the past decade (Hey, 1980; Allen, 1983, 1984; Bridgland, 1988a; Whiteman, 1990) and are now considered to represent various formations within High-level and Low-level Kesgrave Subgroups. The gravel in the vicinity of Little Oakley is attributed to the Oakley Formation, which represents the downstream continuation of the Ardleigh Formation (Fig. 5.2; see above, Ardleigh). The Ardleigh and Oakley Gravels are distinguished on the basis of clast content, the latter deposit containing significantly more material of southern origin (Table 5.2). The change from Ardleigh to Oakley Gravel composition is considered to record the contemporary confluence of the Thames with the Medway (Bridgland, 1988a; Fig. 5.4C).

The Little Oakley channel deposits were not recognized in the Old Series Geological Survey mapping, although a patch of Red Crag is indicated on the map at their approximate location, suggesting that the fluviatile shelly sand may have been mistaken for part of the Crag. In recent investigations, however, a remnant of Red Crag was encountered between the interglacial channel sediments and the London Clay (Bridgland *et al.*, 1990).

No permanent section has ever existed at Little Oakley, the sediments originally being discovered in spoil from sewer trenches (Warren, 1940). Detailed work at the locality has been carried out in recent years (Bridgland *et al.*, 1988, 1990). Using Warren's notes, the fossiliferous channel was relocated by augering near to the site of the original discovery (borehole LOA, Fig. 5.11; TM 233294). Numerous further auger holes were sunk in the area, enabling the form of the sedimentary body, its internal variability and its relation to the neighbouring sediments to be determined. A strip of undeveloped land, including the site of the original discovery, was selected as a potential GCR site. Temporary exposures were excavated mechanically on this land, allowing the examination of sedimentary characteristics and relations and the bulk sampling of sediments.

Detailed mapping of the interglacial deposits at Little Oakley has demonstrated that they fill a WSW–ENE trending channel between 150 m and 175 m wide (Fig. 5.11). The overall geometry and sedimentary facies of these deposits suggest deposition in the channel of a single-thread river flowing under a relatively low-energy regime. They have been given the formal lithostratigraphical name 'Little Oakley Silts and Sands' (Bridgland *et al.*, 1988, 1990). Up to 4 m thick, this member predominantly comprises material in the fine sand, silt and clay grades, with scattered pebbles and occasional thin sand laminae. Mollusca are abundant throughout the deposits (Preece, 1990b; Fig. 5.12, see below), although they are rarer in the upper levels, which are rather poorly bedded and may have suffered some post-depositional decalcification. The dominant species are the gastropods *Valvata naticina* and *Tanousia* cf. *stenostoma* (Nordmann) and the bivalves *Pisidium moitessierianum* Paladilhe and *P. supinum*. The deposits also contain ostracods (Robinson, 1990), pollen (Gibbard and Peglar, 1990) and an important vertebrate assemblage. The vertebrates include two species of early giant deer (*Megaloceros verticornis* Dawkins and *M. dawkinsi* (Newton)), wild boar, horse, spotted hyaena and eight species of small mammal as well as amphibians, reptiles and a large variety of river-dwelling fish (Bridgland *et al.*, 1988; Lister *et al.*, 1990). Particularly significant amongst the fish are records of burbot (*Lota lota*) and carp (*Cyprinus carpio*), the latter being the first recorded from the British Pleistocene. In the area of Newhouse Farm (Fig. 5.11) the channel sediments include a thin bed of coarse, shelly red-orange sand containing abundant reworked Red Crag Mollusca as well as indigenous species. The sand itself resembles the Red Crag, which, since it underlies parts of the channel, is probably its direct source. Sandy and pebbly horizons within the silts indicate occasional higher energy flood events. These were found particularly beneath the eastern end of the village, including the area of the GCR site (Bridgland *et al.*, 1990).

The fluviatile Oakley Gravel and Little Oakley Silts and Sands are capped by a complex, poorly bedded unit of variable thickness. This unit, which predominantly comprises silty or clayey sand with pebbles and calcareous nodules, thickening downslope, has been attributed to solifluction (Bridgland *et al.*, 1990). The sequence at Little Oakley can thus be summarized as follows (for thicknesses, see Fig. 5.11):

3. Colluvium

2. Little Oakley Silts and Sands

1. Oakley Gravel

Pre-Pleistocene strata: London Clay and Waltonian Red Crag

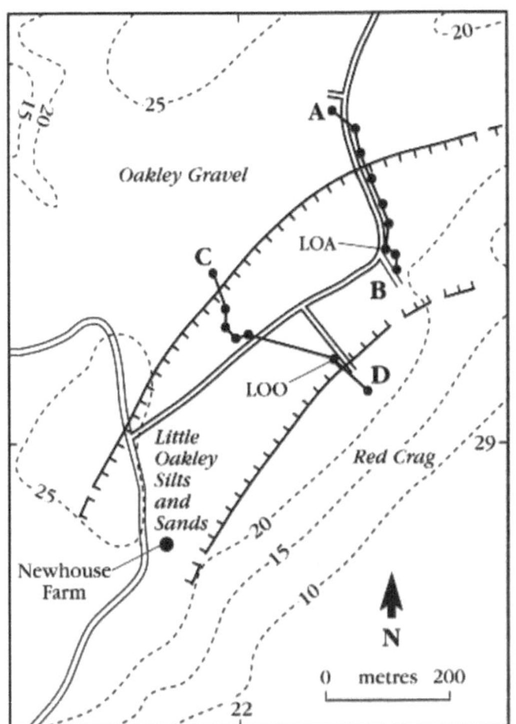

Figure 5.11 Plan and sections showing the distribution and geometry of the Little Oakley Silts and Sands (after Bridgland *et al.*, 1988).

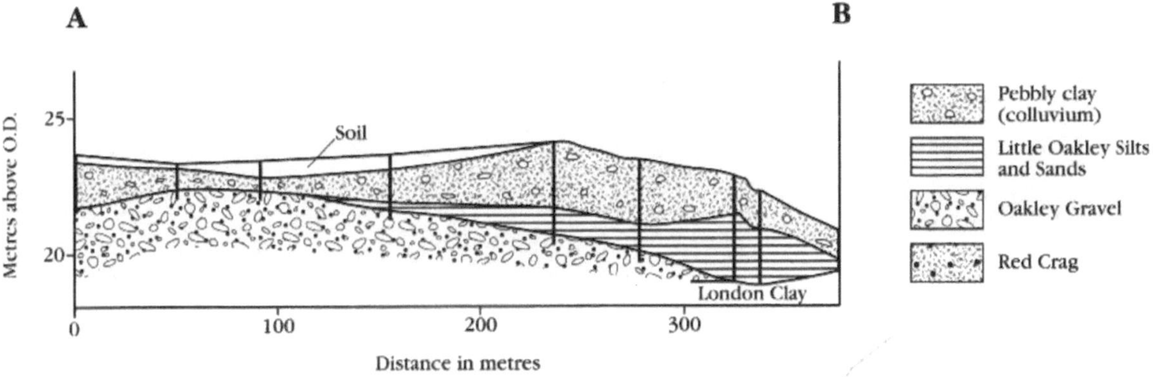

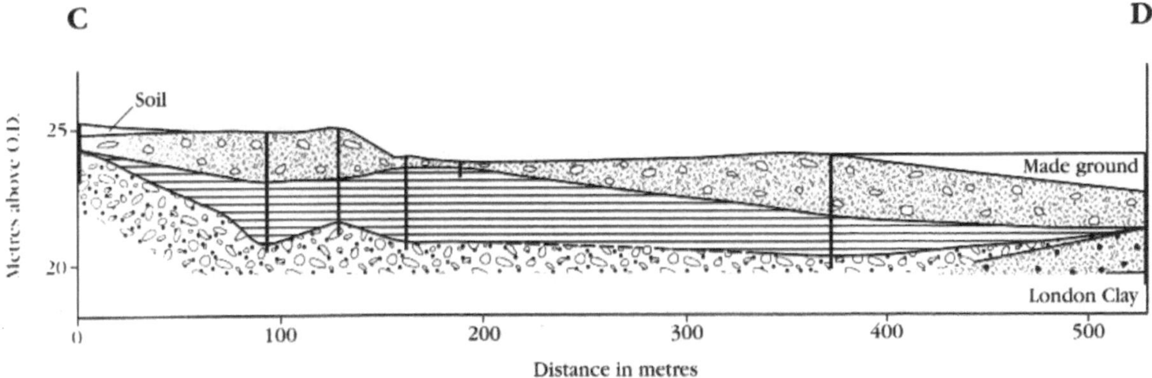

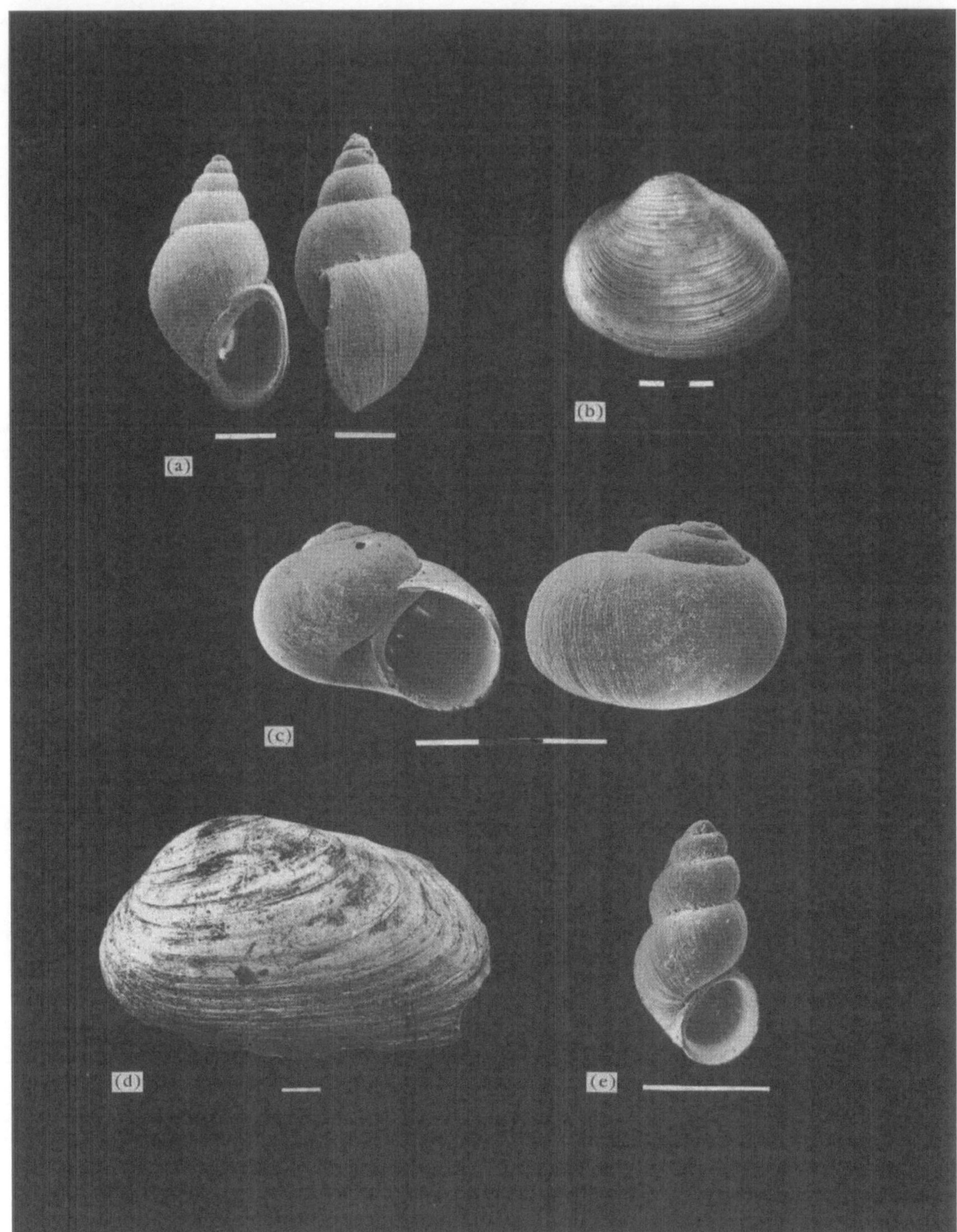

Figure 5.12 Characteristic Mollusca from the Little Oakley Silts and Sands. (A) *Tanousia* cf. *stenostoma* (Nordmann); (B) *Sphaerium solidum* (Normand); (C) *Valvata naticina* Menke; (D) *Unio crassus* Philipsson; (E) *Belgrandia marginata* (Michaud). A, C and E are scanning electron micrographs. Scale bars are graduated in mm. (Photos: Department of Zoology, University of Cambridge.)

Interpretation

The Little Oakley site, after several decades of neglect, has benefited recently from a multidisciplinary appraisal using modern techniques. The sedimentological interpretation has been restricted, however, by the lack of exposure; the reconstruction of the three-dimensional form of the sediment body has relied primarily on augering and occasional temporary excavations. The best evidence for palaeoenvironmental conditions has therefore come from studies of the palaeontology.

The Little Oakley Silts and Sands contain rich assemblages of freshwater molluscs and ostracods, comprising taxa indicative of deposition in the lower reaches of a large, well-oxygenated, calcareous river, upstream from any tidal influence (Preece, 1990b; Robinson, 1990). Terrestrial mollusca are also present, which testifies that the river had wide, open floodplains with fringing marsh habitats (indicated, for example, by *Vertigo antivertigo* (Draparnaud) and *Zonitoides nitidus* (Müller)) and extensive areas of dry, calcareous grassland (indicated by *Trochoidea geyeri* (Soos) and *Truncatellina cylindrica* (Férussac)) (Preece, 1990b). Pollen analyses have confirmed the presence of grassland, but have suggested that woodland also existed in the catchment during most of the time represented (Gibbard and Peglar, 1990). The vertebrate assemblage includes species typical of fluviatile, marsh, grassland and woodland environments (Lister *et al.*, 1990), supporting the evidence from the pollen and molluscs.

A more detailed assessment of the palynology (Gibbard and Peglar, 1990) reveals that early herb-dominated vegetation gave way to boreal forest, with birch and pine dominant, and then to deciduous forest in which oak and particularly elm were major constituents. This vegetational history was reconstructed from the palynological records of several separate boreholes, since deposits in different parts of the channel were found to represent different time periods (Gibbard and Peglar, 1990). The earliest fossiliferous sediments, shown by pollen analyses to date from the transition from cold conditions at the beginning of an interglacial, have been found only on the southern margin of the channel; the basal channel sediments become progressively younger northwards (Gibbard and Peglar, 1990). The Mollusca from borehole LOO, near the southern edge of the channel, record a progressive replacement of aquatic taxa by marsh snails (as above), thus indicating a shallowing sequence (Preece, 1990b). This suggests that this part of the feature was filled as the active channel shifted northwards. The precise width of the active channel must therefore have been somewhat less than the maximum width of the Little Oakley Silts and Sands sediment body.

The vertebrate assemblages from the channel deposits are fully temperate in character. Knowledge of the modern breeding requirements of certain fish species (Cyprinidae and Percidae) recovered from Little Oakley suggests that during the summer months (May–August) water temperatures must have reached a minimum of 15°C and a maximum of 22°C, whereas in winter (December–March) they must have fallen no lower than 0.6°C. Moreover, the occurrence of the pond tortoise (*Emys orbicularis*) implies mean July temperatures well in excess of 18°C if, as seems likely, this represents a breeding population (Lister *et al.*, 1990).

Palaeogeography

Little Oakley lies in an area that is highly important for the study of Pleistocene geology, since it is one in which the Thames and East Anglian stratigraphies can be compared. The results of clast-lithological analysis indicate that the Oakley Gravel, the terrace formation into which the Little Oakley Silts and Sands are channelled, is part of the pre-diversion Thames drainage system (Bridgland, 1988a; see above). Detailed clast compositional data from the Tendring Plateau (Table 5.2) indicates that the early Thames was joined, as it flowed north-eastwards, by an important tributary draining northwards from the Weald, the direct ancestor of the modern River Medway (Bridgland, 1980, 1983a, 1988a; Fig. 5.4). This ancient fluvial confluence is recognized in the gravels of the area by an increase in southern material, predominantly Lower Greensand chert, an important component of Medway gravels. Thus Thames gravels of Kesgrave Group type change downstream into Thames-Medway deposits (Bridgland, 1988a; Figs 5.2 and 5.4).

The ratio of southern material to quartz and quartzites has been employed to demonstrate the change in clast composition resulting from the confluence between the Kesgrave Thames and the Medway (Bridgland, 1988a; Bridgland *et al.*, 1990). In the gravels upstream from the confluence, this ratio is relatively low, ranging from 0.02 to 0.10. A ratio of approximately 0.10 is considered to represent the upstream limit of Medway influence, this being the highest ratio encountered in Kesgrave Group gravels further upstream. Ratios generally increase downstream from this westward limit to 0.50 and above (Bridgland, 1988a; Bridgland *et al.*, 1990; Table 5.2). A gradual compositional change is observed, a phenomenon for which several causes can be envisaged. Firstly, full mixing of the gravel loads of the two rivers may not have occurred for a considerable distance downstream from the confluence. Secondly, the general distribution of gravel deposits in Essex indicates a progressive southward migration of the Thames and an eastward migration of the Medway. This means that the Thames, on its southern flank, would have been reworking west-bank terrace deposits of the Medway, causing an increase in southern material in its bedload gravel several kilometres above the actual confluence.

The composition of the Oakley Gravel in the vicinity of Little Oakley shows it to fall within the Thames-Medway category; in particular, southern to quartz and quartzite ratios in excess of 0.10 were revealed (Table 5.2). However, the analysis of scattered gravel-sized clasts in the Little Oakley Silts and Sands has revealed equivalent ratios close to, but generally below 0.10 (Bridgland *et al.*, 1988, 1990; Table 5.2). There are two possible interpretations of this data. Firstly, the Little Oakley channel deposits may be the product of a tributary river that has reworked the Kesgrave gravels over a wide area, mixing material derived from further upstream, remote from the Medway confluence area, with that from the local Oakley Gravel. Alternatively, the Little Oakley Silts and Sands may have accumulated in the channel of the Thames, immediately upstream from its confluence with the contemporary Medway channel. Therefore the Little Oakley sediments, although attributed to the Thames, might be expected to show some Medway influence, because of reworking either from older Medway terraces (as described above) or from the underlying Oakley Gravel, which represents a wider gravel-covered Thames floodplain that had already coalesced with that of the Medway (Fig. 5.4). The palaeontological and sedimentological evidence for the presence of a large river at Little Oakley, combined with regional stratigraphical evidence indicating that the Thames occupied the area of the Tendring Plateau until the Anglian Stage, provides support for a Thames origin for the Little Oakley Silts and Sands (Bridgland *et al.*, 1988, 1990).

Biostratigraphy and correlation

The palaeontological evidence from the Little Oakley Silts and Sands not only enables detailed palaeoecological reconstructions, but also provides evidence for the relative age of the deposits. The pollen assemblages, from various profiles through the deposits, together represent the early part (biozones I and II) of an early Middle Pleistocene interglacial (Bridgland, 1990b; Gibbard and Peglar, 1990). Hornbeam (*Carpinus*) and 'Tertiary relics' (such as *Tsuga*, *Carya* or *Eucommia*) are absent, suggesting that the deposits are unlikely to be of pre-Cromerian age. A number of features of the pollen record are suggestive of a Cromerian (*sensu* West Runton) age: *Ulmus* (elm) expands early and becomes dominant, *Picea* (spruce) is present throughout, whereas *Quercus* (oak) expands late and is followed by *Corylus* (hazel). The similarity between this sequence of woodland development and that of the Cromerian stratotype at West Runton, Norfolk, has led to suggestions that the two sites may be correlatives (Bridgland *et al.*, 1988, 1990).

The vertebrate fauna is also suggestive of a broadly Cromerian age for the Little Oakley Silts and Sands (Lister *et al.*, 1990). The presence of the giant deer *Megaloceros verticornis* and *M. dawkinsi*, together with the vole *Mimomys savini*, strongly indicate deposition during the early Middle Pleistocene. In western Europe these two deer species are restricted to deposits of Cromerian and early Elsterian (Anglian) ages, whereas *M. savini* extends from the late Early Pleistocene to the type Cromerian, but is replaced in some late pre-Anglian ('late Cromerian') assemblages by its evolutionary descendant, *Arvicola cantiana* (see Chapter 1). The Little Oakley vertebrate fauna is fully temperate in character; it clearly represents an interglacial later than the Pastonian Stage but earlier than those 'late Cromerian' sites with *A. cantiana*, such as Westbury-sub-Mendip (Bishop, 1982).

The rich vertebrate fauna from the West Runton Freshwater Bed (biozones CrIb-IIb) shares with Little Oakley such characteristic extinct taxa as the water vole *Mimomys savini*, the pine vole *Pitymys gregaloides* (Hinton), the Etruscan rhinoceros *Dicerorhinus etruscus* and the giant deer *Megaloceros verticornis* (Stuart, 1975, 1981, 1982a). The much more limited fauna from Little Oakley compares closely with that of West Runton, although pond tortoise (*Emys orbicularis*) has not yet been recorded from the latter. Shrew remains from Little Oakley represent a potentially valuable means of comparison with the fauna from West Runton, in which *Sorex* species of three different sizes occur. There are indications from the collections accumulated to date from Little Oakley that a similar assemblage of shrews occurs, but further sampling is required in order to obtain crucial mandibular remains (Lister *et al.*, 1990).

The molluscan assemblage from Little Oakley includes *Tanousia*, a genus known only from the Cromerian in Britain. *Valvata naticina*, *Bithynia troscheli* and *Unio crassus* are unknown in Britain before this stage. The presence of *B. troscheli* to the exclusion of *Bithynia tentaculata* is a feature that characterizes most British Cromerian sites. These features of the assemblage are consistent with a broadly Cromerian age, although the same taxa are found in earlier sediments in The Netherlands and elsewhere (Preece, 1990b). Recent consideration of the molluscan assemblages from Cromerian (*sensu lato*) sites in Britain and north-west Europe (Meijer and Preece, in press) suggests that the Little Oakley fauna is peculiar, thus far, to this one locality. Meijer and Preece point to significant differences between molluscan faunas that can be regarded as 'early' and others that can be regarded as 'late' within the 'Cromerian Complex'. Significant taxa amongst the 'early' assemblages are *Valvata goldfussiana* and *Tanousia runtoniana*, extinct species that do not seem to survive into the 'late' faunas. The 'late' assemblages have *V. naticina*, *Belgrandia marginata* and *Bithynia tentaculata*, which have yet to be found in the 'early' faunas, and are indistinguishable, malacologically, from Hoxnian (Holsteinian) assemblages. The British sites at West Runton and Sugworth both have molluscan faunas that can be classified, according to these criteria, as 'early'. Other British Cromerian (*sensu lato*) sites such as Sidestrand and Trimmingham have yielded the 'late' type of molluscan assemblage. Little Oakley, uniquely, has a fauna that seems intermediate between these two categories. *Valvata goldfussiana*, an element of the 'early' fauna, is absent, whereas *Bithynia tentaculata* is present. The *Tanousia* from Little Oakley is a different species from that found at West Runton and is close to, but smaller than, *T. stenostoma* a species recognized in Denmark (Preece, 1990b). Meijer and Preece (in press) have suggested that the Little Oakley site represents a later temperate episode within the 'Cromerian Complex' than either West Runton or Sugworth.

Significant amongst the ostracod fauna from Little Oakley are *Candona tricicatricosa* (Diebel and Pietrzeniuk), *Ilyocypris quinculminata* (Sylvester and Bradley), *Sclerocypris clavata prisca* (Diebel and Pietrzeniuk) and *Scottia browniana*. None of these is restricted to the Cromerian, but they are unknown together in Britain after the Hoxnian, thus supporting a broadly Middle Pleistocene age (Robinson, 1990).

Bridgland *et al.* (1990) have cited amino acid ratios from shells from Little Oakley, analysed by two different laboratories. Specimens of both *Valvata piscinalis* and *V. naticina* were analysed in London (London Quaternary Centre; laboratory now relocated to the Institute of Earth Sciences, the University College of Wales, Aberystwyth), and ratios from *V. naticina* were also obtained in Colorado (INSTAAR Laboratory). The D : L ratios from *V. piscinalis* are somewhat higher than those from *V. naticina*, which may indicate that epimerization is faster in the former species. Comparison with *V. piscinalis* ratios from other Middle Pleistocene sites is informative, the following being listed by Bridgland *et al.* (1990) and/or Bowen *et al.* (1989):

Site	Mean D : L ratios		Laboratory
Hoxne	0.243 ± 0.023	(n=3)	INSTAAR
Hoxne	0.261 ± 0.01	(n=4)	London
Swanscombe	0.30 ± 0.016	(n=10)	London
Swanscombe	0.297 ± 0.009	(n=5)	INSTAAR
Clacton	0.299 ± 0.002	(n=3)	London
Little Oakley	0.324 ± 0.004	(n=2)	London
Little Oakley	0.336 ± 0.027	(n=4)	London
West Runton	0.348 ± 0.011	(n=5)	London

These ratios are consistent with a broad correlation between Little Oakley and West Runton and confirm that these sites are older than those at Hoxne, Swanscombe and Clacton, all attributed to the Hoxnian Stage (*sensu lato*). However, ratios from *V. goldfussiana* shells from Sugworth compare closely with those quoted above from Swanscombe and Clacton, despite the convincing biostratigraphical indications that this site is a broad correlative of Little Oakley and West Runton (see Chapter 2, Sugworth).

The stratigraphical record of the lower Middle Pleistocene in Britain, best represented on the Norfolk coast, is now known to be far from complete (Zalasiewicz and Gibbard, 1988). Comparison with the sequence in The Netherlands, in particular, shows that repeated climatic fluctuations occurred in the period following the Matuyama–Brunhes palaeomagnetic reversal and prior to the Elsterian (= Anglian) Stage; four warm/cold climatic cycles are recognized below Elsterian tills in The Netherlands and are collectively termed the 'Cromerian Complex', the magnetic reversal occurring between the peak of the earliest of the four temperate episodes and the trough of the succeeding glacial (Zagwijn *et al.*, 1971; de Jong, 1988; Chapter 1).

Exactly how the Little Oakley interglacial relates to the 'Cromerian Complex' of The Netherlands is difficult to determine, as the four interglacials comprising this complex were distinguished using palynology alone. Correlation with 'Cromerian Complex Interglacial I' (or earlier temperate intervals) is precluded by the absence of Tertiary relics such as *Eucommia* and *Tsuga* (Gibbard and Peglar, 1990). Similarly, the absence of *Taxus* and/or *Carpinus* in the early temperate substage (biozone II) differentiates the sequence in the Little Oakley Silts and Sands from 'Cromerian Complex Interglacials II and III'. There are similarities, however, between the palynology of Little Oakley and Noordbergum, a site in The Netherlands that has been assigned to 'Interglacial IV' (de Jong, 1988). Noordbergum has also been tentatively correlated with the Cromerian stratotype at West Runton (Zagwijn, 1985), a suggestion supported by amino acid ratios from Noordbergum (Miller and Mangerud, 1985). However, the recent recognition of *Arvicola cantiana* amongst collections from Noordbergum (von Kolfschoten, 1988) argues against these correlations, since the earlier vole *Mimomys savini* occurs at both West Runton and Little Oakley, rather than *A. cantiana* (Stuart, 1975, 1981, 1982a; Bridgland *et al.*, 1988). This has led to the suggestion that the Dutch sequence is itself incomplete, that the British Cromerian (*sensu stricto*) is missing in The Netherlands and that Little Oakley and West Runton are broadly of 'late Cromerian Complex' age (Bridgland *et al.*, 1990).

The Matuyama–Brunhes palaeomagnetic boundary is recognized as a highly significant stratigraphical marker within the Pleistocene, widely adopted as the base of the Middle Pleistocene (Richmond and Fullerton, 1986). Since this magnetic reversal approximately coincides, in the sedimentary record, with the end of 'Cromerian Complex Interglacial I' of the Dutch sequence, palaeomagnetic information is of considerable value in the study of sites of 'Cromerian Complex' age. The polarity of the Little Oakley Silts and Sands has been established as normal (Austin, in Bridgland *et al.*, 1990), indicating deposition after the Matuyama–Brunhes reversal and therefore after 'Cromerian Complex Interglacial I' of The Netherlands. This implies an age of somewhat less than 780,000 years BP, the approximate date of the magnetic reversal (Shackleton *et al.*, 1990).

Within the British sequence, the interglacial at Little Oakley has been interpreted as more recent than that at Ardleigh, since the Little Oakley Channel is cut through the local Oakley Gravel, the downstream equivalent of the Ardleigh Gravel (Gibbard, 1988b; Bridgland *et al.*, 1990). It is important to note, however, that the Little Oakley Silts and Sands occupy a position well below the original upper surface level of the Oakley Formation, in an area of considerable dissection (Fig. 5.3). The maximum thickness of the fossiliferous sediments is *c.* 4 m and they do not extend higher than *c.* 24 m O.D. (Fig. 5.11), whereas the Oakley Gravel terrace surface, prior to subsequent erosion, was probably aggraded to around 28–30 m O.D. It is therefore possible that a later aggradation of Oakley Gravel occurred, after the deposition of the Little Oakley Silts and Sands, and was removed by the erosion of the upper 4–5 m of the terrace deposits. Thus it is possible that the interglacials represented at Little Oakley and Ardleigh are one and the same. There is nothing in the palynological records from these two sites to disprove this alternative hypothesis (P.L. Gibbard, pers. comm.), but pollen is the only significant biostratigraphical evidence that the two localities both provide. If the terminology of the climatic model for terrace formation (presented in Chapter 1) is adopted, the two possible interpretations of the relation between the Ardleigh and Little Oakley sequences can be further examined. If the same interglacial is represented at both sites, the gravel into which the Little Oakley channel is incised must be the pre-interglacial (phase 2) part of the Ardleigh/Oakley Formation, with the post-interglacial (phase 4) gravel missing from the immediate area. A single cold-warm-cold cycle would thus be represented in the deposits of the Ardleigh/Oakley Formation. If different interglacials are represented, and the Little Oakley Silts and Sands were deposited by the Thames and not a later tributary, the implication would be that no rejuvenation occurred in this part of the Thames catchment during the cold episode represented by the Ardleigh Upper Gravel.

Conclusions

The Little Oakley interglacial site is clearly a critical locality for Pleistocene palaeontology and stratigraphy. The interglacial sediments, filling a channel cut through the local early Thames-Medway gravel (the Oakley Gravel), are richly fossiliferous, yielding pollen and the remains of molluscs, ostracods, mammals and fish. The combination of the various fossil types present indicates that the sediments represent a period of generally warm climate, comparable to that of the present day. The pollen provides a detailed record of climatic and vegetational change during the first half of an interglacial (temperate) period. The mammals include the extinct vole *Mimomys savini*, the Etruscan rhinoceros and the giant deer *Megaloceros verticornis*. These species, as well as certain extinct molluscs from Little Oakley, are characteristic of several temperate-climate episodes that, alternating with colder periods, preceded the major glaciation during which the Thames was diverted. This period of fluctuating climate, recognized in The Netherlands and called the 'Cromerian Complex', covers a long span of Pleistocene time, between around 750,000 and 450,000 years BP. The name derives from the Cromerian Stage, defined in north Norfolk, which is thought to coincide with part of the 'Cromerian Complex'. The significance of the Little Oakley interglacial deposits is heightened by the fact that they occur within the early terrace deposits of the Thames system. They therefore provide a means for dating the Thames terrace sequence and to enable improved correlation between the Thames Basin and Pleistocene sequences in other parts of Britain and western Europe.

WIVENHOE GRAVEL PIT (TM 005235)

D.R. Bridgland

Highlights

At Wivenhoe, periglacial Thames gravels both underlie and overlie an organic silty clay that contains the remains of temperate-climate plants and beetles. It is believed that this entire sequence pre-dates the Anglian glaciation, which would suggest that the temperate interval represented falls within the 'Cromerian Complex' as recognized in The Netherlands.

Introduction

Wivenhoe Gravel Pit is located near the southern edge of the Tendring Plateau, *c.* 2 km from the estuary of the River Colne. Deposits classified as part of the Kesgrave Sands and Gravels (Rose *et al.*, 1976) have been exploited here for many years, old workings covering around half a square kilometre. The site is now regarded as the type locality for the Wivenhoe Gravel, which is chronologically the third of the four terrace formations (Figs 5.2 and 5.3) that constitute the Low-level Kesgrave Subgroup (Bridgland, 1988a; see above, Introduction to Part 1).

Recent appraisal of the site has revealed fossiliferous sediments apparently interbedded with the Wivenhoe Gravel (Bridgland *et al.*, 1988; Fig. 5.13). These sediments have yielded pollen from temperate-climate trees, together with other plant fossils and beetle remains. The biostratigraphical evidence is as yet insufficiently distinctive to identify any particular temperate episode.

The stratigraphical position of the fossiliferous sediments at Wivenhoe is highly significant. They occur within a terrace formation (the Wivenhoe/Cooks Green Gravel) that is stratigraphically younger than the Ardleigh/Oakley Gravel, the formation that includes the Cromerian deposits at Little Oakley, and stratigraphically older than the St Osyth/Holland Formation, which is correlated with the Anglian Stage glacial (Bridgland, 1988a; Bridgland *et al.*, 1988; Fig. 5.3; see also Little Oakley, and St Osyth and Holland-on-Sea). This makes the Wivenhoe deposits strong candidates for correlation with 'late Cromerian' sites recognized elsewhere in Britain and in north-west Europe. There is controversy at present over whether such sites represent the latter part of the Cromerian Stage (*sensu* West Runton) or an additional temperate episode between the type Cromerian interglacial and the Anglian glacial (see Chapter 1). Further research is required, however, to establish the age of the Wivenhoe deposits satisfactorily.

Description

Peaty deposits containing beetles were first reported from the Wivenhoe pit by McKeown and Samuel (1985), who provided a photographic record of a section showing organic sediments above gravel. These authors suggested that the peaty deposits were of Cromerian age, but cited no supporting evidence. Later investigations at the present GCR site revealed that deposits rich in plant material occur near the top of the Wivenhoe Gravel in a restricted area around TM 050236 (Bridgland *et al.*, 1988). The GCR section at Wivenhoe reveals the following sequence (Fig. 5.13):

		Thickness
4. Silty clay, locally ?organic, cryoturbated		*c.* 1.5 m
3. Gravel and sand, horizontally bedded	(Wivenhoe Upper Gravel)	1.5 m
2. Organic silty clay with scattered pebbles and plant remains, brecciated		*c.* 1 m
1. Medium-coarse sandy gravel interbedded with sand	(Wivenhoe Lower Gravel)	*c.* 5 m

The sequence is disrupted by frost cracks and ice-wedge pseudomorphs. Most of these appear to emanate from various levels within the upper gravel (member 3), but there are clearly some that formed in the sediments below prior to the deposition of this member (Fig. 5.13). These features indicate that the upper gravel was laid down during a periglacial episode that followed the temperate period represented by the organic clay (member 2). Analysis of the clast content of the gravels below and above the organic clay has indicated that they have close similarities and that both are typical of the Low-level Kesgrave Subgroup (Bridgland, 1988a; Bridgland *et al.*, 1988; Table 5.2).

The organic silty clay contains pollen, plant macrofossils and insect remains. A pollen sequence has been determined (Gibbard, in Bridgland *et al.*, 1988), its arboreal component dominated by birch, pine, spruce and alder, with smaller quantities of, for example, silver fir, elm, oak, hornbeam, hazel and willow. A 500 gm sample from this unit has yielded a limited flora, comprising fruits of *Schoenoplectus lacustris* (L.) Palla, trigonal nutlets of *Carex* spp. and seeds of *Menyanthus trifoliata* L. (M.H. Field, pers. comm.). Two flint flakes, evidently formed by percussion, were recovered from the organic silty clay (Fig. 5.13). Their interpretation as Palaeolithic artefacts, claimed by Bridgland *et al.* (1988), remains equivocal (see below).

Thin, involuted lenses of dark grey, possibly organic clay occur at higher levels in the sequence, within the cryoturbated upper part of the upper gravel (Fig. 5.13). No pollen or other fossil material has been recovered from these levels (P.L. Gibbard, pers. comm.), which may represent reworking of organic material from the underlying silty clay.

Interpretation

This site is an important source of evidence for the reconstruction of Thames drainage evolution in the early Middle Pleistocene, shortly before the diversion of the river during the Anglian Stage. Prior to the discovery of various pre-Anglian interglacial sites within the Low-level Kesgrave Subgroup deposits of Essex (Bridgland *et al.*, 1988; see above, Ardleigh and Little Oakley), little was known about this time period. It had been widely accepted that all of

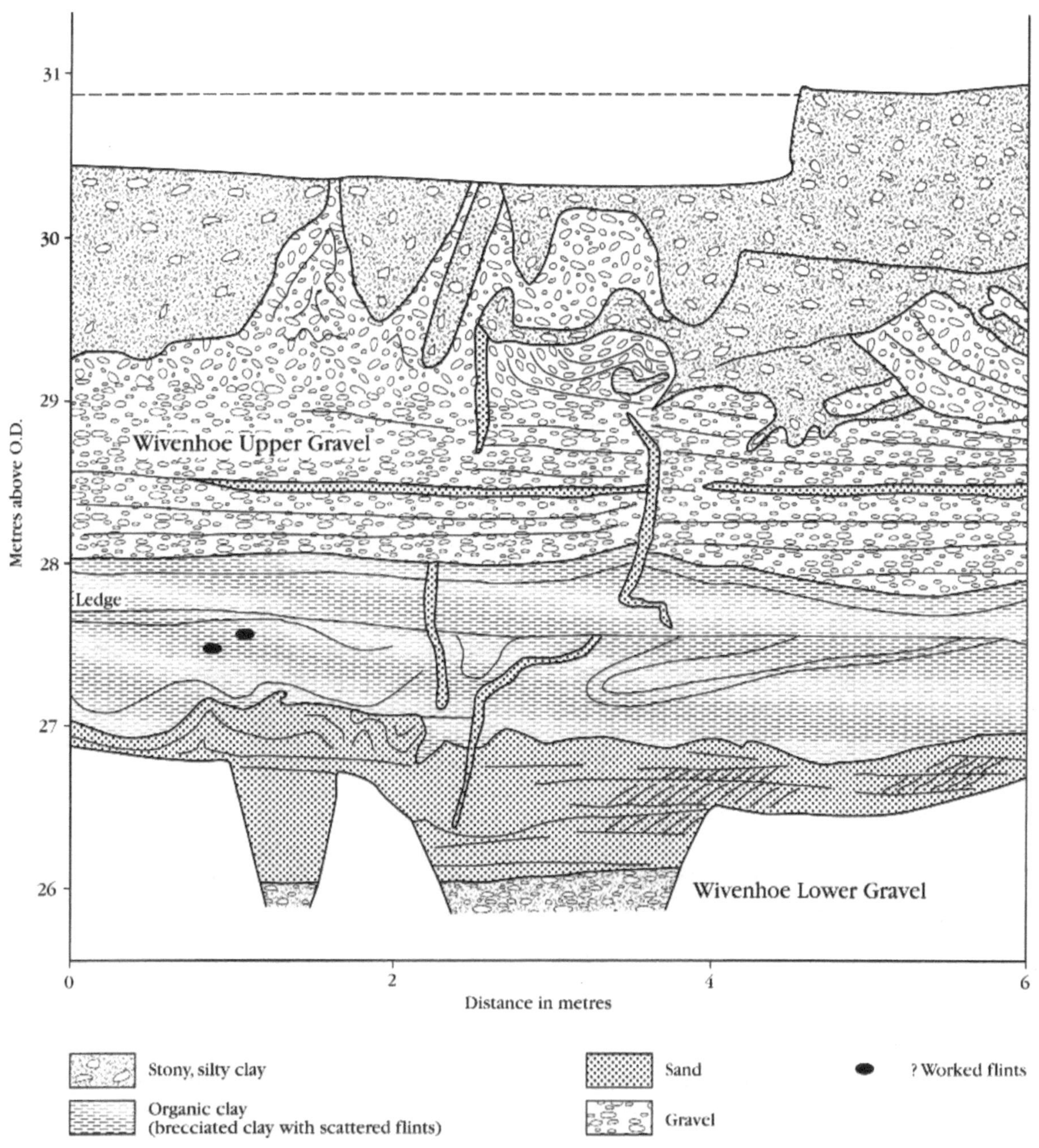

Figure 5.13 Section at Wivenhoe, showing the organic clay (modified from Bridgland *et al.*, 1988)

the Kesgrave Sands and Gravels were aggraded during the Beestonian Stage (Rose *et al.*, 1976; Rose and Allen, 1977), but later studies revealed that a series of distinct terrace formations is represented and that deposition of these early Thames gravels spanned a large proportion of Pleistocene time (Rose, 1983a; Kemp, 1985a; Rose *et al.*, 1985b; see above, Introduction to Part 1).

The temperate-climate Valley Farm Soil and the (superimposed) periglacial Barham Soil were both identified at Wivenhoe by Rose *et al.* (1976), although the stratigraphical control of overlying Anglian glacial sediments is lacking at this site, which lies outside the maximum ice limit (Fig. 5.4F). Kemp (1985a) observed that the Soil Survey of England and Wales had mapped a 'stagnogleyic palaeo-argillic brown earth' soil (their Tendring Association) on undissected remnants of the Waldringfield,

Ardleigh/Oakley and Wivenhoe/Cooks Green Gravels, the oldest three of the four Low-level Kesgrave Subgroup formations. Kemp considered this soil, which is restricted to the unglaciated area of north-east Essex, to contain relict features of the Valley Farm Soil. The temperate palaeosol is not, however, present in the GCR section, although the cryoturbation structures at the top of the sequence there may indicate the presence of the early Anglian Barham Soil.

The stratigraphical position of the Wivenhoe Formation in relation to other parts of the Low-level Kesgrave Subgroup puts considerable constraint on its relative age. It is later than the Ardleigh/Oakley Formation, which incorporates the Ardleigh and Little Oakley interglacial sediments. The Little Oakley channel deposits in particular, if correctly ascribed to the Thames, suggest that the river flowed at the level of the Ardleigh/Oakley Gravel until late in 'Cromerian Complex' times (see above, Little Oakley). The terrace formation immediately (altitudinally) below the Wivenhoe Gravel, the St Osyth/Holland Formation, has been attributed to the Anglian Stage on the basis of its apparent correlation with the diversion of the Thames (Bridgland, 1988a; see above, Introduction to Part 1 and below, St Osyth and Holland-on-Sea). This means that the Wivenhoe Formation must have accumulated during the interval between the Little Oakley interglacial and the downcutting phase that preceded the deposition of the Lower St Osyth and Lower Holland Gravels. If the organic silty clay and the upper gravel at the GCR site (members 2 and 3) were laid down by the Thames, the former must represent a pre-Anglian temperate episode that post-dates the deposition of the Little Oakley interglacial deposits (Bridgland *et al.*, 1988).

The palynology of the Wivenhoe organic sediments is not stratigraphically diagnostic. It merely provides a record of boreal forest vegetation of a type found in many interglacial and interstadial sequences (P.L. Gibbard, pers. comm.). The plant macrofossils are equally undiagnostic. They indicate the presence of a marshy area adjacent to a water body, but the three species listed above occur in cold- and warm-climate sediments throughout the British Pleistocene (M.H. Field, pers. comm.). It is hoped that further details of the palaeontology, when available, will provide information of stratigraphical significance.

The two flint flakes from the organic clay, if they represent a Palaeolithic occupation at Wivenhoe, may point to another aspect of the site's significance. The occurrence of palaeoliths in sediments of pre-Anglian age in Britain was not accepted until recently, but they have now been described from a number of sites of probable late 'Cromerian Complex' age (see, for example, Wymer, 1988; Chapter 1). These include cave deposits at Westbury-sub-Mendip, Somerset (Bishop, 1982), raised-beach deposits at Boxgrove, Sussex (Roberts, 1986), and lacustrine sediments rafted by Anglian Stage ice at High Lodge, Mildenhall, Suffolk (Wymer, 1988). The pre-Anglian age of all these sites is based on their mammalian faunas; all have the rhinoceros *Dicerorhinus etruscus*, which is unknown in Hoxnian and later sediments. The sites at Westbury and Boxgrove have also yielded the stratigraphically significant water vole *Arvicola cantiana*, which replaced the species *Mimomys savini* after Cromerian biozone CrIII, as represented at West Runton and Sugworth. *Arvicola cantiana* is claimed to occur in sediments attributed to the Cromer Forest Bed at Ostend, Norfolk, in pollen biozone CrIV (Stuart, 1982a, 1988). As has been noted in Chapter 1, there is considerable controversy, both in Britain and on the continent, about whether the evolutionary change between these vole species occurred within a single temperate episode or whether an additional temperate interval is represented that was both post-Cromerian (*sensu* West Runton) and pre-Anglian. It is worth noting that there is no convincing record of Palaeolithic artefacts from the Cromer Forest Bed (Wymer, 1988) or from any deposit with *Mimomys savini*, whereas Boxgrove, High Lodge and Westbury all yield palaeoliths (Wymer, 1988).

Deposits attributable to this 'late Cromerian' interval (characterized by *A. cantiana*) have yet to be identified in association with the Thames sequence. They would be anticipated in a stratigraphical position within the terrace sequence between Cromerian *sensu lato* sites that have *Mimomys savini*, such as Little Oakley, and the Anglian Stage St Osyth/Holland Formation. As the Wivenhoe Formation occupies exactly this position, the Wivenhoe organic deposit is a prime candidate for the first record of the Westbury temperate interval in the Thames system.

However, the evidence for assigning the Wivenhoe organic deposit to this interval is at present equivocal. In the absence of definitive palaeontological evidence, the interpretation of the site hinges on the stratigraphical relations of the various Low-level Kesgrave Subgroup formations. Another problem with this correlation is that it relies on the organic clay and the Wivenhoe Upper Gravel being products of the Thames and not of a tributary river. The clast composition of the Wivenhoe Upper Gravel is indistinguishable from the Wivenhoe Lower Gravel and from other Kesgrave Group deposits in the area (Table 5.2), providing some evidence for a Thames origin. It is possible, however, that a tributary stream with a localized catchment might have produced a gravel of identical composition to the various early Thames deposits. Such a river could have laid down the organic deposits and the upper gravel at any time after the deposition of the Wivenhoe Lower Gravel. Thus a pre-Anglian age is only indicated for the organic sediments if the Wivenhoe Upper Gravel is correctly interpreted as part of the Kesgrave Group.

A hypothetical post-Anglian tributary stream would have to have been confined to the western part of the Tendring Plateau, since Medway- and Red Crag-derived material is present in the gravels to the north-east (see above, Ardleigh and Little Oakley), but is not found in the Wivenhoe Upper Gravel. There is also no indication of Anglian glacial erratics in the Wivenhoe Upper Gravel, and so the hypothetical post-Anglian river could not have had a catchment extending to the glacial limit. Given the sparse but widespread occurrence of clast types foreign to the Kesgrave Group in the area, it seems unlikely that a post-Anglian stream of any size could have produced a deposit with the composition of the Wivenhoe Upper Gravel. A pre-Anglian Thames origin for this unit and the underlying biogenic sediments seems, therefore, to be indicated.

Supporting evidence for the occurrence of a further temperate interval between the Little Oakley interglacial and the Anglian Stage is provided by the identification of relict elements of the Valley Farm Soil in the upper levels of the Wivenhoe/Cooks Green Formation (Rose *et al.*, 1976; Kemp, 1985a). In the unglaciated area of north-east Essex, the soil is only present in relict form and there is no upper stratigraphical control on its age. The occurrence of rubified soils similar to the pre-Anglian Valley Farm Soil on post-Anglian deposits in the Chelmsford area (Rose *et al.*, 1978; Sturdy *et al.*, 1978), suggests that caution should be exercised in identifying the Valley Farm Soil outside the Lowestoft Till limit. It may prove possible, however, to relate the palaeosol exposed in other parts of the Wivenhoe workings, recorded by Rose *et al.* (1976), to the stratigraphical level of the organic clay member; an assessment of this relationship is awaited.

Conclusions

The GCR site at Wivenhoe provides sections in sediments formed by the early River Thames at a time when it flowed north-eastwards across East Anglia, before its diversion to its modern course. This is the only locality discovered to date in the Thames system that is likely to represent the temperate interval that immediately preceded the (Anglian) glaciation (this glaciation brought about the river's diversion about 450,000 years ago). Much work is required before this interval, recognized in recent years elsewhere in Britain but as yet undefined, can be fully evaluated. Its status as a full interglacial has yet to be firmly established; it could be that an interstadial (a short-lived temperate-climate event during a predominantly cold period) is represented and it remains possible that the sediments date from the latter part of the type-Cromerian interglacial, as defined at Cromer. More work is required on the fossiliferous sediments and the stratigraphy at Wivenhoe to confirm or deny correlation with this period.

ST OSYTH GRAVEL PIT (TM 120174) and HOLLAND-ON-SEA CLIFF (TM 211166)

D.R. Bridgland

Highlights

Sediments at these two sites record the events immediately prior to and during the glaciation of the Thames valley, leading up to the diversion of the river. At both sites, gravels of the pre-diversion (Kesgrave Group) are overlain by sediments rich in outwash from the ice sheet that blocked the course of the river in central Essex and the Vale of St Albans. At St Osyth the sediments were laid down immediately upstream from the Medway confluence, whereas the gravels at Holland-on-Sea are of Thames-Medway type. The upper gravel at Holland-on-Sea demonstrates that the Medway was unaffected by the glaciation that blocked the Thames.

Introduction

The St Osyth and Holland-on-Sea GCR sites are situated near the southern edge of the Tendring Plateau, in the vicinity of Clacton-on-Sea (Fig. 5.2). The St Osyth pit lies 6 km west of the coast, whereas the Holland cliffs are near the northern end of a 6 km length of erstwhile coastal exposure running north-eastwards from the West Cliff at Clacton, itself part of another GCR site (see Part 2 of this chapter). The St Osyth and Holland sections are both in the lowest of the four Low-level Kesgrave Subgroup formations recognized on the Tendring Plateau (Bridgland, 1988a; Fig. 5.2; see above, Introduction to Part 1). St Osyth lies upstream from the contemporary confluence between the Thames and the Medway, whereas Holland lies within the confluence area (Bridgland *et al.*, 1988, 1990; Fig. 5.4E).

The sites at St Osyth and Holland-on-Sea provide complementary evidence enabling the reconstruction of events in the lower part of the Thames basin during the Anglian glaciation, when the diversion of the river took place. At both sites deposits typical of the Kesgrave Group are overlain by later gravels, believed to have been laid down at the time of the Lowestoft glaciation (Bridgland, 1980, 1983a, 1988a; Bridgland *et al.*, 1988). Considered together, the sites are of considerable stratigraphical significance, since they provide a basis for correlating the terrace sequence in southern East Anglia with the succession in the Middle Thames and the Vale of St Albans, which can also be related to the Anglian glaciation (Table 1.1 and Fig. 1.3).

Description

There have been relatively few descriptions of the Pleistocene deposits in this area, with the notable exception of the Clacton interglacial sediments. Wood (1866b) attributed gravels underlying the coastal district, between St Osyth and Clacton, to his 'East Essex Gravel', equating them with deposits south of the Blackwater estuary (see Part 2 of this chapter). The only available geology map (Old Series, Sheet 48) shows the Tendring Plateau largely covered by 'Glacial Gravel' and 'Glacial Loam', but with the patches of 'Post Glacial' drift fringing the valleys of the Colne and Stour and in the extreme south-east of the area, between St Osyth and the coast. The 'loam' comprises post-Anglian loess mixed with stones from the underlying gravels (Eden, 1980). Misinterpretation of this material has led to the unfounded suggestion that till occurs on the Tendring peninsula (Geological Survey, 1:625,000 Quaternary sheet).

The St Osyth site falls within an outcrop mapped as 'Glacial Gravel', separated from deposits to the east, which were mapped as 'Post Glacial', by the valley of a stream flowing into the St Osyth Creek. The 'Post Glacial' deposits extend to the coast, where they are synonymous with the 'Holland Gravel' of Warren (1923a; 1955). However, it is apparent from clast-lithological studies that the gravels on either side of the above-mentioned stream were formerly continuous (Bridgland, 1983a; Bridgland *et al.*, 1988), despite the fact that they were classified differently by the Geological Survey. Indeed, Oakley and Leakey (1937, fig. 10) classified the St Osyth deposits, in common with all the gravels east of a line from Brightlingsea to Great Oakley, as fluviatile.

Warren (1923a, 1924b, 1933) had already interpreted the gravels of the Tendring Plateau as fluvial deposits and attributed them to the Thames, although at that time he was not aware of their considerable antiquity. The discovery of the Cromerian channel-fill at Little Oakley (see above, Little Oakley) led Warren (1940, 1955, 1957) to realize that these gravels were the products of Thames drainage prior to the diversion of the river into its modern valley. This anticipated the inclusion of these deposits by Rose *et al.* (1976) in the newly-defined Kesgrave Sands and Gravels, which they attributed to the pre-diversion Thames.

There are, in fact, significant differences between the Kesgrave Group deposits at St Osyth and Holland-on-Sea, but these represent downstream compositional changes within a single gravel formation. These early Thames terrace deposits are locally overlain by later gravels of a different type, which prove to be of considerable stratigraphical significance.

St Osyth Gravel Pit

Various commercial workings have exploited the spread of gravel to the north-west of St Osyth over the past few decades. The full thickness of the aggradational sequence is only preserved in a small part of the area, the upper horizons having been widely denuded as a result of later dissection. The GCR site occupies approximately the highest point on the outcrop and appears to preserve the most complete sequence. The lower and major part of the sediments here comprise up to 10 m of typical coarse, predominantly matrix-supported gravels of Kesgrave type. Palaeocurrent data from cross-bedded sandy intercalations indicate flow to the south-east, in keeping with the interpretation of the sediments as products of the early Thames, which flowed from the Colchester area towards Clacton (Bridgland, 1980, 1983a). This, the Lower St Osyth Gravel, is overlain by up to 3 m of sand, into which is channelled 1–2 m of gravel of a quite different character, the Upper St Osyth Gravel (Fig. 5.14). The latter comprises fine gravel material scattered in a matrix of coarse sand. It contains a higher proportion of flint than the lower gravel, but a much smaller proportion of rounded pebbles reworked from the Palaeogene. Furthermore, the Upper St

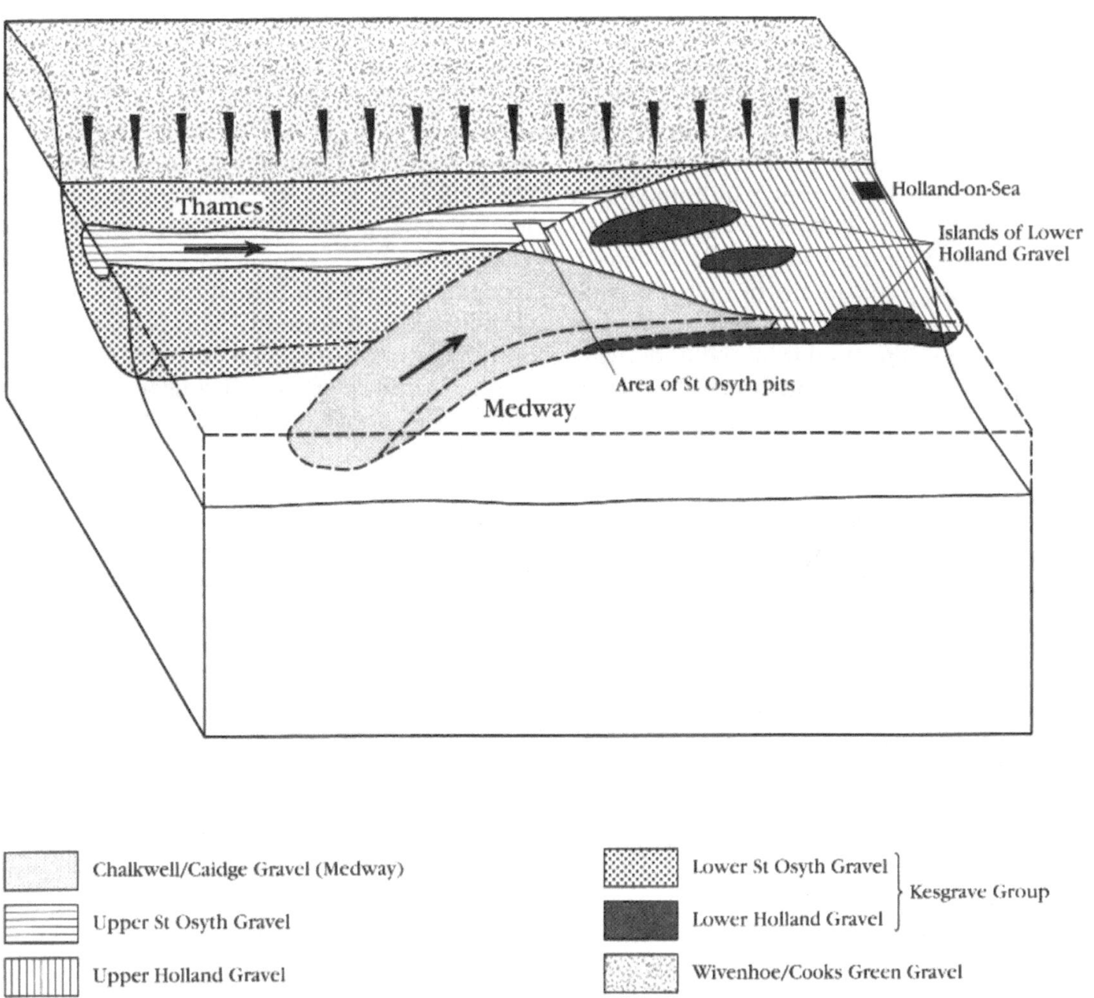

Figure 5.14 Stylized block diagram showing the stratigraphical relations of the Lower and Upper St Osyth and Holland Gravels.

Osyth Gravel contains fewer exotic rock-types than Kesgrave Group gravels (including the Lower St Osyth Gravel) and those present show greater affinities to Anglian glacial gravels than to the underlying Kesgrave Thames deposits. In particular, the exotic suite in the Upper St Osyth Gravel includes *Rhaxella* chert, a rock derived from the Oxfordian of north Yorkshire that, in the London Basin and southern East Anglia (outside the Crag Basin), is present only in Anglian glacial deposits or, reworked, in post-Anglian sediments (Bridgland, 1983a, 1986b). It is uncertain at present whether the intermediate sand has closer affinities with the upper or lower gravel. It is hoped that an analysis of heavy minerals, presently being undertaken, will answer this particular question.

Holland-on-Sea Cliff

The Pleistocene sequence exposed in the Holland-on-Sea cliffs is closely comparable to that at St Osyth. Again there is a dominant lower gravel, *c.* 5 m thick, of Kesgrave type, although with a marked increase in southern clasts (Lower Greensand chert) in comparison with the Lower St Osyth Gravel. Such material constitutes 3–11% of the Lower Holland Gravel, an increase from a maximum of only 2% at St Osyth. A fine-grained, sandy upper gravel is again recognized, separated from the underlying coarser gravel by sands. However, this separation is less well-marked, with alternations of sand and gravel, of which the lower horizons have clast compositions transitional between the Lower and Upper Holland Gravels (Bridgland *et al.*, 1988; Table 5.2, sample 2B), possibly reflecting reworking of the former. The Upper Holland Gravel has been sampled over a wide area inland, as well as at the type section. At some localities it contains large proportions of southern lithologies, one analysis yielding nearly 32% of such material (Bridgland, 1983a; see below). The deposit is otherwise similar in composition and sedimentary characteristics to the Upper St Osyth Gravel.

Interpretation

The principal scientific interest at the St Osyth and Holland-on-Sea GCR sites is stratigraphical. Each is the type site of two lithostratigraphical units: St Osyth pit is the type locality for the Lower St Osyth Gravel and the Upper St Osyth Gravel, whereas Holland Cliffs provide the type section for the Lower Holland Gravel and the Upper Holland Gravel. The interrelations between these deposits and the palaeogeographical interpretation that has been determined from their clast composition provide an illustration, in north-east Essex, of the glacial interruption of Thames drainage that occurred, further upstream, during the Anglian Stage. This provides an important means of correlation with the sequence in the Vale of St Albans (see Fig. 1.3).

The Lower St Osyth and Lower Holland Gravels together represent the lowest (and therefore the youngest) formation within the Kesgrave Group (Bridgland, 1983a, 1988a; Fig. 5.2). They are separately named because they differ from one another in clast-lithological composition; the Lower St Osyth Gravel is a typical Kesgrave Group Thames gravel, but the Lower Holland Gravel contains a higher proportion of southern material. This change is considered to reflect the contemporary confluence of the Kesgrave Thames with the extended River Medway (Fig. 5.4E; see above, Introduction to Part 1). The upstream limit of the contemporary Thames-Medway confluence area has been recognized, for each formation, from changes in clast-lithological content. This is best illustrated by differences in the ratio of southern material to quartz and quartzites (Bridgland *et al.*, 1988, 1990; see above, Little Oakley). Kesgrave Thames gravels upstream from the Medway confluence are characterized by values for this ratio below 0.10, whereas the ratio increases progressively eastwards from the upstream edge of the confluence area (Bridgland *et al.*, 1988; Table 5.2). The Lower St Osyth/Lower Holland Gravel has been sampled at various points between Fingringhoe (TM 042202) and Holland-on-Sea, enabling a more detailed appraisal of these compositional changes than has been possible so far in any of the higher-level formations (Bridgland *et al.*, 1990). This has shown the change from Lower St Osyth to Lower Holland Gravel to occur within the eastern extension of the St Osyth pit (TM 120170; Fig. 5.14), less than 0.5 km to the east of the GCR site (Bridgland *et al.*, 1988).

The fact that the Lower St Osyth/Lower Holland Gravel is the lowest of the pre-diversion (Low-level Kesgrave) Thames formations suggests that its deposition closely preceded the diversion of the river, particularly since temperate sediments of 'Cromerian Complex' age are recognized within higher formations in the area. This led Bridgland (1983a, 1988a) to ascribe this formation to the Anglian Stage and propose a correlation with the Winter Hill Gravel of the Middle Thames. It may be significant that the 'Tendring Association', the soil unit believed to contain relict elements of the Valley Farm Soil (see above, Wivenhoe), has not been mapped on the St Osyth/Holland Formation (Kemp, 1985a), nor were the Valley Farm or Barham Soils recognized in exposures recorded at St Osyth by Rose *et al.* (1976). This may indicate that no temperate-climate interval separated the aggradation of the Lower St Osyth/Lower Holland Gravel and the Anglian glaciation, as is implied by the ascription of the former to the Anglian Stage. However, J. Rose (pers. comm.) has pointed out that palaeo-argillic soils are developed at the land surface on both the Lower and Upper St Osyth and Holland Gravels and that these are difficult to distinguish from the Valley Farm Soil. They are, however, developed on dissection slopes and are independent of the original St Osyth/Holland Formation terrace surface, so they might be expected to be of relatively recent origin. They may be comparable with palaeoargillic and rubified soils developed on the Anglian till of central Essex (Rose *et al.*, 1978; Sturdy *et al.*, 1978).

There is strong stratigraphical support for the correlation of the St Osyth and Holland Gravel with the Winter Hill/Westmill Gravel of the Middle Thames from the interpretation of the two upper units at St Osyth and Holland-on-Sea. The Upper St Osyth Gravel, recognized at the type site and at Fingringhoe, is a fine sandy gravel with closer compositional affinities to Anglian glacial gravel than to Kesgrave Group Thames deposits. In particular, it is characterized by a low Palaeogene (rounded) flint content, a relatively large non-quartzose exotic component and the occurrence of *Rhaxella* chert, all features that typify Anglian outwash gravels but would be unusual in the Kesgrave Group (*Rhaxella* chert is found consistently in Kesgrave Group deposits only within the Crag Basin, where the lithology has been reworked from the Crag). The deposit lacks the non-durable, calcareous component of ice-proximal gravels associated with the Lowestoft Till; calcareous clasts would probably not have survived fluvial transport from the ice front, which lay *c.* 20 km west of St Osyth (Fig. 5.4F). The Upper St Osyth Member is therefore interpreted as a distal outwash gravel.

East of St Osyth, in the former Thames-Medway confluence area, the Lower Holland Gravel is overlain by another fine-grained gravel, similar to that at St Osyth in that it is relatively poor in Palaeogene flint and quartzose exotics. However, this Upper Holland Gravel contains large amounts of southern material (Bridgland, 1983a; Bridgland *et al.*, 1988). In fact, the deposit contains as much Lower Greensand chert as the early Medway gravels to the south of the Blackwater estuary (Tables 5.3 and 5.5; see Part 2 of this chapter). The favoured interpretation of the Upper Holland Gravel is that it represents the confluence of the Medway, not with the Thames, but with the outwash stream, occupying that river's former course, that deposited the Upper St Osyth Gravel. A further observation can be made from the clast composition of these various deposits: whereas Thames-derived sediment considerably dominates the various Thames-Medway Kesgrave Group gravels, such as the Lower Holland Gravel, Medway-derived material completely dominates the Upper Holland Gravel (Bridgland, 1983a, 1988a; Table 5.2). The Upper Holland Gravel has been sampled over a wide area between the St Osyth pits and the type section. Its southern component generally increases in size downstream, although the highest figure, 31.4%, was encountered in a sample from Burrs Road (TM 193173), 2.5 km to the north-west of the Holland-on-Sea GCR site (Table 5.2). The Upper Holland Gravel appears to be much more widely distributed than the Upper St Osyth Gravel, the occurrence of which is extremely localized. This again suggests that the outwash stream responsible for the Upper St Osyth Gravel was much smaller than the contemporary River Medway, into which it flowed to the east of St Osyth.

The Upper St Osyth Gravel, even as far upstream as Fingringhoe, also contains a significant southern component (Table 5.2). This is unexpected in what is believed to be the product of an outwash stream that issued from the Anglian ice sheet. Anglian Stage glacial deposits in Essex generally contain such material in small quantities, reworked from earlier sediments such as the Kesgrave Sands and Gravels, in which Greensand chert has been traced as far

Table 5.5 Clast-lithological composition of gravels described in Chapter 5, Parts 2 and 3.

			Flint			Southern			Exotics								
Gravel	**Site**	**Sample**	**Tertiary**	**Nodular**	**Total**	**Gnsd chert**	**Hastings Beds**	**Total**	**Quartz**	**Quartzite**	**Carb chert**	**Rhax chert**	**Total**	**Ratio (sthrn: q/qtzt)**	**Ratio (qtz:qtzt)**	**Total count**	**National Grid Reference**
Tributary gravels																	
Blackwater	Gt Totham	1	31.7	10.0	80.5	0.2		0.2	8.7	7.2	2.0	0.5	19.0	0.01	1.20	609	TL 865091
Terrace 2	*11.2-16*	*1*	*28.4*	*5.7*	*78.1*	*1.1*		*1.1*	*9.8*	*6.4*	*2.7*	*0.2*	*20.5*	*0.07*	*1.53*	*1092*	
gravel		2	41.2	9.8	78.6				11.2	8.5	0.8	0.6	21.2		1.32	481.0	TL 865091
	11.2-16	*2*	*34.3*	*5.2*	*77.8*	*0.4*		*0.4*	*11.8*	*6.8*	*1.6*	*0.4*	*21.8*	*0.02*	*1.72*	*834*	
E. Mersea	Restrnt site	1	35.4	12.7	85.2	7.4		7.6	4.6	2.0	0.3	0.3	7.1	1.07	2.25	393	TM 0526 1362
Restaurant	*11.2-16*	*1*	*41.5*	*4.2*	*83.1*	*8.3*		*8.4*	*4.3*	*2.3*	*1.0*	*0.1*	*8.4*	*1.26*	*1.86*	*1197*	
Gravel	Hippo site	1	42.2	12.9	83.7	6.2		6.2	4.0	4.0	1.1	0.2	10.2	0.78	1.00	630	TM 0653 1434
Tollesbury	Garlands Fm	1A	37.8	12.9	83.6				9.4	3.5	1.1	0.1	16.2		2.71	805	TL 9467 1059
Gravel		1B	40.4	•	82.6	0.1		0.1	11.6	3.9	0.4		17.3	0.01	3.00	987	
	11.2-16	*1B*	*33.9*	*8.6*	*77.5*	*0.5*		*0.5*	*14.7*	*3.1*	*1.8*		*22.0*	*0.03*	*4.72*	*1475*	
Gravel above	Maldon	1	32.1	18.2	78.8				7.3	8.5	1.2		21.2		0.86	411	TL 8417 0670
Maldon Till	*11.2-16*	*1*	*28.2*	*8.0*	*65.0*	*1.2*		*1.5*	*18.4*	*7.4*	*1.8*	*0.3*	*33.1*	*0.06*	*2.50*	*326*	
	11.2-16	*3*	*25.5*	*7.3*	*74.2*	*0.4*		*0.4*	*11.3*	*8.0*	*1.8*		*25.1*	*0.02*	*1.41*	*275*	
Anglian	Ugley	1	41.9	23.7	87.9				3.5	0.8	1.5	0.4	11.9		0.22	520	TL 516278
glacial		2	3.6	37.6	87.1				2.6	1.7	2.1	1.7	12.6		0.64	420	
gravels (1)																	
Brightlingsea	Bghtlngsea	1	26.4	12.9	80.5	0.3		0.3	11.0	8.2	1.9		21.4	0.01	1.33	364	TM 1282 3125
Gravel	*11.2-16*	*2*	*27.8*	*73.0*	*0.9*			*0.9*	*15.3*	*7.0*	*2.5*	*0.4*	*26.1*	*0.04*	*2.18*	*800*	
East Mersea Hippo site, gravel in brickearth		1	44.4	12.5	84.7	11.6		11.6	1.6	1.9		0.3	3.8	3.36	0.83	320	TM 0652 1434
Low-level East Essex Gravel																	
Barling/Dam-	D. Wick	1	52.0	14.5	88.3	10.6		10.9	0.4	0.4			0.8	14.00	1.00	256	TQ 9614 9268
mer Wick	*11.2-16*	*1*	*46.9*	*2.4*	*87.8*	*9.8*		*10.0*	*0.9*	*0.4*	*0.5*	*0.4*	*2.3*	*7.50*	*2.33*	*752*	
Gravel	Barling	1	33.7	•	80.4	18.6		18.6	1.0				1.0	19.00		306	TQ 9318 9018
Mersea	West Mersea	1	38.6	•	82.4	14.0	0.2	14.2	1.9	0.9		0.3	3.5	5.13	2.20	578	TM 0134 1361
Island		2	44.8	7.9	87.7	9.5		10.0	1.2	0.2	0.5		2.3	7.17	5.00	431	TM 0144 1373
Gravel	Fen Farm	1	47.6	•	87.2	10.7		10.7	1.5	0.4	0.2		2.2	5.50	4.00	553	TM 0590 1444
		2	52.3	2.9	90.0	7.6		7.6	1.2	0.6	0.2	0.4	2.3	4.33	2.00	512	TM 0583 1437
	11.2-16	*2*	*47.5*	*2.9*	*88.2*	*8.7*		*8.7*	*1.7*	*0.5*	*0.4*	*0.1*	*3.1*	*3.86*	*3.38*	*1573*	
	Cudmore Grove	1	47.1	11.6	89.9	6.7		6.8	1.1	1.5	0.3	0.2	3.3	2.57	0.75	1061	TM 0667 1451
		2	45.3	8.8	85.0	11.5		11.6	1.4	1.0	0.7		3.4	4.88	1.29	671	TM 0676 1458
	Point Clear	1	33.1	•	77.3	19.9	0.2	20.1	0.9		0.7	0.4	2.6	22.80		568	TM 1023 1480
Cudmore Grove Channel		1	40.2	11.6	85.0	14.0		14.0	0.7	0.3			1.0	14.00	2.00	301	TM 0664 1447
lag gravel (2)	*11.2-16*	*1*	*48.1*	*4.1*	*86.9*	*9.9*	*0.6*	*10.8*	*1.3*	*0.7*	*0.4*		*2.4*	*5.50*	*1.80*	*715*	
Wigborough	Wigborough	1A	42.9	•	85.0	4.1	0.3	4.4	5.9	3.4	0.3		10.1	0.53	1.75	387	TM 1176 1447
Gravel	Wick	1B	40.4	8.0	79.7	6.7	0.2	7.1	8.0	3.5	0.5		13.1	0.62	2.25	565	
	Jaywick	1	51.0	4.3	81.3	4.8		4.8	7.0	5.3	0.2		13.7	0.39	1.32	416	TM 1502 1419
	11.2-16	*1*	*42.1*	*4.1*	*82.4*	*6.0*		*6.0*	*7.3*	*2.0*	*0.7*	*0.2*	*11.6*	*0.65*	*3.69*	*813*	
Upper gravel	Clacton	4A	45.0	10.5	89.1	8.1		8.1	0.8	0.9	0.9		2.8	4.62	0.86	742	TM 1739 1433
at West Cliff	cliffs	4B	41.0	8.1	83.8	13.4		13.4	0.4	0.9	1.3		2.9	10.17	0.50	456	
	11.2-16	*4B*	*51.4*	*4.6*	*86.6*	*8.9*		*9.0*	*2.7*	*1.4*	*0.2*	*0.2*	*4.4*	*2.18*	*1.94*	*1217*	
Clacton	Lion Point	1	28.2	•	79.2	17.8		17.8	1.2	1.5	0.3		3.1	6.57	0.75	259	TM 1445 1274
Channel		2	42.3	9.8	86.9	8.9	0.3	9.2	2.6	0.7	0.3		3.9	2.80	4.00	305	TM 1445 1274
Gravel	*11.2-16*	*2*	*46.5*	*5.6*	*88.9*	*6.1*		*6.1*	*2.2*	*0.7*	*0.7*	*0.4*	*4.9*	*2.10*	*3.20*	*721*	
	Butlins	1	46.0	9.3	90.1	7.2		7.2	1.5	0.3	0.6		2.4	4.00	4.00	335	TM 1546 1382
	11.2-16	*1*	*39.8*	*4.9*	*85.9*	*8.0*	*0.1*	*8.2*	*2.2*	*2.3*	*0.4*	*0.4*	*5.7*	*1.86*	*0.95*	*973*	
Southchurch	Southend	1	33.6	•	76.2	20.9		21.2	0.7	0.2	0.5	1.1	2.6	26.00	4.00	613	TQ 8962 8750
/Asheldham	Goldsands	1A	41.6	•	84.0	12.8		13.5	0.9		0.5	0.9	2.3	15.00		445	TQ 9609 9901
Gravel	Pit	1B(3)	51.7	•	88.8	9.50	0.2	9.9	0.4	0.4	0.5	0.1	1.4	14.17	1.00	862	
		2(3)	41.3	11.9	88.0	10.0		10.2	1.1	0.4	0.2	0.1	1.8	7.08	3.00	834	TQ 9608 9897

			Flint			Southern			Exotics								
Gravel	**Site**	**Sample**	**Tertiary**	**Nodular**	**Total**	**Gnsd chert**	**Hastings Beds**	**Total**	**Quartz**	**Quartzite**	**Carb chert**	**Rhax chert**	**Total**	**Ratio (sthrn: q/qtzt)**	**Ratio (qtz:qtzt)**	**Total count**	**National Grid Reference**
High-level East Essex Gravel																	
Chalkwell	Caidge Fm	1	44.7	10.3	74.6	23.7	1.3	25.2	0.3				0.3	98.00		389	TQ 9471 9940
/Caidge	*11.2-16*	*1*	*38.4*	*5.9*	*69.1*	*29.4*	*1.2*	*30.9*								*524*	
Gravel	Chalkwell Pk	1	58.9	2.4	55.1	15.2					0.2		0.2			494	TQ 8579 8636
Canewdon/St	St Law.	1	11.2	*	36.3	62.6		63.2		0.2	0.2		0.4	289.00		457	TQ 9677 0408
Lawrence	*11.2-16*	*1*[(4)]	*9.8*	*1.9*	*34.6*	*65.1*		*65.4*								*1069*	
Gravel	Canewdon	1B	45.5	9.8	73.0	26.2	0.5	26.9		0.2			0.2	167.00		622	TQ 8973 9468
Belfairs	Bovill Uplands	1A	60.3	*	85.5	14.0	0.3	14.2		0.3			0.3	54.00		380	TQ 9252 9998
/Mayland		1C	46.6	13.3	74.4	24.1	0.6	25.6								324	
Gravel	*11.2-16*	*1C*	*50.3*	*3.7*	*72.8*	*25.6*	*1.1*	*27.0*								*644*	
	Belfairs Pk	1	39.5	8.4	65.2	34.1	0.3	34.5			0.3		0.3			299	TQ 8336 8764
Ashingdon	Mount View	1	43.2	*	80.3	19.7		19.7								620	TQ 8545 9339
Gravel	*11.2-16*	*1*	*46.2*	*3.3*	*75.2*	*23.9*	*0.1*	*24.3*		*0.1*			*0.3*	*184.00*		*757*	
Oakwood	Oakwood	1	62.9	6.8	73.1	26.7	0.2	26.9								558	TQ 8234 8839
Gravel	*11.2-16*	*1*	*44.6*	*3.1*	*69.0*	*30.2*	*0.4*	*30.9*			*0.2*		*0.2*			*1099*	
Daws Heath	Daws Heath	1	63.5	8.3	86.1	13.2	0.3	13.7			0.2		0.2			613	TQ 8068 8887
Gravel	*11.2-16*	*1*	*52.3*	*2.5*	*76.5*	*22.4*	*0.6*	*23.3*			*0.2*		*0.2*			*1200*	
Claydons	Claydons	1	72.7	2.7	89.9	9.6	0.2	10.0								553	TQ 8017 8896
Gravel	*11.2-16*	*1*	*61.2*	*2.0*	*83.9*	*15.8*		*16.0*	*0.1*				*0.1*	*112.00*		*701*	

* Not separately recorded

(1) For comparison only - non-durables excluded

(2) Feather edge

(3) From the lower gravel at the GCR site

(4) Subsample

See also notes to Table 4.2, page 181

north as Norfolk (Hey, 1980). However, other features of the clast composition of the Upper St Osyth Gravel (the paucity of reworked Palaeogene flint pebbles, for instance) imply that reworking of material from the Kesgrave Group has been insufficient to account for the high southern count. This is an obvious fact, since the amount of southern material is higher in the Upper St Osyth Gravel than in any Kesgrave Group formation upstream from the Medway confluence. The provenance of this extra southern material was probably the area to the south of Colchester, which, in the Middle Pleistocene, would probably have been covered by high-level left-bank terraces of the early Medway. Work in south-eastern Essex and north Kent has indicated that the Medway is a river of considerable antiquity and that throughout its early course from the Medway Towns to the Blackwater estuary it was progressively migrating eastwards (Bridgland, 1980, 1983a, 1988a; Bridgland and Harding, 1985). It is therefore likely that an extensive terrace system existed to the west of the Anglian course of the Medway, which is depicted in Fig. 5.4 (E and F). It is possible that a small river system drained northwards from this area into the old Thames valley. Its contribution would have been insignificant whilst the Thames continued to supply huge quantities of gravel to the area, but, as with the Medway in the case of the Upper Holland Gravel, this contribution made a significant difference to the gravel load of the Upper St Osyth Gravel outwash stream.

Detailed analysis of the clast composition of these various gravels in north-eastern Essex therefore shows that the Thames was replaced, in Upper St Osyth/Upper Holland Gravel times, by an outwash stream. The explanation of this remarkable change lies in events during the Anglian glaciation of the northern London Basin. Gibbard (1977, 1979) showed that, during this glaciation, ice blocked the early Thames course through the Vale of St Albans, leading to the diversion of the river into its modern valley through London (see Chapter 3). The Upper St Osyth and Upper Holland gravels have been interpreted as deposits laid down during this period when the Thames valley was blocked and the river was not reaching the Tendring Plateau. The Medway, however, was unaffected and continued to flow to Clacton and beyond without hindrance from the glaciation, receiving the Upper St Osyth Gravel outwash stream as a west-bank tributary (Fig. 5.4F).

This interpretation of the sequences at St Osyth and Holland-on-Sea is strengthened by the identification of the next (and final) aggradation in the terrace succession of the Tendring Plateau as the product of the post-diversion Thames-Medway (Fig. 5.5B; see Part 2 of this chapter). The Upper St Osyth/Upper Holland Gravel can be correlated with deposits in other parts of the Thames drainage system that were affected by the (Anglian) glacial diversion of the river. It has been considered (Bridgland *et al.*, 1988) to be a time-equivalent of the (lacustrine) Moor Mill Laminated Clay and the (deltaic) Winter Hill Upper Gravel, both laid down in a proglacial lake that formed when the Thames was blocked, immediately prior to its diversion (Gibbard, 1977; see Chapter 3, Moor Mill). It is probable that the time interval during which the Moor Mill lake existed was relatively brief, perhaps only a few centuries (Bridgland *et al.*, 1988). The reinterpretation of events during the glaciation of the Vale of St Albans, by Cheshire (1981, 1986a), suggests that the Thames was diverted following the formation of the Watton Road Lake, near Hertford, rather than at Moor Mill (see Chapter 3, Part 2). This would indicate precise correlation between the Watton Road lacustrine sediments and the Upper St Osyth/ Upper Holland Gravel, making the latter slightly older than is indicated by the correlation with the Moor Mill lake beds. Whichever of these two correlations is correct, the Upper St Osyth/ Upper Holland Gravel is clearly one of the most closely datable gravel units in the Thames basin.

Summary

At St Osyth, an important section reveals the lowest Kesgrave Thames formation, the Lower St Osyth Gravel, overlain by fine-grained sandy gravel (Upper St Osyth Gravel), interpreted as distal outwash laid down while the Thames was blocked further upstream. The palaeogeographical interpretation of this sequence cannot be made without reference to the evidence further downstream in the area of the contemporary Thames-Medway confluence. The sequence there is revealed in equally important cliff sections at Holland-on-Sea, where a comparable sequence to that at St Osyth is exposed, with equivalent Lower Holland Gravel overlain by Upper Holland Gravel. The composition of the former is typical of the Kesgrave Sands and Gravels, but the latter is dominated by Medway-derived material. The only way that gravel so closely resembling the Medway deposits further south could have been deposited at Holland, which was clearly within the contemporary Thames-Medway valley, is for there to have been no contribution from the Thames. The only time when the Thames and Medway have not joined, either before or after the diversion of the former, was during the brief period when the Thames was blocked by the Lowestoft Till ice sheet (immediately prior to its diversion). The Upper Holland Gravel, and its upstream equivalent the Upper St Osyth Gravel, are therefore correlated with this glacial event. These deposits thus provide a key stratigraphical marker within the terrace sequence of north-east Essex, one that assists correlation within the Thames system as a whole and with the Pleistocene sequence in other areas.

Conclusions

Using only the evidence of the rock-types present in the gravels at St Osyth and Holland-on-Sea, it is possible to demonstrate the rapid and catastrophic changes that affected the Thames as a result of the most extensive Pleistocene glaciation, during the Anglian Stage (around 450,000 years ago). Comparison of the gravels at these two sites reveals that the north-eastward-flowing Thames abruptly ceased to reach this area, because it was blocked by ice upstream, in what is now Hertfordshire and central Essex. The lower gravel at St Osyth is a typical pre-diversion Thames deposit. The Holland-on-Sea section is downstream of the contemporary confluence with the Medway, so that the Lower Gravel there is a Thames-Medway deposit, although much dominated by Thames material. The upper (later) gravels at both sites are significantly different. In particular, they contain material carried by meltwater streams from the Anglian ice sheet. The River Medway, lying beyond the direct influence of the ice, continued to flow northwards. This is demonstrated by the composition of the upper gravel at Holland-on-Sea, which is very much dominated by Medway material, in marked contrast to the lower gravel. The implication of this is that the Medway was very much larger than the meltwater river that replaced the Thames at that time. After it was diverted into its modern course, the Thames joined with the Medway in the Southend area, approximately at the location where the estuaries of the two rivers join today.

Part 2:

EASTERN ESSEX
D.R. Bridgland

Introduction

The second part of this chapter deals with sites associated with the sequence of terrace deposits classified as 'East Essex Gravel' by Wood (1866b). These deposits primarily occupy the coastal district of Essex between the estuaries of the Thames and Blackwater, although they are also represented on Mersea Island and in the south-eastern corner of the Tendring Plateau (Fig. 5.1). Bisected by the Crouch estuary, this region is characterized by a series of gravel terraces descending south-eastwards, towards the North Sea (see Figs 5.15 and 5.16). These give way inland to higher, isolated hills capped by Bagshot Beds, Claygate Beds and, frequently, high-level gravel remnants (still part of the East Essex Gravel), the highest of which are found on the Rayleigh Hills, up to a maximum height of 76 m O.D. at Hadleigh. The northward extension of part of this sequence to the Clacton area provides a direct link with the stratigraphical sequence on the Tendring Plateau, described in Part 1 of this chapter.

Wood (1866b) believed the East Essex Gravel to be a dissected spread of marine shingle formed in an embayment of the North Sea. Whitaker (1889) interpreted the deposits as a continuation of the gravels of the Lower Thames valley, laid down by a united Thames-Medway river. Holmes (1896), Gregory (1922) and Coles (1934) all attempted the reconstruction of former fluvial courses across this area, the first proposing a route trending east-north-eastwards to the north of the Langdon and Rayleigh Hills, his 'Romford River'. Of these three early workers, Gregory paid most attention to the deposits, noting abundant material from Kent in the composition of the East Essex Gravel. He envisaged these deposits as accumulations, of considerable antiquity, on the northern slope of an extended Weald, pre-dating the excavation of the modern Thames valley.

Work in the late 1960s by Gruhn *et al.* (1974) demonstrated that the gravels of this area represent the left-bank terrace deposits of a fluvial valley system whose eastern side has been lost to the North Sea. In an early application of clast-lithological analysis, they recognized an abundance of Lower Greensand material in the gravels, combined with a paucity of quartz and other 'exotic' lithologies from the north and west. They therefore attributed the bulk of the gravel to the Medway, although noting (after Whitaker, 1889) that the Thames was probably confluent with the Medway at that time. They suggested that the Medway, with a steeper gradient and a more proximal supply of gravel-forming source materials, provided the major part of the gravel load of the Thames-Medway system.

The work of Gruhn *et al.* pre-dated the publication of New Series geological maps of the area (Sheets 241 and 258/9; Bristow, 1985; Lake *et al.*, 1986; also maps in Hollyer and Simmons, 1978, and Simmons, 1978). The East Essex Gravel is divided on these maps into 'Sand and Gravel of Unknown Age' (high-level deposits south of the Crouch) and four terraces, designated 'Crouch Terraces 1–4', the nomenclature reflecting their distribution either side of the modern Crouch estuary rather than implying deposition by that river. A number of buried channels were recognized beneath these terrace gravels and were attributed to subglacial or partly subglacial streams associated with a hitherto unrecognized ice lobe occupying the southern North Sea (Lake *et al.*, 1977). These authors also suggested that the fourth terrace was formed at approximately the same time, as a kame terrace system at the margin of this ice lobe.

More recent appraisal of these deposits has augmented the Geological Survey mapping, using a lithostratigraphical approach based on detailed analysis of clast types and frequencies (Bridgland, 1980, 1983a, 1983b, 1986a, 1986b, 1988a). Two broad types of gravel have been recognized within Wood's East Essex Gravel, differentiated on both clast lithology (Table 5.5) and altitudinal distribution. They comprise an earlier 'High-level East Essex Gravel', composed almost exclusively of local and southern rocks, and a later type, the 'Low-level East Essex Gravel', containing similar materials but with the important addition of a significant suite of exotic rocks derived from the north and west. Within each of these two types of gravel, a number of separate terrace formations can be distinguished on the basis of geological mapping (Table 5.6); the High- and Low-level East Essex Gravels are therefore classified as subgroups.

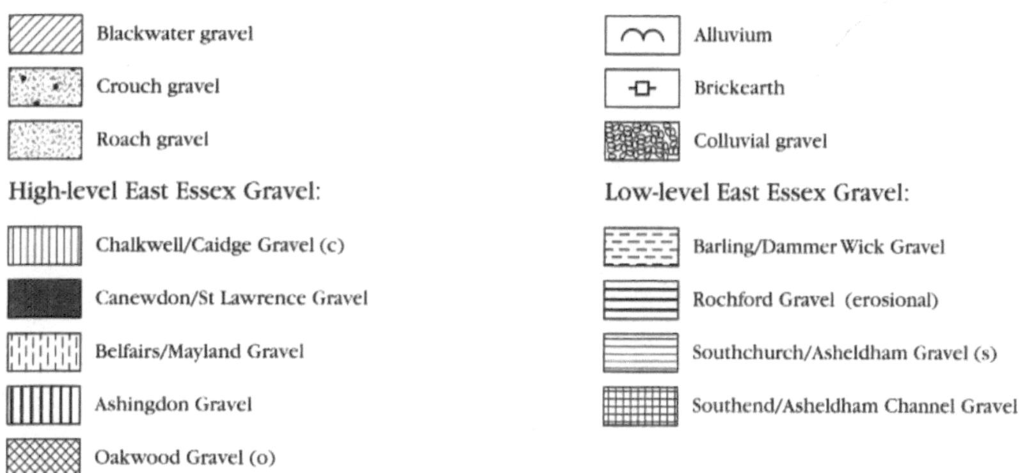

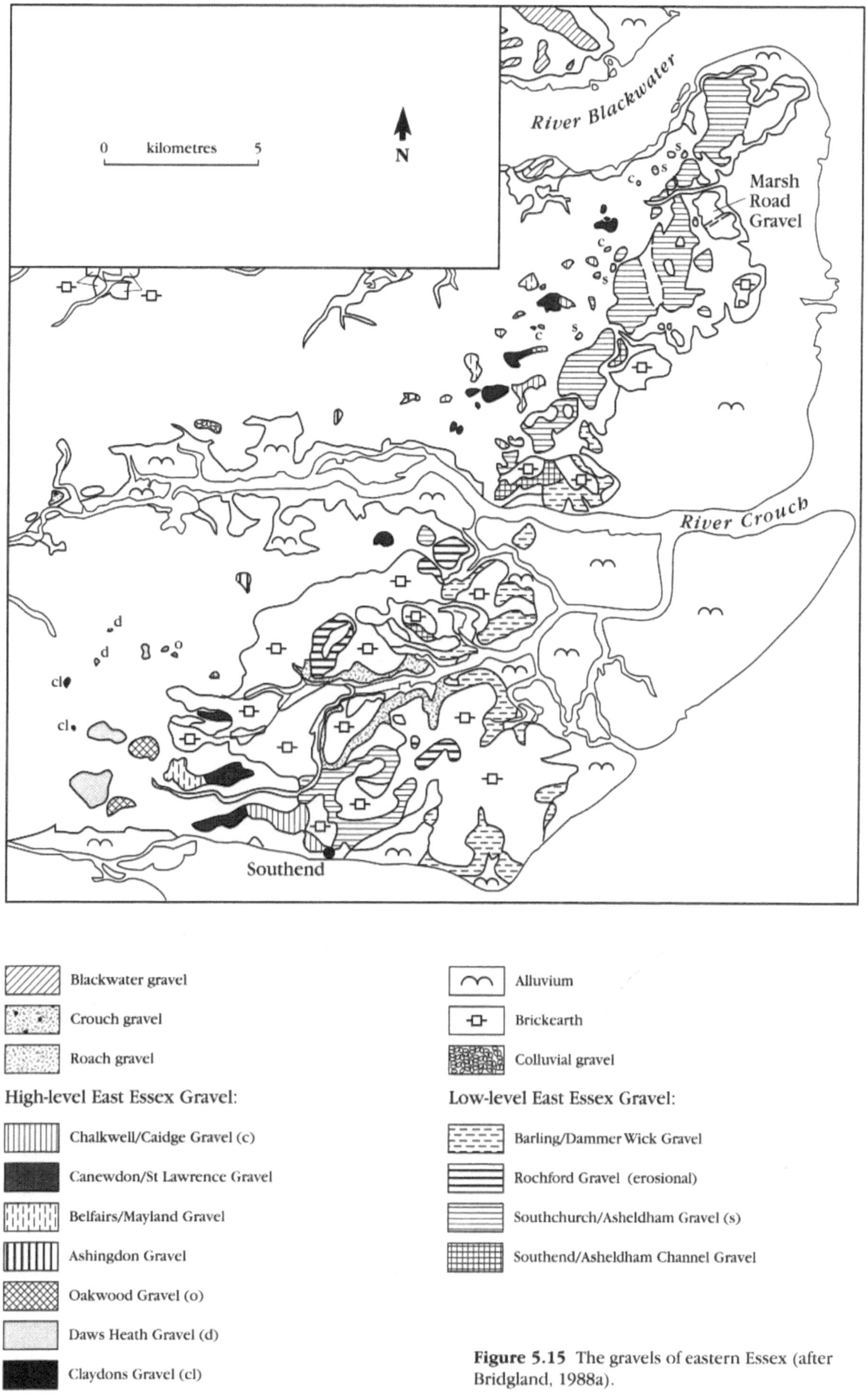

Figure 5.15 The gravels of eastern Essex (after Bridgland, 1988a).

The local gravel component is made up of flint from the Chalk and (in rounded pebble form) from the Palaeogene. Flint accounts for 35–68% of the High-level East Essex Gravel and 74–94% of the Low-level East Essex Gravel. The southern component consists largely of chert from the Folkestone and Hythe Beds (Lower Greensand), supplemented by sandstones, siltstones and ironstones from the Hastings Beds of the central Weald and rare arenaceous lithologies from the Lower Greensand (Bridgland, 1986b). This southern component falls from 13–65% in the High-level East Essex Gravel to 5–25% in the Low-level East Essex Gravel. The most marked difference between these two subgroups, however, is the presence in the Low-level East Essex Gravel of a small but consistent exotic component (0.5–3%), comprising vein quartz, various quartzites, Carboniferous chert, ***Rhaxella*** chert and igneous rocks (Bridgland, 1986b; see Table 5.5).

The High-level East Essex Gravel is believed to be the product of an extended River Medway, laid down by that river when the Thames flowed further north, prior to its diversion by Anglian ice. Gravels of this type can, in fact, be traced upstream to the Hoo Peninsula, north Kent, where they form part of a terrace system of the Medway (Bridgland, 1980, 1983a; Bridgland and Harding, 1985). The exotic rocks recognized in the Low-level East Essex Gravel are identical to those found in the terrace deposits of the Lower Thames (see Chapter 4). The Low-level East Essex Gravel is therefore attributed to the combined Thames-Medway, confirming the interpretation of Whitaker (1889). This type of gravel clearly post-dates the diversion of the Thames. The diversion was the direct result of the Anglian glaciation of the Vale of St Albans, which produced a proglacial lake, the overspill from which is thought to have effected the rerouting of the river (Gibbard, 1977, 1979; see Chapter 3, Part 2), probably by way of a pre-existing left-bank tributary of the 'pre-glacial' Medway (Bridgland, 1988a). The diverted Thames appears to have joined the pre-existing Medway valley in the area of the present Thames estuary and then flowed northwards across east-

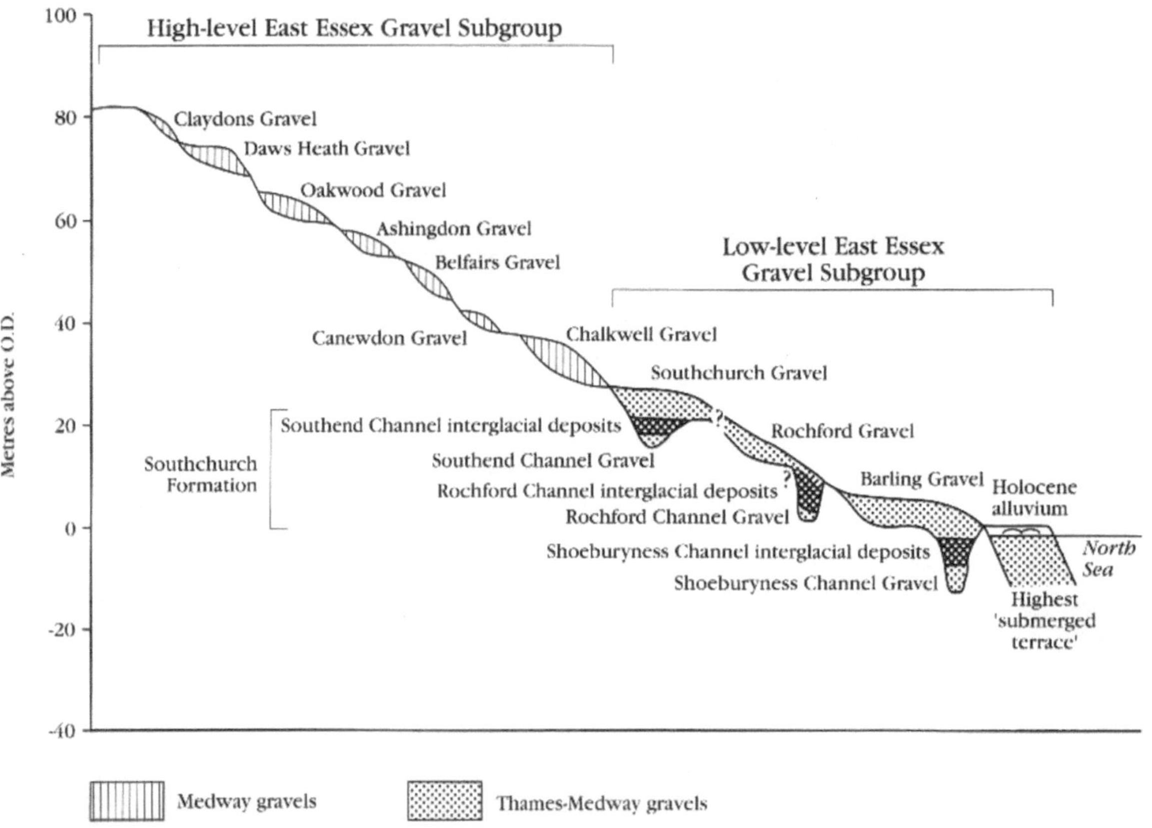

Figure 5.16 Idealized transverse section through the gravels of the Southend area (modified from Bridgland, 1988a).

Table 5.6 Gravel formations in eastern Essex.

Southend area	**Dengie Peninsula**	**Tendring Plateau**	**Middle/Lower Thames equivalent**	**Stage**	**^{18}O**
------------ Offshore ------------			Kempton Park/East Tilbury Marshes	late Saalian - Devensian	6-4 or 2
----------- Below alluvium -----------			Taplow/Mucking	late Saalian	8-6
Barling	Dammer Wick		Lynch Hill/Corbets Tey	mid-Saalian	10-8
Southchurch	Asheldham	Mersea Island/Wigborough	Boyn Hill (and Black Park)	Anglian - early Saalian	12-10
-------------- Thames diversion (stratigraphic marker) -----------				Anglian	12
Chalkwell	Caidge	St Osyth/Holland	Winter Hill	Anglian	12
Canewdon	St Lawrence	Wivenhoe/Cooks Green		early Anglian	12
Belfairs	Mayland	Ardleigh/Oakleigh	?Rassler	'Cromerian Complex'	21-13
Ashingdon		Waldringfield			
Oakwood			?Gerrards Cross		
Daws Heath			?Beaconsfield	Early Pleistocene	Pre-21
Claydons			?Satwell		

ern Essex, eventually rejoining its pre-diversion course in the Clacton area (Bridgland, 1980, 1983a, 1983b, 1988a; Fig. 5.5).

This new route is exemplified by the distribution of the Southchurch Gravel of the Southend area and its downstream equivalents, the Asheldham Gravel of the Dengie Peninsula, the Mersea Island Gravel and the Wigborough Gravel of the Clacton area (see Figs 5.2, 5.5 and 5.15). The northern part of the old Medway course across eastern Essex was abandoned following the deposition of the Southchurch/Asheldham Gravel, the river subsequently turning eastwards towards the North Sea Basin in the region of the modern Crouch estuary (Fig. 5.5). Later gravels in this northern part of the area are products of tributary rivers such as the Chelmer and Blackwater (see Part 3 of this chapter). The Thames-Medway course across the coastal fringe of Essex is reflected by evidence offshore of the submerged (pre-Holocene transgression) valley, which turns northwards off Southend to run parallel to the coastline as far north as the Crouch estuary (Fig. 5.5F), from which a substantial submerged tributary valley emerges (D'Olier, 1975; Bridgland and D'Olier, 1989). The East Essex Gravels essentially form left-bank terraces of this continuation of the Thames-Medway valley, the axis of which is now submerged.

The abrupt compositional change between the High-level and Low-level East Essex Gravels provides an important stratigraphical marker, of great assistance to correlation both within eastern Essex and, thanks to its causal link with the Anglian glaciation and the associated diversion of the Thames, with other areas. Furthermore, the highest of the Low-level East Essex Gravel formations, the Southchurch/Asheldham Gravel, can be traced as far north as Clacton (Fig. 5.5B), thus linking with the Kesgrave Group sequence in southern East Anglia

(Bridgland, 1980, 1983a, 1988a). This stratigraphical marker is the basis for correlation of the lowest three High-level East Essex Gravel Medway formations with the lowest three Kesgrave Group formations on the Tendring Plateau, where the pre-diversion confluence between the two rivers has been recognized (Fig. 5.4; see Part 1 of this chapter).

Since it was the glacial diversion of the Thames that brought the river into its modern lower valley and effected the change to Thames-Medway drainage in eastern Essex, the terrace gravels of the Lower Thames (see Chapter 4) and the Low-level East Essex Gravel must be lateral equivalents. Correlation between the two areas has proved difficult, largely because of a lengthy downstream gap in the terrace record between Stanford-le-Hope and Southend, where the Pleistocene gravels are cut out by the extensive Holocene alluvium of Fobbing Marshes and Canvey Island. Previous correlations have relied primarily on the downstream projection of gravel bodies, although with some palaeontological and archaeological support (Bridgland, 1983a, 1988a). Recently it has become apparent, partly from offshore evidence, that a revision of the correlation scheme suggested by Bridgland (1988a) is necessary (Bridgland *et al.*, 1993; Table 1.1 and Fig. 1.3; Table 5.6). Further discussion of this correlation will appear in the three site reports below.

The buried channels first recognized by Lake *et al.* (1977), and attributed by them to glaciofluvial processes, have been reinterpreted as integral parts of the fluvial stratigraphy of the Low-level East Essex Gravel sequence (Bridgland, 1980, 1983a, 1988a; Bridgland *et al.*, 1993). These channels cover a range of altitudes, a possible reason for the original glaciofluvial interpretation. Bridgland (1983a, 1988a) believed that three separate downcutting events were represented, associated with three Low-level East Essex Gravel formations, his Southchurch, Rochford and Barling Gravels (Fig. 5.15). Each of the channels contains basal gravels overlain by probable interglacial sediments, usually of apparent estuarine character, but these have only been recorded from boreholes and have yet to be described in detail (new work on these channel-fills has been carried out recently by H.M. Roe). They are generally capped by the deposits mapped as terrace gravels, thus showing tripartite sequences that correspond to phases 2, 3 and 4 of the climatic model for terrace formation (see Chapter 1). The oldest of these channels is believed to represent an upstream continuation of the Clacton Channel, traditionally assigned to the Hoxnian (Warren, 1955; Fig. 5.5A; see below, Clacton). It is also thought to be preserved at Cudmore Grove, East Mersea (see Cudmore Grove).

Eastern Essex is an important area for research on the British Pleistocene, because it lies directly between the Lower Thames valley and East Anglia, both of which have well-documented and detailed stratigraphical records. The area is also of considerable significance for studies of the southern North Sea, since many of the deposits in eastern Essex can be traced offshore (Bridgland and D'Olier, 1989). Because it lies at the edge of the North Sea Basin, the gravel sequence in eastern Essex is interbedded with estuarine channel-fills related to high sea-level events. The sequence in this area, although poorly documented through lack of exposure, promises to be more complete than elsewhere in the Thames system and will probably be the source of valuable future contributions to the Pleistocene record. As much of the evidence from this area lies deep below ground level, the coverage of GCR sites is limited. The Low-level East Essex Gravel is represented, however, in GCR sites at Clacton, Cudmore Grove (East Mersea) and Southminster. The Clacton and Cudmore Grove sites include important interglacial sequences; the former is also the internationally recognized type locality for the Clactonian Palaeolithic Industry. It is possible that future research in this area will result in other important sites being recognized and added to this coverage.

CLACTON (CLIFFS, FORESHORE AND GOLF COURSE)
D.R. Bridgland

Highlights

A key locality for studies of Pleistocene stratigraphy and palaeontology, the complex site at Clacton reveals a channel-fill traditionally assigned to the Hoxnian Stage. This series of deposits, attributed to the Thames-Medway, contains faunal and floral remains indicative of temperate-climate conditions. In addition to the considerable stratigraphical, palaeontological and palaeoenvironmental significance of the site, it is famous as the type locality for the Clactonian Palaeolithic Industry. The location of the Clacton deposits is such that a stratigraphical link between the Thames system and the East Anglian Pleistocene succession is provided, making this one of the most important Pleistocene sites in southern Britain. The palaeogeographical position of the site in relation to the regional Thames terrace sequence, together with the stratigraphical evidence it provides, indicates a Hoxnian (*sensu* Swanscombe) age for the interglacial sediments here, immediately post-dating the Anglian diversion of the Thames.

Introduction

The cliffs, foreshore and immediate inland area at Clacton-on-Sea together constitute a complex Pleistocene site of international significance. Clacton lies in the south-eastern corner of the Tendring Plateau. Recent work has shown that the gravels in this area belong mainly to the pre-diversion Thames system (Rose *et al.*, 1976; Bridgland, 1980, 1988a; Green *et al.*, 1982; Bridgland *et al.*, 1988, 1990; Part 1 of this chapter). At Clacton, fossiliferous Pleistocene channel deposits are preserved in an arcuate area to the south of the town centre (Fig. 5.17), intersecting with the present coastline at Lion Point, Jaywick (western end) and to the south of the pier (eastern end). These sediments have yielded many Palaeolithic artefacts, which form a characteristic assemblage of flakes and cores, with no formal tools such as hand-axes (Warren, 1912, 1922, 1933, 1958). Clacton is the type locality of this particular Palaeolithic industry, to which the name Clactonian was first applied by Warren (1926; see below).

Although they were discovered and extensively described in the last century (Brown, 1838,

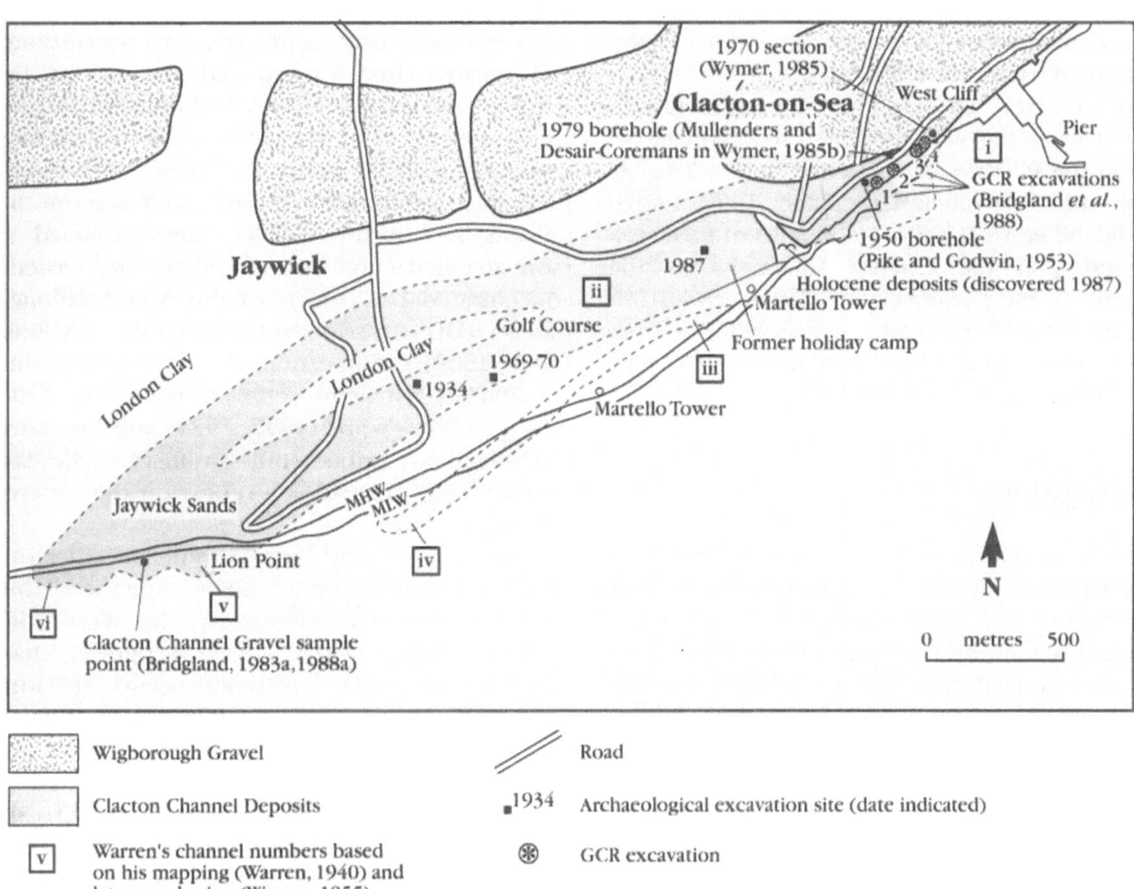

Figure 5.17 Map showing the distribution of Pleistocene deposits in the region of Clacton and the location of the various sites mentioned in the text.

1840, 1841; Fisher, 1868a; Dalton, 1880), much of our present knowledge of the Clacton Channel Deposits can be credited to S.H. Warren, who devoted a considerable proportion of his life's work to the deposits at Clacton and the Clactonian Industry (Warren, 1922, 1923a, 1924b, 1933, 1940, 1951, 1955, 1958). Pollen from these sediments was used to construct one of the first British interglacial pollen diagrams (Pike and Godwin, 1953), subsequently the basis for their ascription to the Hoxnian Stage (West, 1956, 1963; Turner, 1973). The deposits have been widely regarded as downstream correlatives of those at Swanscombe in the Lower Thames (Chapter 4), the two sites having been correlated by a comparison of their molluscan faunas (Kerney, 1971; Turner and Kerney, 1971). Two full-scale archaeological excavations have taken place on the golf course (Oakley and Leakey, 1937; Singer *et al.*, 1973) and a further investigation, as yet unpublished, was recently undertaken of exposures created during the redevelopment of the holiday camp (Wymer, 1988; Fig. 5.17). The most recent summaries of palaeoenvironmental evidence from Clacton were by Wymer (1974, 1985b) and Roe (1981).

The Clacton site has figured prominently in recent work that has attempted to correlate between the Pleistocene sequences in the Thames Basin and East Anglia (Bridgland, 1980, 1988a; Bridgland *et al.*, 1988). The results of this work suggest that the Clacton Channel was the product of the post-diversion Thames and that it represents an early phase of deposition by that river following the adoption of its modern valley through London (see Fig. 5.5A).

Description

The Clacton Channel Deposits have been poorly accessible during the past half century, so that most work on them during this period has relied on temporary exposures, boreholes or museum collections. Earlier workers were able to describe continuous cliff exposures between the site of the holiday camp and Holland-on-Sea, but these disappeared long ago beneath ornamental gardens. Although it has recently been possible to re-excavate fragments of this section (Bridgland *et al.*, 1988), it is necessary to consult the early published descriptions to assess the characteristics and extent of the Clacton sediments.

The deposits were discovered in the late 1830s by John Brown of Stanway, who wrote a number of short papers describing them and their fossil content (Brown, 1838, 1839, 1840, 1841, 1845, 1857). Brown (1840, 1841) noted the occurrence of both marine and freshwater molluscs at Clacton and that only the latter type occurred in the lowest stratum, which also yielded mammalian remains. He divided the sequence into seven separate beds, broadly reflecting a change from a freshwater/lacustrine environment to 'fluvio-marine' (estuarine) conditions. He also recorded a bed with freshwater shells near the top, possibly an early reference to Warren's (1923a, 1955) 'bed l' (see below).

References to the fossiliferous beds at Clacton also appeared in a number of other early publications, notably those of Owen (1846), Wood (1848), who suggested that the 'lacustrine' deposit might be the freshwater equivalent of the Red Crag, and his son (Wood, 1866b), in his original description of the East Essex Gravel (see above, Introduction to Part 2). In addition, Fisher (1868a, 1868b) published descriptions of the sections, partly from his own observations and partly based on an old manuscript by Brown, and Dalton (1880) included an illustrated description of the site in the Colchester memoir. Both Dalton and Fisher reproduced Brown's stratigraphy without significant modification, but Picton (1912) later proposed a more detailed subdivision. Other early writers concentrated on aspects of the palaeontology: Jones (1850) described the ostracods, Dawkins (1868, 1869) and Ransome (1890) described the mammalian remains, Webb (1894, 1900), Kennard and Woodward (1897, 1923) and Kennard (1924) all described the Mollusca, whereas Hinton (1923) provided a report on the rodent remains.

Prior to the work of Warren, the Clacton Channel Deposits were known only from the site of their original discovery, the West Cliff section. Warren (1923a, 1933) recognized that part of the Clacton sequence occurs at Lion Point, Jaywick, exhumed from beneath saltings. He later interpreted the Clacton and Lion Point sites as 'sections across the same fluviatile channel a few miles apart' (Warren, 1933, p. 15). As Warren showed, the channel follows an arcuate course between Jaywick Sands and the West Cliff (Fig. 5.17), but the full sequence of deposits is only preserved at the latter (eastern) end. The channel is excavated in London Clay, but in the cliffs it can also be observed to dissect the Lower Holland Gravel (Fig. 5.18). Its base is reputed to decline to at least 6 m below O.D. (Warren, 1955).

The most complete succession of Clacton Channel Deposits is preserved at the West Cliff locality, where a sequence of fluvial beds overlain by estuarine sediments occurs. Warren (1923a, figs 1 and 2; p. 611) originally proposed a complex subdivision of the sediments here, with an upper series of estuarine clays and sands and a lower series of freshwater gravel, loam and clay. He subsequently found that many of his earlier subdivisions could not be followed laterally for any great distance and adopted the following more generalized sequence for what is widely regarded as the definitive description of the channel deposits (Warren, 1955; Fig. 5.19):

		Thickness
6. Surface soil and colluvium		1–3 m
5. Upper bedded gravel	(Mersea Island/ Wigborough Gravel?)	*c.* 2 m
4. Estuarine sand with shells, passing laterally into estuarine calcareous clay	(Clacton Estuarine Beds)	up to 4 m
3. Estuarine laminated clay ('peaty shale'); contains a localized lens with freshwater fauna, Warren's (1923a) 'bed l'	(Clacton Estuarine Beds)	up to 5 m
2. Loamy sands and clays, with much channelling	(Upper Freshwater Beds)	up to 4 m
1. Clayey gravel and sand	(Lower Freshwater Beds)	up to 7 m

London Clay (or Lower Holland Gravel)

Thicknesses vary considerably as the northern feather-edge of the channel sequence is approached (Fig. 5.18). The basal sand and gravel is typically *c.* 1 m thick, the minimum thickness of the Clacton Freshwater Beds (beds 1 and 2 combined) being just over 2 m. The overlying Estuarine Beds (beds 3 and 4) continue the sequence up to *c.* 10 m O.D. (Figs 5.18 and 5.19). Within the Estuarine Beds, Warren (1923a) recorded a thin (0.3 m) and discontinuous bed ('bed l') containing only non-marine fauna. Molluscs, ostracods, plant macrofossils and pollen have been obtained from both the Freshwater and Estuarine Beds, except where the latter are oxidized, near the modern land surface. The Freshwater Beds have also yielded a rich mammalian fauna and large collections of Clactonian artefacts; the richest concentrations were in the upper part of the Lower Freshwater Beds (Warren, 1923a, 1955). Unfortunately most of the early collections from the different beds have been combined and it is also possible that material from the Lion Point foreshore locality may have been grouped with that from the West Cliff section (Wymer, 1985b).

According to the summary by Wymer (1985b), the mammalian assemblage from the natural exposures includes beaver (*Castor fiber*), the voles *Arvicola cantiana* and *Microtus agrestis*, lion (*Panthera leo*), straight-tusked elephant (*Palaeoloxodon antiquus*), horse (*Equus ferus*), the extinct rhinoceroses *Dicerorhinus kirchbergensis* and *D. hemitoechus*, red deer (*Cervus elaphus*), the large fallow deer *Dama dama clactoniana* and boar (*Sus scrofa*).

Assimilating data from the work and collections of A.S. Kennard, A.G. Davis, the Museum of the Geological Survey and the British Museum (Natural History), Warren recorded *c.* 100 species of land and freshwater Mollusca from the Upper and Lower Freshwater Beds at Clacton. The most common freshwater taxa, all of which persist into at least the basal part of the Estuarine Beds, are: *Bithynia tentaculata*, *Lymnaea peregra*, *L. truncatula*, *Gyraulus albus*, *Armiger crista* (L.), *Valvata piscinalis*, *Pisidium amnicum*, *P. clessini* (Neumayr), *P. henslowanum*, *P. nitidum*, *Sphaerium corneum*, *Potamida littoralis* (Cuvier) and *V. cristata* (Müller). *Vallonia costata* is the most common terrestrial species. Only eight freshwater taxa were listed by Warren as present exclusively in the Freshwater Beds and all are uncommon. All the common freshwater species listed above are also abundant at Swanscombe. Additionally, the Freshwater Beds yielded four species of ostracod, all currently living in rivers or lakes in Europe (Withers, in Warren, 1923a).

The Clacton Estuarine Beds contain a number

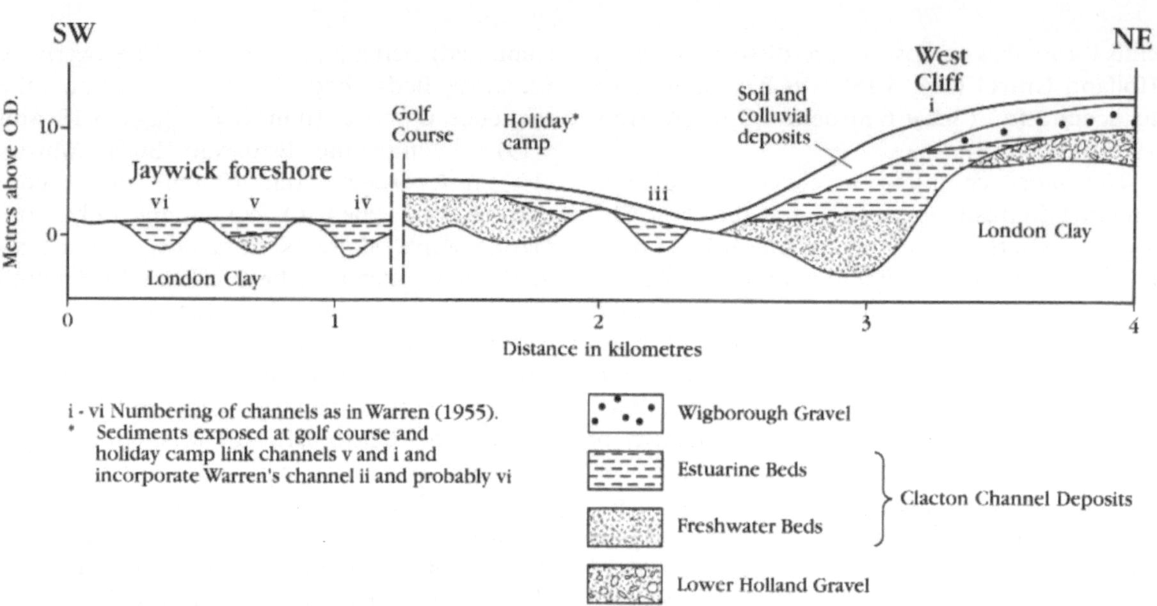

Figure 5.18 Section through the Clacton area, showing the various Clacton Channel occurrences (modified from Warren, 1955).

of freshwater mollusc species that are absent in the Freshwater Beds; for example, Kennard and Woodward (1923) listed *Paladilbia radigueli* (Bourguignat), *Viviparus diluvianus* and *Corbicula fluminalis.* Of these, *P. radigueli*, which is probably part of the *Hydrobia ventrosa* complex (R.C. Preece, pers. comm.), was abundant. Warren (1955) subsequently recorded small numbers of this snail in the Freshwater Beds from the cliff-top borehole and from Jaywick. It is nowadays, however, regarded as a probable brackish-water species (R.C. Preece, pers. comm.). *Paladilbia radigueli* and *V. diluvianus* are also present in small quantities in Warren's (freshwater) 'bed l', which occurs within the Estuarine Beds (see above and Fig. 5.19). Marine species from the Estuarine Beds included *Cerastoderma edule* L., *Hydrobia ulvae* (Pennant), *Littorina littoralis* (L.), *Mytilus edulis* (L.), *Scrobicularia plana* (da Costa), *Macoma balthica* (L.) and *Turritella communis* Risso (Brown, 1841; Dalton, 1880; Baden-Powell, 1955). This assemblage comprises estuarine taxa characteristic of a sandy mud substrate (R.C. Preece, pers. comm.). Other taxa were recorded by Baden-Powell (1955) from estuarine deposits filling Warren's channels iii-iv; these channels should probably be treated separately, as their relation to the main Clacton Channel requires further investigation (see below). A number of species of foraminifera have also been recorded from the Estuarine Beds, the dominant taxon being *Nonion depressula* Walker and Jacob (Ovey, in Warren, 1955; van Voorthuysen, in Baden-Powell, 1955).

Although a wealth of plant remains had been recognized in the deposits several years previously (Reid and Chandler, 1923), it was with the development of pollen analysis that the biostratigraphical significance of the Clacton palaeobotany was first realized. The pioneering palynological study of the channel deposits by Pike and Godwin (1953) was based on a cliff-top borehole (see Figs 5.17 and 5.19). In this borehole, pollen-bearing clays and silts ascribed to the Estuarine Beds overlay organic silty sands with freshwater shells, also polleniferous, which were attributed by Warren (1955) to the Lower Freshwater Beds. The pollen sequence from this borehole showed that the freshwater sediments were laid down during a warm-temperate period, with deciduous woodland established in the region, whereas the overlying Estuarine Beds represent a period of declining warmth, in which coniferous forests became dominant (Pike and Godwin, 1953). The spectra from the Estuarine Beds, which record a marked increase in silver fir (*Abies*) pollen, have subsequently been assigned to biozone IIIb of the Hoxnian interglacial, whereas the underlying freshwater

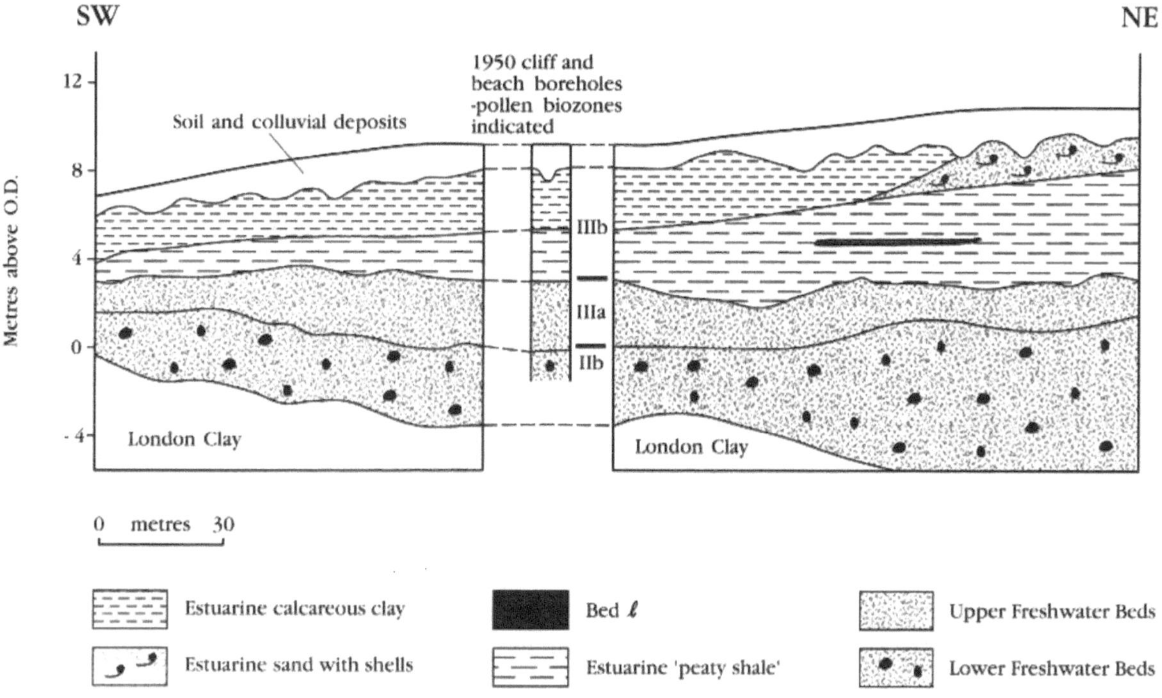

Figure 5.19 Section through the fill of the main Clacton Channel, as exposed at the West Cliff (modified from Warren, 1955).

sediments have been attributed to biozone HoIIIa (Turner and Kerney, 1971). Turner and Kerney (1971) also succeeded in extracting pollen from the Upper Freshwater Beds from a borehole through the modern beach (drilled in the 1950s but not analysed at that time). These sediments proved to contain high levels of oak and alder and were ascribed by Turner and Kerney to biozone Ho IIb, indicating that they pre-date the entire sequence in the cliff-top borehole. Warren's view that the Lower Freshwater Beds were represented in the Pike and Godwin pollen sequence was therefore refuted (Fig. 5.19).

Plant macrofossils, mostly seeds and fruits, were described from both boreholes by Turner and Kerney (1971). These authors broadly confirmed the earlier records of Reid and Chandler (1923), but were also able to relate the new material to the pollen biozones. Plant macrofossils appear to have been distributed throughout the deposits; Reid (in Reid and Chandler, 1923) considered there to be little difference between the assemblages from different beds. Reid and Chandler recorded 137 species, generally indicative of temperate, rather dry conditions. An important new record by Turner and Kerney was that of *Azolla filiculoides* (Lam.), a water fern that is no longer native to Britain and is thought to characterize the Hoxnian Stage.

Sections excavated on the golf course in the early 1930s, in the inland part of the outcrop of the Clacton Channel Deposits (Fig. 5.17), revealed the following sequence (Oakley and Leakey, 1937):

		Thickness
	Pale brown hillwash	0.3–0.6 m
Clacton Channel Deposits	Variegated silty clay ('loam')	0.3–0.6 m
	White or variegated calcareous clay	0.0–0.6 m
	Pebbly silver-sand, cross-bedded, with lenses of silty clay and local seams of calcareous clay ('marl')	0.6–1.2 m
	Red sandy gravel	>0.9 m

In a second excavation at the golf course, 300 m to the east of the first (Figs 5.17 and 5.20), a similar sequence was revealed, although the variegated calcareous clay ('marl') was overlain by a brown fissile, sometimes stony clay, devoid of fossils or artefacts (Singer *et al.*, 1973). Over much of the area the calcareous clay directly overlay the gravel. The deposits were considerably deformed and disturbed by post-depositional periglacial processes (Gladfelter, 1972; Singer *et al.*, 1973; Wymer, 1985b), confirming that the site has experienced at least one period of intense cold since the sediments were laid down. An important new fossil record from the second golf course excavation was that of ***Trogontherium cuvieri*** Fischer, an extinct large beaver unknown in Britain after the Hoxnian Stage (Stuart, 1982a). An additional ostracod species was identified in the calcareous clay (Robinson, in Wymer, 1985b), *Cyclocypris **hucki*** Triebel (synonymous with *Scottia tumida* Kempf – J.E. Robinson, pers. comm.), which is typical of flowing water.

Oakley and Leakey (1937) had considered the calcareous clay to belong to the Estuarine Beds, but shells discovered in less weathered parts of this deposit in the recent holiday camp exposures suggest that it is of fluvial origin (R.C. Preece, pers. comm.). The deposit observed by Warren in his Channel vi on the Jaywick foreshore appears to have been a continuation of this calcareous clay, termed 'marl' by Warren and most subsequent authors. Exposures at the holiday camp revealed remnants of bedding in this material, which is an oxidized silty clay with redeposited calcium carbonate nodules and occasional shells (mostly in its lower part). It is probably equivalent to the Upper Freshwater Beds of the West Cliff section.

Channels iii and iv of Warren (1940, 1955), which actually represent the two ends of a single continuous feature (Fig. 5.17), are entirely separated from the main channel complex by a

Figure 5.20 Photograph of the second archaeological excavation at Clacton, taken in 1970. The Clacton Channel Gravel is clearly seen, beneath calcareous silt (marl). London Clay forms the floor of the excavated area; careful removal of the overlying gravel has revealed undulations in its surface, probably scour features. (Photo: J.J. Wymer.)

ridge of London Clay (Figs 5.17 and 5.18). A recent borehole survey of the holiday camp area (J.J. Wymer, pers. comm.) has confirmed Warren's observation that a separate channel occurs on the seaward side of the main Clacton Channel. Warren (1955) interpreted this as part of the same channel system and considered it to be filled with the Clacton Estuarine Beds. He reported (1955, p. 284) that all the channel fragments seen by him yielded the same fauna, with the elephant *Palaeoloxodon antiquus* and the bivalve *Potamida littoralis*, and the same Clactonian industry. This seems, however, to conflict with his observation that Channels iii, iv and vi contain only the Estuarine Beds, since the elephant bones and artefacts were only supposed to occur in the Freshwater Beds. A sentence in Warren's 1955 paper (p. 288) clarifies this in the case of Channel iii–iv; this contains marine Mollusca of species recorded from the Estuarine Beds in the West Cliff section, together with occasional bones of *Palaeoloxodon antiquus*, but non-marine molluscs are absent. According to Warren's (1940) observations of the Jaywick foreshore and hinterland, the deposits of the main Clacton Channel are themselves divided into two parallel strips by a low ridge of London Clay, his (1955) Channels v and vi (see Figs 5.17 and 5.18). The recognition of estuarine sediments at foreshore level in Channel iii–iv (that on the seaward side of the main channel) raises a number of interesting possibilities, as they are significantly lower than the Estuarine Beds at the West Cliff locality (Fig. 5.19). This may indicate a period of erosion following the deposition of the Upper Freshwater Beds, an explanation that was apparently favoured by Warren (1955) – his 'minor non-sequence'? The recognition at Cudmore Grove, East Mersea (only 9 km upstream from Clacton), of a sequence, also ascribed to the Hoxnian Stage, in which estuarine sediments underlie fluviatile ones (Bridgland *et al.*, 1988), raises the possibility that these low-level estuarine sediments on the Jaywick–Clacton foreshore might pre-date the Freshwater Beds (see below). The possibility that the separate iii–iv channel-fill at Clacton is a later Colne deposit and totally unrelated to the Hoxnian sequence is another alternative that cannot be ruled out. The fauna recorded from this channel is not stratigraphically diagnostic; the marine molluscs are all species that are extant and straight-tusked elephant (even if not reworked into these deposits, as Warren (1955) suspected) is recorded from later interglacials. Detailed investigations of the sediments in this channel and at Cudmore Grove will hopefully resolve these uncertainties in the near future.

The earliest description of Palaeolithic artefacts recovered from the Clacton Channel Deposits was by Warren (1912), although a number of brief notices of earlier discoveries from (or possibly from) these beds had appeared (Evans, 1872, p. 521; Anon., 1906, 1911a, 1911b; Warren, 1911) and J.W. Kenworthy had assembled a small collection of material from the exposures at Clacton that is now in the Passmore Edwards Museum at Stratford (Wymer, 1985b). As well as flint artefacts, the collections from Clacton include a number of bone implements (regarded as questionable by Wymer, 1985b) and the tip of a wooden spear, found by Warren in 1911, which remains unique in the British Palaeolithic (Oakley *et al.*, 1977; McNabb, 1989). The artefacts come from the Lower and Upper Freshwater Beds, the richest concentrations occurring in the gravel of the former, whereas the fine-grained sediments of the Upper Freshwater Beds (including the calcareous clay at the golf course) have yielded the best preserved material, including the celebrated wooden spear.

Interpretation

The Clacton Channel Deposits provide a wealth of palaeontological evidence of considerable value for environmental reconstruction; in addition, some of the taxa recognized are of considerable biostratigraphical significance. The sediments also represent an important element within the sequence of Thames and Thames-Medway deposits that is now recognized in north-east Essex, the interpretation of which has significant implications for the Pleistocene evolution of Thames drainage. All of the above help to evaluate the palaeoenvironmental context of the type Clactonian Industry and to relate this to the British Pleistocene sequence.

The relation of the Clacton sediments to the Thames system has been established only in recent decades. Warren (1923a) originally attributed the Clacton Channel to a small local stream. Later, when he observed that the channel originally exposed in the West Cliff was one of several running side by side, he concluded that 'scoured-out deeps in the bed of a wide river' were represented (Warren, 1955, p. 284) and that the deposits were the product of the main Thames-Medway. Estimates of the age of the Clacton sediments, which were attributed to the 'Great Interglacial' (= Hoxnian) by Pike and Godwin (1953), implied that they belong to the post-diversion Thames system; the subsequent correlation of the Swanscombe and Clacton sediments on the basis of their molluscan faunas by Kerney (1971) reinforced this view. Although alternative interpretations have been proposed (Gladfelter, 1975), the post-diversion Thames origin of the Clacton Channel Deposits has been confirmed by recent work in eastern Essex (Bridgland, 1980, 1983a, 1988a; Fig. 5.5A).

Palaeontology

The wealth of palaeontological data from Clacton is important for both environmental reconstruction and relative dating. Most recent authors have agreed that the deposits accumulated under fully interglacial conditions during the Hoxnian. Many early descriptions included faunal lists, but the most detailed summaries of the palaeontology were by Warren (1923a, 1924b, 1955). Warren's final (1955) summary was revised in the light of Pike and Godwin's (1953) description of an interglacial pollen sequence, later assigned to the Hoxnian Stage (West, 1963; Turner, 1973) .

The molluscan faunas from the various deposits at Clacton are particularly informative. In addition to the most abundant taxa, listed above, there are certain species of biostratigraphical significance. Most notable amongst these are *Belgrandia marginata*, *Valvata piscinalis* forma *antiqua*, *Viviparus diluvianus* and *Corbicula fluminalis*, all of which first appear in the Estuarine Beds or very near the top of the Freshwater Beds (Warren, 1955; Kerney, 1971). These species are part of the so-called 'Rhenish fauna' recognized at Swanscombe, where they appear near the junction between the Lower Loam and the Lower Middle Gravel (Kennard, 1942; Kerney, 1971; see Chapter 4, Swanscombe). According to Kennard (1942), this assemblage was indicative of a connection between the Thames and Rhine at this time. The condition of specimens of *C. fluminalis*, *V. piscinalis* f. *antiqua* and *V. diluvianus* from Clacton suggests reworking from an older deposit (Kennard and Woodward, 1923), possibly a lower bed within the Clacton sequence that was destroyed by intraformational erosion. The appearance of these 'Rhenish' taxa, coupled with other similarities in the molluscan faunas, enabled Kerney (1971) to relate the interglacial pollen sequence recognized at Clacton to the succession at Swanscombe, which lacks a satisfactory palynological record (see Chapter 4, Swanscombe). The Mollusca also provide important palaeoenvironmental information; in particular, they allow the distinction of the freshwater and estuarine sediments at Clacton. Freshwater species dominate the assemblage from the Estuarine Beds, but they are accompanied by marine taxa (Warren, 1955). The most complete summary of the marine Mollusca from Clacton was by Baden-Powell (1955), who noted that the assemblage could not be distinguished from those found in Holocene deposits. His faunal list indicates in which of Warren's six channels the various taxa have been found.

The mammalian remains at Clacton were described in detail by Warren (1923a). He noted that the elephants were of the straight-tusked species (*Palaeoloxodon antiquus*) and that mammoth did not occur, a feature consistent with a Hoxnian age (Stuart, 1982a). The important observation that fallow deer remains from Clacton differ from living examples was made by Dawkins (1868) and Falconer (1868), who both (independently) identified them as a distinct (larger) species, although they are currently interpreted as a subspecies *Dama dama clactoniana* (Stuart, 1974, 1982a; Leonardi and Petronio, 1976; Lister, 1986). The Clacton fallow deer is thought to characterize Hoxnian deposits in Britain; it is an important component of the Swanscombe fauna (see Chapter 4, Swanscombe), supporting the proposed correlation of the two sites (Sutcliffe, 1964).

Two types of extinct rhinoceros occur in the Clacton Channel, *Dicerorhinus hemitoechus* and *D. kirchbergensis* (Sutcliffe, 1964). Rhinoceros teeth from the basal gravel (basal bed 1) were found to have fibrous vegetable matter lodged in crevices, which was interpreted as the remains of the animals' food (Pike and Godwin, 1953; Warren, 1955). This material yielded pollen taxa dominated by non-arboreal types, which led Pike and Godwin to suggest that it represented an earlier (pre-temperate) phase of the interglacial, earlier than any of the sediments encountered in their cliff-top borehole (see above). Warren, on the other hand, suggested that non-arboreal pollen from food remains might have mixed with tree pollen from the containing sediments to give the assemblages obtained. He thought that the remains might provide important evidence for the diet of these animals.

Occasional remains of small mammals have been known to occur in the Clacton Channel Deposits from the time of their earliest description, Brown (1840) having recorded water rat from his original section (see, however, Hinton, 1923). Unfortunately, the assemblage of small vertebrates remains sparse, particularly in comparison with possible correlative sites such as Cudmore Grove and Little Thurrock (Stuart, 1974; Sutcliffe and Kowalski, 1976; Wymer, 1985b; Bridgland *et al.*, 1988; see below, Cudmore Grove). This is despite the careful sieving of the sediments exposed in the second golf course excavation (Singer *et al.*, 1973; Wymer, 1985b), which added only the vole *Clethrionomys* sp. and giant beaver to the assemblage.

Analyses of samples from the cliff-top borehole (see above), as well as older samples collected by Warren, formed the basis for the Clacton pollen diagram published by Pike and Godwin (1953). This has the following characteristics: abundant (but generally declining) alder; oak and elm relatively abundant in the lower part of the sequence; hazel declining throughout the sequence; pine consistently important throughout; and silver fir becoming dominant towards the top. A comparison with other British and continental sites suggested a correlation with the 'Penultimate Interglacial' (Pike and Godwin, 1953), which was later redefined as the Hoxnian (West, 1963; Turner, 1973). Turner and Kerney (1971) subsequently confirmed this interpretation. These authors worked additionally on samples, obtained from the borehole at beach level (see above), of earlier sediments within the Upper Freshwater Beds than those analysed by Pike and Godwin. These sediments, which they attributed to biozone HoIIb, yielded the biostratigraphically significant unnamed palynomorph known as 'Type X', which is thought to be characteristic of the Hoxnian Stage in Britain (Turner and Kerney, 1971).

Wymer (1974) reported that pollen analysis of the calcareous clay ('marl') at the golf course site had revealed spectra dominated by pine, birch and grasses. These spectra, obtained using heavy liquid flotation techniques, were attributed by Mullenders and Desair-Coremans (in Wymer, 1974; preliminary interpretation in Singer *et al.*, 1973) to the early Hoxnian Stage (biozone HoI). This interpretation led to the conclusion that the golf course sequence was older than that at the West Cliff, in which biozones HoIIb, IIIa and IIIb have been identified (Wymer, 1974, 1981, 1985b; Gladfelter, 1975; Fig. 5.19). Supporting evidence for differentiating the sediments at the golf course from those in the West Cliff has been claimed from the recognition of periglacial structures in the basal gravel that pre-date the accumulation of the overlying calcareous clay (Gladfelter, 1972; Singer *et al.*, 1973; Wymer, 1985b). On the basis of this evidence, Wymer (1985b) suggested that the earliest human occupation of the site, represented by abraded artefacts in the gravel (see below), occurred in a pre-Hoxnian (*sensu* Clacton) temperate interval. This interpretation was disputed by West (in Wymer, 1985b), who considered all the cryoturbation structures observed in the golf course sections to have formed after deposition of the calcareous clay. It is possible that some deformation of the gravel prior to burial by the calcareous clay might have been caused by some other process, perhaps related to waterlogging.

The palynological basis for regarding the golf course sediments as earlier than those at the West Cliff has also been challenged. Turner (1975, 1985) has seriously questioned the reliability of the pollen assemblages from the golf course site, because of the very low pollen concentrations in these sediments. He argued that the observed concentration of pine and birch had occurred as a result of the destruction of less durable grains during the weathering to which these oxidized sediments have clearly been subjected. Indeed, it was pointed out by Turner (1985) that the occasional grains of hazel, oak and alder pollen identified by Mullenders and Desair-Coremans imply that these taxa, which are highly susceptible to weathering, were once considerably more common. These deciduous species are not characteristic of the early parts of interglacials; their presence in the calcareous clay suggests instead that it accumulated during biozone II or III (Turner, 1985). It is therefore likely that the golf course sequence is entirely comparable to that in the

West Cliff sections, with the Lower and Upper Freshwater Beds represented, the former by the gravel and the latter by the weathered calcareous clay. The Estuarine Beds are apparently missing at the golf course; in fact the westernmost feather-edge of the outcrop of these beds has been observed recently at the holiday camp (Fig. 5.18). Warren's (1955) observations suggest that these or other estuarine deposits may be present on the foreshore at Jaywick, however (see Fig. 5.18).

Geochronological evidence

Szabo and Collins (1975) obtained a radiometric (uranium-thorium) date of 245,000 (+35,000/ -25,000) years BP from a bone sample from the 1969 excavations at the golf course. As noted by Wymer (1985b), this is considerably younger than many predictions for the Hoxnian, based on an age for the preceding Anglian glaciation of between 400,000 and 470,000 years, as suggested by Kukla (1977).

The most recent evidence, independent of stratigraphy, for the age of the Clacton deposits has come from amino acid analyses (see Chapter 1 for explanation). Early work in this field by Miller *et al.* (1979), using shells of *Corbicula fluminalis*, grouped Clacton with sites in the Lower Thames such as Crayford and Aveley, which are attributed in this volume to Oxygen Isotope Stage 7 (see Chapter 4). However, in more recent analyses using a modified technique, Bowen *et al.* (1989) have obtained higher ratios, indicative of a greater age, using specimens of ***Pisidium*** (0.305 ± 0.001 (n = 2)) and ***Valvata*** (0.299 ± 0.002 (n = 3)) from Clacton. These ratios are comparable with results (mainly using different species) from Swanscombe, which Bowen *et al.* correlated with Oxygen Isotope Stage 11. Since the Anglian Stage is considered to correlate with Oxygen Isotope Stage 12 (see Chapter 1), a Stage 11 age for Clacton would conform with the widely held view that the site represents an immediately post-Anglian (post-diversion) Thames-Medway channel, an interpretation that has been reaffirmed by palaeogeographical reconstructions, based on terrace stratigraphy, of the sequence in eastern Essex (Bridgland, 1988a; Fig. 5.5). The biostratigraphical correlation of the sequences at Clacton and Swanscombe is also upheld by the conclusions of Bowen *et al.*, but the ascription to the Hoxnian interglacial is questioned, since amino acid ratios obtained from shells from the Hoxne type locality are lower and were considered to be indicative of Stage 9 (see Chapter 1). The term Hoxnian, as applied here to the Clacton sequence, should therefore be taken to mean Hoxnian *sensu* Swanscombe, pending confirmation (or otherwise) that the Hoxne sediments represent the same time interval as those at Swanscombe and Clacton.

Evidence for sea levels and possible subsidence

The Clacton site provides important evidence for Hoxnian sea levels, since part of the sequence is of estuarine origin. Palynological studies have revealed that the change from freshwater to estuarine conditions at Clacton occurred in the late-temperate phase of the interglacial, biozone HoIII (see above). There is evidence for a minor non-sequence at this point in the succession, with erosion to the lower levels occupied by the estuarine deposits at Lion Point (see Fig. 5.18; see above). However, the pollen record (Pike and Godwin, 1953) shows little evidence for a lengthy break in deposition. Warren (1955) noted that the estuarine deposits overlap the Freshwater Beds, but considered this to be exactly what would be expected as a result of a marine transgression. There are several reasons for concluding that this transgression did not extend upstream far beyond Clacton and that sea level may have generally declined during the interval represented by the Estuarine Beds. Firstly, Warren (1955, p. 287) noted that land and freshwater taxa greatly predominate over marine species throughout the lower part of the Estuarine Beds, except at their extreme base. Secondly, the occurrence of the freshwater lens, Warren's 'bed l' (see above; Fig. 5.19), suggests that there was a brief break in estuarine conditions, although this may simply represent the transport of fluvial sediment into the estuarine environment during a flood event. Finally, the palaeobotanical record shows that deciduous forest was replaced by coniferous woodland during the deposition of the Estuarine Beds, suggesting a deterioration of climate that might be expected to have been accompanied by a fall in sea level.

A problem encountered in attempts to

determine contemporary sea levels from the Clacton Estuarine Beds is that many workers have considered the area of the southern North Sea coast to have been significantly lowered by subsidence during the late Pleistocene (West, 1963, 1972; Evans, 1971; Kerney, 1971). This interpretation appears to be based largely on the views of Wooldridge (1927b, 1928; Wooldridge and Henderson, 1955), who defined a precise western limit to such downwarping, his 'Braintree line', and the widespread acceptance of a Hoxnian sea-level maximum at *c.* 32 m O.D. (Zeuner, 1945, 1959; West, 1963, 1972), much higher than the Clacton Estuarine Beds. Both these bases have been challenged in recent years. The bulk of the evidence for subsidence of the edge of the North Sea Basin is for differential movement during the Holocene, which has been demonstrated by various authors (Rossiter, 1972; D'Olier, 1975; Devoy, 1977, 1979; Greensmith and Tucker, 1980). The measured rates of differential warping during the Holocene are sufficiently high to have lowered the coastal area of Essex by many metres if this relative movement had occurred continuously throughout the Middle and Late Pleistocene. Differential warping on this scale would have lowered all the Middle Pleistocene terrace deposits in eastern Essex to below modern sea level, so it is apparent that the process could only have operated during a small proportion of this time, perhaps during high sea-level phases when sedimentation was occurring in the present offshore areas (Bridgland, 1983a, 1988a). The association of a 32 m sea level with the Hoxnian interglacial is traditionally based on extrapolation from Mediterranean areas by Zeuner (1945, 1959) and the occurrence of Hoxnian deposits at this elevation in the 'Goodwood raised beach' of West Sussex and at Swanscombe. The Goodwood raised beach deposit has recently been thoroughly reinvestigated at Boxgrove and appears to be pre-Hoxnian (Roberts, 1986), whereas Swanscombe is clearly a fluvial site, with no direct relevance to contemporary sea level. Moreover, the height difference between the deposits at Swanscombe and Clacton, a fall of around 27 m over a distance of approximately 110 km (along the course reconstructed in Fig. 5.5A), implies a downstream gradient of *c.* 1:4000, which is within the range of gradients observed amongst the fluvial terraces of the Middle and Lower Thames. This important fact, which was noticed by a number of previous workers, including Zeuner (1945), Singer *et al.* (1973) and Clayton (1977), allows the reconstruction of a Thames and Thames-Medway course for the Hoxnian Stage (*sensu* Swanscombe) in which the Clacton and Swanscombe sediments are shown to be broadly contemporaneous deposits of the same river system (Bridgland, 1980, 1983a, 1988a; Fig. 5.5A). The problems of reconciling the various evidence for sea levels during the Hoxnian Stage will be further discussed below (see Cudmore Grove).

Palaeolithic evidence

Warren (1912, 1922) was the first to recognize that the flint artefacts from Clacton, a mixture of thick stone-struck flakes and relatively crude cores, did not belong either to the Acheulian Industry, since there were no hand-axes, or the Mousterian Industry, which is characterized by more advanced preparation of cores (Levallois technique; see Chapter 1) than is seen in the Clacton material. The assemblage was initially defined as Mesvinian by Breuil and a detailed description was published under this title by Warren (1922). However, Breuil (in Warren, 1926) subsequently recognized earlier and later divisions of the Mesvinian Industry and recommended confining the term Mesvinian to the later division. Warren (1926) noted that the Clacton assemblage belonged to the earlier of these divisions and was therefore pre-Mesvinian; he proposed the term Clactonian for this industry.

At about the same time, a comparable Clactonian industry was identified in the Lower Gravel at Swanscombe (Chandler, 1930, 1931; see Chapter 4). A detailed description of the Clactonian artefacts from both these sites was published by Breuil (1932b), who considered (in agreement with Chandler) that the artefacts from the Swanscombe Lower Gravel were generally older than those from Clacton. This view, supported by Warren (1933), was probably influential in the formulation of King and Oakley's (1936) model for Lower Thames evolution, in which the Clacton sequence was correlated with a hiatus between the Lower Loam and Lower Middle Gravel at Swanscombe (see Chapter 4, Swanscombe and Globe Pit).

In 1934 the first of two archaeological

excavations was carried out at the Jaywick golf course, from which 190 artefacts were recovered (Oakley and Leakey, 1937). This material was not restricted to definite levels, nor was there any correlation between differences in typology and stratigraphical position. The unabraded condition of many of the artefacts led Oakley and Leakey to suggest the proximity of a working floor. The second series of excavations at the golf course (in 1969 and 1970) yielded over 1200 artefacts (Wymer and Singer, 1970; Singer et al., 1973; Wymer, 1985b; Fig. 5.20). These were most common in the basal gravel, occurring throughout the deposit, but with a slight concentration near its southern edge. The condition of the material varied from mint to very rolled, the former becoming increasingly common towards the top of the gravel. The industry continues sporadically in the overlying calcareous clay, which produced low numbers of well-preserved artefacts. Certain flakes have been refitted on to cores or other flakes, including two examples from the calcareous clay fitting on to pieces from the gravel/clay interface (Wymer, 1985b). The majority of conjoinable material occurred at the latter interface, in an 'accumulation of discarded flint work and broken bones' resting on the gravel (Wymer, 1985b, p. 283). Wymer claimed the mint and conjoinable material from the top of the gravel to represent the debris of human occupation in primary context, a view apparently supported by the successful identification of microwear characteristics on the edges of some artefacts, indicating the type of usage to which they had been put (Keeley, in Roe, 1981; in Wymer, 1985b). Mint artefacts occurred frequently in the upper part of the gravel, whereas abraded ones were evenly distributed throughout the deposit. Wymer regarded the former as part of the primary context debitage that had been incorporated into the gravel from above, perhaps by trampling while the deposit was waterlogged; he considered the abraded material to have been derived from an earlier occupation, pre-dating the deposition of the gravel. A minimum age in the late Anglian or very early Hoxnian Stage is implied for this early occupation. Wymer (1974, 1985b) was inclined to attribute this assemblage to a pre-Hoxnian (*sensu* Clacton) 'mild phase', but this interpretation relied heavily on the pollen evidence from the golf course site, now regarded as unreliable (see above).

A later artefact assemblage, again in derived condition, occurred in a small gravel fan overlying the calcareous clay at the golf course, attributed by Wymer (1985b) to solifluction in a periglacial climate. Wymer considered this deposit, which also contains bone fragments, to pre-date the Freshwater Beds of the West Cliff section. It seems more likely, however, that the calcareous clay is itself part of the Freshwater Beds (see above) and that the gravel fan, if it is really of periglacial origin, is of post-Hoxnian age and incorporates material reworked from the Clacton Channel Deposits.

Until recently the Clactonian Industry was regarded as stratigraphically older in Britain than the earliest appearance of the Acheulian, a view based largely on the sequence at Swanscombe and the absence of hand-axes in the Clactonian gravels at Swanscombe, Clacton and Little Thurrock (Wymer, 1974; Gladfelter and Singer, 1975; see Chapter 1). Flaws in this argument have long been apparent; later Clactonian industries were noted, for example, in the Lynch Hill Gravel of the Reading area (Wymer, 1968), with which the Little Thurrock site would be correlated according to the interpretation presented in Chapter 4 (see Globe Pit). There are also a few records of hand-axes or of evidence for their manufacture from Clacton and the Swanscombe Lower Gravel. Breuil (1932a) intimated that derived 'Chellean' (= Abbevillian or Early Acheulian) implements occurred at the base of the Swanscombe succession, mixed with the indigenous Clactonian material. A hand-axe was apparently recovered from the Lower Gravel during the Waechter excavations (Ohel, 1979; Newcomer, in Ohel, 1979). Wymer (1985b) pointed out that a number of hand-axes had been found on the foreshore at Lion Point and Clacton, but none was located *in situ* in the channel deposits. Furthermore, Singer *et al.* (1973) discovered three possible hand-axe finishing flakes in the 1970 Clacton excavations, all in derived condition. There is now abundant stratigraphical evidence that hand-axes were made in Britain prior to the deposition of the Clacton Channel sediments (see Chapter 1). Industries comprising only primitive hand-axes, classified as Early Acheulian, have also been described in likely pre-Hoxnian gravels at Farnham, Surrey and Fordwich, Kent (Roe, 1964, 1968a, 1975); they are presumably contemporary with or earlier than the Clactonian industries of both Clacton and Swanscombe. Hand-axes have been found in gravels of the Thames system that are now attributed to the late Anglian (see Chapter 3, Highlands Farm Pit and Hamstead Marshall) and in conjunction with a Cromerian mammalian fauna at Boxgrove, West Sussex (Roberts, 1986; Chapter 1). Thus there is no reason why derived hand-axes or characteristic flakes from their manufacture should not occur in the Clacton Channel Deposits, although it is clear that the important industry in the immediate area, which apparently persisted from late Anglian to mid-Hoxnian times, involved no hand-axe manufacture.

Various authors have proposed chronological divisions of the Clactonian Industry over the years, generally based on typology and/or condition, together with stratigraphical considerations (Breuil, 1932b; Warren, 1958; Collins, 1969; for summary, see Roe, 1981). Wymer (1968) argued that such divisions were unjustified and subsequent stratigraphical revisions have revealed that there is little basis for them. Following a recent re-evaluation of the industry, J. McNabb (pers. comm.) found no typological or technological grounds for separating the assemblages from Clactonian sites at Barnham, Suffolk (see Wymer, 1985b), Little Thurrock and Swanscombe in the Lower Thames (see Chapter 4) and Clacton itself. He also considered the essential characteristics of the Clactonian to be present within an assemblage of apparently pre-Anglian artefacts from High Lodge, near Mildenhall, Suffolk (see Wymer, 1985b, 1988; Chapter 1), implying that the industry was practised in Britain before the Anglian Stage.

The occurrence of Palaeolithic assemblages lacking both hand-axes and evidence for the refined flaking techniques employed in the Mousterian industries has puzzled archaeologists since they were first recognized. Warren (1951) compared cores from Clactonian sites to the early pebble and flake tool industries of Africa and Asia, considering the Clactonian to be an offshoot of the tradition of these primitive industries, an idea first mooted by Oakley (1949). In this industry flint working was carried out using bold strokes with a hard (stone) hammer, producing characteristically pronounced bulbs of percussion (Baden-Powell, 1949). Wymer (1985b) has emphasized, however, that there is no such thing as a 'typical Clactonian flake'; hard-hammer flakes of the type making up the bulk of Clactonian assemblages are also found in other industries, including post-Palaeolithic ones, although in these they tend to represent a smaller proportion of the total material. This fact has led to the suggestion that the Clactonian is merely a 'facies' of the Acheulian, assemblages without hand-axes (Clactonian) representing debitage from locations where preliminary working of raw material took place, with tool manufacture occurring elsewhere (Ohel, 1977, 1979). This suggestion scarcely seems feasible, as hand-axe manufacture is not a lengthy process; it is inconceivable, if the Clactonian knappers were hand-axe makers, that no evidence of this would be found amongst many thousands of artefacts at a site such as Clacton (excepting the three questionable and abraded finishing flakes from the Singer/Wymer excavation – see above). The above suggestion also fails to explain the geographical distribution of flake and hand-axe industries. Clacton lies at the western extremity of a 'province' of flake-core industries, which dominate the eastern European and Asian Palaeolithic, whereas hand-axe industries are dominant in the south-western and southern part of Europe (see Roe, 1981). Britain lies, according to some authorities, on the frontier between these two 'provinces', in an area where both Clactonian and hand-axe industries are found, possibly resulting from separate cultural groupings.

Stratigraphical relations of the Clacton deposits

Doubt has existed for many years about the precise stratigraphical relations between the Clacton Channel Deposits and the Pleistocene gravels of the Clacton area. The early descriptions and illustrations of the cliff sections at Clacton (Brown, 1840; Wood, 1866b; Fisher, 1868a; Dalton, 1880; Prestwich, 1890b) indicate that the channel deposits cut through earlier stratified gravel, which was later termed the Holland Gravel (Warren, 1923a; 1955). Fisher (1868a) claimed (in the caption to an illustration of the cliff section) that the channel deposits and the Holland Gravel were both overlain by an 'obliquely bedded gravel of unascertained relation', newer than the channel deposits but older than the overlying colluvial 'trail' (Fisher, 1868a, p. 214). Fisher's figure shows what appears to be cross-bedding, inclined towards the northern end of the section. Most later descriptions either ignored any gravel overlying the channel deposits or described it as a poorly bedded unit that might have resulted from solifluction. Fisher's claim, coupled with Warren's consistent caution regarding the matter (Warren never saw the critical area exposed), led to controversy over the relation between the channel deposits and the Holland Gravel (summarized by Wymer (1985b)). Clarification of this relationship has only been achieved by recent trial excavations in the West Cliff (Bridgland *et al.*, 1988; see below).

Most early workers considered the Holland Gravel to belong to the 'Glacial Series' of Wood and Harmer (1868), but Warren (1923a, 1924b, 1933) recognized that the gravels covering the Tendring Plateau, including that in the Clacton district, were terrace deposits of the Thames. In these early papers he suggested a correlation of these gravels with the Boyn Hill Terrace, regarding the Clacton Channel Deposits as a possible equivalent, laid down by a smaller stream, of the Taplow aggradation of the Thames. Subsequently, Warren (1951) cited the opinion of Solomon (in Oakley and Leakey, 1937), that the Holland Gravel was not of Thames but of Colne origin. He later changed his opinion again, realizing that the channel deposits were the product of a river of sufficient size to be the main Thames, and suggested that the Holland Gravel was a pre-diversion Thames deposit (Warren, 1955), a view that has been confirmed by recent work (Rose *et al.*, 1976; Bridgland, 1983a, 1988a).

A number of separate deposits have recently been identified within what was formerly classified as Holland Gravel. Within the main sheet of Holland Gravel, which stretches from the cliffs at Clacton and Holland to the St Osyth gravel pits, lower and upper divisions have been recognized (see above, St Osyth and Holland-on-Sea). On the basis of clast composition, it has been demonstrated that the Lower Holland Gravel is a pre-diversion Thames-Medway deposit, whereas the Upper Holland Gravel, overlying and apparently channelled into the top of the Lower Holland Gravel, appears to date from the brief period during the Anglian glaciation when the Thames was blocked by ice and no longer reached the Clacton area (Bridgland, 1983a, 1988a; Bridgland *et al.*, 1988, 1990). This interpretation is again based on the clast content of the Upper Holland Gravel, which is predominantly indicative of a Medway provenance, supplemented by small amounts of the type of material introduced by the Anglian glaciation. On the basis of these observations, the Holland Gravel Formation has been ascribed to the Anglian Stage and correlated with the Winter Hill Formation of the Middle Thames (Table 1.1 and Fig. 1.2; Table 5.3 and see above, St Osyth and Holland-on-Sea). Thus the Clacton Channel Deposits cut through and overlie the downstream equivalent of the Winter Hill Gravel, the last Thames formation to be aggraded prior to the diversion of the river.

A later gravel formation, the Mersea Island/Wigborough Gravel, has also been recognized in the Clacton area within the broad definition of the Holland Gravel applied by earlier workers (Bridgland, 1983a, 1988a; Bridgland *et al.*, 1990). This deposit is represented by a string of remnants along the southern fringe of the Tendring Plateau, running from Point Clear to Jaywick. These appear, from their distribution and altitude, to be a continuation of the gravels of Mersea Island (Figs 5.2 and 5.5B). This has been confirmed by clast-lithological analysis (Table 5.5), which indicates that the deposits are part of the Low-level East Essex Gravel Subgroup, interpreted as the product of the post-diversion Thames-Medway (Bridgland, 1980, 1983a, 1988a; Bridgland *et al.*, 1988; see above, Introduction to Part 2). The Mersea Island Gravel (Fig. 5.2) contains only a few percent of quartzose exotic clasts, in contrast to the 12–20% present in pre-diversion Thames-Medway gravels, and also contains significantly more Greensand chert from the Medway catchment (Table 5.5). However, between Point Clear and Jaywick the clast content of this formation changes markedly over a short lateral (downstream) distance. In this transition, from gravel of Mersea Island to Wigborough type, quartzose exotics become more important at the expense of Greensand chert, so that the deposit closely resembles the Lower Holland Gravel. A possible explanation for this change is an influx of material reworked from the pre-diversion gravels covering the Tendring Plateau, possibly as a result of a confluence between the Thames-Medway and an early River Colne, which would have formed in the eastern end of the abandoned pre-diversion Thames valley (Fig. 5.5B). It is possible that the upper gravel in the West

Cliff at Clacton, illustrated by Fisher (1868a), represents a downstream continuation of this post-diversion formation (see below).

Analysis of the clast content of the Clacton Channel Gravel (basal Lower Freshwater Beds) indicates that it too is the product of the post-diversion Thames-Medway (Bridgland, 1980, 1983a, 1988a; Bridgland *et al.*, 1988; Table 5.5). However, the relation between the Clacton Channel Gravel and the Wigborough Gravel is unclear from the mapped distribution of these deposits. The former is benched into the London Clay at the extreme south-eastern edge of the Tendring Plateau, at a height of *c.* -6 m O.D., and was aggraded to at least +10 m O.D. (Fig. 5.18), whereas the latter has a surface level falling from 15 m O.D. at Point Clear to 11 m O.D. at Jaywick. In the latter area the London Clay rises to the land surface between the outcrops of the Wigborough Gravel and the Clacton Channel Deposits (Fig. 5.2). The geomorphology would therefore seem to suggest that the excavation of the Clacton Channel resulted from post-Wigborough Gravel rejuvenation. It has been suggested, however, that the Wigborough Gravel formerly extended further south, covering the channel deposits (Bridgland, 1988a; Bridgland *et al.*, 1988). Only a single post-diversion Thames-Medway terrace formation has been recognized north of the Crouch, the Asheldham Gravel/Mersea Island/Wigborough Gravel (Bridgland, 1983a, 1988a; Fig. 5.5B). On the Dengie Peninsula (between the Crouch and Blackwater estuaries), this formation fills and overlies a buried channel, the Asheldham Channel (see below, Southminster and Fig. 5.28). An upstream equivalent of this channel has been identified south of the Crouch, the Southend Channel (Fig. 5.5A), and it has been suggested that the Swanscombe Lower Gravel channel is its correlative (Bridgland, 1980, 1983a, 1988a). Downstream the same channel is believed to continue across Mersea Island, where it contains estuarine and fluvial deposits that are exposed, beneath the Mersea Island Gravel, at Cudmore Grove (Bridgland *et al.*, 1988; Fig. 5.5A). This has been interpreted as an upstream equivalent of the Clacton Channel (Bridgland, 1980, 1983a, 1988a). According to this regional interpretation, the Wigborough Gravel must once have covered the Clacton Channel Deposits, but has since been removed by erosion, except for the series of remnants between Point Clear and Jaywick.

In order to clarify the relations between the gravels of the Clacton cliffs and the Clacton Channel Deposits, temporary sections were excavated at four points along the West Cliff during April 1987 by P. Harding and the author. The first three sections were entirely within the Clacton Estuarine Beds, although augering at the base of section 3 may have reached London Clay. The fourth section was of considerable interest, however, in that it revealed the feather-edge of the interglacial beds, in the form of a blue-grey clay, interstratified between bedded gravels (Bridgland *et al.*, 1988). A large flake was discovered ***in situ*** at the intersection between the top of the lower gravel and the base of this clay (Fig. 5.21); a small flake was also found in the upper gravel. The lower deposit is considered to be the Lower Holland Gravel, the last terrace formation of the pre-diversion Thames-Medway. The gravel above the channel deposits, presumably that observed by Brown and illustrated by Fisher, is clearly a bedded fluvial deposit, albeit with a rather high silt/clay content and perhaps somewhat disturbed (both of which could be the result of proximity to the land surface, within the zone of cryoturbation and pedogenesis). This upper gravel has a clast-content similar to the Mersea Island and Clacton Channel Gravels, suggesting that it too represents the post-diversion Thames-Medway (Low-level East Essex Gravel Subgroup). There is a surprising similarity in clast composition between the Clacton Channel Gravel and the upper gravel in the West Cliff section, on the one hand, and the Mersea Island Gravel, on the other (Table 5.5). The Clacton site is well downstream of the supposed confluence with the Colne, so reworked quartzose material from the pre-diversion gravels, as seen in the Wigborough Gravel at Jaywick, should be present there. An eastward projection of the series of Wigborough Gravel remnants indicates that the formation should intersect with the present coast in the area of the West Cliff section. This poses a problem of interpretation, in that the characteristics of the gravel above the channel deposits do not support its identification as part of the Wigborough Gravel, although it occupies the geographical and stratigraphical position in which that formation would be expected. A possible explanation is that the upper gravel in the cliffs is more closely allied to the channel deposits than to the Wigborough Gravel, which is assumed to represent

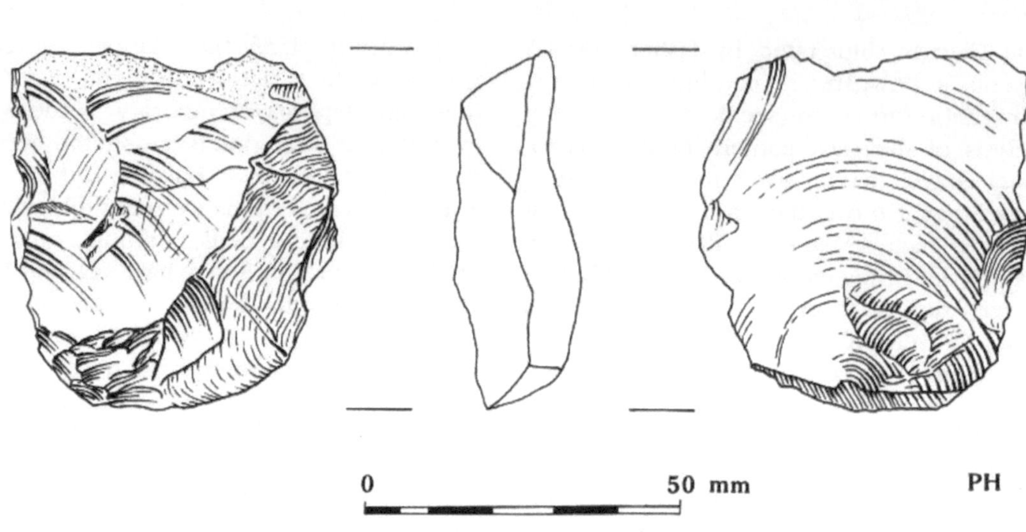

Figure 5.21 Flint flake from the West Cliff at Clacton, found *in situ* in GCR Section 4 in April 1987, at *c.* 9.9 m O.D. This flake was lying immediately below the wedge of blue-grey clay, interpreted as the feather-edge of the Clacton Channel Deposits. This is probably the highest point at which an artefact has been found in the Clacton deposits, although stratigraphically it was at the same level as the earlier Palaeolithic finds. (Drawing by P. Harding.)

a later cold episode. The Colne tributary may not have operated as a major source of gravel (reworked from the Kesgrave Group deposits to the north and west) until after the Hoxnian Stage (*sensu* Clacton). Alternatively, mixing of Thames-Medway- and Colne-derived gravel may have been incomplete, allowing deposits of Mersea Island Gravel affinities to be deposited downstream from the confluence with the Colne.

The possibility that the regional sequence is more complex than has yet been established, and that the Clacton Channel post-dates the Wigborough Gravel, cannot be ruled out however, since no gravel of Wigborough type has yet been found overlying the channel deposits. Indeed, since the Wigborough Gravel and the Lower Holland Gravel are very difficult to separate on the basis of clast-lithological content (see Tables 5.2 and 5.5), it is possible that the gravel underlying the channel deposits in the cliff section could be the Wigborough Gravel, rather than the Lower Holland Gravel. At present the interpretation of this cliff section still relies heavily on the regional stratigraphical framework derived from studies of the terrace sequence in eastern Essex as a whole. Further consideration will be given to this question below, in the report on the closely related site at Cudmore Grove, East Mersea.

The correlation of the Clacton Channel Deposits with the lower part of the Swanscombe sequence (Table 1.1 and Fig. 1.2), implied by the above interpretation, confirms many previously published opinions, notably those of Singer *et al.* (1973), on the basis of the Palaeolithic industries at the two sites. In fact a small collection of Clactonian artefacts has also been obtained from equivalent deposits (Asheldham Channel Gravel) at Burnham-on-Crouch (TQ 945972), further establishing a link between this particular phase of Thames evolution and the Clactonian Industry (Bridgland, 1988a). Bridgland (1980, 1983a, 1988a) suggested a correlation between these early post-diversion deposits and both the Black Park and Boyn Hill Gravels of the Middle Thames. He believed that the steeper Black Park aggradation had fallen below the level of the Boyn Hill in eastern Essex, so that the Black Park Gravel is represented within the lower part of the Boyn Hill/Mersea Island/Wigborough Formation (no rejuvenation can be recognized between the Black Park and Boyn Hill Formations downstream from London). This implies that the erosion of the Clacton Channel, prior to its infilling during the Hoxnian Stage, may have resulted from downcutting associated with the pre-Black Park rejuvenation; this incision has been directly attributed elsewhere in this volume to the diversion of the Thames (see Chapters 1, 3 and 4). It is also possible that any basal, pre-Hoxnian sediments in the Clacton Channel may equate with the Black Park Formation of the Middle Thames.

Summary

The complex GCR site at Clacton has been shown to be a Pleistocene locality of international significance. It provides evidence from many of the various disciplines involved in Quaternary studies, in particular from Palaeolithic archaeology and palaeontology. The site has yielded a range of faunal material seldom bettered in others of this age, including highly significant large-mammal and mollusc faunas. It is extremely likely that modern sampling of the freshwater beds at the West Cliff site, which have not been exposed for many years, would produce an assemblage of smaller vertebrates to match those from other sites more recently investigated. The site also has a particularly rich palaeobotanical record.

The wealth of palaeontological information from Clacton is of increased value since the site can be directly related to the Thames terrace sequence. This is not the case with the Hoxnian type site, which represents a lacustrine infill and is therefore more difficult to relate to an integrated regional stratigraphy. Swanscombe, the only other well-known Hoxnian site within the Thames system, lacks a convincing palaeobotanical record; indeed, the Clacton sequence has been used to calibrate that at Swanscombe, by a comparison of their molluscan faunas.

Conclusions

The Clacton locality, with its complex fossil record and stratigraphy, has been central to the debate about the interglacial (temperate-climate) period immediately following the glaciation that brought about the diversion of the Thames. The site is internationally famous for the occurrence there, in channel deposits of the early Thames-Medway, of abundant Palaeolithic (early Stone Age) artefacts. There are no recognizable tools amongst these artefacts; instead the 'industry' comprises a characteristic mixture of crudely worked flint. Clacton was established as the type locality for this particular type of industry in 1926. The Clacton sections also yield the remains of terrestrial, fluvial and marine faunas and floras. These illustrate life in and around the river during the early part of this temperate interval and, higher in the sequence, an influx of marine species marks a change to estuarine conditions as the sea level rose. The Thames at this time flowed in its modern valley, through what is now London, but turned northwards to rejoin its old pre-diversion valley just upstream from Clacton. There is widespread agreement that the Clacton sequence can be correlated with sediments at Swanscombe, in the Lower Thames valley, both being attributed to the interglacial (around 400,000 years BP) that immediately followed the major glaciation during the Anglian Stage. The Clacton sediments have also been equated, on the basis of similarities in pollen content, with lake sediments at Hoxne in Suffolk, the type locality for the Hoxnian interglacial.

CUDMORE GROVE (EAST MERSEA) CLIFFS AND FORESHORE (TM 068146)

D.R. Bridgland

Highlights

A highly fossiliferous Middle Pleistocene channel-fill, underlying gravels of post-diversion Thames-Medway origin, has been revealed here in recent years by coastal erosion. The deposits at Cudmore Grove include estuarine and freshwater beds. They are probably equivalent in age to the fossiliferous deposits at Clacton and Swanscombe, which have been interpreted as part of the same interglacial Thames channel-fill. The tracing of this channel from the Lower Thames valley through eastern Essex charts the course of the Thames during the period that followed the Anglian glaciation. The Cudmore Grove site has produced the most extensive small-vertebrate fauna from the British Pleistocene.

Introduction

The Pleistocene interest of Mersea Island has long been overshadowed by the richly fossiliferous Middle Pleistocene channel deposits at

Clacton-on-Sea, situated only 9 km to the east. However, the gravels covering parts of the island were the subject of discussion by early workers (Wood, 1866b; Dalton, 1908; Anon., 1913). Furthermore, records of Pleistocene fossils discovered at East Mersea can be traced back for at least 80 years (Dalton, 1908; Warren, 1917, 1924b, 1933; Cornwall, 1958; Zeuner, 1958; Sutcliffe, 1964; Spencer, 1966). Today three separate Pleistocene localities are recognized here (Bridgland *et al.*, 1988) within a single, complex GCR site. Pleistocene channel-fills occur at all three sites (Figs 5.22, 5.23 and 5.24), exposed by present-day marine erosion. At Cudmore Grove two channels of different ages are recognized, the Cudmore Grove Channel, which is attributed to the Thames-Medway and is the subject of the present report, and a later feature, the sediments in which yield hippopotamus. The latter, at the Cudmore Grove Hippopotamus Site, is closely associated with the third East Mersea locality, the Restaurant Site, 1 km to the west. The sediments at the Hippopotamus Site and the Restaurant Site are attributed to a later tributary river, rather than the main Thames-Medway, and are thus described below in Part 3 of this chapter.

Most early records of fossils from East Mersea seem to be related to the later tributary deposits, but there is one probable reference to the Cudmore Grove Thames-Medway sediments. A Pleistocene estuarine deposit with shells of *Cerastoderma* and *Scrobicularia* was encountered in 1906 in an excavation for a well in a small pit 'one mile east by north of East Mersea church' (Dalton, 1908, p. 136).

At that time there was, to the south of Cudmore Grove, a tract of well-established salt-marsh, still remembered by local inhabitants, between the high, gravel-covered land and the sea. Subsequent destruction of the salt-marsh has led to the formation of cliffs at the edge of the higher land. These had reached a height of just over 3 m in 1971, when they were described by Tucker and Greensmith (1973). Greatly accelerated erosion in recent years has doubled their height and given rise to fine cliff and foreshore exposures in the fossiliferous deposits and the overlying gravel. It seems likely that the pit referred to by Dalton was situated within the present Country Park at Cudmore Grove, where there is a small overgrown hollow (TM 065148). This lies *c.* 300 m from the present cliffs, so, if the channel deposits are continuous between the two, they are of considerable extent.

Description

The cliffs at Cudmore Grove expose fossiliferous estuarine deposits beneath up to 4 m of

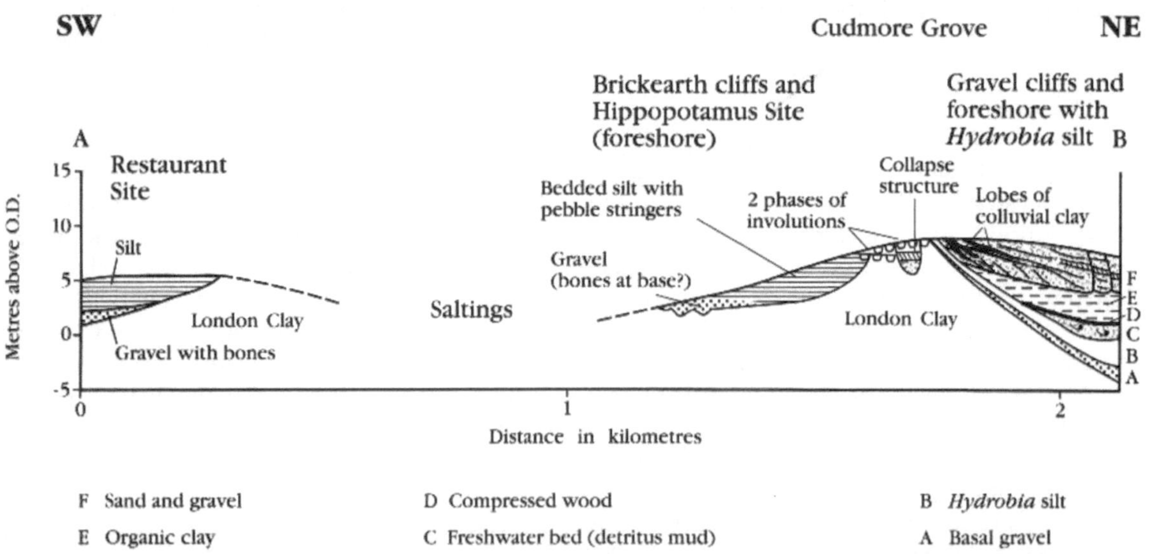

Figure 5.22 SW–NE section through the deposits at East Mersea, showing the relations of the Cudmore Grove Channel to the Blackwater deposits at the Hippopotamus and Restaurant Sites. Points A and B are indicated on Fig. 5.23.

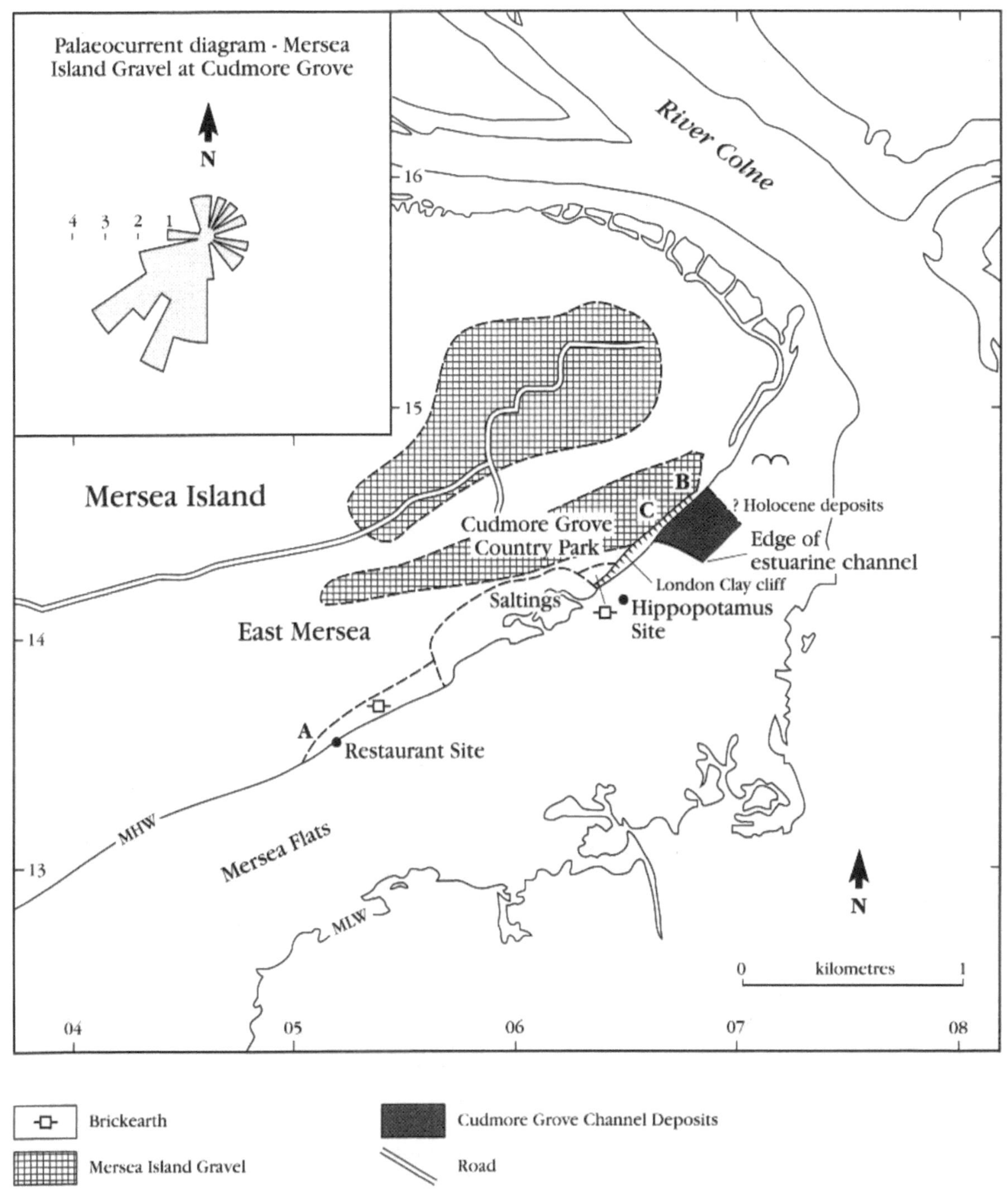

Figure 5.23 Map showing the Pleistocene deposits of East Mersea (after Bridgland *et al.*, 1988). The points A and B refer to the ends of the section in Fig. 5.22. Point C is the location of the section in Fig. 5.24.

well-bedded gravel and sand (Fig. 5.24). These deposits occupy a channel deeply excavated into the London Clay, the southern edge of which can be traced across the foreshore (Fig. 5.23). The base of the channel has been reached in a borehole in the central part of the outcrop at -11 m O.D., where some 3 m of gravel underlie the fossiliferous sediments (Roe, in Bridgland *et al.*, 1988). The sequence can be summarized as follows:

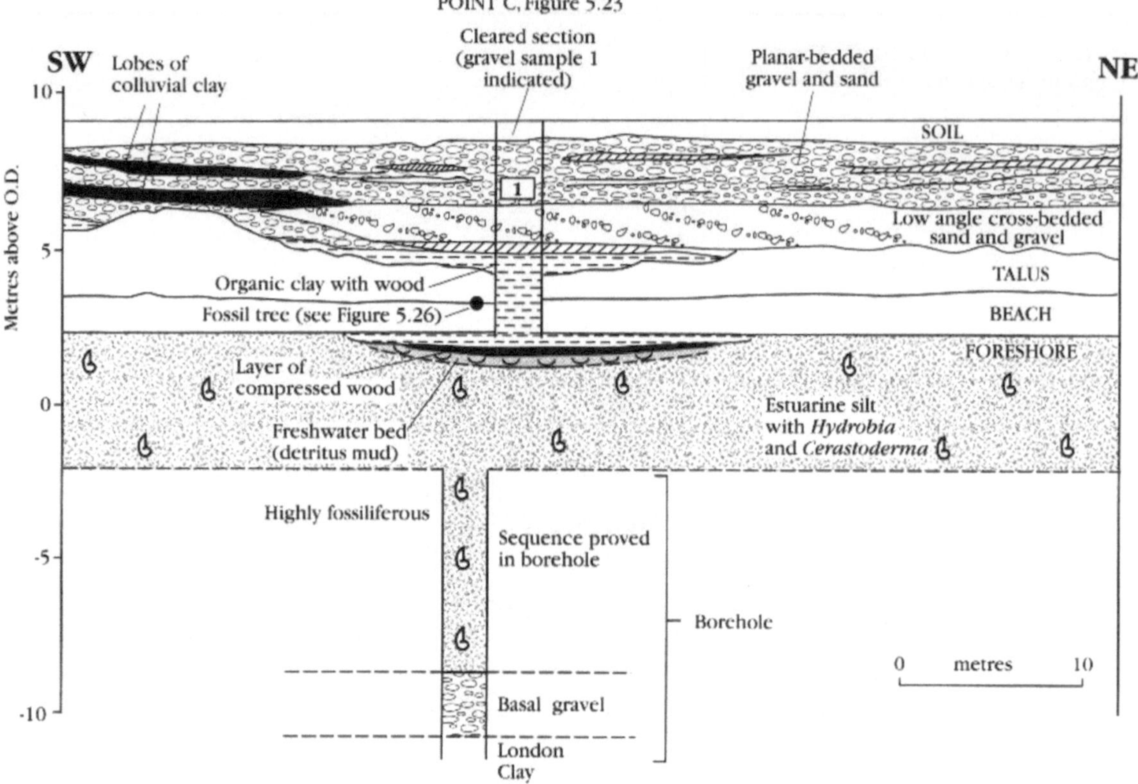

Figure 5.24 Section through the Cudmore Grove Channel Deposits, located at Point C on Fig. 5.23.

		Thickness
5. Gravel and sand, well-bedded	(Mersea Island Gravel)	up to 4 m
4. Organic clay with wood fragments, especially at its base	(Cudmore Grove Channel Deposits)	up to 3 m
3. Shelly 'detritus mud', richly fossiliferous		0–0.3 m
2. Homogeneous clayey silt with estuarine Mollusca		up to 10 m
1. Gravel	(Cudmore Grove Channel Gravel)	3 m
London Clay		

The channel deposits (1–4) and the overlying gravel (5) thin rapidly to the south-west and are replaced in the cliffs by London Clay (Fig. 5.22). This allows an attenuated representation of the channel sequence to be observed in exposure; the rapid coastal erosion and the periodic cliff collapses that it causes have provided opportunities to study different parts of the succession at different times. At the boundary between the channel deposits and the London Clay is a 'lag gravel', only 0.2–0.3 m thick, composed largely of rounded London Clay pebbles set in a clay matrix. Scattered durable clasts also occur, however, which resemble the material in the gravel overlying the channel sequence (Table 5.5). A Palaeolithic flake in very fresh condition has been recovered from this basal deposit (Bridgland *et al.*, 1988; see below).

The homogeneous clayey silt (bed 2) constitutes the bulk of the channel-fill (Fig. 5.24). It contains abundant molluscs, mostly hydrobiids and *Cerastoderma*, indicative of brackish or intertidal conditions. From its 'feather-edge', in the cliffs, this bed yields a rich ostracod

Figure 5.25 GCR excavation at Cudmore Grove, May 1987, which coincided with a field excursion of the Geologists' Association (Bridgland *et al.*, 1988). A pit has been dug into the fossiliferous sequence beneath the beach. The natural exposure of the layer of compressed wood is visible in the foreground and the Mersea Island Gravel is superbly exposed in the cliffs. (Photo: A.J. Sutcliffe.)

assemblage dominated by the brackish species *Cyprideis torosa*, the valves of which have well-formed nodes, thought to indicate low salinity. Other ostracod taxa recorded (J.E. Robinson, pers. comm.) are *Candona neglecta* Sars, *C. caudata* Kaufmann, *C. marchica* Hartwig, *Loxoconcha elliptica* (Brady), *Cythere lutea* (Müller), *Darwinula stevensoni* (Brady and Robertson), *Herpetocypris reptans* (Baird), *Cypridopsis vidua* (Müller), *Cytheromorpha fuscata* (Brady), *Ilyocypris gibba* (Ramdohr) and *Limnocythere inopinata* (Baird). The majority of the above are freshwater species, but *L. elliptica*, *C. lutea* and *C. fuscata* live in marine or brackish conditions. The deposit also contains pollen from trees such as oak, elm and hazel, implying fully temperate conditions (Roe, in Bridgland *et al.*, 1988).

Above the estuarine silt, but apparently of restricted lateral extent, occurs the most richly fossiliferous of the Cudmore Grove deposits, the shelly 'detritus mud' (bed 3). This bed is usually only *c.* 0.3 m thick, but it is packed with shell fragments and is very rich in small vertebrate remains. The molluscan fauna is dominated by *Corbicula fluminalis* and a hydrobiid referred to in the British literature as *Paladilhia radigueli*. Other freshwater taxa occur, including *Bithynia*, *Valvata*, *Pisidium* and fragments of freshwater mussels. These taxa occur in proportions suggestive of brackish conditions, but with a lower salinity than the environment indicated by the underlying homogeneous silt (Bridgland *et al.*, 1988).

Thanks to the efforts of a number of collectors (notably J.D. Clayden, D. Harrison, M. Warren and R. Wrayton), who have sieved huge volumes of sediment from the foreshore exposures, bed 3 has produced a wealth of small vertebrate remains that is unparalleled in the British Middle Pleistocene. The rodents include *Apodemus sylvaticus* (L.) (wood mouse), *Clethrionomys glareolus* (bank vole) and *Microtus agrestis* (short-tailed field vole); the insectivores *Neomys fodiens* (water shrew) and *Crocidura* sp. (white-toothed shrew) are also present, the former in some abundance. Larger mammals include quite common remains of *Castor fiber* (beaver), much rarer *Ursus* sp. (bear), and a tooth of *Macaca sylvanus* (macaque), a species

Figure 5.26 Part of a fossil tree protrudes from beneath the beach at Cudmore Grove. An interesting analogue is provided by the modern tree, which has fallen over the cliffs from the rapidly diminishing grove, a victim of the rapid coastal erosion. (Photo: A.J. Sutcliffe.)

known only from pre-Ipswichian interglacial deposits in Britain. Other vertebrate remains discovered in this bed include *Emys orbicularis* (pond tortoise), *Bufo bufo* (L.) (toad), *Rana* sp. (frog), *Eptesicus* sp. (bat) and several species of birds and fish. Several elements of the herpetofauna are new to the British Pleistocene (Holman *et al.*, 1990). The palynological sequence from the site places this bed in the late temperate phase of the interglacial; it has been attributed to Hoxnian biozone HoIIIb (Roe, in Holman *et al.*, 1990).

The overlying organic clay (bed 4) lacks calcareous fossils, but contains, except in the oxidized upper few centimetres, abundant wood fragments and well-preserved pollen. The latter is similar to that in the homogeneous silt (bed 2), with the addition of substantial hornbeam and alder, small amounts of silver fir (*Abies*) and occasional records of the palynomorph 'Type X' (Roe, in Bridgland *et al.*, 1988). At the base of this bed is a layer of compressed wood (Figs 5.24 and 5.25) that includes large fragments. Two tree-trunks, one with roots attached, have been observed at different times protruding from this bed and forming prominent features on the modern beach and foreshore (Fig. 5.26).

The gravel that overlies the channel deposits contains a mixture of local, southern and exotic clasts similar to those in the Clacton Channel Gravel and in the Low-level East Essex Gravel south of the Blackwater (Bridgland, 1980, 1983a, 1983b, 1988a; Bridgland *et al.*, 1988; Table 5.5). Excellent exposures in this deposit are provided by the cliffs at Cudmore Grove, allowing sedimentary and post-depositional structures to be observed. At the western end of its exposure in the cliffs, as it thins against rising London Clay bedrock, the gravel is interbedded with steeply dipping beds of redeposited clay (Fig. 5.24), thought to represent lobes of colluvial material at the edge of the contemporary floodplain. Towards the opposite end of the exposure, the gravel is disrupted by numerous near-vertical structures resembling ice-wedge casts, but often with a slight downthrow on their eastern side. The underlying organic clay is diapirically uplifted beneath a number of these features, which might result from cambering rather than (or as well as) ground-ice development (Bridgland *et al.*, 1988).

Interpretation

Although research on the Cudmore Grove site is still in progress, sufficient information has already come to light to indicate that this is an extremely important Middle Pleistocene locality. Much of the site's significance results from its exceptional wealth of faunal evidence, the interest of which is enhanced (as at Clacton) by the fact that the deposits are a component part of the Thames-Medway terrace sequence.

The palaeontology of the channel deposits provides information about the palaeoenvironment as well as evidence for relative dating and correlation with other sites. The sequence records a change from a fluvial channel-fill, as evidenced by the thick basal gravel, to an estuarine environment, indicated by the characteristics and faunal content of the homogeneous silt (bed 2). Later deposits suggest a tendency towards marine regression, with a decrease in salinity indicated by the fauna of the 'detritus mud'. The overlying gravel (bed 5) presumably marks a return to a fluvial environment, very possibly coupled with a climatic deterioration, major gravel units normally being regarded as cold-climate deposits (see Chapter 1). However, there is nothing to suggest a lengthy hiatus between the organic clay and the overlying gravel. In fact the lower part of the gravel comprises a series of low-angle cross sets (Fig. 5.24), dipping south-eastwards, that may represent deltaic progradation over the estuarine sequence, perhaps in response to an increased supply of sediment, prior to any major fall in sea level (Whiteman, in Bridgland *et al.*, 1988).

The channel sediments above the basal gravel are of clear interglacial character. The rodent assemblage in the 'detritus mud' (bed 3), for example, is typical of the temperate woodland phases of the Middle and Late Pleistocene (Currant, 1986; Bridgland *et al.*, 1988). The pollen and fossil wood further indicate that deciduous woodland was established in the region. The fact that sea level was sufficiently high to allow estuarine deposition within the present land area is also indicative of an interglacial, as sea levels during cold episodes were many metres below that at the present time.

The overlying Mersea Island Gravel is of sedimentological interest, comprising a varied sequence displaying a number of different types of bedding structure. These range from small-scale cross-stratification to the large foresets mentioned above. Several of the sandier horizons show evidence of penecontemporaneous deformation, suggesting that they were highly saturated when the overlying beds were laid down above them. Palaeocurrent measurements from small-scale foresets are widely distributed (Fig. 5.23, inset), giving rise to a radial distribution that is somewhat difficult to interpret. A south-south-east mean direction can be calculated, if the largest gap in the distribution is taken as a 'false origin', but this is of dubious value. It fails to take into account the strong double-peaked concentration of dips to the south-west. Double peaks of this type are typical in braided river gravels, in which they are believed to represent deposition on either side of lozenge-shaped bars (see Reading, 1978). These peaks appear to indicate that a river flowing towards the south-west deposited at least some parts of the gravel sequence.

The Mersea Island Gravel was excluded by Wood (1866b) from his original definition of the East Essex Gravel and was mapped as 'Glacial' by the early officers of the Geological Survey (Old Series, Sheet 48SW). However, Dalton (1908) decided that the gravel must be 'Post-glacial', following the discovery of fossiliferous deposits beneath it (see above), and it was subsequently described as a Thames-Medway deposit (Anon., 1913). Wooldridge (1927b), perhaps swayed by the Geological Survey mapping, interpreted the gravel as glacial outwash. The palaeocurrent evidence, which suggests that all or part of the gravel sequence at Cudmore Grove represents a braided river floodplain trending towards the south or south-west, would appear to support its interpretation as outwash. The composition of the gravel, however, provides important evidence precluding a provenance from the north. The deposit bears no resemblance to outwash, even of distal type, and it contains little of the quartzose exotic material that a river from the north or north-east would have reworked from the pre-diversion

Thames gravels of the Tendring Plateau (Bridgland *et al.*, 1988; Table 5.2). Instead, its clast composition implies parity with the Low-level East Essex Gravel deposits to the south of the Blackwater estuary (Table 5.5), which are interpreted as the products of the post-diversion Thames-Medway. Thus the south-westward palaeocurrents from the Cudmore Grove cliffs, if correctly interpreted, must be a localized trend, possibly representative of a braided floodplain that followed a sinuous course across the area of Mersea Island. The gravel is considered to be part of a single, thick aggradational unit, the Mersea Island Gravel, that also includes the much higher remnants at West Mersea and on the eastern side of the Colne estuary at Point Clear (Bridgland, 1983a, 1988a; Figs 5.2 and 5.27).

Correlation

In what appears to be the first record of their existence, Dalton (1908, pp 136–7) compared the Cudmore Grove deposits with those at Clacton: 'the silt much resembles the unweathered condition of the Clacton Postglacial deposit, as found below the beach; it is there full of land and freshwater shells and plant remains, though of estuarine character at a higher level, as seen in the cliffs. Possibly a similar sequence obtains in East Mersea, the estuarine passing down into a lacustrine deposit'.

Dalton's suggestion, made before the complexity of climatic fluctuations during the Pleistocene was widely accepted and based largely on the character of the sediments, is supported by much of the scientific evidence gathered in recent years. Most palaeontological indications of the age of the deposits suggest that they, like the Clacton Channel sequence, are of Hoxnian (*sensu* Swanscombe) age. Amongst the rich vertebrate fauna from Cudmore Grove, the vole *Arvicola cantiana* has a restricted occurrence within the Pleistocene between the late 'Cromerian Complex', where it

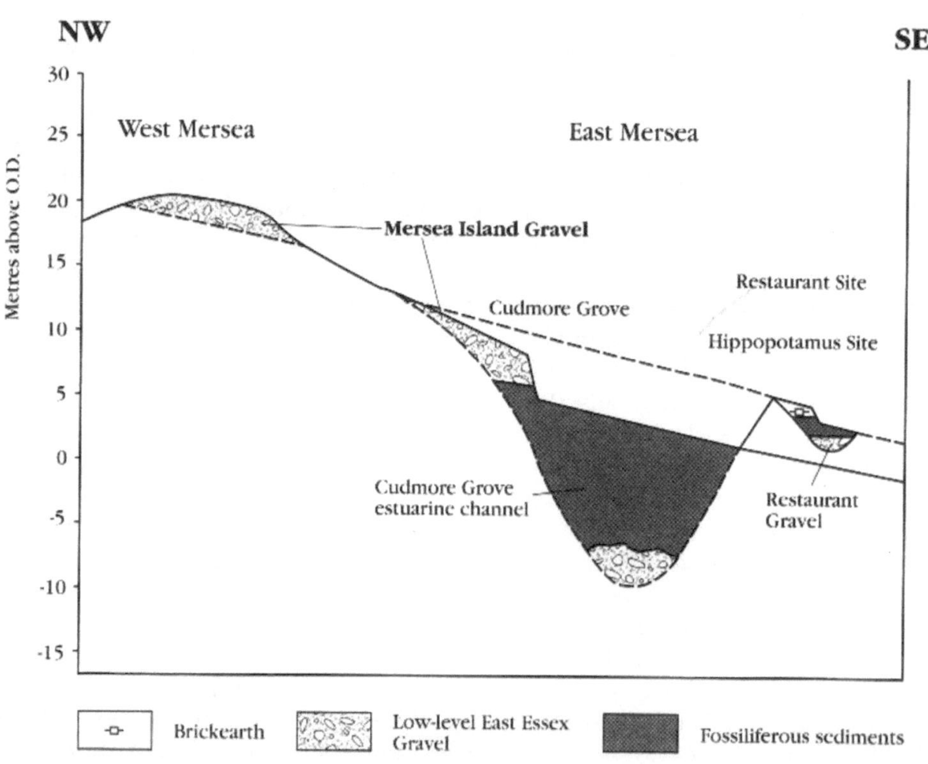

Figure 5.27 Idealized transverse section through Mersea Island (after Bridgland *et al.*, 1988).

replaces the earlier form *Mimomys savini* (see Chapter 1), and the Ipswichian (*sensu* Trafalgar Square), in which it in turn is replaced by the modern water vole *Arvicola terrestris* (Hinton) (Sutcliffe and Kowalski, 1976).

The pollen record from the Cudmore Grove channel sequence is also of particular value for relative dating, in that it resembles those from Hoxnian sites in the general area, such as Marks Tey (Turner, 1970) and Clacton (Pike and Godwin, 1953; Turner and Kerney, 1971; see above, Clacton). The detailed analysis of pollen and spores from Cudmore Grove is as yet unpublished, but preliminary observations have already proved informative (Roe, in Bridgland *et al.*, 1988; in Holman *et al.*, 1990). Of considerable significance is the occurrence of 'Type X', which is generally regarded as an indicator of the Hoxnian Stage (see above, Clacton). Correlation of the Cudmore Grove sequence with the Hoxnian (*sensu* Swanscombe) would support recent palaeogeographical reconstructions of the various Pleistocene deposits in eastern Essex, based on terrace stratigraphy (Bridgland, 1988a; Figs 5.5A and 5.5B). This reconstruction involves a correlation of the Cudmore Grove and the Clacton channels. There are, however, a number of uncertainties to overcome before detailed correlation between the sequences at the two sites can be established.

The estuarine character of the deposits at Cudmore Grove would seem to indicate that they post-date the most fossiliferous part of the Clacton sequence, which is of freshwater origin (see above, Clacton). The Clacton Estuarine Beds were regarded by Warren (1955) as representing a period of declining sea level following a marine transgression. The palynological record from Clacton places this transgression between biozones HoIIIa and IIIb, the Estuarine Beds being ascribed to IIIb (see above, Clacton and Fig. 5.19). The fossiliferous sequence at Cudmore Grove also appears to record a period of declining sea level, as suggested by the molluscan fauna of the detritus mud. The pollen from Cudmore Grove indicates that biozone HoIIIb is also represented there. The fact that deposits of this age occur at a lower elevation at Cudmore Grove than at the Clacton West Cliff site (the transgression is represented at *c.* 3 m in the West Cliff section – see Fig. 5.19) lends support to the suggestion that there was a period of erosion between the deposition of the Freshwater and Estuarine Beds at Clacton (see above, Clacton). Thus the lower part of the estuarine sequence at Cudmore Grove probably has no equivalent in the Clacton West Cliff section, but may correlate with the lower-level estuarine deposits in other parts of the Clacton site, such as Warren's Channel vi at Lion Point (Fig. 5.18).

The basal gravel and the peripheral 'lag' gravel at Cudmore Grove are probably fluviatile, but these have yet to yield any palaeontological evidence and therefore cannot be related to the standard Hoxnian pollen sequence. The flake found in the 'lag gravel' may, however, indicate an upstream continuation of the industry at Clacton, where artefacts are concentrated in the Freshwater Beds. The flake was found *in situ* during sampling for clast-lithological analysis; this involved removing only a very small volume of the deposit, so a rich Palaeolithic content may be indicated. Wymer (1985b) recorded an earlier discovery of a flake at the site. This was found in 1978 'in the top of an orange clay with large flint pebbles', at TM 067144 (Vincent and George, in Wymer, 1985b, p. 258). The grid reference closely coincides with the outcrop of the 'lag gravel' at the western edge of the channel and the description also suggests that this artefact (now in the Passmore Edwards Museum) came from this deposit. Two further flakes, again in fresh condition, have been found on the beach in close proximity to the exposure of the 'lag gravel'. It is likely that these specimens, one found by D. Maddy during the Geologists' Association excursion to East Mersea in May 1987 and the other by P. Spencer in February 1989, were derived from the Cudmore Grove deposits, possibly the basal gravel. Unfortunately, because of the location of these deposits, at (and below) foreshore level beneath actively eroding cliffs, it will be difficult to verify that an important Palaeolithic industry is present.

The reconstruction of post-Anglian palaeodrainage in eastern Essex (Bridgland, 1980, 1983a, 1983b, 1988a; Fig. 5.5A and 5.5B) also supports the correlation of the Cudmore Grove and Clacton sediments. This reconstruction, based on terrace stratigraphy and aided by gravel clast analyses, suggests that the Thames-Medway river only flowed as far north as Mersea Island and Clacton for the period of a single terrace cycle, that during which the Asheldham/Mersea Island/Wigborough Formation was deposited (see Fig. 5.5). Subsequent formations are only found to the south of the Blackwater, their distribution reflecting the progressive south-eastward migration of the river. The ages attributed to these various formations (Tables 1.1 and 5.3) imply that the course across Mersea Island to Clacton persisted, following the Anglian Stage diversion of the Thames, until Oxygen Isotope Stage 10. During the latter stage, rejuvenation to the Barling Formation level coincided with a southward shift in the course of the Thames-Medway. This means that, unless this interpretation is incorrect, all interglacial deposits of post-diversion Thames-Medway origin to the north of the Dengie Peninsula must date from Oxygen Isotope Stage 11. Such deposits, which include the sediments filling parts of the Southend/Asheldham Channel (Fig. 5.5A) as well as those at Cudmore Grove and Clacton, are therefore seen as downstream equivalents of the deposits at Swanscombe (Bridgland, 1983a, 1988a; see Chapter 4).

A view contradictory to this interpretation has, however, been expressed by Currant (1989), in an appraisal of small-mammal faunas in Britain. Currant noted that there are many similarities between the rich assemblage of small mammals at Cudmore Grove and that from Grays in the Lower Thames (see Chapter 4, Globe Pit). The shrew *Crocidura*, several specimens of which have been recovered from the 'detritus mud' (bed 3) at Cudmore Grove, is a very rare element of the British Pleistocene mammalian fauna, but it has also been recorded from the Orsett Road section at Grays (Hinton, 1901; Bridgland *et al.*, 1988). An abundance of water shrew (*Neomys fodiens*) is another rare phenomenon, but this species is also well-represented at Grays (Hinton, 1911). Currant placed both Cudmore Grove and Grays within his Group 2 assemblages, along with Aveley (see Chapter 4). He regarded this group as intermediate in age between Hoxnian stage assemblages (his 'Group 3'), in which he included Swanscombe and Clacton, and the last interglacial '*Hippopotamus* fauna', his 'Group 1'. There is a recognizable weakness in Currant's groupings, however; this stems from the fact that the assemblages he discusses have very different levels of richness. Thus the Grays and Cudmore Grove faunas are both from prolific sites that have yielded large numbers of specimens. The assemblages from Clacton and Swanscombe, differentiated from Cudmore Grove by Currant, are sparse by comparison. It is possible, therefore, that the strong similarities between the Grays and Cudmore Grove assemblages, and their distinction from Currant's 'Group 3 assemblages', merely result from the relative richness of these two faunas and are of no stratigraphical significance.

The interpretation of the Lower Thames sequence proposed in this volume (Chapter 4) also fails to conform with Currant's groupings. Sites from two aggradational formations (the Corbets Tey and Mucking Formations) appear within his Group 2 assemblages, representing, according to the evidence presented in Chapter 4, warm Oxygen Isotope stages 9 and 7. This suggests that evidence from small-mammal assemblages (in common with other biostratigraphical evidence – see Chapter 4) does not allow at present a distinction to be made between faunas from Stages 9 and 7. It therefore seems wise to give precedence to the evidence from terrace stratigraphy and palaeodrainage reconstruction, which provide the most reliable framework for interpreting the fluvial record in the lower reaches of the Thames Basin and for correlation with the deep-sea record (see Chapter 1 and Table 1.1). In the absence, thus far, of a complete analysis of the Mollusca and pollen from Cudmore Grove, or of amino acid ratios from the former, the principal evidence for correlation with other sites comes from the reconstruction of terrace formations in eastern Essex.

Conclusions

Evidence from the sequence at Cudmore Grove, only revealed by coastal erosion during the last decade, shows this to be an extremely important Pleistocene locality. The site yields a wealth of palaeontological information that is of great importance for reconstructing the contemporary environment and for dating the deposits. This evidence indicates warm (interglacial) conditions and a Middle Pleistocene age. The plant and animal remains in the Cudmore Grove channel sediments provide a picture of life during this interglacial period. A huge range of fossils are included – molluscs, ostracods (microscopic crustaceans), pollen, other plant fossils, mammals (including monkey, beaver, bear, bat and extinct voles), reptiles, birds, amphibians and fish. In fact the fauna of small vertebrates is considered to be the richest ever found in Britain.

Pollen preserved in the fine-grained sediments suggests a correlation with the nearby (Hoxnian Stage) fossiliferous sediments at Clacton. The latter suggestion conforms with reconstructions of the course taken by the Thames-Medway after the Thames was diverted by ice during the most severe Middle Pleistocene glaciation. This course took the river through both East Mersea and Clacton. The Cudmore Grove sediments are thus interpreted as part of an interglacial channel-fill that can be recognized widely within the lower reaches of the Thames system. This channel-fill, which includes the well-known fossiliferous and artefact-bearing deposits at Swanscombe as well as at Clacton, is believed to have been laid down during Oxygen Isotope Stage 11 (Hoxnian *sensu* Swanscombe), *c.* 400,000 years ago.

SOUTHMINSTER, GOLDSANDS ROAD PIT (TQ 961991)

D.R. Bridgland

Highlights

This pit exposes typical gravel of the post-diversion Thames-Medway, deposited by the Thames, downstream from its confluence with the Medway, as it flowed north-eastwards across this part of Essex. This stretch of the river's course was formerly part of the Medway valley but was adopted by the Thames upon its diversion. The Asheldham Gravel at Southminster is believed to equate with both the Black Park Gravel and the Boyn Hill Gravel of the Middle Thames (the former underlying the latter in the Southminster area), the first two formations to be deposited by the river after its diversion during the Anglian Stage.

Introduction

Exposures at Goldsands Road Pit, Southminster, reveal the Asheldham Gravel, the highest and oldest formation of the Low-level East Essex Gravel Subgroup, which is attributed to the post-diversion Thames-Medway (Bridgland, 1983a, 1983b, 1988a; Figs 5.5B and 5.15). This formation is broadly equivalent to both the Southminster and Asheldham Terrace gravels of Gruhn *et al.* (1974) and to the '3rd Terrace' of the Geological Survey (Lake *et al.*, 1977, 1986).

Southminster lies in the south-eastern corner of the Dengie Peninsula, which separates the Crouch and Blackwater estuaries. The Asheldham Gravel is the most extensively preserved terrace formation on this peninsula, largely because later Thames-Medway deposition seems to have been confined, onshore, to the area further south. Higher deposits to the west belong to the High-level East Essex Gravel Subgroup, the product of the tributary Medway system (Figs 5.15 and 5.16; see above, Introduction to Part 2). The Asheldham Gravel forms an almost continuous sheet, averaging approximately 1 km in width, between Burnham-on-Crouch and Bradwell, dissected only by small streams flowing eastwards into the North Sea (Fig. 5.15). Geological Survey borehole data and subdrift contour mapping (Lake *et al.*, 1977; Simmons, 1978) show that over much of the area the Asheldham Gravel overlies a substantial buried channel, the Asheldham Channel (Fig. 5.28). Boreholes have revealed that this channel contains localized fossiliferous sediments, but scientific evaluation of these has only recently been undertaken (H.M. Roe, pers. comm.) and results have yet to be published.

Description

The exposures in the Goldsands Road Pit show mainly matrix-supported, massive and cross-stratified sandy gravel, interbedded with sands and clayey sands (Fig. 5.29). These sediments essentially comprise upper and lower gravel units, separated by ripple-drift-laminated and cross-bedded sands (Bridgland, 1983a, 1983b). Palaeocurrent measurements from the sands indicate flow to the east-north-east. In a large part of the pit the sands and gravels are overlain by 1.5 m of silty clay (brickearth), containing scattered pebbles. This may be either a flood-plain (overbank) deposit or a colluvial accumulation and, except for the pebbles, has an appearance similar to weathered London Clay. These deposits, with the exception of the clay, are typical of the products of a braided-river environment (Miall, 1977), the range of sediment types suggesting deposition on

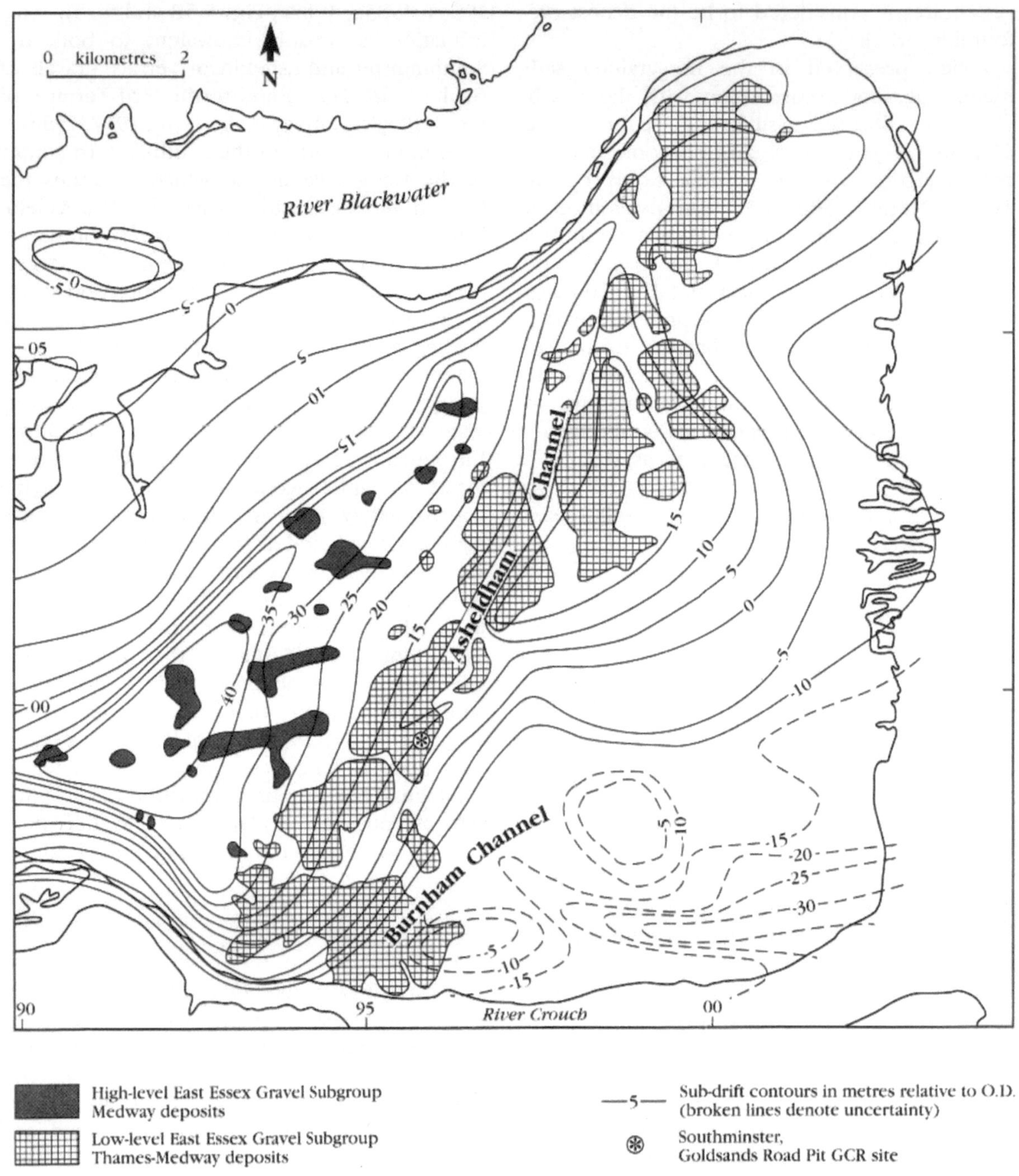

Figure 5.28 Map showing the outcrop of the Asheldham Gravel and bedrock surface contours, revealing the form of the Asheldham Channel (modified from Lake *et al.*, 1977).

longitudinal and linguoid bars. The original land surface in the vicinity of the pit was probably between 20 m and 21 m O.D., rising north-westwards to 25 m O.D. in the middle of Southminster (TQ 956998), well within the mapped range of the Asheldham Gravel.

A total thickness of over 4.5 m of Pleistocene sediments overlies London Clay at 15.5 m O.D. in the GCR site. Gruhn and Bryan, who worked here at a time of more extensive quarrying, reported a sloping 'bench' beneath the gravel of the area, ranging between 9.7 and 10.7 m O.D.,

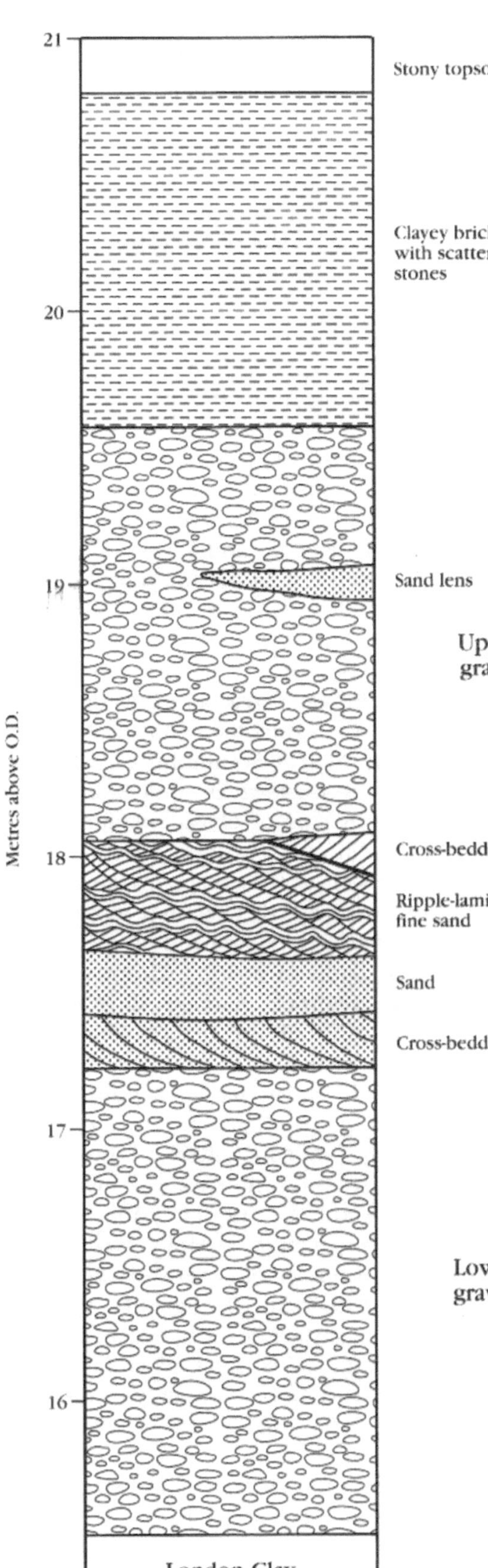

with the highest bedrock level in the north-east (Gruhn *et al.*, 1974, unpublished appendix). Three Geological Survey boreholes in the vicinity add to the general picture of variable relief. A borehole to the north-west of Goldsands Road Pit showed 3.8 m of sandy, silty clay and soil, overlying 2.4 m of gravel, the London Clay being reached at 14.4 m O.D. Another, 350–400 m to the south-west, showed only 3 m of Pleistocene sediments overlying the London Clay at 16.3 m O.D. The third borehole, near Newmoor (TL 9964 0035), to the north of Southminster, revealed 10.5 m of gravel overlying the London Clay at 10.8 m O.D. (Simmons, 1978). This bedrock surface information suggests that the deposits fill a channel eroded into the London Clay, the eastern side of which appears to be preserved beneath the eastern edge of the Asheldham Gravel outcrop (Fig. 5.28). This contradicts the subdrift contour maps published by the Geological Survey (Lake *et al.*, 1977; Simmons, 1978), which suggest that the eastern side of the channel has been removed by later erosion.

Interpretation

Clast-lithological analysis of the Asheldham Gravel at Southminster reveals the combination of local, southern and exotic lithologies that characterizes the Low-level East Essex Gravel Subgroup (Bridgland, 1983a, 1983b, 1988a; Table 5.5). In comparison with the High-level East Essex Gravel deposits, which lie to the west and at a greater elevation, the Asheldham Gravel contains rather more local material (84–91%) and significantly less southern material (7.8–13.5%). The occurrence of a number of pebbles of Ightham Stone and Hastings Beds lithologies (Bridgland, 1983a, 1986b) indicates a continued Medway influence, these being derived from north Kent and the central Weald respectively. Most significant, however, is the appearance of an exotic component (1.4–2.25%) in the Asheldham Gravel, including all the types characteristic of the Lower Thames gravels of the Tilbury area (see Chapter 4): predominantly

Figure 5.29 Section at Goldsands Pit. This shows the division into upper and lower gravels, separated by cross-bedded and ripple-laminated sands.

quartz, quartzites, Carboniferous chert and *Rhaxella* chert. The appearance of this characteristic exotic suite in the Asheldham Gravel marks the initiation of Thames drainage in this part of eastern Essex, which resulted from the Anglian diversion of the river (see above, Introduction to Part 2).

The buried channel underlying the Asheldham Gravel, revealed by various bedrock surface data, is part of a complex feature recognized by Lake *et al.* (1977) as their 'Burnham Buried Channel'. However, the feature they described incorporates a deep channel, eroded to well below ordnance datum, beneath 'First Crouch Terrace' (Barling/Dammer Wick Gravel) deposits in the Burnham-on-Crouch area (see Figs 5.15 and 5.28). This deep channel is here attributed to later incision, which has partly dissected the earlier deposits. Although the name Burnham Channel has previously been used for the older channel beneath the Asheldham Gravel (Bridgland, 1983a, 1983b), it is now restricted to the later, deeply incised feature. The Burnham Channel has been interpreted as a downstream continuation of the Shoeburyness Channel of the Southend area (Bridgland, 1988a; Fig. 5.5C). Little is known of the sediments filling this feature, but the bivalve *Corbicula fluminalis* is known both from Shoeburyness (Whitaker, 1889; Kennard and Woodward, 1907) and from a borehole at East Wick, Burnham-on-Crouch (Warren, 1951), possibly from the channel(s) in question. The occurrence of this bivalve might be an indication of deposition in a temperate episode pre-dating the Ipswichian Stage (*sensu* Trafalgar Square); in other words, pre-Oxygen Isotope Stage 5 (see Chapter 2, Stanton Harcourt and Magdalen Grove). A correlation between the Shoeburyness/Burnham Channel Deposits and the Mucking Formation of the Lower Thames was suggested by Bridgland (1988a). However, following a recent revision of terrace correlation between the Lower Thames and the Southend area (Bridgland *et al.*, 1993), the Shoeburyness/Burnham Channel is thought to be a downstream equivalent of the interglacial Thames channel recognized within the Corbets Tey Formation in the Ockendon–Purfleet area, which suggests correlation with the Oxygen Isotope Stage 9 temperate episode (see Chapter 4, Purfleet; Tables 1.1 and 4.1).

The older channel, underlying the Asheldham Gravel, has been redefined as the Asheldham Channel (Bridgland, 1988a). The deposits filling this channel comprise a basal gravel (Asheldham Channel Gravel) and an overlying sequence of fine-grained, fossiliferous deposits (Asheldham Channel interglacial deposits). The type locality for these units (and for the Asheldham Gravel) is a gravel pit at Asheldham (TQ 971917), which overlies the central part of the channel (Bridgland, 1983a). Again, little is known about the fossiliferous channel sediments; clays, silts and sands recorded in boreholes near Bradwell at TL 9872 0581 and 9971 0657 revealed silty deposits with 'shell and reed beds' and 'carbonaceous material' respectively (Simmons, 1978). The distribution and elevation of these channel deposits strongly suggest correlation with those to the north of the Blackwater, at Cudmore Grove and Clacton (Bridgland, 1988a; Figs 1.3 and 5.5A).

The deposits to the west of Southminster and at Asheldham, which, at up to 25 m O.D., are the highest within the Asheldham Gravel (Bridgland, 1983a), were assigned by Gruhn *et al.* to their Asheldham Terrace. Other deposits now included in the Asheldham Gravel, those at Burnham, to the east of Southminster, west of Tillingham and south-west of Bradwell, have considerably lower surface elevations and were included by Gruhn *et al.* in their Southminster Terrace. Their long-profile diagram (Gruhn *et al.*, 1974, fig. 10) showed these two terraces as vertically overlapping aggradations with an altitudinal separation of 3–5 m. However, the more detailed bedrock surface information available as a result of the recent Geological Survey borehole programme indicates that the differences in bedrock surface level, interpreted by Gruhn *et al.* as evidence for two distinct terraces, merely reflect different positions relative to the cross-profile of the Asheldham Channel. Indeed, the upper part of the Asheldham Gravel extends laterally away from the channel in a number of areas and overlies a separate, higher 'bench' (Bridgland, 1983a). The difference in surface level between gravels assigned by Gruhn *et al.* to their Asheldham and Southminster Terraces probably results, therefore, from differential erosion. Thus the deposits underlying both terraces can be variously reinterpreted as Asheldham Gravel or Asheldham Channel Gravel (Fig. 5.15). It is apparent that the Asheldham Terrace represents the maximum aggradational level of the Asheldham Formation, whereas areas attributed by Gruhn *et al.* to their Southminster Terrace have been lowered by later erosion; the latter term should no longer be used, therefore.

Consideration of its elevation suggests that the lower part of the sequence observed at Southminster may belong to the Asheldham Channel Gravel rather than the Asheldham Gravel. The interglacial sediments that separate the Asheldham Channel Gravel and the Asheldham Gravel are not present throughout the area; where they are absent, distinction between the two gravel units is extremely difficult. Fine-grained sediments occur between lower and upper gravels at Goldsands Road Pit, but it is impossible to ascertain whether these occupy the stratigraphical position of the Asheldham Channel interglacial deposits. There is nothing in the clast composition of samples collected from the lower and upper gravels (Table 5.5) to support any distinction between the units on this basis, but no lithological separation is generally possible between the various formations of the Low-level East Essex Gravel Subgroup (Bridgland, 1983a, 1988a).

A further piece of evidence of possible relevance to the identity of the lower gravel unit is the discovery in it of the butt-half of a rolled hand-axe by P. Harding (Bridgland, 1983a, p. 227; Wymer, 1985b). Two such broken artefacts were in fact discovered at the site during the cleaning of the sections for the visit of the Quaternary Research Association in April 1983, a broken point of a hand-axe, less rolled, being recovered from the upper gravel on the same occasion. The occurrence of hand-axes (Acheulian Industry) in the Asheldham Gravel is no surprise; numerous examples are recorded from its upstream equivalents, the Southchurch Gravel of the Southend area (Bridgland, 1983a; Wymer, 1985b), the Orsett Heath Gravel of the Lower Thames (Chapter 4) and the Boyn Hill Gravel of the Middle Thames (Chapter 3). The Asheldham Channel Gravel, on the other hand, is believed to correlate (Bridgland, 1988a; Table 1.1 and Fig. 1.3) with the Swanscombe Lower Gravel and the basal gravel of the Clacton Channel (Lower Freshwater Beds), both of which contain abundant Clactonian artefacts but no hand-axes (excluding a very few, possibly unreliable records – see above, Clacton). The Asheldham Channel Gravel has itself yielded a small assemblage of Clactonian artefacts, collected by Warren (1933) from a site at Burnham-on-Crouch (Wymer, 1985b; Bridgland, 1988a). The apparent association of Clactonian material with the pre-interglacial and early interglacial parts of the Asheldham Formation (phase 2 and early phase 3 of the climatic terrace model – see Chapter 1) may indicate that the lower gravel at Southminster, which has yielded a hand-axe, post-dates the interglacial. However, there is good evidence to suggest that hand-axe makers occupied the Thames valley prior to the Clactonian occupation represented by the Swanscombe and Clacton industries, as has been discussed above (see Clacton). Three rolled, probable hand-axe finishing flakes were, in fact, found amongst the material collected in the second golf course excavation at Clacton (Singer *et al.*, 1973; Wymer, 1985b), implying that derived Acheulian material is to be expected in the Asheldham/Clacton Channel Gravel. The hand-axes from Southminster may therefore have no stratigraphical significance, other than indicating that the gravel post-dates the earliest occupation of southern Britain by Palaeolithic Man.

The correlations proposed in this volume, based on terrace stratigraphy, imply that aggradation of the Asheldham Formation spanned the period from the late Anglian (late Oxygen Isotope Stage 12) to early Oxygen Isotope Stage 10, when rejuvenation to the level of the Barling Formation occurred (Chapter 1). The Asheldham Formation and its upstream correlative in the Lower Thames, the Orsett Heath Formation, are considered to correlate with the Boyn Hill Formation of the Middle Thames (see Chapter 4). They are also believed to incorporate, in their lower parts, downstream equivalents of the late Anglian Black Park Gravel of the Middle Thames, the earliest post-diversion formation, which appears to have been graded to a very low base level (see Chapter 4, Wansunt Pit). It must be emphasized that the degree of complexity implied by this interpretation is indicated by regional stratigraphical evidence (summarized in Table 1.1) and cannot, as yet, be determined from the sediments of the Asheldham Formation at Southminster or elsewhere.

Conclusions

Fluvial gravels occurring at this locality contain a mixture of rocks from Kent, to the south, and from the north-west, carried down the main Thames valley. This is because they were deposited by the combined Thames-Medway river, formed by the confluence of the Medway and the Thames in the area south of Southend. Older deposits in the Southminster area show that this part of Essex was formerly in the Medway valley, at a time when that river extended from Kent to the Clacton area, where it joined the old (pre-diversion) Thames. When diverted, the Thames adopted the old Medway valley between Southend and Clacton, depositing gravels of the type found at Southminster. The GCR site at Goldsands Road Pit provides exposures in the Asheldham Gravel and, possibly, in the Asheldham Channel Gravel. The study of these deposits is of considerable importance in reconstructing the evolution of the river system in this area during the Middle Pleistocene. This area of eastern Essex provides an important link between the Lower Thames sequence, with its abundance of fossiliferous and Palaeolithic sites, and the Tendring Plateau, where a comparable wealth of information also exists.

Part 3:

DEPOSITS OF LOCAL RIVERS

D.R. Bridgland

Introduction

It has been shown earlier in this chapter that a large part of Essex was, at different times during the Early and Middle Pleistocene, drained by the lower reaches of the Thames. In the Early Pleistocene the Thames flowed across the north-western part of the county towards Suffolk, and much of Essex was drained northwards towards it (although not necessarily directly into it), the main north-flowing river being the Medway. By the Middle Pleistocene the south-eastward migration of the Thames, possibly aided by diversion or capture by a Medway tributary (Bridgland, 1988a), had resulted in a north-eastward course for the river across Essex to the Tendring Plateau, where it was joined by the Medway (see Part 1 of this chapter). During the Anglian Stage the Thames was diverted, by way of its present valley through London, into the Medway, thus bringing into being the Thames drainage of south-western and eastern Essex. Subsequent migration caused the river to abandon the north-eastern part of this course, although the northward trend is still represented in the off-shore continuation of the Late Pleistocene Thames-Medway valley (D'Olier, 1975; Bridgland and D'Olier, 1989; Fig. 5.5F).

In the latter part of the Pleistocene, as the Thames moved further towards the south and east, new tributary streams were initiated, draining the areas once occupied by the main river. In particular, the Colne appears to have formed in the old beheaded valley of the Kesgrave (Lower St Osyth Gravel) Thames; this former river is presumably the 'misfit' remnant of the pre-diversion Thames itself (see Fig. 5.5B). The Chelmer and Blackwater also appear to drain parts of the old Thames valley in central Essex (Fig. 5.4E), the latter flowing in the opposite direction to the Thames. This part of the old valley was apparently modified by glacial activity between its occupation by the Thames and by the later rivers, since considerable thicknesses of till occur in overdeepened sections of it (Bristow, 1985).

The tributary rivers of the northern and western parts of Essex thus appear to have been initiated immediately following the Anglian glaciation. It is therefore not surprising that they have extensive terrace systems of their own (see Geological Survey Sheet 241, Chelmsford), although these have received comparatively little attention from geologists. A number of important Pleistocene sites have come to light within these terrace systems over the years, but it has not always been easy to distinguish the deposits of these rivers from those of the Thames; there has, for example, been uncertainty about whether the deposits in the Clacton area are the products of the Thames or the local River Colne (see above, Clacton). The richest source of hand-axes in Essex, a gravel at Upper Dovercourt (TM 240313) (Underwood, 1913; Warren, 1933; Wymer, 1985b), appears to be a Stour terrace deposit banked against the much earlier Oakley Gravel, of (pre-diversion) Thames-Medway origin (Bridgland *et al.*, 1990). Most of the Palaeolithic discoveries in the Chelmsford, Maldon and Colchester areas, carefully catalogued by Wymer (1985b), are probably from the terraces of the Chelmer, Blackwater and Colne. Occasionally collections of mammalian bones have also been made from these deposits (Wymer, 1985b). There are a few sites in areas where the gravels are generally believed to be pre-diversion (Kesgrave Group) Thames deposits, but where important fossil or Palaeolithic discoveries raise doubts about this interpretation. One such is near Thorpe-le-Soken (Daking's Pit – TM 155233), where a rich assemblage of artefacts has been recovered from deposits mapped as part of the Cooks Green Gravel (Warren, 1933; Oakley and Leakey, 1937; Wymer, 1985b). A recent reinvestigation has confirmed the presence of abundant worked flakes, but has also shown the gravel to be perceptibly different to the local Cooks Green Formation. In particular, it contains ***Rhaxella*** chert, which is very rare in the Kesgrave Group gravels upstream of the Crag basin (see Part 1 of this chapter), but is present in the Red Crag and in Anglian Stage glacial deposits. This site lies in the valley of the Holland Brook, which suggests that the gravel may be a post-Anglian deposit laid down by that river.

Three GCR sites are included in this part of the chapter, covering very different areas of interest, although all three are associated with the River Blackwater. The first, Maldon Railway Cutting, is an important site for stratigraphical evaluation of the deposits of the Anglian glaciation, since it is the type locality of the controversial Maldon Till. The other two sites are of Late Pleistocene age, dating from the last interglacial/glacial cycle; they therefore represent a part of the Pleistocene for which no Thames deposits are known in Essex (if the various Lower Thames interglacial sites are correctly interpreted as pre-Ipswichian in Chapter 4). The Ipswichian Stage (*sensu* Trafalgar Square) is represented by hippopotamus-bearing deposits at East Mersea, while a site in a low terrace of the Blackwater at Great Totham has yielded an abundance of palaeontological data that suggests deposition during the Devensian Stage.

MALDON RAILWAY CUTTING (TQ 842067)

D.R. Bridgland

Highlights

This is the type locality of the Maldon Till, deposited here during the Anglian Stage at a position close to the maximum extent of Lowestoft glaciation. The till at Maldon has previously been interpreted as stratigraphically earlier than the main Lowestoft Till, separated from the latter by an intermediate glacial gravel. Gravels overlying the till at Maldon were hitherto regarded as part of this intermediate deposit, but are now interpreted as fluvial terrace sediments. These are taken to be the product of the Blackwater–Chelmer river system, which came into being, following the Anglian glaciation, as the new drainage of the area once occupied by the Thames.

Introduction

Maldon Railway Cutting is of importance to Pleistocene studies as the type locality of the Maldon Till, which has been claimed to represent an early ice advance into southern East Anglia. It was formerly held that this represented the second of three separate glacial advances into the Chelmsford area (Clayton, 1957, 1960). However, following the recognition that glaciation occurred in Essex only during the Anglian Stage (Turner, 1970; Baker, 1971; Bristow and Cox, 1973; Perrin *et al.*, 1973), the Maldon Till was later attributed to the earlier of two advances of the Lowestoft Till ice into the region (Baker, 1971, 1983; Ambrose, 1973; Baker and Jones, 1980; Bristow, 1985). These various interpretations of the evidence at Maldon were based largely on the original description of the railway cutting sections by Whitaker (1889), the exposures having been obscured since that date. However, till was mapped at the locality in the late 1960s by the Geological Survey (Ambrose, 1973; Sheet 241).

A recent re-excavation of the site, soon to be part of the Maldon by-pass road scheme (the railway closed many years ago), confirmed the presence of till beneath coarse gravel. However, analysis of the clast composition of the gravel (Table 5.5) and the discovery of a hand-axe, apparently from the deposit, suggest that it is not glacial outwash, as was previously assumed. This reappraisal has undermined previous interpretations of Anglian glacial stratigraphy, which held the Maldon Till to be earlier than the main Anglian ice advance into the region. The till at Maldon has subsequently been interpreted as an isolated outlier of the widespread sheet of Lowestoft Till that covers much of central and north-western Essex (Whiteman, 1987; in Allen *et al.*, 1991). The Maldon outlier, in fact, falls marginally outside most published reconstructions of the Anglian ice limit (see, for example, Rose, 1983b; Bowen *et al.*, 1986a). It is one of several small till remnants along the line of the Danbury–Tiptree Ridge, regarded by many authors as the maximum south-eastward extent of the Lowestoft Till (see Geological Survey, New Series Sheet 241; Bristow, 1985).

Description

Whitaker (1889) described sections created in August 1887 at Maldon, during the construction of the railway to Wickford. Whitaker recorded detailed variations that he observed at various points along the cutting. He also listed the following generalized succession (Whitaker, 1889, p. 317):

(e) 'Gravel. ?At one place becoming a gravelly loam. Overlying, or ?locally replaced by

(d) Brown bedded loam and sand, with gravelly layers, beneath which there occurs, also locally

(c) Bedded gravel and loam.

(b) Grey boulder clay, or stony loam, with, at the base,

(a) Irregular gravelly bed.'

London Clay was exposed beneath the glacial deposits and also occurred as a lenticular mass up to 2.5 m thick within the Pleistocene sequence. Whitaker regarded the latter as a 'boulder', although emplacement as a result of diapirism or glaciotectonic deformation may also be envisaged.

In March 1984 two small sections were reopened (as part of the Geological Conservation Review) in the side of the disused cutting by P. Allen, C.A. Whiteman and the author, located in the steep face of an old landslip-scar. The sections broadly confirmed Whitaker's observations, revealing the following sequence (see Fig. 5.30):

	Thickness
4. Sand and silt, poorly exposed, to land surface (39 m O.D.)	4.0 m
3. Sandy silt, with gravel stringers and sand lenses	1.0 m
2. Gravel, silty and poorly bedded	2.3 m
1. Chalky till, sandy, fresh grey (weathered brown near top)	>1.0 m

London Clay (seen lower on the cutting side)

The basal gravel described by Whitaker (see above, a) was not seen; this does not necessarily indicate its absence, since the base of the till could not be exposed because of waterlogging. The till itself contains conspicuous Chalk clasts, ensuring that it is readily distinguished from the London Clay bedrock. The deposit appeared unusually well-bedded in the GCR section, but was otherwise of typical Lowestoft Till appearance (Whiteman, 1990).

The gravel (bed 2) was found to be similar, in terms of clast composition (Table 5.5), to the Kesgrave Group Thames deposits that cover much of central Essex (see Part 1 of this chapter and Table 5.2). Material of the type associated with Anglian glacial deposits, such as *Rhaxella* chert (Bridgland, 1986b), is only present in small quantities. The deposit comprises only durable clasts, indicating that it is not an ice-proximal outwash gravel. A hand-axe was discovered whilst removing talus from the section. This is a rolled and patinated specimen (Fig. 5.31), which, judging from its condition and its location in pebbly talus, almost certainly came from the gravel.

The uppermost 4 m of the sequence (bed 4) was observed in a narrow trench cut in the sloping cutting-side above the face illustrated in Fig. 5.30. This sandy silty deposit, which has a reddish brown colour, may equate with Whitaker's brown loam and sand (above, d), which he described as locally replacing the upper part of the gravel.

Interpretation

The interpretation of the till at Maldon as a 'lower boulder clay' dates back to the original description by Whitaker, who suggested that it correlated with similar deposits occurring beneath 'glacial gravel' in Suffolk and Norfolk (Whitaker, 1889, pp. 299 and 316). The site achieved the status of a type section three-quarters of a century later, as a result of the work of Clayton (1957, 1960, 1964), despite having been obscured by talus and vegetation throughout the intervening period. Clayton (1957) recognized three separate tills in central Essex, (1) an older, dissected 'Hanningfield Till', confined to high ground, (2) a lower till within a 'sandwich' of deposits filling valleys, his 'Maldon Till' and (3) a later 'Springfield Till', forming the upper leaf of the sequence in the valleys and separated from the Maldon Till by gravel. This last deposit, which he termed the 'Chelmsford Gravels', was interpreted as glacial outwash. Clayton (1957) suggested a correlation between the Hanningfield Till and the continental Elsterian (= Anglian) glaciation and between the later tripartite sequence (Maldon Till, Chelmsford Gravels and Springfield Till) and the Saalian Stage. Within a few years Clayton had modified his views, suggesting that the Hanningfield and Maldon Tills were both Anglian, with the Chelmsford Gravels representing the Hoxnian and the Springfield Till representing the Saalian (Clayton, 1960).

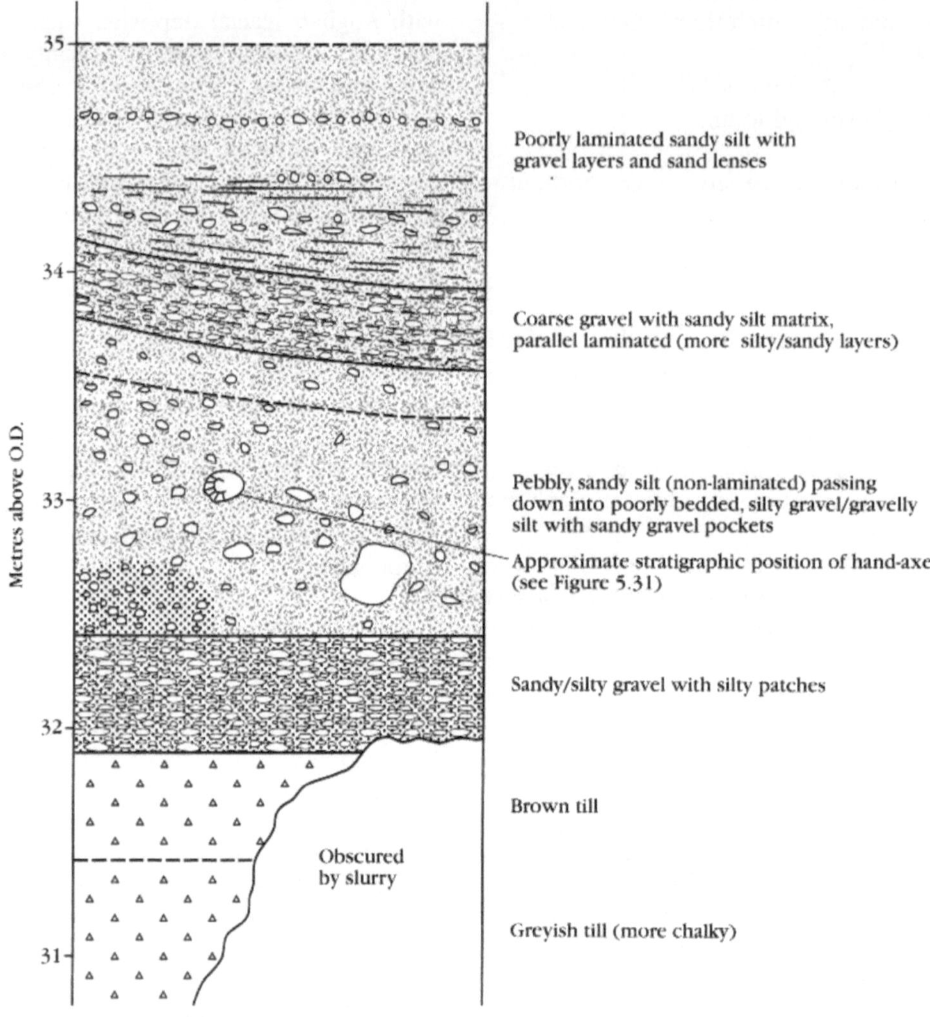

Figure 5.30 Section excavated at the Maldon GCR site in March 1984. Beds 1–3 are illustrated (see Description).

Later workers demonstrated that no distinction could be made between the till occurring on plateaux and that in valleys, leading to the conclusion that only a single glaciation has occurred in southern East Anglia (Turner, 1970; Baker, 1971; Bristow and Cox, 1973; Perrin *et al.*, 1973), during the Anglian Stage (Turner, 1973). The Maldon Till was therefore attributed to an early advance of the Anglian ice and the Chelmsford Gravels were, once again, regarded as outwash (Baker and Jones, 1980; Baker, 1983).

Bristow (1985) expressed doubt as to whether the till described at Maldon was an *in situ* glacial deposit, observing that other records of a lower till in the Chelmsford area could instead be interpreted as London Clay, glacial lake deposits or colluvium. A lower till was found during recent Geological Survey mapping, however, in the Witham area (Bristow, 1985). The name Maldon Till was applied to this deposit, despite reservations about the type locality, which had not been seen in section since the construction of the railway.

Wooldridge (1957) had suggested a correlation between the Maldon Till of Clayton and the till at Hornchurch, which he believed to represent the glaciation that diverted the Thames into its modern valley (see Chapter 4, Hornchurch). This correlation was later supported by Clayton (1960, 1964). A summary of the progression of views on the glacial history of southern Essex was provided by Baker and Jones (1980). They suggested that the Maldon

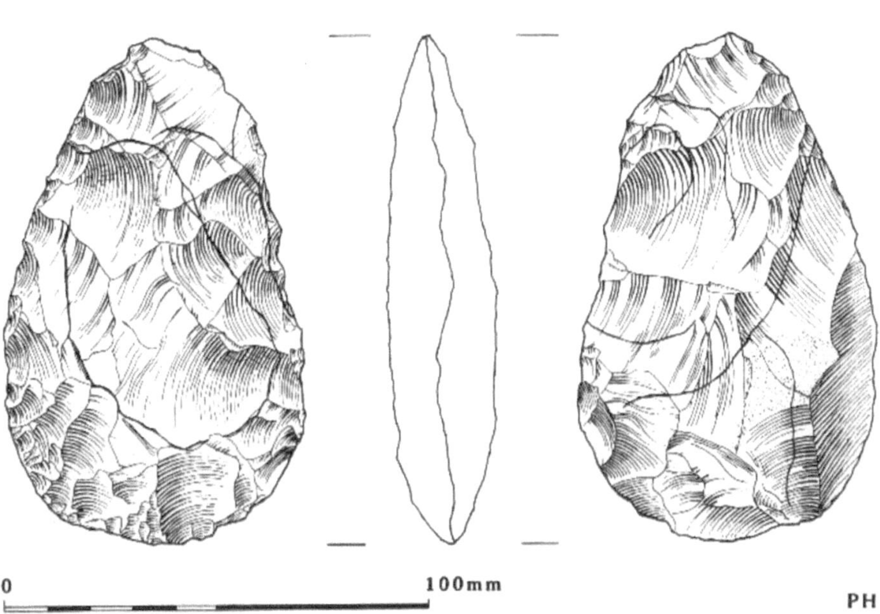

Figure 5.31 Flint hand-axe from Maldon Railway Cutting, found during the GCR excavations. The artefact is a cordate hand-axe of Wymer's (1968) type J, having a symmetrical shape with a cutting edge around the entire circumference. The implement has a white surface patina and is slightly rolled, although with rather more severe damage of the edges. Both sides show a network of incipient thermal fractures. (Drawing and description by P. Harding).

Till could also be correlated with the Ware Till of the Vale of St Albans (see Chapter 3) and that the early Anglian glacial advance responsible for this deposit was also responsible for the diversion of the Thames from its valley across central Essex. They envisaged that a temporary route was used by the river, carrying it from the Vale of St Albans by way of the modern Lower Lea valley into its modern course downstream of London (see Chapter 3, Part 2), a hypothesis supported by Cheshire (1981) and Baker (1983). Baker and Jones (1980) cited evidence from proglacial lake deposits in the Newport area of north-west Essex for the duration of the interval between lower and upper Anglian tills. In that area they recognized a lower, Quendon Till, separated from the main Lowestoft Till sheet. They correlated this lower till with the Maldon Till of south-east Essex and the Ware Till of Hertfordshire. The counting of supposed annual varves in the above-mentioned lacustrine deposits suggested to Baker and Jones that the Quendon/Maldon Till ice advance and the readvance that resulted in the accumulation of the main Lowestoft Till were separated by an interval of 5400 years. During this interval the ice-front apparently stabilized just to the north of Newport.

The most recent re-evaluation of the till at Maldon is that by Whiteman (in Allen *et al.*, 1991), who, following the 1984 re-excavation, equated it with his Newney Green Member of the Lowestoft Till Formation. Although the Newney Green Member is the lower of two divisions of the Lowestoft Till recognized by Whiteman in central Essex, he considered it to be part of the single till sheet that covers southern East Anglia. Whiteman found no evidence in central Essex for the tripartite sequence (Maldon Till–Chelmsford Gravels–Springfield Till) proposed by Clayton. He pointed out that in many instances the deposits ascribed to the middle part of this sequence, the Chelmsford Gravels, are in fact occurrences of Kesgrave Sands and Gravels and therefore pre-date the glacial sequence altogether (Whiteman, in Allen *et al.*, 1991). Whiteman's work supports the correlation between the tills at Maldon and Hornchurch, since he also assigned the latter to his Newney Green Member. Cheshire (1986a) suggested that the till exposed in 1984 at Maldon might have been geli-flucted, which could account for its stratified appearance.

The recent reinvestigation of the Maldon site shows that the interpretation of the till there as the lowest part of a tripartite sequence, underlying an outwash gravel, is unfounded. The gravel overlying the till at Maldon is not an outwash deposit; it contains none of the non-durable material that characterizes proximal outwash gravels (Bridgland, 1986b) and a distal outwash origin appears also to be precluded by the paucity of 'Anglian erratics' such as *Rhaxella* chert (Table 5.5). The gravel contains similar proportions of local and exotic material to the Kesgrave Sands and Gravels of central Essex, from which it is presumably largely derived. In this it resembles Blackwater terrace gravels at Tollesbury (Bridgland, 1983a) and Great Totham (see below). Thus it is likely that the gravel at Maldon is a terrace deposit within the Blackwater/Chelmer system. The occurrence of a hand-axe in this gravel at Maldon would appear to support this interpretation; although such palaeoliths are no longer regarded as indicative of a post-Anglian age (see Chapters 1 and 3), they are unknown from outwash gravels in this area.

The elevation of the gravel at Maldon, with a base level of 32 m and a maximum surface height of 39 m O.D. (the upper part is replaced by sands and silts in the section recorded here), suggests a correlation with the Tollesbury Gravel (type locality: TL 947106) of Bridgland (1983a). The latter formation, aggraded to *c.* 26 m O.D. in its type area, was correlated by Bridgland with the Mersea Island Gravel, implying a further correlation with the earliest post-diversion formation of the Thames-Medway in eastern Essex and with the Boyn Hill Gravel of the Thames system (see Part 2 of this chapter). This aggradation is believed to have been initiated late in the Anglian Stage and to have been completed in the early Saalian (Bridgland, 1988a; Table 5.6). If the gradient required to trace the Tollesbury Gravel downstream into the Mersea Island Gravel, *c.* 0.7 m per kilometre, is projected upstream to Maldon it would take the top of the formation to *c.* 35 m O.D. Allowing for a slight upstream increase in gradient, this strongly supports the correlation of the Maldon and Tollesbury aggradations (Fig. 5.5B and 5.32). The Maldon deposit can therefore be regarded as an upstream outlier of the Tollesbury Gravel. Its geographical location suggests that it may have been deposited by the Chelmer, the southern branch of the Blackwater system (see Fig. 5.1).

If the gravel overlying the till at Maldon is correctly interpreted as a terrace deposit laid down after the Anglian glaciation, there remains no stratigraphical basis for its interpretation as the 'lower till' within the tripartite sequence described by Clayton (1957). Even if a 'lower till' can be demonstrated elsewhere in the area (Bristow, 1985), the name Maldon Till would not be appropriate for it unless correlation with the till at Maldon can be established.

Summary

The railway cutting section at Maldon is significant in that it has provided a basis for a number of complex stratigraphical models that have been widely used to explain the glacial sequence in southern East Anglia. Doubts about the occurrence of ***in situ*** till at Maldon have been allayed by a recent reinvestigation, but this has itself raised doubts about the status of the deposit as an early 'lower till' within the glacial sequence. This status depends upon gravel overlying the till being a glacial outwash deposit, part of a tripartite sequence with a later 'upper till' (not present in the Maldon area). The discovery that the gravel at Maldon is probably a terrace deposit of the Blackwater/Chelmer system means that the underlying till can no longer be placed at the base of the tripartite sequence of Clayton (1957) and may question the validity of that sequence.

Conclusions

The historic section at Maldon, showing glacial sediments (Maldon Till) overlain by water-lain gravels, has previously been cited as evidence for of a complex regional story of alternating deposition by ice sheets and meltwater streams. The Maldon Till has been widely interpreted as the lower element in a three-part sequence of till, glacial gravel and till, the gravel at Maldon being regarded as the intermediate meltwater deposit. The recognition, presented here for the first time, that this gravel is the product of the Blackwater/Chelmer river system, casts doubt upon the validity of the three-part sequence. The Maldon section is now more simply interpreted as showing till, deposited by East Anglian ice around 450,000 years ago, overlain by river gravels deposited subsequent to the Anglian glaciation.

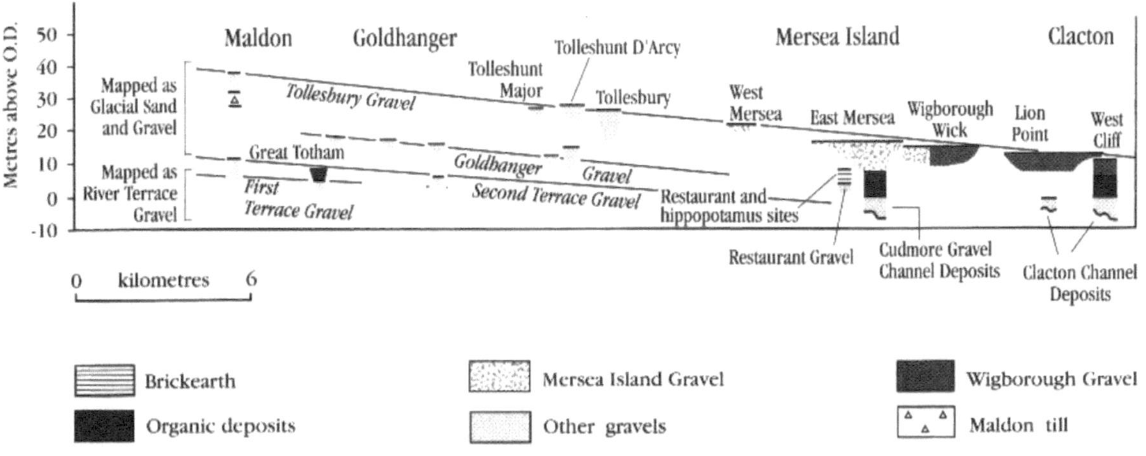

Figure 5.32 Longitudinal profiles of Blackwater terrace gravels.

EAST MERSEA RESTAURANT SITE (TM 053136) and HIPPOPOTAMUS SITE (TM 065142)

D.R. Bridgland

Highlights

Fossiliferous sediments here appear to provide a rare record of the Ipswichian Stage in fluvial deposits within the Thames catchment downstream from London. The site has one of the richest interglacial bone-beds in southern Britain, deposited by an erstwhile Thames tributary, the Essex Blackwater.

Introduction

The East Mersea Restaurant Site and Hippopotamus Site both reveal Late Pleistocene fossiliferous deposits attributed to the River Blackwater. They are of considerable stratigraphical importance because of their close proximity to the Middle Pleistocene Thames-Medway locality at Cudmore Grove (see Part 2 of this chapter). The three sites collectively constitute a single large and complex GCR site, which also includes an area of cryoturbated London Clay between the two sets of Pleistocene deposits (see Figs 5.22 and 5.23).

The Restaurant Site, *c.* 1.5 km to the south-west of Cudmore Grove, appears to coincide with most if not all the early records of fossil bone-bearing deposits at East Mersea (Warren, 1917, 1924b, 1933; Cornwall, 1958; Zeuner, 1958). The sequence at the Restaurant Site has produced Mollusca and ostracods, as well as abundant mammalian remains. Warren (1933) considered it the richest bone-bed he had seen. The occurrence of hippopotamus suggests deposition during the Ipswichian Stage (*sensu* Trafalgar Square) (Sutcliffe, 1964; see Chapter 1).

There is a further occurrence of hippopotamus-bearing sediments at East Mersea, at the Hippopotamus Site, which lies between the Restaurant Site and the outcrop of the Cudmore Grove Channel (Fig. 5.23). The Hippopotamus Site was discovered more recently and has so far only produced a small assemblage of mammalian bones, found on the foreshore in pockets of Pleistocene sediment in the surface of a London Clay platform. Pleistocene deposits that presumably have been stripped from this platform by recent marine erosion are exposed in the nearby cliffs. This fauna consists entirely of taxa already known from the Restaurant Site. Similar sequences of deposits are found at both localities, comprising a basal gravel overlain by finer grained sediments. The (basal) gravel is quite different to that overlying the Cudmore Grove estuarine deposits and provides the basis for attributing these Upper Pleistocene sediments to the River Blackwater.

Description

Casual observation of the Restaurant Site reveals only a shallow cliff of brickearth above a modern sandy, shelly and pebbly beach, below which the foreshore is cut in London Clay. Excavation through the modern beach (Fig. 5.33) reveals fossiliferous Pleistocene beds, however, as part of the following sequence (Fig. 5.34A):

	Thickness
3. Clayey silt, weathered (brown)	1.0 m
2. Sandy silt with bones and shells (grey)	0.2 m
1. Gravel with mammal bones	0.4 m
London Clay	

This sequence appears to fill a channel excavated in London Clay, the latter rising to cut out the later deposits beneath the eastern end of the beach. The gravel contains considerably more quartzose exotic material than the Mersea Island Gravel, but also includes a large proportion of southern rocks (predominantly Greensand chert) of the type characteristic of the East Essex Gravels (Table 5.5 and see Part 2 of this Chapter). It has been termed the (East Mersea) Restaurant Gravel (Bridgland *et al*., 1988).

Better exposures of the sequence were available in the early post-war years, when the Restaurant Gravel was preserved at the landward edge of the foreshore (Cornwall, 1958). Cornwall interpreted this deposit as filling the channel of a 'considerable stream', trending towards 110° east of north. He interpreted the overlying sequence as a 'floodloam' overlain by marine silt and considered there to be a buried soil at the top of the floodloam. In addition to the published descriptions, the Restaurant Site was excavated in 1934 by D. Bate and J. Reach (MS notes in the Natural History Museum) and in the late 1960s by R. Gruhn and A.L. Bryan (MS notes in the Institute of Archaeology, London University) and H.E.P. Spencer. The mammalian remains from those excavations are preserved in the Natural History Museum, except for Spencer's collection, which is in the Ipswich Museum.

The Hippopotamus Site is broadly similar to the Restaurant Site, but the cliff section at the former is considerably higher and exposes over 2 m of bedded silts with sand and gravel stringers (Fig. 5.34B). Again, gravel cannot be seen in the cliffs or on the foreshore, but is present beneath the beach. This gravel is indistinguishable, on the basis of sedimentary characteristics and clast-lithological composition, from that at the Restaurant Site (Table 5.5) and is presumed to be a continuation of the Restaurant Gravel (Bridgland *et al*., 1988). As at the Restaurant Site, the foreshore comprises a wave-cut platform in London Clay. However, all the mammalian bones from the Hippopotamus Site have been obtained from a small area of foreshore (TM 066143), where they were found protruding from pockets of silty material in the bedrock surface (Fig. 5.35). These probably represent scour hollows at the base of the Restaurant Gravel. The Pleistocene deposits at the Hippopotamus Site are separated from the Cudmore Grove Channel by London Clay, which rises to the full height of the cliffs between the two Pleistocene channels (Figs 5.22 and 5.23).

The Restaurant Site has yielded a mammalian fauna of ten species (Bridgland *et al*., 1988): in addition to the important indicator species *Hippopotamus amphibius*, the assemblage includes *Palaeoloxodon antiquus* (straight-tusked elephant), *Dicerorhinus hemitoechus* (narrow-nosed rhinoceros), *Bison priscus*, *Megaloceros giganteus* (giant deer) and *Crocuta crocuta* (Enxleben) (spotted hyaena). Another species of stratigraphical significance to occur is the modern water vole *Arvicola terrestris*, which is known only from the Late Pleistocene (Sutcliffe and Kowalski, 1976). As the bulk of the collections is from early investigations, no attempt has been made to separate assemblages from the gravel and silt (beds 1 and 2). The Hippopotamus Site has produced, in addition to hippopotamus, only giant deer and indeterminate bovine and elephant bones (Bridgland *et al.*, 1988).

The molluscan fauna, which occurs only in the sandy silt (bed 2) at the Restaurant Site, is dominated by freshwater bivalves, including *Pisidium supinum* and *P. moitessierianum*, which suggest deposition in a sizeable stream. Of particular interest is the occurrence of *Sphaerium rivicola* (Lamarck), which is rare in the British Pleistocene (Bridgland *et al.*, 1988). Five ostracod taxa have also been recovered from this same deposit: *Candona neglecta*, *Ilyocypris bradyi*, *I. schwarzbachi* Kempf, *Herpetocypris* sp. and *Cyprideis torosa*. The

Figure 5.33 Excavations at the East Mersea Restaurant Site. This view, looking north-east, shows bones being collected from the channel deposits temporarily exposed in a trench dug through the beach. Only London Clay is exposed on the foreshore. (Photo: A.J. Sutcliffe.)

first two, which dominate the fauna, require a low-energy freshwater environment, but *C. torosa* is a brackish water species. However, only a few specimens of the last-mentioned species were encountered, the poor preservation of which may suggest derivation from an earlier estuarine deposit, such as that at Cudmore Grove, in which the species is common (see above).

Interpretation

Warren (1917, 1924b, 1933) was the first to recognize an 'elephant bed' at East Mersea and suggested that it was equivalent to the Clacton Channel Deposits, which also yield elephant remains (at both sites all identifiable elephant remains are attributable to ***Palaeoloxodon antiquus***). The Clacton deposits were sub-

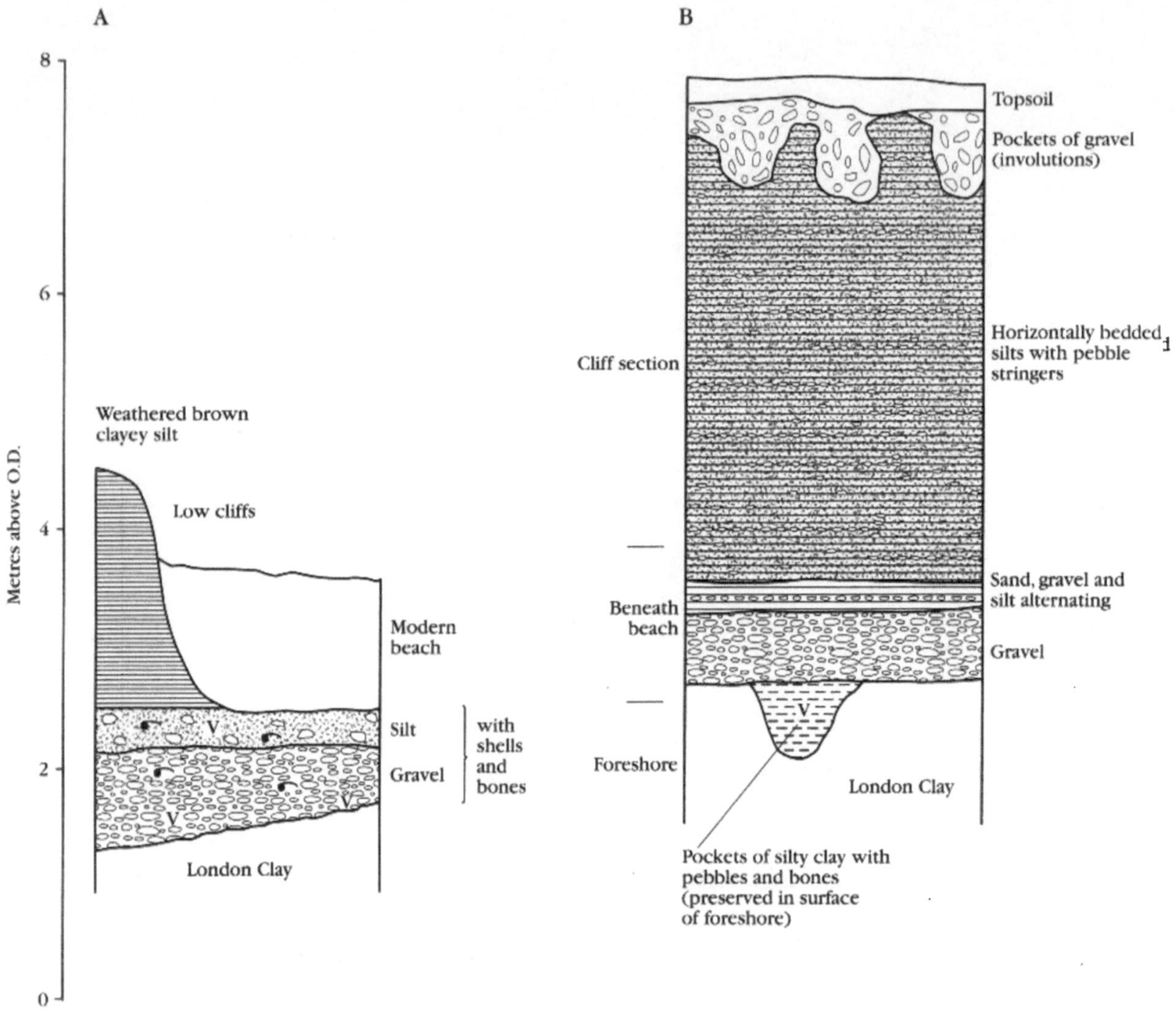

Figure 5.34 Sections at: (A) the East Mersea Restaurant Site; and (B) the Cudmore Grove Hippopotamus Site.

sequently assigned to the 'Great Interglacial' (Hoxnian Stage) (Pike and Godwin, 1953), which led to a similar interpretation of the sediments at East Mersea (Cornwall, 1958; Zeuner, 1958; Spencer, 1966). However, *Hippopotamus* is prominent amongst collections from East Mersea and, since this species is believed to have been absent from Britain during the Hoxnian, the site had been attributed to the Ipswichian Stage (*sensu* Trafalgar Square) (Sutcliffe, 1964) well in advance of the recent investigations, which have confirmed that it is quite unrelated to the Clacton deposits (Bridgland *et al.*, 1988).

The Restaurant and Hippopotamus Sites have closely similar sedimentary sequences at comparable elevations, the clast-lithological composition of the gravel at both sites matches (Table 5.5), and the limited fauna from the Hippopotamus Site is entirely coincident with taxa within the assemblage from the Restaurant Site. All these factors lead to the conclusion that a single set of deposits is represented at the two sites. These deposits are not, however, continuous between the two outcrops, but are cut out by Holocene saltings (Fig. 5.23).

The most important biostratigraphical evidence from this sequence derives from the mammalian fauna. This is of fully interglacial character and, unlike that from the Clacton and Cudmore Grove channels, includes hippopotamus. It also includes straight-tusked

Figure 5.35 East Mersea Hippopotamus Site: an elephant tooth is shown protruding from a silty pocket in the London Clay foreshore. All the faunal remains from this site have been recovered from similar situations, thought to represent pockets at the base of the Restaurant Gravel (see text). (Photo: A.J. Sutcliffe.)

elephant and narrow-nosed rhinoceros, but lacks horse. Its closest match is with mammalian assemblages from Joint Mitnor Cave (Devon), Trafalgar Square (London), Barrington (Cambridgeshire) and Victoria Cave (North Yorkshire), all ascribed to the Ipswichian Stage. The last of these has yielded a uranium-series date of *c.* 120,000 years BP, confirming a last interglacial age (Gascoyne *et al.*, 1981). There is little in the molluscan or ostracod faunas to confirm or deny the correlation of the deposits at East Mersea with the Ipswichian, although they do imply temperate conditions, and the two dominant ostracod species (*C. neglecta* and *I. bradyi*) survive in the Holocene. *Ilyocypris schwarzbachi*, however, was first described from a Holsteinian interglacial site at Karlich, near Koblenz.

If the fossiliferous deposits are of Ipswichian age, it is likely that the overlying unfossiliferous bedded silts and sands at the Hippopotamus Site represent aggradation under cold conditions during the Devensian Stage. These contain 'stringers' of gravel, the contents of which were analysed along with the underlying Restaurant Gravel (Table 5.5). The clast composition of the Restaurant Gravel is of considerable significance to palaeogeographical reconstruction. It differs in this respect from the Mersea Island Gravel (see above, Cudmore Grove) in that it contains a higher exotic fraction, dominated by the quartzose lithologies that characterize the gravels of the pre-diversion Thames. This composition is consistent with a Blackwater origin, in which case this exotic material is presumed to have been reworked from Kesgrave Group deposits in central Essex. Other exotic lithologies, in particular *Rhaxella* chert, are probably derived from Anglian Stage glacial deposits in the same area, although they could have been reworked from older Blackwater terrace gravels or from the Mersea Island Gravel. The latter has probably contributed much of the Greensand chert, but the High- and Low-level East Essex Gravels to the south of the Blackwater are also likely to have provided this rock type (see Part 2 of this chapter). The gravel stringers within the bedded silts and sands at the Hippopotamus Site contain fewer exotic rocks and more Greensand chert, having a composition intermediate between that of the Restaurant Gravel and the Mersea Island Gravel, although closer to the latter (Table 5.5). This suggests that much of the material was derived from the immediate valley side, from the Mersea Island Gravel (Bridgland *et al.*, 1988). Such a process might be considered more likely under cold-climate conditions, thus supporting the interpretation of these upper deposits as of Devensian age.

The conclusion that the fluvial sequence at the Restaurant and Hippopotamus Sites is the product of the River Blackwater and not the Thames-Medway is fully consistent with its apparent Ipswichian age. It is thought that the Thames-Medway had migrated a considerable distance to the south of Mersea Island by the Ipswichian, leaving the area to be drained by the Blackwater (Bridgland, 1988a; Fig. 5.5). The plotting of long-profiles of River Blackwater gravel formations (Fig. 5.32) suggests that the Restaurant Gravel lies higher in the sequence than the mapped terraces in the Maldon area (Geological Survey, New Series, Sheet 241). To the north of the terrace gravels, however, a belt of 'Glacial Sand and Gravel' has been mapped, extending from Great Totham to Tollesbury. These have been identified as older terrace gravels of the Blackwater system (Bridgland, 1983a; Bridgland *et al.*, 1990). Two formations can be recognized within these higher terrace deposits, the Tollesbury Gravel, already mentioned (see above, Maldon), and the Goldhanger Gravel, which is well-represented around the village of Goldhanger (TL 905090). The gradient of the Goldhanger Gravel suggests possible correlation with the Restaurant Gravel (Fig. 5.32). This would imply that all the mapped terrace deposits of the Maldon area are later in age than the Ipswichian Stage (*sensu* Trafalgar Square) and, conversely, that the deposits mapped as 'Glacial Sand and Gravel' account for all Blackwater deposition between the Anglian Stage (when the river was formed – see Maldon) and the Late Pleistocene. Comparison with the fluvial record in the Lower Thames valley (Chapter 4), in which four formations are recognized representing this time interval, suggests that the sequence thus far established in the Blackwater is unlikely to be complete.

The London Clay exposures between the Cudmore Grove Channel–Mersea Island Gravel section and the silts (brickearth) at the Hippopotamus Site show evidence of severe and repeated periglacial activity. Involutions, filled with gravel similar to that above the Cudmore Grove Channel, have been observed in the upper levels of the cliff and extend into the upper part of the silt sequence above the Restaurant Gravel near its eastern edge. Further west, the height of the cliffs declines and the uppermost, cryoturbated silts have probably been removed by erosion. Immediately east of the Blackwater sequence at the Hippopotamus Site, two sets of gravel-filled involutions occur at the top of the London Clay cliffs (Fig. 5.22). The lower of these, which is formed in the top of the London Clay, is overlain by a sheet of redeposited (colluvial) clay, indistinguishable from the bedrock and probably derived directly from it. The higher set of involutions is developed in this remobilized clay. Also of interest are large dislocated pockets of gravel that occur in the London Clay between the silts and Mersea Island Gravel exposures. These apparently comprise relatively undisturbed Mersea Island Gravel, so much so that palaeocurrent measurements have been obtained from foresets in one of them. The latter are inclined directly towards the near-vertical wall of London Clay at the edge of the pocket, suggesting that the contact is not erosional, but has been produced by post-depositional deformation. The clay itself is highly fissile and appears to have been subjected to considerable disturbance. It seems likely that the gravel pockets represent the feather-edge of the Mersea Island Gravel, perhaps higher parts of the unit than are now preserved, which have been let down undisturbed into the London Clay, probably when the latter was itself highly mobile and rising diapirically around the gravel pockets. Such processes are likely to have operated during periods of summer melting in a periglacial environment. Since the Restaurant Gravel is attributed to the Ipswichian, the periglacial episode during which this occurred must belong within the Devensian Stage.

The deposits at these two localities are

therefore interpreted as products of an Ipswichian to early Devensian Blackwater aggradation, laid down long after the Thames-Medway had ceased to flow further north (within the present land area of south-eastern Essex) than the immediate vicinity of its present estuary (Bridgland, 1988a; Fig. 5.5).

Relation to the regional sequence

The Restaurant and Hippopotamus Sites at East Mersea, together with Cudmore Grove, form a highly significant locality spanning a large part of the Middle and Late Pleistocene. Whereas the Cudmore Grove site exposes a succession ascribed to the Hoxnian (*sensu* Swanscombe), the Restaurant and Hippopotamus Sites represent the Ipswichian (*sensu* Trafalgar Square) and Devensian Stages, the last interglacial and glacial episodes. The sediments at the Restaurant and Hippopotamus Sites are contained in the later and smaller of two juxtaposed channels (Fig. 5.27), attributed in this case to the River Blackwater, whereas the earlier, deeper channel at Cudmore Grove represents the Thames-Medway. The coastal sections at East Mersea thus reveal sediments from both of the post-Anglian interglacials defined by Mitchell *et al.* (1973). However, evidence from other parts of the Thames Basin now suggests that this sequence is an oversimplification, so the preservation at this locality of sediments representing the Hoxnian and Ipswichian Stages is regarded as coincidental; there were two further post-Anglian temperate intervals that are not represented at East Mersea, but which occurred between the deposition there of the Thames-Medway (Cudmore Grove Channel) and Blackwater (Restaurant Gravel) deposits. It is clear that major changes in palaeogeography took place between the emplacement of the two sets of deposits. The Thames-Medway, represented at Cudmore Grove, had migrated a considerable distance to the south by the Ipswichian, leaving the Blackwater as the main drainage line in the area of Mersea Island (as it is today).

The Blackwater deposits at the East Mersea Restaurant and Hippopotamus Sites are therefore of considerable stratigraphical significance. Notwithstanding this, they are also of importance as a source of palaeoenvironmental evidence for the Ipswichian Stage. If the various Corbets Tey and Mucking Formation sites in the Lower Thames are correctly interpreted as of pre-Ipswichian (*sensu* Trafalgar Square) age (see Chapter 4), East Mersea is the only true 'last interglacial' site within the Thames system downstream of London. The fact that it represents a tributary is no accident. It is apparent from the elevation of the Trafalgar Square site that the projected Ipswichian thalweg level of the main river falls below the valley floor in the Lower Thames and probably below ordnance datum before the modern coast is reached. Therefore Thames-Medway Ipswichian deposits are more likely to be found offshore from the Essex coast or beneath coastal alluvium than within the onshore terrace system. Whether this is because of post-Ipswichian subsidence or whether it indicates that Ipswichian sea level was much lower than is generally believed is at present uncertain.

Conclusions

These two related localities provide exposures in fluvial sediments containing an assemblage of mammal remains typical of the last Pleistocene interglacial episode (the Ipswichian). These include such characteristic elements as hippopotamus, hyaena and straight-tusked elephant. The remains of molluscs and ostracods (small crustaceans) are also found at the Restaurant Site. During the last interglacial, the main channel of the Thames, downstream from London, lay well below the level of the present river floodplain, below modern sea level. It is because the interglacial sediments at East Mersea were deposited in a former tributary of the Thames, the Blackwater, that they lie at a sufficiently high level to be studied at the surface – the steeper upstream gradient of the Blackwater brings its last interglacial floodplain level above modern sea level at East Mersea. Because of this, the two East Mersea localities expose the only unequivocal last interglacial (Ipswichian) deposits (dated at around 125,000 years BP) yet recognized in the Thames catchment downstream from London. They are therefore comparable to the famous but inaccessible fossiliferous site at Trafalgar Square.

GREAT TOTHAM (LOFTS FARM PIT; TL 866092)

D.R. Bridgland, T. Allen, G.R. Coope, P.L. Gibbard and R. Wrayton

Highlights

A view, rare in the Thames catchment, of presumed Devensian fossiliferous sediments is here afforded by a section in a gravel terrace of the River Blackwater, a former tributary of the Thames. These deposits contain a typical cold-climate mammalian fauna, with reindeer, woolly mammoth, woolly rhinoceros and hyaena. They also contain pollen and the remains of insects and ostracods.

Introduction

A gravel pit in the 2nd Terrace of the Blackwater at Great Totham exploited what appear to be the most recent sediments included within the GCR Thames coverage. This site is located 2 km NNE of Maldon, on the northern side of the River Blackwater, at the upstream limit of its present-day estuary (Fig. 5.1). Organic sediments interbedded with the gravel near Lofts Farm have yielded a large collection of mammalian bones, as well as pollen, other plant remains, insects and ostracods. The deposits are believed to represent the Devensian Stage, on the bases both of their faunal contents and of regional terrace stratigraphy.

Devensian sediments are relatively common in the valley bottom (floodplain) gravels of the Thames upstream from London and in many of its tributaries (Coope and Angus, 1975; Sutcliffe and Kowalski, 1976; Gibbard *et al.*, 1982; Kerney *et al.*, 1982; Gibbard, 1985). In these situations, however, they invariably lie below the water table and are difficult to study except where accessible in active gravel pits or temporary exposures. Downstream from London, Holocene subsidence of the North Sea Basin, coupled with the low Devensian base level, ensures that sediments dating from the last glacial are well below floodplain level, within the buried channel. In Essex such sediments, like the Ipswichian deposits at East Mersea (see above), can therefore be studied only in the valleys of minor rivers such as the Blackwater, where steeper upstream gradients bring them more rapidly above floodplain level than is the case with the Thames (Bridgland, 1988a).

At present the Blackwater flows out to sea east of Maldon, but offshore bedrock surface contour mapping indicates that before the Holocene marine transgression it flowed into the now submerged valley of the Thames-Medway, *c.* 20 km to the south of Clacton (D'Olier, 1975; Bridgland and D'Olier, 1989; Fig. 5.5F). The Blackwater may therefore be regarded as a tributary of the Thames-Medway system.

The Great Totham locality has not been described hitherto, so the present report is the first documentation of evidence collected at the site. An excavation was carried out at Lofts Farm Pit by the GCR Unit in the autumn of 1985, when the organic sediments were sampled in detail. Analyses of the samples collected on this occasion are still incomplete.

Description

The former gravel workings at Great Totham exposed deposits attributed by the Geological Survey (Sheet 241) to the 2nd Terrace of the Blackwater-Chelmer system, immediately downstream of the confluence between these two rivers (Fig. 5.1). In a small area to the north-west of Lofts Farm (TL 866092) the following section was recorded:

	Thickness
Surface stripped prior to quarrying (the original surface level was *c.* 9 m O.D.)	
3. Disturbed clayey gravel occurs within ice-wedge casts and involutions in (2)	*c.* 3 m
2. Organic clays and silts, oxidized near top	2.6 m
1. Gravel, horizontally bedded	*c.* 2 m
London Clay	

Involutions and ice-wedge pseudomorphs, which occur in the upper part of the sequence, indicate that permafrost conditions have prevailed at Great Totham at some time since the deposition of the fossiliferous sediments. Clast-lithological analysis of the gravel (bed 1) reveals a close similarity to other deposits attributed to deposition by the River Blackwater (Table 5.5).

The most spectacular discoveries at Great Totham were a collection of vertebrate bones (Fig. 5.36) amassed by R. Wrayton, who visited the site frequently while it was in operation (1982–4) and discovered the organic sediments. The mammalian assemblage consists of *Canis lupus* (wolf), *Crocuta crocuta* (spotted hyaena), *Rangifer tarandus* L. (reindeer), *Megaloceros* sp. (giant deer), *Bison* and/or *Bos* (bovid), *Coelodonta antiquitatis* (woolly rhinoceros), *Equus ferus* (horse) and *Mammuthus primigenius* (woolly mammoth). Freshwater fish and Amphibia are also represented, namely *Perca fluviatilis* L. (perch), *Esox lucius* L. (pike), *Gasterosteus aculeatus* L. (three-spined stickleback), a cyprinid (carp family), frog (*Rana* sp.) and indeterminate newt (Fig. 5.37). The bones are well preserved, although many show evidence of damage during transport by the river. The remains of small mammals have been discovered but have yet to be identified. During the 1985 excavation an ulna of woolly mammoth was discovered *in situ* in the organic silty clay, the only addition to the assemblage of large bones. The organic deposits also contain pollen and the remains of plants, insects and ostracods.

The majority of the insect remains recovered at Great Totham came from a bulk sample from bed 2, the organic clays and silts. These were supplemented by a similar but smaller assemblage from a sample collected in 1984, which has not been related to the stratigraphical sequence described above, but which can be assumed to come from bed 2. Altogether 45 taxa of Coleoptera were recorded, of which 37 could be identified to species or species group. Six are no longer living in Britain and are distinguished by an asterisk in the following list:

Taxon	Number
Carabidae	
Notiophilus aquaticus (L.)	23
**Diacheila polita* (Fald.)	3
Elaphrus cupreus Duft.	1
Loricera pilicornis (F.)	1
Clivina fossor (L.)	2
Dyschirius globosus (Hbst.)	2
Trechus secalis (Payk.)	1
Bembidion (Metallina) properans (Steph.)	3
Bembidion (Princidium) bipunctatum (L.)	6
Bembidion (Notaphus) obliquum Sturm	1
Bembidion (Blepharoplataphus) virens Gyll.	1
Bembidion (Plataphodes) sp.	1
**Bembidion (Plataphus) hasti* Sahlb.	1
Bembidion (Philochthus) aeneum Germ.	4
Pterostichus strenuus (Panz.)	1
Pterostichus melanarius (Ill.)	1
Calathus melanocephalus (L.)	1
Agonum ericeti (Panz.)	1
**Amara municipalis* (Duft.)	5
**Amara torrida* (Panz.)	2
Dytiscidae	
Ilybius sp.	1
Gyrinidae	
Gyrinius aeratus Steph.	1
Hydraenidae	
Helophorus aquaticus (L.) type	1
Hydrophilidae	
Cercyon melanocephalus (L.)	2
Cercyon tristis (Ill.)	1
Cercyon analis (Payk.)	2
Cryptopleurum minutum (F.)	1
Hydrobius fuscipes (L.)	1
Silphidae	
Thanatophilus dispar (Hbst.)	1
Liodidae	
Agathidium marginatum Sturm	1
Staphylinidae	
Stenus sp.	1
Xantholinus sp.	1
Tachyporus sp.	1
Tachinus sp.	1
Scarabaeidae	
Aegialia sabuleti (Panz.)	3
Aphodius fimetarius (L.)	5
Aphodius sp.	27
Chrysomelidae	
**Phaedon segnis* Weise	1
Curculionidae	
Apion sp.	2
Otiorhynchus fuscipes (Ol.)	7
Otiorhynchus arcticus (F.)	3
Otiorhynchus ligneus (Ol.)	3
Otiorhynchus rugifrons (Gyll.)	5
Notaris aethiops (F.)	3
Alophus triguttatus (F.)	12

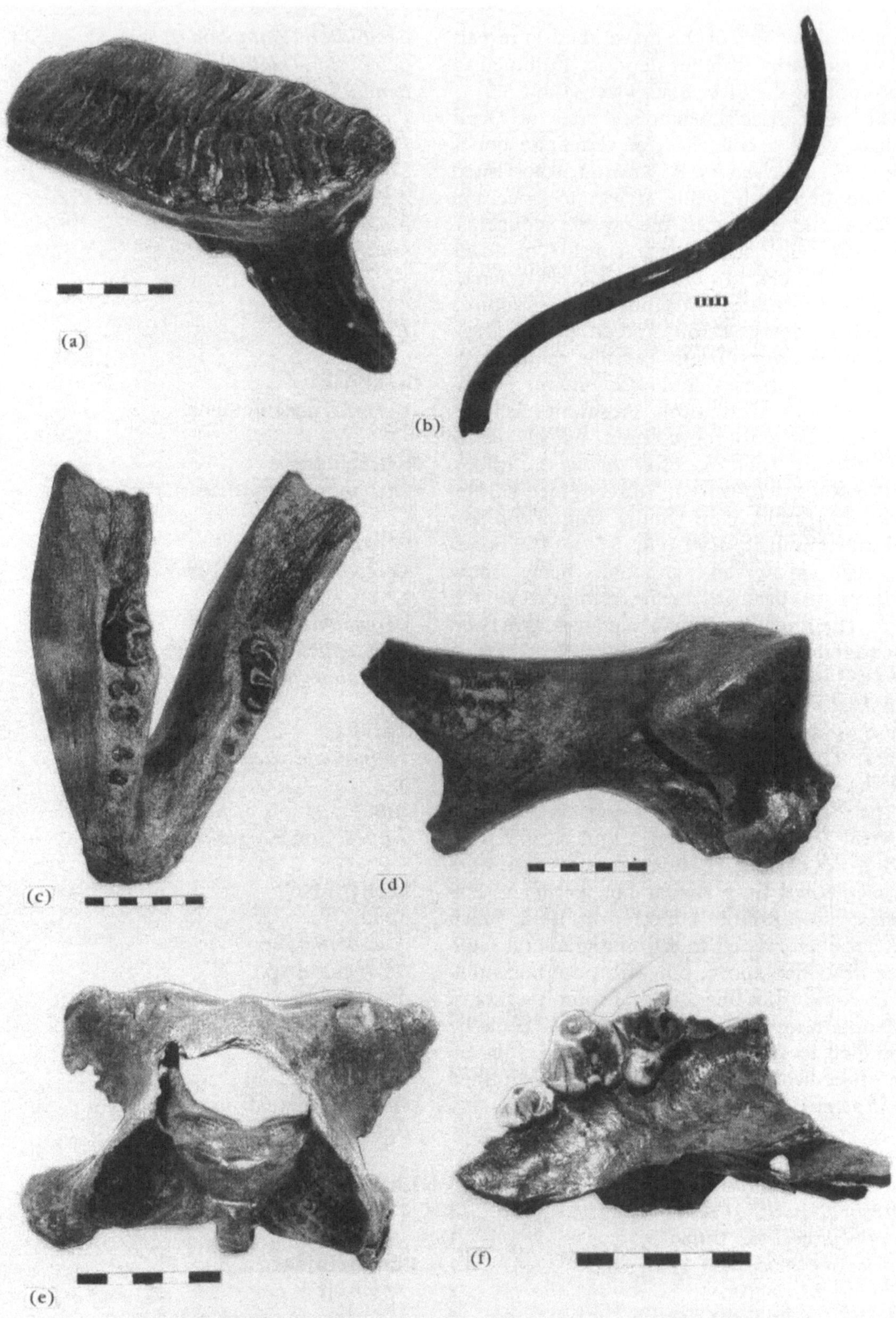

Figure 5.36 Mammalian bones from Great Totham (R. Wrayton collection). (A) Molar tooth of mammoth (*Mammuthus primigenius*). (B) Tusk of mammoth (*Mammuthus primigenius*). (C) Lower jaw of woolly rhinoceros (*Coelodonta antiquitatis*). (D) Humerus of woolly rhinoceros (*Coelodonta antiquitatis*). (E) Vertebra of horse (*Equus ferus*). (F) Fragment of jaw of spotted hyaena (*Crocuta crocuta*). Scale bars are graduated in cm. (Photos: R. Wrayton).

In this faunal list the nomenclature follows that of Lucht (1987). The number opposite each taxon indicates the minimum number of individuals present in the sample.

During the period when R. Wrayton was collecting fossil bones from the Great Totham site, he also sieved a large amount of the biogenic sediment of bed 2 and found it to contain a rich fossil fauna and flora. At this time a number of broken valves of the bivalve *Pisidium* sp. and an apex of the gastropod *Bithynia tentaculata* were recovered. These remain the only Mollusca recorded from the site. Also collected at this time (1984) was the preliminary sample from bed 2 that produced significant numbers of beetles and ostracods; it was the assessment of this preliminary sample that prompted the more detailed investigation of the site that was undertaken by the GCR Unit the following year.

A study has been made of the cladoceran and ostracod remains from five serial bulk samples that were taken from the lower 0.9 m of bed 2 during the 1985 GCR excavation. The uppermost of these samples was barren, but the other four, from the lowest 0.70 m of bed 2, all contained small crustacea (Table 5.7). Between 0.15 and 0.50 m from the base of the unit, cladoceran ephippia and ostracod valves were present in profusion, particularly in the 0.15–0.30 m sample. As a similar assemblage of small crustacea was present in all four fossiliferous samples, the 0.15–0.30 m sample can be regarded as representative of the bed 2 fauna at its optimum. In this sample, cladoceran ephippia, the saddle-shaped coverings of 'winter' or 'resting' eggs, were common. Both the elongate ephippia of the *Daphnia magna* group and more triangulate ephippia belonging to the *D. pulex* or *D. longispina* groups were

Figure 5.37 Small vertebrate remains from Great Totham. The identifiable species represented are pike (*Esox lucius*), perch (*Perca fluviatilis*), stickleback (*Gasterosteus aculeatus*) and frog (*Rana* sp.; probably *R. temporaria*, common frog). Remains of a cyprinid fish (carp family) and indeterminate newt are also present. Identifications by B. Clarke (Amphibia) and A. Wheeler (fish) of the Natural History Museum. Scale bar is graduated in mm, numbered in cm. (Photo: Paul Douthwaite.)

Table 5.7 Ostracods from Great Totham (identifications by T. Allen).

Species	Group	Height (cm) above base of organic clays and silts				
		0 - 15	**15 - 30**	**30 - 50**	**50 - 70**	**70 - 90**
Cyclocypris serena (Koch)	1	-	Common	Very rare	-	-
Candona candida (Müller)	2	Very rare	Common	Common	Very rare	-
Candona neglecta Sars	2	Very rare	Abundant	Abundant	Very rare	-
Ilyocypris bradyi Sars	2	-	Common	Common	-	-
Cypria ophthalmica (Jurine)	3	-	Common	Very rare	-	-
Ilyocypris gibba (Ramdohr)	4	Very rare	Very common	Common	-	-
Eucypris zenkeri (Chyzer)	4	-	Rare	-	-	-
Limnocythere inopinata (Baird)	4	-	Rare	Very rare	-	-
Herpetocypris sp.	?	-	Rare	Very rare	-	-
Potamocypris sp.	?	-	Very rare	Very rare	-	-
Overall frequency		**Very rare**	**Very common**	**Common**	**Very rare**	**Absent**

present in equal quantities. Ten species of ostracods were recorded from this sample (Table 5.7). The most abundant species was *Candona neglecta*, which was represented by many adult male and female valves and large numbers of instars. Valves of *Ilyocypris gibba*, often strongly tuberculate, were also very common (Table 5.7). Four other species were found in significant numbers: *Candona candida*, *Cyclocypris serena*, *Cypria ophthalmica* and *Ilyocypris bradyi*. Adult valves and instars of the remaining species (Table 5.7) were present only in small numbers. Single valves of *Candona protzi* Hartwig and *Candona weltneri* Hartwig were found in the 1984 preliminary sample, but no examples of these species were encountered in the material collected from the 1985 GCR excavation.

Only a preliminary assessment of the palaeobotany of the organic deposits can be given here. Three samples from bed 2 have been analysed, from 2 m, 2.5 m and 3 m (approximately) above the London Clay. The second of these was from a particularly fossiliferous level, rich in macroscopic plant remains and ostracods (although collected from an earlier exposure, this was probably broadly equivalent to the richest ostracod-bearing levels sampled in 1985 – see above). All three samples yielded a similar pollen assemblage, dominated by herbs, grasses and sedges (Fig. 5.38). The plant macrofossil counts from the three levels are also closely comparable (Table 5.8).

Interpretation

The organic sediments (bed 2) at Great Totham have provided considerable palaeontological evidence, but this is predominantly of value to environmental reconstruction rather than stratigraphy. Although further assessment of the flora and fauna is in progress, the fossil assemblage recognized thus far is not stratigraphically diagnostic. It is necessary to look to the terrace record of the River Blackwater, and to the

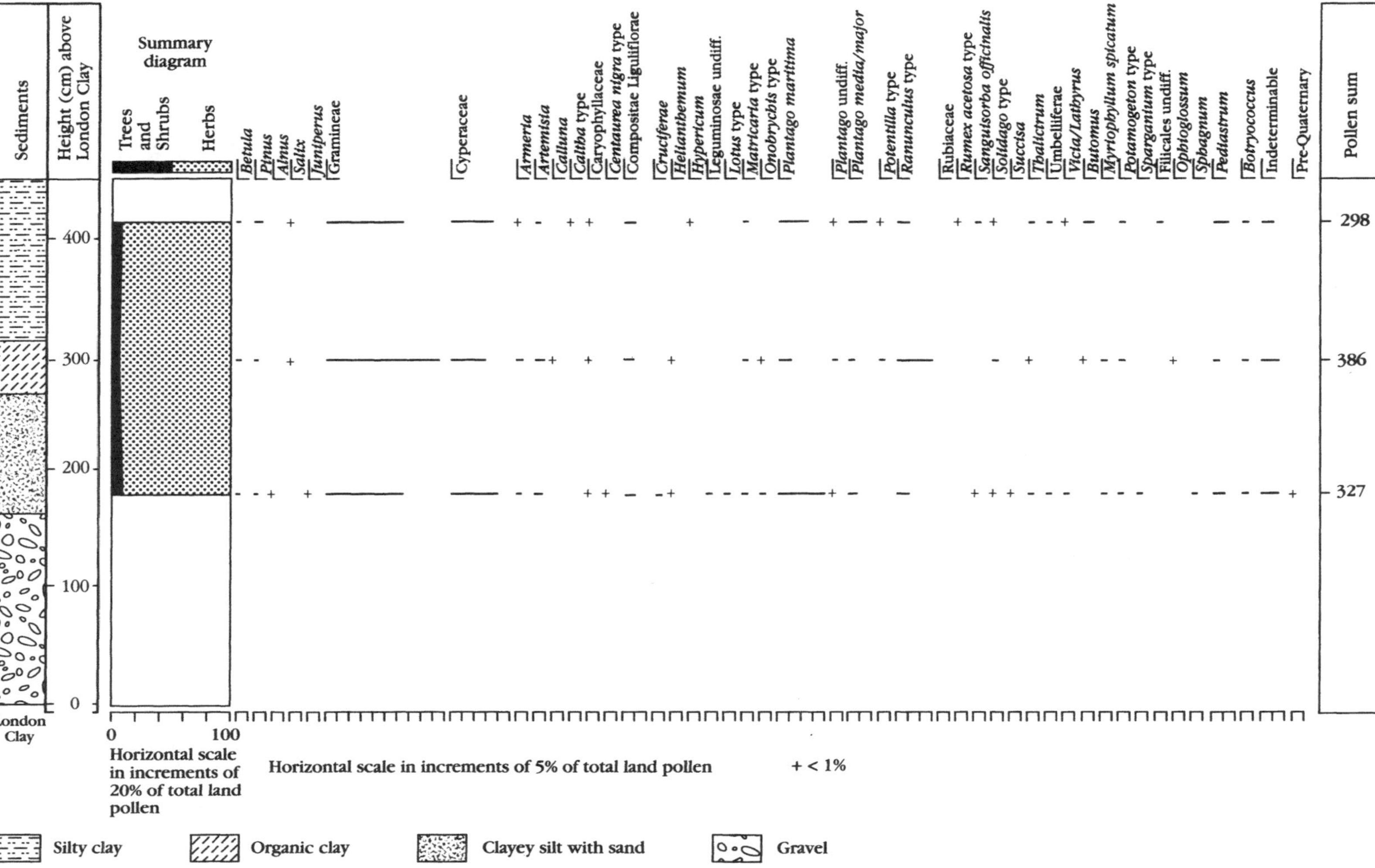

Figure 5.38 Pollen from the organic deposits at Great Totham.

Table 5.8 Plant macrofossils from Lofts Farm Pit, Great Totham (identifications by M. Pettit).

	Height above London Clay		
	2m	2.5m	3m
Ranunculus subgenus *Ranunculus*	1a	-	1a
Ranunculus subgenus *Batrachium*	8.5a	9.5a	5a
Potentilla anserina	1a	-	-
Potentilla sp.	2a	2a	5a
Viola sp.	-	1s	-
Silene vulgaris cf. subspecies (*maritima*)	1s	1s	-
Carex sp.	7n	6n	3n
Scirpus lacustris	1n	-	-
Eleocharis palustris	2n	-	-
Potamegeton sp.	20fst	20fst	20fst
Zannichellia palustris	16a	22a	3a
Hippuris vulgaris	-	4a	-
Linum perennes subspecies (*anglicum*)	-	1s	1s
Moss fragment	-	+	-

a: achene; n: nut; fst: fruitstone; s: seed.

position of the deposits at Great Totham within that sequence, to gain an indication of their relative age.

The mammalian assemblage from Great Totham comprises species that are all well known from the Devensian Stage in Britain. All have been obtained from Devensian sediments that have been dated using the radiocarbon technique (see Stuart, 1982a, 1991). A similar mammalian assemblage occurs, for example, in the Cave Earth of Kent's Cavern, which is regarded as Devensian in age (Sutcliffe, 1974). However, all the species present at Great Totham have been recorded from earlier Middle and Late Pleistocene deposits, so their occurrence together at Great Totham does not necessarily indicate a Devensian age. For this reason a horse metatarsal has been submitted to the Godwin Laboratory at Cambridge for radiocarbon dating. Although some of the species in this assemblage also occur in interglacials (giant deer, mammoth and horse, for example), the assemblage as a whole is suggestive of a cold episode. It fully conforms with the environmental indications from the palaeobotany, from which a treeless, open habitat can be envisaged.

The insect assemblage provides a considerable insight into conditions prevailing at the time when the organic deposits were laid down. Viewed as a whole, the assemblage is characteristic of open ground, sparsely or patchily vegetated, on a substrate that must have included both clay and gravel. ***Notiophilus aquaticus***, for example, prefers rather open dry ground with short heaths and grasses, sometimes living in apparently sterile places. ***Bembidion bipunctatum*** is often found in the company of ***B. virens*** and ***B. hasti*** on sterile stony banks. ***Amara municipalis*** also lives in sandy or gravelly habitats with sparse vegetation. ***Diacheila polita*** is today one of the characteristic species of the tundra, where it is usually found on dry peaty soil. ***Agonum ericeti***, a surface dweller that likes sunlight, is a stenotopic species requiring acid soils. Clay substrates are required by ***Bembidion aeneum***, which is believed to prefer somewhat saline soils. ***Clivina fossor*** burrows in clayrich soils, avoiding pure sand

(Lindroth, 1985).

There are very few Coleoptera present that require fully aquatic habitats. Single specimens of *Ilybius* and *Gyrinus* indicate that some open water must have been available, while *Hydrobius fuscipes* occurs in stagnant water. The remainder of the Hydrophilidae in the assemblage are mainly species of damp, decomposing vegetation, or are dung-dwellers. The relatively high numbers of the scarabaeid *Aphodius* also suggests the presence of large quantities of dung, although some members of this genus can live in well-rotted vegetable material. *Thanatophilus dispar* is a carcass beetle often found under rotting fish in northern Europe (Strand, 1946).

The beetle assemblage provides little information about the specific composition of the flora at the time of deposition. All the species of *Otiorhynhus* are polyphagous herbaceous plant eaters. Their abundance in the assemblage indicates that such plants were readily available. *Notaris aethiops* is usually said to prefer *Sparganium*, but its ubiquity in arctic Eurasia suggests that it can feed on a variety of different Cyperacaea.

The beetle assemblage is strongly indicative of a climate substantially colder than that of today. With the exception of *Phaedon segenis*, which lives at the present time in the mountains of eastern Europe, all the species occur today in the far north of Europe. *Diacheila polita* is especially significant in this respect. Its nearest modern habitat to Britain is on the Kola peninsula, in northern Russia, from where it ranges across the tundra of Eurasia as far as northwestern Alaska. Only occasionally is it found below the tree line. *Bembidion hasti* is an exclusively north-palaeoarctic species that lives today in the mountains of Fennoscandia, extending eastwards into western Siberia.

An interesting feature of the Great Totham insect assemblage is that it lacks the important group of exclusively Asiatic species that characterizes faunas from Devensian interstadial sites (Coope, 1968, 1987). The implication of this particular faunal difference is that the Great Totham deposits were laid down under climatic conditions that were less continental than those prevailing during the major part of the Upton Warren Interstadial Complex.

One species in this insect assemblage, *Phaedon segnis*, has up to now been found only in Devensian interstadial contexts in Britain. This was found (incorrectly recorded under the name *P. pyritosus*) at Upton Warren (Coope *et al.*, 1961) and at Marlow (Coope, in Gibbard, 1985), in deposits that have been attributed to the Middle Devensian. The absence at Great Totham of the group of exclusively eastern Asiatic species makes correlation with the major part of the Upton Warren Interstadial Complex unlikely. The Devensian interstadial fauna from Isleworth (Coope and Angus, 1975) also lacks far-eastern species, but in that case the insect assemblage is indicative of temperate conditions, quite different to those inferred at Great Totham. The Great Totham assemblage may thus represent a hitherto unrecognized episode within the Devensian.

The ostracod assemblage provides further information about the palaeoclimate at the time of the deposition of bed 2. The species present can be divided into four groups based on differences in their capacity to tolerate the annual cycle of seasonal changes in water temperature (Table 5.7), as follows:

Group 1: *Cyclocypris serena*

This species has been classed as a cold stenothermal (confined to low-temperature environments) form by Diebel and Wolfschläger (1975). The two species encountered only in the 1984 sample, *Candona protzi* and *C. weltneri* (see above, Description), have also been listed as cold stenothermal forms (Hiller, 1972) and therefore belong within this group, although not featuring in Table 5.7.

Group 2: *Candona candida, Candona neglecta* and *Ilyocypris bradyi*

These three species thrive in cool water and can survive quite cold aquatic temperatures, but they may decline in numbers or become absent as water temperatures reach their peak in mid- and late summer. *Candona candida* is found only rarely when water temperatures exceed 18°C (Hiller, 1972).

Group 3: *Cypria ophthalmica*

Cypria ophthalmica, the only species in this group, is tolerant of a wide range of water temperatures, living not only in cold and cool water, but also thriving in the warmer water of summer.

Group 4: *Ilyocypris gibba*, *Eucypris zenkeri* and *Limnocythere inopinata*

These species flourish in the warm aquatic conditions of the summer months and, while able to tolerate the cooler water temperatures of spring and autumn, they would be absent during the coldest winter months.

Overall the ostracod assemblage contains a preponderance of species adapted to cool water conditions. As aquatic temperatures started to rise in late spring to early summer, the cold stenothermal forms (Group 1) would have become scarce and then failed, leaving the cool water ostracods (Group 2) to dominate the assemblage. A subsidiary fauna of summer species (Group 4) would have begun to appear at this time, becoming increasingly important as temperatures increased during the early summer. By mid- to late summer the Group 4 species would have formed a substantial part of the ostracod fauna, perhaps even achieving a short period of dominance if water temperature peaked at above 18–20°C. As aquatic temperature decreased with the onset of autumn, the summer species would have declined, leaving the cool-water species again predominant. *Cypria ophthalmica*, the single Group 3 species, would have been a constant member of the fauna throughout the year.

This ostracod assemblage lived in a permanent body of fresh or oligohaline (slightly brackish) moving water with at least moderate summer weed growth. *Eucypris zenkeri*, in particular, is characteristic of slowly moving, shallow, plant-rich waters (Klie, 1938). The assemblage is similar to faunas described from the Devensian interstadial sites at Fladbury, Isleworth and Upton Warren (Siddiqui, 1971). The presence of numerous Daphniidae ephippia, such as are found in the silts and clays at Great Totham, may be an indication of seasonality, as these egg covers are produced to enable the survival of periods of inhospitable conditions; however, the climatic significance of these is limited, as ephippia are produced by the Daphniidae at the present time over a wide range of latitudes, from arctic to subtropical (Shotton and Osborne, 1965).

The preliminary pollen analysis of the organic sediments (Fig. 5.38) indicates that vegetation was sparse at the time of deposition. Tree and shrub pollen are so rare as to preclude the growth of such plants locally, the small quantities present probably resulting from either long-distance transport or reworking. Amongst the herb pollen, several distinct plant communities are indicated. It is clear that the channel and adjacent wet ground supported communities of the aquatics *Thalictrum*, *Myriophylum spicatum* L., *Sparganium* and *Butomus*. The abundance of grass pollen, together with a range of dry-ground herbs, points to the occurrence of grassland further away from the depositional site. The pollen of Compositae (Liguliflorae), *Artemisia*, *Plantago majorimedia*, Umbelliferae, Caryophyllaceae, *Helianthenmum*, *Centaurea nigra*, *Vicia/Lathyrus* and *Matricaria* type were all probably derived from this habitat. Damp grassland and meadow environments also occurred in the vicinity, to judge by the abundance of pollen from Cyperaceae, *Sanguisorba officinalis* L. and *Caltha* type.

Of particular interest is the repeated occurrence of the pollen of halophytic plants such as *Plantago maritima* L. and *Armeria*. These records, reinforced by the finds of *Silene vulgaris maritima* (Withering) (A. and D. Löve) in the plant macrofossil assemblage (see below), are indicative of high soil salinity. This phenomenon has been frequently attributed to high evaporation rates under cold-climate conditions (West, 1988).

The limited plant macrofossil assemblage (Table 5.8) compares closely with that of the pollen. It is dominated by the remains of aquatics such as *Potamogeton* spp., *Hippuris vulgaris* (L.), *Ranunculus* (*Batrachium*), *Myriophyllum* cf. *spicatum* and *Zannichelia palustris* (L.). Marsh plants (*Eleocharis palustris* (L.) Roemer and Schultes, *Potentilla* sp.) and dry grassland taxa (*Linum perenne anglicum* Ockendon, *Viola* sp. and *Ranunculus* (*Ranunculus*)) occur in all three samples (Table 5.8), confirming that similar conditions prevailed throughout the period of the deposition of the organic clays and silts. The palaeobotanical evidence is typical of cold episodes during the Middle and Late Pleistocene, intervals that are often termed 'full-glacial'. Such floras recur repeatedly and are not therefore characteristic of any particular period.

As indicated above, the terrace record of the Blackwater provides some stratigraphical indication of the likely age of the Great Totham deposits. The long-profiles of the terraces in both branches of the Blackwater system, the Blackwater itself and the River Chelmer, have been illustrated by Bristow (1985). In the Maldon area and further downstream, Bristow recognized no well-marked formations above the 2nd Terrace. A minor spread, mapped as 3rd Terrace, to the north of Chigborough Farm (around TL 878091) appears to be poorly differentiated from the 2nd Terrace (Bristow, 1985, figure 21). In the present volume the deposits on the northern side of the Blackwater estuary, which were mapped as 'Glacial Sand and Gravel', are recognized as higher, older terraces of the Blackwater system (see above, Maldon and East Mersea). Figure 5.32 shows projections of the long-profiles of the terrace gravels of the Blackwater system downstream from Maldon. The downstream gradient of the 2nd Terrace indicates that it falls below ordnance datum between Tollesbury and Mersea Island. As has been noted previously, this places the 2nd Terrace lower in the Blackwater terrace sequence than the fossiliferous deposits at the East Mersea Restaurant and Hippopotamus sites (Fig. 5.32). Since the latter have been attributed to the Ipswichian Stage, this would appear to provide important confirmatory evidence for the Devensian age of the 2nd Terrace and, therefore, the deposits at Great Totham.

In terms of the scheme for Thames terrace stratigraphy promoted in this volume, however, the occurrence of Ipswichian and Devensian deposits in different terraces of the Blackwater system is somewhat surprising. In the London area Ipswichian sediments and deposits dating from the mid-Devensian Upton Warren Interstadial both occur within the Kempton Park Formation of the Thames (see Chapter 3, Fern House Pit). No terrace rejuvenation appears to separate the Ipswichian and the Middle Devensian in this area, the downcutting to the Shepperton Gravel level occurring in the Late Devensian (Gibbard, 1985). It is possible that a rejuvenation occurred in the early Devensian in the Upper Thames, however; recent indications from a site in the Northmoor Formation at Cassington (see Chapter 2, Stanton Harcourt and Magdalen Grove) suggest that sediments of Chelford Interstadial or Upton Warren Interstadial age occur there (D. Maddy, pers. comm.), which would require there to have been a downcutting event soon after the deposition of the Ipswichian Eynsham Gravel at the Summertown-Radley Terrace level.

Conclusions

The Great Totham site provides a rare opportunity for the study of last-glacial fossiliferous sediments in the lower reaches of the Thames system, in this case situated in the valley of the River Blackwater (a former Thames tributary). Organic sediments here have yielded an impressive assemblage of large mammal remains, as well as smaller vertebrates, pollen, other plant remains, insects and ostracods. The mammal fauna has many of the elements regarded as typical of cold climatic episodes: wolf, giant deer, reindeer, woolly rhinoceros and mammoth are all represented. A rich assemblage of fossil beetles has been collected from the organic sediments. It too points to conditions much colder than today, with several species present that now live in the far north of Europe and/or Eurasia. The last glacial (Devensian) age cannot be ascertained with certainty from the palaeontology, however. It is necessary to compare the height of the Great Totham sediments within the sequence of Blackwater terraces with that of the last interglacial site at East Mersea, which appears to be in a higher (and therefore older) terrace, implying that the Great Totham terrace post-dates the last interglacial.

References

Abbott, W.J.L. (1890) Notes on some Pleistocene sections in and near London. *Proceedings of the Geologists' Association*, **11**, 473–80.

Abbott, W.J.L. (1911) On the classification of the British Stone Age industries and some new, and little known well marked horizons and cultures. Journal *of the Royal Anthropological Institute*, **41**, 458–81.

Allen, P. (1983) Middle Pleistocene stratigraphy and landform development in south-east Suffolk. Unpublished Ph.D. thesis, University of London.

Allen, P. (1984) *Field Guide to the Gipping and Waveney Valleys*, Quaternary Research Association, Cambridge, 116 pp.

Allen, P. (1991) Deformation structures in British Pleistocene sediments. In *Glacial Deposits in Great Britain and Ireland* (eds J. Ehlers, P.L. Gibbard and J. Rose), A.A. Balkema, Rotterdam, pp. 455–69.

Allen, P., Cheshire, D.A. and Whiteman, C.A. (1991) Glacial deposits of southern East Anglia. In *Glacial Deposits in Great Britain and Ireland* (eds J. Ehlers, P.L. Gibbard and J. Rose), A.A. Balkema, Rotterdam, pp. 255–78.

Allen, T. (1977) Interglacial sea-level change: evidence for brackish water sedimentation at Purfleet, Essex. *Quaternary Newsletter*, **22**, 1–3.

Allen, T.J. (1978) Disposition of the terraces of the River Thames in the vicinity of Yiewsley. In *Early Man in West Middlesex* (ed. D. Collins), HMSO, London, pp. 5–10.

Almaine, H.G.W.D. (1922) Palaeolithic gravel near Abingdon. *Antiquaries Journal*, **2**, 257–8.

Ambrose, J.D. (1973) The sand and gravel resources of the country around Maldon, Essex. *Report of the Institute of Geological Sciences*, 73/1.

Anon. (1906) Flint implements and fossils from Clacton. *Essex Naturalist*, **14**, 164.

Anon. (1908) Palaeolithic flint implement from a gravel pit, Handborough, Oxon. *Oxford University Gazette*, **38**, 752.

Anon. (1911a) Visit to Clacton-on-Sea, and 301st ordinary meeting. Saturday, 30th September 1911. *Essex Naturalist*, **16**, 322–4.

Anon. (1911b) Exhibition by S.H. Warren of plaster casts of Palaeolithic wooden spear (?) and some flint-flakes from a Pleistocene deposit at Clacton-on-Sea. *Essex Naturalist*, **16**, 326.

Anon. (1913) Excursion to Mersea Island (the 427th Meeting), Saturday, 20th September 1913. *Essex Naturalist*, **17**, 229–34.

Anon. (1931) The Newton Collection. *Antiquaries Journal*, **11**, 420–1.

Anon. (1966) The Aveley elephants. *Report of the British Museum (Natural History)* [for 1963–1965], pp. 30–1.

Anon. (1982a) Waltham Cross, Hertfordshire. *Earth Science Conservation*, **19**, 35.

Anon. (1982b) Hornchurch, Essex. *Earth Science Conservation*, **19**, 35.

Anon. (1984a) Hornchurch railway cutting. *Earth Science Conservation*, **21**, 42.

Anon. (1984b) Globe Pit SSSI, Essex. *Earth Science Conservation*, **21**, 39–40.

Arkell, W.J. (1943) The Pleistocene rocks at Trebetherick Point, north Cornwall; their interpretation and correlation. *Proceedings of the Geologists' Association*, **54**, 41–170.

References

Arkell, W.J. (1945) Three Oxfordshire palaeoliths and their significance for Pleistocene correlation. *Proceedings of the Prehistoric Society*, **2**, 20–31.

Arkell, W.J. (1947a) *The Geology of Oxford*, Clarendon Press, Oxford, 268 pp.

Arkell, W.J. (1947b) The geology of the Evenlode Gorge, Oxfordshire. *Proceedings of the Geologists' Association*, **58**, 87–113.

Arkell, W.J. (1947c) A palaeolith from the Hanborough Terrace. *Oxoniensia*, **11–12**, 1–4.

Arkell, W.J. and Oakley, K.P. (1948) The implements in the Treacher Collection. *In* On the ancient channel between Caversham and Henley, Oxfordshire, and its contained flint implements. *Proceedings of the Prehistoric Society*, **14**, 126–54.

Avery, B.W. and Catt, J.A. (1983) Northaw Great Wood. In *The Diversion of the Thames* (ed. J. Rose), Field Guide, Quaternary Research Association, Cambridge, pp. 96–101.

Baden-Powell, D.F.W. (1948) The chalky boulder clays of Norfolk and Suffolk. *Geological Magazine*, **85**, 279–96.

Baden-Powell, D.F.W. (1949) Experimental Clactonian technique. *Proceedings of the Prehistoric Society*, **15**, 38–41.

Baden-Powell, D.F.W. (1950) The Pliocene-Pleistocene boundary in the British deposits. In *The Pliocene-Pleistocene boundary* (ed. K.P. Oakley), International Geological Congress 18th session [G.B., 1948], Vol. 9, pp. 8–10.

Baden-Powell, D.F.W. (1951) The age of interglacial deposits at Swanscombe. *Geological Magazine*, **88**, 344–56.

Baden-Powell, D.F.W. (1955) Appendix B: Report on the marine fauna of the Clacton Channels. In Warren, S.H., The Clacton (Essex) channel deposits. *Quarterly Journal of the Geological Society of London*, **111**, 301–5.

Baker, C.A. (1971) A contribution to the glacial stratigraphy of west Essex. *Essex Naturalist*, **32**, 318–30.

Baker, C.A. (1977) Quaternary stratigraphy and environments in the Upper Cam valley. Unpublished Ph.D. thesis, University of London.

Baker, C.A. (1983) Glaciation and Thames diversion in the Mid-Essex Depression. In *The Diversion of the Thames* (ed. J. Rose), Field Guide, Quaternary Research Association, Cambridge, pp. 39–49.

Baker, C.A. and Jones, D.K.C. (1980) Glaciation of the London Basin and its influence on the drainage pattern: a review and appraisal. In *The Shaping of Southern England* (ed. D.K.C. Jones), Institute of British Geographers Special Publication 11, Academic Press, London, pp. 131–76.

Barrow, G. (1919a) Some future work for the Geologists' Association. *Proceedings of the Geologists' Association*, **30**, 1–48.

Barrow, G. (1919b) Notes on the correlation of the deposits described in Mr C. J. Gilbert's paper with the high-level gravels of the south of England (or the London Basin). *Quarterly Journal of the Geological Society of London*, **75**, 44–50.

Barrow, G. (1919c) Excursion to Stanmore Hill and Bushey Heath. *Proceedings of the Geologists' Association*, **30**, 122–6.

Bell, A.M. (1894a) Palaeolithic remains at Wolvercote, Oxfordshire, I and II. *Antiquary*, **30**, 148–52 and 192–8.

Bell, A.M. (1894b) On the Pleistocene gravels at Wolvercote near Oxford. *Report of the British Association, Oxford*, pp. 663–4.

Bell, A.M. (1904) Implementiferous sections at Wolvercote (Oxfordshire). *Quarterly Journal of the Geological Society of London*, **60**, 120–32.

Bell, F.G. (1969) The occurrence of southern, steppe and halophyte elements in Weichselian (last glacial) floras from southern Britain. *New Phytologist*, **68**, 913–22.

Bennett, K.D., Peglar, S.M. and Sharp, M.J. (1991) Holocene lake sediments in central East Anglia. In *Central East Anglia and the Fen Basin* (eds S.G. Lewis, C.A. Whiteman and D.R. Bridgland), Field Guide, Quaternary Research Association, London, pp. 111–118.

Berckhemer, F. (1933) Ein Menschen-Schädel aus den diluvialen Schottern von Steinheim a.d. Murr. *Anthropologische Anzeiger*, **10**, 318–21.

Bishop, M.J. (1982) The mammal fauna of the early Middle Pleistocene cavern infill site of Westbury-sub-Mendip, Somerset. *Special Papers in Palaeontology*, **28**, 1–108.

Bishop, W.W. (1958) The Pleistocene geology and geomorphology of three gaps in the Middle Jurassic escarpment. *Philosophical Transactions of the Royal Society of London*, **B241**, 255–306.

Blair, K.G. (1923) Some coleopterous remains from the peat-bed at Wolvercote, Oxfordshire. *Transactions of the Royal Entomological Society of London*, **71**, 558–63.

Blake, J.H. (1891) Excursion to Henley-on-Thames and Nettlebed. *Proceedings of the Geologists' Association*, **12**, 204–6.

Blake, J.H. (1900) Excursion to Silchester. *Proceedings of the Geologists' Association*, **16**, 513–6.

Blake, J.H. (1903) *The Geology of the Country around Reading*. Memoir of the Geological Survey of Great Britain, 91 pp.

Blezard, R.G. (1966) Field meeting at Aveley and West Thurrock. *Proceedings of the Geologists' Association*, **77**, 273–6.

Blezard, R.G. (1973) South Essex. In *The Estuarine Region of Suffolk and Essex* (eds J.T. Greensmith, R.G. Blezard, C.R. Bristow *et al.*), Geologists' Association Guide. Benham, Colchester, pp. 35–41.

Boswell, P.G.H. (1940) Climates of the past: a review of the geological evidence. *Quarterly Journal of the Royal Meteorological Society of London*, **66**, 249–74.

Boswell, P.G.H. (1952) The Pliocene-Pleistocene boundary in the east of England. *Proceedings of the Geologists' Association*, **63**, 301–12.

Bowen, D.Q. (1978) *Quaternary Geology: A Stratigraphic Framework for Multidisciplinary Work*, Pergamon Press, Oxford, 237 pp.

Bowen, D.Q. (1989) The last interglacial-glacial cycle in the British Isles. *Quaternary International*, **3/4**, 41–7.

Bowen, D.Q. (1991) Amino acid geochronology. In *Central East Anglia and the Fen Basin* (eds S.G. Lewis, C.A. Whiteman and D.R. Bridgland), Field Guide, Quaternary Research Association, London, pp. 21–4.

Bowen, D.Q., Sykes, G.A., Reeves, A., *et al.* (1985) Amino acid geochronology of raised beaches in south west Britain. *Quaternary Science Reviews*, **4**, 279–318.

Bowen, D.Q., Hughes, S.A., Sykes, G.A., *et al.* (1989) Land-sea correlations in the Pleistocene based on isoleucine epimerization in non-marine molluscs. *Nature, London*, **340**, 49–51.

Bowen, D.Q., Rose, J., McCabe, A.M., *et al.* (1986a) Correlation of Quaternary glaciations in England, Ireland, Scotland and Wales. *Quaternary Science Reviews*, **5**, 299–340.

Bowen, D.Q., Richmond, G.M., Fullerton, D.S., *et al.* (1986b) Correlation of Quaternary glaciations in the Northern Hemisphere. *Quaternary Science Reviews*, **5**, 509–10 + loose figures.

Bowen, D.Q. and Sykes, G.A. (1988) Correlation of the marine events and glaciations on the north-east Atlantic margin. *Philosophical Transactions of the Royal Society of London*, **B318**, 619–35.

Breitinger, E. (1952) Zur Morphologie und systematischen Stellung des Schädelfragmentes von Swanscombe. *Homo*, **3**, 131–3.

Breitinger, E. (1955) Das Schädelfragment von Swanscombe und das 'Praesapiensproblem'. *Mitteilungen der Anthropologischen Gesellschaft Wien*, **84/85**, 27–38.

Breitinger, E. (1964) Reconstruction of the Swanscombe skull. In *The Swanscombe Skull: a Survey of Research on a Pleistocene Site*, (ed. C.D. Ovey), Royal Anthropological Institute, Occasional Paper No. 20, 161–72. Translated by D.M. Watson from: Das Schädelfragment von Swanscombe und das 'Praesapiensproblem'. *Mitteilunger der Anthropologischen Gesellschaft Wien*, **84/85**, 27–38.

Breuil, H. (1932a) Appendix in Sandford, K.S., The Pleistocene succession in England. *Geological Magazine*, **69**, 17–18.

Breuil, H. (1932b) Les industries à éclats du Palaéolithique ancien, I: Le Clactonien. *Préhistoire, Paris*, **1**, 148–57.

Breuil, H. (1934) De l'importance de la solifluction dans l'étude des terrains Quaternaires de la France et des pays voisins. *Revue de Géographie Physique et de Géologie Dynamique*, **7**, 269–331.

Breuil, H. (1947) Age of the Baker's Hole Coombe Rock, Northfleet, Kent. *Nature, London*, **160**, 831.

Bridgland, D.R. (1980) A reappraisal of Pleistocene stratigraphy in north Kent and eastern Essex, and new evidence concerning the former courses of the Thames and Medway. *Quaternary Newsletter*, **32**, 15–24.

Bridgland, D.R. (1983a) The Quaternary fluvial deposits of north Kent and eastern Essex. Unpublished Ph.D. thesis, City of London Polytechnic, 2 volumes.

Bridgland, D.R. (1983b) Eastern Essex. In *Diversion of the Thames* (ed. J. Rose), Field Guide, Quaternary Reasearch Association, Cambridge, pp. 170–84.

Bridgland, D.R. (1985a) Pleistocene sites in the Thames-Avon system. *Earth Science Conservation*, **22**, 36–9.

Bridgland, D.R. (1985b) Uniclinal shifting; a speculative reappraisal based on terrace distribution in the London Basin. *Quaternary Newsletter*, **47**, 26–33.

Bridgland, D.R. (1986a) Discussion of procedures and recommendations. In *Clast Lithological Analysis* (ed. D.R. Bridgland), Technical Guide No. 3, Quaternary Research Association, Cambridge, pp. 1–33.

Bridgland, D.R. (1986b) The rudaceous components of the East Essex Gravels; their characteristics and provenance. *Quaternary Studies*, **2**, 34–44.

Bridgland, D.R. (1986c) The provenance of gravel at Great Fanton Hall, near Wickford, Essex. In *Clast Lithological Analysis* (ed. D.R. Bridgland), Technical Guide No. 3, Quaternary Research Association, Cambridge, pp. 147–52.

Bridgland, D.R. (1988a) The Pleistocene fluvial stratigraphy and palaeogeography of Essex. *Proceedings of the Geologists' Association*, **99**, 291–314.

Bridgland, D.R. (1988b) Problems in the application of lithostratigraphic classification to Pleistocene terrace deposits. *Quaternary Newsletter*, **55**, 1–8.

Bridgland, D.R. (1988c) The Quaternary derivation of quartzites used by Palaeolithic Man in the Thames Basin for tool manufacture. *In* Non-flint stone tools and the Palaeolithic occupation of Britain, *British Archaeological Report, British Series*, **189**, 187–98.

Bridgland, D.R. (1990a) Pleistocene stratigraphy and river basin sediments: a reply to D. Maddy and C.P. Green. *Quaternary Newsletter*, **60**, 10–2.

Bridgland, D.R. (1990b) Little Oakley (TM 223294), In *The Cromer Symposium Field Excursion Guidebook* (ed. C. Turner), Symposium of European Quaternary Stratigraphy/Quaternary Research Association, Cambridge, pp. 48–57.

Bridgland, D.R. and D'Olier, B. (1987) Attempted correlation of onshore and offshore Thames channels and terraces in the eastern London Basin and the southern North Sea. *Programme and Abstracts XII INQUA Congress* [July 1987], pp. 136.

Bridgland, D.R. and D'Olier, B. (1989) A preliminary correlation of the onshore and offshore courses of the Rivers Thames and Medway during the Middle and Upper Pleistocene. In *Quaternary and Tertiary Geology of the Southern Bight, North Sea* (eds J.P. Henriet and G. De Moor), Belgian Ministry of Economic Affairs, Geological Survey, 161–72.

Bridgland, D.R. and Gibbard, P.L. (1990) Ardleigh (Martell's Quarry) TM053280. In *The Cromer Symposium Field Excursion Guidebook* (ed. C. Turner), Symposium of European Quaternary Stratigraphy/Quaternary Research Association, Cambridge, 57–62.

Bridgland, D.R. and Harding, P. (1985) Palaeolithic artifacts from the gravels of the Hoo Peninsula. *Archaeologia Cantiana*, **101**, 41–55.

Bridgland, D.R. and Harding, P. (1986) An attempt to locate the 'Wolvercote Channel' in the railway cutting adjacent to Wolvercote Brick Pit. *Quaternary Newsletter*, **48**, 12–6.

Bridgland, D.R. and Lewis, S.G. (1991) Introduction to the Pleistocene geology and drainage history of the Lark valley. In *Central East Anglia and the Fen Basin* (eds S.G. Lewis, C.A. Whiteman and D.R. Bridgland), Field Guide, Quaternary Research Association, London, pp. 37–44.

Bridgland, D.R., Gibbard, P.L., Harding, P., *et al.* (1985) New information and results from recent excavations at Barnfield Pit, Swanscombe. *Quaternary Newsletter*, **46**, 25–39.

Bridgland, D.R., Allen, P., Currant, A.P., *et al.* (1988) Report of the Geologists' Association field meeting in north-east Essex, May 22nd–24th, 1987. *Proceedings of the Geologists' Association*, **99**, 315–33.

Bridgland, D.R., Keen, D.H. and Maddy, D. (1989) The Avon Terraces: Cropthorne, Ailstone and Eckington. In *West Midlands* (ed. D.H. Keen), Field Guide, Quaternary Research Association, Coventry, pp. 51–67.

Bridgland, D.R., Gibbard, P.L. and Preece, R.C. (1990) The geology and significance of the interglacial sediments at Little Oakley, Essex. *Philosophical Transactions of the Royal Society of London*, **B328**, 307–39.

Bridgland, D.R., D'Olier, B., Gibbard, P.L. and Roe, H.M. (1993) Correlation of Thames terrace deposits between the lower Thames, eastern Essex and the submerged offshore continuation of the Thames-Medway valley. *Proceedings of the Geologists' Association*, **104**, 51–58.

Briggs, D.J. (1973) Quaternary deposits of the Evenlode valley and adjacent areas. Unpublished Ph.D. thesis, University of Bristol.

Briggs, D.J. (1976a) River terraces of the Oxford area. In *Field Guide to the Oxford Region* (ed. D. Roe), Quaternary Research Association, Oxford, pp. 8–15.

Briggs, D.J. (1976b) Some Quaternary problems in the Oxford area. In *Field Guide to the Oxford Region* (ed. D. Roe), Quaternary Research Association, Oxford, pp. 6–7.

Briggs, D.J. (1988) The environmental background to human occupation in the Upper Thames valley during the Quaternary Period. *In* Non-flint stone tools and the Palaeolithic occupation of Britain (eds R.J. MacRae and N. Moloney), *British Archaeological Report, British Series*, **189**, 167–86.

Briggs, D.J. and Gilbertson, D.D. (1973) The age of the Hanborough Terrace of the River Evenlode, Oxfordshire. *Proceedings of the Geologists' Association*, **84**, 155–73.

Briggs, D.J. and Gilbertson, D.D. (1974) Recent studies of Pleistocene deposits in the Evenlode valley and adjacent areas of the Cotswolds. *Sound [Journal of the Plymouth Polytechnic Geological Society*], **3**, 7–22.

Briggs, D.J. and Gilbertson, D.D. (1980) Quaternary processes and environments in the Upper Thames basin. *Transactions of the Institute of British Geographers*, **5**, 53–65.

Briggs, D.J., Coope, G.R. and Gilbertson, D.D. (1975a) Late Pleistocene terrace deposits at Beckford, Worcestershire, England. *Geological Journal*, **10**, 1–16.

Briggs, D.J., Gilbertson, D.D., Goudie, A.S., *et al.* (1975) New interglacial site at Sugworth. *Nature, London*, **257**, 477–9.

Briggs, D.J., Coope, G.R. and Gilbertson, D.D. (1985) The chronology and environmental framework of early Man in the Upper Thames Valley: a new model. *British Archaeological Report, British Series*, **137**, 176 pp.

Bristow, C.R. (1985) *The Geology of the Country around Chelmsford*. Memoir of the Geological Survey of Great Britain, 108 pp.

Bristow, C.R. and Cox, F.C. (1973) The Gipping Till: a reappraisal of East Anglian glacial stratigraphy. *Journal of the Geological Society of London*, **129**, 1–37.

Bromehead, C.E.N. (1912) On diversions of the Bourne near Chertsey. *Summary of Progress, Geological Survey of Great Britain* [for 1911], pp. 74–7.

Bromehead, C.E.N. (1925) *The Geology of North London*. Memoir of the Geological Survey of Great Britain, 63 pp.

Brown, E.H. (1975) The Quaternary terraces of the River Thames. In *L'évolution Quaternaire des Bassins Fluviaux de la Mer du Nord Méridionale* (ed. P. Macar), Société Géologique de Belgique, Liege, 318 pp.

Brown, J. (1838) Discovery of a large pair of fossil horns in Essex. *Magazine of Natural History, Series 2*, **2**, 163–4.

Brown, J. (1839) Fossil bones at Clacton. *Essex Literary Journal* [for 1839], 29.

Brown, J. (1840) Notice of a fluvio-marine deposit containing mammalian-remains occurring in the parish of Little Clacton on the Essex coast. *Magazine of Natural History, Series 2*, **4**, 197–201.

Brown, J. (1841) A list of the fossil shells found in a fluvio-marine deposit at Clacton in Essex. *Annals and Magazine of Natural History, Series 1*, **7**, 427–9.

Brown, J. (1845) On certain conditions and appearances of the strata on the coast of Essex near Walton. *Quarterly Journal of the Geological Society of London*, **1**, 341–2.

Brown, J. (1857) Note on bovine remains, lately found at Clacton, Essex. *Annals and Magazine of Natural History, Series 2*, **20**, 397–8.

Brunnacker, K. (1986) Quaternary stratigraphy in the Lower Rhine area and northern Alpine foothills. *Quaternary Science Reviews*, **5**, 373–9.

Brunnacker, K., Löscher, M., Tillmanns, W., *et al.* (1982) Correlation of the Quaternary terrace sequences in the Lower Rhine valley and northern Alpine foothills of central Europe. *Quaternary Research*, **18**, 152–73.

Bryant, I.D. (1983) Facies sequences associated with some braided river deposits of late-Pleistocene age from southern Britain. In *Modern and Ancient Fluvial Systems: Sedimentology and Processes* (eds J.D. Collinson and J. Lewin), International Association of Sedimentologists, Special Publication, **No. 6**, pp. 267–75.

Bryant, I.D. and Holyoak, D.T. (1980) Devensian deposits at Brimpton, Berkshire. *Quaternary Newsletter*, **30**, 17.

Bryant, I.D., Holyoak, D.T. and Moseley, K.A. (1983) Late Pleistocene deposits at Brimpton, Berkshire, England. *Proceedings of the Geologists' Association*, **94**, 321–43.

Buckland, W. (1823) *Reliqiae Diluvianae: or Observation on the Organic Remains Contained in Caves, Fissures and Diluvial Gravel and on Other Geological Pheno-*

mena, Attesting the Action of a Universal Deluge. John Murray, London, 303 pp.

Buckman, S.S. (1897) Deposits of the Bajocian age in the northern Cotteswolds: The Cleeve Hill Plateau. *Quarterly Journal of the Geological Society of London*, **53**, 607–29.

Buckman, S.S. (1899a) Gravel at Moreton-in-Marsh, Gloucestershire. *Quarterly Journal of the Geological Society of London*, **55**, 220–3.

Buckman, S.S. (1899b) The development of rivers; and particularly the genesis of the Severn. *Natural Science*, **14**, 273–89.

Buckman, S.S. (1900) Excursion notes: chiefly on river features. Salisbury meeting. *Proceedings of the Cotteswold Naturalists Field Club*, **13**, 175–92.

Bull, A.J. (1942) Pleistocene chronology. *Proceedings of the Geologists' Association*, **53**, 1–45.

Burchell, J.P.T. (1931) Early Neanthropic Man and his relation to the Ice Age. *Proceedings of the Prehistoric Society of East Anglia*, **6**, 253–303.

Burchell, J.P.T. (1933) The Northfleet 50-foot submergence later than the coombe rock of the post-Early Mousterian times. *Archaeologia*, **83**, 67–91.

Burchell, J.P.T. (1934a) The Middle Mousterian culture and its relation to the coombe rock of post-early Mousterian times. *Antiquaries Journal*, **14**, 33–9.

Burchell, J.P.T. (1934b) Fresh facts relating to the Boyn Hill Terrace of the Lower Thames valley. *Antiquaries Journal*, **14**, 163–6.

Burchell, J.P.T. (1935a) Evidence of a further glacial episode within the valley of the Lower Thames. *Geological Magazine*, **72**, 90–1.

Burchell, J.P.T. (1935b) Some Pleistocene deposits at Kirmington and Crayford. *Geological Magazine*, **72**, 327–31.

Burchell, J.P.T. (1936a) A final note on the Ebbsfleet Channel series. *Geological Magazine*, **73**, 550–4.

Burchell, J.P.T. (1936b) Hand-axes later than the Main Coombe Rock of the Lower Thames valley. *Antiquaries Journal*, **16**, 260–4.

Burchell, J.P.T. (1936c) Evidence of a Late Glacial episode within the valley of the Lower Thames. *Geological Magazine*, **73**, 91–2.

Burchell, J.P.T. (1954) Loessic deposits in the fifty-foot terrace post-dating the Main Coombe Rock of Baker's Hole, Northfleet, Kent. *Proceedings of the Geologists' Association*, **65**, 256–61.

Burchell, J.P.T. (1957) A temperate bed of the last interglacial period at Northfleet, Kent. *Geological Magazine*, **94**, 212–14.

Callaway, C. (1905) The occurrence of glacial clay on the Cotteswold Plateau. *Geological Magazine*, **2**, 216–9.

Cambridge, P.G. (1977) Whatever happened to the Boytonian? A review of the marine Plio–Pleistocene of the southern North Sea Basin. *Bulletin of the Geological Society of Norfolk*, **29**, 23–45.

Campbell, S. and Bowen, D.Q. (1989) *Quaternary of Wales*, Geological Conservation Review Series, Nature Conservancy Council, 240 pp.

Carreck, J.N. (1972) Chronology of the Quaternary deposits of south-east England, with special reference to their vertebrate faunas. Unpublished M.Phil. thesis, University of London.

Carreck, J.N. (1976) Pleistocene mammalian and molluscan remains from 'Taplow' Terrace deposits at West Thurrock, near Grays, Essex. *Proceedings of the Geologists' Association*, **87**, 83–92.

Case, H.J. and Kirk, J.R. (1952) Notes and news: Henley-on-Thames. *Oxoniensia*, **15**, 107.

Case, H.J. and Kirk, J.R. (1955) Notes and news: Rotherfield Peppard. *Oxoniensia*, **19**, 118.

Castell, C.P. (1964) The non-marine Mollusca. In *The Swanscombe Skull: a Survey of Research on a Pleistocene Site*, (ed. C.D. Ovey), Royal Anthropological Institute of London, pp. 77–83.

Catt, J.A. (1977) Loess and coversands. In *British Quaternary Studies: Recent Advances* (ed. F.W. Shotton), Clarendon Press, Oxford, pp. 221–9.

Catt, J.A. (1978) The contribution of loess to soils in lowland Britain. In *The Effect of Man on the Landscape: the Lowland Zone* (eds S. Imbrey and J.G. Evans), Council for British Archaeological Resources, Report No. 21, pp. 12–20.

Catt, J.A. (1979) Soils and Quaternary geology in Britain. *Journal of Soil Science*, **30**, 607–42.

Catt, J.A. and Hodgson, J.M. (1976) Soils and geomorphology of the chalk in south-east England. *Earth Surface Processes*, **1**, 181–93.

Catt, J.A., Corbett, W.M., Hodge, C.A.H., *et al.* (1971) Soils of north Norfolk. *Journal of Soil Science*, **22**, 444–52.

Catt, J.A., Weir, R.A. and Madgett, P.A. (1974) The loess of eastern Yorkshire and Lincolnshire. *Proceedings of the Yorkshire Geological Society*, **40**, 23–34.

Cepek, A.G. (1986) Quaternary stratigraphy of the German Democratic Republic. *Quaternary Science Reviews*, **5**, 359–64.

Cepek, A.G. and Erd, K. (1982) Classification and stratigraphy of the Holsteinian and Saalian complex in the Quaternary of the German Democratic Republic. In *Quaternary Glaciations in the Northern Hemisphere* (eds D.J. Easterbrook, P. Hansliêk, K-D. Jäger and F. W. Shotton), UNESCO – International Geological Correlation Programme, Project 73/1/24 Report 7, Prague 1981, pp. 50–7.

Chandler, R.H. (1914) The Pleistocene deposits of Crayford. *Proceedings of the Geologists' Association*, **25**, 61–70.

Chandler, R.H. (1916) The implements and cores of Crayford. *Proceedings of the Prehistoric Society of East Anglia*, **2**, 240–8.

Chandler, R.H. (1930) On the Clactonian Industry at Swanscombe. *Proceedings of the Prehistoric Society of East Anglia*, **6**, 79–116.

Chandler, R.H. (1931) On the Clactonian Industry and report of field meeting at Swanscombe. *Proceedings of the Geologists' Association*, **42**, 175–7.

Chandler, R.H. (1932a) Notes on types of Clactonian implements at Swanscombe. *Proceedings of the Prehistoric Society of East Anglia*, **6**, 377–8.

Chandler, R.H. (1932b) The Clactonian industry and report of field meeting at Swanscombe (II), *Proceedings of the Geologists' Association*, **43**, 70–2.

Chandler, R.H. and Leach, A.L. (1907) Excursion to Crayford and Dartford Heath. *Proceedings of the Geologists' Association*, **20**, 122–6.

Chandler, R.H. and Leach, A.L. (1911) Excursion to Dartford Heath. *Proceedings of the Geologists' Association*, **22**, 171–5.

Chandler, R.H. and Leach, A.L. (1912) On the Dartford Heath Gravel and on a Palaeolithic implement factory. *Proceedings of the Geologists' Association*, **23**, 102–11.

Chartres, C.J. (1975) Soil development on the terraces of the River Kennet. Unpublished Ph.D. thesis, University of Reading.

Chartres, C.J. (1980), A Quaternary soil sequence in the Kennet valley, central southern England. *Geoderma*, **23**, 125–146.

Chartres, C.J. (1981) The mineralogy of Quaternary deposits in the Kennet valley, Berkshire. *Proceedings of the Geologists' Association*, **92**, 93–103.

Chartres, C.J. (1984) The micromorphology of Quaternary river terrace deposits in the Kennet valley, Berkshire, England. *Earth Surface Processes and Landforms*, **9**, 343–55.

Chartres, C.J., Cheetham, G.H. and Fenwick, I.M. (1976) Excursion to the Kennet valley. In *Field Guide to the Oxford Region* (ed. D. Roe), Quaternary Research Association, Oxford, pp. 23–31.

Chatwin C.P. (1927) Fossils from the ironsands on Netley Heath (Surrey), *Summary of Progress, Geological Survey of Great Britain* [for 1926], pp. 154–7.

Cheetham, G.H. (1980) Late Quaternary palaeohydrology: the Kennet valley case study. In *The Shaping of Southern England* (ed. D.K.C. Jones), Institute of British Geographers Special Publication 11, Academic Press, London, pp. 203–23.

Cheshire, D.A. (1978) The Glaciation of the Lea valley between Hertford and Enfield. Unpublished M.Sc. thesis, City of London Polytechnic and Polytechnic of North London.

Cheshire, D.A. (1981) A contribution towards a glacial stratigraphy of the lower Lea valley, and implications for the Anglian Thames. *Quaternary Studies*, **1**, 27–69.

Cheshire, D.A. (1983a) Till lithology in Hertfordshire and west Essex. In *Diversion of the Thames* (ed. J. Rose), Field Guide, Quaternary Research Association, Cambridge, pp. 50–9.

Cheshire, D.A. (1983b) Westmill. In *Diversion of the Thames* (ed. J. Rose), Field Guide, Quaternary Research Association, Cambridge, pp. 120–32.

Cheshire, D.A. (1983c) Hoddesdon, St Albans Sand and Gravel Co. Quarry and Hoddesdon, Nursery Grove Pits. In *Diversion of the Thames* (ed. J. Rose), Field Guide, Quaternary Research Association, Cambridge, pp. 140–8.

Cheshire, D.A. (1986a) The lithology and stratigraphy of the Anglian deposits of the Lea Basin. Unpublished Ph.D. thesis, Hatfield Polytechnic.

Cheshire, D.A. (1986b) The use of small clast counts as a means of till differentiation in Hertfordshire and western Essex. In *Clast Lithological Analysis* (ed. D.R. Bridgland), Technical Guide 3, Quaternary Research Association, Cambridge, 129–43.

References

Cheshire, D.A. and Gibbard, P.L. (1983) Harper Lane. In *Diversion of the Thames* (ed. J. Rose), Field Guide, Quaternary Research Association, Cambridge, 102–9.

Clark, W.E.LeG. (1955) *The Fossil Evidence for Human Evolution*. Chicago University Press.

Clarke, M.R. and Dixon, A.J. (1981) The Pleistocene braided river deposits in the Blackwater area of Berkshire and Hampshire, England. *Proceedings of the Geologists' Association*, **92**, 139–57.

Clayton, K.M. (1957) Some aspects of the glacial deposits of Essex. *Proceedings of the Geologists' Association*, **68**, 1–19.

Clayton, K.M. (1960) The landforms of parts of southern Essex. *Transactions of the Institute of British Geographers*, **28**, 55–74.

Clayton, K.M. (1964) The glacial geomorphology of southern Essex. In *Guide to London Excursions* (ed. K.M. Clayton), International Geographical Congress [London 1964], **20**, 123–8.

Clayton, K.M. (1977) River terraces. In *British Quaternary Studies: Recent Advances* (ed. F.W. Shotton), Clarendon Press, Oxford, pp. 153–68.

Clayton, K.M. and Brown, J.C. (1958) The glacial deposits around Hertford. *Proceedings of the Geologists' Association*, **69**, 103–19

Clinch, G. (1908) Early Man. In *The Victoria History of the County of Kent* (ed. W. Page), Vol. 1, Archibald Constable Ltd., Westminster, pp. 307–38.

Coles, R. (1934) The evolution of the coastal drainage of Essex. *Essex Naturalist*, **25**, 36–49 and 65–70.

Collins, D. (1969) Culture, traditions and environment of early Man. *Current Anthropologist*, **10**, 267–316.

Conway, B.W. (1969) Preliminary geological investigation of Boyn Hill Terrace deposits at Barnfield Pit, Swanscombe, Kent during 1968. *Proceedings of the Royal Anthropological Institute* [for 1968], 59–61.

Conway, B.W. (1970a) Geological investigation of Boyn Hill Terrace deposits at Barnfield Pit, Swanscombe, Kent, during 1969. *Proceedings of the Royal Anthropological Institute* [for 1969], 90–3.

Conway, B.W. (1970b) Written discussions on R.G. West 1969. Pollen analyses from interglacial deposits at Aveley and Grays, Essex. *Proceedings of the Geologists' Association*, **81**, 177–9.

Conway, B.W. (1971) Geological investigation of Boyn Hill Terrace deposits at Barnfield Pit, Swanscombe, Kent during 1970. *Proceedings of the Royal Anthropological Institute* [for 1970], 60–4.

Conway, B.W. (1972) Geological investigation of Boyn Hill Terrace deposits at Barnfield Pit, Swanscombe, Kent during 1971. *Proceedings of the Royal Anthropological Institute* [for 1971], 80–5.

Conway, B.W. (1985) Research history and geology of Barnfield Pit. In *The Story of Swanscombe Man* (ed. K.L. Duff), Kent County Council and Nature Conservancy Council, pp. 6–13.

Conway B.W. and Waechter, J. d'A. (1977) Lower Thames and Medway valleys – Barnfield Pit, Swanscombe. In *South East England and the Thames Valley* (eds E.R. Shephard-Thorn and J.J. Wymer), Guide Book for Excursion A5, X INQUA Congress, Birmingham, Geoabstracts, Norwich, pp. 38–44.

Cook, J., Stringer, C.B., Currant, A.P., *et al.* (1982) A review of the chronology of the European Middle Pleistocene hominid record. *Yearbook of Physical Anthropology*, **25**, 19–65.

Coope, G.R. (1968) An insect fauna from the Mid-Weichselian deposits at Brandon, Warwickshire. *Philosophical Transactions of the Royal Society of London*, **B254**, 425–56.

Coope, G.R. (1987) The response of late Quaternary insect communities to sudden climatic changes. In *Organisation of Communities, Past and Present* (eds J.H.R. Gee and P.S. Giller), Blackwell Scientific, Oxford, pp. 421–38.

Coope, G.R. and Angus, R.B. (1975) An ecological study of a temperate interlude in the middle of the last glaciation, based on fossil Coleoptera from Isleworth, Middlesex. *Journal of Animal Ecology*, **44**, 365–91.

Coope, G.R., Shotton, F.W. and Strachan, I. (1961) A Late Pleistocene fauna and flora from Upton Warren, Worcestershire. *Philosophical Transactions of the Royal Society of London*, **B244**, 379–421.

Cooper, J. (1972) Last interglacial (Ipswichian) non-marine Mollusca from Aveley, Essex. *Essex Naturalist*, **33**, 9–14.

Cornwall, I.W. (1950) Pleistocene and Holocene sections in deposits of the Lower Thames. *University of London Institute of Archaeology, 6th Annual Report*, 34–43.

Cornwall, I.W. (1958) *Soils for the Archaeologist*, Phoenix House, London. 230 pp.

Cotton, R.P. (1847) On the Pliocene deposits of the valley of the Thames at Ilford. *Annals and Magazine of Natural History, Series 1*, **20**, 164–9.

Cranshaw, S. (1983) Handaxes and cleavers: selected English Acheulian industries. *British Archaeological Report, British Series*, **113**, 283 pp.

Currant, A.P. (1986) Man and Quaternary interglacial faunas in Britain. In *The Palaeolithic of Britain and its Nearest Neighbours; Recent Trends* (ed. S.N. Collcutt), J.R. Collis Publications, Department of Archaeology and Prehistory, Sheffield University, pp. 50–2.

Currant, A.P. (1989) The Quaternary origins of the modern British mammal fauna. *Biological Journal of the Linnean Society,* **38**, 23–30.

Curry, D., Adams, C.G., Boulter, M.C., *et al.* (1978) *A Correlation of Tertiary Rocks in the British Isles*, Special Report of the Geological Society of London, No. 12, 72 pp.

Dalrymple, J.B. (1958) The application of soil micromorphology to fossil soils and other deposits from archaeological sites. *Journal of Soil Science*, **9**, 199–209.

Dalton, W.H. (1880) *The Geology of the Neighbourhood of Colchester*. Memoir of the Geological Survey of Great Britain, 24 pp.

Dalton, W.H. (1890) Note on the Upminster brickyard. *Essex Naturalist*, **4**, 186–7.

Dalton, W.H. (1908) Post-glacial beds in Mersea, Essex. *Essex Naturalist*, **15**, 136–7.

Davies, G.M. (1915) The rocks and minerals of the Croydon Regional Survey area. *Proceedings and Transactions of the Croydon Natural History and Scientific Society*, **8**, 53–96.

Davies, G.M. (1917) Excursion to Netley Heath, Newlands Corner and the Silent Pool. *Proceedings of the Geologists' Association*, **28**, 48–51.

Davis, A.G. (1953) On the geological history of some of our snails illustrated by some Pleistocene and Holocene deposits in Kent and Surrey. *Journal of Conchology*, **23**, 355–64.

Davis, W.M. (1895) The development of certain English rivers. *Geographical Journal*, **5**, 127–46.

Davis, W.M. (1899) The drainage of cuestas. *Proceedings of the Geologists' Association*, **16**, 87–93.

Davis, W.M. (1909) The valleys of the Cotswold Hills. *Proceedings of the Geologists' Association*, **21**, 150–2.

Dawkins, W.B. (1867) On the age of the lower brickearth of the Thames valley. *Quarterly Journal of the Geological Society of London*, **23**, 91–109.

Dawkins, W.B. (1868) On a new species of fossil deer from Clacton. *Quarterly Journal of the Geological Society of London*, **24**, 511–13.

Dawkins, W.B. (1869) On the distribution of the British post-glacial mammals. *Quarterly Journal of the Geological Society of London*, **25**, 192–217.

Day, M.H. (1977) *Guide to Fossil Man: A Handbook of Human Palaeontology*, 3rd edn, Cassell, London, 346 pp.

Devoy, R.J.N. (1977) Flandrian sea level changes in the Thames estuary and the implications for land subsidence in England and Wales. *Nature, London*, **270**, 712–15.

Devoy, R.J.N. (1979) Flandrian sea level changes and vegetational history of the Lower Thames estuary. *Philosophical Transactions of the Royal Society of London*, **B285**, 355–407.

Dewey, H. (1919) On some Palaeolithic flake implements from the high level terraces of the Thames valley. *Geological Magazine*, **6**, 49–57.

Dewey, H. (1930) Palaeolithic Thames deposits. *Proceedings of the Prehistoric Society of East Anglia*, **6**, 147–155.

Dewey, H. (1932) The Palaeolithic deposits of the Lower Thames valley. *Quarterly Journal of the Geological Society of London*, **88**, 35–56.

Dewey, H. (1934) The excursion to the 100-foot terrace of the Thames at Swanscombe, Kent (4th August), *International Congress of Prehistoric and Protohistoric Science* (London), 70–2.

Dewey, H. (1959) Palaeolithic deposits of the Thames at Dartford Heath and Swanscombe, north Kent. Unpublished, edited text of Henry Stopes memorial lecture, Geologists' Association, 1959.

Dewey, H. and Bromehead, C.E.N. (1915) *The Geology of the Country around Windsor and Chertsey*. Memoir of the Geological Survey of Great Britain, 123 pp.

Dewey, H. and Bromehead, C.E.N. (1921) *The Geology of South London*. Memoir of the Geological Survey of Great Britain, 92 pp.

Dewey, H. and Smith, R.A. (1914) The Palaeolithic sequence at Swanscombe, Kent. *Proceedings of the Geologists' Association*, **25**, 90–7.

Dewey, H., Bromehead, C.E.N., Chatwin, C.P., *et al.* (1924) *The Geology of the Country around Dartford*, Memoir of the Geological Survey of Great Britain, 136 pp.

Dibley, G.E. and Kennard, A.S. (1916) Excursion to Grays. *Proceedings of the Geologists' Association*, **27**, 103–5.

Diebel, K. and Wolfschläger, H. (1975) Ostracoden aus dem junpleistozänen Travertin von Ehringsdorf bei Weimar. *Abhandlungen des Zentralen Geologischen Instituts* [Berlin], **23**, 91–136.

Dines, H.G. (1928) On the glaciation of the north Cotteswold area. *Summary of Progress, Geological Survey of Great Britain* [for 1927], pp. 66–71.

Dines, H.G. (1946) Pleistocene and recent deposits. In *The Geology of the Country around Witney* (eds L.S. Richardson, W.J. Arkell and H.G. Dines), Memoir of the Geological Survey of Great Britain, 105–29.

Dines, H.G. and Chatwin, C.P. (1930) Pliocene sandstone from Rothamstead (Hertfordshire), *Summary of Progress, Geological Survey of Great Britain* [for 1929], pp. 1–7.

Dines, H.G. and Edmunds, F.H. (1925) *The Geology of the Country around Romford*, Memoir of the Geological Survey of Great Britain, 53 pp.

Dines, H.G. and Edmunds, F.H. (1929) *The Geology of the Country around Aldershot and Guildford*, Memoir of the Geological Survey of Great Britain, 182 pp.

Dines, H.G., King, W.B.R. and Oakley, K.P. (1938) A general account of the 100 ft terrace gravels of the Barnfield Pit, Swanscombe. *Journal of the Royal Anthropological Institute*, **68**, 21–7.

Docherty, J. (1967) The exhumed sub-Tertiary surface in north-west Kent. *South East Naturalist*, **70**, 19–31.

Docherty, J. (1971) Chalk karst: a synthesis of C.C. Faggs' theories of chalkland morphology in the light of recent hydrological research. *Proceedings of the Croydon Natural History and Scientific Society*, **15**, 21–34.

D'Olier, B. (1975) Some aspects of late Pleistocene-Holocene drainage of the River Thames in the eastern part of the London Basin. *Philosophical Transactions of the Royal Society of London*, **A279**, 269–77.

Duff, K.L. (1985) (ed.) *The Story of Swanscombe Man*, Kent County Council and Nature Conservancy Council, 40 pp.

Duigan, S.L. (1955) Plant remains from the gravels of the Summertown-Radley Terrace near Dorchester, Oxfordshire. *Quarterly Journal of the Geological Society of London*, **111**, 225–38.

Duigan, S.L. (1956) Interglacial plant remains from the Wolvercote channel, Oxford. *Quarterly Journal of the Geological Society of London*, **112**, 363–72.

Duphörn, K., Grube, F., Meyer, K.D., *et al.* (1973) Pleistocene and Holocene. *Eiszeitalter und Gegenwart*, **23/24**, 222–50.

Eden, D.N. (1980) The loess of north-east Essex, England. *Boreas*; **9**, 165–77.

Ehlers, J. (1981) Problems of the Saalian stratigraphy in the Hamburg area. *Mededelingen Rijks Geologische Dienst*, **34**, 26–9.

Ellis, T.S. (1882) On some features in the formation of the Severn valley as seen near Gloucester. *Transactions of the School of Science Philosophical Society, Gloucester* [for 1882], pp. 3–15.

Emiliani, C. (1955) Pleistocene temperatures. *Journal of Geology*, **63**, 538–78.

Emiliani, C. (1957) Temperature and age analysis of deep-sea cores. *Science, New York*, **125**, 383–7.

Evans, J. (1860) On the occurrence of flint implements in undisturbed beds of gravel, sand and clay. *Archaeologia*, **38**, 280–307.

Evans, J. (1872) *The Ancient Stone Implements, Weapons and Ornaments of Great Britain*, 1st edn, Longmans, Green and Co., London, 640 pp.

Evans, J. (1897) *The Ancient Stone Implements, Weapons and Ornaments of Great Britain*, 2nd edn, Longmans, Green, and Co, London, 747 pp.

Evans, P. (1954) Field meeting in the Vale of St Albans. *Proceedings of the Geologists' Association*, **65**, 18–22.

Evans, P. (1971) Towards a Pleistocene timescale. Part 2 of *The Phanerozoic Time-scale – A Supplement*, Special Publication of the Geological Society of London, No. 5, pp. 123–356.

Falconer, H. (1868) *Palaeontological Memoirs and Notes, Compiled and Edited by Charles Murchison*, Vol. 2, R. Hardwicke, London, 675 pp.

Federoff, N. (1971) Caractères micromorphologiques des pédogénèses quaternaire en France. *Bulletin de l'Association Française pour l'Étude du Quaternaire*, Supplément, 4, 341–9.

Fisher, O. (1868a) A few notes on Clacton, Essex. *Geological Magazine*, **5**, 213–5.

Fisher, O. (1868b) The boulder clay at Witham and the Thames valley. *Geological Magazine*, **5**, 98–100.

Fisher, P.F. (1982) A study of the plateau gravels in the western part of the London Basin. Unpublished Ph.D. thesis, Kingston Polytechnic, 2 volumes.

Franks, J.W. (1960) Interglacial deposits at Trafalgar Square, London. *New Phytologist*, **59**, 145–52.

Franks, J.W., Sutcliffe, A.J., Kerney, M.P., *et al.* (1958) Haunt of the elephant and rhinoceros: the Trafalgar Square of 100,000 years ago – new discoveries. *Illustrated London News*, 14th June, Vol. 232, pp. 1011–3.

French, H.H. (1888) Excursion to Gomshall, Netley Heath, and Clandon. *Proceedings of the Geologists' Association*, **10**, 182–6.

French, J. (1891) On the occurrence of Westleton Beds in part of north-western Essex. *Essex Naturalist*, **5**, 210–18.

Friedman, G.M. (1967) Dynamic processes and statistical parameters compared for size frequency distribution of beach and river sands. *Journal of Sedimentary Petrology*, **37**, 327–54.

Gascoyne, M., Currant, A.P. and Lord, T.C. (1981) Ipswichian fauna of Victoria Cave and the marine palaeoclimatic record. *Nature, London*, **294**, 652–4.

Geikie, A. and Reid, C. (1866) The Pliocene deposits of north-western Europe. *Nature, London*, **34**, 341–3.

Gibbard, P.L. (1974) Pleistocene stratigraphy and vegetational history of Hertfordshire. Unpublished Ph.D. thesis, University of Cambridge.

Gibbard, P.L. (1977) Pleistocene history of the Vale of St Albans. *Philosophical Transactions of the Royal Society of London*, **B280**, 445–83.

Gibbard, P.L. (1978a) Quaternary geology and landform development in the Vale of St Albans. In *Field Guide to the Vale of St Albans* (eds J. Rose and P.L. Gibbard), Quaternary Research Association, London, pp. 9–29.

Gibbard, P.L. (1978b) Westmill. In *Field Guide to the Vale of St Albans* (eds J. Rose and P.L. Gibbard), Quaternary Research Association, London, pp. 63–7.

Gibbard, P.L. (1978c) Hatfield Polytechnic. In *Field Guide to the Vale of St Albans* (eds J. Rose and P.L. Gibbard), Quaternary Research Association, London, pp. 79–85.

Gibbard, P.L. (1978d) Moor Mill. In *Field Guide to the Vale of St Albans* (eds J. Rose and P.L. Gibbard), Quaternary Research Association, London, pp. 87–90.

Gibbard, P.L. (1979) Middle Pleistocene drainage in the Thames valley. *Geological Magazine*, **116**, 35–44.

Gibbard, P.L. (1982) Terrace stratigraphy and drainage history of the plateau gravels of north Surrey, south Berkshire, and north Hampshire, England. *Proceedings of the Geologists' Association*, **93**, 369–84.

Gibbard, P.L. (1983) The diversion of the Thames – a review. In *Diversion of the Thames* (ed. J. Rose), Field Guide, Quaternary Research Association, Cambridge, pp. 8–23.

Gibbard,P.L. (1985) *The Pleistocene History of the Middle Thames Valley*, Cambridge University Press, 155 pp.

Gibbard, P.L. (1986) Comparison of the clast lithological composition of the gravels in the middle Thames using canonical variates analysis and principal components analysis. In *Clast Lithological Analysis* (ed. D.R. Bridgland), Technical Guide No. 3, Quaternary Research Association, Cambridge, pp. 153–64.

Gibbard, P.L. (1988a) The history of the great northwest European rivers during the past three million years. *Philosophical Transactions of the Royal Society of London*, **B318**, 559–602.

Gibbard, P.L. (1988b) Palynological problems and the vegetational sequence of the Pliocene–preglacial Pleistocene of East Anglia. In *Pliocene–Middle Pleistocene of East Anglia* (eds P.L. Gibbard and J.A. Zalasiewicz), Field Guide, Quaternary Research Association, Cambridge, pp. 42–9.

Gibbard, P.L. (1989) The geomorphology of a part of the Middle Thames forty years on: a reappraisal of the work of F. Kenneth Hare. *Proceedings of the Geologists' Association*, **100**, 481–503.

Gibbard, P.L. and Cheshire, D.A. (1983) Hatfield

Polytechnic (Roe Hyde Pit), In *Diversion of the Thames* (ed. J. Rose), Field Guide, Quaternary Research Association, Cambridge, pp. 110–9.

Gibbard, P.L. and Peglar, S.M. (1990) Palynology of the interglacial deposits at Little Oakley, Essex, and their correlation. *Philosophical Transactions of the Royal Society of London*, **B328**, 341–57.

Gibbard, P.L. and Wymer, J.J. (1983) Highlands Farm. In *Diversion of the Thames* (ed. J. Rose) Field Guide, Quaternary Research Association, Cambridge, pp. 69–76.

Gibbard, P.L. and Pettit, M. (1978) The palaeobotany of the interglacial deposits at Sugworth, Berkshire. *New Phytologist*, **81**, 465–77.

Gibbard, P.L. and Stuart, A.J. (1974) Trace fossils from pro-glacial lake sediments. *Boreas*, **3**, 69–74.

Gibbard, P.L., Coope, G.R., Hall, A.R., *et al.* (1982) Middle Devensian river deposits beneath the 'Upper Floodplain' terrace of the River Thames at Kempton Park, Sunbury, Surrey, England. *Proceedings of the Geologists' Association*, **93**, 275–90.

Gibbard, P.L., Wintle, A.G. and Catt, J.A. (1987) Age and origin of clayey silt 'brickearth' in West London, England. *Journal of Quaternary Science*, **2**, 3–9.

Gibbard, P.L., Whiteman, C.A. and Bridgland, D.R. (1988) A preliminary report on the stratigraphy of the Lower Thames valley. *Quaternary Newsletter*, **56**, 1–8.

Gibbard, P.L., West, R.G., Zagwijn, W.H., *et al.* (1991) Early and early Middle Pleistocene correlations in the southern North Sea Basin. *Quaternary Science Reviews*, **10**, 23–52.

Gilbert, C.J. (1919a) On the occurrence of the extensive deposits of high-level sands and gravels resting upon the chalk at Little Heath, near Berkhampstead. *Quarterly Journal of the Geological Society of London*, **75**, 32–43.

Gilbert, C.J. (1919b) Excursion to Berkhamstead and Little Heath. *Proceedings of the Geologists' Association*, **30**, 87–91.

Gilbertson, D.D. (1976) Non-marine molluscan faunas of terrace gravels in the Upper Thames Basin. In *Field Guide to the Oxford Region* (ed. D.A. Roe), Quaternary Research Association, Oxford, pp. 16–9.

Gilbertson, D.D. (1980) The palaeoecology of the Middle Pleistocene Mollusca from Sugworth, Oxfordshire. *Philosophical Transactions of the Royal Society of London*, **B289**, 107–18.

Gladfelter, B.G. (1972) Cold-climate features in the vicinity of Clacton-on-Sea, Essex (England), *Quaternaria*, **16**, 121–35.

Gladfelter, B.G. (1975) Middle Pleistocene sedimentary sequences in East Anglia (UK), In *After the Australopithecines: Stratigraphy, Ecology and Culture Change in the Middle Pleistocene* (eds K.W. Butzer and G.L. Isaac), Mouton, The Hague, pp. 225–58.

Gladfelter, B.G. and Singer, R. (1975) Implications of East Anglian glacial stratigraphy for the British Lower Palaeolithic. *In* Quaternary Studies (eds R.P. Suggate and M.M. Cresswell), Selected papers from IX INQUA Congress, Christchurch, New Zealand, 2–10 December 1973. *Bulletin of the Royal Society of New Zealand*, **13**, 139–45.

Goudie, A.S. (1976) The Oxford region. In *Field Guide to the Oxford Region* (ed. D.A. Roe), Quaternary Research Association, Oxford, pp. 1–5.

Goudie, A.S. and Hart, M.G. (1975) Pleistocene events and forms in the Oxford region. In *Oxford and its Region* (eds C.G. Smith and D.I. Scargill), Oxford University Press, pp. 3–13.

Gray, J.W. (1911) The north and mid Cotteswolds and the Vale of Moreton during the Glacial Epoch. *Proceedings of the Cotteswolds Naturalists Field Club*, **17**, 257–74.

Green, A.H. (1864) *The Geology of Banbury, Woodstock, Bicester and Buckingham*, Memoir of the Geological Survey of Great Britain, 62 pp.

Green, C.P. and McGregor, D.F.M. (1978a) Pleistocene gravel trains of the River Thames. *Proceedings of the Geologists' Association*, **89**, 143–56.

Green, C.P. and McGregor, D.F.M. (1978b) Pleistocene gravel deposits of the Vale of St Albans and the Middle Thames. In *Field Guide to the Vale of St Albans* (eds J. Rose and P.L. Gibbard), Quaternary Research Association, London, pp. 31–7.

Green, C.P. and McGregor, D.F.M. (1978c) Westwood. In *Field Guide to the Vale of St Albans* (eds J. Rose and P.L. Gibbard), Quaternary Research Association, London, p. 91.

Green, C.P. and McGregor, D.F.M. (1980) Quaternary evolution of the River Thames. In *The Shaping of Southern England* (ed. D.K.C. Jones), Institute of British Geographers

Special Publication 11, Academic Press, London, pp. 177–202.

Green, C.P. and McGregor, D.F.M. (1983) Lithology of the Thames gravels. In *Diversion of the Thames* (ed. J. Rose), Field Guide, Quaternary Research Association, Cambridge, pp. 24–8.

Green, C.P. and McGregor, D.F.M. (1986) The utility of intercomponent ratios in the interpretation of stone count data. In *Clast Lithological Analysis* (ed. D.R. Bridgland), Technical Guide No.3, Quaternary Research Association, Cambridge, pp. 83–93.

Green, C.P. and McGregor, D.F.M. (1987) River terraces: a stratigraphic record of environmental change. In *International Geomorphology 1986 Part 1* (ed. V. Gardiner), Wiley, Chichester, pp. 977–87.

Green, C.P., Hey, R.W. and McGregor, D.F.M. (1980) Volcanic pebbles in Pleistocene gravels of the Thames in Buckinghamshire and Hertfordshire. *Geological Magazine*, **117**, 59–64.

Green, C.P., McGregor, D.F.M. and Evans, A. (1982) Development of the Thames drainage system in Early and Middle Pleistocene times. *Geological Magazine*, **119**, 281–90.

Green, C.P., Coope, G.R., Currant, A.P., *et al.*, (1984) Evidence for two temperate episodes in late Pleistocene deposits at Marsworth, Buckinghamshire. *Nature, London*, **309**, 778–81.

Green, H.S. (1984) *Pontnewydd Cave. A Lower Palaeolithic Hominid Site in Wales: the First Report*, National Museum of Wales, Cardiff, 227 pp.

Greensmith, J.T. and Tucker, E.V. (1980) Evidence for differential subsidence on the Essex coast. *Proceedings of the Geologists' Association*, **91**, 169–75.

Gregory, J.W. (1894) Evolution of the Thames. *Natural Science*, **5**, 97–108.

Gregory, J.W. (1922) *Evolution of the Essex Rivers and of the Lower Thames*, Benham, Colchester, 64 pp.

Grube, F., Christensen, S. and Vollmer, T. (1986) Glaciations in north west Germany. *Quaternary Science Reviews*, **5**, 347–57.

Gruhn, R., Bryan, A.L. and Moss, A.J. (1974) A contribution to Pleistocene chronology in south east Essex, England. *Quaternary Research*, **4**, 53–71.

Grün, R., Schwarcz, H.P. and Chadwin, J. (1988) ESR dating of tooth enamel: coupled correction for U-uptake and U-series disequilibrium. *Nuclear Tracks and Radiation Measures*, **14**, 237–41.

Harding, P. and Gibbard, P.L. (1984) Excavations at Northwold Road, Stoke Newington, north east London, 1981. *Transactions of the Middlesex Archaeological Society*, **34**, 1–18.

Harding, P., Bridgland, D.R., Madgett, P.A. *et al.* (1991) Recent investigations of Pleistocene sediments near Maidenhead, Berkshire, and their archaeological content. *Proceedings of the Geologists' Association*, **102**, 25–53.

Hare, F.K. (1947) The geomorphology of a part of the Middle Thames. *Proceedings of the Geologists' Association*, **58**, 294–339.

Harmer, F.W. (1902) A sketch of the later Tertiary history of East Anglia. *Proceedings of the Geologists' Association*, **17**, 416–79.

Harmer, F.W. (1907) On the origin of certain cañon-like valleys associated with lake-like areas of depression. *Quarterly Journal of the Geological Society of London*, **63**, 470–514.

Harries, W.J.R. (1977) *The Sand and Gravel Resources of the Country around Eynsham, Oxfordshire*. Mineral Assessment Report of the Institute of Geological Sciences 28, 88 pp.

Hart, J. McA. (1960), Field meeting at Grays Thurrock. *Proceedings of the Geologists' Association*, **71**, 242–4.

Hawkins, H.L. (1922) The relation of the River Thames to the London Basin. *Report of the British Association for the Advancement of Science* [for 1922], 365–6.

Hawkins, H.L (1928) Excursion to Kingsclere. *Proceedings of the Geological Society of London*, **39**, 98–102.

Hedberg, H.D. (1976) *International Stratigraphic Guide*. Wiley, New York, 200 pp.

Hey, R.W. (1965) Highly quartzose pebble gravels in the London Basin. *Proceedings of the Geologists' Association*, **76**, 403–20.

Hey, R.W. (1967) The Westleton Beds reconsidered. *Proceedings of the Geologists' Association*, **78**, 427–45.

Hey, R.W. (1976a) The terraces of the Middle and Lower Thames. *Studia Societatis Scientiarum Torunensis*. Torun-Polonia, **8C**, 115–22.

Hey, R.W. (1976b) Provenance of far-travelled pebbles in the pre-Anglian Pleistocene of East Anglia. *Proceedings of the Geologists' Association*, **87**, 69–82.

Hey, R.W. (1980) Equivalents of the Westland

Green Gravels in Essex and East Anglia. *Proceedings of the Geologists' Association*, **91**, 279–90.

Hey, R.W. (1982) Composition of Pre-Anglian gravels in Norfolk. *Bulletin of the Geological Society of Norfolk*, **32**, 51–9.

Hey, R.W. (1983) Ferneux Pelham. In *Diversion of the Thames* (ed. J. Rose), Field Guide, Quaternary Research Association, Cambridge, pp. 94–5.

Hey, R.W. (1986) A re-examination of the Northern Drift of Oxfordshire. *Proceedings of the Geologists' Association*, **97**, 291–302.

Hey, R.W. and Auton, C.A. (1988) Compositions of pebble-beds in the Neogene and pre-Anglian Pleistocene of East Anglia. In *Pliocene-Middle Pleistocene of East Anglia* (eds P.L. Gibbard and J.A. Zalasiewicz), Field Guide, Quaternary Research Association, Cambridge, pp. 35–41.

Hey, R.W. and Brenchley, P.J. (1977) Volcanic pebbles from Pleistocene gravels in Norfolk and Essex. *Geological Magazine*, **114**, 219–25.

Hey, R.W., Krinsley, D.H. and Hyde, P.J.W. (1971) Surface textures of sand grains from the Hertfordshire pebble gravels. *Geological Magazine*, **108**, 377–82.

Hiller, D. (1972) Untersuchungen zur Biologie und zur Ökologie limnischer Ostracoden aus der Umbebung von Hamburg. *Archiv für Hydrobiologie, Supplementband 40* (Stuttgart), **Heft 4**, 400–97.

Hinton, M.A.C. (1900a) The Pleistocene deposits of the Ilford and Wanstead district, Essex. *Essex Naturalist*, **11**, 161–5.

Hinton, M.A.C. (1900b) The Pleistocene deposits of the Ilford and Wanstead district. *Proceedings of the Geologists' Association*, **16**, 271–81.

Hinton, M.A.C. (1901) Excursion to Grays Thurrock. *Proceedings of the Geologists' Association*, **17**, 141–4.

Hinton, M.A.C. (1910) A preliminary account of the British voles and lemmings; with some remarks on the Pleistocene climate and geography. *Proceedings of the Geologists' Association*, **21**, 489–507.

Hinton, M.A.C. (1911) The British fossil shrews. *Geological Magazine*, **8**, 529–39.

Hinton, M.A.C. (1923) Note on the rodent remains from Clacton-on-Sea. *Quarterly Journal of the Geological Society of London*, **79**, 626.

Hinton, M.A.C. (1926a) The Pleistocene mammalia of the British Isles and their bearing upon the date of the Glacial Period. *Proceedings of the Yorkshire Geological Society New Series*, **20**, 325–48.

Hinton, M.A.C. (1926b) *Monograph of the Voles and Lemmings (Microtinae), Living and Extinct*, Volume 1 [Volume 2 not published], British Museum, London, 488 pp.

Hinton, M.A.C. and Kennard, A.S. (1900) Contributions to the Pleistocene geology of the Thames valley, I. The Grays Thurrock area, part I. *Essex Naturalist*, **11**, 336–70.

Hinton, M.A.C. and Kennard, A.S. (1905) The relative ages of the stone implements of the Lower Thames valley. *Proceedings of the Geologists' Association*, **19**, 76–100.

Hinton, M.A.C. and Kennard, A.S. (1907) Contributions to the Pleistocene geology of the Thames valley I. The Grays Thurrock area, Part II (Revised), *Essex Naturalist*, **15**, 56–88.

Holland, C.H., Audley-Charles, M.G., Bassett, M.G., *et al.* (1978) *A Guide to Stratigraphic Procedure*, Special Report for the Geological Society of London, No. 10, 18 pp.

Hollin, J.T. (1971) Ice-sheet surges and interglacial sea levels. Unpublished Ph.D. thesis, Princeton University, 179 pp.

Hollin, J.T. (1977) Thames interglacial sites, Ipswichian sea levels and Antarctic ice surges. *Boreas*, **6**, 33–52.

Hollyer, S.E. and Simmons, M.B. (1978) *The Sand and Gravel Resources of the Country around Southend-on-Sea, Essex*. Mineral Assessment Report of the Institute of Geological Sciences 36, 212 pp.

Holman, J.A. (1987) Middle Pleistocene herpetological records from interglacial deposits at Sugworth near Oxford. *British Herpetological Society Bulletin*, **21**, 5–7.

Holman, J.A. and Clayden, J.D. (1988) Pleistocene interglacial herpetofauna from the Greenlands Pit, Purfleet, Essex. *British Herpetological Society Bulletin*, **26**, 26–7.

Holman, J.A., Stuart, A.J. and Clayden, J.D. (1990) A Middle Pleistocene herpetofauna from Cudmore Grove, Essex, England, and its paleogeographic and paleoclimatic implications. *Journal of Vertebrate Paleontology*, **10**, 86–94.

Holmes, T.V. (1890) Some sections between West Thurrock and Stifford on the Grays and Upminster railway. *Essex Naturalist*, **4**, 143–9.

Holmes, T.V. (1892a) The new railway from Grays Thurrock to Romford: sections between Upminster and Romford. *Quarterly Journal of the Geological Society of London*, **48**, 365–72.

Holmes, T.V. (1892b) Excursion to the cuttings on the new railway between Upminster and Romford. *Proceedings of the Geologists' Association*, **12**, 316–9.

Holmes, T.V. (1892c) Recent excursions of the Geologists' Association in Essex. Upminster and Hornchurch. *Essex Naturalist*, **6**, 96–7.

Holmes, T.V. (1893) The new railway between Upminster and Romford. Boulder Clay beneath old river gravel at Hornchurch. Conclusions therefrom. *Essex Naturalist*, **7**, 1–14.

Holmes, T.V. (1894) Further notes on some sections of the new railway from Romford to Upminster, and on the relations of the Thames valley beds to the boulder clay. *Quarterly Journal of the Geological Society of London*, **50**, 443–52.

Holmes, T.V. (1896) Notes on the ancient physiography of south Essex. *Essex Naturalist*, **9**, 193–200.

Holyoak, D.T. (1983) A late Pleistocene interglacial flora and molluscan fauna from Thatcham, Berkshire, with notes on the Mollusca from the interglacial deposits at Aveley, Essex. *Geological Magazine*, **120**, 623–9.

Hopson, P.M. (1981) *The Sand and Gravel Resources of the Country around Stansted Mountfitchet, Essex*, Mineral Assessment Report of the Institute of Geological Sciences, **104**, 110 pp.

Horton, A. (1977) Nettlebed. In *South East England and the Thames Valley* (eds E.R. Shephard-Thorn and J.J. Wymer), Guide Book for Excursion A5, X INQUA Congress, Birmingham, Geoabstracts, Norwich, pp. 16–8.

Horton, A. (1983) Nettlebed. In *Diversion of the Thames* (ed. J. Rose), Field Guide, Quaternary Research Association, Cambridge, pp. 63–5.

Howell, F.C. (1960) European and northwest African Middle Pleistocene hominids. *Current Anthropologist*, **1**, 195–232.

Hubbard, R.N.L.B. (1972) An interim report of the pollen record at Swanscombe. *Proceedings of the Royal Anthropological Institute* [for 1971], p. 79.

Hubbard, R.N.L.B. (1982) The environmental evidence from Swanscombe and its implications for Palaeolithic archaeology. In *Archaeology in Kent to AD 1500* (ed. P.E. Leach), Council for British Archaeology, Research Report 48, pp. 3–7.

Hughes, T.McK. (1868) On the two plains of Hertfordshire and their gravels. *Quarterly Journal of the Geological Society of London*, **24**, 283–7.

Hull, E. (1855) On the physical geography and Pleistocene phenomena of the Cotteswold Hills. *Quarterly Journal of the Geological Society of London*, **11**, 475–96.

Hull, E. (1859) *The Geology of the Country around Woodstock, Oxfordshire*, Memoir of the Geological Survey of Great Britain, 30 pp.

Hull, E. and Whitaker, W. (1861) *The Geology of Parts of Oxfordshire and Berkshire*, Memoir of the Geological Survey of Great Britain, 57 pp.

Hunt, C.O. (1985) Pollen from the Eynsham Gravel at Magdalen College, Oxford. *In* The chronology and environmental framework of Early Man in the Upper Thames Valley: a new model (eds D.J. Briggs, G.R. Coope and D.D. Gilbertson), *British Archaeological Report, British Series*, **137**, 85–7.

Janossy, D. (1975) Mid-Pleistocene microfauna of Continental Europe. In *After the Australopithecines: Stratigraphy, Ecology and Culture Change in the Middle Pleistocene* (eds K.W. Butzer and G.L. Isaac), Mouton, The Hague, pp. 375–397.

Janossy, D. (1987) *Pleistocene Vertebrate Faunas of Hungary*, Elsevier, Amsterdam, 208 pp.

Jessop, R.F. (1930) *The Archaeology of Kent*. Methuen, London, 272 pp.

John, D.T. (1980) The soils and superficial deposits on the North Downs of Surrey. In *The Shaping of Southern England* (ed. D.K.C. Jones), Institute of British Geographers Special Publication 11, Academic Press, London, pp. 101–30.

John, D.T. and Fisher, D.F. (1984) The stratigraphical and geomorphological significance of the Red Crag fossils at Netley Heath, Surrey: a review and re-appraisal. *Proceedings of the Geologists' Association*, **95**, 235–47.

Johnson, J.P. (1901) Additions to the Palaeolithic fauna of the Uphall Brickyard, Ilford, Essex. *Essex Naturalist*, **11**, 209–12.

Jones, D.K.C. (1974) The influence of the

Calabrian transgression on the drainage evolution of south-east England. In *Progress in Geomorphology* (eds E.H. Brown and R.S. Waters), Institute of British Geographers, Special Publication 7, Academic Press, London, pp. 139–158.

Jones, D.K.C. (1981) *The Geomorphology of the British Isles: Southeast and Southern England*. Methuen, London and New York, 332 pp.

Jones, T.R. (1850) Description of the Entomostraca of the Pleistocene beds of Newbury, Copford, Clacton and Grays. *Annals and Magazine of Natural History, Series 2*, **6**, 25–71.

De Jong, J. (1988) Climatic variability during the past three million years, as indicated by vegetational evolution in northwest Europe and with emphasis on data from the Netherlands. *Philosophical Transactions of the Royal Society of London*, **B318**, 603–17.

Jukes-Browne, A.J. and White, H.J.O. (1908) *The Geology of the Country around Henley-on-Thames and Wallingford*, Memoir of the Geological Survey of Great Britain, 113 pp.

Kahlke, H.D. (ed.) (1965) Das Pleistozän von Voigtstedt. *Paläeontologische Abhandlungen*, **A11 2/3**, 227–692.

Kahlke, H.D. (1969) Das Pleistozän von Süssenborn. *Paläeontologische Abhandlungen*, **A111 3/4**, 367–788.

Kahlke, H.D. (1975) The macrofaunas of continental Europe during the Middle Pleistocene: stratigraphic sequence and problems of intercorrelation. In *After the Australopithecines: Stratigraphy, Ecology and Culture Change in the Middle Pleistocene* (eds K.W. Butzer and G.L. Isaac), Mouton, The Hague, pp. 309–74.

Keen, D.H. (1990) Significance of the record provided by Pleistocene fluvial deposits and their included molluscan faunas for palaeoenvironmental reconstruction and stratigraphy: case study from the English Midlands. *Palaeogeography, Palaeoclimatology, Palaeoecology*, **80**, 25–34.

Keith, A. (1939) A resurvey of the anatomical features of the Piltdown Skull with some observations on the recently discovered Swanscombe Skull. Parts I and II. *Journal of Anatomy, London*, **73**, 155–85 and 234–54.

Kellaway, G.A., Horton, A. and Poole, G. (1971) The development of some Pleistocene structures in the Cotswolds and Upper Thames Basin. *Bulletin of the Geological Survey of Great Britain*, **37**, 1–28.

Kelly, M.R. (1964) The Middle Pleistocene of north Birmingham. *Philosophical Transactions of the Royal Society of London*, **B247**, 533–92.

Kemp, R.A. (1983) Stebbing: the Valley Farm palaeosols layer. In *Diversion of the Thames* (ed. J. Rose), Field Guide, Quaternary Research Association, Cambridge, pp. 154–8.

Kemp, R.A. (1984) Quaternary soils in southern East Anglia and the Lower Thames Basin. Unpublished Ph.D. thesis, University of London.

Kemp, R.A. (1985a) The Valley Farm Soil in southern East Anglia. In *Soils and Quaternary Landscape Evolution* (ed. J. Boardman), Wiley, Chichester, pp. 179–96.

Kemp, R.A. (1985b) The decalcified Lower Loam at Swanscombe, Kent: a buried Quaternary soil. *Proceedings of the Geologists' Association*, **96**, 343–55.

Kemp, R.A. (1987a) Genesis and environmental significance of a buried Middle Pleistocene soil in eastern England. *Geoderma*, **41**, 49–77.

Kemp, R.A. (1987b) The interpretation and environmental significance of a buried soil near Ipswich airport, Suffolk. *Philosophical Transactions of the Royal Society of London*, **B317**, 365–91.

Kemp, R.A. (1991) Micromorphology of the buried Quaternary soil within Burchell's 'Ebbsfleet Channel', Kent. *Proceedings of the Geologists' Association*, **102**, 275–87.

Kennard, A.S. (1904) Notes on a palaeolith from Grays, Essex. *Essex Naturalist*, **13**, 112–13.

Kennard, A.S. (1916) The Pleistocene succession in England. *Proceedings of the Prehistoric Society of East Anglia*, **2**, 249–67.

Kennard, A.S. (1924) The Pleistocene non-marine Mollusca of England. *Proceedings of the Malacological Society of London*, **16**, 84–97.

Kennard, A.S. (1938) Report on the non-marine Mollusca from the Middle Gravels of the Barnfield Pit. *Journal of the Royal Anthropological Institute of London*, **68**, 28–30.

Kennard, A.S. (1942) Faunas of the High Terrace at Swanscombe. *Proceedings of the Geologists' Association*, **53**, 105.

Kennard, A.S. (1944) The Crayford Brickearths. *Proceedings of the Geologists' Association*, **55**, 121–69.

Kennard, A.S. and Woodward, B.B. (1897) The post-Pliocene non-marine Mollusca of Essex. *Essex Naturalist*, **10**, 87–109.

Kennard, A.S. and Woodward, B.B. (1900) The Pleistocene non-marine Mollusca of Ilford. *Proceedings of the Geologists' Association*, **16**, 282–6.

Kennard, A.S. and Woodward, B.B. (1907) Notes on the post-Pliocene Mollusca of the Milne collection. *Proceedings of the Malacological Society of London*, **7**, 261–3.

Kennard, A.S. and Woodward, B.B. (1923) On the non-marine Mollusca of Clacton-on-Sea. *Quarterly Journal of the Geological Society of London*, **79**, 629–34.

Kennard, A.S. and Woodward, B.B. (1924) Appendix 3: The Pleistocene non-marine Mollusca. *In* Sandford, K.S., The river gravels of the Oxford district, *Quarterly Journal of the Geological Society of London*, **80**, 170–5.

Kerney, M.P. (1959a) An interglacial tufa near Hitchin, Hertfordshire. *Proceedings of the Geologists' Association*, **70**, 322-37.

Kerney, M.P. (1959b) Pleistocene non-marine Mollusca of the English interglacial deposits. Unpublished Ph.D. thesis, University of London.

Kerney, M.P. (1971) Interglacial deposits at Barnfield Pit, Swanscombe, and their molluscan fauna. *Journal of the Geological Society of London*, **127**, 69–86.

Kerney, M.P. and Sieveking, G. deG. (1977) Northfleet. In *South East England and the Thames Valley* (eds E.R. Shephard-Thorn and J.J. Wymer), Guide Book for excursion A5, X INQUA Congress, Birmingham, Geoabstracts, Norwich, pp. 44–6.

Kerney, M.P., Gibbard, P.L., Hall, A.R., *et al.* (1982) Middle Devensian river deposits beneath the 'Upper Floodplain' terrace of the River Thames at Isleworth, West London. *Proceedings of the Geologists' Association*, **93**, 385–93.

King, W.B.R. and Oakley, K.P. (1936) The Pleistocene succession in the lower part of the Thames valley. *Proceedings of the Prehistoric Society*, **1**, 52–76.

Klie, W. (1938) Krebstiere oder Crustacea III; Ostracoda, Muschelkrebse. In *Die Tierwelt Deutschlands* (ed. F. Dahl), **Band 34**, Jena, pp. 1–130.

Koenigswald, W. von. (1973) Veranderungen in der Kleinäugerfauna von Mitteleuropa zwischen Cromer und Eem (pleistozaen), *Eiszeitalter und Gegenwart*, **23–24**, 159–67.

Van Kolfschoten, T. (1988) The Pleistocene mammalian faunas from Zuurland boreholes at Brielle, The Netherlands. *Mededelingen Werkgroep Tertiaire and Kwartaire Geologie*, **25**, 73–86.

Kukla, G.J. (1975) Loess stratigraphy of Central Europe. In *After the Australopithecines: Stratigraphy, Ecology and Culture Change in the Middle Pleistocene* (eds K.W. Butzer and G.L. Isaac), Mouton, The Hague, pp. 99–188.

Kukla, G.J. (1977) Pleistocene land-sea correlations. I. Europe. *Earth Science Reviews*, **13**, 307–74.

Kurtén, B. (1959) On the bears of the Holsteinian Interglacial. *Acta Universitatis Stockholmiensis, Stockholm Contributions in Geology*, **2**, 73–102.

Lacaille, A.D. (1940) The palaeoliths from the gravels of the Lower Boyn Hill Terrace around Maidenhead. *Antiquaries Journal*, **20**, 245–71.

Lacaille, A.D. (1960) On Palaeolithic choppers and cleavers (notes suggested by some Buckinghamshire examples) *Records of Bucks*, **16**, 330–41.

Lake, R.D., Ellison, R.A., Hollyer, S.E., *et al.* (1977) Buried channel deposits in the south-east Essex area: their bearing on Pleistocene palaeogeography. *Report of the Institute of Geological Sciences*, 77/21.

Lake, R.D., Ellison, R.A., Henson, M.R., *et al.* (1986) *The Geology of the Country around Southend and Foulness*, Memoir of the Geological Survey of Great Britain, 85 pp.

Lautridou, J-P. (1982) *The Quaternary of Normandy*. Field Guide, Quaternary Research Association, Cambridge, 88 pp.

Lautridou, J-P., Masson, M., Paepe, R., *et al.* (1974) Loess, nappes aluviales et tuf de St-Pierre-les-Elbeuf, près de Rouen; les terraces de la Seine de Muids à Caudebec. *Bulletin de l'Association Francaise pour l'Etude Quaternaire*, Supplément, **3–4**, 193–201.

Lautridou, J-P., Monnier, J-L., Mortazec-Kerfourn, M-T., *et al.* (1983) Les subdivisions du Pléistocène de la France septentrionale: stratigraphie et paléolithique. In *Quaternary Glaciations in the Northern Hemisphere* (eds O. Billards, O. Conchon and F.W. Shotton), UNESCO – International Geological Correlation Programme, Project 73/1/24 Report 9, Paris 1982, pp. 148–70.

Leach, A.L. (1913) On buried channels in the

Dartford Heath Gravel. *Proceedings of the Geologists' Association*, **24**, 337–44.

Leakey, L.S.B. (1972) Homo sapiens in the Middle Pleistocene and the evidence of Homo sapiens evolution. In *The Origin of Homosapiens* (ed. F. Bordes), UNESCO, Paris, pp. 25–9.

Leonardi, G. and Petronio, C. (1976) The fallow deer of European Pleistocene. *Geologica Roma*, **15**, 1–67.

Lindroth, C.H. (1985) The Carabidae (Coleoptera) of Fennoscandia and Denmark. *Fauna Entomologica Scandinavica* (Copenhagen), **15**, 1–228.

Linke, G., Katzenberg, O. and Grün, R. (1985) Description and ESR dating of the Holsteinian interglaciation. *Quaternary Science Reviews*, **4**, 319–31.

Lister, A.M. (1986) New results on deer from Swanscombe, and the stratigraphical significance of deer in the Middle and Upper Pleistocene of Europe. *Journal of Archaeological Science*, **13**, 319–38.

Lister, A.M. (1989) Mammalian faunas and the Wolstonian debate. In *West Midlands* (ed. D.H. Keen), Field Guide, Quaternary Research Association, Cambridge, pp. 5–11.

Lister, A.M., McGlade, J.M. and Stuart, A.J. (1990) The early Middle Pleistocene vertebrate fauna from Little Oakley, Essex. *Philosophical Transactions of the Royal Society of London*, **B328**, 359–85.

Lonsdale, C.A. (1978) A sedimentological investigation of a supposed Ipswichian Interglacial site at Purfleet, Essex. Unpublished M.Sc. thesis, City of London Polytechnic and Polytechnic of North London.

Lucht, W.H. (1987) *Die Käfer Mitteleuropas, Katalog,* Geoke and Evers, Kreleid, 342 pp.

Lyell, C. (1865) *Elements of Geology*, 6th edn, John Murray, London, 294 pp.

McGregor, D.F.M. and Green, C.P. (1978) Gravels of the River Thames as a guide to Pleistocene catchment changes. *Boreas*, **7**, 197–203.

McGregor, D.F.M. and Green, C.P. (1983a) Post-depositional modification of Pleistocene terraces of the River Thames. *Boreas*, **12**, 23–33.

McGregor, D.F.M. and Green, C.P. (1983b) Gerrards Cross. In *Diversion of the Thames* (ed. J. Rose), Field Guide, Quaternary Research Association, Cambridge, pp. 80–4.

McGregor, D.F.M. and Green, C.P. (1983c) Lithostratigraphic subdivisions in the gravels of the proto-Thames between Hemel Hempstead and Watford. *Proceedings of the Geologists' Association*, **94**, 83–5.

McGregor, D.F.M. and Green, C.P. (1986) Early and Middle Pleistocene gravel deposits of the Thames – development of a lithostratigraphic model. In *Clast Lithological Analysis* (ed. D.R. Bridgland), Technical Guide No. 3, Quaternary Research Association, Cambridge, pp. 95–115.

McKeown, M.C. and Samuel, M.D.A. (1985) *Regional Study of the Sand and Gravel Resources of Essex and South Suffolk*. British Geological Survey, Keyworth.

McNabb J. (1989) Sticks and stones: a possible experimental solution to the question of how the Clacton spear point was made. *Proceedings of the Prehistoric Society,* **55**, 251–71.

MacRae, R.J. (1982) Palaeolithic artefacts from Berinsfield, Oxfordshire. *Oxoniensa*, **47**, 1–11.

MacRae, R.J. (1985) Palaeolithic archaeology of the Upper Thames Basin. *In* The chronology and environmental framework of early Man in the Upper Thames Valley (eds D.J. Briggs, G.R. Coope and D.D. Gilbertson), *British Archaeological Reports, British Series*, **137**, 8–25.

MacRae, R.J. (1987) The great giant handaxe stakes. *Lithics*, **8**, 15–7.

MacRae, R.J. (1988) The Palaeolithic of the Upper Thames and its quartzite implements. *In* Non-flint stone tools and the Palaeolithic occupation of Britain (eds R.J. MacRae and N. Moloney), *British Archaeological Reports, British Series*, **189**, 123–54.

MacRae, R.J. (1989) Belt, shoulder-bag or basket? An enquiry into handaxe transport and flint sources. *Lithics*, **10**, 2–8.

MacRae, R.J. (1991) New finds and old problems in the Lower Palaeolithic of the Upper Thames valley. *Lithics*, **11**, 3–15.

MacRae, R.J. and Moloney, N. (1988) Gazetteer of Lower Palaeolithic non-flint artefacts in Great Britain. *In* Non-flint stone tools and the Palaeolithic occupation of Britain (eds R.J. MacRae and N. Moloney), *British Archaeological Reports, British Series*, **189**, 243–63.

Maddy, D. (1989) The Middle Pleistocene of the Rivers Severn and Avon. Unpublished Ph.D. thesis, University of London.

Maddy, D., Keen, D.H., Bridgland, D.R., *et al.* (1991a) A revised model for the Pleistocene development of the River Avon, Warwickshire. *Journal of the Geological Society of London*, **148**, 473–84.

Maddy, D., Lewis, S.G. and Green, C.P. (1991b) A review of the stratigraphic significance of the Wolvercote Terrace of the Upper Thames valley. *Proceedings of the Geologists' Association*, **102**, 217–25.

Manning, P. and Leeds, E.T. (1921) An archaeological survey of Oxfordshire. *Archaeologia*, **71**, 227–65.

Marston, A.T. (1937) The Swanscombe Skull. *Journal of the Royal Anthropological Institute*, **67**, 339–406.

Marston, A.T. (1942) Flint industries of the High Terrace at Swanscombe. *Proceedings of the Geologists' Association*, **53**, 106.

Martinson, D.G., Pisias, N.J., Hayes, J.D., *et al.* (1987) Age dating and the orbital theory of the ice ages, development of a high resolution nought to 300,000 year chronostratigraphy. *Quaternary Research*, **27**, 1–29.

Meijer, T. (1985) Maastricht-Belvedere: stratigraphy, palaeoenvironment and archaeology of the Middle and Late Pleistocene deposits. *Mededelingen Rijks Geologische Dienst*, **39**, 76–103.

Meijer, T. and Preece, R.C. (in press) Malacological evidence relating to the stratigraphical position of the Cromerian. In *The Early Middle Pleistocene of Europe* (eds P.L. Gibbard and C. Turner), A.A. Balkema, Rotterdam.

Miall, A.D. (1977) A review of the braided river depositional environment. *Earth Science Reviews*, **13**, 1–62.

Miller, G.H. and Mangerud, J. (1985) Aminostratigraphy of European marine interglacial deposits. *Quaternary Science Reviews*, **4**, 215–78.

Miller, G.H., Hollin, J.T. and Andrews, J. (1979) Aminostratigraphy of UK Pleistocene deposits. *Nature, London*, **281**, 539–43.

Mitchell, G.F., Penny, L.F., Shotton, F.W., *et al.* (1973) *A Correlation of Quaternary Deposits in the British Isles*. Geological Society of London Special Report, No. 4, 99 pp.

Moffat, A.J. (1980) The Plio-Pleistocene transgression in the northern part of the London Basin – a re-examination. Unpublished Ph.D. thesis, University of London.

Moffat, A.J. (1986) Quartz signatures in Plio-Pleistocene gravels in the northern part of the London Basin. In *Clast Lithological Analysis* (ed. D.R. Bridgland), Technical Guide No. 3, Quaternary Research Association, Cambridge, pp. 117–28.

Moffat, A.J. and Catt, J.A. (1982) The nature of the pebbly clay drift at Epping Green, southeast Hertfordshire. *Transactions of the Hertfordshire Natural History Society*, **28**, 16–24.

Moffat, A.J. and Catt, J.A. (1983) A new excavation in Plio-Pleistocene deposits at Little Heath. *Transactions of the Hertfordshire Natural History Society*, **29**, 5–10.

Moffat, A.J. and Catt, J.A. (1986a) A re-examination of the evidence for a Plio-Pleistocene marine transgression on the Chiltern Hills. II. Drainage patterns. *Earth Surface Processes and Landforms*, **11**, 169–80.

Moffat, A.J. and Catt, J.A. (1986b), A re-examination of the evidence for a Plio-Pleistocene marine transgression on the Chiltern Hills, III. Deposits. *Earth Surface Processes and Landforms*, **11**, 233–47.

Moffat, A.J., Catt, J.A., Webster, R. *et al.* (1986) A re-examination of the evidence for a Plio-Pleistocene marine transgression on the Chiltern hills, I. Structures and surfaces. *Earth Surface Processes and Landforms*, **11**, 95–106.

Monckton, H.W. (1892) On the gravels south of the Thames from Guildford to Newbury. *Quarterly Journal of the Geological Society of London*, **48**, 29–59.

Monckton, H.W. and Herries, R.S. (1891) On some hill gravels north of the Thames. *Proceedings of the Geologists' Association*, **12**, 108–14.

Montagu, M.F.A. (1960) *An Introduction to Physical Anthropology*, 3rd edn, Charles G. Thomas, Springfield, Illinois.

Morant, G.M. (1938) The form of the Swanscombe Skull. *Journal of the Royal Anthropological Institute of London*, **68**, 67–97.

Morgan, A. (1969) A Pleistocene fauna and flora from Great Billing, Northamptonshire, England. *Opuscula Entomologica, Lund*, **34**, 109–29.

Morgan, A.V. (1973) Late Pleistocene environmental changes indicated by fossil insect faunas of the English Midlands. *Boreas*, **2**, 173–212.

Morris, J. (1836) On a freshwater deposit containing mammalian remains, recently discovered at Grays, Essex. *Magazine of Natural History, Series 1*, **9**, 261–4.

Newcomer, M.H. (1971) Conjoined flakes from

the Lower Loam, Barnfield Pit, Swanscombe. *Proceedings of the Royal Anthropological Institute of London* [for 1970], pp. 51-59.

Newton, E.T. (1895) On a human skull and limb bones found in the Palaeolithic terrace gravel at Galley Hill, Kent. *Quarterly Journal of the Geological Society of London*, **51**, 505–27.

Newton, E.T. (1907) Note on specimens of 'Rhaxella-chert' or 'Arngrove stone' from Dartford Heath. *Proceedings of the Geologists' Association*, **20**, 127–8.

Newton, R.B. (1916) On the conchological features of the Lenham Sandstone of Kent. *Journal of Conchology, London*, **15**, 56–84, 97–118 and 137–49.

Newton, W.M. (1901) Kent: flint implements. The occurrence in a very limited area of the rudest with the finest forms of worked stones. *Man*, **1**, 81–2.

Newton, W.M. (1930) A remarkable gravel pit. *Man*, **30**, 41–4.

Oakley, K.P. (1937) Field meeting at Taplow, Burnham and Iver, Bucks. *Proceedings of the Geologists' Association*, **48**, 276–9.

Oakley, K.P. (1939) *A Survey of the Prehistory of the Farnham District (Surrey), Part 1 Geology and Palaeolithic Studies*, Surrey Archaeological Society, Guildford, pp. 3–58.

Oakley, K.P. (1943) The future of Quaternary research in Britain. *South East Naturalist*, **48**, 25–32.

Oakley, K.P. (1949) *Man the Toolmaker*, 1st edn, British Museum (Natural History), London, 98 pp.

Oakley, K.P. (1952) Swanscombe Man. *Proceedings of the Geologists' Association*, **63**, 271–300.

Oakley, K.P. (1964) The evidence of fire at Swanscombe. In *The Swanscombe Skull: a Survey of Research on a Pleistocene Site* (ed. C.D. Ovey), Royal Anthropological Institute, Occasional Paper No. 20, pp. 63–6.

Oakley, K.P. and Gardiner, E. (1964) Analytical data on the Swanscombe bones. In *The Swanscombe Skull: A Survey of Research on a Pleistocene Site* (ed. C.D. Ovey), Royal Anthropological Institute, Occasional Paper No. 20, pp. 117–23.

Oakley, K.P. and King, W.B.R. (1945) Age of the Baker's Hole Coombe Rock, Northfleet, Kent. *Nature, London*, **155**, 51–2.

Oakley, K.P. and Leakey, M. (1937) Report on excavations at Jaywick Sands, Essex (1934), with some observations on the Clactonian industry, and on the fauna and geological significance of the Clacton channel. *Proceedings of the Prehistoric Society*, **3**, 217–60.

Oakley, K.P., Andrews, P., Keeley, L.H., *et al.* (1977) A reappraisal of the Clacton spearpoint. *Proceedings of the Prehistoric Society*, **43**, 13–30.

Ohel, M.Y. (1977) On the Clactonian reexamined, redefined, reinterpreted. *Current Anthropologist*, **18:2**, 329–31.

Ohel, M.Y. (1979) The Clactonian: an independent complex or an integral part of the Acheulian? *Current Anthropologist*, **20:4**, 685–744.

Osborne, P.J. (1980) The insect fauna of the organic deposits at Sugworth and its environmental and stratigraphic implications. *In* Shotton, F.W., Goudie, A.S., Briggs, D.J. *et al.* Cromerian interglacial deposits at Sugworth, near Oxford, England, and their relation to the Plateau Drift of the Cotswolds and the terrace sequence of the Upper and Middle Thames. *Philosophical Transactions of the Royal Society of London*, **B289**, 119–33.

Ovey, C.D. (ed.) (1964) *The Swanscombe Skull: a Survey of Research on a Pleistocene Site*, Royal Anthropological Institute, Occasional Paper No. 20, 215 pp.

Owen, R. (1846) *History of the British Fossil Mammals and Birds*, John Van Voorst, London, 560 pp.

Owen, R. (1855) Description of a fossil cranium of the musk-buffalo from the 'lower level drift' at Maidenhead, Bucks. *Quarterly Journal of the Geological Society of London*, **12**, 124–31.

Palmer, S. (1975) A Palaeolithic site at North Road, Purfleet, Essex. *Transactions of the Essex Archaeological Society*, **7**, 1–13.

Parks, D.A. and Rendell, H.M. (1988) TL dating of brickearths from SE England. *Quaternary Science Reviews*, **7**, 305–8.

Paterson, T.T. (1940) The Swanscombe Skull: a defence. *Proceedings of the Prehistoric Society*, **6**, 166–9.

Penck, A and Brückner, E. (1909) *Die Alpen im Eiszeitalter*, Tauchmitz, Leipzig, 3 vols, 1199 pp.

Perrin, R.M.S., Davies, H. and Fysh, M.D. (1973) Lithology of the Chalky Boulder Clay. *Nature, London*, **245**, 101–4.

Perrin, R.M.S., Rose, J. and Davies, H. (1979) The distribution, variation and origins of pre-

Devensian tills in eastern England. *Philosophical Transactions of the Royal Society of London*, **B287**, 535–70.

Pettit, M. and Gibbard, P.L. (1980) Palaeobotany. *In* Shotton, F.W., Goudie, A.S., Briggs, D.J. *et al.* Cromerian interglacial deposits at Sugworth, near Oxford, England, and their relation to the Plateau Drift of the Cotswolds and the terrace sequence of the Upper and Middle Thames, *Philosophical Transactions of the Royal Society of London*, **B289**, 63.

Phillips, J. (1871) *The Geology of Oxford and the Valley of the Thames*, Clarendon Press, Oxford, 523 pp.

Phillips, L.M. (1976) Pleistocene vegetational history and geology in Norfolk. *Philosophical Transactions of the Royal Society of London*, **B275**, 215–86.

Picton, H. (1912) Observations on the bone bed at Clacton. *Proceedings of the Prehistoric Society*, **1**, 158–9.

Pike, K. and Godwin, H. (1953) The interglacial at Clacton-on-Sea, Essex. *Quarterly Journal of the Geological Society of London*, **108**, 261–72.

Pinchemel, P. H. (1954) *Les Plaines de Craie du Nord-ouest du Bassin Parisien et du Sud-est du Bassin de Londres et leurs Bordures*. Librairie Armand Collin, Paris, 502 pp.

Pocock, T.I. (1903) On the drifts of the Thames valley near London. *Summary of Progress, Geological Survey of Great Britain*, for 1902, 199–207.

Pocock, T.I. (1908) *The Geology of the Country around Oxford*. Memoir of the Geological Survey of Great Britain, 142 pp.

Podmore, J.A. (1976) The geomorphology of a selected archaeological site in South Essex; Botany Pit, Purfleet, Essex, Unpublished B.Sc. thesis, City of London Polytechnic, 34 pp.

Porrenga, D.H. (1967) Glauconite and chamosite as depth indicators in the marine environment. *Marine Geology*, **5**, 495–501.

Poulton, E.B. (1880) On mammalian remains and tree trunks in Quaternary sands and gravels at Reading. *Quarterly Journal of the Geological Society of London*, **36**, 296–306.

Preece, R.C. (1988) A second British interglacial record of *Margitifera auricularia*. *Journal of Conchology*, **33**, 50–1.

Preece, R.C. (1989) Additions to the molluscan fauna of the early Middle Pleistocene deposits at Sugworth, near Oxford, including the first British Quaternary record of *Perforatella bidentata* (Gmelin), *Journal of Conchology*, **33**, 179–82.

Preece, R.C. (1990a) Alfred Santer Kennard (1870–1948): his contribution to malacology, Quaternary research and to the Geologists' Association. *Proceedings of the Geologists' Association*, **101**, 239–58.

Preece, R.C. (1990b) The molluscan fauna of the Middle Pleistocene interglacial deposits at Little Oakley, Essex, and its environmental and stratigraphic implications. *Philosophical Transactions of the Royal Society of London*, **B328**, 387–407.

Prestwich, J. (1854) On the structure of the strata between the London Clay and the Chalk in the London and Hampshire Tertiary systems. Part II. The Woolwich and Reading Series. *Quarterly Journal of the Geological Society of London*, **10**, 75–170.

Prestwich, J. (1855) Note on the gravel near Maidenhead in which the skull of the musk-buffalo was found. *Quarterly Journal of the Geological Society of London*, **12**, 131–3.

Prestwich, J. (1858a) On the age of the sands and iron-sandstones on the North Downs. *Quarterly Journal of the Geological Society of London*, **14**, 322–35.

Prestwich, J. (1858b) On the occurrence of the boulder clay or Northern Clay Drift, at Brick-et Wood, near Watford. *Geologist*, **1**, 241.

Prestwich, J. (1881) On the extension into Essex, Middlesex and other inland counties, of the Mundesley and Westleton Beds, in relation to the age of certain hill-gravels of some of the valleys of the south of England. *Geological Magazine*, **8**, 466–8.

Prestwich, J. (1882) The occurrence of *Cyrena fluminalis* at Summertown near Oxford. *Geological Magazine*, **9**, 49–51.

Prestwich, J. (1890a) On the relation of the Westleton Beds, or pebbly sands of Suffolk, to those of Norfolk, and on their extension inland; with some observations on the period of the final elevation and denudation of the Weald and of the Thames valley. Part I. *Quarterly Journal of the Geological Society of London*, **46**, 84–119.

Prestwich, J. (1890b) On the relation of the Westleton Beds, or pebbly sands of Suffolk, to those of Norfolk, and on their extension inland; with some observations on the period of the final elevation and denudation of the Weald and of the Thames valley. Part II.

Quarterly Journal of the Geological Society of London, **46**, 120–54.

Prestwich, J. (1890c) On the relation of the Westleton Beds, or pebbly sands of Suffolk, to those of Norfolk, and on their extension inland; with some observations on the period of the final elevation and denudation of the Weald and of the Thames valley. Part II. *Quarterly Journal of the Geological Society of London*, **46**, 155–81.

Prestwich, J. (1891) On the age, formation and successive drift-stages of the valley of the Darent; with remarks on the Palaeolithic implements of the district and on the origin of its Chalk escarpment. *Quarterly Journal of the Geological Society of London*, **47**, 126–63.

Ransome, E.R. (1890) Fossil Mammalia at Clacton-on-Sea. *Essex Naturalist*, **4**, 201.

Reading, H.G. (1978) *Sedimentary Environments and Facies*. Blackwell Scientific, Oxford, 557 pp.

Reid, C. (1890) *The Pliocene Deposits of Britain*, Memoir of the Geological Survey of England and Wales, 326 pp.

Reid, C. (1897) On Pleistocene plants from Casewick, Shacklewell and Grays. *Quarterly Journal of the Geological Society of London*, **53**, 463–4.

Reid, C. (1899) The origin of the British flora. Dalau, London, 191 pp.

Reid, C. (1900) [untitled comments on the high-level gravel at Stanmore]. *In* Anon (ed.) Field work, Tertiary, London Basin, *Summary of Progress, Geological Survey of Great Britain* [for 1899], p. 140.

Reid, C. and Chandler, M.E.J. (1923) The fossil flora of Clacton-on-Sea. *Quarterly Journal of the Geological Society of London*, **79**, 619–23.

Reineck, H-E. and Singh, I.B. (1975) *Depositional Sedimentary Environments*, Springer-Verlag, Berlin, 439 pp.

Richardson, L. (1935) Weekend field meeting in the Witney district. *Proceedings of the Geologists' Association*, **46**, 403–11.

Richardson, L. and Sandford, K.S. (1963) Ditchford Gravel Pit near Stretton-on-Fosse, Gloucestershire and the occurrence of a mammoth tooth. *Proceedings of the Cotteswolds Naturalists Field Club*, **33**, 172–6.

Richmond, G.M. and Fullerton, D.S. (1986) Introduction to Quaternary glaciations in the United States of America. *Quaternary Science Reviews*, **5**, 3–10.

Roberts, M.B. (1986) Excavation of the Lower Palaeolithic site at Amey's Eartham Pit, Boxgrove, West Sussex: a preliminary report. *Proceedings of the Prehistoric Society*, **52**, 215–45.

Robinson, J.E. (1978) Ostracods from deposits in the Vale of St Albans. *Quaternary Newsletter*, **2**, 8–9.

Robinson, J.E. (1980) The ostracod fauna of the interglacial deposits at Sugworth, Oxfordshire. *Philosophical Transactions of the Royal Society of London*, **B289**, 99–106.

Robinson, J.E. (1983) Ostracods from the Westmill Lower Gravels, Westmill. In *Diversion of the Thames* (ed. J. Rose), Field Guide, Quaternary Research Association, Cambridge, pp. 132.

Robinson, J.E. (1990) The Ostracod fauna of the interglacial deposits at Little Oakley, Essex. *Philosophical Transactions of the Royal Society of London*, **B328**, 409–23.

Robson, P. (1976) *The Sand and Gravel Resources of the Thames Valley, the Country between Lechlade and Standlake*, Mineral Assessment Report of the Institute of Geological Sciences 23, 141 pp.

Roe, D.A. (1964) The British Lower and Middle Palaeolithic: some problems, methods of study and preliminary results. *Proceedings of the Prehistoric Society*, **30**, 245–67.

Roe, D.A. (1968a) British Lower and Middle Palaeolithic hand-axe groups. *Proceedings of the Prehistoric Society*, **34**, 1–82.

Roe, D.A. (1968b) *A Gazetteer of British Lower and Middle Palaeolithic Sites*, Council for British Archaeology Research Report No. 8, 355 pp.

Roe, D.A. (1975) Some Hampshire and Dorset hand-axes and the question of Early Acheulian in Britain. *Proceedings of the Prehistoric Society*, **41**, 1–9.

Roe, D.A. (1976) Palaeolithic industries in the Oxford Region. In *Field Guide to the Oxford Region* (ed. D.A. Roe), Quaternary Research Association, Oxford, pp. 36–43.

Roe, D.A. (1977) Fordwich and Sturry. In *South East England and the Thames Valley* (eds E.R. Shephard-Thorn and J.J. Wymer), Guide Book for Excursion A5, X INQUA Congress, Birmingham, Geoabstracts, Norwich, pp. 53–4.

Roe, D.A. (1981) *The Lower and Middle Palaeolithic Periods in Britain*. Routledge and Kegan Paul, London, 324 pp.

Rolfe, W.D.I. (1958) A recent temporary section through Pleistocene deposits at Ilford. *Essex Naturalist*, **30**, 93–103.

Rose, J. (1974) Small scale variability of some sedimentary properties of lodgement and slumped till. *Proceedings of the Geologists' Association*, **85**, 223–37.

Rose, J. (1979) River terraces and sea level change. *Brighton Polytechnic Geographic Society Magazine*, **3**, 13–30.

Rose, J. (1983a) Early and Middle Pleistocene sediments and palaeosols in west and central Essex. In *Diversion of the Thames* (ed. J. Rose), Field Guide, Quaternary Research Association, Cambridge, pp. 135–9.

Rose, J. (1983b) Introduction. In *Diversion of the Thames* (ed. J. Rose), Field Guide, Quaternary Research Association, Cambridge, pp. 1–7.

Rose, J. (1987) Status of the Wolstonian glaciation in the British Quaternary. *Quaternary Newsletter*, **53**, 1–9.

Rose, J. (1988) Stratigraphic nomenclature for the British Middle Pleistocene – procedural dogma or stratigraphic common sense. *Quaternary Newsletter*, **54**, 15–20.

Rose, J. (1989) Tracing the Baginton–Lillington Sands and Gravels from the West Midlands to East Anglia. In *West Midlands* (ed. D.H. Keen), Field Guide, Quaternary Research Association, Cambridge, pp. 102–10.

Rose, J. (1991) Stratigraphic basis of the 'Wolstonian glaciation', and retention of the term 'Wolstonian' as a chronostratigraphic stage name – a discussion. In *Central East Anglia and the Fen Basin* (eds S.G. Lewis, C.A. Whiteman and D.R. Bridgland), Field Guide, Quaternary Research Association, London, pp. 15–20.

Rose, J. and Allen, P. (1977) Middle Pleistocene stratigraphy in south-east Suffolk. *Journal of the Geological Society of London*, **133**, 83–102.

Rose, J., Allen, P. and Hey, R.W. (1976) Middle Pleistocene stratigraphy in southern East Anglia. *Nature, London*, **263**, 492–4.

Rose, J., Sturdy, R.G., Allen, P., *et al.* (1978) Middle Pleistocene sediments and palaeosols near Chelmsford, Essex. *Proceedings of the Geologists' Association*, **89**, 91–6.

Rose, J., Allen, P., Kemp, R.A., *et al.* (1985a) The early Anglian Barham Soil of eastern England. In *Soils and Quaternary Landscape Evolution* (ed. J. Boardman), Wiley, Chichester, pp. 197–229.

Rose, J., Boardman, J., Kemp, R.A., *et al.* (1985b) Palaeosols and the interpretation of the British Quaternary stratigraphy. In *Geomorphology and Soils* (eds K. Richards, R. Arnett and S. Ellis), Allen and Unwin, London, p. 348–75.

Ross, B.R.M. (1932) The physiographic evolution of the Kennet-Thames. *Report of the British Association, London* [for 1931], p. 368.

Rossiter, J.R. (1972) Sea level observations and their secular variations. *Philosophical Transactions of the Royal Society of London*, **A272**, 131–9.

Ruddiman, W.F., Raymo, M.E., Martinson, D.G., *et al.* (1989) Pleistocene evolution: Northern Hemisphere ice sheets and North Atlantic Ocean. *Palaeoceanography*, **4**, 353–412.

Salter, A.E. (1896) 'Pebbly gravel' from the Goring Gap to the Norfolk Coast. *Proceedings of the Geologists' Association*, **14**, 389–404.

Salter, A.E. (1898) Pebbly and other gravels in southern England. *Proceedings of the Geologists' Association*, **15**, 264–86.

Salter, A.E. (1901) Excursion to Stanmore. *Proceedings of the Geologists' Association*, **17**, 175–6.

Salter, A.E. (1903) Excursion to Erith and Crayford. *Proceedings of the Geologists' Association*, **18**, 165–6.

Salter, A.E. (1905) On the superficial deposits of central and parts of southern England. *Proceedings of the Geologists' Association*, **19**, 1–56.

Sandford, K.S. (1924) The river gravels of the Oxford district. *Quarterly Journal of the Geological Society of London*, **80**, 113–79.

Sandford, K.S. (1925) The fossil elephants of the Upper Thames basin. *Quarterly Journal of the Geological Society of London*, **81**, 62–86.

Sandford, K.S. (1926) Pleistocene deposits. In *The Geology of the Country around Oxford* (ed. J. Pringle), Memoir of the Geological Survey of Great Britain, pp. 104–172.

Sandford, K.S. (1932) Some recent contributions to the Pleistocene succession in England. *Geological Magazine*, **69**, 1–18.

Sandford, K.S. (1939) Early Man. The Quaternary geology of Oxfordshire with reference to Palaeolithic Man. In *The Victoria County History of Oxfordshire*, **1**, 223–38.

Sandford, K.S. (1954) River development and

superficial deposits. In *The Oxford Region. A Scientific and Historical Survey* (eds A.F. Martin and R.W. Steel), University Press, Oxford, pp. 21–4.

Sandford, K.S. (1965) Notes on the gravels of the Upper Thames floodplain between Lechdale and Dorchester. *Proceedings of the Geologists' Association*, **76**, 61–75.

Saner, B.R.M. and Wooldridge, S.W. (1929) River development in Essex. *Essex Naturalist*, **22**, 244–50.

Sarnthein, M., Stremme, H.E. and Mangini, A. (1986) The Holstein interglaciation: time stratigraphic position and correlation to stable-isotope stratigraphy of deep-sea sediments. *Quaternary Research*, **26**, 283–96.

Schreuder, A. (1950) Microtinae from the Middle Gravels of Swanscombe. *Annals and Magazine of Natural History, London*, **Series 12**, **3**, 629–35.

Schwarcz, H.P. and Grün, R. (1988) Comment on Sarnthein, M., Stremme, H.E. and Mangini, A. 'The Holstein interglaciation: time stratigraphic position and correlation to stable-isotope stratigraphy of deep-sea sediments'. *Quaternary Research*, **29**, 75–9.

Schwertmann, U., Murad, E. and Schulze, D.G. (1982) Is there Holocene reddening (hematite formation) in soils of axeric temperate areas? *Geoderma*, **27**, 209–23.

Sealy, K.R. and Sealy, C.E. (1956) The terraces of the Middle Thames. *Proceedings of the Geologists' Association*, **67**, 369–92.

Seddon, M.B. and Holyoak, D.T. (1985) Evidence of sustained regional permafrost during deposition of fossiliferous Late Pleistocene sediments at Stanton Harcourt, Oxfordshire, England. *Proceedings of the Geologists' Association*, **96**, 53–73.

Shackleton, N.J. (1969) The last interglacial in the marine and terrestrial records. *Proceedings of the Royal Society of London*, **B174**, 135–54.

Shackleton, N.J. (1987) Oxygen isotopes, ice volume and sea level. *Quaternary Science Reviews*, **6**, 1835–90.

Shackleton, N.J. and Opdyke, N.D. (1973) Oxygen Isotope and palaeomagnetic stratigraphy of Equatorial Pacific Core V28–238, Oxygen Isotope temperatures and ice volumes on a 10^5 year – 10^6 year scale. *Quaternary Research*, **3**, 39–55.

Shackleton, N.J. and Opdyke, N.D. (1976) Oxygen-isotope and palaeomagnetic stratigraphy of Pacific core V28–239: Late Pliocene to latest Pleistocene. *Geological Society of America Memoir*, **145**, 449–64.

Shackleton, N.J., Berger, A. and Peltier, W.R. (1990) An alternative astronomical calibration of the lower Pleistocene time scale based on ODP site 677. *Transactions of the Royal Society of Edinburgh: Earth Sciences*, **81**, 251–61.

Shephard, R.W. (1976) The geomorphology of a part of the Taplow and Boyn Hill Terrace sequence in south Essex. Unpublished B.Sc. thesis, City of London Polytechnic, 35 pp.

Sherlock, R.L. (1919) Discussion of the two foregoing papers [Barrow (1919b) and Gilbert (1919a)]. *Quarterly Journal of the Geological Society of London*, **75**, 46–8.

Sherlock, R.L. (1922) *The Geology of the Country around Aylesbury and Hemel Hempstead*, Memoir of the Geological Survey of Great Britain, 66 pp.

Sherlock, R.L. (1924) The superficial deposits of south Buckinghamshire and south Hertfordshire and the old course of the Thames. *Proceedings of the Geologists' Association*, **35**, 1–28.

Sherlock, R.L. (1929) Discussion on the alleged Pliocene of Buckinghamshire and Hertfordshire. *Proceedings of the Geologists' Association*, **40**, 357–70.

Sherlock, R.L. and Noble, A.H. (1912) On the glacial origin of the clay-with-flints of Buckinghamshire, and on the former course of the Thames. *Quarterly Journal of the Geological Society of London*, **68**, 199–212.

Sherlock, R.L. and Noble, A.H. (1922) *The Geology of the Country around Beaconsfield*, Memoir of the Geological Survey of Great Britain, 59 pp.

Sherlock, R.L. and Pocock, R.M. (1924) *The Geology of the Country around Hertford*, Memoir of the Geological Survey of Great Britain, 66 pp.

Shotton, F.W. (1953) The Pleistocene deposits of the area between Coventry, Rugby and Leamington and their bearing upon the topographic development of the Midlands. *Philosophical Transactions of the Royal Society of London*, **B237**, 209–60.

Shotton, F.W. (1968) The Pleistocene succession around Brandon, Warwickshire. *Philosophical Transactions of the Royal Society of London*, **B254**, 387–400.

Shotton, F.W. (1973a) The English Midlands. In

A Correlation of Quaternary Deposits in the British Isles (eds G.F. Mitchell, L.F. Penny, F.W. Shotton and R.G. West), Geological Society of London Special Report, No.4, pp. 18–22.

Shotton, F.W. (1973b) General principles governing the subdivision of the Quaternary system. In *A Correlation of Quaternary Deposits in the British Isles* (eds G.F. Mitchell, L.F. Penny, F.W. Shotton and R.G. West), Geological Society of London Special Report, No.4, pp. 1–7.

Shotton, F.W. (1977) Chronology, climate and marine record, the Devensian stage: its development, limits and substages. *Philosophical Transactions of the Royal Society of London*, **B280**, 107–18.

Shotton, F.W. (1981) A Lower Pleistocene glaciation in England. In *Quaternary Glaciations in the Northern Hemisphere* (eds D.J. Easterbrook, P. Hanslièk, K-D. Jäger and F.W. Shotton), UNESCO – International Geological Correlation Programme, Project 73/1/24, Report 7, Prague 1981, pp. 203–13.

Shotton, F.W. (1983) Interglacials after the Hoxnian in Britain. In *Quaternary Glaciations in the Northern Hemisphere* (eds O. Billards, O. Conchon and F.W. Shotton), UNESCO – International Geological Correlation Programme, Project 73/1/24, Report 9, Paris 1982, pp. 109–15. Reproduced in *Quaternary Newsletter*, **39**, 20–5.

Shotton, F.W. (1986) Glaciations in the United Kingdom. *Quaternary Science Reviews*, **5**, 293–7.

Shotton, F.W. and Osborne, P.J. (1965) The fauna of the Hoxnian interglacial deposits of Nechells, Birmingham. *Philosophical Transactions of the Royal Sociey of London*, **B248**, 353–78.

Shotton, F.W., Goudie, A.S., Briggs, D.J., *et al.* (1980) Cromerian interglacial deposits at Sugworth near Oxford, England, and their relation to the Plateau Drift of the Cotswolds and the terrace sequence of the Upper and Middle Thames. *Philosophical Transactions of the Royal Society of London*, **B289**, 55–86.

Shrubsole, O.A. (1898) On some high level gravels in Berkshire and Oxfordshire. *Quarterly Journal of the Geological Society of London*, **54**, 585–600.

Shrubsole, O.A. (1906) Early Man – the Palaeolithic age. In *A History of the County of Berkshire* (ed. W. Page), Victoria History of the Counties of England, Vol. 1, Archibald Constable, Westminster, pp. 173–80.

Shrubsole, O.A. and Whitaker, W. (1902) Excursion to Reading. *Proceedings of the Geologists' Association*, **17**, 381–3.

Sibrava, V. (1986a) Scandinavian glaciations in the Bohemian Massif and Carpathian foredeep and their relation to the extraglacial areas. *Quaternary Science Reviews*, **5**, 381–6.

Sibrava, V. (1986b) Correlations of European glaciations and their relation to the deep sea record. *Quaternary Science Reviews*, **5**, 433–42.

Siddiqui, Q.A. (1971) The palaeoecology of non-marine Pleistocene Ostracoda from Fladbury, Worcestershire and Isleworth, Middlesex. In *Colloque sur la Paléoécologie des Ostracodes* (ed. H.J. Oertli), Bulletin du Centre de Recherches, Société Nationale des Pétroles d'Aquitaine, Pau, Supplément 5, pp. 331–9.

Simmons, M.B. (1978) *The Sand and Gravel Resources of the Dengie Peninsula*. Mineral Assessment Report of the Institute of Geological Sciences 34, 90 pp.

Simpson, I.M. and West, R.G. (1958) On the stratigraphy and palaeobotany of a Late Pleistocene organic deposit at Chelford, Cheshire. *New Phytologist*, **57**, 239–50.

Singer, R., Wymer, J.J., Gladfelter, B.G., *et al.* (1973) Excavation of the Clactonian Industry at the golf course, Clacton-on-Sea, Essex. *Proceedings of the Prehistoric Society*, **39**, 6–74.

Smith, R.A. (1911) A Palaeolithic industry at Northfleet, Kent. *Archaeologia*, **62**, 515–32.

Smith, R.A. (1915) Prehistoric problems in geology. *Proceedings of the Geologists' Association*, **26**, 1–20.

Smith R.A. (1917) Plateau deposits and implements. *Proceedings of the Prehistoric Society of East Anglia*, **2**, 392–408.

Smith, R.A. (1922) Flint implements of special interest. *Archaeologia*, **72**, 25–40.

Smith, R.A. (1923) Prehistoric Man in Kent. *South East Naturalist*, **28**, 32–7.

Smith, R.A. (1926) *A Guide to Antiquities of the Stone Age*. 3rd edn, British Museum, London, 204 pp.

Smith, R.A. (1933) Implements from high-level gravels near Canterbury. *Proceedings of the Prehistoric Society of East Anglia*, **7**, 165–70.

Smith, R.H. and Dewey, H. (1913) Stratification at Swanscombe: report on excavation made on behalf of the British Museum and HM Geological Survey. *Archaeologia*, **64**, 177–204.

Smith, R.H. and Dewey, H. (1914) The High Terrace of the Thames: report on investigations made on behalf of the British Museum and H.M. Geological Survey in 1913. *Archaeologia*, **65**, 187–212.

Smith, W.G. (1883) On a Palaeolithic floor at North East London. *Journal of the Anthropological Institute*, **13**, 357–84.

Smith, W.G. (1894) *Man the Primaeval Savage: His Haunts and Relics from the Hill Tops of Bedfordshire to Blackwall*. E. Stanford, London, 349 pp.

Snelling, A.J.R. (1964) Excavations at the Globe Pit, Little Thurrock, Grays, Essex 1961. *Essex Naturalist*, **31**, 199–208.

Snelling, A.J.R. (1975) A fossil molluscan fauna at Purfleet, Essex. *Essex Naturalist*, **33**, 104–8.

Solomon, J.D. (1935) The Westleton Series of East Anglia; its age, distribution and relations. *Quarterly Journal of the Geological Society of London*, **91**, 216–38.

Sparks, B.W., West, R.G., Williams, R.B.G., *et al.* (1969) Hoxnian interglacial deposits near Hatfield, Herts. *Proceedings of the Geologists' Association*, **80**, 243–67.

Spencer, H.E.P. (1966) An Essex fossil ziphoid whale and its implications of the geographical changes in geological times. *Essex Naturalist*, **31**, 348–53.

Spurrell, F.J.C. (1880) On the discovery of the place where Palaeolithic implements were made at Crayford. *Quarterly Journal of the Geological Society of London*, **36**, 544–8.

Spurrell, F.J.C. (1883a) Palaeolithic knapping tools and modes of using them. *Journal of the Anthropological Institute*, **13**, 109–18.

Spurrell, F.J.C. (1883b) Palaeolithic implements found in West Kent. *Archaeologia Cantiana*, **15**, 89–103.

Spurrell, F.J.C. (1886) A sketch of the history of the rivers and denudation of West Kent. *Report of the West Kent Natural History Society* [for 1886], 53–104.

Spurrell, F.J.C. (1892) Excursion to Grays Thurrock, Essex. *Proceedings of the Geologists' Association*, **12**, 194.

Spurrell, F.J.C. (1893) Excursion to Dartford Heath. *Proceedings of the Geologists' Association*, **13**, 70.

Squirrell, H.C. (1978) *The Sand and Gravel Resources of the Country around Sonning and Henley. Berkshire, Oxfordshire and Buckinghamshire*. Mineral Assessment Report of the Institute of Geological Sciences 32, 98 pp.

Stebbing, W.P.D. (1900) Excursion to Netley Heath and Newlands Corner. *Proceedings of the Geologists' Association*, **16**, 524–6.

Stopes, C. (1903) Palaeolithic implements from the shelly gravel pit at Swanscombe, Kent. *Report of the British Association for the Advancement of Science, Southport* [1903], pp. 803–4.

Stopes, H. (1900) On the discovery of *Neritina fluviatalis* with a Pleistocene fauna and worked flints in high terrace gravels of the Thames valley. *Journal of the Anthropological Institute*, **29**, 302–3.

Strand, A. (1946) Nord-Norges Coleoptera. *Tromsø Museums Årshefter Naturhistorisk*, **67**, 1–699.

Straw, A. (1979) The geomorphological significance of the Wolstonian glaciation in Eastern England. *Transactions of the Institute of British Geographers*, **4**, 540–9.

Straw, A. (1983) Pre-Devensian glaciation of Lincolnshire (eastern England) and adjacent areas. *Quaternary Science Reviews*, **2**, 239–60.

Stringer, C.B. (1974) Population relationships of later Pleistocene hominids: a multivariate study of available crania. *Journal of Archaeological Science*, **1**, 317–42.

Stringer, C.B. (1978) Some problems in Middle and Upper Pleistocene hominid relationships. In *Recent Advances in Primatology. Volume 3. Evolution* (eds D.J. Chivers and K. Joysen), Academic Press, London, pp. 395–418.

Stringer, C.B. (1983) *Our Fossil Relatives – More About Man's Place in Evolution*, British Museum (Natural History), London, 23 pp.

Stringer, C.B. (1985) The Swanscombe fossil skull. In *The Story of Swanscombe Man* (ed. K.L. Duff), Kent County Council and Nature Conservancy Council, pp. 14–19.

Stringer, C.B. (1986) The British fossil hominid record. In *Recent Studies in the Palaeolithic of Britain and its Nearest Neighbours* (ed. S.N. Collcutt), J.R. Collis Publications, Department of Archaeology and Prehistory, Sheffield University, pp. 59–61.

Stringer, C.B., Currant, A.P., Schwarcz, H.P., *et*

al. (1986) Age of Pleistocene faunas from Bacon Hole, Wales. *Nature, London*, **320**, 59–62.

Stringer, C.B., Hublin, J.J. and Vandermeersch, B.V. (1984) The origin of anatomically modern humans. In *The Origins of Modern Humans* (eds F.H. Smith and F. Spencer), Alan Liss, New York, pp. 51–135.

Stuart, A.J. (1974) Pleistocene history of the British vertebrate fauna. *Biological Reviews*, **49**, 225–66.

Stuart, A.J. (1975) The vertebrate fauna of the type Cromerian. *Boreas*, **4**, 63–76.

Stuart, A.J. (1976) The history of the mammal fauna during the Ipswichian/Last Interglacial in England. *Philosophical Transactions of the Royal Society of London*, **B276**, 221–50.

Stuart, A.J. (1980) The vertebrate fauna from the interglacial deposits at Sugworth, near Oxford. *Philosophical Transactions of the Royal Society of London*, **B289**, 87–97.

Stuart, A.J. (1981) A comparison of the middle Pleistocene mammal faunas of Voigtstedt (Thüringia, German Democratic Republic) and West Runton (Norfolk, England), *Quartärpaläeontologie, Berlin*, **4**, 155–63.

Stuart, A.J. (1982a) *Pleistocene Vertebrates of the British Isles*, Longman, London, 212 pp.

Stuart, A.J. (1982b) Pleistocene occurrences of hippopotamus in Britain. *Quartärpaläeontologie, Berlin*, **6**, 209–18.

Stuart, A.J. (1988) Preglacial Pleistocene vertebrate faunas in East Anglia. In *Pliocene–Middle Pleistocene of East Anglia* (eds P.L. Gibbard and J.A. Zalasiewicz), Field Guide, Quaternary Research Association, Cambridge, pp. 57–64.

Stuart, A.J. (1991) Mammalian extinctions in the Late Pleistocene of northern Eurasia and North America. *Biological Reviews*, 66, 453–562.

Stuart, A.J. and West, R.G. (1976) Late Cromerian fauna and flora at Ostend, Norfolk. *Geological Magazine*, **113**, 469–73.

Sturdy, R.G., Allen, R.H., Bullock, P., *et al.* (1978) Palaeosols developed on Chalky Boulder Clay. *Journal of Soil Science*, **30**, 117–37.

Sumbler, M.G. (1983a) A new look at the type Wolstonian glacial deposits of Central England. *Proceedings of the Geologists' Association*, **94**, 23–31.

Sumbler, M.G. (1983b) The type Wolstonian sequence – some further comments. *Quaternary Newsletter*, **40**, 36–9.

Sutcliffe, A.J. (1960) Joint Mitnor Cave, Buckfastleigh. *Transactions of the Torquay Natural History Society*, **13**, 1–26.

Sutcliffe, A.J. (1964) The mammalian fauna. In *The Swanscombe Skull: a Survey of Research on a Pleistocene Site* (ed. C.D. Ovey), Royal Anthropological Institute, Occasional Paper No. 20, pp. 85–111.

Sutcliffe, A.J. (1974) The caves of south Devon. In *Exeter Field Guide* (ed. A. Straw), Quaternary Research Association, Cambridge, pp. 8–10.

Sutcliffe, A.J. (1975) A hazard in the interpretation of glacial-interglacial sequences. *Quaternary Newsletter*, **17**, 1–3.

Sutcliffe, A.J. (1976) The British glacial-interglacial sequence: a reply. *Quaternary Newsletter*, **18**, 1–7.

Sutcliffe, A.J. (1985) *On the Track of Ice Age Mammals*, British Museum (Natural History), London, 224 pp.

Sutcliffe, A.J. and Bowen, D.Q. (1973) Preliminary report on excavations in Minchin Hole, April–May 1973. *Newsletter of the William Pengelly Cave Studies Trust*, **21**, 12–25.

Sutcliffe, A.J. and Kowalski, K. (1976) Pleistocene rodents of the British Isles. *Bulletin of the British Museum of Natural History (Geology)*, **27**, 33–147.

Sutcliffe, A.J., Currant, A.P. and Oakley, K.P. (1979) Some little known and potentially important Middle and Upper Pleistocene mammalian localities in Essex. *Quaternary Newsletter*, **29**, 5–12.

Swanscombe Committee (1938) Report on the Swanscombe skull. *Journal of the Royal Anthropological Institute*, **68**, 17–98.

Szabo, B.J. and Collins, D. (1975) Ages of fossil bones from British interglacial sites. *Nature, London*, **254**, 680–2.

Tester, P.J. (1951) Palaeolithic flint implements from Bowman's Lodge Gravel Pit, Dartford Heath. *Archaeologia Cantiana*, **63**, 122–34.

Tester, P.J. (1953) The discovery of Acheulian implements in the deposits of the Dartford Heath terrace. *Archaeologia Cantiana*, **66**, 72–6.

Tester, P.J. (1955) Destruction of Rickson's Pit, Swanscombe. *Archaeologia Cantiana*, **69**, 216–7.

Tester, P.J. (1958) The age of the Baker's Hole industry. *Archaeological Newsletter*, **6**, 123–5.

Tester, P.J. (1975) Further consideration of the Bowman's Lodge industry. *Archaeologia Cantiana*, **91**, 29–39.

Thomas, M.F. (1961) River terraces and drainage development in the Reading area. *Proceedings of the Geologists' Association*, **72**, 415–36.

Tomlinson, M.E. (1929) The drifts of the Stour-Evenlode watershed and their extension into the valleys of the Warwickshire Stour and Upper Evenlode. *Proceedings of the Birmingham Natural History and Philosophical Society*, **15**, 157–96.

Tomlinson, M.E. (1963) The Pleistocene chronology of the Midlands. *Proceedings of the Geologists' Association*, **74**, 187–202.

Treacher, L. (1896) Palaeolithic Man in east Berkshire. *Berks., Bucks. and Oxon. Archaeological Journal, New Series*, **2**, 16–8 and 39–43.

Treacher, L. (1904) On the occurrence of stone implements in the Thames valley between Reading and Maidenhead. *Man*, **4**, 17–20.

Treacher, L. (1909) Excursion to Maidenhead. *Proceedings of the Geologists' Association*, **21**, 198–201.

Treacher, L. (1916) Excursion to Bourne End. *Proceedings of the Geologists' Association*, **27**, 107–9.

Treacher, L. (1926) Excursion to Shiplake. *Proceedings of the Geologists' Association*, **37**, 440–1.

Treacher, L. (1934) Field meeting in the Marlow district. *Proceedings of the Geologists' Association*, **45**, 107–8.

Treacher, M.S., Arkell, W.J. and Oakley, K.P. (1948) On the ancient channel between Caversham and Henley, Oxfordshire, and its contained flint implements. *Proceedings of the Prehistoric Society,* **14**, 126–54.

Trimmer, J. (1853) On the origin of the soils which cover the Chalk of Kent. Part 3. *Quarterly Journal of the Geological Society of London*, **9**, 286–96.

Trimmer, W.K. (1813) An account of some organic remains found near Brentford, Middlesex. *Philosophical Transactions of the Royal Society of London*, **53**, 131–7.

Tucker, E.V. and Greensmith, J.T. (1973) South East Essex. A. East Mersea. In *The Estuarine Region of Suffolk and Essex* (eds J.T. Greensmith, R.G. Blezard, C.R. Bristow, *et al.*), Geologists' Association Guide 12, pp. 12–7.

Turner, C. (1970) The Middle Pleistocene deposits at Marks Tey, Essex. *Philosophical Transactions of the Royal Society of London*, **B257**, 373–440.

Turner, C. (1973) Eastern England. In *A Correlation of Quaternary Deposits in the British Isles* (eds G.F. Mitchell, L.F. Penny, F.W. Shotton and R.G. West), Geological Society of London Special Report, No.4, pp. 8–18.

Turner, C. (1975) The correlation and duration of the Middle Pleistocene interglacial periods in North-west Europe. In *After the Australopithecines: Stratigraphy, Ecology and Culture Change in the Middle Pleistocene* (eds K.W. Butzer and G.L. Isaac), Mouton, The Hague, pp. 259–308.

Turner, C. (1983) Nettlebed interglacial deposits. In *Diversion of the Thames* (ed. J. Rose), Field Guide, Quaternary Research Association, Cambridge, pp. 66–8.

Turner, C. (1985) Problems and pitfalls in the application of palynology to Pleistocene archaeological sites in western Europe. In *Palynologie Archéologique* (eds J. Renault-Miskovsky, Bui-Thi-Mai and M. Girard), Actes des Journées du 25-26-27 janvier 1984, Éditions du Centre National de la Recherche Scientifique, Paris, pp. 347–73.

Turner, C. and Kerney, M.P. (1971) The age of the freshwater beds of the Clacton Channel. *Journal of the Geological Society of London*, **127**, 87–93.

Tyldesley, J.A. (1986a) The Wolvercote Channel handaxe assemblage: a comparative study. *British Archaeological Report, British Series*, **153**, 211 pp.

Tyldesley, J.A. (1986b) A re-assessment of the handaxe assemblage recovered from the Wolvercote Channel, Oxford. In *Recent Studies in the Palaeolithic of Britain and its Nearest Neighbours* (ed. S.N. Collcutt), J.R. Collis Publications, Department of Archaeology and Prehistory, Sheffield University, pp. 23–5.

Tyldesley, J.A. (1988) Quartzite implements recovered from the Wolvercote Channel, Oxfordshire. *In* Non-flint stone tools and the Palaeolithic occupation of Britain (eds R.J. MacRae and N. Moloney), *British Archaeological Report, British Series*, **189**, 159–66.

Tylor, A. (1869) On Quaternary gravels. *Quarterly Journal of the Geological Society of London*, **25**, 57–100.

Underwood, W. (1913) A discovery of Pleistocene bones and flint implements in a gravel pit at Dovercourt, Essex. *Proceedings of the Prehistoric Society of East Anglia*, **1**, 360–8.

Vallois, H.V. (1954) Neandertals and Praesapiens. *Journal of the Royal Anthropological Institute*, **84**, 111–30.

Vallois, H.V. (1958) La Grotte de Fontécherade: Part 2, Anthropologie. *Archives de l'Institut de Paléontologie Humaine, Paris*, Memoire No. 29, pp. 157–64.

Waechter, J. d'A (1969) Swanscombe 1968. *Proceedings of the Royal Anthropological Institute* [for 1968], pp. 53–8.

Waechter, J. d'A. (1970) Swanscombe 1969. *Proceedings of the Royal Anthropological Institute* [for 1969], pp. 83–5.

Waechter, J. d'A. (1971) Swanscombe 1970. *Proceedings of the Royal Anthropological Institute* [for 1970], pp. 43–9.

Waechter, J. d'A. (1972) Swanscombe 1971. *Proceedings of the Royal Anthropological Institute* [for 1971], pp. 73–8.

Waechter, J. d'A. (1973) The Late Middle Acheulian industries in the Swanscombe area. In *Archaeological Theory and Practice* (ed. D.E. Strong), Seminar Press, London and New York, pp. 67–86.

Walder, P.S. (1967) The composition of the Thames gravels near Reading, Berkshire. *Proceedings of the Geologists' Association*, **78**, 107–19.

Walker, H. (1871) On the glacial drifts of North London. *Proceedings of the Geologists' Association*, **2**, 289–93.

Ward, G.R. (1984) Interglacial fossils from Upminster, Essex. *London Naturalist*, **3**, 24–6.

Warren, S.H. (1911) Palaeolithic wooden spear from Clacton. *Quarterly Journal of the Geological Society of London*, **67**, cxix.

Warren, S.H. (1912) Palaeolithic remains from Clacton-on-Sea, Essex. *Essex Naturalist*, **17**, 15.

Warren, S.H. (1917) The study of pre-history in Essex as recorded in the publications of the Essex Field Club. *Essex Naturalist*, **18**, 145–52.

Warren, S.H. (1922) The Mesvinian industry of Clacton-on-Sea. *Proceedings of the Prehistoric Society of East Anglia*, **3**, 597–602.

Warren, S.H. (1923a) The *Elephas-antiquus* bed of Clacton-on-Sea (Essex) and its flora and fauna. *Quarterly Journal of the Geological Society of London*, **79**, 606–36.

Warren, S.H. (1923b) The sub-soil flint flaking sites at Grays. *Proceedings of the Geologists' Association*, **34**, 38–42.

Warren, S.H. (1923c) Sub-soil pressure flaking. *Proceedings of the Geologists' Association*, **34**, 153–75.

Warren, S.H. (1924a) Pleistocene classifications. *Proceedings of the Geologists' Association*, **35**, 265–82.

Warren, S.H. (1924b) The elephant-bed of Clacton-on-Sea. *Essex Naturalist*, **21**, 32–40.

Warren, S.H. (1926) The classification of the Lower Palaeolithic with especial reference to Essex. *South East Naturalist*, **31**, 38–50.

Warren, S.H. (1933) The Palaeolithic industries of the Clacton and Dovercourt districts. *Essex Naturalist*, **24**, 1–29.

Warren, S.H. (1940) Geological and prehistoric traps. *Essex Naturalist*, **27**, 2–19.

Warren, S.H. (1942) The drifts of south-western Essex. Parts I and II. *Essex Naturalist*, **27**, 154–79.

Warren, S.H. (1945) Some geological and prehistoric records on the north-west border of Essex. *Essex Naturalist*, **27**, 273–80.

Warren, S.H. (1951) The Clacton flint industry: a new interpretation. *Proceedings of the Geologists' Association*, **62**, 107–35.

Warren, S.H. (1955) The Clacton (Essex) channel deposits. *Quarterly Journal of the Geological Society of London*, **111**, 283–307.

Warren, S.H. (1957) On the early pebble gravels of the Thames Basin from the Hertfordshire-Essex border to Clacton-on-Sea. *Geological Magazine*, **94**, 40–6.

Warren, S.H. (1958) The Clacton flint industry: A supplementary note. *Proceedings of the Geologists' Association*, **69**, 123–9.

Webb, W.M. (1894) Museum notes: Pleistocene non-marine Mollusca from Walton-on-the-Naze. *Essex Naturalist*, **8**, 160–2.

Webb, W.M. (1900) Pleistocene non-marine Mollusca from Clacton-on-Sea, Essex. *Essex Naturalist*, **11**, 225–9.

Wehmiller, J.F. (1982) A review of amino acid racemization studies in Quaternary molluscs: stratigraphic and chronological applications in coastal and interglacial sites. Pacific and Atlantic coasts, United States, United Kingdom, Baffin Island and tropical islands. *Quaternary Science Reviews*, **1**, 83–120.

Weidenreich, F. (1940) The *torus occipitalis* and related structures and their transformations in the course of human evolution. *Bulletin of the Geological Society of China*, **19**, 480–558.

Weidenreich, F. (1943) The skull of *Sinanthropus pekinensis*: a comparative study on a primitive hominid skull. *Palaeontologia Sinica, New Series D*, **10**, 1–485.

Weiner, J.S. and Campbell, B.G. (1964) The taxonomic status of the Swanscombe skull. In *The Swanscombe Skull: a Survey of Research on a Pleistocene Site* (ed. C.D. Ovey), Royal Anthropological Institute, Occasional Paper No. 20, pp. 175–209.

Weir, A.H., Catt, J.A. and Madgett, P.A. (1971) Postglacial soil formation in the loess of Pegwell Bay, Kent (England), *Geoderma*, **5**, 131–49.

Wenban-Smith, F.F. (1990) The location of Baker's Hole. *Proceedings of the Prehistoric Society*, **56**, 11–14.

West, R.G. (1956) The Quaternary deposits at Hoxne, Suffolk. *Philosophical Transactions of the Royal Society of London*, **B239**, 265–356.

West, R.G. (1963) Problems of the British Quaternary. *Proceedings of the Geologists' Association*, **74**, 147–86.

West, R.G. (1968) *Pleistocene Geology and Biology*, 1st edn, Longman, London, 379 pp.

West, R.G. (1969) Pollen analyses from interglacial deposits at Aveley and Grays, Essex. *Proceedings of the Geologists' Association*, **80**, 271–82.

West, R.G. (1972) Relative land-sea-level changes in south eastern England during the Pleistocene. *Philosophical Transactions of the Royal Society of London*, **A272**, 87–98.

West, R.G. (1977) *Pleistocene Geology and Biology*, 2nd edn, Longman, London, 440 pp.

West, R.G. (1980) *The Pre-glacial Pleistocene of the Norfolk and Suffolk Coasts*, Cambridge University Press, 203 pp.

West, R.G. (1988) The record of the cold stages. *Philosophical Transactions of the Royal Society of London*, **B318**, 505–22.

West, R.G. and Donner, J.J. (1956) The glaciations of East Anglia and the East Midlands: a differentiation based on stone orientation measurements of the tills. *Quarterly Journal of the Geological Society of London*, **112**, 69–91.

West, R.G., Lambert, C.A. and Sparks, B.W. (1964) Interglacial deposits at Ilford, Essex. *Philosophical Transactions of the Royal Society of London*, **B247**, 185–212.

West, R.G., Dickson, C.A., Catt, J.A., *et al.* (1974) Late Pleistocene deposits at Wretton, Norfolk. II. Devensian deposits. *Philosophical Transactions of the Royal Society of London*, **B267**, 337–420.

Whitaker, W. (1862) On the western end of the London Basin; on the westerly thinning of the Lower Eocene beds in that basin; and of the Grey Wethers of Wiltshire. *Quarterly Journal of the Geological Society of London*, **18**, 258–74.

Whitaker, W. (1864) *The Geology of Parts of Middlesex, Hertfordshire, Buckinghamshire, Berkshire and Surrey*, Memoir of the Geological Survey of Great Britain, 112 pp.

Whitaker, W. (1875) *Guide to the Geology of London and the Neighbourhood*, Memoir of the Geological Survey of Great Britain, 72 pp.

Whitaker, W. (1877) *The Geology of the Eastern End of Essex*. Memoir of the Geological Survey of Great Britain, 32 pp.

Whitaker, W. (1884) *Guide to the Geology of London and the Neighbourhood*, 4th edn, Memoir of the Geological Survey of Great Britain, 98 pp.

Whitaker, W. (1889) *The Geology of London and Parts of the Thames Valley*, Volume 1, Memoir of the Geological Survey of Great Britain, 556 pp.

White, H.J.O. (1892) Notes on the Westleton Beds near Henley-on-Thames. *Proceedings of the Geologists' Association*, **12**, 379–84.

White, H.J.O. (1895) On the distribution and relations of the Westleton and glacial gravels in Oxfordshire and Berkshire. *Proceedings of the Geologists' Association*, **14**, 11–23.

White, H.J.O. (1897) On the origin of the high-level gravel with Triassic debris adjoining the valley of the Upper Thames. *Proceedings of the Geologists' Association*, **15**, 157–74.

White, H.J.O. (1902) On the peculiarity in the course of certain streams in the London and Hampshire Basins. *Proceedings of the Geologists' Association*, **17**, 399–413.

White, H.J.O. (1906) On the occurrence of quartzose gravel in the Reading Beds at Lane End, Bucks. *Proceedings of the Geologists' Association*, **19**, 371–7.

White, H.J.O. (1907) *The Geology of the Country around Hungerford and Newbury*. Memoir of the Geological Survey of Great Britain, 150 pp.

White, H.J.O. (1908a) Eocene. In *The Geology of the Country around Henley on Thames and Wallingford* (eds A.J. Jukes-Browne and H.J.O. White), Memoir of the Geological Survey of Great Britain, pp. 58–76.

White, H.J.O. (1908b) Scenery and superficial deposits. In *The Geology of the Country around Henley on Thames and Wallingford* (eds A.J. Jukes-Browne and H.J.O. White), Memoir of the Geological Survey of Great Britain, pp. 77–103.

Whiteman, C.A. (1983) Great Waltham. In *Diversion of the Thames* (ed. J. Rose), Field guide, Quaternary Research Association, Cambridge, pp. 163–9.

Whiteman, C.A. (1987) Till lithology and genesis near the southern margin of the Anglian ice sheet in Essex, England. In *Tills and Glaciotectonics* (ed. J.J.M. Van der Meer), A.A. Balkema, Rotterdam, pp. 55–66.

Whiteman, C.A. (1990) Early and Middle Pleistocene stratigraphy and soils in central Essex, England. Unpublished Ph.D. thesis, University of London.

Whiteman, C.A. and Kemp, R.A. (1990) Pleistocene sediments, soils and landscape evolution at Stebbing, Essex. *Journal of Quaternary Science*, **5**, 145–61.

Wiegank, K. Von.F. (1972) Ekologische Analyse quartärer Foraminiferen. *Geologie 21*, 7, 1–111.

Wilson, D. and Lake, R.D. (1983) Field meeting to north Essex and west Suffolk, 20–22 June, 1980. *Proceedings of the Geologists' Association*, **94**, 75–9.

Wiseman, C.R. (1978) A palaeoenvironmental reconstruction of part of the Lower Thames terrace sequence based on sedimentological studies from Aveley, Essex. Unpublished M.Sc. thesis, City of London Polytechnic and Polytechnic of North London.

Wolpoff, M.H. (1971) Is Vértessozöllös an occipital of *Homo erectus*? *Nature, London*, **232**, 867–8.

Wood, S.V. (1848) Introduction, v-xii. In *A Monograph of the Crag Mollusca, or Descriptions of Shells from the Middle and Upper Tertiaries of the East of England, Part 1 Univalves* (S.V. Wood), Monograph of the Palaeontographical Society, London, 208 pp.

Wood, S.V., Jun. (1866a) On the structure of the Thames valley and its contained deposits. I and II. *Geological Magazine*, **3**, 57–63 and 99–107.

Wood, S.V., Jun. (1866b) On the structure of the valleys of the Blackwater and the Crouch and of the East Essex Gravel, and on the relation of this gravel to the denudation of the Weald. *Geological Magazine*, **3**, 348–54 and 398–406.

Wood, S.V., Jun. (1867) On the structure of the Postglacial deposits of the south-east of England. *Quarterly Journal of the Geological Society of London*, **23**, 394–417.

Wood, S.V., Jun. (1868) On the pebble-beds of Middlesex, Essex and Herts. *Quarterly Journal of the Geological Society of London*, **24**, 464–72.

Wood, S.V., Jun. (1870) Observations on the sequence of the glacial beds. *Geological Magazine*, **7**, 17–22 and 61–8.

Wood, S.V., Jun. (1872) On the climate of the Post-Glacial Period. *Geological Magazine*, **9**, 153–61.

Wood, S.V., Jun. and Harmer, F.W. (1868) On the Glacial and Post-Glacial structure of Norfolk and Suffolk. *Geological Magazine*, **5**, 452.

Wood, S.V., Jun. and Harmer, F.W. (1872) An outline of the geology of the Upper Tertiaries of East Anglia. *In* Supplement to the Monograph of the Crag Mollusca (ed. S.V. Wood), *Monograph of the Palaeontolographical Society*, **3**, 2–31.

Woodland, A.W. (1970) The buried tunnel-valleys of East Anglia. *Proceedings of the Yorkshire Geological Society*, **37**, 521–78.

Woodward, B.B. (1890) On the Pleistocene (non-marine) Mollusca of the London district. *Proceedings of the Geologists' Association*, **11**, 335–87.

Woodward, H.B. (1904) Excursion to Upminster, Great Warley and Brentwood. *Proceedings of the Geologists' Association*, **18**, 479–86.

Woodward, H.B. (1909) *The Geology of the London District*, 1st edn, Memoir of the Geological Survey of Great Britain, 142 pp.

Woodward, H.B. and Davies, W. (1874) Note on the Pleistocene deposits yielding mammalian remains in the vicinity of Ilford, Essex. *Geological Magazine*, **1**, 390–8.

Woodward, H.B., Bromehead, C.E.N. and Chatwin, C.P. (1922) *The Geology of the London District*, 2nd edn, Memoir of the Geological Survey of Great Britain, 99 pp.

Wooldridge, S.W. (1927a) The Pliocene history of the London Basin. *Proceedings of the Geologists' Association*, **38**, 49–132.

Wooldridge, S.W. (1927b) The Pliocene Period in western Essex and the Pre-glacial topography of the district. *Essex Naturalist*, **21**, 247–68.

Wooldridge, S.W. (1928) The 200-foot platform in the London Basin. *Proceedings of the Geologists' Association*, **39**, 1–26.

Wooldridge, S.W. (1938) The glaciation of the London Basin, and the evolution of the Lower Thames drainage system. *Quarterly Journal of the Geological Society of London*, **94**, 627–64.

Wooldridge, S.W. (1957) Some aspects of the physiography of the Thames valley in relation to the Ice Age and early Man. *Proceedings of the Prehistoric Society*, **23**, 1–19.

Wooldridge, S.W. (1960) The Pleistocene succession in the London Basin. *Proceedings of the Geologists' Association*, **71**, 113–29.

Wooldridge, S.W. and Ewing, C.J.C. (1935) The Eocene and Pleistocene deposits of Lane End, Bucks. *Quarterly Journal of the Geological Society of London*, **41**, 293– 317.

Wooldridge, S.W. and Gill, D.M.C. (1925) The Reading Beds of Lane End, Bucks., and their bearing on some unsolved questions of London geology. *Proceedings of the Geologists' Association*, **36**, 146–73.

Wooldridge, S.W. and Henderson, H.C.K. (1955) Some aspects of the physiography of the eastern part of the London Basin. *Transactions of the Institute of British Geographers*, **21**, 19–31.

Wooldridge, S.W. and Linton, D.L. (1939) *Structure, Surface and Drainage in South-east England*. Transactions of the Institute of British Geographers, No. **10**, 124 pp.

Wooldridge, S.W. and Linton, D.L. (1955) *Structure, Surface and Drainage in South-east England*, 2nd edn, G. Phillip, London, 176 pp.

Wright, W.B. (1937) *The Quaternary Ice Age*, Macmillan, London, 478 pp.

Wymer, B.O. (1955) The discovery of the right pariental bone at Swanscombe. *Man*, **55**, 124.

Wymer, J.J. (1956) Palaeoliths from the gravel of the Ancient Channel between Caversham and Henley at Highlands, near Henley. *Proceedings of the Prehistoric Society*, **22**, 29–36.

Wymer, J.J. (1957) A Clactonian flint industry at Little Thurrock, Grays, Essex. *Proceedings of the Geologists' Association*, **68**, 159–77.

Wymer, J.J. (1958) Archaeological notes from Reading Museum: Highlands, Henley. *Berkshire Archaeological Journal*, **56**, 56–7.

Wymer, J.J. (1959) Archaeological notes from the Reading Museum: Henley. *Berkshire Archaeological Journal*, **57**, 121–2.

Wymer, J.J. (1960) Archaeological notes from the Reading Museum: Henley. *Berkshire Archaeological Journal*, **58**, 52–8.

Wymer, J.J. (1961) The Lower Palaeolithic succession in the Thames valley and the date of the Ancient Channel between Caversham and Henley, Oxfordshire. *Proceedings of the Prehistoric Society*, **27**, 1–27.

Wymer, J.J. (1962) Archaeological notes from the Reading Museum: Rotherfield Peppard, Oxon. *Berkshire Archaeological Journal*, **60**, 114–5.

Wymer, J.J. (1964a) Archaeological notes from the Reading Museum: Rotherfield Peppard, Oxon. *Berkshire Archaeological Journal*, **61**, 96–7.

Wymer, J.J. (1964b) Excavations at Barnfield Pit, 1955–1960. In *The Swanscombe Skull: a Survey of Research on a Pleistocene Site* (ed. C.D. Ovey), Royal Anthropological Institute, Occasional Paper No. 20, pp. 19–60.

Wymer, J.J. (1968) *Lower Palaeolithic Archaeology in Britain, as Represented by the Thames Valley*, John Baker, London, 429 pp.

Wymer, J.J. (1974) Clactonian and Acheulian industries in Britain – their chronology and significance. *Proceedings of the Geologists' Association*, **85**, 391–421.

Wymer, J.J. (1976) Highlands Farm Pit, Rotherfield Peppard. In *Field Guide to the Oxford Region* (ed. D.A. Roe), Quaternary Research Association, Oxford, pp. 48–9.

Wymer, J.J. (1977a) Highlands Farm, Rotherfield Peppard. In *South East England and the Thames Valley* (eds E.R. Shephard-Thorn and J.J. Wymer), Guide Book for excursion A5, X INQUA Congress, Birmingham, Geoabstracts, Norwich, pp. 24–8.

Wymer, J.J. (1977b) Sulhamstead. In *South East England and the Thames Valley* (eds E.R. Shephard-Thorn and J.J. Wymer), Guide Book for excursion A5, X INQUA Congress, Birmingham, Geoabstracts, Norwich, pp. 11–2.

Wymer, J.J. (1977c) Furze Platt. In *South East England and the Thames Valley* (eds E.R. Shephard-Thorn and J.J. Wymer), Guide Book for excursion A5, X INQUA Congress, Birmingham, Geoabstracts, Norwich, pp. 30–4.

Wymer, J.J. (1981) The Palaeolithic. In *The Environment in British Prehistory* (eds I.G. Simmons and M.J. Tooley), Duckworth, pp. 49–81.

Wymer, J.J. (1985a) Early Man in Britain – time and change. *Modern Geology*, 9, 261–72.

Wymer, J.J. (1985b) *The Palaeolithic Sites of East Anglia*, Geobooks, Norwich, 440 pp.

Wymer, J.J. (1988) Palaeolithic archaeology and the British Quaternary sequence. *Quaternary Science Reviews*, 7, 79–98.

Wymer, J.J. and Singer, R. (1970) The first season of excavations at Clacton-on-Sea, Essex, England: a brief report. *World Archaeology*, **2**, 12–16.

Zagwijn, W.H. (1973) Pollen analytic studies of Holsteinian and Saalian Beds in the northern Netherlands. *Mededelingen Rijks Geologische Dienst, New Series*, **24**, 139–56.

Zagwijn, W.H. (1978) A macroflora of Holsteinian age from the northern part of the Netherlands. *Review of Paleobotany and Palynology*, **26**, 243–8.

Zagwijn, W.H. (1985) An outline of the Quaternary Stratigraphy of the Netherlands. *Geologie en Mijnbouw*, **50**, 41–58.

Zagwijn, W.H. (1986) The Pleistocene of the Netherlands with special reference to glaciation and terrace formation. *Quaternary Science Reviews*, **5**, 341–6.

Zagwijn, W.H., Montfrans, H.M. Van and Zandrsta, J.G. (1971) Subdivision of the 'Cromerian' in the Netherlands; pollen analysis, palaeomagnetism and sedimentary petrology. *Geologie en Mijnbouw*, **50**, 41–58.

Zalasiewicz, J.A. and Gibbard, P.L. (1988) The Pliocene to early Middle Pleistocene of East Anglia: an overview. In *Pliocene–Middle Pleistocene of East Anglia* (eds P.L. Gibbard and J.A. Zalasiewicz), Field Guide, Quaternary Research Association, Cambridge, pp. 1–31.

Zeuner, F.E. (1945) *The Pleistocene Period: its Climate, Chronology and Faunal Successions*, 1st edn, Ray Society, Publication No. 130, London, 322 pp.

Zeuner, F.E. (1946) *Dating the Past: an Introduction to Geochronology*. Methuen, London, 444 pp.

Zeuner, F.E. (1954) Riss or Würm? *Eiszeitalter und Gegenwart*, 4, 98–105.

Zeuner, F.E. (1955) Loess and Palaeolithic chronology. *Proceedings of the Prehistoric Society*, **21**, 51–64.

Zeuner, F.E. (1958) *Dating the Past: An Introduction to Geochronology*, 4th edn, Methuen, London, 516 pp.

Zeuner, F.E. (1959) *The Pleistocene Period: its Climate, Chronology and Faunal Successions*, 2nd edn, Hutchinson, London, 447 pp.

Index

Page numbers in **bold** type refer to figures and page numbers in *italic* type refer to tables.

Index

Index

Index

Index

Index

Index

Index

Index

Index